TA
1800
.067
V.1
0 0 2 6 2 5 7 4 4
C0-AYR-766

Optical Fiber Technology

OTHER IEEE PRESS BOOKS

Selected Papers in Digital Signal Processing, II
Edited by the Digital Signal Processing Committee
A Guide for Better Technical Presentations
Edited by R. M. Woelfle
Career Management: A Guide to Combating Obsolescence
Edited by H. G. Kaufman
Energy and Man: Technical and Social Aspects of Energy
Edited by G. Morgan
Magnetic Bubble Technology: Integrated-Circuit Magnetics for Digital Storage and Processing
Edited by H. Chang
Frequency Synthesis: Techniques and Applications
Edited by J. Gorski-Popiel
Literature in Digital Signal Processing: Author and Permuted Title Index (Revised and Expanded Edition)
Edited by H. D. Helms, J. F. Kaiser, and L. R. Rabiner
Data Communications via Fading Channels
Edited by K. Brayer
Nonlinear Networks: Theory and Analysis
Edited by A. N. Willson, Jr.
Computer Communications
Edited by P. E. Green, Jr. and R. W. Lucky
Stability of Large Electric Power Systems
Edited by R. T. Byerly and E. W. Kimbark
Automatic Test Equipment: Hardware, Software, and Management
Edited by F. Liguori
Key Papers in the Development of Coding Theory
Edited by E. R. Berlekamp
Technology and Social Institutions
Edited by K. Chen
Key papers in the Development of Information Theory
Edited by D. Slepian
Computer-Aided Filter Design
Edited by G. Szentirmai
Laser Devices and Applications
Edited by I. P. Kaminow and A. E. Siegman
Integrated Optics
Edited by D. Marcuse
Laser Theory
Edited by F. S. Barnes
Digital Signal Processing
Edited by L. R. Rabiner and C. M. Rader
Minicomputers: Hardware, Software, and Applications
Edited by J. D. Schoeffler and R. H. Temple
Semiconductor Memories
Edited by D. A. Hodges
Power Semiconductor Applications, Volume I: General Considerations
Edited by J. D. Harnden, Jr., and F. B. Golden
Power Semiconductor Applications, Volume II: Equipment and Systems
Edited by J. D. Harnden, Jr., and F. B. Golden
A Practical Guide to Minicomputer Applications
Edited by F. F. Coury
Active Inductorless Filters
Edited by S. K. Mitra
Clearing the Air: The Impact of the Clean Air Act on Technology
Edited by J. C. Redmond, J. C. Cook, and A. A. J. Hoffman

Optical Fiber Technology

Edited by

Detlef Gloge

Research Head

Bell Laboratories

A volume in the IEEE PRESS Selected
Reprint Series, prepared under the sponsorship
of the IEEE Electron Devices Society and
the IEEE Microwave Theory and Techniques Society.

The Institute of Electrical and Electronics Engineers, Inc. New York

International Standard Book Numbers: Clothbound: 0-87942-061-8
Paperbound: 0-87942-062-6

Library of Congress Catalog Card Number 75-23777

PRINTED IN THE UNITED STATES OF AMERICA

Contents

Introduction

This book considers the prospects of optical fibers in the communications field. Unlike the older and established applications of fibers in endoscopic devices and image processors, communications applications are still in an exploratory stage. Although feasibility is demonstrable in principle, detailed systems requirements and conditions imposed by manufacturing and maintenance procedures must still be met. A reprint book seems useful at this time because it fills the need for a compendium covering the present state of the art until the time when the prevailing concepts, tested by fabrication and field trials, will be collected in permanent textbooks.

On the other hand, presenting the principal ideas of this new and evolving technology by way of a reprint collection is a delicate matter indeed. For among the close to 400 publications covering the subject, there are many equally well suited to represent the essential concepts. Out of this number, the Editor must choose a painfully small fraction, selected not on the basis of priority, originality, or technical excellence, but with the objective of the book in mind; he must fit all parts together in the most lucid and educational way; he must cover all aspects as completely as the limited space permits; and he must include controversial or futuristic views if they have any chance of realization.

These standards of selection have inevitably a subjective component. When viewed in isolation, individual cases may appear arbitrary. Optical signal processing, for example, and the exploration of related components receive little coverage in this book solely because a comprehensive book on the subject (D. Marcuse, *Integrated Optics,* IEEE Press, 1974) is available. Because of the limited space, few review papers are included, in spite of a rapid sequence of excellent reviews that has appeared in recent years (see the appended list). These reviews, as well as expert counsel obtained from colleagues all over the world, were an appreciated help in preparing this book.

Bibliography

Review Articles

[1] K. C. Kao and G. A. Hockham, "Dielectric fiber surface waveguides for optical frequencies," *Proc. Inst. Elec. Eng.,* vol. 113, pp. 1151-1158, July 1966.

[2] D. Gloge, "Optical waveguide transmission," *Proc. IEEE,* vol. 58 pp. 1516-1522, Oct. 1970.

[3] T. Li and E. A. J. Marcatili, "Research on optical fiber transmission," *Bell Lab Rec.,* pp. 331-337, Dec. 1971.

[4] R. Kompfner, "Optics at Bell Laboratories—Optical communications," *Appl. Opt.* vol. 11, pp. 2412-2425, Nov. 1972.

[5] M. M. Ramsay, "Fiber optical communications within the United Kingdom," *Opto-Electron.,* vol. 5, pp. 261-274, July 1973.

[6] S. Maslowski, "Activities in fiber optical communications in Germany," *Opto-Electron.,* vol. 5, pp. 275-284, July 1973.

[7] C. P. Sandbank, "The challenge of fiber optical communication systems," *Radio Electron. Eng.,* vol. 43, pp. 665-674, Nov. 1973.

[8] D. Marcuse, "Optical fibers for communications," *Radio Electron. Eng.,* vol. 43, pp. 655-665, Nov. 1973.

[9] H. Ohnesorge, "Neue Moglichkeiten fur Nachrichtensysteme auf der Basis des Laser-Glasfaserkanals," NTZ Rep. 17, Berlin, Germany, 1973.

[10] R. A. Andrews, A. F. Milton, and T. G. Giallorenzi, "Military applications of fiber optics and integrated optics," *IEEE Trans. Microwave Theory Tech.,* vol. MTT-21, pp. 763-769, Dec. 1973.

[11] O. Krumpholz and S. Maslowski, "Nachrichtenubertragung uber Lichtleitfasern," *Ver. Deut. Ing. Z.,* vol. 116, pp. 230-234, Feb. 1974.

[12] D. Gloge, "Optical fibers for communication," *Appl. Opt.,* vol. 13, pp. 249-254, Feb. 1974.

[13] F. F. Roberts, "Optical-fiber telecommunication transmission systems," *Post Office Elec. Eng. J.,* vol. 67, pp. 32-36, Apr. 1974.

[14] W. B. Bielawaski, "Low-loss optical waveguide," *Electron Eng.,* pp. 59-66, May 1974.

[15] M. Borner and D. Rosenberger, "Laser communication technology in Germany," *IEEE Trans. Commun.* (*Special Issue on Communications in Europe*), vol. COM-22, pp. 1305-1309, Sept. 1974.

Books and Conference Proceedings

[1] N. S. Kapany, *Fiber Optics.* New York: Academic, 1967.

[2] D. Marcuse, *Light Transmission Optics.* New York: Van Nostrand, 1972.

[3] M. P. Lisitsa, L. I. Berezhinskii, and M. Ya. Valakh, *Fiber Optics* (translated from Russian). New York: Halsted, 1972.

[4] N. S. Kapany and J. J. Burke, *Optical Waveguides.* New York: Academic, 1972.

[5] W. B. Allen, *Fiber Optics: Theory and Practice.* London: Plenum, 1973.

[6] D. Marcuse, *Theory of Dielectric Optical Waveguides.* New York: Academic, 1974.

[7] *Proc. Conf. on Trunk Telecommun. by Guided Waves,* London, England, Sept. 1970.

[8] *Proc. Topical Meeting on Optical Fiber Transmission,* Williamsburg, Va., 1975.

Part I
Introductory Survey

Research Toward Optical-Fiber Transmission Systems

Part I: The Transmission Medium

STEWARD E. MILLER, ENRIQUE A. J. MARCATILI, AND TINGYE LI

Part II: Devices and Systems Considerations

STEWARD E. MILLER, TINGYE LI, AND ENRIQUE A. J. MARCATILI

Invited Paper

Abstract—The fundementals of optical fiber transmission systems including the fiber transmission medium, sources suitable for use as a carrier, modulation and detection techniques, and some system design considerations are reviewed.

The advent of low-loss optical fibers brings new dimensions to optical communication prospects. Fibers may soon be used much as wire pairs of coaxial cable are now used in communication systems. Transmission losses as low as 2 dB/km have been achieved. Experimental repeaters for fiber systems with 10^{-9} error rate at about 300-Mb/s pulse rate have been reported.

Fiber cabling and splicing are among the problems requiring new ideas in order to make feasible an operable system.

TABLE OF CONTENTS

PART I: THE TRANSMISSION MEDIUM

PART II: DEVICES AND SYSTEMS CONSIDERATIONS

This invited paper is one of a series planned on topics of general interest—The Editor.

Manuscript received August 2, 1973; revised August 27, 1973.

The authors are with the Crawford Hill Laboratory, Bell Laboratories, Holmdel, N. J. 07733.

Reprinted from *Proc. IEEE*, vol. 61, pp. 1703–1751, Dec. 1973.

I. Introduction

OPTICAL-FIBER transmission systems will find application in the communications industry. The question is *when* will this occur, not *whether* it will occur. Whereas the idea of using light for the transmission of information dates back at least to the 1870's [1], it was not until the invention of the laser that a major research effort appeared warranted [2]. There followed several years of exploration of materials and laser structures, modulation and detection fundamentals, and techniques of guidance. The prospect appeared bright for systems carrying very large volumes of communications, but the anticipated date when very large systems would be needed appeared to be after 1980 [3], [4].

The advent of low-loss optical fibers added new dimensions to optical communication prospects. No longer is it necessary to have a large number of customers on a single transmission path in order to reduce the cost of the transmission medium to an attractive per-channel value. Fibers may soon be used much as we now use wire pairs or coaxial cables and may be applied to existing communications needs. Moreover, the complexity of the optical apparatus associated with multichannel multicarrier systems, such as would be needed on high-capacity lens-guided systems, is largely eliminated in fiber systems.

Fiber system advantages will be developed more fully later in this paper. The remarks here serve as introductory background for the choice of material in the body of the paper, and serve also to alert the reader to the fact that the prospects for optical communication systems are much more immediate and pervasive than previously viewed.

A brief historical note may be interesting to those new to the field. The first communications expert to direct serious attention to glass fibers for long-distance communications was Kao, then at Standard Telecommunications Laboratories in England [5]. At that time, 1968, typical fiber losses were above 1000 dB/km, but Kao suggested that purer materials should permit much lower losses. There followed a British Post Office sponsored effort to purify glass and to explore fiber transmission problems, both in industry and in the British universities as well as in the Post Office research laboratory. Interest in fibers soon became serious elsewhere—in the United States at Bell Laboratories and at Corning Glass Works, in Japan at Nippon Electric Company and at Nippon Sheet Glass Company, and in Germany at AEG-Telefunken and at Siemens and Halske. The breakthrough came in 1970 when Kapron, Keck, and Maurer of Corning Glass Works announced the achievement of losses under 20 dB/km in single-mode fibers hundreds of meters long [6]. Thereafter, progress in the science and technology of fiber transmission has developed along a broad front and continues vigorously now—as will be apparent in the following pages.

Section II treats the transmission properties of fiber waveguides. The various types of single-mode or multimode fibers are described, a description is given for the fields, and a review of new analytical techniques is presented. Descriptions of modes as well as ray-optic approaches are given. The mechanisms responsible for loss and dispersion are reviewed, and the techniques available for reducing dispersion are outlined. After a brief indication of the techniques for measuring loss and dispersion, the state of the art for the various fiber types is presented.

Section III discusses sources or carrier generators suitable for use in fiber transmission systems. The types of sources, the choice of carrier wavelength, the spectral width of their outputs, their brightness, their modulation capabilities, and their efficiencies are reviewed.

Section IV discusses modulation techniques and indicates their relation to fiber systems and to the particular carrier source. Direct modulation of the sources is considered in some detail and various optical modulators suitable for fiber systems are reviewed.

Section V contains a discussion of detectors in relation to fiber systems, with references to recent review papers.

Section VI gives an indication of the relation between the work on integrated optics and fiber transmission systems.

Section VII discusses some systems design considerations. The effects of modulation format, noise, and dispersion on system signal-to-noise ratio (or error rate) and repeater spacing are discussed, and the factors affecting choice of fiber design are reviewed.

Section VIII lists the advantages and disadvantages of fiber transmission systems and gives the general form of potential applications.

The paper is divided into two parts: Part I: The Transmission Medium, contains Section II; Part II: Devices and Systems Considerations, contains Sections III through VIII.

Part I: The Transmission Medium

II. Transmission Properties of Fiber Waveguides

A. Types of Fiber

In Fig. 1 we show the cross sections of important fiber types, and above each cross section we present a typical profile of the index of refraction for that fiber. A few typical dimensions are also shown to provide a physical feel for the scale.

The simplest fiber is the single-dielectric or unclad fiber shown in Fig. 1(a). A discontinuous change in index of refraction occurs at the fiber surface, from the dielectric's value n_1

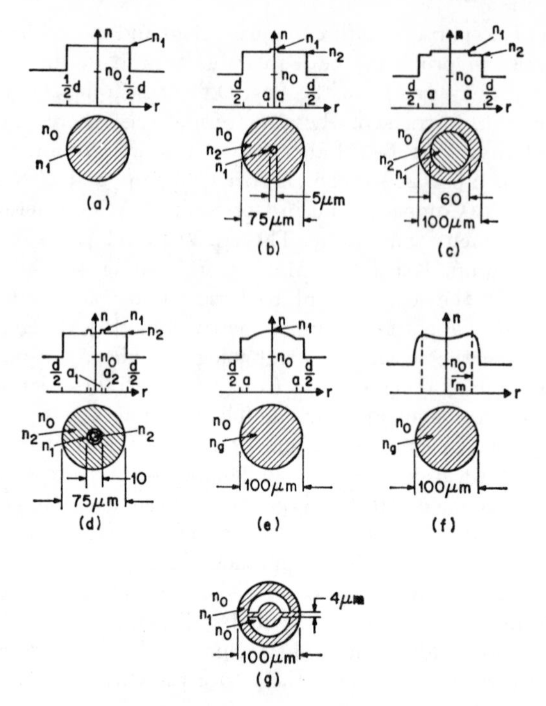

Fig. 1. Cross-sectional drawings for various fiber types, and their associated index of refraction profiles. (a) Unclad fiber. (b) Single-mode clad fiber. (c) Multimode clad fiber. (d) Dielectric-tube fiber. (e) Parabolic-index fiber. (f) Graded-index fiber. (g) Single-material fiber.

to the ambient index n_0. Typical values are $n_1 \simeq 1.5$ for glass and $n_0 = 1$ for an air ambient. Theoretical studies of this type go back to 1910 (Hondros and Debye [7]) and more recently were carried out by Prof. R. E. Beam and his students at Northwestern University around 1950 [8].[1]

The field of the guided wave is appreciable at the fiber surface and some field extends into the surrounding air. Thus any support for the unclad fiber disturbs its transmission properties. The unclad fiber is therefore unsuitable for practical usage. Its importance now is as a test vehicle for evaluating materials and fiber drawing processes. Unclad fibers have been measured in unsupported lengths up to 60 m and individual material losses and loss mechanisms determined [9], [10].

Fig. 1(b) and (c) shows viable practical forms of fiber. In both cases, the core of radius a has an index n_1 and is surrounded by a cladding of slightly lower index n_2, where

$$n_2 = n_1(1 - \Delta) \tag{1}$$

and $\Delta \ll 1$. Under these conditions a single mode of propagation can be supported with $a \simeq 5$ μm [Fig. 1(b)], which is readily achievable in practice. The single-mode fiber is used with single-mode sources—lasers—and provides the ultimate in transmission bandwidth. For many uses it may be desirable to utilize an incoherent source to carry the signal, and for these applications the multimode fiber [Fig. 1(c)] is essential. For both clad fibers, the guided-wave field is mainly confined to the core region of radius a; some field does extend into the cladding, but it decays almost exponentially with radial distance from the core boundary and can be made negligibly small at the outer fiber surface of diameter d. Thus the fiber can be supported and handled without affecting the transmission characteristics. Note that typical fiber diameters are 75–100 μm (3–4 mils), dictated primarily by strength requirements in the single-mode case and by electrical transmission requirements in the multimode case.

Closely related to the multimode fiber of Fig. 1(c) is the liquid-core fiber that was independently conceived by Stone [11] and Ogilvie [12] and later also explored by Gambling and his associates [13]. Instead of using a glass for both the light-guiding core and the cladding, the liquid-core fiber employs a liquid such as tetrachloroethylene in the center with a glass tube as the lower index cladding. The combination is fabricated by first drawing a hollow capillary tube of either fused silica or a more common glass, and then filling the tube with a low-loss liquid. Because the index of a liquid is quite sensitive to temperature, the number of modes could vary appreciably in a field environment. Consequently, it is feasible to use the liquid-core fiber only in multimode form for long-distance transmission. Liquid-core fibers, using active liquids, may also be used as components such as modulators or oscillators.

In all-glass fibers the size of the guided-wave field can be enlarged somewhat, within the single-mode design objective, by using the configuration of Fig. 1(d). A dielectric tube of index n_1 is embedded in a region of index n_2, with (1) describing the relation between n_1 and n_2. A larger single-mode field is advantageous because that eases the problem of fiber splicing.

Fig. 1(e) shows diagrammatically an important fiber type in which the composition of the glass is varied in such a way as to develop an index of refraction which changes parabolically with a maximum on the axis:

$$n = n_1[1 - \Delta(r/a)^2], \qquad 0 \leq r \leq a \tag{2}$$

where r is the distance from the axis and a is defined in Fig. 1(e). The practical value of Δ is on the order of 0.01 to 0.02 and n_1 is in the vicinity of 1.5. A fiber of this form can transmit an optical image (in focus at periodic positions along the length of the fiber) for a few meters [14]. The value of the parabolic-index fiber for communication lies in reduced dispersion as compared with the discontinuous-index multimode fiber of Fig. 1(c). The parabolic-index fiber, called Selfoc by the group from Nippon Electric Company and Nippon Sheet Glass Company, supports many modes with nearly the same group velocity [15]–[19].

The fiber of Fig. 1(f) has an axially symmetric index distribution with a maximum near $r = r_m$. The index variation across the fiber's cross section is again only a percent or less. Although not an optimum design choice, a fiber of this type has exhibited simultaneously very low loss (less than 6 dB/km) and very low dispersion (less than 2 ns/km) and is cited here for that reason [20], [21].

Fig. 1(g) shows a new fiber form that creates a viable and handleable fiber using only one material, such as pure fused silica. The difficulties of fabricating multicomponent glasses lead to the desirability of using just one material. In the structure of Fig. 1(g), the usefully guided modes are carried in the central enlarged member which may be of any convenient shape. The thin supporting membrane provides for exponential decay of the field with increasing radial distance from the central enlargement and the guided field can be made negligibly small at the outer protective sheath with no change in material composition. Either single-mode or multimode guid-

[1] Further contributions to this field can also be found in the following books. N. S. Kapany, *Fiber Optics, Principles and Applications*. New York: Academic Press, 1967; N. S. Kapany and J. J. Burke, *Optical Waveguides*. New York: Academic Press, 1972; D. Marcuse, *Light Transmission Optics*. New York: Van Nostrand-Reinhold, 1972.

TABLE I

Range of v	Additional Modes	Total Number of Propagating Modes[a]
0–2.4048	HE_{11}	2
2.4048–3.8317	TE_{01}, TM_{01}, HE_{21}	6
3.8317–5.1356	HE_{12}, EH_{11}, HE_{31}	12
5.1356–5.5201	EH_{21}, HE_{41}	16
5.5201–6.3802	TE_{02}, TM_{02}, HE_{22}	20
6.3802–7.0156	EH_{31}, HE_{51}	24
7.0156–7.5883	HE_{13}, EH_{12}, HE_{32}	30
7.5883–8.4172	EH_{41}, HE_{61}	34

[a] Accounts for two polarizations of one mode type where applicable.

ance can be provided with a suitable choice of dimensions for the supporting sheet and for the central enlargement [22].

All of these fibers may be coated with a thin film of optically lossy material whose function is to absorb radiation that is scattered from the core into the cladding. This reduces the possibility of crosstalk between adjacent fibers in a multifiber cable.

As we have implied, these fibers are all readily realizable in practice and will be discussed further in the following sections. We are primarily concerned in this paper with the fundamentals of wave propagation in the fibers. The science and technology of materials and fabrication form the subject of other papers [23], [26].

B. Field and Modal Description

We will have frequent need to refer to the clad-fiber type and present here a few expressions that are most useful. To obtain manageable relations we will follow the lead of Snyder [24a] and make use of the practically important condition

$$\Delta \ll 1 \tag{3}$$

where Δ is as defined in (1). Then the transverse field is polarized in essentially one direction and typical components are

$$E_y = H_x \begin{Bmatrix} z_0/n_1 \\ \\ z_0/n_2 \end{Bmatrix} = \begin{Bmatrix} AJ_l\left(u\dfrac{r}{a}\right) \\ \\ BK_l\left(w\dfrac{r}{a}\right) \end{Bmatrix} \cos l\phi \tag{4}$$

with

$$u = a(k^2 n_1^2 - \beta^2)^{1/2} \tag{5}$$

$$w = a(\beta^2 - k^2 n_2^2)^{1/2} \tag{6}$$

$$v = (u^2 + w^2)^{1/2} = ka(n_1^2 - n_2^2)^{1/2} \tag{7}$$

$$k = \frac{2\pi}{\lambda} \tag{8}$$

and with

- β longitudinal propagation constant,
- z_0 impedance of free space,
- λ free-space wavelength,

and where [in (4)] the upper line refers to the core and the lower line refers to the cladding. An exact solution of the characteristic equation leads to three E and three H components for all modes except the circularly symmetric series that are transverse, TE_{0n} and TM_{0n}. The noncircularly symmetric modes have been designated EH_{lm} and HE_{lm}. Table I shows the order of appearance of the modes as the size of the fiber increases relative to wavelength [8], [24b]–[24d]. The various groups (for example, TE_{01}, TM_{01}, and HE_{21}) have different field configurations but very nearly the same propagation constant.

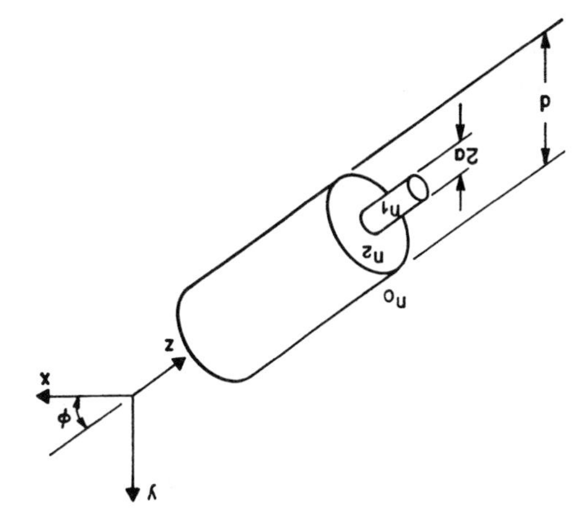

Fig. 2. Sketch of fiber, showing core (of index n_1 and diameter $2a$) and cladding (of index n_2 and outer diameter d), surrounded by a region of index n_0.

Using the small-index-difference approximation, from (3), Snyder [24a] and Gloge have developed a simplified characteristic equation for linearly polarized (LP) modes:[2]

$$\frac{uJ_{l-1}(u)}{J_l(u)} = -w\frac{K_{l-1}(w)}{K_l(w)}. \tag{9}$$

Solutions from this equation are designated as LP_{lm} by Gloge [25]. It is assumed in the foregoing that the cladding diameter d (Fig. 2) is so large that the outer region of index n_0 has no effect on the wave propagation.

Gloge has developed an approximate analytic expression for the propagation constant, with $\Delta \ll 1$:

$$\beta = n_1 k(b\Delta + 1) \tag{10}$$

where

$$b = 1 - (u^2/v^2) \tag{11}$$

is plotted in Fig. 3 and u is obtained from a solution of (9).

The quantity b can be understood as a normalized propagation constant, which has zero value at cutoff and approaches unity as λ approaches zero. Fig. 3 (taken from Gloge's paper [25]) shows $b(v)$ for 18 LP modes.

Another quantity of importance is the group delay for energy propagating in the various modes. Gloge's formulation of this is

$$\tau_{gr} = \frac{L}{c}\left\{\frac{d(n_1 k)}{dk} + n_2\Delta\frac{d(vb)}{dv}\right\} \tag{13}$$

in which L is fiber length and c is velocity of light in vacuum.

The first term of (13) is the dispersion due to the bulk material of which the fiber is made; it is the same for all modes. The second term of (13) is group delay associated with wave guidance; the derivative may be expressed [25] as

$$\frac{d(vb)}{dv} = 1 - \left(\frac{u}{v}\right)^2(1 - 2\kappa) \tag{14}$$

[2] These have been called pseudomodes because they are not the exact solutions [25].

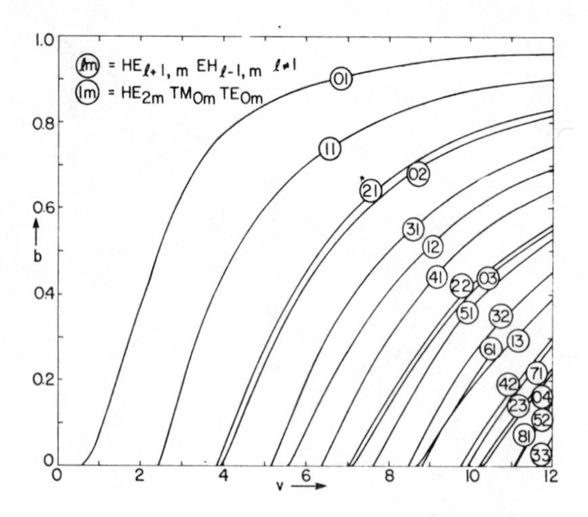

Fig. 3. Normalized propagation parameter b as a function of normalized frequency v (after D. Gloge, [25]).

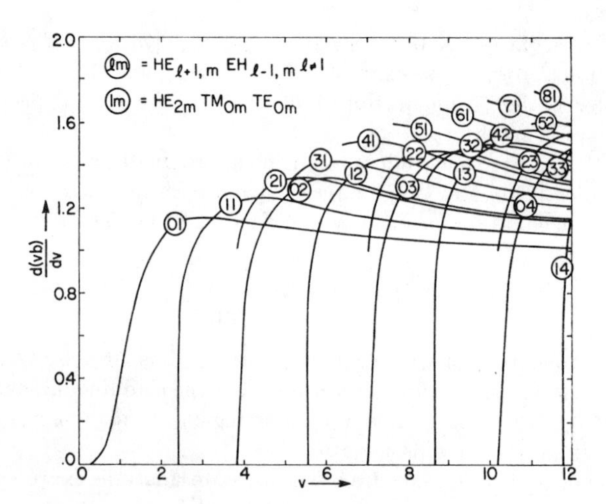

Fig. 4. Normalized group delay as a function of v (after D. Gloge, [25]).

where

$$\kappa = \frac{K_l^2(w)}{K_{l-1}(w)K_{l+1}(w)}. \tag{15}$$

For λ approaching zero, (14) approaches unity and at cutoff (14) approaches $2\kappa(w=0)$. Fig. 4 (from Gloge's paper) shows $d(vb)/dv$ for a series of modes. Note that the LP_{21}, LP_{31}, and other modes have group delays at cutoff of more than n_1L/c rather than n_2L/c. This result, first found by Gloge, shows an important difference between round-fiber theory and two-dimensional slab-waveguide theory, which leads one to expect all modes to have delays characteristic of the cladding near cutoff.

The spread of group delay for the modes of a multimode fiber with $\Delta \ll 1$ and $v \gg 1$ is approximately [25]

$$\Delta\tau = n_1 \frac{\Delta L}{c}\left(1 - \frac{2}{v}\right). \tag{16}$$

The distribution of the power in the guide is important and we record here Gloge's formulation [25]. For total power carried in the mode equal to P, the power in the core P_{core} and the power in the cladding P_{clad} are

$$\frac{P_{\text{core}}}{P} = 1 - \left(\frac{u}{v}\right)^2(1-\kappa) \tag{17}$$

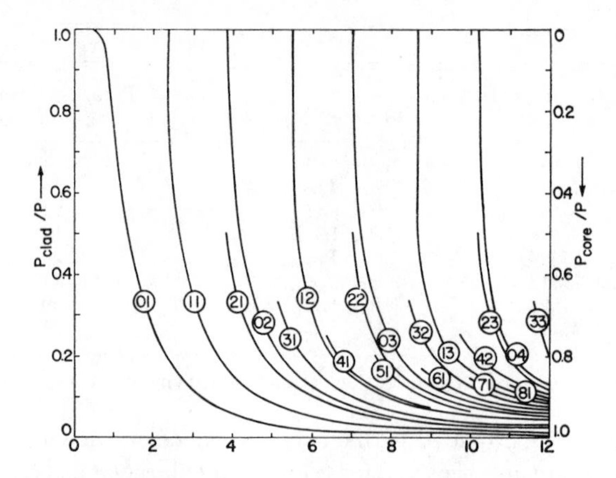

Fig. 5. Portion of the mode power which propagates in the cladding plotted versus v (after D. Gloge, [25]).

$$\frac{P_{\text{clad}}}{P} = \left(\frac{u}{v}\right)^2(1-\kappa). \tag{18}$$

The quantities are plotted in Fig. 5. The power density $\bar{p}(a)$ averaged over ϕ at $r=a$ is [25]

$$\bar{p}(a) = \kappa\left(\frac{u}{v}\right)^2 \frac{P}{\pi a^2} \tag{19}$$

which is plotted in Fig. 5. For the lowest order mode, the maximum power density occurs at $v \simeq 1.8$; for higher order modes, the maximum power density assumes larger values and occurs at larger values of v.

The power density in the cladding is given by

$$\bar{p}(r) \approx \kappa\left(\frac{u}{v}\right)^2 \frac{P}{\pi a r} \exp\left[-2\frac{w(r-a)}{a}\right] \tag{20}$$

for $r \gg a$. For all modes except the lowest azimuthal order, the power in the cladding decreases with distance from the axis even at cutoff [25].

Ray theory is useful in appraising the characteristics of multimode fibers. An incident ray at the end of a fiber that is off-axis by a maximum angle θ will be accepted by the fiber; this is related to the numerical aperture (NA), where

$$\text{NA} = \sin\theta = (n_1^2 - n_2^2)^{1/2} \simeq n_1\sqrt{2\Delta}. \tag{21}$$

Internal to the fiber, this limiting ray makes an angle to the axis $\sqrt{2\Delta}$ which is the approximate critical angle for the interface at the core–cladding boundary. There are N free-space modes accepted by the same number of modes in the fiber, where

$$N = 2\left(\frac{\theta}{\delta}\right)^2 \tag{22}$$

and

$$\delta = \frac{\lambda}{\pi a}. \tag{23}$$

Wave theory yields the same expression:

$$N \simeq \frac{v^2}{2} \simeq (kan_1)^2\Delta. \tag{24}$$

For the parabolic-index fiber [Fig. 1(e)], with an index profile defined by

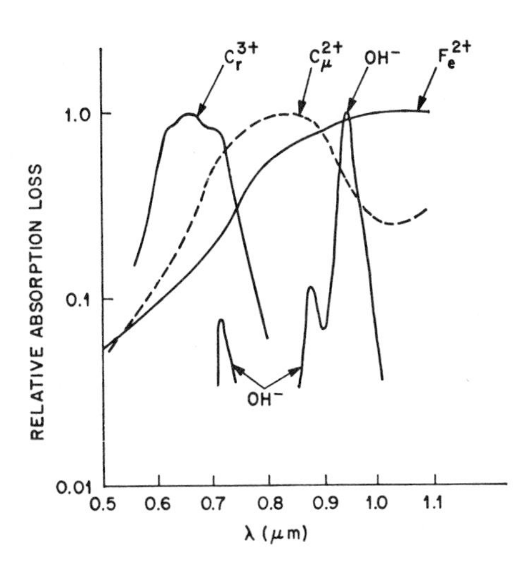

Fig. 6. Relative absorption loss versus wavelength for certain ions in glass.

$$n = n_1\left(1 - \Delta\frac{r^2}{a^2}\right) \qquad (25)$$

Gloge has shown that the number of modes contained within the diameter $2a$ is

$$N_p = (kan_1)^2\frac{\Delta}{2}\cdot \qquad (26)$$

If the parabolic-index fiber is given a Δ twice as large as that for the simple clad fiber, then the two fibers propagate the same number of modes.

C. Loss Mechanisms

There are several mechanisms through which transmission loss may occur in fiber waveguides: 1) material absorption; 2) material scattering, active and/or passive; 3) waveguide scattering; 4) radiation due to curvature; and 5) cladding design effects.

1) Material Absorption: Up to the present the most important of these mechanisms has been material absorption. When serious work was initiated to reduce glass fiber losses, the state of the art on loss was 1000–4000 dB/km and study revealed that almost all of this loss was due to absorption. It appears that a variety of glass compositions would yield very small material absorption if no foreign elements were present, but there are numerous metallic ions which have electronic transitions in the 0.5–1-μm region and which therefore cause absorption bands. Fig. 6 shows these bands for the Cr^{3+}, Cu^{2+}, and Fe^{2+} ions. It is apparent that the absorption bands are of varying width depending on the element. It is also true that the peak absorption wavelength and the width of the band may be somewhat different for a given ion in different glass compositions and Fig. 6 should be regarded as an illustration rather than precise definition. Also, the valence state of a given ion influences the peak absorption wavelength and absorption bandwidth; for example, iron in the plus-three state causes peak absorption below 0.4 μm. Thus the details of the glass-making process have profound effects on the transmission-loss versus wavelength plot.

Another important absorbing ion in glass is OH^-. Fig. 6 also shows the absorption bands due to OH^- in fused silica glass in the 0.5–1.0-μm region; the fundamental absorption peak for OH^- is near 2.7 μm and the peaks at about 0.95 μm and 0.72 μm are the second and third overtones of that vibrational absorption. (It is common for the overtones not to be at exact harmonics of the fundamental.) The process for making fused silica influences whether or not there is much OH^- in the glass.

TABLE II

Ion	Concentration for 1-dB/km Peak Loss in Glass (fractions by weight)
OH^-	1.25 parts in 10^6
Cu^{2+}	2.5 parts in 10^9
Fe^{2+}	1 part in 10^9
Cr^{3+}	1 part in 10^9

The absorptions at the wavelength of peak absorption in Fig. 6 are related to ion concentration approximately as presented in Table II. Other transition-metal ions also cause important losses.

It is clear that high purity is necessary in order to achieve low losses. However, unclad fibers have been made showing total losses of from 2 to 3 dB/km in the 1-μm region [10]. Even lower losses were measured by Pinnow and his associates on bulk crystalline quartz [26]. Further discussion of materials science and technology is to be found in other papers [23], [26].

2) Material Scattering: There is a series of scattering mechanisms that can cause loss, including Rayleigh scattering, Mie scattering, stimulated Raman scattering, and stimulated Brillouin scattering.

Rayleigh scattering will always be present and the coefficient is independent of the light-wave field strength. It may be caused by thermal fluctuations, compositional fluctuations, phase separations, etc., on a scale small compared to the wavelength [23]. A well-made glass will cause Rayleigh scattering loss of about 0.9 dB/km at 1.0 μm; this loss varies as $1/\lambda^4$. The Rayleigh scattered energy appears both in the cladding (where it can be absorbed) and in the core as a backward-scattered guided wave. The latter has been analyzed, with the conclusion that no serious limitation in transmission bandwidth results [27], [28].

Mie scattering is caused by inhomogeneities comparable in size to the wavelength, and causes predominantly forward scatter; it has been observed in fibers [29].

Stimulated Raman and stimulated Brillouin scattering are both nonlinear effects. Very little effect on the transmission will be noted below a threshold power-density level. Beginning at a critical power-density level, power will be shifted in wavelength in the nonlinear interaction between the traveling wave and the material. Since fibers concentrate the power into a small cross section, the large fields required to make these effects significant can be observed at modest absolute power levels; long interaction lengths can also be achieved. These effects can be useful in scientific studies and in device design [30]. On the other hand, for long-distance transmission these nonlinear effects constitute an upper limit on the power level that can be transmitted [31], [32].

For Raman scattering in the forward direction in vitreous silica at $\lambda = 1$ μm, the maximum usable light-wave power is related to fiber transmission loss by [31], [32a]

$$P_{\max} \simeq 4 \times 10^{-2}\alpha w^2 \quad \text{W} \qquad (27)$$

where α is transmission loss of the fiber, in decibels per kilometer, and w is full width at half-maximum power density for the guided light wave in the fiber, expressed in microns. This formula is based on data taken at 0.53 μm by Stolen and Ippen [32a] and extrapolated to 1 μm. The field width w is used instead of the core size since in the single-mode fibers, where nonlinear effects are more important, the field is not confined to the core. Equation (27) is valid for either single-frequency or multifrequency sources, provided the spectral width is equal to or less than 0.01 μm. With light-emitting diode (LED) sources (spectral width$\simeq$0.03 μm), P_{max} will be from two to four times greater than given by (27).

As an example, for a guided field diameter w of 3 μm and a total fiber (passive) loss of 4 dB/km, $P_{max} \simeq 1.5$ W at 1-μm wavelength. At other wavelengths, P_{max} can be estimated with the relation $P_{max} \sim \lambda$.

Stimulated Brillouin scattering is predominantly in the backward direction. The critical power is different depending on whether the source is single-frequency or multifrequency. For the single-frequency source,

$$P_{max} \simeq 8 \times 10^{-5} \alpha w^2 \qquad \text{W} \tag{28}$$

which is taken from Smith [31] and from Ippen and Stolen [32b], assuming that the stimulated gain coefficient does not vary with wavelength. For example, with $\alpha = 4$ dB/km and $w = 3$ μm, P_{max} from (28) is about 3 mW.

For the multifrequency source of bandwidth equal to $\Delta\nu$ (MHz) or $\Delta\lambda$ (Å) (either a laser or a LED)

$$P_{max} \simeq 3 \times 10^{-6} \alpha w^2 \Delta\nu \qquad \text{W} \tag{29}$$

or

$$P_{max} = 8 \times 10^{-2} \alpha w^2 \Delta\lambda \qquad \text{W.} \tag{30}$$

For example, with $\alpha = 4$ dB/km, $w = 3$ μm, and $\Delta\lambda = 1$ Å, P_{max} from (30) is about 3 W.

All of the foregoing is for long fibers, of such length L that $\alpha L \gg 1$. Also, the example, $w = 3$ μm, is selected to exaggerate the effects. A single-mode fiber can be designed with a w of at least 6–12 μm, which leads to a factor of 4–16 larger permitted signal powers. In multimode fibers designed for LED sources, the field width is typically on the order of 50 μm; although the higher order modes have somewhat larger field strength in the fiber than does the lowest order wave assumed previously, the nonlinear effects in multimode fibers are vastly smaller than in the examples cited.

3) Waveguide Scattering: Geometrical variations in the size of the fiber core as a function of z will cause transfer of power between guided modes and/or from a guided mode to the radiation field. This is a form of scattering with no ***immediate*** absorptive effects, although absorption changes may later occur by virtue of differing absorption coefficients for radiation fields and for the various propagating modes.

Consider first two guided modes having propagation constants β_p and β_q. It is known that any mechanism which causes periodic coupling between these waves and which has the form

$$C(z) = A \sin \theta z \tag{31}$$

will result in a complete transfer of power from one mode to the other, provided [33]

$$\theta = \beta_p - \beta_q. \tag{32}$$

In fibers, a more typical mode coupling would be a spectrum

$$C(z) = \sum_n A_n \sin \theta_n z. \tag{33}$$

The thickness of the core, for example, could have a spectrum of variations (versus z) according to (33). For any pair of modes having propagating constants β_p and β_q there will be a coupling parameter $\theta_{pq} = (\beta_p - \beta_q)$ to which that mode pair will be extremely sensitive.

Marcuse has studied these effects in two-dimensional slab waveguides and in round-clad fibers [34]–[46]. Fortunately, for small (perturbation) coupling magnitudes, the round-fiber theory and slab-guide theory give comparable results. (A word of caution, however; for large steps, the round-guide radiation loss is much larger than slab-guide radiation loss [35].) When the driven wave has a phase constant β_p and

$$\beta_p - \beta_2 < \theta < \beta_p + \beta_2 \tag{34}$$

where $\beta_2 = 2\pi n_2/\lambda$ is the cladding phase constant, then power is radiated into the cladding, assumed for the moment to be infinitely thick. In the following section we will describe how the finite cladding thickness is accommodated.

Marcuse has calculated the power distribution among guided modes and the radiation losses in multimode dielectric slab waveguides [37]. With the coupling written

$$C_{\nu\mu} = K_{\nu\mu} f(z) \tag{35}$$

and with the assumption that the correlation function of $f(z)$ is Gaussian

$$\langle f(z) f(z-u) \rangle = \bar{\sigma}^2 e^{-(u^2/L_c^2)} \tag{36}$$

his work yields a scattering power-loss coefficient for the νth mode:

$$\alpha_\nu = \frac{(\pi)^{7/2}}{4} \frac{(n_1^2 - n_2^2)}{n_1} \frac{(\nu+1)^2}{\left(1 + \frac{1}{\gamma_\nu t}\right)} \frac{\bar{\sigma}^2 L_c}{\lambda t^3} \tag{37}$$

where

- t slab-core thickness,
- γ_ν $= (\beta_\nu^2 - n_2 k^2)^{1/2}$,
- β_ν propagation constant of mode ν,
- n_1, n_2 index of core and cladding, respectively,
- L_c correlation length,
- σ $= (n_1^2 k^2 - \beta^2)^{1/2}$,
- β propagation constant in the z direction for the radiation modes,

and subject to the restriction

$$\frac{L_c}{\lambda} \ll 1. \tag{38}$$

Marcuse presents other expressions for L_c large [37]. In the absence of loss, this short-correlation-length coupling produces a steady-state distribution of power primarily in the lowest order mode. Coupling with very long correlation length yields a steady-state distribution of power in all modes equally. Dissipation loss that is not equal for all modes will prevent these distributions from occurring.

We conclude with a numerical example. Assuming a systematic deviation of the core wall of 0.1 percent of the core width, total exchange of energy between the lowest order mode and the next mode can occur in a fiber length of approximately 5 cm. An rms deviation of one of the waveguide walls of 9 Å will cause radiation loss of 10 dB/km (core–cladding index difference 1 percent, guide width 5 μm), pro-

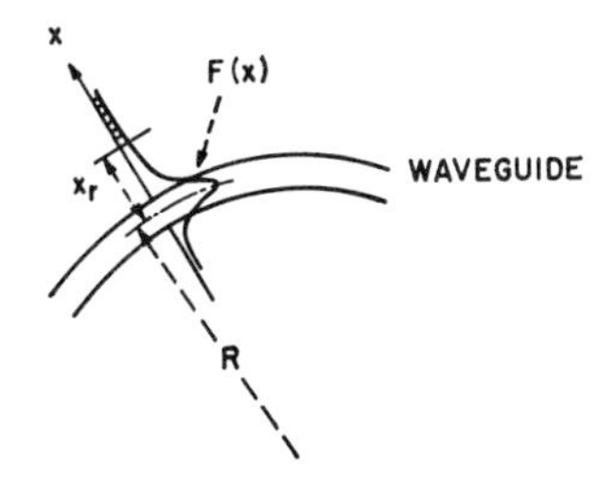

Fig. 7. Sketch of curved dielectric waveguide to illustrate radiation effects (after E. A. J. Marcatili and S. E. Miller, [47]).

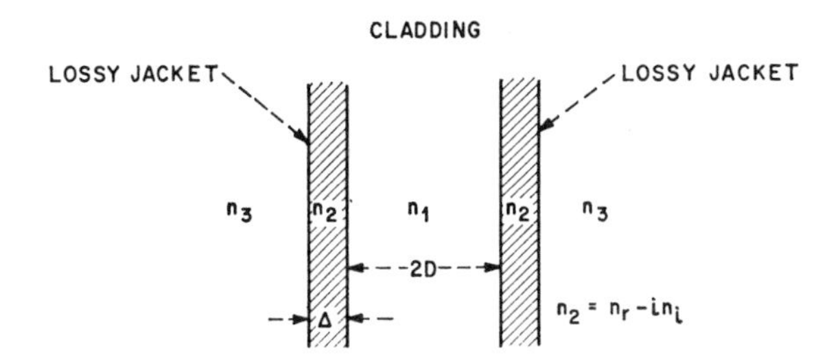

Fig. 8. Sketch of slab dielectric waveguide for use in analysis of cladding effects (after D. Marcuse, [45]).

vided the correlation length is near its worst possible value, on the order of the guide width.

Scattering effects of these magnitudes do not occur in well-made drawn glass fibers. (These effects are observed, however, in other dielectric waveguides, such as those made for integrated optics.) Avoidance of waveguide scattering losses is evidently achieved by 1) maintaining a uniform-fiber cross section, aided by surface tension in the flowing glass as it is drawn; and 2) having residual core size changes occur with very long periods, $\theta \ll (\beta_p - \beta_2)$.

Mixing of power among the guided modes can have important effects on multimode dispersion, as will be described.

4) Radiation Due to Curvature: Any dielectric waveguide will radiate if it is not absolutely straight. Why this is so can be seen with reference to Fig. 7. For gradual bends the transverse field in the radial (x-direction) planes will differ little from the normal-mode field for the straight guide. Since the field in the cladding (unshaded region of Fig. 7) extends to x indefinitely large, the introduction of a radius of curvature R implies energy propagating at greater than the velocity of light. One can use this concept to calculate the radiation loss associated with losing the energy at $x > x_r$ (Fig. 7), where x_r is the value of x where the velocity of light is reached along the propagation path [47], [48]. The radiation attenuation coefficient has the form

$$\alpha_r = c_1 \exp(-c_2 R) \tag{39}$$

in which c_1 and c_2 are independent of R. The important thing is the exponential dependence of α_r on bend radius R. This means in practice that α_r can go from a negligibly small value to a prohibitively large value within a range of about 2:1 in R. We illustrate this with calculations from a slab-waveguide model [47], [48]. Assume a slab width of 1.18 μm, an index difference $\Delta = 0.001$, and $\lambda = 0.63$ μm; this is a severe case for curvature since the field extends well into the cladding. (This is a single-mode guide; the second slab mode occurs at a slab width of 4×1.18 μm.) Then the value of $x_r \simeq 16$ μm, c_1 is of order 10^4, c_2 is of order 100, and α_r is of order 8.68 dB/m at $R = 0.18$ m. Doubling R decrease α_r by the factor $e^{200} \simeq 6.5 \times 10^7$, which leaves α_r negligible.

Typical values of Δ are larger than 0.001 leading to smaller curvature losses. Experimental values of fiber bend losses are in agreement with the preceding theory [6]. Tolerable bend radii of order 1 cm can be achieved.

The number of modes guided by a multimode fiber is reduced by the bending of its axis. As a matter of fact, for step-index fibers and for parabolic-index fibers, Gloge has shown [49] that the number of guided modes is

$$N(R) \simeq N\left(1 - \frac{a}{R\Delta}\right)$$

and

$$N_p(R) = N_p\left(1 - \frac{2a}{R\Delta}\right)$$

respectively. The explicit values of the number of modes N and N_p of the straight guides are given in (24) and (26). Both curved guides support the same number of modes, that is, $N(R) = N_p(R)$, if the parabolic-index fiber is given a Δ twice as large as that of the step-index fiber.

5) Cladding Design Effects: It is inevitable that some signal power will be scattered into the cladding; a significant part of the Rayleigh scattering is captured by the cladding and some waveguide scattering may occur. The cladding with air or other ambient as the surrounding medium acts as a second dielectric waveguide. Reconversion of power from cladding modes into core-guided modes would have undesirable effects on the group-delay characteristic for the signal mode. This is an important reason for providing *loss* in the cladding which is large for cladding modes but negligible for core-guided modes. This may be done by having a lossy jacket on the outside of the cladding. However, a lossy jacket will also introduce some loss for the core-guided modes. Moreover, if no lossy jacket is used and if the index of the medium outside the cladding (n_0 of Fig. 1) is larger than n_2, then there will also be loss for the core-guided modes. These cladding-design effects are interrelated with the topic of crosstalk between parallel fibers in a multifiber cable. Consequently both will be considered together in the next section.

D. Cladding Loss and Fiber Crosstalk

We consider here the factors that influence the cladding dimensions and lossy jacket as related to core-guided mode losses and to crosstalk between parallel fibers of a multifiber cable.

First consider the losses that can be achieved for the cladding-guided modes by the addition of a lossy jacket. Fig. 8 is a sketch of the slab-guide model used by Marcuse [45]. The indexes of the cladding, lossy jacket, and surrounding region are n_1, n_2, and n_3; the cladding diameter is $2D$; and the lossy-jacket thickness is Δ.[3] The lowest order even mode has the least loss and is used for study. With a thin jacket, interference effects are found on the cladding-mode loss, as illustrated in Fig. 9; this is plotted for $kD = 50$, $n_1 = 1.6$, $n_r = 1.65$ (real part of n_2), and $n_3 = 1.0$. The amplitude attenuation coefficient of the bulk jacket material is α_j; Fig. 9 is plotted for $2\alpha_j D = 1$ dB. We note that for the abscissa above $\Delta/D = 0.1$, $2\alpha D$ is larger than about 10^{-3}; using $D = 35$ μm, this amounts to a cladding-mode loss α of 30 dB/m with a lossy-jacket film loss $2\alpha_j\Delta$ of

[3] In this paper we have often used symbols to conform to the original publication. This results in some duplication of symbols.

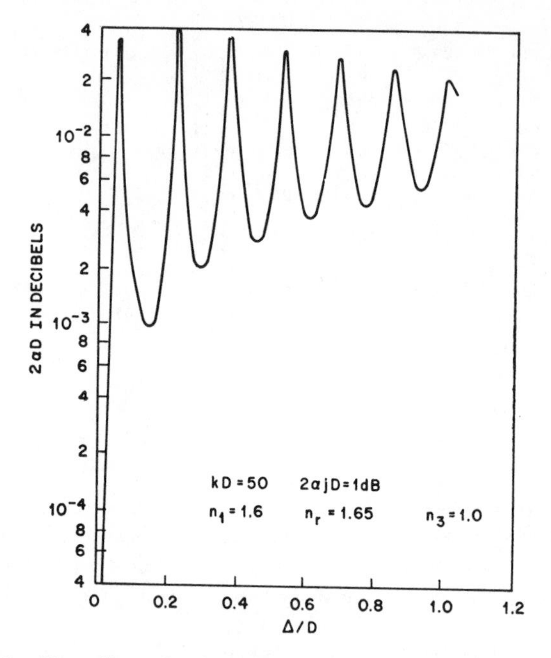

Fig. 9. Normalized cladding-guided-mode loss α versus normalized jacket thickness Δ. $2D$ is the width of the cladding (after D. Marcuse, [45])

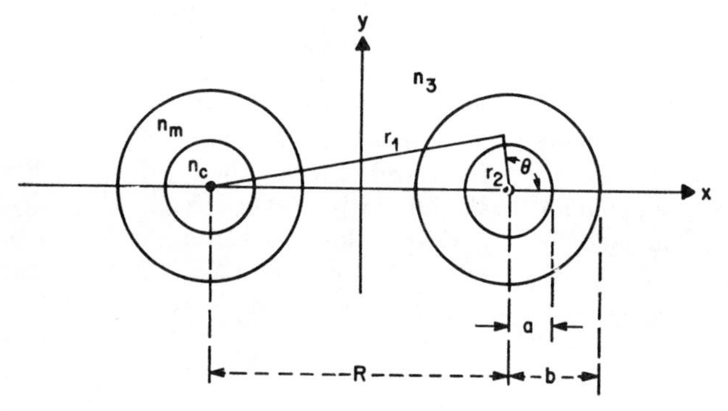

Fig. 10. Cross section of two coupled clad fibers (after D. Marcuse, [50]).

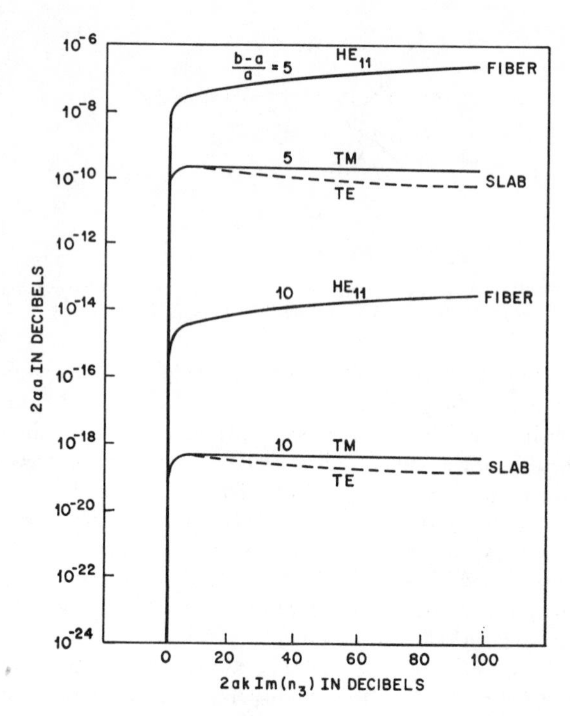

Fig. 11. Additional mode loss caused by the lossy surrounding medium. The abscissa indicates the loss (in decibels) that a plane wave would suffer by traveling a distance equal to the core radius in the surrounding medium. $n_m ka=16$, $n_c/n_m=1.01$ (Fig. 10) (after D. Marcuse, [50]).

0.1 dB in $\Delta=3.5$ μm. Larger cladding-mode losses should be readily achievable if desired. Preferentially, the real part of n_2 should approximate n_1, which can be done perhaps by doping the outer layer of the cladding with an absorbing ion.

Some undesired loss for the core-guided modes will be introduced by the lossy jacket. This loss and the capabilities of the lossy jacket for reducing crosstalk between parallel fibers have been analyzed by Marcuse in another paper [50]. Fig. 10 shows Marcuse's model and definition of symbols; the lossy jacket occupies the gap between the two fibers and has the thickness $(R-2b)$ along the x axis. Figs. 11 and 12 illustrate the results for identical single-mode guides. Fig. 11 shows the normalized loss to the core-guided mode (ordinate) versus normalized lossy-jacket loss (abscissa), the latter measured as the loss that a plane wave would suffer by traveling a distance equal to the core radius in the lossy-jacket type of material (abscissa $=8.68 \text{ Im}(n_3)ka$). Results are given for both the round fiber and the slab guide for cladding thicknesses of five and ten times the core half-width or radius. We note that a cladding thickness of five times the core radius gives $2\alpha a \simeq 8\times 10^{-7}$ for abscissa ~ 1 dB. For $a=2$ μm this yields $\alpha \simeq 0.2$ dB/m or 200 dB/km for the added HE_{11} mode loss due to the lossy jacket. Thus the cladding thickness of five times the core radius is not sufficiently large; however, a cladding thickness

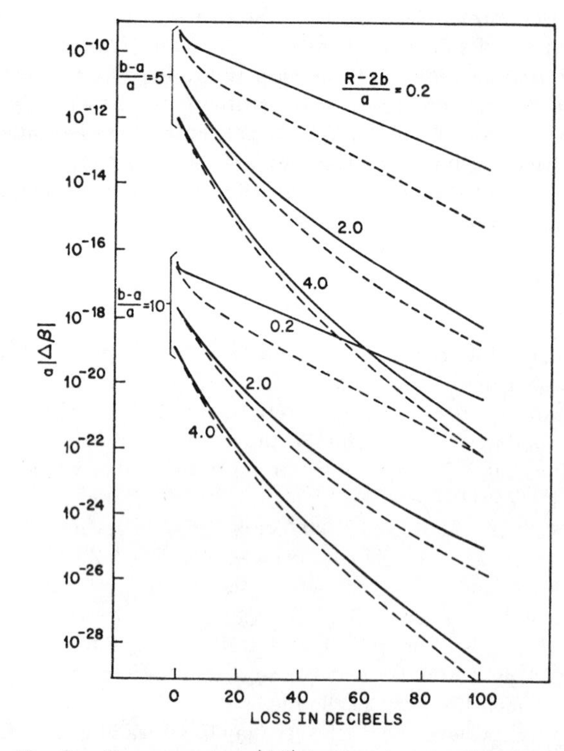

Fig. 12. Coupling parameter $a|\Delta\beta|$ (ordinate) versus loss that a plane wave would experience in going from the boundary of one fiber to the boundary of the other fiber (Fig. 10). The solid lines hold for horizontal polarization while the dotted lines apply to the case of vertical polarization. $n_m ka=16$, $\gamma a=1.594$, $n_c/n_m=1.01$, $\text{Re}(n_3)=n_m$ (after D. Marcuse, [50]).

of ten times the core radius yields an HE_{11} mode loss smaller by 10^6 as shown in Fig. 11.

From Fig. 12 we can infer the crosstalk reduction obtainable from the lossy jacket; the abscissa is the same quantity as in Fig. 11, as already described. The ordinate $a|\Delta\beta|$ can be

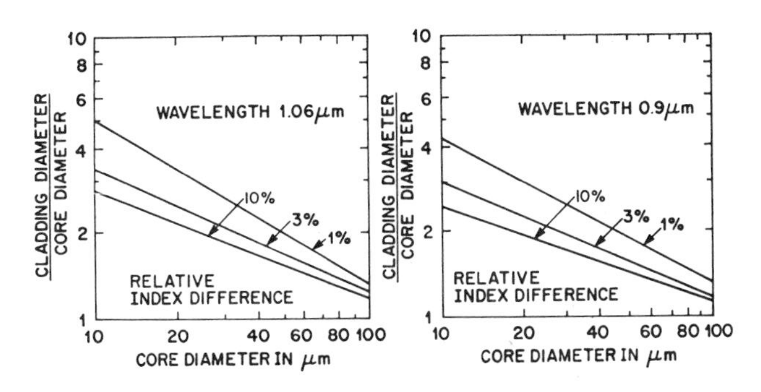

Fig. 13. Diameter ratio required for 90 percent of the modes to have less than 1-dB/km extra loss, due to absorption or radiation, at the cladding surface (after D. Gloge, [51]).

interpreted into crosstalk ratio V, using the relation [50]

$$L = \frac{V^{1/2}}{|\Delta\beta|} \tag{40}$$

in which $\Delta\beta$ is the change of propagation constant due to the crosstalk coupling between two otherwise degenerate modes [50], and L is the distance at which unit power incident in one fiber yields a power of V in the other fiber. Fig. 12 was calculated for $n_m ka = 16$ (note that n_m of Fig. 10 is our n_2) and $n_c/n_m = 1.01$ (core-cladding index ratio). To illustrate the use of Fig. 12, let us assume that a crosstalk ratio of 60 dB is tolerable; hence $V = 10^{-6}$. Then for $2a = 4$-μm core diameter and $L = 1$ km we find from (40) that $|\Delta\beta|$ may not exceed 10^{-9}/m and $a|\Delta\beta|$ may not exceed 2×10^{-15}. We see from Fig. 12 that to achieve this with a cladding thickness of five times the core radius requires at least 15 dB of abscissa loss. However, by doubling the cladding thickness we drop the crosstalk an additional 70 dB with no lossy-jacket loss.

In summary, it has been shown that it is easy to obtain large losses for cladding-guided modes, thus 1) assuring that appreciable dispersion will not be caused by conversion and reconversion between core-guided and cladding-guided modes, and 2) minimizing crosstalk due to cladding-guided modes. Moreover, it is easy to prevent appreciable core-guided mode loss due to the lossy jacket. However, the lossy jacket is not very effective in reducing direct crosstalk between core-guided modes; a cladding of appropriate thickness is a better approach.

We have discussed the choice of cladding size for the single-mode guide. For multimode guides the work of Gloge [51] gives us a quantitative indication for the choice of the ratio of cladding diameter to core diameter, recorded in Fig. 13. The ordinate is cladding/core diameter ratio and the abscissa is core diameter, with core-cladding index difference as the parameter. Plots for a $\lambda = 1.06$ and $\lambda = 0.9$ μm are given. The criterion is setting the cladding diameter such that 90 percent of the core-guided modes have less than 1-dB/km extra loss due to absorption or radiation from the cladding surface. For a cladding diameter of 100 μm and an index difference $\Delta = 0.01$, we find from Fig. 13 a cladding/core ratio of 1.33 at $\lambda = 0.9$ μm; the cladding thickness thus is 12 μm and the core diameter is 75 μm.

Finally, we note a cladding loss that may result from differences between the intrinsic material-loss properties of the cladding and the core. In conventional clad fibers [Fig. 1(b)], the core and the cladding must have different indexes and therefore must differ in composition. The result may be two different intrinsic attenuation coefficients α_{core} and α_{clad}. Gloge has shown that the clad-fiber loss is given by

$$\alpha_{\text{fiber}} = \alpha_{\text{core}} + \frac{\alpha_{\text{clad}} - \alpha_{\text{core}}}{2ak(\text{NA})} \tag{41}$$

where the NA is given by (21). Because most of the guided energy is confined to the core, the fiber loss is relatively insensitive to α_{clad}. To cite an example for NA $= 0.1$ and $a = 50\lambda$, an excess loss coefficient $(\alpha_{\text{clad}} - \alpha_{\text{core}}) = 100$ dB/km contributes only 1.6 dB/km to the fiber loss. This cladding loss affects higher order modes more than low-order modes, with the result that the power distribution in the modes varies with length. We will discuss this further in the section under dispersion.

E. Group Delay in Lossless Uniform Fibers

In this section we consider the group delay associated with various fiber types, assuming the structure is uniform along its length, yielding no mode conversion, and assuming negligible losses—or equivalently, the same loss for all modes.

Group-delay distortion sets a limit on the information rate that may be used in a fiber transmission system. Contributors to group-delay distortion are the following:

1) material dispersion—the variation of index of refraction of the glass as a function of wavelength;
2) waveguide-delay distortion—the delay versus wavelength effect in each propagating mode;
3) multimode group-delay spread—the variation in group delay among the propagating modes at a single frequency.

All three effects must be considered. With broad-band sources (LED's), material dispersion and multimode delay distortion are limiting. With narrow-band laser sources and a single-mode fiber,[4] material dispersion tends to control, but the possible use of a several-mode fiber to ease splicing problems makes it important to evaluate where the waveguide effects begin to contribute.

We will present the rms impulse response associated with each of the three delay-distortion contributors and will compare several fiber types.

Most work up to now has been done using round-clad fibers [Fig. 1(b) or (c)] with small index difference. Gloge has provided a comprehensive analysis of group delay in such structures. Using the approximation (not always exactly true) that the material dispersion characteristics of core and cladding are very similar, Gloge derives the following expression for the specific group delay (seconds per meter):

$$\tau = \frac{1}{c}\left\{N_1 + (N_1 - N_2)\frac{m}{M}\right\} \tag{42}$$

where $N_1 = d(kn_1)/dk$ and $N_2 = d(kn_2)/dk$ are called *group indexes* by Gloge; they differ only a small amount from the indexes n_1 and n_2 [52]. The symbol m is a designation of mode-group number, increasing as v of (7) increases, and M is the maximum value that m can assume. Equation (42) shows that the specific group delay for the lowest order modes is nearly N_1/c, the specific group delay in an infinite medium, and for the highest order modes increases by $(N_1 - N_2)/c$ which is approximately $n_1\Delta/c$. For $\Delta = 0.01$ and $n_1 = 1.458$ (fused silica), this multimode delay distortion is about 48.6 ns/km.

[4] Waveguide dispersion in a single-mode fiber and its compensation by material dispersion have been studied by Kapron and Keck [51].

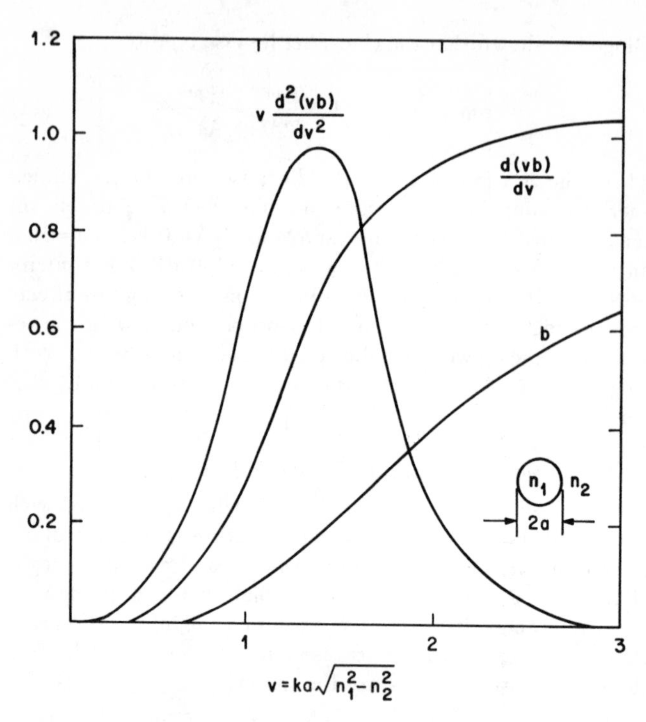

Fig. 14. Waveguide parameter b, (11), and its derivatives $d(vb)/dv$ and $vd^2(vb)/dv^2$ as a function of the normalized frequency v (after D. Gloge, [52]).

For the lowest order wave HE_{11} Gloge shows that

$$\tau = \frac{1}{c}\left\{N_1 + (N_1 - N_2)\frac{d(vb)}{dv}\right\} \tag{43}$$

where v and b are as given by (7) and (11), respectively. Fig. 14 shows the functions $d(vb)/dv$ and $vd^2(vb)/dv^2$ for HE_{11}.

The frequency dependence of group delay is given by $d\tau/df$ and the specific time difference (seconds per meter) between output power components separated by δf cycles is

$$s = \delta f \frac{d\tau}{df} \simeq \frac{\delta f}{f_0}\,\frac{1}{c}\left\{\frac{kdN_1}{dk} + (N_1 - N_2)\frac{vd^2(vb)}{dv^2}\right\} \tag{44}$$

in which f_0 is the center frequency about which δf is measured and at which the derivatives are evaluated. The first term of (44) represents material dispersion, and the second term represents the waveguide effect. For the HE_{11} mode in the practical part of the single-mode region ($v = 2.0$ to 2.4), $vd^2(vb)/dv^2$ runs from 0.2 to 0.1. For glasses the value of kdN_1/dk (or $kd^2(kn_1)/dk^2$) is on the order of 0.05 or more [52]. Hence for $(N_1 - N_2) \simeq \Delta n_1$ on the order of 0.015, material dispersion dominates the output pulse spreading due to wavelength dependence in a single-mode guide. More quantitative relations will be given in a subsequent paragraph.

For the higher order modes, (44) can be expressed as

$$s \simeq \frac{\delta f}{f_0}\,\frac{1}{c}\left\{\frac{kdN_1}{dk} - 2(N_1 - N_2)\frac{m}{M}\right\}. \tag{45}$$

The second term of (45) gives the multimode variation in frequency dependence of output. This term can be significant, as will be stated quantitatively later.

We next provide quantitative expressions for the delay distortion effects in fibers having a parabolic index distribution [Fig. 1(e)], and for other transverse index distributions that may appear in practical fibers [17], [19], [53]–[55]. It has recently been found possible to produce low-loss fibers having very much reduced delay distortion using graded indexes [Fig. 1(f)], so we are addressing an important practical problem [20], [21].

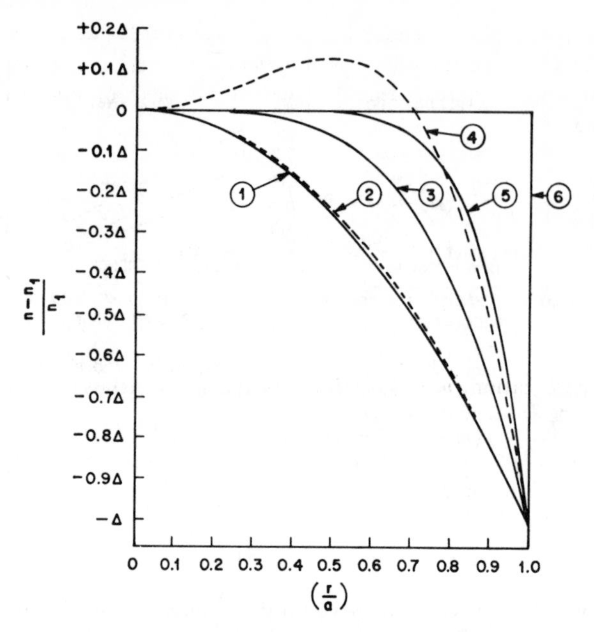

Fig. 15. Index of refraction versus transverse position (r/a) for several graded-index fibers. ①, pure parabolic $u=2$ in (46); ②, perturbed $\delta=0.05$ in (66); ③, pure fourth order; ④, fourth order perturbed by a second-order term; $\delta \sim 2$ in (66); ⑤, pure eighth order; ⑥, step-index, $u=\infty$ in (46).

Fig. 15 shows a series of fiber index distributions for which there are analytical results available. The transverse parameter is r/a, where r is the radial distance from the axis and a is the radius of the core for a round fiber or the half-width of the core for a square-core fiber. The index varies from a value n_1 at the core axis to a value $n_1(1-\Delta)$ at $r=a$, and remains constant at $n_1(1-\Delta)$ for $r>a$. The model is realistic, in that a small amount of loss in the region $r>a$ would render the wave propagation insensitive to the radial extent of the region at $r>a$. Consider the family of index distributions

$$n(r) = n_1\left[1 - \Delta\left(\frac{r}{a}\right)^u\right]. \tag{46}$$

The total number of propagating modes is

$$M_T = \mathcal{C}_m(n_1ka)^2\Delta \tag{47}$$

where $\mathcal{C}_m$ for the round fiber is [19]

$$\mathcal{C}_m^{\circ} = \frac{u}{u+2} \tag{48}$$

and $\mathcal{C}_m$ for the square fiber is [53]

$$\mathcal{C}_m^{\square} = \frac{4}{\pi}\,0.6^{1/u}. \tag{49}$$

For the step-index case, $u=\infty$ and the ratio of the number of modes in the square fiber to the number of modes in the round fiber is

$$\frac{\mathcal{C}_m^{\square}}{\mathcal{C}_m^{\circ}} = \frac{4}{\pi}, \qquad \text{for } u = \infty \tag{50}$$

as one might expect from the relative cross-sectional areas. Thus the two theories, based on grossly different approaches, do compare quite reasonably and we will draw on both in this discussion.

The specific group delay for each mode of the *square* fiber is

$$\tau = \frac{n_1}{c}\left\{1 + \frac{(u-2)}{(u+2)}\Delta\left[\frac{(m+1)}{(m+1)_{\max}}\right]^{2u/u+2}\right\} \quad \text{s/m} \tag{51}$$

in which a simplification has been introduced which gives a slight error for the lowest order modes. The ratio $(m+1)/(m+1)_{\max}$ is the equivalent of $\sqrt{m/M}$ in (42). The mode identification $(m+1)$ is related to the fields within the fiber by

$$(m+1)^2 = (m_x+1)^2 + (m_y+1)^2 \tag{52}$$

and there are $(m_{x,y}+1)$ field extrema in the x or y directions, respectively. The highest propagating mode group is $(m+1)_{\max}$ for which

$$(m+1)_{\max} = \left(\frac{2}{\pi}\mathcal{C}_m{}^{\square}\right)^{1/2} n_1 k a \Delta^{1/2}. \tag{53}$$

When the square fiber is excited by an impulse in all modes at the input, the output is spread over the time interval $\tau n_1 t/c$, where

$$0 < t \leq \frac{n_1}{c}\left(\frac{u-2}{u+2}\right)\Delta. \tag{54}$$

The round-fiber output is spread over an identical interval.

The impulse response for equal input power in all modes is

$$P(t) = \frac{\pi}{2A_u{}^{(u+2)/u}}\left(\frac{u+2}{2u}\right)t^{2/u} \tag{55}$$

where

$$A_u{}^{(u+2)/u} = \left[\frac{n_1}{c}\left(\frac{u-2}{u+2}\right)\right]^{(u+2)/u}\frac{1}{(m+1)_{\max}{}^2} \tag{56}$$

$$0 < t < t_{\max} \tag{57}$$

$$t_{\max} = \frac{n_1}{c}\left(\frac{u-2}{u+2}\right)\Delta \tag{58}$$

for the square fiber, and the round fiber has the identical time dependence $t^{2/u}$. The shape of the impulse response is plotted in Fig. 16 for several index distributions of interest.

A quantity that limits the information rate of transmission[5] over the fiber is the rms value of the width of the impulse response σ, where σ is defined by

$$\sigma^2 = \frac{1}{A}\int_{-\infty}^{\infty} P(t)t^2 dt - T^2 \tag{59}$$

$$T = \frac{1}{A}\int_{-\infty}^{\infty} P(t)t\,dt \tag{60}$$

$$A = \int_{-\infty}^{\infty} P(t)dt. \tag{61}$$

For the index distribution given by (46), σ is

[5] For a more complete discussion, see the text in connection with Figs. 65 and 66.

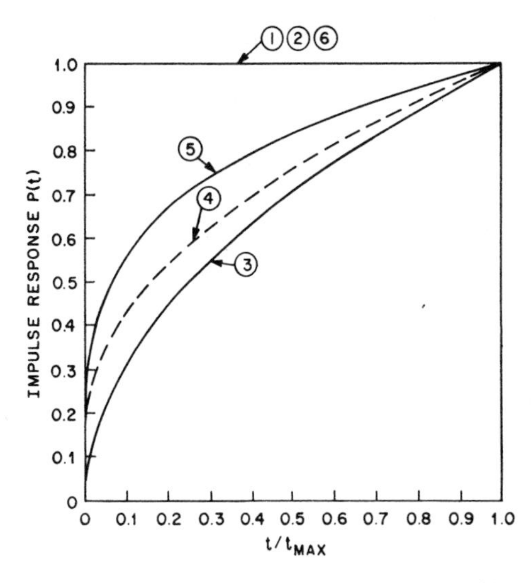

Fig. 16. Impulse response versus normalized time for fibers with index distributions given in Fig. 15.

TABLE III

NORMALIZED ROOT-MEAN-SQUARE WIDTH OF IMPULSE RESPONSE FOR SEVERAL INDEX DISTRIBUTIONS

Index Distribution	$\frac{\sigma c}{n_1 \Delta}$
Step-index ($u = \infty$)	$1/\sqrt{12} = 0.289$
Pure eighth order ($u = 8$)	0.165
Pure fourth order ($u = 4$)	0.0873
Pure second order ($u = 2$)	0.15Δ
"Ideal" near-parabolic (round fiber)	0.037Δ
Practical (5% error) parabolic	0.00591

$$\sigma = \frac{n_1}{c}\Delta(u-2)\left[\frac{1}{3u+2} - \frac{u+2}{(2u+2)^2}\right] \quad \text{s/m}. \tag{62}$$

Values of σ are listed in Table III for step-index, pure eighth-order, and pure fourth-order index distributions. Very significant narrowing of the impulse response occurs with just a rough rounding of the index curve.

Both (54) and (58) indicate no modal dispersion when $u = 2$, the parabolic profile. This is not quite true, and more precise analysis indicates a delay spread in the impulse response of

$$t_{\max} = \frac{n_1}{c}\frac{\Delta^2}{2} \tag{63}$$

for a round fiber [55]; however, for an "ideal" near-parabolic profile in a round fiber, $|t_{\max}|$ is a factor of four smaller [19]. The delay spread in the impulse response for the square fiber with an approximate solution for the near-parabolic profile is [53]

$$t_{\max} = \frac{n_1}{c}0.26\Delta^2. \tag{64}$$

The rms width of the impulse response for the "ideal" round fiber is

$$\sigma_i = 0.037\frac{n_1}{c}\Delta^2. \tag{65}$$

This is remarkably smaller than for the other distributions of index already discussed since σ_i varies as Δ^2 and the others vary as Δ; as the reader will recall, Δ is a small number, typically of order 10^{-2}.

The variation of impulse response as Δ^2 in (65) is a form of singularity in behavior. Consider what happens if the index distribution is

$$n(r) = n_1\left[1 - (1-\delta)\Delta\left(\frac{r}{a}\right)^2 - \delta\Delta\left(\frac{r}{a}\right)^4\right]. \quad (66)$$

Then for the square fiber it has been shown [53] that the spread in output for impulse excitation is, when $\delta \ll 1$,

$$t_{\max} = 0.389\frac{n_1}{c}\frac{\delta}{(1-\delta)}\Delta \quad (67)$$

and the rms width of the impulse response is [53]

$$\sigma_p = 0.1123\frac{n_1}{c}\frac{\delta}{(1-\delta)}\Delta. \quad (68)$$

We now see that for a fixed percentage error in the fabrication of the index profile, i.e., for a fixed δ, the σ varies as Δ. For $\delta = 0.05$, a small error,

$$\sigma_p = 0.00591\frac{n_1}{c}\Delta \quad \text{s/m} \quad (69)$$

which is still far better than fourth-order or less graded distributions (per Table III). However, (69) is larger than (65) by a factor of 16 for a $\Delta = 10^{-2}$.

The question comes to mind: Is the delay distortion equally sensitive to small changes in index at other nominal values of index distribution? To answer this question, consider the impulse response for (66), where $(1-\delta) \ll 1$; the solution has been given [53] but is somewhat cumbersome. However, it is found that for $|1-\delta| \leq 0.1$, the rms width of the impulse response varies less than 2 percent from the value given in Table III for $\delta = 1$. The impulse response for perturbations of other distributions also changes very little. Hence the impulse response is sensitive to small variations in index distribution *only* in the vicinity of the "ideal" near-parabolic distribution.

The preceding discussion has described the group-delay differences between the various modes of multimode fibers. We return to a discussion of delay distortion within each mode due to the dependence of group delay on wavelength, $d\tau/d\lambda$. The specific time difference between output power components separated by $\delta\lambda$ is for the step-index fiber [53]

$$s = \delta\lambda\frac{d\tau}{d\lambda} = \frac{\lambda_0\delta\lambda}{c}\left\{-\frac{d^2n_1}{d\lambda^2}\left[1 + \Delta\frac{(m+1)^2}{(m+1)_{\max}{}^2}\right] + \frac{(m+1)^2}{16n_1a^2} - \frac{dn_1}{d\lambda}\frac{(m+1)^2\lambda_0}{8n_1{}^2a^2} + \left(\frac{dn_1}{d\lambda}\right)^2\frac{\lambda_0{}^2(m+1)^2}{16n_1{}^3a^2}\right\} \text{ s/m.} \quad (70)$$

In practice, it is essential to evaluate $d^2n_1/d\lambda^2$ and $dn_1/d\lambda$ in a particular case to know which effect will dominate. The best available data for n_1, $dn_1/d\lambda$, and $d^2n_1/d\lambda^2$ versus λ are given in Figs. 17–19. Using these data and the parameters $\lambda_0 = 0.8\ \mu$m, $\Delta = 0.01$, and $(m+1) = (m+1)_{\max}$, the value of a drops out and we find that the four terms within the brackets have the values 4×10^{10} m^{-2}, 4.5×10^{10} m^{-2}, 2.2×10^8 m^{-2}, and 4.4×10^6 m^{-2}, respectively. Hence for fused silica represented by Figs. 17–19,

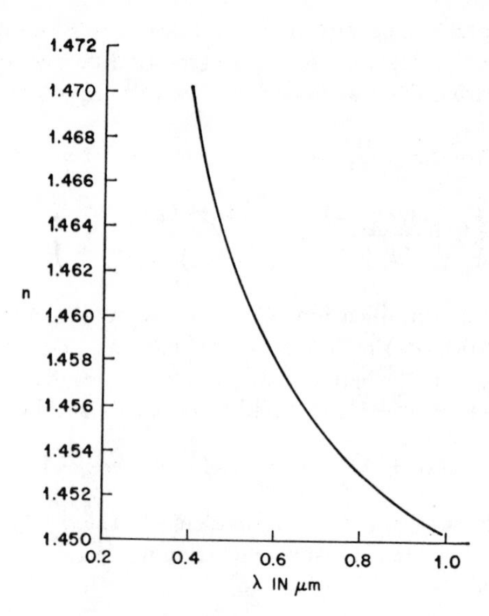

Fig. 17. Index n versus wavelength λ for fused silica (after I. H. Malitson, *J. Opt. Soc. Amer.*, vol. 55, p. 1205, 1965).

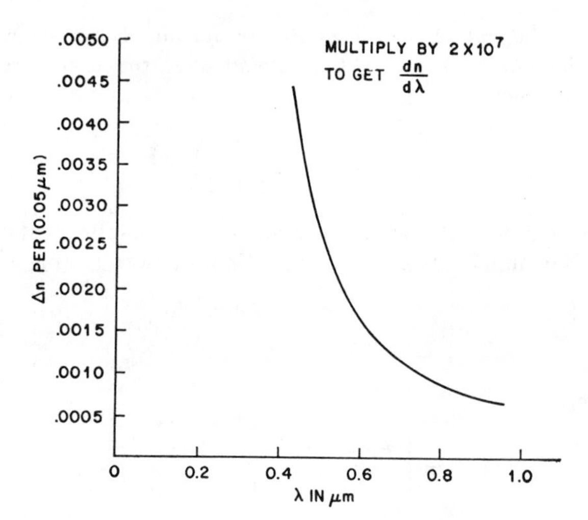

Fig. 18. $dn/d\lambda$ versus λ for fused silica.

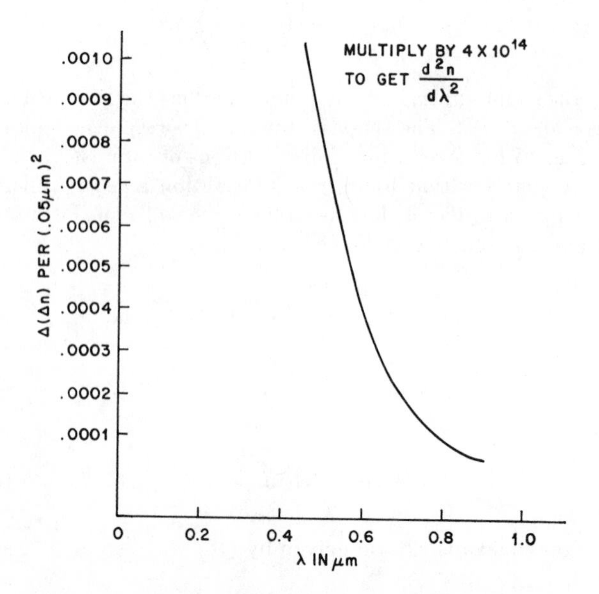

Fig. 19. $d^2n/d\lambda^2$ versus λ for fused silica.

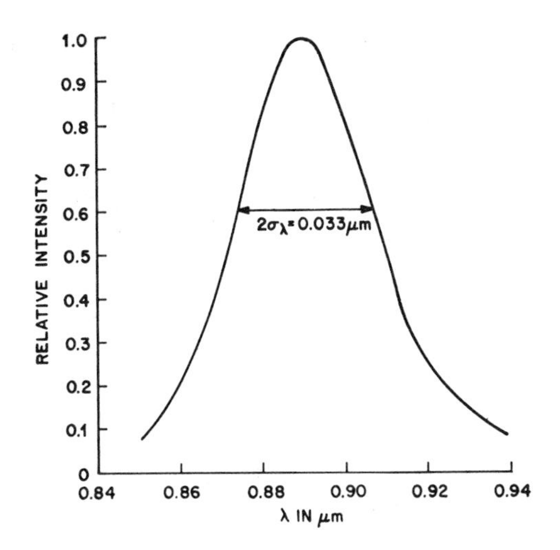

Fig. 20. Spectral distribution of the output of a typical Burrus-type AlGaAs LED.

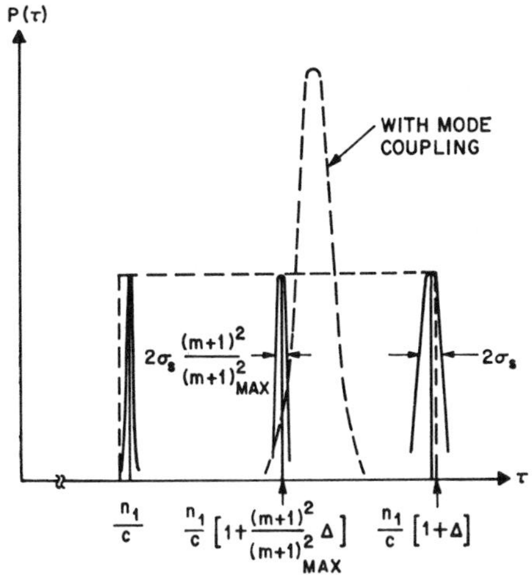

we can accurately simplify (70) to

$$s = -\frac{\lambda_0 \delta\lambda}{c} \frac{d^2 n_1}{d\lambda^2} + \frac{2n_1}{c} \frac{\delta\lambda}{\lambda_0} \frac{(m+1)^2}{(m+1)_{\max}^2} \Delta. \tag{71}$$

The first term shows the effect of material dispersion and the second term gives the output pulse spreading due to wavelength dependence of group delay for each of the modes when $dn_1/d\lambda$ and $d^2n_1/d\lambda^2$ are zero. For clad fibers, the numbers given indicate that the highest modes have about equal wavelength dependence due to material effect and waveguide mode effect.

We can state the output pulse spread for an AlGaAs LED in combination with fused silica fibers as represented by Figs. 17–19. The spectrum of the output of an AlGaAs LED is shown in Fig. 20. The spread of the output pulse maps into the time domain from the spectrum of Fig. 20, giving a similar pulse shape. The full rms width of the LED, representing its output centered at λ_0 as $\exp[-0.5(\lambda-\lambda_0)^2/\sigma_\lambda^2]$, is $2\sigma_\lambda = 330$ Å or 3.3×10^{-8} m. With this value, the material contribution to the rms impulse response of the combination of the AlGaAs LED and fused silica fiber is obtained from the first term of (71) as 1.75×10^{-12} s/m or 1.75 ns/km. From the work of Personick,[6] a choice of PCM pulse rate yielding 1-dB signal-power penalty due to pulse spreading is $1/(4\sigma L)$; this yields for the LED-fused silica wavelength dispersion limitation a pulse rate of 142 Mb in 1-km length or a pulse rate of $142/L$ megabits with a fiber length of L kilometers.

The second term of (71) also results in a mapping of the LED spectrum into the time domain—the time spread varying with mode number. The wavelength-dependent spreading superimposes on the difference in specific delay for each mode as sketched in Fig. 21. For the highest mode, the value of the second term of (71)

$$\frac{2n_1}{c} \frac{\delta\lambda}{\lambda_0} \Delta$$

is 0.128Δ ns/m for the LED of Fig. 20 having $2\sigma_\lambda = 330$ Å, $\lambda_0 = 0.8$ μm, and $n = 1.458$ (fused silica). This compares with 1.40Δ ns/m, from (62), with $u = \infty$ for the rms width of impulse response due to multimode-delay distortion at a single frequency. Hence wavelength-dependent delay distortion of individual modes adds only a small component to the delay difference between modes at a single frequency in multimode glass fibers.

[6] See Section VII-B.

Fig. 21. Sketch of details of the impulse response of a multimode fiber, without and with mode coupling.

F. Mechanisms for Reduction of Delay Distortion

The preceding section described the delay distortion to be expected if there were no coupling between the modes and if there were no *differences* between the attenuation coefficients for the various modes. We show here the reduction in delay distortion which can result from loss differences or from modal coupling.

Consider first the effects of attenuation alone. If the attenuation is the same for all modes, the fiber output is reduced but the relative magnitude of the powers in the various modes is unchanged. However, if there is more loss for higher order modes than for low-order modes, and if all modes are excited equally at the fiber input, then the fiber output will contain more power in low-order modes than in higher order modes. Because the group delay varies with modal number m [see (51), for example], the width of the impulse response is reduced. Consider the simplified case of zero differential loss for $(m+1)$ less than some value $(m+1)_c$ and infinite differential loss for $(m+1)>(m+1)_c$. Then we can relate the loss of output power to the reduction in width of impulse response through (51) for the step-index fiber:

$$\text{signal loss for step-index fiber} = 10 \log_{10} \frac{(m+1)_c^2}{(m+1)_{\max}^2} \cdot \tag{72}$$

For half the modes lost in a square or round fiber, there is a 3-dB signal loss, and the width of the impulse response is reduced by a factor of two. Fig. 22 presents a plot of the normalized rms impulse-response width versus signal loss for this loss model.

For index distributions other than the step change, the specific group delay varies more slowly with signal loss and the reduction in width of the impulse response is not as large. As given by (51), the modal-delay spread varies as $(m+1)^{4/3}$ for the fourth-order index distribution and as $(m+1)^{1.6}$ for the eighth-order index distribution. Fig. 22 shows the resulting signal loss versus impulse-response width. For the lowest 30

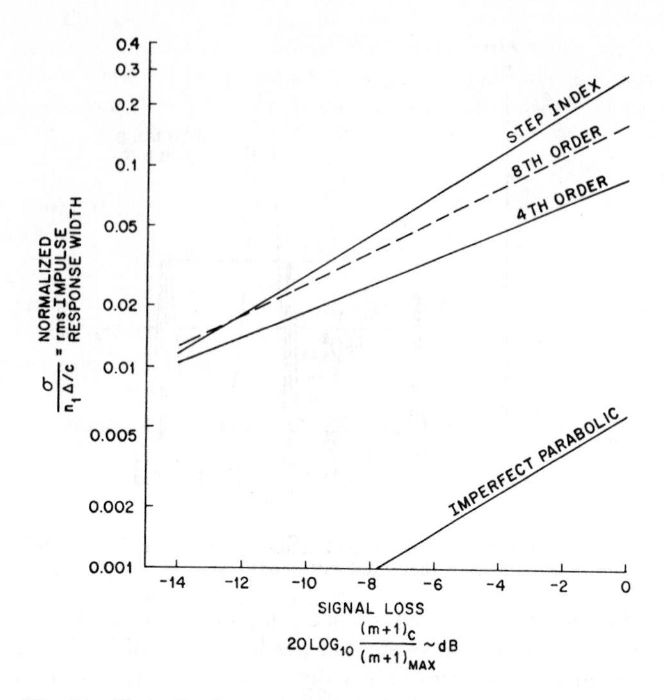

Fig. 22. Normalized rms width of the fiber impulse response versus signal loss for low-pass distribution of modal loss.

percent of the modes (signal loss greater than 10 dB), the impulse response width is similar for the fourth-order, eighth-order, and step-change index distributions. This might be expected since the lower modes are less influenced by the index conditions farther from the axis of the fiber.

In Fig. 22 the plot for the "imperfect parabolic" relates to the parabolic with 5-percent fourth order [$\delta = 0.05$ in (66)]. The resulting modal-delay spread varies as $(m+1)_c^2$ and the signal-loss versus impulse-width slope is the same as for a step-index change.

The preceding discussion related to "low-pass" modal transmission with infinite loss to all modes for which $(m+1) > (m+1)_c$. If the loss varies continuously with increasing mode index, the resulting reduction in width of impulse response can be computed using expressions for specific impulse response from (62). The result will be to modify the *shape* as well as the width of the impulse response and the rms width will have to be calculated for each loss function and fiber length.

Turning now to the effects of mode coupling, let us consider first coupling without modal loss differences. The solid lines of Fig. 21 indicate the series of responses from the various modes at the fiber output when there is no mode coupling. When there is strong mode coupling, the output can be as shown by the dotted line in Fig. 21; the figure is not drawn to scale, but is meant to illustrate a single response for the entire set of modes.

Personick was the first to point out that mode conversion in a multimode waveguide could reduce the width of the impulse response [56]. He showed that the shape of the impulse response in any given mode (for "large" mode conversion, to be defined) is the same regardless of the mode observed and has a shape that approaches a Gaussian for random coupling. One can view the energy propagation as follows: a packet of energy travels first in one mode, then in another, followed by still another—with specific delays characteristic of the uncoupled modes while the packet is in a particular mode. The result is that the energy arrives at the far end of the line with a delay that is a weighted average of the delays of all the modes.

The rms width of the impulse response for a lossless two-mode system is [56]

$$\sigma = \left\{\left\langle\left[t - \left(\frac{\tau_1+\tau_2}{2}\right)L\right]^2\right\rangle_{\text{av}}\right\}^{1/2} = \left(\frac{\tau_1-\tau_2}{2}\right)\sqrt{Ll_c} \tag{73}$$

in which

t	running time variable,
τ_1, τ_2	specific delay coefficients in the absence of coupling for modes 1 and 2, respectively,
L	waveguide length,
l_c	average distance for energy transfer by coupling from one mode to the other.

The output pulse in the absence of loss has the form

$$P_0(t) \sim \exp\left\{-\frac{1}{2}\left[\frac{t - \dfrac{(\tau_1+\tau_2)}{2}L}{\sigma}\right]^2\right\}. \tag{74}$$

The effect of dissipative loss in the modes is to multiply the lossless Gaussian response by the factor P:

$$P = \exp\left[-\left\{\frac{(\alpha_1-\alpha_2)}{(\tau_1-\tau_2)}\left[t - \left(\frac{\tau_1+\tau_2}{2}\right)L\right] + \left(\frac{\alpha_1+\alpha_2}{2}\right)L\right\}\right] \tag{75}$$

and α_1 and α_2 are the attenuation coefficients in modes 1 and 2, respectively. The exponential in time—the part of the exponent proportional to $(\alpha_1 - \alpha_2)$—has the effect of shifting the delay for the mean of the pulse and it alters the pulse magnitude; however, in the currently discussed region where $(\alpha_1 - \alpha_2)$ is small, loss does not alter appreciably the rms width of the impulse response given by (73).

Increased coupling reduces l_c and the impulse response width; the line length L enters (73) as the square root rather than linearly as it would in uniform waveguides. We observe that, in the absence of coupling, the output responses of the two modes would be separated by $L(\tau_1 - \tau_2)$; with strong coupling, the impulse-response width is reduced by the factor $\sqrt{l_c/L}$.

It is important that the coupling be large only between guided modes and not between guided modes and the radiation field [57]. The latter would introduce signal loss.

In an important series of papers [46], [40], [43], [58], Marcuse has provided an analysis for the interrelation between random mode coupling, reduction in width of impulse response, and radiation losses that may occur as an undesired consequence of the perturbations introduced to cause the desired coupling between guided modes. Marcuse derived a set of steady-state power-flow equations [46]:

$$\frac{\partial P_\nu}{\partial z} + \frac{1}{v_\nu}\frac{\partial P_\nu}{\partial t} = -(\alpha_\nu + b_\nu)P_\nu + \sum_{\mu=1}^{N} h_{\nu\mu}P_\mu \tag{76}$$

with

α_ν attenuation coefficient of νth mode in the absence of coupling,

$b_\nu = \sum_{\mu=1}^{N} h_{\nu\mu}$

v_ν group velocity in mode ν,

$h_{\nu\mu}$ coefficient of coupling, mode μ to mode ν.

This set of equations represents the effects of coupling between the modes, exhibiting both the spatial variation and the variation with time in a coordinate system moving at velocity v_ν. Implicit in the analysis is the requirement that the correlation length of the coupling as a function of z be short compared to the distance over which the mode power P_ν changes appreciably; this is easily met in practice.

It is known [33] that a complete transfer of power can occur between modes ν and μ in a two-mode system if the spatial variation in coupling occurs at frequency $\Omega_{\mu\nu}$, where

$$\Omega_{\mu\nu} = \frac{2\pi}{\lambda_B} = (\beta_\mu - \beta_\nu) \tag{77}$$

and λ_B is commonly referred to as the beat wavelength. For coupling components outside a band $\delta\Omega_{\mu\nu}$ centered on $\Omega_{\mu\nu}$, very little transfer of power can occur [33]. The magnitude of the significant spectral band $\delta\Omega_{\mu\nu}$ is approximately

$$\delta\Omega_{\mu\nu} = \begin{bmatrix} 2h_{\text{av}} \\ \dfrac{0.9\pi}{L}, \quad \text{if } h_{\text{av}} \to 0 \end{bmatrix} \tag{78}$$

where L is the distance between repeaters and h_{av} is the average coupling component in the band centered at $\Omega_{\mu\nu} = (\beta_\mu - \beta_\nu)$. The larger of the two right-hand sides of (78) prevails. It is assumed that the value of h_{av} over bands of the order of $\delta\Omega_{\mu\nu}$ does not change much in the vicinity of $\Omega_{\mu\nu}$.

For a many-mode fiber there will be a series of bands in the coupling spectrum, each of width $\delta\Omega_{\mu\nu}$ and centered at $\Omega_{\mu\nu}$, which cause significant energy transfer. Consider the numbers for a typical case. Assume a signal wavelength $\lambda_0 = 1.0$ μm, a fiber core 50 μm wide, an index $n = 1.5$, and a core-cladding index difference $\Delta = 0.01$. Then the beat wavelength, which is identical to the coupling period for maximum modal coupling, is 10 mm for the adjacent lowest order mode groups and 0.7 mm for the adjacent mode groups near cutoff. Also, the value of $\delta\Omega_{\mu\nu}/\Omega_{\mu\nu}$, the fractional width of the significant coupling spectrum centered on those mechanical periods, is 5 percent for the lowest order modes and 0.35 percent for the modes near cutoff on the assumption of a coupling strength h_{av} sufficient to cause complete energy exchange in 10 m. For other coupling strengths, the values of $\delta\Omega_{\mu\nu}/\Omega_{\mu\nu}$ vary directly as the average coupling strength h_{av} and inversely as the average energy exchange distance. We thus see that the significant bands of coupling are very narrow indeed, centered on a series of discrete periods given by (77).

Marcuse has given explicit solutions to (76) for the two-mode case. He notes that when $h \ll |\alpha_2 - \alpha_1|$, the rms output pulsewidth is

$$\sigma = \sqrt{2}\,(\tau_2 - \tau_1) h \sqrt{\frac{L}{(\alpha_2 - \alpha_1)^3}} \tag{79}$$

where increasing the coupling strength h increases the width of the impulse response. This regime is not beneficial—mode coupling does not average the modal velocities since power coupled to another mode is dissipated before it can couple back.

For $h \gg |\alpha_2 - \alpha_1|$,

$$\sigma = \tfrac{1}{2}(\tau_1 - \tau_2)\sqrt{\frac{L}{h}}. \tag{80}$$

This h may be interpreted as the average value h_{av} over the band $\delta\Omega_{1,2}$ centered at $\Omega_{1,2}$ as discussed in connection with (77) and (78). Comparing (80) with Personick's result (73), we see that $l_c = 1/h_{\text{av}}$.

Marcuse has found that (80) holds for a four-mode case with an added multiplier 0.79 and with μ and ν in $(\tau_\mu - \tau_\nu)$ being the slowest and the fastest mode, respectively. It is thus possible that (80) may hold for any multimode fiber.

Marcuse has solved (76) for the case of a Gaussian input pulse. We give here the solution for the steady state when L is large. Then the output $P_\nu(L, t)$ is also a Gaussian,

$$P_\nu(L, t) = \frac{2\sigma_i}{\sigma_o} k_1 B_{\nu 0}^{(1)} \exp\left[-\alpha_0^{(1)} L\right] \cdot \exp\left[-\left\{\frac{\left[t - \dfrac{L}{v_0}\right]^2}{2\sigma_o^2}\right\}\right] \tag{81}$$

where

σ_i rms width of the input pulse, given by

$$P_\nu(0, t) = G_\nu \exp\left[t^2/2\sigma_i^2\right] \tag{82}$$

σ_o rms width of the output pulse:

$$\sigma_o = \frac{1}{\sqrt{2}}\left(2\sigma_i^2 + 4\alpha_2^{(1)} L\right)^{1/2} \tag{83}$$

$k_1 = \sum_{\nu=1}^{N} G_\nu B_\nu^{(1)}$

v_0 weighted average group velocity.

The constant $B_{\nu 0}^{(1)}$ describes the distribution of power among the modes, and $\alpha_2^{(1)}$ characterizes the output pulsewidth σ_o. Unfortunately, it is not possible to give an explicit solution for $\alpha_2^{(1)}$ and we will resort to computed results based on a slab-guide model.

The output in the absence of coupling occurs over a time interval

$$(\tau_{\max} - \tau_0) L$$

where $\tau_{\max}$ and τ_0 are the specific delays for the slowest and the fastest mode, respectively; for step-index fibers we showed in the preceding section that this results in a σ [per (59)]:

$$\sigma_{oo} = \frac{1}{\sqrt{12}}(\tau_{\max} - \tau_0) L. \tag{84}$$

In the presence of coupling we have the output width σ_o given by (83). Marcuse has defined and plotted a pulsewidth improvement factor R, which is

$$R = \frac{2\sqrt{2}\,\sigma_o}{(\tau_{\max} - \tau_0) L} = \sqrt{\frac{2}{3}}\,\frac{\sigma_o}{\sigma_{oo}}. \tag{85}$$

Thus R is a constant times the ratio of rms width of output pulse with and without coupling.

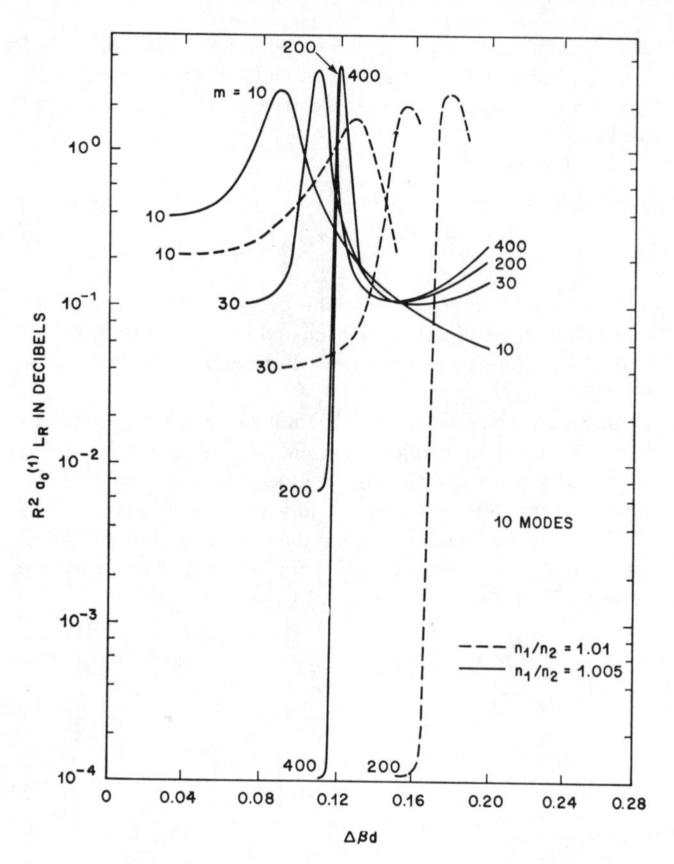

Fig. 23. Radiation-loss penalty $\alpha_0^{(1)}L_R$ times the square of pulsewidth improvement factor versus spectral width $\Delta\beta d$, for the spectrum according to (87) (after D. Marcuse, [46]).

If the coupling effects in the fiber have a broader spectral range than desired, the components in the range

$$\beta_\nu - n_2 k \leqq \Omega_{\mu\nu} \leqq \beta_\nu + n_2 k \tag{86}$$

will couple guided modes to the radiation field causing excess loss, represented within the term exp $[-\alpha_0^{(1)}L]$ in (81). We give one example from Marcuse's paper [46]. Assume a coupling spectrum

$$F(\Omega) = \frac{m\sigma_{cc}^2 \sin(\pi/m)}{\Delta\beta\left[1+\left|\frac{\Omega}{\Delta\beta}\right|^m\right]} \tag{87}$$

in which σ_{cc} is the rms deviation of the core–cladding interface in the slab waveguide. The quantity $\Delta\beta$ measures the width of the coupling spectrum; coupling between modes μ and ν becomes small when $|\Omega/\Delta\beta|^m$ becomes large compared to unity and $\Delta\beta = \beta_\mu - \beta_\nu$. The exponent m controls the sharpness of cutoff of the coupling spectrum. Fig. 23 shows Marcuse's computed relation between the radiation loss $\alpha_0^{(1)}L_R$, the pulsewidth improvement factor R, and the spectral width $\Delta\beta d$ for a slab width d; the figure relates to a 10-mode slab guide (approximately 100-mode square fiber) and includes data for values of m from 10 to 400, and values of $n_1/n_2 = 1.01$ (dotted curves) and $n_1/n_2 = 1.005$ (solid curves). We observe that the ordinate is

$$R^2\alpha_0^{(1)}L_R$$

indicating the tradeoff between pulse narrowing and radiation loss. This is a general result. Small ordinates are desired since this permits small R, corresponding to small-output pulsewidth, for a given loss $\alpha_0^{(1)}L_R$. From Fig. 23, if $m = 30$, $\Delta\beta d = 0.08$, and $n_1/n_2 = 1.005$, the ordinate is about

$$R^2\alpha_0^{(1)}L_R = 0.1 \text{ dB}.$$

Then for radiation loss $\alpha_0^{(1)}L_R = 10$ dB, $R^2 = 0.01$, indicating a $10\sqrt{3/2}$ reduction in rms output pulsewidth [using (85)] due to mode coupling. The effects of the spectral cutoff parameter m and index ratio n_1/n_2 can be found by further study of Fig. 23.

In another analysis, Gloge found closed-form solutions [59], [60] on the basis of several simplifying assumptions: 1) coupling exists only between modes that are adjacent in the spectrum, 2) loss due to coupling between guided modes and the radiation field increases as the square of the mode order [the $A\theta^2$ term of (88)], and 3) the radiation loss is so large that the highest order guided modes do not reach the output. The results relate well to experiments on some liquid-core fibers.

Gloge's starting equation for the output power $P(t, z)$ is

$$\frac{\partial P}{\partial z} + \frac{dt}{dz}\frac{\partial P}{\partial t} = -A\theta^2 P + \frac{1}{\theta}\frac{\partial}{\partial\theta}\left(\theta D \frac{\partial P}{\partial \theta}\right) \tag{88}$$

in which A is the assumed loss constant, and θ represents the mode order, with

$$\theta = \frac{m\lambda}{4an_1} \tag{89}$$

and

- a half-width of the guide,
- D coupling parameter, $= (\lambda/4an_1)^2 d_0$,
- d_0 zero-order coupling coefficient [60].

The output pulse response is a Gaussian in modal distribution $P \sim \exp[\theta^2/\theta^2(z)]$ at large z; this steady state results from mode-coupling feeding power between the modes and the radiation loss shaping the modal spectrum to leave small power in higher order modes. With an input having a modal distribution to match the asymptotic value of the fiber response, the solution shows that the mean delay (in excess of $n_1 z/c$) for each mode characterized by θ is

$$\delta_P = \frac{T}{2}\left[\gamma_\infty z + \left(\frac{\theta^2}{\theta_\infty^2} - \frac{1}{2}\right)(1 - \exp[-2\gamma_\infty z])\right] \tag{90}$$

in which

$$\theta_\infty = (4D/A)^{1/4} \tag{91}$$

$$\gamma_\infty = (4DA)^{1/2} \tag{92}$$

$$T = \frac{n_1}{2cA} = \frac{n_1}{2c}\frac{\theta_\infty^2}{\gamma_\infty}. \tag{93}$$

The parameter θ_∞ is the asymptotic mode-distribution parameter already noted and γ_∞ is the corresponding power-loss parameter [60]. The half-width of the output pulse in each mode (again characterized by θ) is

$$\tau_p = \frac{T}{2}\left[\gamma_\infty z + \left(\frac{\theta^2}{\theta_\infty^2} - \frac{5}{4}\right) - 2\gamma_\infty z\left(2\frac{\theta^2}{\theta_\infty^2} - 1\right)\exp[-2\gamma_\infty z] + \exp[-2\gamma_\infty z] - \left(\frac{\theta^2}{\theta_\infty^2} - \frac{1}{4}\right)\exp[-4\gamma_\infty z]\right]^{1/2}. \tag{94}$$

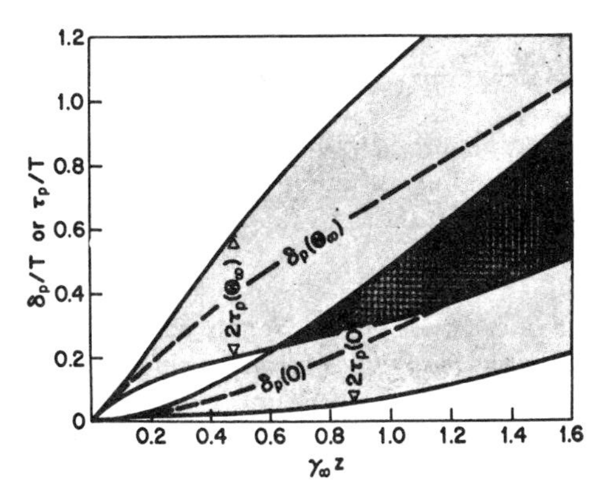

Fig. 24. Normalized mean delay and half-width of the output pulse versus normalized fiber length for an input pulse having a modal distribution to match the asymptotic value of the fiber response.

At large z, both the mean delay from (90) and the width from (94) vary by the fractional amount $1/\gamma_\infty z$ for the lowest mode compared to the highest mode; this clearly is small when $\gamma_\infty z \gg 1$, in agreement with Personick's results. Fig. 24 shows the mean delay and half-width of the output pulse, plotted versus normalized distance $\gamma_\infty z$, for the lowest mode $\theta=0$ and for the mode $\theta=\theta_\infty$ at the self-induced cutoff controlled by A and D [(91) and (92)]. For small z [with input $\theta(z)$ matching $\theta_\infty(z)$] the pulse broadening is small; at larger z the lowest order and high-order modes overlap in time and the mode $\theta=\theta_\infty$ is spreading less than linearly with z. Gloge finds the pulse spreading to be inversely proportional to the square of the output NA; half the NA permits a fourfold increase of the pulse rate for comparable interfering effects [60].

G. Material and Fiber Measuring Techniques

The search and advent of materials and fibers, in general, in small quantities, and with overall losses in the order of 10 dB/km, have taxed the ingenuity of researchers looking for techniques to evaluate the properties of bulk materials, as well as the guiding characteristics of the fibers made with them. Five important types of measurements have evolved which we proceed to review: 1) bulk material absorption, 2) bulk material scattering, 3) fiber absorption, 4) fiber scattering, and 5) fiber dispersion.

1) Bulk Material Absorption: Essentially three techniques have been developed to evaluate the absorption loss in bulk materials. The first is a two-beam bridge-balance technique [5], [61], [62], exemplified in Fig. 25. The properly collimated light from the incoherent source is split into two beams that are alternately chopped and then collected by a narrow-band phase-sensitive detector and amplifier that passes only the first harmonic. Without the samples the bridge is balanced by reducing the amplitude of the periodic signal to zero with the help of the attenuator. Next, samples of the same material, but of about 20-cm different lengths, introduced into the arms of the bridge, produce an imbalance which, properly calibrated, measures the differential loss between the samples.

The use of two samples compensates the end effects. Comparable elimination of end effects and substantial simplification have been achieved by placing one single sample about 2 cm thick at the Brewster angle [63] in one of the beams.

A balance stability of about $\pm 1\times10^{-5}$ and measuring capability of a few decibels per kilometer have been achieved with these instruments. As a matter of fact, loss as low as 22 dB/km in samples of Suprasil I has been measured by Tynes [9] at 0.6328 μm.

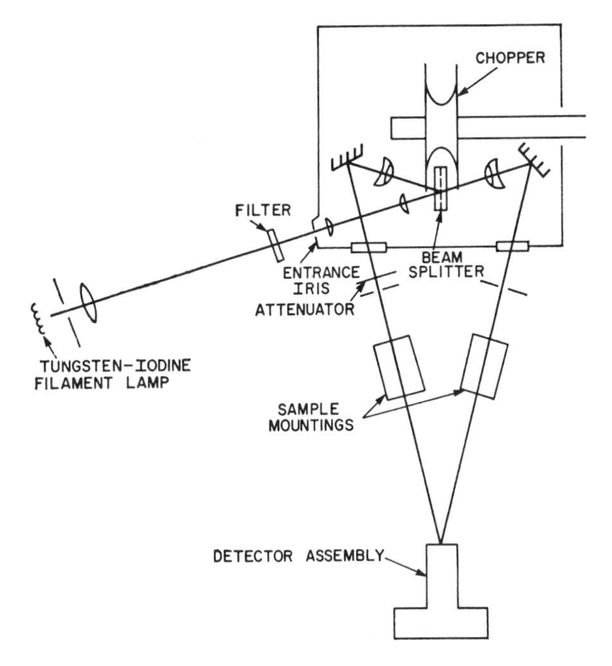

Fig. 25. Bulk loss measurement via a two-beam bridge (after [5], [61], [62], [63]).

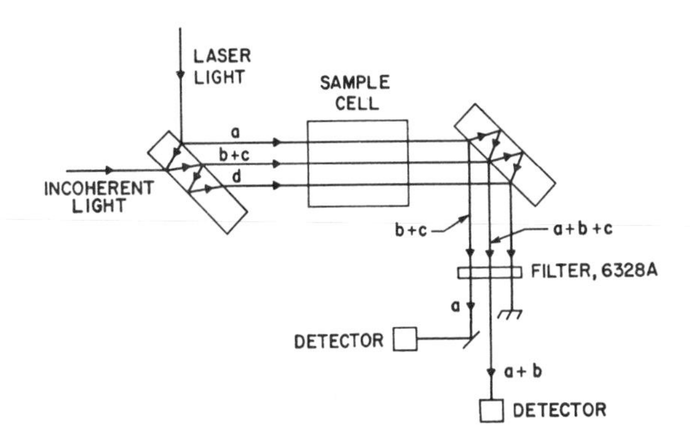

Fig. 26. Technique for the measurement of absorption losses in liquids (after [64]).

These techniques are used to measure all the losses in the samples; consequently, the scattering losses, which are small in general, must be independently measured and subtracted to get the absorption losses.

A sensitive technique to measure the absorption in liquids [64] over a wide spectral range is schematized in Fig. 26. It is based on the idea that two beams of different intensity heat the liquid unequally, creating an unequal change of index and an electrical path length differential that can be measured interferometrically.

Let us ignore for the time being the incoherent beam. The laser beam impinging on the parallel plate launches many parallel beams into the liquid cell, of which we will consider those labeled a, b, and d. After reflections in the second plate, the beam $a+b$ reaching the detector is made of the superposition of two beams of equal intensity (two reflections and two transmissions at the plates for each beam) and easily controllable phase difference by a small rotation of one of the plates.

The intensity of the laser beams in the liquid sample is so small that practically no heat effect is introduced. However, the added incoherent light beam of a few milliwatts passing through the liquid, as beam c, changes the electrical path of

TABLE IV

EXPERIMENTAL RESULTS OF ABSORPTIVITY α AND PUBLISHED DATA OF SCATTERING LOSS α_S

	Density (g/cm^3)	Specific Heat (cal/g °C)	Refractive Index	$-dn/dT$ ($10^4 °C^{-1}$)	α (10^4 cm^{-1})	α_S (10^4 cm^{-1}) at 6328 Å	$\alpha+\alpha_S$ (10^4 cm^{-1})
Carbon tetrachloride	1.575	0.206	1.4601	5.7	<0.1	0.5	<0.6
Chloroform	1.480	0.230	1.446	5.8	0.17	0.5	0.67
Bromobenzene	1.495	0.254	1.560	5.36[a]	0.19		
Benzene	0.875	0.413	1.501	5.45[a]	0.21	1.5	1.71
Fluorobenzene	1.023	0.365	1.468	4.9	≤0.23		
Chlorobenzene	1.106	0.310	1.524	5.53	0.23	2.0	2.23
Methylene chloride	1.327	0.300	1.424	5.4	0.23	0.6	0.83
o-dichlorobenzene	1.30	0.27	1.549	5[a]	0.31		
Trichloroethylene	1.462	0.220	1.478	5[a]	0.41		
Heavy water	1.10	1.00	1.328	1.04[a]	<0.47	0.1	<0.57
Dibromomethane	2.497	0.25[a]	1.542	5.4	0.47	1.5	1.97
Pyridine	0.982	0.400	1.510	5.1	0.50		
Carbon disulfide	1.26	0.25	1.62	7.9	1.4	4.6	6.0
Nitromethane	1.14	0.42	1.38	4.2	2.3	0.13	2.43
Bromoform	2.89	0.127	1.60	6.0	3.3	1.3	4.6
Toluene	0.87	0.40	1.50	5.8	3.3	1.1	4.4
Acetone	0.78	0.519	1.36	4.9	4.1	0.4	4.5
Tetra-hydro-naphthalene	0.98[a]	0.40	1.48	5.0[a]	5.5		
Xylene[b]	0.88	0.40	1.51	4.8	6.5	1.6	8.1
n-hexane	0.655	0.545	1.38	5.6	6.9	0.5	7.4
n-heptane[b]	0.68	0.50	1.39	4.6	7.0		
Cyclohexane[b]	0.66	0.59	1.38	5.2	7.6	0.25	7.8
Iso-octane	0.703	0.530	1.397	4.8	9.8	0.5	10.3
1-propanol	0.803	0.590	1.385	3.8	10.2	0.5	10.7
Methyl-ethyl-ketone[b]	0.81	0.52[a]	1.38	5.0[a]	12.1	0.4	12.5
Iso-butanol	0.817	0.630	1.398	3.9	14.3	0.25	14.5
Sec-butanol	0.808	0.600[a]	1.395	3.9[a]	14.3		
Ethanol	0.79	0.57	1.36	3.6	15.5	0.25	15.8
Methanol[b]	0.79	0.59	1.33	3.90	19.5	0.2	19.7
Water	1.00	1.00	1.33	1.04	29	0.10	29
Nitrobenzene	1.20	0.33	1.55	4.6	58	4.6	62.6

[a] Data not available; estimated from similar liquids.
[b] Not distilled.

beam b but does not reach the detector because of the narrow filter at 0.6328 μm. Any change in the intensity or the wavelength of the incoherent light changes the relative phase of laser beams a and b as they interact interferometrically at the detector. Absorptions as small as 8 dB/km have been measured with 2–3 mW of incoherent light and cells 10 cm long. Table IV contains typical results.

The second technique for measuring the attenuation of light in solid materials is calorimetric. The calibrated increase of temperature between the sample traversed by the light and the reference is used to measure the light absorption [65], [66].

Typically, 0.0007°C of differential temperature is observed with 0.02 W in the light beam if the absorption is 50 dB/km.

By thermally clamping both ends of the sample, heating effects due to surface losses are short-circuited and a sensitivity of a few decibels per kilometer with a 100-mW power laser has been reported [67].

The calorimetric sensitivity can be increased by increasing the beam intensity. For that purpose the sample can be incorporated within the cavity of a laser [68] such as Nd:YAG operating at 1.06 μm. Optical absorption within the glass causes its temperature to rise until it is compensated by heat leakage, which occurs mostly via air convection. For a laser power of 115 W passing through a Suprasil W1 sample, its temperature rise was 0.56°C and the calculated absorption was 2.3±0.5 dB/km. The lowest reported loss of 0.8 dB/km has been measured in a fused silica sample [68].

In all the measuring techniques described in the foregoing, either the limits of sensitivity, or the necessity of measuring

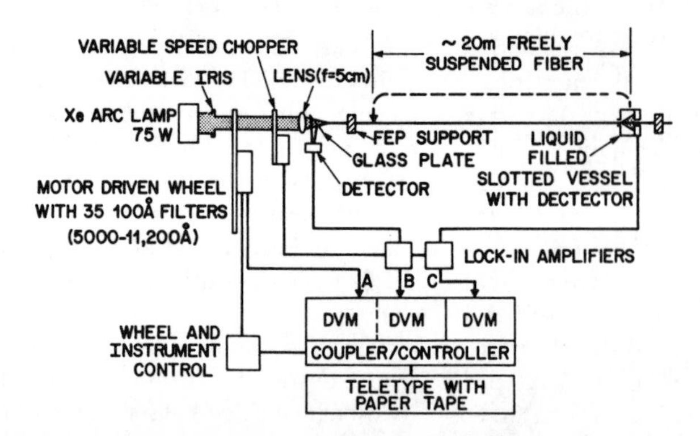

Fig. 27. Automated loss-measuring apparatus applied to unclad optical fibers. Recorded signals: A, voltage proportional to wavelength; B, reference; C, signal in fiber (after Kaiser *et al.* [10], [69], [70]).

at a few discrete wavelengths for which adequate sources are available, arise mostly from the short length of the measured sample. This observation is the starting point for the third method to measure absorption. It consists in pulling the sample into a long unclad fiber (up to 60 m in the case of [10], [69], [70]) supported at the ends with fluorinated ethylene propylene (FEP) clamps, and in measuring with a nondestructive technique the decaying power level of light traveling along such a fiber. The apparatus is shown in Fig. 27. Light from a Xe arc lamp is selectively filtered, chopped, and launched into the fiber. The nondestructive power extraction is achieved by contacting the fiber with a higher refractive index liquid contained in a vessel inside of which the fiber can slide through adequate slots. The extracted light is detected compared to a reference and automatically processed for the

100-Å-wide 200-Å-spaced filters of the frequency-selective wheel and for the different positions of the power-extracting liquid cell along the guide.

The measured loss comprises absorption and scattering. However, as will be shown later, the scattering can be measured independently and can be small if the surface of the fiber is properly cleaned.

The loss spectra of many unclad fibers have been compared with bulk loss measurement of the same materials and their coincidence has shown that the fiber-pulling process does not usually alter the absorption properties of the bulk material. There is one exception: low-OH-content silica, when pulled into fibers, develops a high-loss band centered at 0.625 μm [70].

Loss spectra for various grades of vitreous silica and soda-lime-silicate glasses are shown in Figs. 28 and 29. The lowest overall losses of about 2.5 dB/km at 1.06 μm and 4 dB/km at 0.9 μm have been measured in Spectrosil WF.

2) Bulk Material Scattering: Measurement of the angular distribution of light scattering from a high-optical-quality glass shows a Rayleigh distribution, [66], [71], characteristic of a medium whose refractive index fluctuates over distances small compared to the wavelength of the impinging light and which decays according to λ^{-4}. This Rayleigh loss can be deduced from a single measurement of the light scattered at 90° from the incident beam [72].

Another technique for evaluating Rayleigh scattering [68] consists in measuring, with the aid of a pressure-scanned Fabry–Perot interferometer, the spectrum of the light from a tightly focused argon-ion laser beam traversing the glass sample. The spectrum has spontaneous Brillouin lines that are Doppler shifted from the major Rayleigh line centered at the laser frequency, because of phonon interaction. The level corresponding to the longitudinal acoustic mode can be calculated from first principles and consequently can be used to calibrate the loss of the Rayleigh line. Thus at 1.06 μm the expected Rayleigh scattering in Suprasil W1 is 0.64 ± 0.04 dB/km.

3) Attenuation in Fibers Due to Absorption: The method used to evaluate the absorption in fibers is similar to some of the methods used for bulk material; the absorption is found as the difference between the overall attenuation and the attenuation due to scattering.

The overall loss is evaluated by measuring the power emerging at the end of a fiber successively shortened by known amounts.

As depicted in Fig. 30, light is mostly launched in the core of the fiber and, after stripping the cladding modes, the power flowing in the core is fed into a detector immersed in liquid matching the core index, thus avoiding random directivity of the fiber's broken end. The stripper, consisting of a pool of liquid with refractive index slightly larger than that of the cladding, must be used to eliminate cladding modes since their power level, highly dependent on the fiber's surface condition and on the core–cladding coupling, fluctuates substantially along the fiber.

Only if the measured power levels are exponentially related to fiber length do the results reflect a steady-state condition of the propagation from which reliable values of attenuation can be derived. This exponential dependence decreases as the length of the fiber becomes shorter, and the results become more and more sensitive to launching conditions.

4) Attenuation in Fibers Due to Scattering: Light can be coupled out of the fiber either by Rayleigh scattering or by guide imperfections such as dielectric strain, chemical or physical discontinuities in the core or cladding, irregularities and extraneous inclusions in the core–cladding interface, curvature of the guide axis, tapering, etc.

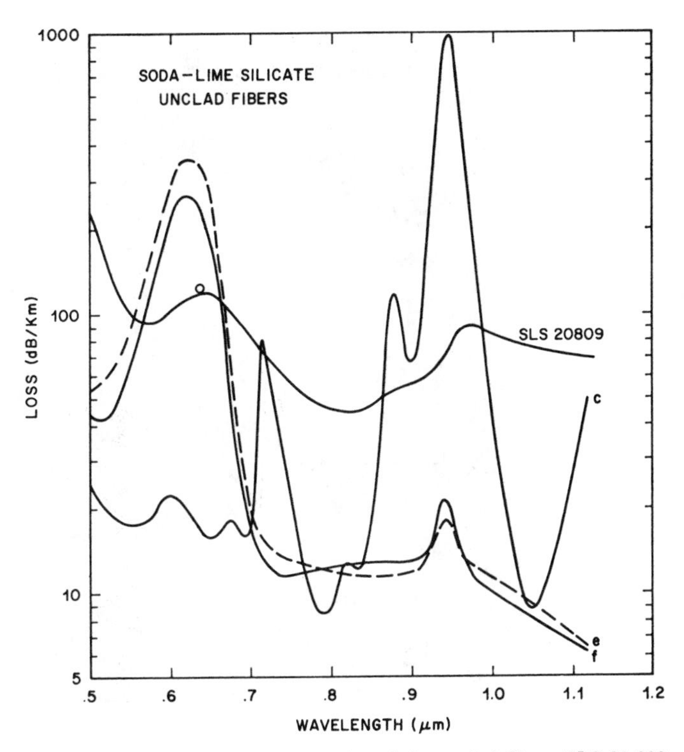

Fig. 28. Measured loss versus wavelength for unclad fibers: SLS 20 809 represents soda-lime silicate; *c* represents Suprasil I; *e* represents Suprasil W2; *f* represents Suprasil W1 (after [10]).

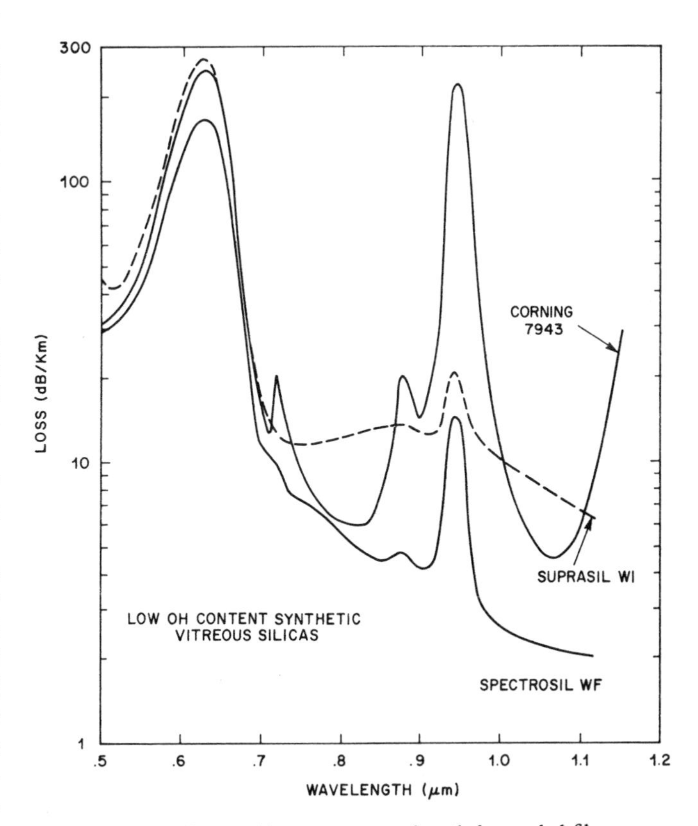

Fig. 29. Measured loss versus wavelength for unclad fibers (after Kaiser, [70]).

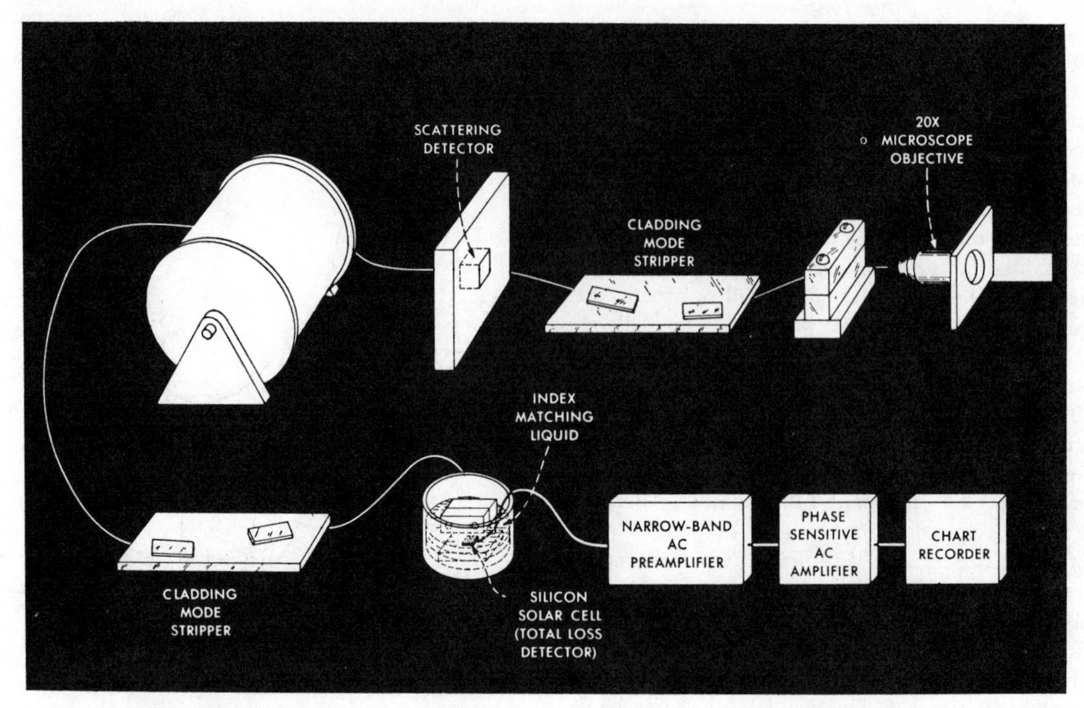

Fig. 30. Configuration for the measurement of total attenuation in fibers (courtesy the editor of *Bell Lab. Rec.*; after Tynes, [66]).

Both types of scattering can in principle be distinguished by their radiation pattern and by their wavelength dependence [73], [34]. However, scanning the radiation especially in directions close to the guide axis, measuring its frequency dependence, and interpreting the data constitute a formidable task, particularly since all the information varies along the guide axis as a function of local imperfections.

It is more useful to measure the light scattered in all directions as an integrated effect. This is achieved by threading the fiber through an integrating detector, such as the one shown in Fig. 30, consisting of a cube made of six 1-cm square silicon solar cells properly balanced [66], [74], or through a similar integrating device, such as that of Fig. 31, made of three 4×1-cm silicon photodetectors arranged in a triangular prism geometry and mounted on an x, y, z micropositioner [69].

These detectors integrate the light scattered over lengths of 1 and 4 cm, respectively, and, by sliding them along the fiber, the light scattered from the full length of the guide can be measured.

5) *Dispersion in Fibers:* Dispersion of a pulse as it travels along a fiber largely limits the information-carrying capacity of the guide.

Of the three mechanisms contributing to dispersion—1) material dispersion, due to the frequency dependence of the refractive index; 2) modal dispersion, caused by the different group velocities of the different modes; and 3) waveguide dispersion of a mode due to the frequency dependence of the propagation constant of that mode—only the first two have practical significance for moderate pulse repetition rates and we proceed to describe the techniques devised to evaluate them.

Fig. 31. Triangular cell for measuring fiber scattering losses (after Kaiser and Astle, [69]).

Fig. 32(a) gives a measuring set which compares, in the time domain, an input pulse into a fiber and the output. A periodic sequence of 200-ps pulses generated in an acousto-optic mode-locked krypton-ion laser is transmitted either directly or through the fiber into matched detectors [75].

Pulses with carriers at 0.647 μm and 0.568 μm, selected by a tuning prism retroreflector and shown in Fig. 32(b) with their peaks coinciding in time, were fed into a 20-m-long fiber with a 13-mm focal length lens (small NA) and the emerging pulses are shown in Fig. 32(c). They are 470-ps displaced replicas of the input pulses which measure the material dispersion of 0.03 ps/mÅ at about 0.6 μm.

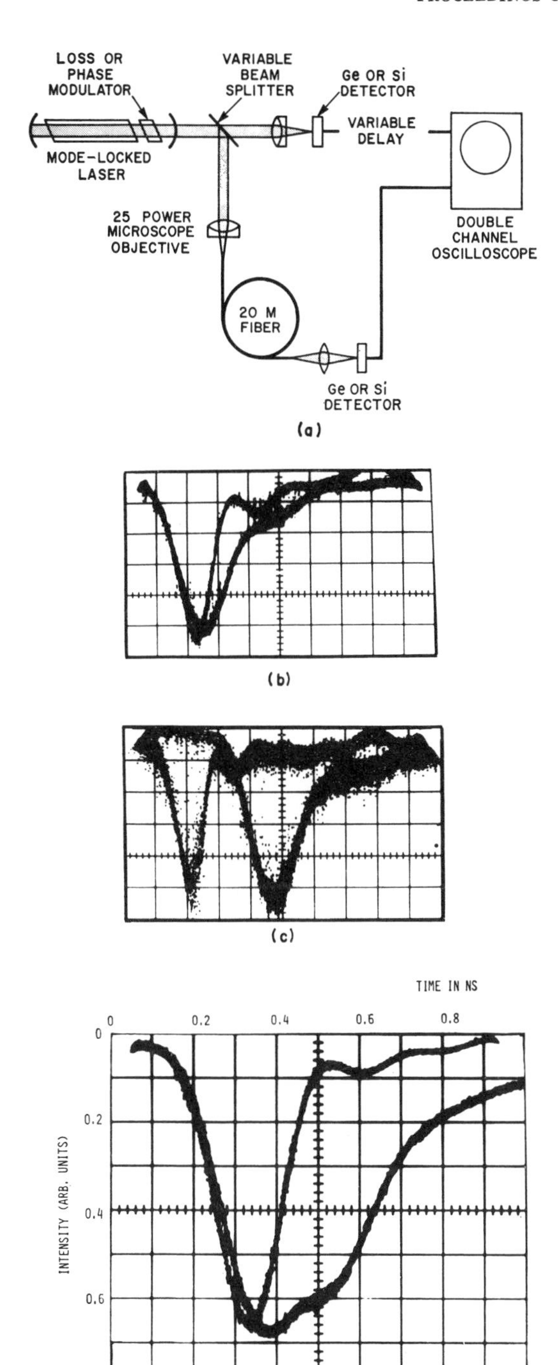

Fig. 32. Fiber dispersion measurement using the time domain. (a) Schematic of equipment layout. (b) Oscilloscope traces of input pulses at 0.647 and 0.568 μm coincident in time; horizontal scale 0.2 ns/div. (c) Output traces separated in time by 0.470 ns because of material dispersion. (d) Input pulse superimposed with output pulse widened by modal spread (after Gloge, [75]).

Injected pulses of a 0.647-μm carrier via a 6-mm focal length lens (large NA) become widened at the output due to modal spread, as shown in Fig. 32(d).

Different implementations of these basic ideas are reported in many publications [76].

Another method for evaluating the modal dispersion in multimode fibers operates in the frequency domain rather than in the time domain [77]. The baseband frequency response of the fiber is measured by comparing the beat spectra

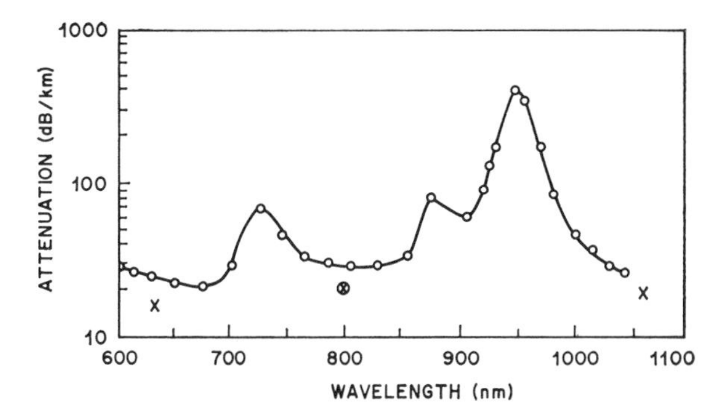

Fig. 33. Single-mode fiber loss versus wavelength (after Keck and Tynes, [78]).

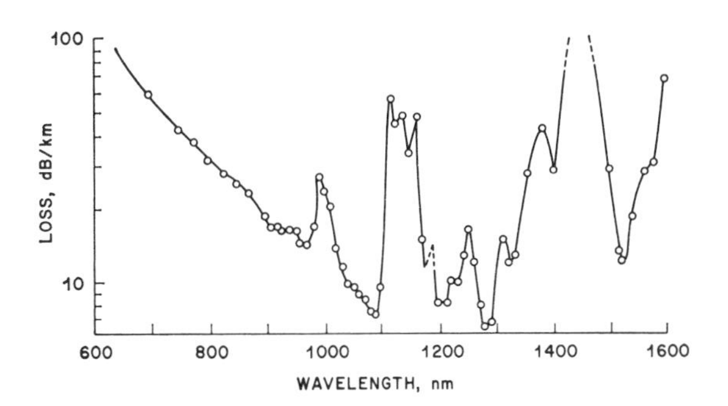

Fig. 34. Liquid-core (C_2Cl_4) fiber loss versus wavelength (after Ogilvie, Esdaile, and Kidd, [12]).

from a free-running laser before and after transmission through a fiber.

H. State of the Art

Since the momentous announcement by Corning in 1970 of their single-mode fiber [6], [78] with ~20-dB/km loss at 0.6328 μm (Fig. 33), a wide variety of fibers with ever-improving properties have been reported. As a matter of fact, the turnover is gaining momentum fast and consequently the following list of outstanding fibers representing the current state of the art may soon be outdated.

1) Liquid-Core Fibers: Toward the end of 1971 and the beginning of 1972, two almost simultaneous disclosures were made of liquid-core quartz fibers [11], [12] filled with tetrachloroethylene and exhibiting losses as low as 13 dB/km at 1.06 μm. Less than one year later, kilometer-long fibers [12], drawn from Heralux silica tubing having an internal diameter between 70 and 100 μm and filled with "chromato-quality" dehydrated tetrachloroethylene, exhibit attenuation troughs lower than 8 dB/km at 1.090, 1.205, and 1.280 μm (Fig. 34).

Similarly, kilometer-long capillary tubes [13], drawn from Chance–Pilkington ME1 glass tubing (10 000-dB/km attenuation) with an internal diameter of 50 μm and filled with hexachlorobuta-1,3-diene, have a minimum loss of 7.3 dB/km at 1.08 μm. The measured dispersion reported by Gambling, Payne, and Matsumura [13] on 200 m of this fiber is shown in Fig. 35(a) as a function of input laser beam angle and radius of curvature of the fiber. With only the lowest order modes excited (input beam half-width = 0.3°), dispersion as low as 320 ps in 200 m was reported. Larger input beams resulted in more dispersion, as would be expected, but only a small fraction of the propagating modes were launched (largest input beam used = 9°, acceptance angle = 27°). Fig. 35(b) indicates that mode mixing occurred as a function of fiber bend radius.

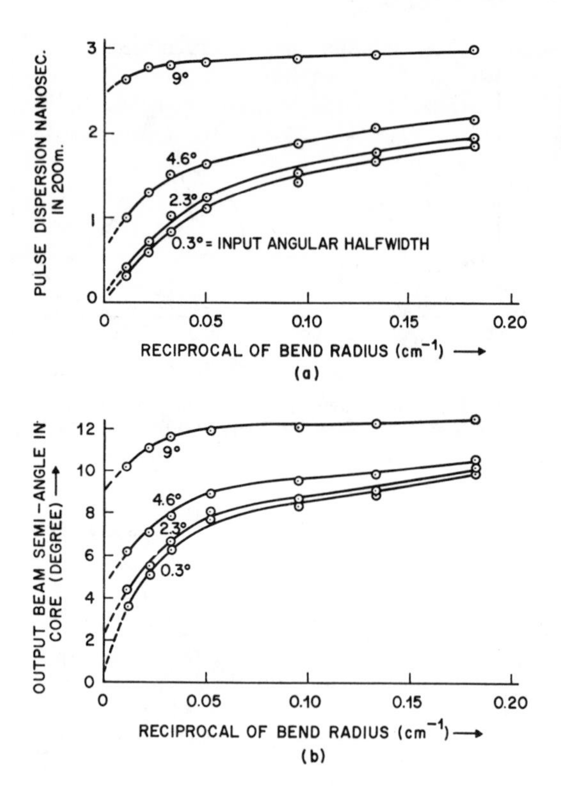

Fig. 35. Liquid-core (hexachlorobuta 1,3-diene) fiber. (a) Dispersion versus bend radius (lowest order modes at input). (b) Fiber output beam angle versus bend radius (lowest order modes at input) (after Gambling *et al.*, [13]).

The authors indicate that 400-m fiber lengths have similar results, but uncontrolled mode-mixing effects at some longer length would yield increased dispersion due to excitation of the higher order modes.

2) Conventional Clad Fibers with Uniform Core Index: The lowest loss optical guide reported in the literature is a multimode Corning fiber [79] that has an NA of 0.14, and consists of a doped fused silica core and fused silica cladding whose diameters are 91 and 125 μm, respectively. The total attenuation (Fig. 36) is 4 dB/km between 0.8 and 0.85 μm and at 1.05 μm.

Corning has announced [79] a new fused silica fiber made by vapor deposition with a loss characteristic similar to that depicted in Fig. 36, except for the important fact that it goes down to about 2 dB/km at 1.06 μm.

At least in principle lower loss should be achieved if the core were made of undoped fused silica. Such a potentially attractive fiber has been reported [80a]. The fused silica core is surrounded by a slightly lower refractive index Vycor glass cladding; their diameters are ~90 μm and 140 μm, respectively, while the NA and attenuation are 0.0512 and 20 dB/km at 0.6328 μm. Similar fibers made with pure silica core and borosilicate claddings have been studied at Bell Laboratories and the lowest reported attenuation [80b]–[80e] is about 7 dB/km at 0.63 and 0.8 μm.

3) Graded-Index Fibers: The effect of graded-index profile upon modal dispersion has been discussed in Section II-E. Fibers with graded-index profile were proposed as a transmission medium in a patent issued in 1969 [18] and fibers with parabolic-index profile (Selfoc) were announced in 1969 [14]. The lowest reported loss [81] is around 20 dB/km between 0.75 and 0.9 μm.

The modal dispersion measured in a relatively short Selfoc guide has been found to be remarkably small [82]. A

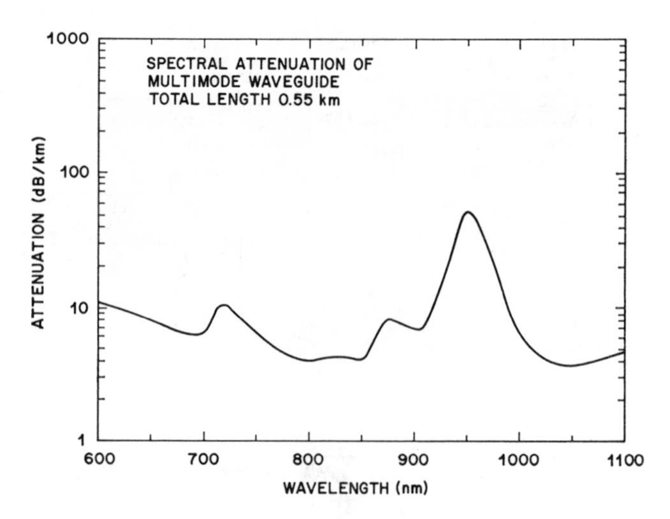

Fig. 36. Fused silica multimode fiber loss versus wavelength (after Keck, Maurer, and Schultz, [79]).

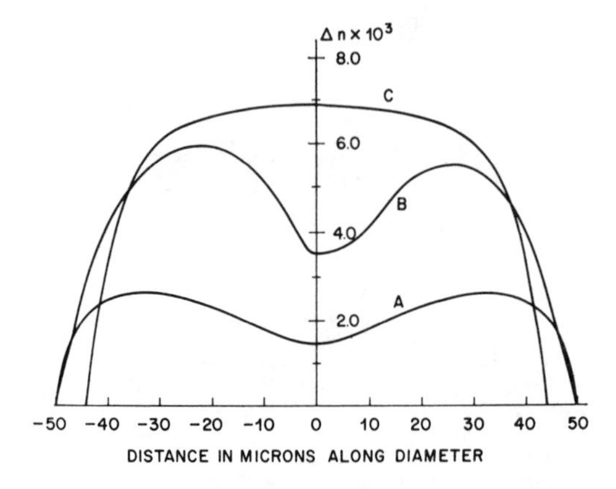

Fig. 37. Index of refraction versus radial position for fibers reported in [20] and [21].

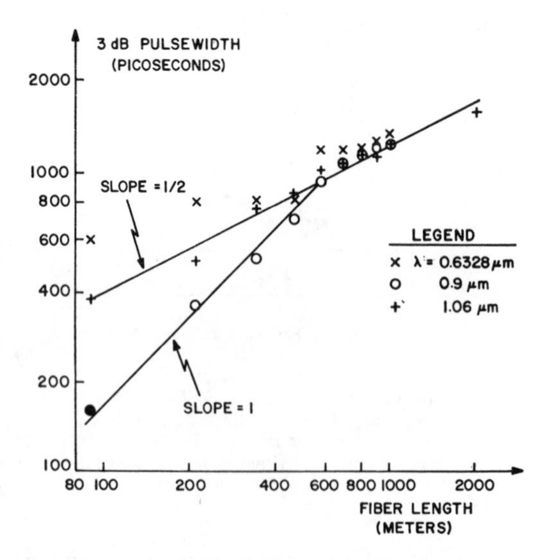

Fig. 38. Fiber output pulsewidth versus length for a Corning Glass Works fiber reported in [21].

100-ps gallium-arsenide laser pulse became only twice as wide after traversing a 70-m-long fiber having a 0.1-mm diameter and a refractive index 1.562 on-axis and 1.54 near the periphery.

Modal dispersion in three lengths of low-loss multimode fibers made by Corning was measured and all showed similar

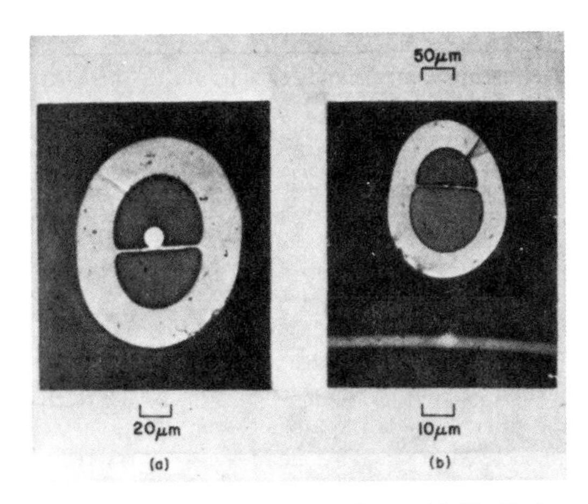

Fig. 39. Photographs of single material fibers with He–Ne laser excitation. (a) Multimode guide. (b) Single-mode guide (after Kaiser, Marcatili, and Miller, [22]).

behavior at the wavelengths of 0.6328 μm, 0.9 μm, and 1.06 μm [20], [21]. The index profiles of these fibers, as plotted in Fig. 37, were graded more or less in a nonuniform manner. Fibers A and B, each 1-km long, exhibited low pulse dispersion of about 1.5 ns; while fiber C, 226-m long, gave 3.5 ns. Subsequently, pulse spreading as a function of length was investigated in fiber A and a square-root-of-the-length dependence was found [21] (Fig. 38). It was apparent that mode-mixing effects reduced the modal dispersion in this fiber, although the index profile might have played a part in its initial reduction.

4) Single-Material Fibers: The low loss of about 2.5 dB/km measured in low-OH-content fused silica both in bulk [68] and in unclad fibers [70] has stimulated the development of an optical guide made exclusively of a single low-loss material [22]. Because of the fiber cross section (Fig. 39), the electromagnetic energy travels mostly in the enlarged central portion and decays exponentially in the thin supporting slabs.

The best reported results [83] of this potentially low-loss guide were 15, 20, and 5 dB/km at 0.7, 0.9, and 1.1 μm. In single-mode single-material fiber, the lowest reported loss [22] is 55 dB/km at 1.06 μm.

Note: The references and perspective comments on the advantages that fibers present for future transmission systems are given at the end of Part II of this paper.

Part II: Devices and Systems Considerations

Part I of this paper has presented the fundamentals for optical-fiber transmission media. This portion, Part II, presents a discussion of carrier generators (Section III), modulation techniques (Section IV), detectors (Section V), integrated optics (Section VI), fiber transmission system considerations (Section VII), general form of potential applications (Section VIII), concluding remarks (Section IX), and References.

III. Carrier Generators (Optical Sources)

A. Requirements and Types

Broadly speaking an optical power source or carrier generator that is suitable for optical-fiber transmission must be reliable, economically viable, and compatible with the fiber. The first two requirements are usually set by past engineering experiences and economic considerations and are beyond the scope of the present discussion. The last requirement, compatibility, is intimately tied with properties of the fiber, which include factors such as fiber geometry, its loss spectrum, group-delay distortion, modal characteristics, etc. The question of simple and reliable source–fiber coupling configuration versus high coupling efficiency will involve compromises which depend on both fibers and sources. Also, the characteristics of the source will have important influences on the choice of fiber types (e.g., single-mode versus multimode, uniform-index profile versus graded-index profile, etc.), and the ultimate design of fibers (e.g., core size, core–cladding index difference, etc.). Hence in selecting carrier generators for fiber systems, the interplay of the above factors with each other and with system requirements must all be taken into consideration.

Both coherent and incoherent sources can be used in fiber systems; while either single-mode or multimode fibers can be used with coherent sources, only the multimode fiber can accept enough power from an incoherent source to be useful. At present, four existing types of solid-state optical sources show promise as carrier generators for fiber transmission; they are the LED, the superluminescent diode (SLD), the injection laser, and the neodymium-doped yttrium-aluminum-garnet (Nd:YAG) laser pumped by LED's. Each of the above types is discussed in the following. Optical sources of a more speculative nature will be considered in Section VI.

1) Light-Emitting Diodes (LED's): The electroluminescent radiation from a LED is incoherent and arises from the process of recombination of carriers injected across the p-n junction of a semiconductor diode. A basic measure of the usefulness of such an incoherent source for optical communication applications is its radiance (or brightness) as measured in watts of optical power radiated into a unit solid angle per unit area of the emitting surface. Another important consideration is its emission response time, which limits the bandwidth with which the source can be modulated directly by variation of the injected current. To be useful for fiber transmission applications, a LED must have a high radiance and a fast response time.

An asset of the LED is the ease with which it can be directly modulated by an analog signal as compared with lasers. This is because the relationship between the output power and the input current is linear over a usefully large range. For digital modulation the LED is usually driven much harder, to the point where thermal effects begin to limit.

One of the first LED's suitable for communications use was fabricated by Keyes and Quist in 1962 from a single crystal of n-type GaAs with a junction formed by zinc diffusion [84]. The diode had a junction area of 7.5×10^{-4} cm^2 and

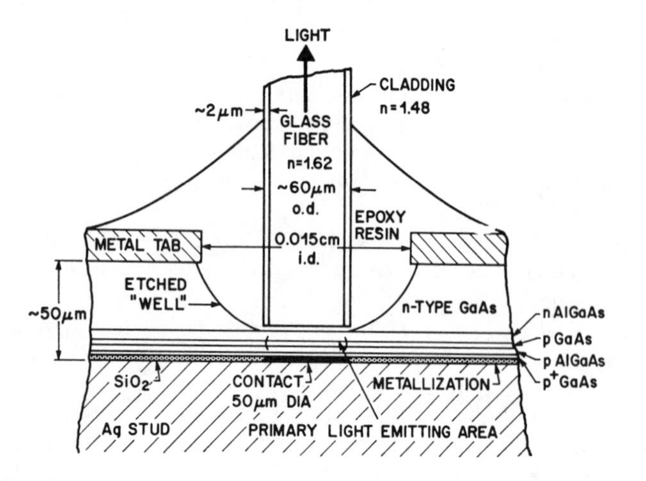

Fig. 40. Cross-sectional drawing of the small-area high-radiance GaAs–$Al_xGa_{1-x}As$ double-heterostructure (DH) light-emitting diode (LED) coupled to an optical fiber (after Burrus and Miller, [90]).

emitted at the wavelength of 0.93 μm with a half-power width of 350 Å at room temperature. Switching time measurements indicated that the diode could be modulated in excess of 100 MHz.

A GaAs LED with a hemispherical dome to enhance coupling was fabricated and measured by Carr [85]. The diode had a junction area of 3.3×10^{-4} cm² and a room-temperature spectral width of 250 A centered at 0.9 μm. An external quantum efficiency of 7.3 percent and an emission rise time of 1.6 ns were measured at room temperature.

A GaAs LED with its emitting surface perpendicular to the p-n junction was made by Zargar'yants and his co-workers for range-finder applications [86]. The area of the junction was about 7×10^{-4} cm² and the thickness of the active layer was 11 μm. A radiance value of 25 W/sr/cm² was obtained with a continuous drive of 600 mA at 300 K.

Recently, a small-area high-current-density LED of high radiance was developed and fabricated by Burrus specifically for application in optical-fiber transmission systems [87], [90]. Earlier units of this type had diffused junctions in GaAs and radiance values of 20–35 W/sr/cm² at about 300-mA drive [87]. More recently LED's of the same design but of higher efficiency and radiance were fabricated using the double-heterostructure aluminum-gallium-arsenide material ($Al_xGa_{1-x}As$) grown by liquid-phase epitaxy for low-threshold injection lasers that operated continuously at room temperature [88], [89]. A desirable property of this ternary compound material system is that the peak of the emission can be made to fall anywhere within the wavelength region of 0.75 to 0.9 μm, a region of low fiber loss, by controlling the mole fraction x or the amount of aluminum in the compound material. A cross-sectional drawing of a LED of this type is shown in Fig. 40 [90]. The radiance of these units with an emission area of 2×10^{-5} cm² could be as high as 100 W/sr/cm² for a drive of 150 mA; about 2 mW of the emitted power could be coupled into a fiber of 10-percent relative index difference between core and cladding. The half-power spectral width of these diodes has been measured to be between 350 to 400 Å, their emission response time found to be about 1–2 ns, and their external quantum efficiency observed to be 2–3 percent [90]. Various characteristics, including pulsing behavior and noise, of LED's of similar design made with homostructure (diffused junctions) single- and double-heterostructure materials have been studied extensively by Burrus and his co-workers [90]–[93]. Although the current density in these devices is extremely high—tens of kiloamperes per square centimeter—their operating life is found to be surprisingly good—over 10^4 h; many units have been used in system experiments.

In order to couple as much power as possible from a LED into a fiber, it is necessary to use a multimode fiber that has a large v number and therefore propagates many modes [see (24) and (26)]; this, of course, is because the emission from a LED, being incoherent, radiates in many spatial modes, and the number of spatial modes that can be coupled into a fiber is equal to the number of modes of that fiber. A simple and reliable method of coupling is that shown in Fig. 40 where the end of the fiber is butted against the emitting surface of the LED and secured in place by means of an epoxy resin. In this case the power P_0 that is coupled into the fiber is given by

$$P_0 = B \cdot A_s \cdot \Omega \simeq 2\pi B A_s n_1^2 \Delta \qquad (95)$$

where B is the radiance (or brightness) of the source in free space, A_s is the area of emission, and Ω is the external solid acceptance angle of the fiber and is approximately equal to $2\pi n_1^2 \Delta$ for small Δ. The above equation assumes that the angular distribution of power is uniform and that the fiber core area is at least as large as the LED emission area. It can be shown that (95) holds in general for any dielectric material that fills in the space between the fiber and the LED, subject to a correction factor due to Fresnel reflection loss, which is neglected in the above formulation. As an example, consider the Burrus-type LED shown in Fig. 40. If B is assumed to be 100 W/sr/cm², P_0, as given by (95), is 2.8 mW, where the experimentally observed value is 1.7 mW [90]. The discrepancy is probably due to the fact that the reflection losses, the falling-off of B with angle, and the distance between the fiber and the emitting surface are not taken into account in (95).

Since the input power P_0 is proportional to the solid acceptance angle of the fiber, one should use a fiber with a Δ as large as is consistent with the limitations imposed by the system requirements of group-delay spread (which increases with Δ). On the other hand, more power could be coupled into a fiber of *larger* core area by using a lens or a conical taper which increases the acceptance angle of the fiber; the lens or the conical taper merely serves as a mode converter so that the brightness in the core remains the same. In practice, the improvement in the increased power would have to be weighed against the complications of the additional optics.

For systems considerations in later sections, it is convenient to assume a value of P_0 between 10Δ mW to 20Δ mW, which is presently attainable with Burrus-type diodes. The rms spectral width of these diodes is about 165 Å, which gives an rms delay spread due to material dispersion in silica fibers of about 1.7 ns/km (see discussion associated with Fig. 20). If the repeater spacing is determined by this material effect rather than by the limitation of optical power, it is possible to use a filter to narrow the source spectral width, at the expense of reduced input power, until maximum repeater spacing is obtained.

Since silica fibers exhibit low loss also in the 1.0–1.1-μm wavelength region, LED's that emit in this region are also of interest. An example is the $In_xGa_{1-x}As$ LED which emits in the neighborhood of 1.05 μm [94]. A general review of LED's is given by Bergh and Dean [95].

2) Superluminescent Diodes (SLD's): When both spontaneous and stimulated emission occur in a LED, its output can be of narrower spectral width and higher radiance than if spontaneous emission alone exists. GaAs LED's operated in

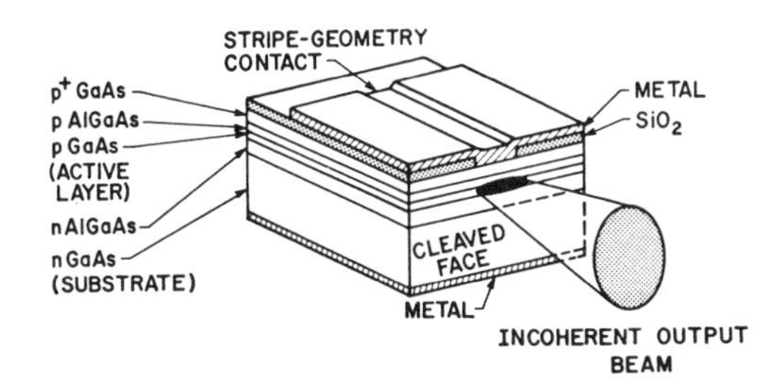

Fig. 41. Sketch of the GaAs–$Al_xGa_{1-x}As$ DH superluminescent diode (SLD) (after Lee, Burrus, and Miller, [97]).

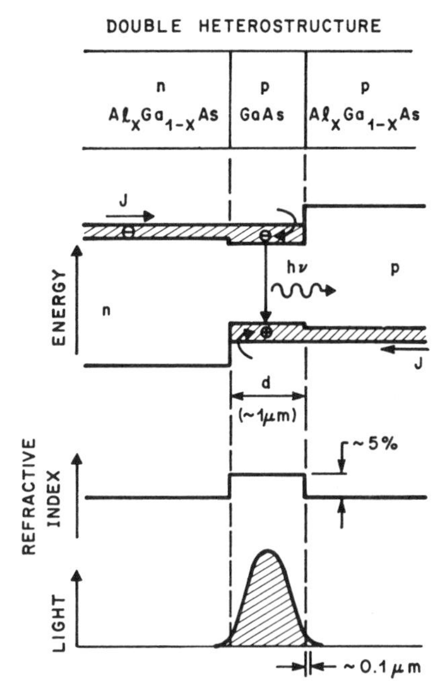

Fig. 42. Schematic representation of the band edge with forward bias, refractive index change, and optical field distribution in GaAs–$Al_xGa_{1-x}As$ DH injection laser (after Hayashi, Panish, and Reinhart, [103]).

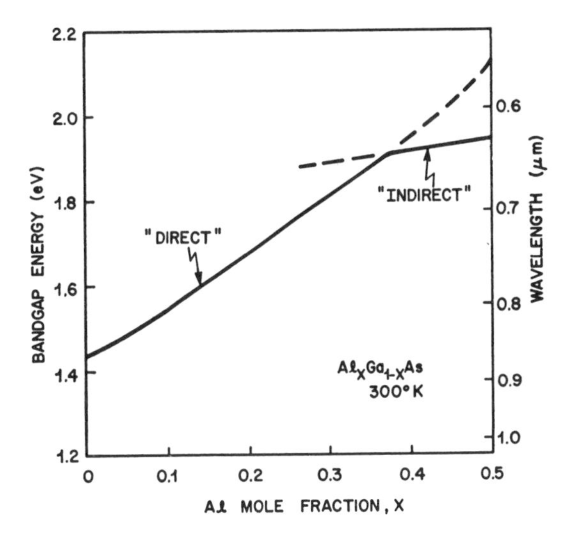

Fig. 43. Bandgap energy as a function of aluminum mole fraction x for $Al_xGa_{1-x}As$ at room temperature (after Berolo and Woolley, [105], and Panish and Hayashi, [106]).

this way were first investigated by Kurbatov and his co-workers who named them superluminescent diodes (SLD's) [96]. The geometry of their diodes was similar to that of stripe-contact laser diodes except that the stripe was inclined at 10° with respect to the normal to the emitting surface to eliminate feedback. The output, therefore, consisted of a single-pass amplified spontaneous emission which was incoherent, but was observed to be 90-percent polarized. Under pulsed operation, a spectral width of 20 Å was obtained with very high pulsed driving current.

Similar characteristics were obtained by Lee and his co-workers with SLD's made from the double-heterostructure $Al_xGa_{1-x}As$ material used to make lasers and LED's [97]. The device configuration, shown in Fig. 41, was similar to that of stripe-geometry double-heterostructure injection laser, except that optical feedback was suppressed by eliminating one of the mirrors and providing absorption for the backward waves in the cavity, thus resulting in single-pass gain only. Unlike lasers which tended to oscillate in many modes at high driving currents, the SLD structure favored low-order modes for large output powers; as much as 50-mW (pulsed) peak power has been efficiently coupled into a fiber of NA = 0.63. The spectral width was measured to be 50–80 Å and some preference for TE-mode polarization was observed.

The SLD in its present experimental form is inefficient compared to the laser and requires rather high driving currents. Nevertheless, the phenomenon of superluminescence remains as a potential means to obtain higher radiance and narrower spectral width from incoherent LED's.

3) Injection Lasers: Amongst lasers of all types the semiconductor injection laser is exceptionally well-suited for optical-fiber transmission; some of the desirable attributes are that it is physically small, inherently rugged, highly efficient, and can be pumped and modulated simply and directly by means of the injected current. Semiconductor lasers date back to 1962 when stimulated emission was first observed in GaAs [98]–[100]. A great deal of progress has been made since then: lasers have been made from many materials, diverse methods of pumping explored, and the wavelength of radiation extended from the ultraviolet to the far infrared. By far the most promising semiconductor laser is that of the double-heterostructure (DH) design consisting of layers of $Al_xGa_{1-x}As$ (with different values of x) grown by liquid-phase epitaxy [88], [89] and with which low-threshold continuous operation at room temperature has been achieved [101], [102]. Fig. 42 shows the energy bands and the sandwich structure by means of which low-threshold operation is obtained through the confinement of the carriers and of the optical field in the central active layer of narrower energy gap (lower Al content) [103]. Threshold current density for pulsed operation at room temperature has now been reduced to below 700 A/cm² using five-layer heterostructures for separate optical and carrier confinement [104a], [104b]; still lower values are in prospect.

As mentioned earlier, an important advantage of the $Al_xGa_{1-x}As$ ternary compound material system is that its bandgap energy is a function of x and hence the wavelength of emission can be made to fall anywhere within the wavelength region of 0.75 to 0.9 μm, a region of low fiber loss. This feature of being able to control the wavelength opens up the possibility of frequency-multiplexing a number of carriers into a single fiber. Fig. 43 shows how the bandgap energy varies as a function of x, the aluminum mole fraction [105], [106]. Plotted in Fig. 44 is the variation of index of refraction with x [107a], [107b].

A state-of-the-art discussion of material and laser characteristics of the $Al_xGa_{1-x}As$ heterostructure semiconductor

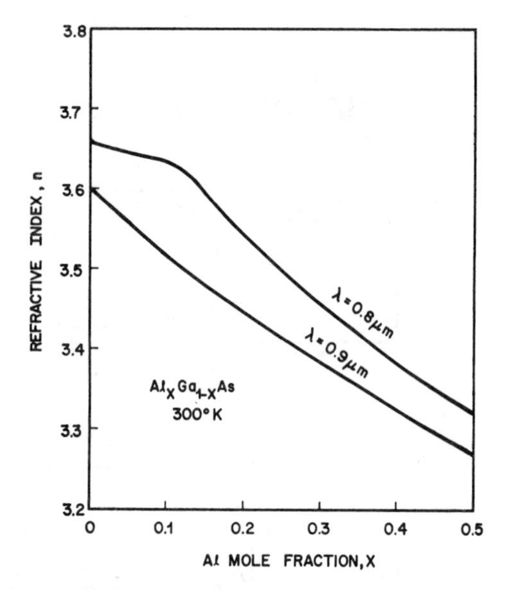

Fig. 44. Refractive index versus aluminum mole fraction x for $Al_xGa_{1-x}As$ at room temperature (after Casey, Sell, and Panish, [107b]).

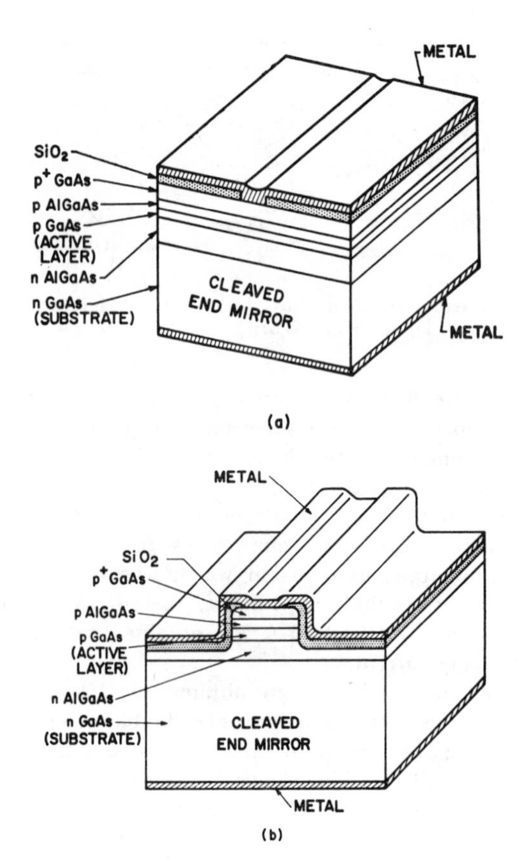

Fig. 45. Sketches of GaAs–$Al_xGa_{1-x}As$ injection lasers with lateral confinement. (a) Oxide-masked stripe-geometry structure. (b) Mesa-stripe structure (after D'Asaro, [113]).

laser is given in a comprehensive review by Panish and Hayashi [106].

One of the difficult problems with injection lasers is the attainment of single-mode operation. Because the gain profile is over 200 Å wide, many longitudinal orders can oscillate. Moreover, inhomogeneities in the material cause emission to occur in filaments [108] which tend to be unstable and cause noise. The stripe geometry has been used to alleviate this problem by defining a narrow active volume in which stable low-order modes can oscillate. Both stripe-contact [109] and proton-bombardment isolations [110] have been used with DH lasers. In each case, thermal conductance is also improved, which is important for low-threshold continuous operation at room temperature. Another structure which provides lateral confinement and low threshold is the mesa-stripe structure [111], [112]. Both the stripe-geometry DH laser and the mesa-stripe structure DH laser are illustrated in Fig. 45. A review of the properties of stripe-geometry GaAs injection lasers is given by D'Asaro [113].

Unlike the situation with the LED, power coupling from injection lasers into fibers is not beset by fundamental limitations. High coupling efficiency into multimode fibers can be accomplished with relative ease by merely placing the fiber end next to the output end of the laser [114a], or by using a fiber with a spherical end [114b]. Efficient single-mode coupling, on the other hand, is not as easy because the laser mode is elliptical in cross section whereas the usual fiber has a circular core. A mode transformer in the form of a cylindrical lens [115] or a section of fiber with suitable taper could be used to effect the mode matching and to increase the coupling efficiency. An alternative solution is to use lasers with large optical cavities which have a thick guiding layer for optical confinement [104], [116]–[118].

Because the radiative recombination times of the carriers in GaAs and in AlGaAs are very short (about 1 ns for GaAs and a few nanoseconds for AlGaAs), and are further shortened by the action of stimulated emission, injection lasers can easily be directly modulated at high speed by variation of the injected current. Much work has been done in the investigation of the properties of the laser for direct modulation and of the various methods of modulation. A review of the work prior to October 1970 is given by Paoli and Ripper [119]. Recent studies on $Al_xGa_{1-x}As$ DH lasers will be treated in the following section on modulation.

The problem of reliability has plagued the GaAs injection laser ever since its inception; lasers would degrade, either catastrophically or gradually, no matter what the environment or the mode of operation. Catastrophic failure has been found to be the result of mechanical damage of the facets and shown to be related to the optical flux density [120], [121]. Understanding and avoidance of this type of failure is important in the case of high-power lasers. Lower power lasers degrade gradually but surely over a period ranging from seconds to hours. Understanding of this gradual mode of failure is incomplete, but a great deal of progress has been made recently. Kressel and Byer [122] report that 1) the degradation process is a bulk phenomenon not amenable to control by surface passivation; 2) gradual degradation may occur with no evidence of facet damage; 3) the formation of nonradiative recombination centers in the junction vicinity is responsible for the reduction in the internal differential quantum efficiency and for the increase in the threshold; 4) the degradation rate depends superlinearly on current density; and 5) the uniformity of emission (which is related to material imperfection) has an important bearing on laser life. Newman and Ritchie [123] have observed that, as in the case of diffused-junction lasers, optical flux is not the cause of degradation in DH lasers, but the formation of nonradiative recombination centers is. More recently, Hartman and Hartman [124] have shown that strain fields introduced during the bonding procedure by differential thermal contraction are related to the degradation of stripe-geometry DH lasers operated continuously at room temperature; they have developed a new bonding procedure which does not produce significant strains. DeLoach *et al.* [125] have identified a major failure mechanism which consists of the formation of localized "dark lines" in the optical cavity of stripe-geometry proton-bombarded DH lasers. These dark lines are thought to be dislocations "decorated" by impurities or natural lattice defects, which trap carriers and eventually lead to optical absorption [126]. Therefore, it appears that reduction of lattice mismatch and cleanliness of fabrication may contribute significantly to the amelioration of laser life. Indeed, lasers have been operated continuously at room temperature over 2000 h [127], [128] and the prognosis is very encouraging.

When driven to yield appreciable power (tens of milliwatts), present-day stripe-geometry DH lasers that operate continuously at room temperatures tend to oscillate in several transverse and longitudinal modes which span a spectral width of several angstroms. However, stable single-frequency single-mode output of about 10 mW has been observed from rather long-lived lasers [126]. Thus it is reasonable to assume that DH lasers can provide 10 mW or more of single-mode or multimode power. In either case, the spectral width is sufficiently narrow to render material dispersion unimportant in long-distance fiber transmission.

4) Neodymium-Doped Yttrium-Aluminum-Garnet (Nd:YAG) Lasers: The Nd:YAG laser is suitable for optical-fiber transmission for the following reasons: 1) the wavelength of emission is 1.064 μm, which coincides with one of the low-loss regions of silica fibers; 2) its longer wavelength of operation (compared to the 0.8–0.9-μm region of the injection laser) is intrinsically more advantageous since the ultimate loss mechanism in glass is Rayleigh scattering, which has a $1/\lambda^4$ dependence; 3) its emission spectral width is narrower than that of the injection laser, thus giving rise to less material dispersion in the fibers; 4) it is relatively easy to obtain single-frequency single-mode output from an Nd:YAG laser; and 5) LED pumping is expected to yield a long life.

The Nd:YAG laser was first operated continuously at room temperature in 1964 using an elliptical-cylinder pumping cavity with a tungsten-halogen lamp [129]. This combination is highly developed as a research instrument and is used in industry as a manufacturing tool. Multimode output power of 10 W or more is easily obtained with an overall efficiency of 1 percent [130]. Bulkiness, larger power consumption, and the necessity of water cooling render this form of the Nd:YAG laser unsuitable for fiber system use.

A compact and simple laser consuming orders of magnitude less power and requiring no forced cooling can be realized by using efficient incoherent electroluminescent diodes (LED's) to pump the Nd:YAG rod. One of the main pump bands is near 0.81 μm, so either $GaAs_{1-x}P_x$ or $Al_xGa_{1-x}As$ diodes can be utilized. At first, such lasers used an array of GaAsP diodes cooled to 77 K in order to shift the spectral output into the desired 0.81-μm pump band [131], [132]. Later, room-temperature CW operation was obtained with an array of sixty-four GaAsP diodes in a semielliptical-cylinder pumping cavity [133]. An electrical input power of 30 W to the diode array produced a laser output power of 1.4 mW. Recently, an output of 52 mW in TEM_{00} modes was obtained from a Nd:YAG laser operating at 269 K and pumped by an array of AlGaAs diodes [134].

A further reduction in size and power consumption has been realized by end-pumping with a single LED. The feasibility of this method of pumping was demonstrated in an experiment where a pulsed heterostructure GaAs laser was temperature-tuned to end-pump a 1.5-mm-diameter by 25.4-mm-long laser rod [135]. Most recently, a miniaturized Nd:YAG laser using a 0.45-mm by 5-mm laser rod end-pumped by a single GaAsP LED was reported (see Fig. 46) [136]. Laser threshold was achieved at room temperature in a pulsed mode of operation. Single-frequency TEM_{00}-mode output power of a few milliwatts is expected from such a laser end-pumped by a single AlGaAs LED.

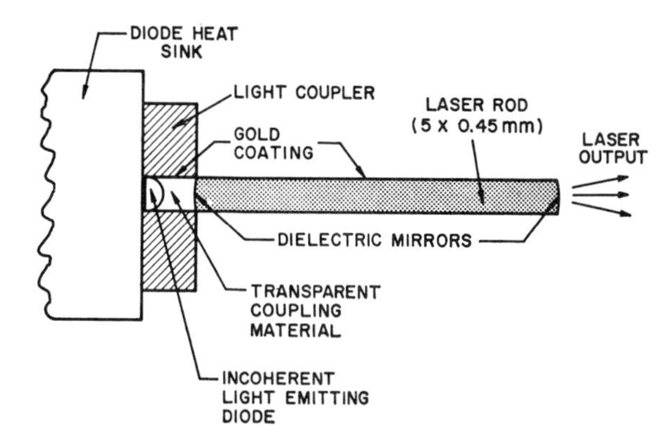

Fig. 46. Schematic configuration of the miniature Nd:YAG laser end-pumped by a single LED (after Chesler and Draegert, [136]).

Also very recently, short lengths (1–2 cm) of fused silica fibers with Nd-doped cores have been operated as end-pumped lasers [137]. The core diameters (a few tens of micrometers) and the measured lasing threshold (1–2 mW of absorbed power) are compatible with present small-area high-radiance LED's [92], and the combination may lead to a truly miniature LED-pumped Nd laser.

The Nd:YAG laser is not suitable for direct modulation by varying either the pump or the cavity loss as in the case of the injection laser. This is because the fluorescence lifetime of the upper laser level is about 230 μs [138], so that even with further shortening by stimulated emission, direct modulation at megahertz rates is difficult to achieve. An external optical modulator is therefore a necessary companion of the Nd:YAG laser for communication applications.

B. Summary and Comparison

Table V summarizes and compares the various pertinent characteristics of the four optical sources discussed earlier, specifically for potential application in optical-fiber transmission systems. The approximate numerical values given are believed to be reasonable estimates of expected performance with presently available data. Continuing research progress may render these estimates conservative in the future.

IV. Modulation Techniques

A. Approach

As discussed in the previous section on carrier generators, the method and approach to modulation of optical signals are highly dependent upon the optical source to be used. The semiconductor sources—the LED, the SLD, and the injection laser—can all be directly modulated at high speed by variation of the pumping current. The Nd:YAG laser, on the other hand, requires an external modulator. An optical modulator can also serve as a switch or a time-division multiplexer. Recent and pertinent work on direct modulation of the LED and of the injection laser and on optical modulators which are suitable for applications in optical-fiber systems are discussed in the following subsections.

TABLE V

POTENTIAL CARRIER GENERATORS FOR OPTICAL-FIBER TRANSMISSION SYSTEMS

	Incoherent		Coherent	
Source	electroluminescent diode (LED)	superluminescent diode (SLD)	semiconductor laser	solid-state ion laser
Material	$Al_xGa_{1-x}As$ double heterostructure (DH)	$Al_xGa_{1-x}As$ double heterostructure (DH)	$Al_xGa_{1-x}As$ double heterostructure (DH)	Nd:YAG
Pump	direct current (dc)	direct current (dc)	direct current (dc)	(AlGaAs) LED
Input electrical power	0.2–0.5 W	3–5 W	0.2–0.5 W	1–2 W
Output optical power	5 mW	~50 mW	~10 mW (single mode) ~50 mW (multimode)	~2 mW (single mode) ~5 mW (multimode)
Wavelength of emission	0.75–0.9 μm	0.75–0.9 μm	0.75–0.9 μm	1.06 μm
Spectral width	~350 Å	~50 Å	≲20 Å	≲1 Å
Coupling efficiency into fiber	$n^2\Delta$	≳50%	≳50%	~100%
Modulation	direct	direct	direct	external modulator
Modulation bandwidth	a few hundred MHz	a few hundred MHz	a few GHz	that of the modulator

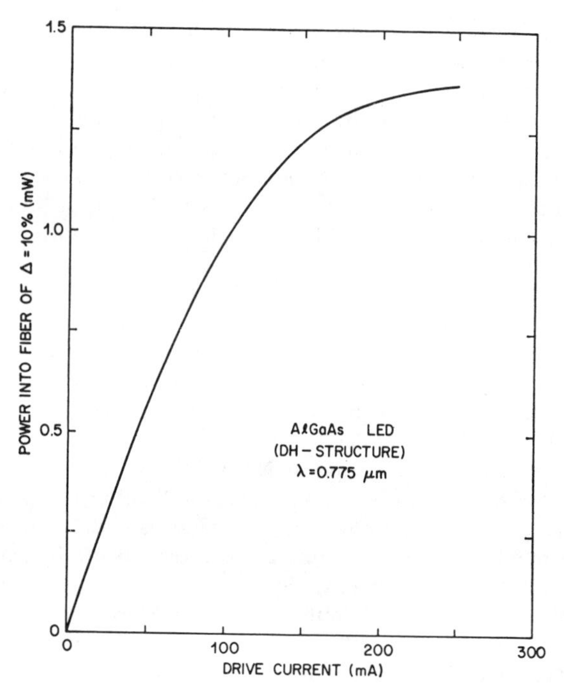

Fig. 47. Output characteristics of a typical Burrus-type AlGaAs double-heterostructure LED (after Burrus, private communication).

B. Direct Modulation of Sources

1) The LED: Shown in Fig. 47 is a typical plot of the output power versus the driving current for a Burrus-type LED. Such a characteristic amply demonstrates the suitability of the LED for direct modulation by either analog or digital signals. The modulation bandwidth can be several hundred megahertz since its emission response time is of the order of a nanosecond [90], [91].

Although simple demonstrations of analog modulation of telephone and television signals are easily done in the laboratory, field applications of analog modulation that must meet certain system standards will require a thorough evaluation of the characteristics of the type of LED to be used. For example, nonlinear distortion in the output characteristics of the LED is important and needs to be measured carefully. Variations of the output characteristics from unit to unit may require some control.

The LED has also been used in many experiments involving digital modulation of PCM signals. In particular, the Burrus-type LED has been employed in repeater and systems studies at rather low speeds—6 and 12 Mb/s [139], [140].

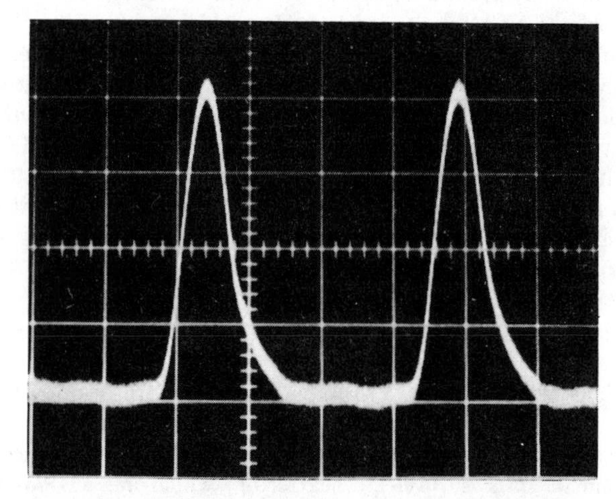

Fig. 48. Detected optical pulses from a Burrus-type GaAs LED driven at 280 Mb/s. Scale: 1 ns/div (after Dawson, [143]).

It has also been successfully and efficiently driven at the higher rate of 100 Mb/s in an optical-fiber communication experiment [141], [142]. Direct modulation at still higher bit rate is possible but becomes increasingly difficult, because the diode presents a nonlinear low impedance of a few ohms to the driver. Nevertheless, a rise time of the order of 1 ns has been measured when the source impedance is matched to that of the diode [90], [91]. Fig. 48 shows the detected pulses from a 25-μm-diameter diffused-junction GaAs LED which is driven at 280 Mb/s by means of a step-recovery diode pulser [143]. The pulsewidth at half maximum in this case is less than 1 ns.

By driving the diode with short pulses of high peak current and low duty cycle (e.g., 100-ns wide pulses at 10-kHz rate) it is possible to obtain a peak output power which is about ten times the dc rating [91]. However, the repetition rate must be restricted to 50 kHz or less in order to avoid the deleterious effects of heating. This low-duty-cycle high-peak-power mode of operation is attractive for pulse-position modulation of low bandwidth signals where high peak power is an advantage [144].

2) The SLD: The behavior of the SLD under direct modulation has not been investigated as yet, but it is not expected to be too different from that of the LED.

3) The Injection Laser: The power output as a function of the driving current for a typical stripe-geometry AlGaAs DH laser that has a life of a few hundred hours under continuous operation at room temperature [126] is given in Fig. 49. The curve shows that besides digital modulation, the injection

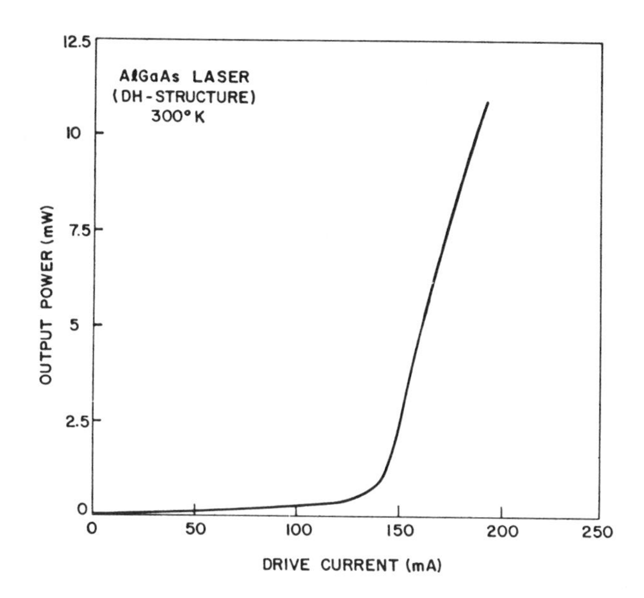

Fig. 49. Output characteristics of a typical stripe-geometry AlGaAs DH injection laser operating continuously at room temperature (after Hartman, Dyment, Hwang, and Kuhn, [128]).

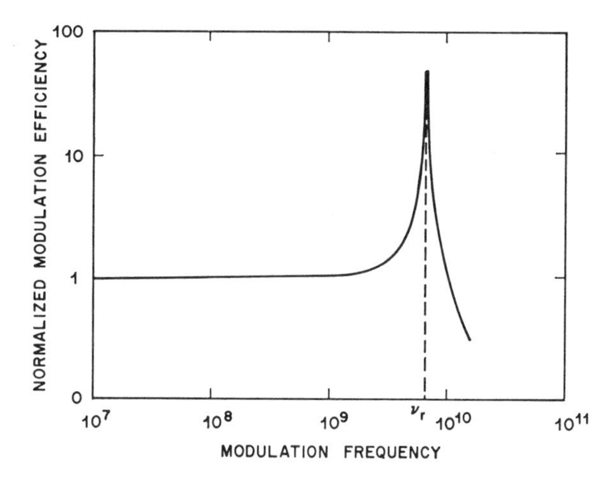

Fig. 50. Normalized modulation efficiency of the GaAs injection laser as a function of modulation frequency (after Ikegami and Suematsu, [146]).

laser is also suitable for direct analog modulation as its output characteristic has a linear portion above the threshold level. The spontaneous radiative recombination time of the carriers in the DH laser is observed to be a few nanoseconds [145] and is shortened further by the laser action of stimulated emission. Large modulation bandwidths are therefore possible with direct modulation.

An important effect is the resonance phenomenon occurring in the laser that strongly affects its transient behavior and, hence, the modulation response or efficiency [119], [146]–[148]. This resonance effect manifests itself as a peaked high-frequency noise, as "spiking" oscillations in the output, as damped relaxation oscillations when the laser is suddenly switched on, or as regular intensity pulsations at some fixed repetition rate. The phenomenon is attributed to the interplay of the optical field in the cavity and the injected electron-density distribution, resulting in oscillations of both the light output and the electron density in the laser. The resonance frequency ν_r for a single mode is approximately given by [119], [146], [148]

$$\nu_r \simeq \frac{m}{2\pi}\left\{\frac{1}{\tau_s\tau_p}\left(\frac{I}{I_t}-1\right)\right\}^{1/2}$$

where τ_s is the electron lifetime for spontaneous recombination, τ_p is the photon lifetime which is inversely proportional to the losses in the cavity, I is the drive current, I_t is the threshold current, and the factor m is the exponent of gain (gain $\propto I^m$) which depends on the band-tail structure [149]. The experimentally inferred value of $m \simeq 3$ is in good agreement with theoretical results [149], [150]. Typical values of ν_r lie in the range of a few tenths of a gigahertz to several gigahertz.

Because of this resonance phenomenon, the modulation efficiency of the injection laser exhibits a peak at the frequency ν_r as illustrated in Fig. 50. Modulation with *narrow-band* signals near ν_r therefore requires little energy, while modulation with wide-band signals must be restricted to the flat portion of the curve below ν_r to avoid signal distortion. Above ν_r, the modulation efficiency falls off very rapidly.

The time delay of a few nanoseconds associated with the building-up of population inversion in the diode when the current is abruptly switched on can be reduced to negligible values by prebiasing the laser just below its threshold. Once this delay is overcome, it is possible to modulate the laser directly with digital signals at bit rates up to the resonance frequency. Thus direct modulation experiments have been done at 200 Mb/s [151], [152], and at 300 Mb/s [153] with prebiased DH lasers operating at room temperature. Chown and his co-workers [154] in similar modulation experiments with a dc-prebiased mesa-stripe DH laser operating at room temperature have found the expected distortion and ringing effects associated with the resonance phenomenon to be suppressed over a range of bias levels for bit rates up to 1 Gb/s. Thim *et al.* [155] have used a Gunn-effect diode to modulate a stripe-geometry DH laser with 1.2-Gb/s pulses superimposed on prebiasing pedestal pulses set slightly above the laser threshold. Microwave TRAPATT oscillator diodes capable of driving short current pulses into low-impedance loads have been used by Carroll and Farrington [156] to produce a pulse train of 1-GHz repetition rate and of pulsewidth less than 350 ps from a DH injection laser. Direct modulation experiments have also been done by Russer and Schulz [157] using a step-recovery diode to drive a prebiased laser at 2.3 Gb/s.

Most recently, longer lived DH lasers that operate continuously at room temperature have been used in optical communication and repeater system experiments in which direct modulation of PCM signals at 200 Mb/s [156] and at 274 Mb/s [158] was successfully implemented. In the latter experiment [158], an error rate of 10^{-9} was achieved with a minimum signal level which was predictable by theory [159].

A laser structure which is of interest as a source of short pulses of high repetition rate is that of the double-diode structure shown in Fig. 51 in which the two sections of the laser are electrically isolated but optically coupled in tandem [160], [161]. The current in one section of the diode is maintained at a level high enough to provide gain while the current in the other section is adjusted to give saturable loss with a suitable relaxation time. Under the appropriate conditions predictable by theory, the laser produces sustained self-induced pulsations of about 100-ps duration at a repetition rate which is dependent on the driving currents and the ratio of the lengths of the two sections and which ranges from a few hundred

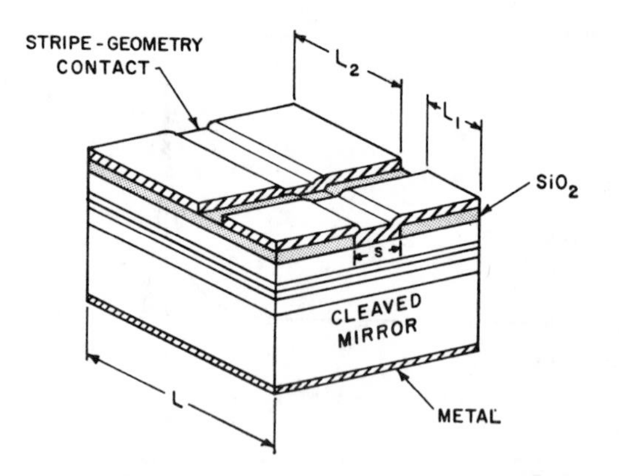

Fig. 51. Sketch of the AlGaAs stripe-geometry injection laser with double-diode structure (after Lee and Roldan, [161]).

megahertz to several gigahertz [160]–[162]. Sustained self-induced pulsations of several hundred megahertz to a gigahertz have also been observed in the output of conventional single-section stripe-geometry injection lasers [163a], [163b]. The repetition rate of these self-pulsing lasers can be synchronized to a stable microwave source by current injection [136c] and so these diodes, together with external modulators, are suitable for application in a high bit-rate time-multiplexed PCM system. However, if the external microwave source is angle-modulated, the output pulses become position-modulated [163c]. Such a technique of deriving PPM optical pulses from an angle-modulated signal can provide appreciable bandwidth with very little modulating power.

Just as in the case of the LED, PPM of low bandwidth signals is possible today with commercially available large-optical-cavity (LOC) type lasers that yield high peak power at low duty cycle [116]. A PPM system using 100-ns pulses at 10-kHz pulse repetition rate has been built by Rao and his co-workers [164]. Pulse repetition rate of one megahertz or higher should be possible in future with improved lasers.

C. External Optical Modulators

1) Types and Approach: Optical modulators can either be reactive type or absorptive type, and can take either the bulk form in which the optical beam is freely propagating or the waveguide form where the optical beam is confined in a guiding structure as in a thin film. Historically, the reactive-type modulator in the bulk form has received the most attention since many materials exist that can be used in this type of modulator. Modulators using miniature guiding structures can operate with much less modulating power and, with the advent of integrated optics [165], have drawn increasing interest. A review of high-speed small-aperture bulk-type modulators is given by Chen [166].

The field-induced interactions upon which reactive-type modulators depend for their function are the electrooptic, the acoustooptic, and the magnetooptic effects. The electrooptic effect has been the most widely exploited in high-speed light modulators [166], [167], while the acoustooptic effect has experienced only moderate attention [168]. The successful application of the magnetooptic effect for light modulators has been hampered by the limited range of optical transparency of the magnetooptic materials [169].

Very little work has been done on the absorptive-type modulator. One of the field-induced interactions investigated for this type of modulator is the electro-absorption near the band edge of the AlGaAs heterojunction material (Franz–Keldysh effect) [170].

There are many performance characteristics to be considered in evaluating a modulator for a specific optical communication system. An important figure of merit is the modulator power dissipation per unit bandwidth ($P/\Delta f$), usually expressed in milliwatts per megahertz required to achieve a given degree of modulation at a particular wavelength.[7] Other factors include optical insertion loss, bandwidth, extinction ratio, temperature sensitivity, etc. (Effects of modulator impairments such as imperfect extinction ratio, reduced drive, and optical bias drift on an optical digital modulation system have been considered by Chen [171].) The physical compatibility of the modulator to the rest of the system, the ease of coupling the modulator to the optical source and to the transmission medium, and, of course, the cost are additional but extremely important considerations.

In the following, modulators of each interaction type are discussed. Emphasis is placed upon recent results pertinent to optical-fiber transmission systems.

2) Electrooptic Modulators: The electric field-induced variation of electronic polarizability in an electrooptic material produces a change in the refractive index which, in turn, impresses an angle or phase modulation upon the light beam. The angle modulation is easily converted into polarization modulation by suitably orienting the modulator crystal with respect to the polarization vector of the incident beam, or, into intensity modulation by the addition of polarizers or optical circuitry. Many such electrooptic modulators have been proposed, built, and evaluated; reviews of the work are given by Kaminow and Turner [167], and by Chen [166].

Among the many electrooptic materials that have been investigated for modulators, lithium niobate ($LiNbO_3$) and lithium tantalate ($LiTaO_3$) have emerged as most practical and useful in the 0.4- to 4-μm wavelength range. Lumped-element modulators of very wide bandwidth have been constructed and tested (at visible wavelengths) using these materials. Denton *et al.* [172] have reported a $LiTaO_3$ baseband modulator which requires 1.1 mW/MHz of modulating power for 100-percent modulation[8] at $\lambda = 0.633$ μm over a bandwidth of 1.3 GHz. A $LiNbO_3$ microwave modulator has been reported by Chow and Leonard [173] to require 3.3 mW/MHz of power for 30-percent modulation at $\lambda = 0.633$ μm over a 1.5-GHz band. Most recently, a baseband intensity modulator using $LiNbO_3$ has been developed by Chen and Benson [174] specifically for optical-fiber applications at the wavelength of 1.06 μm with emphasis placed on low cost, compactness, and simplicity. A photograph of the modulator is shown in Fig. 52. The modulator has been operated at 70 Mb/s with extinction ratio better than 40 to 1 and 1-dB optical insertion loss, and requiring 20-mW/MHz drive power for 100-percent modulation. It has also been tested at 300 MHz and 100-percent modulation by by-passing the built-in transistor amplifier with no degradation of performance.

In contrast to lumped modulators where the required modulating power is independent of the length of the modulator crystal, the power required by traveling-wave modulators can be reduced by increasing the length. Even if the length is insufficient to reduce the required power, the traveling-wave structure affords a means of matching impedance

[7] ($P/\Delta f$) is proportional to (degree of modulation)^2 $\times$ (wavelength)^3.
[8] See [166] for definition of percent modulation in optical modulators.

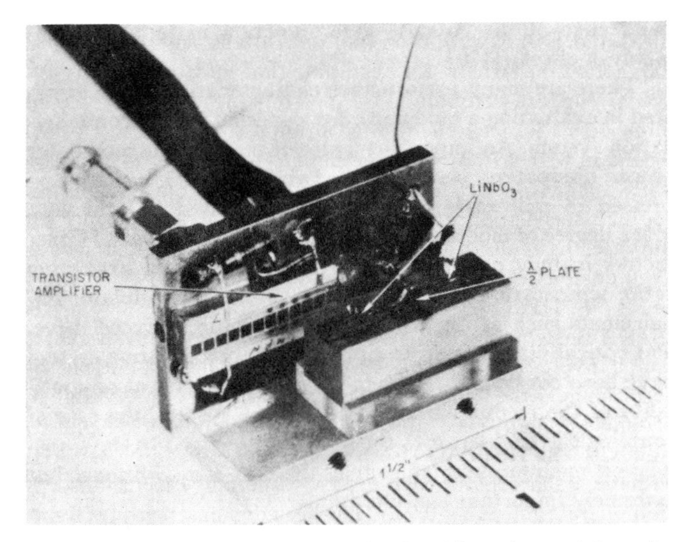

Fig. 52. Photograph of the $LiNbO_3$ baseband intensity modulator developed for optical-fiber applications (after Chen and Benson, [174]).

levels over a large bandwidth. Thus White and Chin [175] have constructed a baseband $LiTaO_3$ modulator using the traveling-wave structure for impedance matching and realizing a $P/\Delta f$ of 0.1 mW/MHz for 30-percent modulation depth at $\lambda = 0.496$ μm over a 1.5-GHz band.

Diffraction modulation by means of field-induced phase gratings is also possible using electrooptic materials. The operation is similar to that of an acoustooptic modulator except that the field-induced phase grating is created by the voltage applied to an interdigital electrode structure deposited on the electrooptic crystal as shown in Fig. 53. Such modulators have been demonstrated by St. Ledger and Ash [176] and by Hammer [177] and have been investigated for high-speed baseband operation by de Barros and Wilson [178] using thin $LiNbO_3$ crystals. The latter have obtained a $P/\Delta f$ figure of 7.5 mW/MHz for 100-percent modulation with a rise time better than 1.5 ns [178]. Instead of diffracting into many higher orders, it is possible to direct most of the power into a single higher order by using a number of harmonically related gratings in series on the crystal to simulate a "blazed" effect as demonstrated by Wright and Wilson [179a] or by using a Bragg phase grating in a single-crystal waveguide film as demonstrated by Hammer, Channin, and Duffy [179b]. High-speed deflection modulators are suitable for implementing time-division multiplex. Electrooptic diffraction and deflection modulators of the above-discussed forms lend themselves particularly to inclusion in thin film or integrated optical circuits.

A thin-film broad-band electrooptic phase modulator has been demonstrated by Kaminow and his co-workers [180] using an out-diffusion technique [235] for producing guiding layers of higher index on $LiNbO_3$ and $LiTaO_3$. The experimental modulator, as shown in Fig. 54, has a calculated bandwidth of 1.6 GHz and has been measured from 50 to 500 MHz to yield a $P/\Delta f$ figure of 0.4 mW/MHz for a modulation index of one radian at $\lambda = 0.633$ μm. The $P/\Delta f$ figure of the thin-film waveguide modulator can be improved dramatically if diffraction spread in the plane of the film is eliminated by confining the optical beam completely in a three-dimensional waveguide.

A field-induced waveguide created in an electrooptic material can be operated as a simple intensity modulator. Such guides have been observed in $LiNbO_3$ by Channin [181].

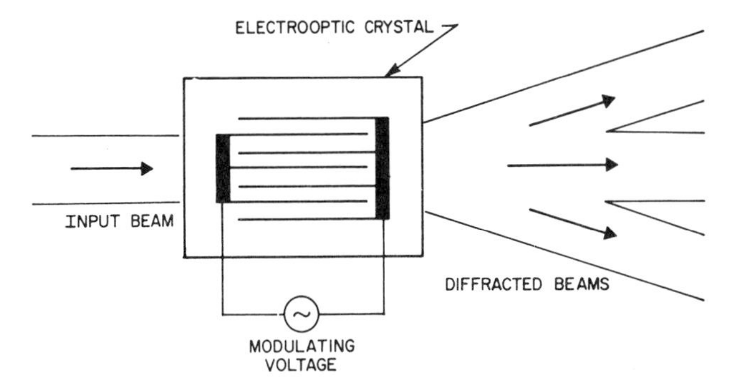

Fig. 53. Schematic configuration of electrooptic modulator with an interdigital electrode structure to create a field-induced phase grating (after deBarros and Wilson, [178]).

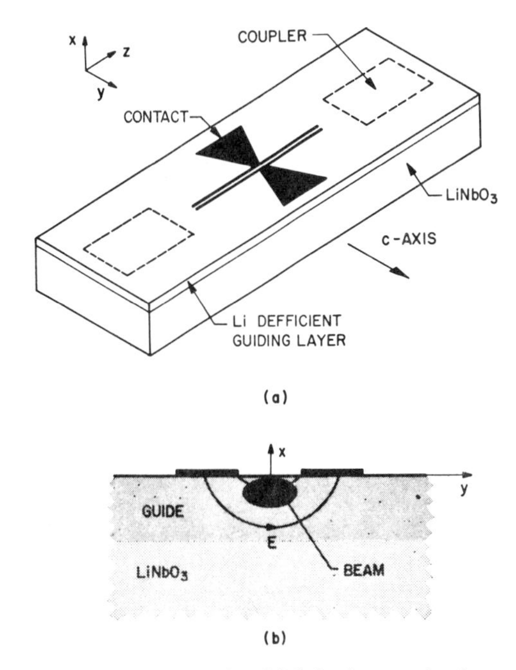

Fig. 54. Sketch of the thin-film $LiNbO_3$ electrooptic phase modulator. (a) Modulator configuration. (b) Field configuration (after Kaminow, Carruthers, Turner, and Stulz, [180]).

The depletion layer of a reverse-biased semiconductor diode exhibits waveguiding as well as electrooptic effects and therefore can function as a high-speed modulator with a potentially low $P/\Delta f$ figure. Such waveguide modulators using reverse-biased p-n junctions in GaP [182] and reverse-biased metal–semiconductor Schottky barriers in GaAs [183] have been reported. An efficient GaAs-$Al_xGa_{1-x}As$ DH modulator has been realized by Reinhart and Miller [184] with a cutoff frequency of 4 GHz and a $P/\Delta f$ figure of 0.1 mW/MHz with a modulation index of one radian at $\lambda = 1.15$ μm. The diode waveguide modulator shares with other thin-film modulators the difficult problem of coupling with other components of the system, which has to be resolved before they can be practical.

3) Acoustooptic Modulators: The phase grating created by an acoustic wave through the photoelastic effect can either diffract a light beam into many orders as in the Raman–Nath regime of operation or deflect a light beam into a single order as in the Bragg regime. For an acoustic wave of width L and wavelength Λ interacting with an optical wave of wavelength λ in a medium of index n, the Raman–Nath regime prevails when $(2\pi\lambda L/n\Lambda^2) < 1$ and the Bragg regime prevails when $(2\pi\lambda L/n\Lambda^2) > 10$ [185]. In either regime, intensity modulation

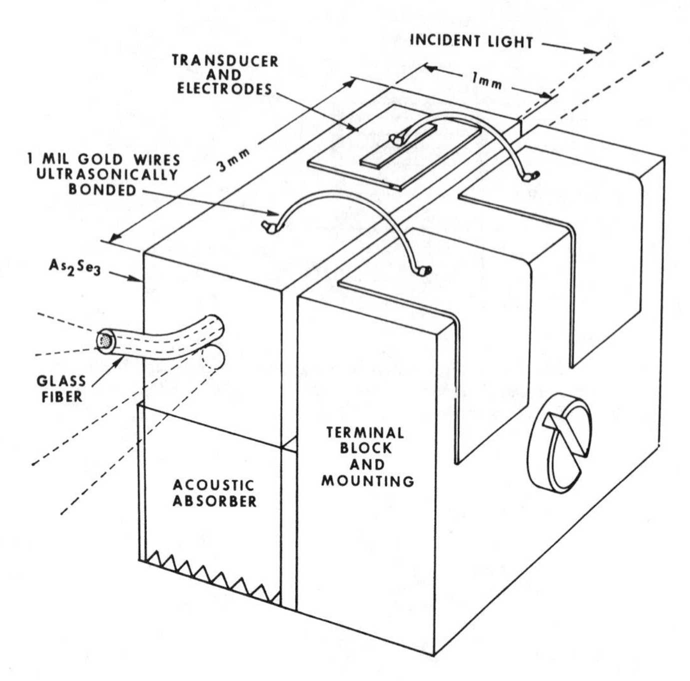

Fig. 55. Sketch of the As_2Se_3 50-Mb/s acoustooptic modulator developed for optical-fiber applications (after Warner and Pinnow, [191]).

of moderate bandwidth is easily accomplished without regard to the polarization of the incident light. At present, the bandwidth of acoustooptic modulators is limited to about a few hundred megahertz by practical considerations of the high-frequency transducer design [186], [187]. Guidelines for the selection of acoustooptic materials for device applications are given by Pinnow [188]. A review of acoustooptic materials and techniques for light deflection is presented by Uchida and Niizeki [189].

A pulse modulator has been built by Maydan [190] using As_2S_3 glass and ZnO transducer. Over 70 percent of the light at $\lambda = 0.633$ μm is deflected with a rise time of 6 ns and 600 mW of RF drive at 350 MHz. Recently, Warner and Pinnow [191] have developed a broad-band modulator specifically for optical-fiber communication at $\lambda = 1.06$ μm. Three acoustooptic materials have been investigated: the $Ge_{33}As_{12}Se_{55}$ and As_2Se_3 glasses, and the crystalline GaAs. The required power for the As_2Se_3 50-Mb/s modulator is slightly less than 1 mW/Mb/s for an acoustic subcarrier frequency of 200 MHz, and the size of the acoustooptic medium is 1 mm³. A sketch of the modulator is shown in Fig. 55.

As is true with the electrooptic interaction, the surface-wave acoustooptic interaction has potential for application in thin-film or guided-wave devices. Deflection and diffraction of optical film-guided waves by surface acoustic waves have been demonstrated by Kuhn *et al.* [192a] in a glass film on a quartz crystal, and, more recently, by Schmidt *et al.* [192b] in a $LiNbO_3$ thin-film guide produced by the out-diffusion technique mentioned earlier. High-speed acoustooptic deflectors are particularly suitable for time-division multiplexing and demultiplexing of optical digital channels.

4) *Magnetooptic Modulators:* The large Faraday rotation obtainable from iron garnet crystals makes these materials interesting candidates for efficient magnetooptic modulators. Although such modulators are quite feasible at $\lambda = 1.52$ μm, for example [169], the extremely high optical loss at $\lambda = 1.06$ μm at room temperature makes most iron garnets impractical for use with the Nd:YAG laser. Wemple *et al.* [169] have measured Faraday rotation and optical absorption in several magnetic iron garnets; they estimate that certain compositions containing praseodymium (or possibly neodymium) should allow operation at 1.06 μm with a signal absorption loss below 3 dB and with a $P/\Delta f$ figure perhaps as low as 0.01 mW/MHz. Thus magnetooptic modulators offer interesting possibilities for high-speed applications. Switching and modulation of 1.15-μm laser light in an epitaxially grown iron garnet film have been demonstrated by Tien and his co-workers [193] at rates up to 80 MHz.

5) *Absorptive-Type Modulator:* The Franz–Keldysh effect, which involves the electric-field induced changes of optical absorption near the band edge of a material, offers the possibility of efficient high-speed absorptive-type modulators. Application with the Nd:YAG laser requires the modulator material to have a sharp band edge near $\lambda = 1.06$ μm. Electro-absorption modulation at $\lambda = 0.9$ μm has been demonstrated by Reinhart [170] using $Al_xGa_{1-x}As$ material; a $P/\Delta f$ figure of 0.2 mW/MHz has been achieved for 90-percent intensity modulation.

D. *Summary and Future Development*

Semiconductor LED's and injection lasers can all be directly modulated at high speed by variation of the driving current. Digital system experiments have been performed using LED's driven at 100 Mb/s and lasers at 300 Mb/s. Direct modulation of lasers with a few gigahertz bandwidth appears feasible and is presently being pursued.

External optical modulators are required in a system using Nd:YAG lasers as the optical source. The acoustooptic bulk-type modulator is well-suited for operation at modulation rates below 100 Mb/s. For higher speed operation the electrooptic modulator is a prime candidate, and efficient electrooptic modulators with gigahertz bandwidth have been built. Presently available magnetooptic materials have too much loss at 1.06 μm to be useful, but new garnet materials show promise, especially for high-speed applications. Thin-film waveguide-type modulators offer advantages of efficiency and speed and are, therefore, being vigorously explored by workers in the field of integrated optics.

The fact that optical modulators can also be made to operate as fast switches and deflectors extends their arena of application well beyond that just for modulating the Nd:YAG lasers.

V. Detectors

A. *Requirements*

The photodetectors [194]–[198] which are needed to demodulate optical signals in optical-fiber transmission systems must satisfy certain requirements regarding performance, compatibility, and cost. The following are the important performance requirements.

1) High response or sensitivity at the wavelength of emission of the prospective sources, viz., the AlGaAs LED and injection laser ($\lambda = 0.75$–0.9 μm), and the Nd:YAG laser ($\lambda = 1.06$ μm).

2) Sufficient bandwidth or speed of response to accommodate the information rate.

3) Minimum additional noise introduced by the detector.

4) Low susceptibility of performance characteristics to changes in ambient conditions.

Compatibility requirements involve considerations of the physical size of the detector, the coupling to the fiber and to the ensuing electronics, and power-supply requirements. Recently developed photomultipliers with III–V compound photocathodes and dynodes have substantially improved sensitivity near $\lambda = 1\ \mu m$ [197], [198], but their relative bulkiness and the requirement of high voltage render them unsuitable for fiber-system use. Solid-state photodiodes can satisfy almost all the requirements of performance, compatibility, and (potentially) low cost. Their miniature size makes coupling to fibers and electronic circuits simple, and they require only low to moderate voltage for bias. This section[9] is devoted to the consideration of semiconductor photodiodes for specific use in optical-fiber communication systems.

Several excellent review papers on photodetectors are found in the literature; they are by Anderson and McMurtry [194] (1966), Anderson, DiDomenico, and Fisher [195] (1970), Melchior, Fisher, and Arams [196] (1970), and Melchior [197], [198] (1972 and 1973).

B. Noise Considerations

The ultimate performance of a communication system is usually set by noise fluctuations that are present at the input to the receiver. Noise degrades the signal and impairs the system performance. In an optical receiver, several sources of noise exist that are associated with the detection and amplification processes. Understanding of the origin, the characteristics, and the interplay of the various noise sources is essential to the design and evaluation of any optical communication system.

The relative importance and the interplay of the receiver noise depend very much on the method of demodulation. The two basic methods of demodulating an optical signal are 1) direct detection, where the output current of the photodetector is a linear function of the incident optical power, and 2) heterodyne detection, in which the incoming optical signal is mixed in the detector with that from a coherent local oscillator to produce a difference frequency from which information is extracted. Heterodyne detection, which is applicable only to a single-mode transmission system, does not appear practical at present for fiber systems because stable single-frequency lasers are required for both carrier generators and local oscillators. Direct detection is simple to implement; the incoming signal can be either incoherent or coherent, and the performance is independent of the polarization state or the modal content. Direct detection is therefore the preferred method of detection for optical-fiber communication systems.

Fig. 56 depicts the various sources of noise associated with the detection and amplification processes in an optical receiver employing direct detection. The background-radiation noise, which is important in an atmospheric propagation system, is negligible in a fiber system. The beat noise, generated in the detector from the various spectral components of an incoherent carrier such as that from a LED, is expected to be insignificant when a large number of modes is transmitted and received [144]. The quantum noise, the dark-current noise, and the surface-leakage-current noise all manifest themselves as shot noise which is characterized by Poisson statistics. The dark-current noise and the surface-leakage-current noise can be reduced by careful design and fabrication of the detector and of the devices in the amplifier. The quantum noise, which arises from the intrinsic fluctuations in the photo-excitation of carriers, is fundamental in nature and sets the ultimate limit

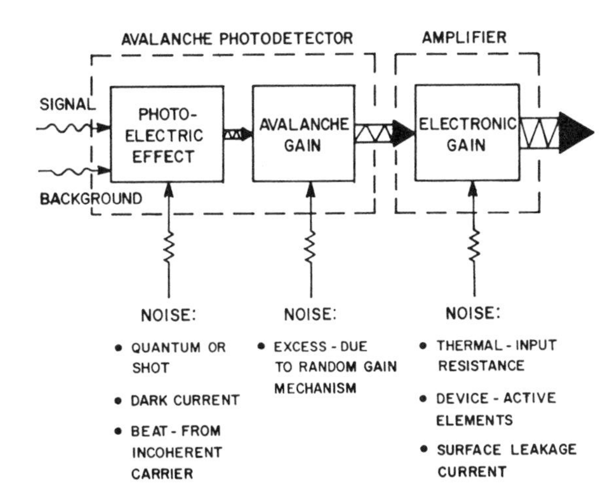

Fig. 56. Diagram depicting the various noise sources at the front end of an optical receiver employing direct detection.

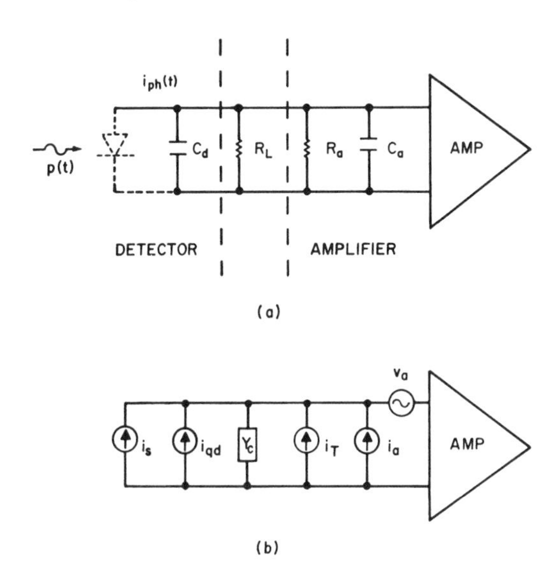

Fig. 57. (a) Equivalent circuit of the front end of an optical receiver. (b) Reduced equivalent circuit showing the signal-current source and the various noise sources.

in receiver sensitivity. Normally, when the photodiode is without internal avalanche gain, thermal noise arising from the detector load resistance and from the active elements of the electronic amplifier dominate. When internal gain is employed, the relative significance of thermal noise is reduced. However, carrier multiplication or avalanche gain is a random process which introduces excess noise into the receiver [199]–[201]. The excess noise manifests itself in the form of increased quantum and dark current noise above the level resulting from amplification of the primary shot noise. Despite the excess noise, avalanche multiplication of photocarriers does provide a useful way to improve significantly the sensitivity of optical receivers which are otherwise limited by the thermal noise of the amplifier [194]–[198], [206].

To illustrate the relative importance and the interplay of the various noise sources, consider the equivalent circuit of the front end of an optical receiver shown in Fig. 57(a). The optical power $p(t)$ is incident on the avalanche photodetector which has a junction capacitance C_d and a bias or load resistance R_L. The input capacitance and resistance of the ensuing amplifier are C_a and R_a, respectively. The primary photocurrent generated in the detector is

[9] Much of the material in this section is based on two review papers by Melchior [197], [198].

$$i_{ph}(t) = \frac{\eta e p(t)}{h\nu} \quad (96a)$$

which η is the quantum efficiency of the detector, e is the electronic charge, and $h\nu$ is the photon energy. This primary photocurrent has a dc component I_p and a signal component i_p. The average value of the signal current after internal avalanche gain is

$$\langle i_s \rangle = i_p \langle g \rangle \quad (96b)$$

where $\langle g \rangle$ is the average internal gain. The dc component I_p determines quantum or shot noise whose mean-square current after avalanche gain is given by

$$\langle i_q^2 \rangle = 2eI_p b \langle g^2 \rangle \quad (97)$$

where b is the signal bandwidth and $\langle g^2 \rangle$ is the mean-square value of the internal gain. The shot noise due to the multiplied part of the dark current I_d behaves in the same way and so can be combined with the quantum noise to give a total shot-noise contribution which is[10]

$$\langle i_{qd}^2 \rangle = 2e(I_p + I_d) b \langle g^2 \rangle. \quad (98)$$

The mean-square gain $\langle g^2 \rangle$ is a function of the statistics of the avalanche multiplication process. For a large class of avalanche photodiodes, it has been found both in theory and in experiment that [197], [199]

$$\langle g^2 \rangle = \langle g \rangle^2 F(g) \simeq \langle g \rangle^{2+x} \quad (99)$$

where $x>0$ and $F(g)$ can be considered as the excess noise factor which is gain-dependent. For germanium photodiodes x is approximately unity, whereas for properly designed silicon photodiodes x is approximately 0.5 or less.

The thermal noise due to the load resistance R_L is given by

$$\langle i_T^2 \rangle = \frac{4k\theta b}{R_L} \quad (100)$$

where k is Boltzmann's constant and θ is the absolute temperature. The noise sources associated with the active elements of the amplifier can be represented by a series voltage noise source $\langle v_a^2 \rangle$ and a shunt current noise source $\langle i_a^2 \rangle$. The latter includes the thermal noise associated with the input resistance of the amplifier.

The various noise sources previously discussed are shown in Fig. 57(b), where the signal current and the shunt admittance Y_c which combines the shunt resistances and capacitances of Fig. 57(a) are also shown. The amplifier is now assumed to be noiseless. Several important conclusions can be drawn from Fig. 57(b) and from (96)–(100):

1) The thermal noise $\langle i_T^2 \rangle$ can be reduced by using a large load resistance R_L.[11] The value of R_L need not be limited by the bandwidth consideration that the time constant $R_L C_d$ be less than the reciprocal of the signaling rate, as the signal which is integrated by the large time constant at the receiver front end can be restored by differentiation at a later stage [159], [198]. This high-impedance integrating front-end approach is a commonly recognized practice in the field of nuclear-particle counters for combating thermal noise and increasing receiver sensitivity [202]; thus, less avalanche gain is required to override the thermal noise. In fact, it may not be necessary to use avalanche gain at all for bandwidths below 1 MHz, provided that the $1/f$ noise and the surface leakage currents are sufficiently small.

2) Since the average signal power is proportional to $\langle g \rangle^2$, the shot-noise power approximately proportional to $\langle g \rangle^{2+x}$, and thermal-noise power independent of $\langle g \rangle$, it can be seen that for a given situation there is an optimal value of $\langle g \rangle$ which maximizes the signal-to-noise ratio. For the high-impedance front-end design, typical values of $\langle g \rangle$ range from about 25 for a bandwidth of a few megahertz to about 250 for a bandwidth of a few gigahertz [159].

3) The required optimal gain is reduced and the receiver sensitivity increased when the shunt admittance Y_c is reduced in value, provided, of course, neither the amplifier noise nor the dark-current noise is dominant.

[10] "Mean-square current or voltage" is understood and is henceforth omitted when we refer to expressions for noise.

[11] We refer here to the passive impedance which appears at the detector terminals, as distinguished from the active impedance presented at the terminals of a network containing feedback. Within Bell Laboratories credit is due to G. L. Miller for pointing out the advantage of using a high passive impedance (or an integrating amplifier) following the photodiode in the optical receiver.

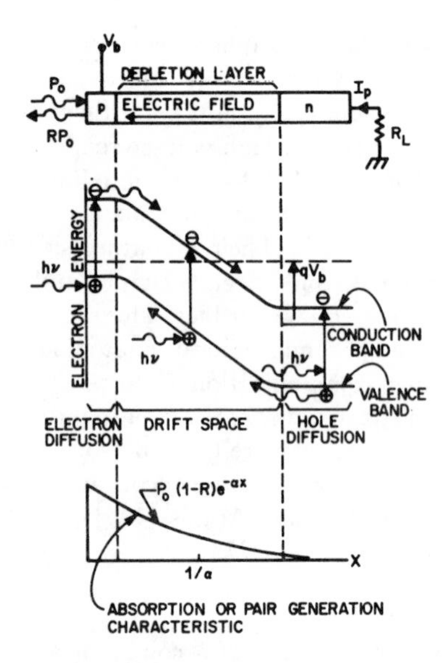

Fig. 58. Schematic representation illustrating the principle of operation of a solid-state photodiode. Cross section of the p-i-n diode, energy-band diagram under reverse bias, and carrier pair generation characteristic are shown (after Melchior, [197]).

C. Detection Process of Photodiodes

Semiconductor photodiodes, with or without avalanche gain, all have in common as an essential element a depleted region with a high electric field between two semiconducting regions. The absorbed photons generate, through band-to-band transitions, electron–hole pairs which are separated in the high-field depletion region and are collected across the junction. While the carriers are moving through the high-field region, photocurrents are induced in the load circuit. For low-level light detection, photodiodes are usually operated with reverse bias where the output current varies linearly with incident light intensity. A relatively large reverse bias (of a few tens of volts) helps to reduce carrier drift time and diode capacitance. Carrier drift time in the depleted region and minority carrier diffusion time in the p and n bulk regions need to be kept short for high-speed operation. The principle of operation of a p-i-n diode is illustrated in Fig. 58 [197], [198].

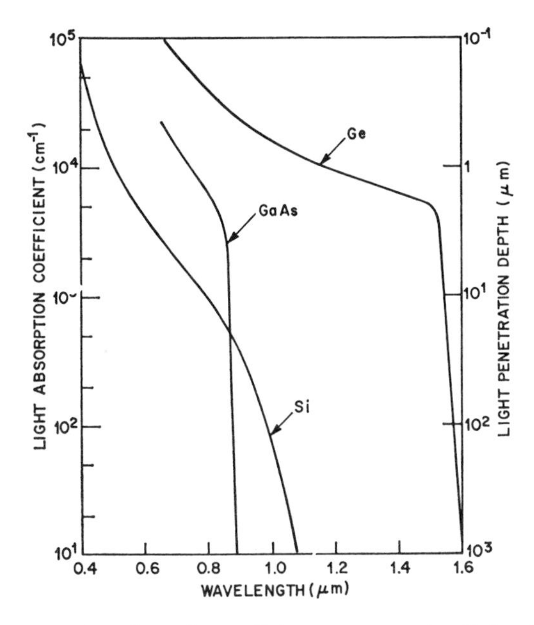

Fig. 59. Optical absorption coefficient versus wavelength for silicon, gallium arsenide, and germanium (after Melchior, [198]).

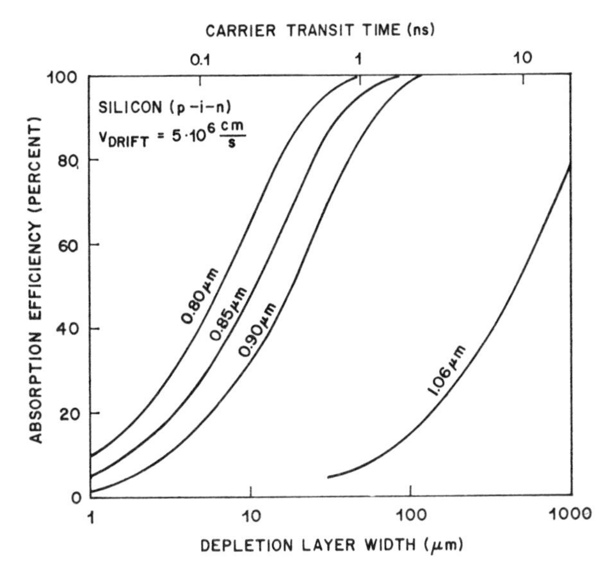

Fig. 60. Tradeoff between speed of response and quantum efficiency for silicon p-i-n photodiode (after Melchior, [203]).

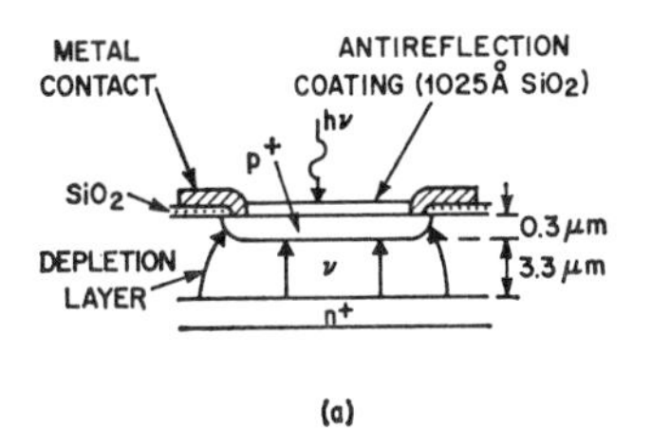

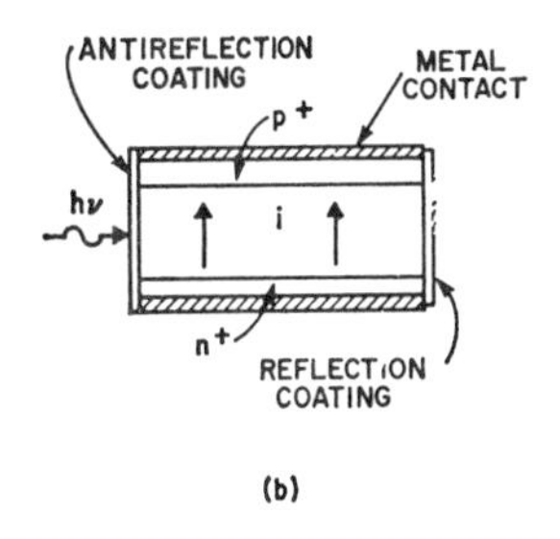

Fig. 61. Sketches showing the construction of nonavalanching p-i-n photodiodes. (a) Front-illuminated silicon photodiode optimized for $\lambda = 0.633\ \mu m$. (b) Side-illuminated silicon photodiode (after Melchior, [197]).

D. Photodiode Materials

Since the absorption coefficients of semiconductor materials are strongly dependent on the wavelength, the required thickness of the carrier collection region to absorb a large percent of the incident light is a function of the material and the wavelength of operation. Absorption coefficients and penetration depths of three commonly used semiconductor materials, gallium arsenide, silicon, and germanium, are given in Fig. 59 for the wavelength range 0.4 to 1.6 μm. It is seen that while all three materials can be used for diodes operating in the visible range, only germanium has response to 1.6 μm. However, due to the narrower bandgap, germanium photodiodes tend to have larger dark currents.

In the wavelength region from the near ultraviolet to the near infrared of about 1 μm, silicon is the preferred material as silicon technology is highly developed. Being an indirect bandgap material, its absorption coefficient changes gradually with wavelength; this allows silicon photodiodes to be optimized for particular combinations of wavelength-of-operation and speed-of-response. The tradeoffs between response speed and absorption efficiency for silicon p-i-n diodes as given by Melchior [203] are shown in Fig. 60 for the wavelengths of operation of the prospective sources for fiber systems.

E. Nonavalanching Photodiodes

Front-illuminated p-i-n structures are commonly used in silicon photodiodes of the nonavalanche type providing fast response and high quantum efficiency. Optimized diodes with quantum efficiency greater than 90 percent and rise time less than 100 ps have been built for application at $\lambda = 0.633\ \mu m$ [see Fig. 61(a) and Table VI] [197], [198]. For application in the wavelength region 0.75–0.9 μm the required depletion layer width would be about 20 to 40 μm, with which a quantum efficiency of 70 percent and a bandwidth of several hundred megahertz could be achieved. An example of a diode with high efficiency at $\lambda = 0.9\ \mu m$ is given in Table VI [204]. At $\lambda = 1.06\ \mu m$, however, the required thickness of the depletion layer (for high efficiency) is 500 μm, thus giving rise to long carrier transit time which leads to limited bandwidth. A compromise between quantum efficiency and speed of response is realized if the light is allowed to enter the depletion region from the side as shown in Fig. 61(b). An example of a side-illuminated silicon photodiode is given in Table VI [205].

Both front-illuminated and side-illuminated germanium photodiodes have been built with individual response spanning the wavelength range from 0.6 to 1.6 μm [206], [207]; the characteristics of a front-illuminated diode are given in Table VI [206].

F. Avalanche Photodiodes

Avalanche photodiodes combine the detection of optical signals with internal amplification of the photocurrent. The internal gain is realized through avalanche multiplication of carriers in the high-field region of a highly reverse biased junction where photocarriers gain sufficient energy to create new electron–hole pairs through impact ionization. High gain is

TABLE VI
CHARACTERISTICS OF PHOTODIODES

Diode Type	Material and Structure	Wavelength Range (μm)	Quantum Efficiency[a] (%)	Gain $\times$ Band-width (GHz)	Response Time (ns)	Sensitive Area (cm²)	Break-down Voltage (V)	Dark Current (nA)	Capac-itance (pF)	Refer-ence
Front-illuminated (no gain)	silicon p-i-n	0.5–0.7	>90 (0.633 μm)		0.1 ($R_L=50\ \Omega$)	2×10^{-5}		<1 @ −40 V	<1 @ −5 V	[197]
Front-illuminated (no gain)	silicon p-i-n	0.4–1.1	90 (0.9 μm)		3 ($R_L=50\ \Omega$)	2×10^{-2}		50 @ −45 V	3 @ −45 V	[204]
Side-illuminated (no gain)	silicon p^+-n-n^+	0.4–1.1	90 (0.8–1 μm)		<0.5			10 @ −100 V	1.8 @ −300 V	[205]
Front-illuminated avalanche	silicon p^+-π-p-n^+	0.6–1.1	85 (0.9 μm)	~30	2 ($R_L=50\ \Omega$)	51×10^{-3}	~400	100 @ −300 V	2 @ −300 V	[209]
Side-illuminated avalanche	silicon p^+-n-n^+	0.4–1.1	85 (0.6–1 μm)	$\lesssim$100	0.3		~210	20 @ −200 V	<1 @ −200 V	[210]
Front-illuminated avalanche	germanium n^+-p	0.6–1.6	>85 (0.7–1.5 μm)	~60	0.12 ($R_L=50\ \Omega$)	2×10^{-5}	16.3	20 @ −16 V	0.8 @ −16 V	[206]

[a] Internal quantum efficiency—with antireflection coating or assumed to have antireflection coating.

achieved by taking special precautions to assure uniformity of carrier multiplication and to avoid microplasmas and excessive leakage currents. Gain–bandwidth products between 20 and 100 GHz have been reported for various silicon and germanium avalanche photodiodes [196]–[198], [206].

The current gain of the avalanche photodiode fluctuates due to the fundamental statistical nature of the avalanche multiplication process. The gain fluctuation gives rise to excess noise which can be accounted for by means of an excess noise factor $F(g)$ which is a function of the average gain $\langle g\rangle$ and of the electron-to-hole ionization ratio (α/β) [199]. In silicon avalanche photodiodes with $\alpha/\beta\simeq10$, electron injection leads to $F(g)\simeq\langle g\rangle^{0.5}$. For germanium diodes with $\alpha/\beta\simeq1$, $F(g)\simeq\langle g\rangle$.

Front-illuminated silicon avalanche photodiodes with fast response and high quantum efficiency at $\lambda=0.9\ \mu$m have been built using p^+-π-p-n^+ structures [208], [209]. As shown in Fig. 62(a) the multiplication is confined to the narrow n^+-p region; the wide π region serves as the collection region for the photo-generated electrons. For application at $\lambda=1.06\ \mu$m side-illuminated silicon avalanche diodes promise to show superior overall response [210], [211], construction of such a diode is shown in Fig. 62(b). A front-illuminated germanium avalanche photodiode employing n^+-p structure which exhibits the fastest response in the wavelength region 0.9 to 1.6 μm is shown in Fig. 62(c) [206]. Characteristics of the various avalanche photodiodes previously discussed are given in Table VI.

Avalanche diodes require power supplies of moderately high voltage (a few hundred volts) and stabilization of gain against temperature variations. These facts plus the added cost of the avalanching device lead one to consider carefully their use in a system.

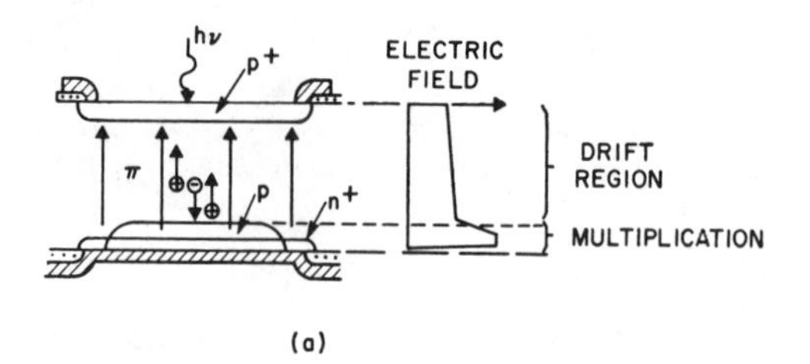

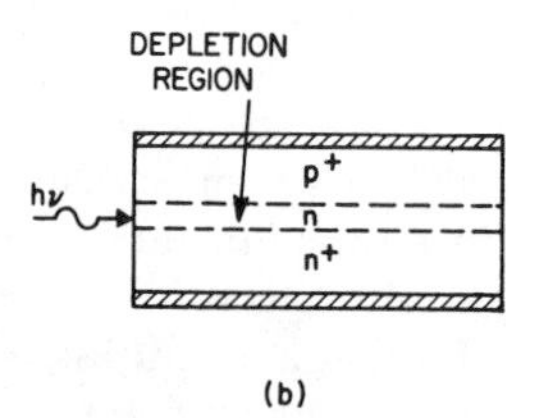

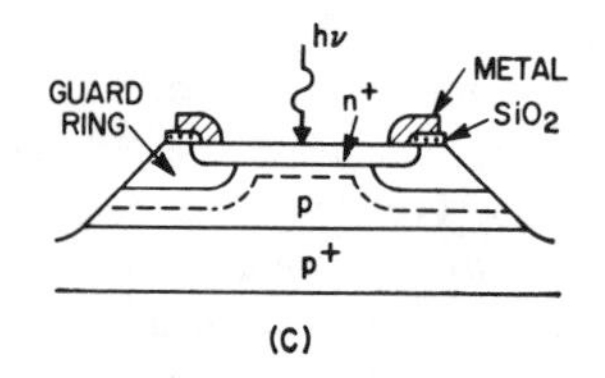

Fig. 62. Sketches showing the construction of avalanche photodiodes. (a) Front-illuminated silicon avalanche photodiode—the electric fields in the drift and multiplication regions are also shown. (b) Side-illuminated silicon avalanche photodiode. (c) Front-illuminated germanium avalanche photodiode (after Melchior, [203]).

G. Detectors for Delay Equalization

Delay distortion due to the spread of the group velocities of the various modes in a multimode fiber which does not couple modes can be equalized by means of specially designed detectors or detector arrays which take advantage of the separability of the modes at the output end of the fiber. One proposal is a detector array consisting of concentric rings of photodiodes which conform to the radiation patterns of the various groups of modes at the end of the fiber [212]; another is a linear array with each photodiode receiving power from groups of modes which have been spatially separated by a mode stripper [213]. Equalization of mode delay is accomplished by delaying the output signal from each detector by an appropriate amount before being combined in time coincidence in a common output circuit. The detection, delay, and combining can also be effected in a single specially designed detector which makes use of the drift time of the photo-generated carriers for delay equalization [214].

H. Future Development

The performance characteristics of presently available silicon photodiodes are adequate in the wavelength region below 0.9 μm. More work needs to be done on side-illuminated diodes for application at 1.06 μm. Continuing progress in material technology and further refinement in device design will bring the expected improvement in quantum efficiency and speed, and reduction of dark current and capacitance. GaAs avalanche photodiodes [215] with somewhat faster response than silicon devices may be useful in the 0.75- to 0.88-μm wavelength range. Several ternary compound semiconductor materials offer interesting possibilities for detector application in the near-infrared region [216]–[218]. Some of these are: $In_xGa_{1-x}As$, $GaAs_{1-x}Sb_x$, and $CdSnP_2$–InP. Much work will have to be done before these materials can be used for photodetectors in the way that silicon is used today.

VI. Integrated Optics

A. General

Integrated optics is a burgeoning new field of research in which novel optical devices and components in miniature waveguide form on planar substrates are being explored for possible applications in the general area of optical signal processing which includes optical communications. One of the goals of integrated optics is to achieve integrated assemblages of miniature optical devices and components that can perform useful functions and which are rugged in construction, insensitive to vibrations, and sparing of power consumption. Another is to explore two-dimensional and three-dimensional thin-film structures to realize new devices and components which are not feasible otherwise. Yet another is to exploit the techniques of thin-film waveguides for the investigation of new scientific phenomena. Since the advent of integrated optics five years ago [165], many novel devices have been studied, several new fabrication techniques explored, a few scientific phenomena investigated, but the practical integration of devices and components still remains a major challenge.

Because thin-film waveguide optics deals with highly coherent wavefronts, the application of integrated optics in optical-fiber transmission is mainly in the area of single-mode systems with laser sources. Thin-film waveguide lasers, modulators, switches, directional couplers, filters, etc., are all useful, provided they can be coupled easily with fibers or with the rest of the fiber system. This section is devoted to a brief discussion of materials and processing techniques for integrated optics work and of devices and components of the integrated-optics type which appear to have potential application in future optical-fiber communication systems.

Several review papers on integrated optics are found in the literature; they are by Goell and Standley [219] (1970), Tien [220] (1971), Miller [221] (1972), and Pole, Miller, Harris, and Tien [222] (1972).

B. Materials

Materials which show promise for *integrated*-optics work are those from which both active and passive elements could be fabricated on a single chip. One of these is the familiar ternary semiconductor compound $Al_xGa_{1-x}As$. It is conceivable that, by using selective-doping, epitaxial-regrowth, chemical-etching, and ion-beam-etching techniques, one could build a whole class of active and passive devices on a single chip [223]. For example, a laser, a modulator, a coupler, and a detector (for monitoring) all connected by waveguides on a single chip could be made by the techniques of selective etching and epitaxial regrowth with different amounts of aluminum. Work in this area is being vigorously pursued at present [224]–[229].

Neodymium-bearing materials are candidates for application at $\lambda = 1.06$ μm; some of these materials being explored at present are NdLa-pentaphosphate (or Nd-ultraphosphate) [230]–[232], Nd:YAG epitaxial films grown on YAG [233] and Sapphire [234], etc.

Other materials of interest include those exhibiting electrooptic, electroacoustic, and magnetooptic effects, and from which modulators, switches, multiplexers, and filters could be built in an integrated form on a single crystal. Some of these are $LiNbO_3$–$LiTaO_3$ [235a], [235b], ZnO [236a], II–VI compounds of Cd, Zn, S, and Se [236b], certain iron and gallium garnets [237], and piezoelectric sillenites (bismuth gallate, titanate, and germanate) [238].

C. Processing and Fabrication Techniques

Many techniques, both new and old, are being explored for processing and fabricating thin-film waveguide elements and devices. Some of these include selective chemical etching, epitaxial growth and regrowth [227], [229], diffusion [235], [236], ion-beam etching [226], [239], [240], sputtering [241], ion implantation [224], electron-beam exposure and writing [242], [243], etc.

D. Lasers

Thin-film waveguide lasers with distributed feedback [244], [245] are capable of single-frequency operation which is stabilized and controlled by the choice of the grating constant. Such lasers would be useful in a communication system which exploits frequency multiplexing or heterodyne detection. A distributed feedback GaAs laser operating at liquid-nitrogen temperature and pumped by a pulsed dye laser has been reported recently [246], [247]. Much work needs to be done before room-temperature operation and dc pumping can be achieved.

Neodymium lasers of the integrated optics form could be competitive with other prospective sources if they were pumped by LED's. The neodymium-bearing film must therefore be at least several wavelengths thick in order to accept the incoherent pumping radiation. Neodymium-bearing materials mentioned previously in Section VI-B are possible candidates.

E. Modulators, Switches, and Multiplexers

The principal advantage of a modulator in the thin-film waveguide configuration is that it requires very little drive power. Thin-film modulators, switches, and multiplexers employing electrooptic, electroacoustic, and magnetooptic effects have been discussed in Section IV-C. In particular the grating-type interaction utilizing any one of the three effects appears to be a versatile approach to realizing practical devices.

F. Passive Components

Useful passive components for optical-fiber communication include directional couplers and frequency-selective filters. Successful couplers have been fabricated in glass [242], [243], and in gallium arsenide [225]. Frequency-selective filters or couplers which have passbands a few angstroms to a few hundred angstroms wide are useful for frequency multiplexing or demultiplexing applications. Dielectric gratings or periodic waveguides could serve as the frequency-selective element [248], [249]. Etched gratings of periodicities as short as 1100 Å have been achieved [240]. A high-frequency cutoff periodic dielectric waveguide with a periodicity of 3640 Å has been demonstrated at $\lambda = 1.06\ \mu m$ [250].

VII. Fiber Transmission System Considerations

In this section we discuss the features of system design selected to achieve the overall transmission objective. The topics covered include choice of carrier wavelength, modulation method, fiber design, and some comparison with prior-art systems.[12]

A. Considerations in the Choice of Carrier Wavelength

As discussed in Section II, typical fiber losses are a gross function of wavelength. The OH^- overtone absorption at 0.95 μm should be avoided. The best fibers approach at discrete wavelengths the Rayleigh scattering limit, which falls off as $1/\lambda^4$; this is a strong system consideration leading toward use of longer wavelengths. There are several broad low-loss regions: between 0.75 and 0.9 μm, which can be met with AlGaAs LED's or lasers, and from 1 to 1.1 μm which includes the Nd-emission region near 1.06 μm. For detectors, existing silicon technology provides an excellent solution out to about 1.1 μm. Thus systems in the 0.75–0.9-μm or 1.0–1.1-μm regions are optimal now. In the future, research on new source and detector materials may permit use of somewhat longer wavelengths where the fiber may provide even lower losses [10], [70], [79a], and relaxed mechanical tolerances.

B. Modulation Format

The type of modulation used on a fiber transmission system can be selected to be compatible with almost any associated system, possibly at some penalty in signal-to-noise ratio (SNR) or repeater spacing, or may be chosen to optimize the performance of the optical fiber link. We review here the use of simple analog-intensity modulation, pulse-position modulation (PPM), and coded-digital modulation.

1) Analog-Intensity Modulation: Perhaps the simplest technique for interfacing between an optical-fiber transmission system and either another analog or digital system is intensity modulation of the source and envelope detection of the optical signal. Both GaAs LED's and injection lasers have output versus bias current characteristics which are sufficiently linear for many simple applications, and the use of baseband feedback from a detected sample of the optical output can provide a high degree of linearity if it is desired.

The optical receiver can be an avalanche photodiode with quantum efficiency η and avalanche current gain $G = \langle g \rangle$, followed by a baseband amplifier which has a noise figure F_t and which presents an equivalent load resistance R to the photodiode.

The SNR for a sinusoidally modulated signal is [194]–[196], [144]

$$\mathrm{SNR} = \frac{\frac{1}{2}\left(\eta \frac{e}{h\nu} G m p_0\right)^2}{\langle i_n^2 \rangle} \tag{101}$$

where

$$\langle i_n^2 \rangle = \langle i_Q^2 \rangle + \langle i_T^2 \rangle + \langle i_D^2 \rangle + \langle i_L^2 \rangle + \langle i_B^2 \rangle \tag{102}$$

$$\langle i_Q^2 \rangle = 2e \frac{e}{h\nu} \eta p_0 G^2 F_d b \qquad \text{(quantum noise)} \tag{103}$$

$$\langle i_T^2 \rangle = \frac{4k\theta}{R} b F_t \qquad \text{(thermal noise)} \tag{104}$$

$$\langle i_D^2 \rangle = 2e I_d G^2 F_d b \qquad \text{(dark-current plus background noise)} \tag{105}$$

$$\langle i_L^2 \rangle = 2e I_L b \qquad \text{(leakage current)} \tag{106}$$

$$\langle i_B^2 \rangle = 2\left(\frac{e}{h\nu} G \eta p_0\right)^2 \frac{b}{JW}\left(1 - \frac{1}{2}\frac{b}{W}\right) \qquad \text{(beat noise)} \tag{107}$$

$$I_d = I_d' + \frac{e}{h\nu} \eta p_G \tag{108}$$

where

- m modulation index of a sinusoidally modulated carrier;
- e electronic charge;
- $h\nu$ energy per photon;
- p_0 average received optical power;
- b bandwidth of the information source, taken to be equal to the channel bandwidth for analog intensity modulation;
- G average avalanche gain of the photodetector ($G = \langle g \rangle$);
- F_d excess noise factor associated with the random nature of the avalanche process; approximately $F_d \simeq \sqrt{G}$ for silicon;
- $k\theta$ Boltzmann's constant times absolute temperature;
- R equivalent load resistor;
- F_t noise figure of the baseband amplifier;
- I_d' primary detector dark current;
- I_L detector leakage current;
- p_G average power of background radiation incident on the detector;
- W spectral width of the source;
- J number of spatial modes reaching the detector (fiber output modes).

In each case $\langle i_x^2 \rangle$ is the mean-square value of the corresponding noise current after avalanche gain.

We consider an illustrative example based on an AlGaAs LED source and a silicon avalanche photodetector [144]:

$$\lambda = 0.85\ \mu m \qquad \frac{R}{F_t} = 1000\ \Omega$$
$$\eta = 0.5 \qquad b = 4\ \text{kHz}$$
$$m = 0.85 \qquad WJ = 10^{17}\ \text{Hz}$$
$$I_d = 10^{-10}\ \text{A} \qquad G = 17.$$
$$I_L = 10^{-9}\ \text{A}$$

First let us note that the numerator of (101) varies as p_0^2, the quantum noise (103) varies as p_0, and thermal noise (104) is independent of p_0. For the above numbers, Hubbard [144] calculates that the SNR is set by quantum noise $\langle i_Q^2 \rangle$ for in-

[12] Further background material on optical communication systems can be found in the following books: W. K. Pratt, *Laser Communication Systems.* New York: Wiley, 1969; M. Ross, *Laser Receivers.* New York: Wiley, 1966.

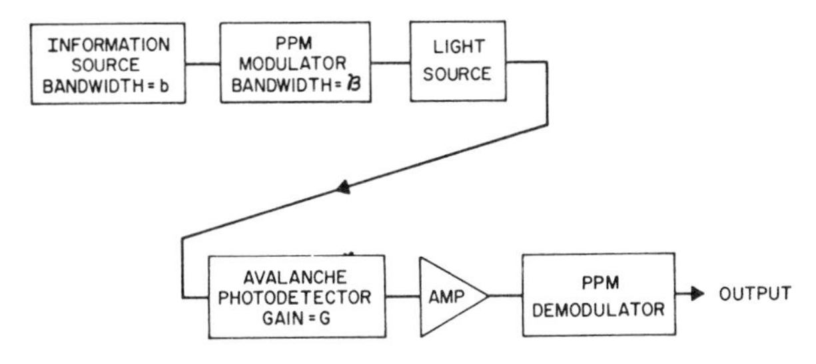

Fig. 63. Block diagram of a pulse-position modulation (PPM) transmission channel (after Hubbard, [144]).

put power p_0 above -38 dBm, at which point $10 \log_{10}$ (SNR) is 60 dB. For p_0 below -38 dBm, SNR is controlled by thermal noise and falls to 20 dB for $p_0 = -62$ dBm. Using higher avalanche gain improves performance mainly in the region of low input power ($p_0 < -40$ dBm). For example, it is possible to achieve SNR = 20 dB for $p_0 = -70$ dBm using an optimal gain of 130.

The value for R/F_t in the above example is very low. By using a high-impedance detector load (integrating front-end approach—see Section V-B) better performance could be achieved with lower or even unity avalanche gain.

This example and (101)–(108) illustrate a number of features characteristic of optical-fiber systems that differ from microwave or lower frequency systems:

1) Ultimate receiver sensitivity is set by quantum noise, which is proportional to signal power.

2) Thermal noise can be reduced to a level below that of the quantum noise by using avalanche gain and/or large detector load impedance.

3) Since avalanche gain introduces excess noise which is gain-dependent there exists an optimal value of gain for a given set of operating conditions.

4) With the signal output dependent on p_0^2 and quantum noise dependent on p_0, there is advantage in using highly peaked signals (for a given average power); that is, bandwidth-expansion schemes may be advantageous for low-bandwidth signals.

For the above example, the beat noise (107) does not contribute importantly for p_0 up to 0 dBm. However, if fewer modes were present at the detector input, beat noise would set a limit to SNR for large p_0.

2) Pulse-Position Modulation (PPM): Improvement in noise immunity can be achieved by using highly peaked pulses, as noted above, and PPM has this character [144]. It does require more bandwidth, but if the signal does not in its baseband form require the band available on the fiber, then bandwidth expansion is desirable.

A block diagram of the PPM channel is shown in Fig. 63. Analysis by Hubbard [144] assumes a sequence of light pulses of the form

$$p(t) = \frac{p_m}{2}\left[1 + \cos\left(\pi \frac{t}{\mathfrak{I}}\right)\right], \qquad -\mathfrak{I} < t < \mathfrak{I} \tag{109}$$

in which p_m is the peak pulse power at the receiver and $2\mathfrak{I}$ is the total pulse duration. The average power p_0 is

$$p_0 = \frac{\mathfrak{I}}{T} p_m = \frac{p_m}{2\kappa} \tag{110}$$

where T is the sampling interval. Noise affects the SNR of a PPM signal in two ways: 1) by perturbing the threshold crossing of the received signal and effectively shifting the position of the pulse, and 2) by causing the received current to exceed the threshold in the absence of a pulse, thereby initiating an extraneous pulse or "false alarm" in the receiving circuits. On the assumption that the reciprocal pulsewidth $1/\mathfrak{I}$ is equal to the noise bandwidth $\mathcal{B}$, Hubbard finds

$$\text{SNR} = \frac{\dfrac{\pi^2}{2}(\kappa - 1)^2\left(\eta \dfrac{e}{h\nu}\kappa p_0\right)^2 G^2}{\langle i_n^2\rangle + \dfrac{8\pi^2}{3\sqrt{3}}(\kappa - 1)^3 e^{-XNR/2}} \tag{111}$$

where

$$XNR = \frac{i_m^2}{4\langle i_n^2\rangle} \tag{112}$$

$$i_m^2 = \eta\left(\frac{e}{h\nu}\right) G p_m. \tag{113}$$

For comparison with the analog-intensity modulation system, the following relationships are established:

$$T = \frac{1}{2b} \qquad \kappa = \frac{T}{2\mathfrak{I}} = \frac{\mathcal{B}}{4b}. \tag{114}$$

Then, for signal power p_0 above the threshold violation point the SNR for PPM is formally the same as for analog intensity modulation provided κp_0 replaces p_0 and $(\pi/2)[(\kappa-1)/\sqrt{\kappa}]$ replaces m in (101). Thus the PPM system has an effective power κ times that of the analog-intensity modulation system and the PPM system has an effective modulation index $\sqrt{\kappa}\,\pi/2$ for $\kappa \gg 1$. Using the same numerical parameters given for the illustrative example for analog-intensity modulation and using $\kappa = 250$, Hubbard finds the following [144].

1) The SNR (in decibels) varies linearly with p_0 in dBm from SNR = 143 dB at $p_0 = -20$ dBm to SNR = 70 dB at $p_0 = -80$ dBm; and 2) threshold violation occurs at $p_0 = -84$ dBm for which SNR = 65 dB. It is to be noted again that the preceding example does not make use of the integrating front-end approach which may yield improved performance. The high SNR may be difficult to obtain in practice where coding and decoding circuitry may set the ultimate performance obtainable.

It is evident that PPM is very advantageous for the transmission of low-bandwidth signals in optical fiber systems. The cost is the coding and decoding equipment, and bandwidth expansion on the fiber.

3) Digital Modulation: We consider here digital systems such as PCM and delta modulation which require only the identification of one of two states in each sampling interval, with a clean noise-free signal (i.e., a regenerated signal) sent out from each repeater. Further we concern ourselves only with those aspects which are different in the optical-fiber system compared with previous conventional systems. Previous publications cover digital transmission more generally [251].

The end-result of theoretical studies is to specify the receiver input power p_0 required for a given objective error rate. In addition to accounting for the various noise contributions previously discussed, one must evaluate the errors made due to intersymbol interference as a result of pulse-delay distortion—i.e., the overlapping of received pulses in successive time slots.

In an important study, Personick [252] has shown that a

fiber system that utilizes many fiber modes and/or has a carrier bandwidth many times the modulation bandwidth can be considered as linear in power; this allows baseband equalization and use of the guide at higher bit rates (with noise penalties) than would be associated with the criterion that the output pulses must not overlap.

As previously noted, it is now recognized that the circuit following the optical envelope detector should have a high impedance as has been recognized in the field of nuclear-particle counters [202]. After the signal is amplified above circuit noise, an equalizer is then used to provide adequately resolved pulses so that a conventional pulse regenerator can make the necessary decision on presence or absence of a mark in each sampling interval. Personick [159] has recently published a comprehensive study of optical receiver design using the high-impedance integrating input-circuit design. In this approach the lowest received power for a given error rate is obtained when the parallel resistance R_T at the detector output is made as large as possible and the parallel capacitance C_T is made as small as possible. Assuming no avalanche gain Personick [159] found that the use of $C_T R_T = T$, the signal-pulse spacing, required 4 dB more signal power than would a very large parallel resistance in a 25-Mb/s receiver; the advantage of large R_T increases at lower bit rates and decreases at higher bit rates.

Personick's work, based on statistical detection theory, showed other important results which we now summarize. The reader is referred to his paper for a full discussion [159]. The results are for the case in which the dominant noise source of the amplifier is the series-voltage noise source $\langle v_a^2 \rangle$ whose spectral height is frequency independent (see Fig. 57), such as in the case of a field-effect transistor (FET). In this case the required receiver input power is, for no avalanche gain and with a high input impedance,

$$p_0 \sim \frac{h\nu}{\eta} \cdot \frac{Q}{T^{3/2}} \tag{115}$$

where $(1/T)$ is the pulse rate and Q, the sole factor which determines the error rate, is shown in Fig. 64. Thus for no avalanche gain, the required receive input power increases at 4.5 dB per octave of bit rate. For optimal avalanche gain in silicon detectors

$$p_0 \sim \frac{Q^{5/3}}{T^{7/6}} \tag{116}$$

which shows 3.5 dB more signal power required for an octave increase in bit rate.

We note in Fig. 64 the weak dependence of Q on error rate. Thus either with or without avalanche gain, the error rate varies by a factor of 10 for a fraction of a decibel change in received signal power.

The optimal avalanche gain varies with bit rate according to

$$G_{\text{optimal}} \sim \frac{1}{T^{1/3}}. \tag{117}$$

Moreover, the use of high input impedance (rather than setting the impedance for faithful pulse reproduction in that portion of the receiver) can bring the optimal avalanche gain down by large factors for systems running at low bit rates ($1/T < 50$–100 Mb/s). For very low bit rates or bandwidth, it may not be necessary to use avalanche gain; this is an important economic consideration in detector and repeater design.

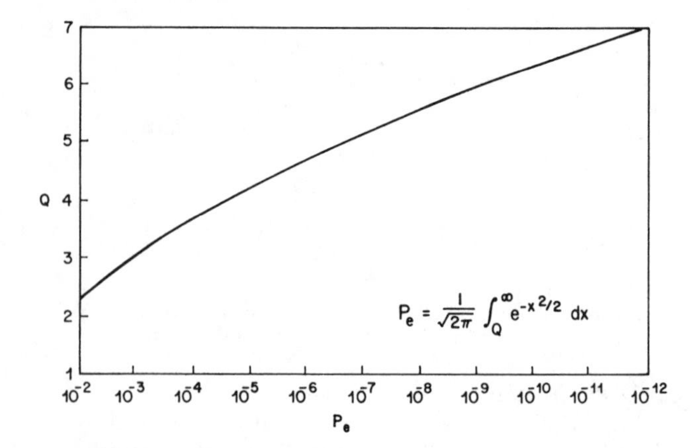

Fig. 64. Q versus error rate P_e for a digital receiver with additive Gaussian noise. Q is the ratio of the required baseband signal to the rms noise at the sampling time (after Personick, [159]).

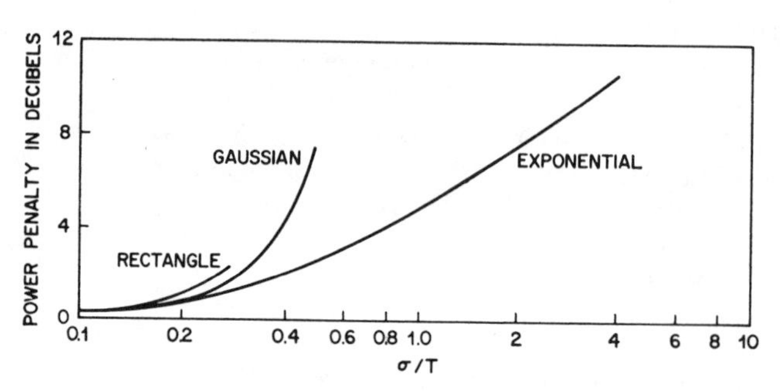

Fig. 65. Additional signal-power required (ordinate) versus normalized rms width of the input pulse; no avalanche gain in the detector (after Personick, [159]).

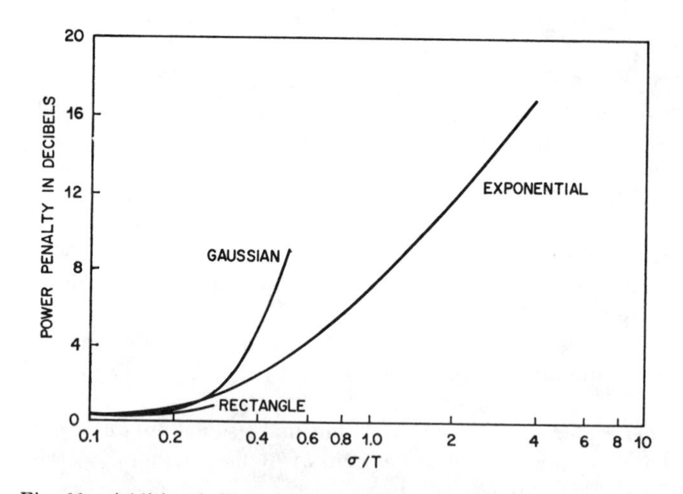

Fig. 66. Additional signal-power required (ordinate) versus normalized rms width of the input pulse; with optimal avalanche gain in the detector (after Personick, [159]).

The shape of the received pulse influences the intersymbol interference effect at large pulse overlap. This is shown in Figs. 65 and 66 for the cases of no avalanche gain and optimal avalanche gain, respectively [159].[13] These curves, which apply to an amplifier with a FET first stage, show the added power needed to hold the error rate constant at 10^{-9} as a function of σ, the rms width of the received pulse. Note, however,

[13] Personick's σ, plotted in Figs. 65 and 66, is equivalent to our σL, with σ defined in (59)–(61).

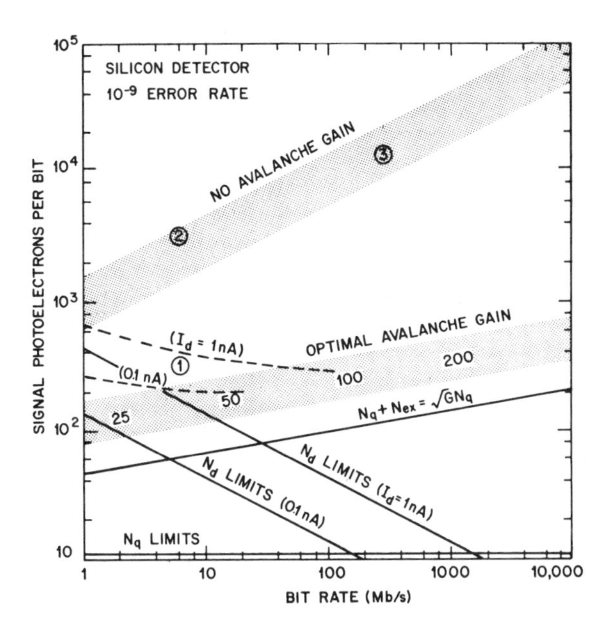

Fig. 67. Average primary signal photoelectrons per bit interval required to achieve 10^{-9} error rate versus bit rate for high-impedance integrating front-end optical receivers employing direct detection. Silicon photodetectors and FET first-stage amplifiers are assumed. The ordinate is also proportional to the required minimum received average power divided by the bit rate, the proportionality constant being $(\eta/h\nu)$. The dotted bands indicate the expected performance based on current device parameters. The numbers (25, 50, 100, 200) shown in the middle of the lower band are the calculated values of optimal avalanche gain for the respective bit rates. The dotted lines show degradation of performance due to primary dark current (I_d) in the presence of optimal avalanche gain. The solid horizontal line gives the quantum noise (N_q) limit without gain and represents the ultimate in receiver performance. The two solid lines with negative slope set the limit of performance for no gain if dark current noise (N_d) were the only noise source. The solid line with positive slope represents the limit due to quantum and excess noise (N_q+N_{ex}) with optimal gain but negligible dark current. The circles represent experimentally achieved values by Goell [139], [158]: ① corresponds to −62 dBm, ② corresponds to −52 dBm, and ③ corresponds to −29 dBm. The curves are based on calculations of Personick [159] for a 25-Mb/s system and extrapolated to other bit rates using (115)–(117).

that for $\sigma/T \leq 0.25$ there is not much dependence on pulse shape. Therefore, for about 1-dB penalty in the received power requirement, one can specify an allowable pulse rate of

$$\frac{1}{T} = \frac{1}{4\sigma L} \tag{118}$$

for a fiber of length L and with a specific impulse response characterized by σ, regardless of shape. Acceptance of larger power penalties requires attention to the particular shape of the impulse response.

Fig. 67 summarizes the state of the art, for a FET first-stage amplifier, relating the received signal energy expressed as photoelectrons per bit to system bit rate for 10^{-9} error rate. The two bands represent the range of calculated performance relevant to existing devices extrapolated to higher bit rates [159]. N_q represents the quantum-noise limit and N_d the dark-current limits. The line marked (N_q+N_{ex}) represents the shot noise from both excess avalanche-detector noise and quantum noise. Three points represent published results on laboratory-model repeaters. Points ① and ② represent a 6-Mb unit using a GaAs LED source, reported by Goell [139]. Point ③ represents a 274-Mb experimental repeater using a GaAs injection laser, also due to Goell [158]. Both repeaters used the high-impedance integrating front-end approach and both employed direct modulation of the current through the carrier generator to produce pseudorandom optical test signals.

Finally, we may note a system arrangement suggested by the very small size of fibers. To simplify the circuitry, one might use two fibers to send a single channel in one direction—one fiber for the "mark" and the other for the "space." An analysis for this configuration has been provided [253]; there is little to be gained performance-wise over conventional single-channel transmission, but the scheme may be worthy of consideration for easing the problems associated with signaling and control.

C. Choice of Fiber Type

It is fortunate that the total attenuation of the best optical fibers is not dependent on whether they are single-mode or multimode. This simplifies the fiber choice. There are strong interdependencies with other system considerations:

1) If a LED is to be used as a carrier generator, the fiber must be multimode and more power is captured by the fiber if it has a larger relative index difference Δ.

2) If a laser is to be used, either a single-mode or multimode fiber may be used.

3) The index distribution for the multimode fiber controls the modal delay distortion. This is under design control, and economy is the only reason to design for anything other than the near-parabolic distribution which minimizes delay distortion [17], [19], [53].

4) Fiber splicing is greatly eased if the transverse size of the guided light wave is large. The most obvious way to achieve this is by using the multimode fiber. The single-material fiber in single-mode form may provide a second way of providing a large field cross section. The use of small index difference in conventional fibers seems to be limited to a $\Delta \simeq 0.002$ which yields a field size around 8–10 μm for single-mode transmission.

Let us consider typical numerical values for system information rate as related to fiber choice.

The bulk material delay distortion is given by the first term of (45) or (71), and for distance L between repeaters this becomes

$$sL = -\frac{\lambda_0 \delta\lambda}{c} \frac{d^2 n}{d\lambda^2} L. \tag{119}$$

This term is independent of the number of fiber modes, but is proportional to the spectral width of the source $\delta\lambda$, the carrier wavelength λ_0, the second derivative of fiber material index $d^2n/d\lambda^2$, and the distance L between repeaters. For the GaAs LED emission spectrum, Fig. 20, and for fused silica fibers the rms width of the impulse response due to material dispersion is $\sigma = 1.75 \times 10^{-12}$ s·m^{-1}, which [using (118)] translates into an allowable bit-rate-length product of 142 Mb·km/s. It is noteworthy that this contribution is always linear with distance between repeaters. It can only be reduced by narrowing the source spectral width or by reducing $d^2n/d\lambda^2$.

For a multimode fiber, the modal delay distortion depends on the index difference Δ and the degree of perfection employed in grading the transverse index distribution. The rms width of the impulse response is given in Section II-E. We tabulate here (see Table VII) the allowed bit rate according

TABLE VII

Fiber Index Distribution[a]	Bit-Rate-Length Product (Mb·km)/s
Step-index ($u=\infty$)	18
Pure eighth order ($u=8$)	31
Pure fourth order ($u=4$)	59
"Ideal" near-parabolic [19]	13 800
Imperfect (5% error) parabolic [53]	870

[a] See Section II-E for expressions for the rms width of the impulse response for the various fibers.

to (118) for $\Delta=0.01$ and negligible mode mixing—which will be valid in many fibers for lengths up to a few hundred meters, and may be valid for very uniformly made fibers of longer lengths. The root-sum-square σ of both material dispersion and modal dispersion may be used when both effects contribute simultaneously.

For lowest multimode dispersion the effects of mode mixing are important [20], [21]. In the desirable case of not too strong mode coupling (in order to avoid radiation losses), the modal delay distortion will increase linearly up to some fiber length L_l and thereafter increase as the square root of length. Then an effective length for delay distortion calculation is

$$L_{\text{eff}} = \sqrt{L_l L}, \qquad \text{if } L > L_l$$
$$L_{\text{eff}} = L, \qquad \text{if } L < L_l \qquad (120)$$

where L is the total fiber length. (Note the similarity to the factor $\sqrt{Ll_c}$ in (73) of Section II-F. The above gives an alternate interpretation of Personick's l_c.) The allowed bit rate is then, from (118),

$$\text{bit rate} = \frac{1}{T} = \frac{1}{4\sigma L_{\text{eff}}}. \qquad (121)$$

We take for illustration a desired bit rate of 100 Mb/s, assume a total repeater spacing $L=10$ km, assume the fiber has $L_l=1$ km, and has an imperfect parabolic distribution represented by $\sigma=0.0059n\Delta/c$ (last line of Table III). We ask what Δ is the largest acceptable value, and find from (121) and the above parameters that $\Delta=0.0275$. Thus a laser (narrow-band) source exciting all modes at the input of this fiber could be used up to 100 Mb/s, whereas the GaAs LED would be limited in the same fiber to a bit rate of $142/10=14$ Mb/s due to material dispersion. By using a smaller Δ the laser bit rate would be increased but no change would occur for the LED carrier. Note that the bit rate is not dependent on the fiber diameter or on the number of propagating modes.

Finally, we note the small delay distortion associated with the single-mode fiber. For $2\sigma_\lambda=1$ Å carrier spectral width (a laser) the material dispersion (fused silica) allows a bit-rate-length product of 47 000 Mb·km/s. The waveguide mode dispersion is negligible.

D. *Transmission Line Reflections*

The nature of wave propagation on dielectric waveguides is such that imperfections in cross section or misalignment at splices cause little reflection. No detailed study has been published because those effects are so small. The intrinsic Rayleigh scattering which continuously sends power both forward and backward from the signal results in a far-end received wave that has suffered two such scatterings—i.e., "double-reverse" scatter. This has been analyzed and found small [27], [28].

E. *Comparison with Coaxial Transmission Systems*

Transmission via optical fibers requires a carrier generator, modulation means, and a detector. Some wire-pair or coaxial systems do not require a carrier and in those cases the above optical elements represent added complexity. High-speed coaxial systems do require a carrier and demodulation means, and there the cost comparison is one of degree.

There are several features of fiber transmission which render it simpler than copper wire systems.

1) The loss of fibers is essentially flat over a very useful modulation band, thereby avoiding the $\sqrt{f}$ loss equalizers needed on coaxial systems. (For long or very wide-band systems, dispersion is important in fibers, necessitating equalization.)

2) For the region where dispersion is not limiting, the repeater spacing is relatively insensitive to bit rate or modulation bandwidth.

3) The loss of fibers is essentially independent of ambient temperatures [255], over the range encountered in practice (i.e., −40°C to +100°C). This again simplifies the repeater circuitry, since coaxial systems require variable $\sqrt{f}$ loss equalizers to compensate for line temperature changes.

F. *Fiber Splicing and Cable Design*

It is recognized that the handling of glass-fiber cables in the field may be the most formidable educational task associated with new fiber systems' use.

A connector design for joining individual single-mode fibers has been described by Krumpholz of AEG-Telefunken [255]. System-research work based on single-mode fibers has been reported by Börner [256].

Studies of splicing single-mode fibers by Someda yielded average losses as low as 0.5 dB [257].

Splicing of multimode fibers should be easier, due to the far larger transverse field width, and a technique for fusing two fibers into one continuous fiber was described by Bisbee [258] and by Dyott *et al.* [259]. The core maintained its integrity across the junction and losses of ½ dB were achieved.

More work will be needed. One view is to use a multiplicity of fibers to carry the same signal, and then to accept partial loss due to fiber breakage along the cable or at splices. Another view is to provide design for low probability of failure on every fiber and to use only one for each channel.

Some have expressed the view that fibers should not be handled individually in making joints, but rather should be bundled for handling in groups.

This branch of the art is at such an early stage that further review is not now possible.

VIII. General Form of Potential Applications

We have developed the thesis that glass fibers are candidates for use wherever wire pairs or coaxials are now used. Their advantages are:

1) *Small size*—overall fiber diameter in the range 75 to 125 μm.

2) *Small bending radius*—tolerable radii as small as a few centimeters feasible with design care, and of order 10 cm for almost any fiber.

3) *Nonconductive and noninductive*—a feature of value when many complex electronic assemblies are being interconnected.

4) *Wide bandwidth*—with flexibility in the bandwidth utilized according to the electronics associated with it. As noted in the example above, the same fiber could be used at rates up to 14 Mb over 10-km length with a LED carrier generator, or up to 100 Mb over 10-km length with a laser source.

5) *Small crosstalk*—with proper attention to design.

6) *Graceful growth*—as noted in item 4), the electronics can be changed on an installed fiber to increase its communication capacity. Moreover, the investment in facilities for producing new optical elements is very limited. One LED design, one laser design, and one detector design might furnish the building blocks for a large volume of applications.

7) *Potentially, low cost*—if the cost of purification of starting materials is not unreasonable, the economics should prove attractive, since the glass constituents are commonly available and only small amounts are needed.

Weighing against these features are the following disadvantages.

1) The new problems of forming fibers into cables, and of making splices must be solved, first by producing viable designs and second by training installers to use new techniques.

2) The fiber cannot carry the dc signaling currents now used in subscriber loops, and hence carrier methods (which technically are feasible) or an auxiliary system of some kind will be needed for this type of application.

The work reviewed above leads to the conclusion that it will soon be technically feasible to build digital fiber links at rates near 300 Mb with roughly 30–40-dB loss between repeaters. At 5-dB/km fiber loss, this yields a repeater spacing of 6–8 km, which is notably larger than the 1.6-km coaxial repeater spacing of L-5. Laser life for such a fiber system is now only several thousand hours; considerable improvement is needed to provide economic feasibility for this potential intercity application. In the longer range future, it seems clear that gigabit rates on individual fibers will become technically feasible; the economic factors must be evaluated in comparison with alternate systems, including systems employing several fibers carrying lower bit-rate signals.

Other potential applications that can be based on LED carrier generators (which now have long life) include 1) intracity links, between central offices, for example, or 2) distribution links within a building. It is evident that many of the obstacles for fiber system use—device life, fiber loss, installer training—become progressively easier to overcome as the length of the link becomes shortened to within one building. One might expect to find an early telecommunication application there.

IX. Concluding Remarks

We have attempted to provide a review of fiber transmission research useful both to one new to the field, and to specialists in one or another parts of the field. We are both reporters and inevitably interpreters for the rather vast individual contributions that have been made. We hope the comprehensive reference list will prove useful and facilitate study of the original contributor's presentation. Readers are urged to do so. A special issue of *Opto-Electronics* devoted to optical transmission has been prepared concurrently with this article and this provides another view [260].

References

[1] R. Kompfner, "Optics at Bell Laboratories—Optical communications," *Appl. Opt.*, vol. 11, pp. 2412–2425, Nov. 1972.

[2] ——, "Optical communications," *Science*, vol. 150, pp. 149–155, Oct. 8, 1965.

[3] S. E. Miller and L. C. Tillotson, "Optical transmission research," *Appl. Opt.*, vol. 5, pp. 1538–1549, Oct. 1966.

[4] S. E. Miller, "Optical communications research progress," *Science*, vol. 170, pp. 685–695, Nov. 13, 1970.

[5] K. C. Kao and T. W. Davies, "Spectrophotometric studies of ultra low loss optical glasses 1: Single beam method," *J. Sci. Instrum.*, vol. 1 (ser. 2), pp. 1063–1068, 1968.

[6] F. P. Kapron, D. B. Keck, and R. D. Maurer, "Radiation losses in glass optical waveguides," *Appl. Phys. Lett.*, vol. 17, pp. 423–425, Nov. 15, 1970.

[7] D. Hondros and P. Debye, "Elektromagnetische Wellen an Dielektrischen Drahten," *Ann. Phys.*, vol. 32, pp. 465–476, 1910.

[8] R. E. Beam, M. M. Astrahan, W. C. Jakes, H. M. Wachowski, and W. L. Firestone, "Dielectric tube waveguides," Army Signal Corps, Final Rep. under Contract W36-039 sc-38240, Microwave Lab., Northwestern Univ., Evanston, Ill., pp. 184–269, 1951–1952.

[9] P. Kaiser, A. R. Tynes, A. H. Cherin, and A. D. Pearson, "Loss measurements of unclad optical fibers," presented at the Topical Meeting on Integrated Optics-Guided Waves, Materials and Devices, Feb. 7–10, 1972.

[10] P. Kaiser *et al.*, "Spectral losses of unclad vitreous silica and soda-lime-silicate fibers," *J. Opt. Soc. Amer.*, vol. 63, pp. 1141–1148, Sept. 1973.

[11] J. Stone, "Optical transmission in liquid-core quartz fibers," *Appl. Phys. Lett.*, vol. 20, pp. 239–240, Apr. 1972.

[12] G. J. Ogilvie, R. J. Esdaile, and G. P. Kidd, "Transmission loss of tetrachloroethylene-filled liquid-core-fibre light guide," *Electron. Lett.*, vol. 8, pp. 533–534, Nov. 2, 1972; also G. J. Ogilvie, Australian provisional patent PA 7211/71, 1971.

[13] W. A. Gambling, D. N. Payne, and H. Matsumura, "Gigahertz bandwidths in multimode, liquid core, optical fibre waveguide," *Opt. Commun.*, vol. 6, pp. 317–322, Dec. 1972.

[14] T. Uchida, M. Furukawa, I. Kitano, K. Koizumi, and H. Matsumura, "A light-focusing fiber guide," *IEEE J. Quantum Electron.* (Abstract), vol. QE-5, p. 331, June 1969.

[15] E. A. J. Marcatili, "Modes in a sequence of thick astigmatic lens-like focusers," *Bell Syst. Tech. J.*, vol. 43, pp. 2887–2904, Nov. 1964.

[16] S. E. Miller, "Light propagation in generalized lens-like media," *Bell Syst. Tech. J.*, vol. 44, pp. 2017–2063, Nov. 1965.

[17] S. Kawakami and J. Nishizawa, "An optical waveguide with the optimum distribution of refractive index with reference to waveform distortion," *IEEE Trans. Microwave Theory Tech.*, vol. MTT-16, pp. 814–818, Oct. 1968.

[18] S. E. Miller, "Waveguide for millimeter and optical waves," U. S. Patent 3 434 774, Mar. 25, 1969.

[19] D. Gloge and E. A. J. Marcatili, "Multimode theory of graded-core fibers," *Bell Syst. Tech. J.*, vol. 52, pp. 1563–1578, Nov. 1973.

[20] C. A. Burrus *et al.*, "Pulse dispersion and refractive-index profiles of some low-loss multi-mode optical fibers," *Proc. IEEE* (Lett.), vol. 61, pp. 1498–1499, Oct. 1973.

[21] E. L. Chinnock, L. G. Cohen, W. S. Holden, R. D. Standley, and D. B. Keck, "The length dependence of pulse spreading in CGW-Bell-10 optical fiber," *Proc. IEEE* (Lett.), vol. 61, pp. 1499–1500, Oct. 1973.

[22] P. Kaiser, E. A. J. Marcatili, and S. E. Miller, "A new optical fiber," *Bell Syst. Tech. J.*, vol. 52, pp. 265–269, Feb. 1973.

[23] R. D. Maurer, "Glass fibers for optical communications," *Proc. IEEE*, vol. 61, pp. 452–462, Apr. 1973.

[24] a) A. W. Snyder, "Asymptotic expressions for eigenfunctions and eigenvalues of a dielectric or optical waveguide," *IEEE Trans. Microwave Theory Tech.*, vol. MTT-17, pp. 1310–1138, Dec. 1969.
——, "Excitation and scattering of modes on a dielectric or optical fiber," *IEEE Trans. Microwave Theory Tech.*, vol. MTT-17, pp. 1138–1144, Dec. 1969.
——, "Power loss on optical fibers," *Proc. IEEE* (Lett.), vol. 60, pp. 757–758, June 1972.
——, "Radiation losses due to variations of radius on dielectric or optical fibers," *IEEE Trans. Microwave Theory Tech.*, vol. MTT-18, pp. 608–615, Sept. 1970.
b) S. A. Schelkunoff, *Electromagnetic Waves.* New York: Van Nostrand, 1943.
c) E. Snitzer, "Cylindrical dielectric waveguide modes," *J. Opt. Soc. Amer.*, vol. 51, pp. 491–498, May 1961.
E. Snitzer and H. Osterberg, "Observed dielectric waveguide modes in the visible spectrum," *J. Opt. Soc. Amer.*, vol. 51, pp. 499–504, May 1961.
d) G. Biernson and D. J. Kinsley, "Generalized plots of mode patterns in a cylindrical dielectric waveguide applied to retinal cones," *IEEE Trans. Microwav Tehory Tech.*, vol. MTT-13, pp. 345–356, May 1965.

[25] D. Gloge, "Weakly guiding fibers," *Appl. Opt.*, vol. 10, pp. 2252–2258, Oct. 1971.

[26] E. D. Kolb, D. A. Pinnow, T. C. Rich, R. A. Laudise, and A. R. Tynes, "Low optical loss synthetic quartz," *Nat. Res. Bull.*, vol. 7, pp. 397–406, 1972.
D. A. Pinnow and T. C. Rich, "Development of a caloric method for making precision optical absorption measurements," *Appl. Opt.*, vol. 12, pp. 984–992, May 1973.
D. B. Keck, R. D. Maurer, and P. C. Schultz, "On the ultimate lower limit of attenuation in glass optical waveguides," *Appl. Phys. Lett.*, vol. 22, pp. 307–309, Apr. 1973.
D. A. Pinnow, T. C. Rich, F. W. Ostermeyer, and M. DiDomenico, "Fundamental optical attenuation limits in the liquid and glassy state with application to fiber optical waveguide materials," *Appl. Phys. Lett.*, vol. 22, pp. 527–529, May 1973.

[27] F. P. Kapron, R. D. Maurer, and M. P. Teter, "Theory of backscattering effects in waveguides," *Appl. Opt.*, vol. 11, pp. 1352–1356, June 1972.

[28] W. M. Hubbard, "Double-reverse-scatter interference in optical fiber communication systems," *Appl. Opt.*, vol. 11, pp. 2495–2501, Nov. 1972.

[29] E. G. Rawson, "Calculation of Mie scattering by spherical particles in low-loss glasses for optical waveguides," *Appl. Opt.*, vol. 10, pp. 2778–2779, Dec. 1971.
——, "Measurement of the angular distribution of light scattered from a glass fiber optical waveguide," *Appl. Opt.*, vol. 11, pp. 2477–2481, Nov. 1969.

[30] R. H. Stolen, E. P. Ippen, and A. R. Tynes, "Raman oscillation in glass optical waveguides," *Appl. Phys. Lett.*, vol. 20, pp. 62–64, Jan. 1972.

[31] R. G. Smith, "Optical power handling capacity of low-loss optical fibers as determined by stimulated Raman and Brillouin scattering," *Appl. Opt.*, vol. 11, pp. 2489–2494, Nov. 1972.

[32] a) R. H. Stolen and E. P. Ippen, "Raman gain in glass optical waveguides," *Appl. Phys. Lett.*, vol. 22, pp. 276–278, Mar. 1973.
b) E. P. Ippen and R. H. Stolen, "Stimulated Brillouin scattering in optical fibers," *Appl. Phys. Lett.*, vol. 21, pp. 539–541, Dec. 1972.

[33] a) S. E. Miller, "Some theory and applications of periodically coupled waves," *Bell Syst. Tech. J.*, vol. 48, pp. 2189–2219, Sept. 1969.
b) A. L. Jones, "Coupling of optical fibers and scattering in fibers," *J. Opt. Soc. Amer.*, vol. 56, pp. 261–267, Mar. 1965.

[34] D. Marcuse, "Radiation losses of dielectric waveguides in terms of the power spectrum of the wall distortion function," *Bell Syst. Tech. J.*, vol. 48, pp. 3233–3242, Dec. 1969.

[35] ——, "Radiation losses of the dominant mode in round dielectric waveguides," *Bell Syst. Tech. J.*, vol. 49, pp. 1665–1693, Oct. 1970.

[36] ——, "Crosstalk by scattering in slab waveguides," *Bell Syst. Tech. J.*, vol. 50, pp. 1817–1832, July–Aug. 1971.

[37] ——, "Power distribution and radiation losses in multimode dielectric slab waveguides," *Bell Syst. Tech. J.*, vol. 51, pp. 429–454, Feb. 1972.

[38] ——, "Mode conversion caused by surface imperfections of a dielectric slab waveguide," *Bell Syst. Tech. J.*, vol. 48, pp. 3187–3215, Dec. 1969.

[39] ——, "Higher-order scattering losses in dielectric waveguides," *Bell Syst. Tech. J.*, vol. 51, pp. 1801–1817, Oct. 1972.

[40] ——, "Derivation of coupled power equations," *Bell Syst. Tech. J.*, vol. 51, pp. 229–237, Jan. 1972.

[41] ——, "Radiation losses of tapered dielectric slab waveguides," *Bell Syst. Tech. J.*, vol. 49, pp. 273–290, Feb. 1970.

[42] ——, "Coupling coefficients of the asymmetric slab waveguide," *Bell Syst. Tech. J.*, vol. 52, pp. 62–81, Jan. 1973.

[43] ——, "Fluctuations of the power of coupled modes," *Bell Syst. Tech. J.*, vol. 51, pp. 1793–1800, Oct. 1972.

[44] ——, "Excitation of the dominant mode of a round fiber by a Gaussian beam," *Bell Syst. Tech. J.*, vol. 49, pp. 1695–1703, Oct. 1970.

[45] ——, "Attenuation of unwanted cladding modes," *Bell Syst. Tech. J.*, vol. 50, pp. 2565–2583, Oct. 1971.

[46] ——, "Pulse propagation in multimode dielectric waveguides," *Bell Syst. Tech. J.*, vol. 51, pp. 1199–1232, July–Aug. 1972.

[47] E. A. J. Marcatili and S. E. Miller, "Improved relations describing directional control in electromagnetic wave guidance," *Bell Syst. Tech. J.*, vol. 48, pp. 2161–2187, Sept. 1969.

[48] E. A. J. Marcatili, "Bends in optical dielectric guides," *Bell Syst. Tech. J.*, vol. 48, pp. 2103–2132, Sept. 1969.

[49] D. Gloge, "Bending loss in multimode fibers with graded and ungraded core," *Appl. Opt.*, vol. 11, pp. 2506–2513, Nov. 1972.

[50] D. Marcuse, "The coupling of degenerate modes in two parallel dielectric waveguides," *Bell Syst. Tech. J.*, vol. 50, pp. 1791–1816, July–Aug. 1971.

[51] F. P. Kapron and D. B. Keck, "Pulse transmission through a dielectric optical waveguide," *Appl. Opt.*, vol. 10, pp. 1519–1523, July 1971.

[52] D. Gloge, "Dispersion in weakly guiding fibers," *Appl. Opt.*, vol. 10, pp. 2442–2445, Nov. 1971.

[53] S. E. Miller, "Delay distortion in generalized lens-like media," *Bell Syst. Tech. J.*, vol. 53, Feb. 1974.

[54] D. Gloge and E. A. J. Marcatili, "Impulse response of fibers with ring-shaped parabolic index distribution," *Bell Syst. Tech. J.*, vol. 52, pp. 1161–1168, Sept. 1973.

[55] D. Marcuse, "The impulse response of an optical fiber with parabolic index profile," *Bell Syst. Tech. J.*, vol. 52, Sept. 1973.

[56] S. D. Personick, "Time dispersion in dielectric waveguide," *Bell Syst. Tech. J.*, vol. 50, pp. 843–859, Mar. 1971.

[57] S. E. Miller and S. D. Personick, "Reduction of dispersion in multimode waveguide," U. S. Patent 3 687 514, Aug. 29, 1972.

[58] D. Marcuse, "Pulse propagation in a two-mode waveguide," *Bell Syst. Tech. J.*, vol. 51, pp. 1785–1791, Oct. 1972.

[59] D. Gloge, "Optical power flow in multimode fibers," *Bell Syst. Tech. J.*, vol. 51, pp. 1767–1783, Oct. 1972.

[60] ——, "Impulse response of clad optical multimode fibers," *Bell Syst. Tech. J.*, vol. 52, pp. 801–815, July–Aug. 1973.

[61] M. W. Jones and K. C. Kao, "Spectrophotometric studies of ultra low loss optical glasses II: Double beam method," *J. Sci. Instrum.* (*J. Phys. E*), vol. 2, pp. 331–335, 1969.

[62] P. J. Laybourn, W. A. Gambling, and D. T. Jones, "Measurements of attenuation in low loss optical glass," *Opto-Electron.*, vol. 3, pp. 137–144, 1971.

[63] A. R. Tynes and D. L. Bisbee, "Balanced two-beam optical loss measuring apparatus," presented at the Spring Meet. of the Opt. Soc., Denver, Colo., Mar. 13–16, 1973.

[64] J. Stone, "A thermo-optical technique for the measurement of absorption loss spectrum in liquids," to be published in *Appl. Opt.*, Sept. 1973.

[65] H. N. Daglish and J. C. North, "Watching light waves decline," *New Scientist*, vol. 49, pp. 14–15, Oct. 1970.

[66] A. R. Tynes, "Measuring loss in optical fibers," *Bell Lab. Rec.*, pp. 303–310, Nov. 1972.

[67] K. I. White and J. E. Midwinter, "An improved technique for the measurement of low optical absorption losses in bulk glass," *Opto-Electron.*, vol. 5, July 1973.

[68] a) T. C. Rich and D. A. Pinnow, "Total optical attenuation in bulk fused silica," *Appl. Phys. Lett.*, vol. 20, pp. 264–266, Apr. 1972.
b) ——, "Optical absorption in fused silica and fused quartz at 1.06 μm," *Appl. Opt.*, vol. 12, p. 2234, Oct. 1973.

[69] P. Kaiser and H. W. Astle, "Measurement of spectral total and scattering losses in unclad optical fibers," *J. Opt. Soc. Amer.*, to be published.

[70] P. Kaiser, "Spectral losses of unclad fibers made from high grade vitreous silica," *Appl. Phys. Lett.*, vol. 21, July 1, 1973.

[71] J. P. Dakin and W. A. Gambling, "Angular distribution of light scattering in bulk glass and fibre waveguides," *Opt. Commun.*, in press.

[72] M. Kerker, *The Scattering of Light and Other Electromagnetic Radiation.* New York and London: Academic Press, 1969, pp. 31–39.

[73] E. G. Rawson, "Measurement of the angular distribution of light scattered from a glass fiber optical waveguide," *Appl. Opt.*, vol. 11, pp. 2477–2481, Nov. 1969.

[74] A. R. Tynes, "Integrating cube scattering detector," *Appl. Opt.*, vol. 9, pp. 2706–2710, Dec. 1970.

[75] D. Gloge, E. L. Chinnock, R. D. Standley, and W. S. Holden, "Dispersion in a low-loss multimode fibre measured at three wavelengths," *Electron. Lett.*, vol. 8, pp. 527–529, Oct. 19, 1972.

[76] S. D. Personick *et al.*, "Measurement of material dispersion in an optical fiber," presented at the CLEA Meeting, Washington, D. C., May 30–June 1, 1973.
D. Gloge, A. R. Tynes, M. A. Duguay, and J. W. Hansen, "Picosecond pulse distortion in optical fibers," *IEEE J. Quantum. Electron.*, vol. QE-8, pp. 217–221, Feb. 1972.
W. A. Gambling, D. N. Payne, and H. Matsumura, "Dispersion in low-loss liquid-core optical fibres," *Electron. Lett.*, vol. 8, pp. 568–569, Nov. 16, 1972.
R. Bouillie and J. K. Andrews, "Measurement of broadening of pulses in glass fibers," *Electron. Lett.*, vol. 8, p. 309, June 15, 1972.
R. Kahn, H. H. Witte, and D. Schicketanz, "Pulse propagation through multimode fibers," *Opt. Commun.*, vol. 4, pp. 352–353, Jan. 1972.

[77] D. Gloge, E. L. Chinnock and D. H. Ring. "Direct measurement of the (baseband) frequency response of multimode fibers," *Appl. Opt.*, vol. 11, pp. 1534–1538, July 1972.

[78] D. B. Keck and A. R. Tynes, "Spectral response of low-loss optical waveguides," *Appl. Opt.*, vol. 11, pp. 1502–1506, July 1972.

[79] a) D. B. Keck, R. D. Maurer, P. C. Schultz, "On the ultimate lower limit of attenuation in glass optical waveguides," *Appl. Phys. Lett.*, vol. 22, pp. 307–309, Apr. 1, 1973.
b) P. C. Schultz, "Preparation of very low-loss optical waveguides," presented at the 1973 Amer. Ceramic Soc. Meet., Cincinnati, Ohio (abstract in *Amer. Ceramic Soc. Bull.*, Apr. 1973).

[80] a) D. Kato, "Fused-silica-core glass fiber as a low-loss optical waveguide," *Appl. Phys. Lett.*, vol. 22, p. 34, Jan. 1, 1973.
b) W. G. French, A. D. Pearson, G. W. Tasker, and J. B. Mac-

Chesney, "A low-loss fused silica optical waveguide with borosilicate cladding," *Appl. Phys. Lett.*, vol. 23, pp. 338–339, Sept. 1973.
c) L. G. van Uitert *et al.*, "Borosilicate glasses for fiber-optical waveguides," *Mat. Res. Bull.*, vol. 8, p. 469, 1973.
d) J. B. MacChesney *et al.*, "Low-loss silica core-borosilicate clad fiber optical waveguide," *Appl. Phys. Lett.*, vol. 23, pp. 340–341, Sept. 1973.
e) S. H. Wemple, D. A. Pinnow, T. C. Rich, R. E. Jaeger, and L. G. Van Uitert, "The binary SiO_2–B_2O_3 glass system: Refractive index behavior and energy gap considerations," *J. Appl. Phys.*, to be published.

[81] K. Koizumi, Post Deadline paper at the Topical Meet. on Integrated Optics, Las Vegas, Nev., Feb. 7–10, 1972.
K. Koizumi, Y. Ikeda, I. Kitano, M. Furukawa, and T. Sumimoto, "New light-focusing glass fibers made by a continuous process," presented at the 1973 Conf. on Laser Engineering and Applications, Washington, D. C. (abstract in *IEEE J. Quantum Electron.*, vol. QE-9, p. 639, June 1973).

[82] D. Gloge, E. L. Chinnock, and K. Koizumi, "Study of pulse distortion in selfoc fibres," *Electron. Lett.*, vol. 8, pp. 856–857, Oct. 19, 1972.

[83] P. Kaiser, "Ultra-low-loss synthetic vitreous silica, their spectral losses and their application in single-material fibers," presented at the 1973 Conf. on Laser Engineering and Applications, Washington, D. C.

[84] R. J. Keyes and T. M. Quist, "Recombination radiation emitted by gallium arsenide," *Proc. IRE* (Corresp.), vol. 50, pp. 1822–1823, Aug. 1962.

[85] W. N. Carr, "Characteristics of a GaAs spontaneous infrared source with 40 percent efficiency," *IEEE Trans. Electron Devices*, vol. ED-12, pp. 531–535, Oct. 1965.

[86] M. N. Zargar'yants, Yu. S. Mezin, and S. I. Kolonenkova, "Electroluminescent diode with a flat surface emitting continuously 25 $W \cdot cm^{-2} \cdot sr^{-1}$ at 300°K," *Fiz. Tekh. Poluprov.*, vol. 4, pp. 1596–1598, Aug. 1970. (*Sov. Phys.—Semicond.*, vol. 4, pp. 1371–1372, Feb. 1971.)

[87] C. A. Burrus and R. W. Dawson, "Small-area, high-current-density GaAs electroluminescent diodes and a method of operation for improved degradation characteristics," *Appl. Phys. Lett.*, vol. 17, pp. 97–99, Aug. 1970.

[88] I. Hayashi, "High power fundamental transverse-mode operation in double-heterostructure lasers," U. S. Patent 3 733 561, issue date May 15, 1973.
M. B. Panish, I. Hayashi, and S. Sumski, "Double-heterostructure injection lasers with room-temperature thresholds as low as 2300 A/cm^2," *Appl. Phys. Lett.*, vol. 16, pp. 326–327, Apr. 1970.

[89] Zh. I. Alferov, V. M. Andreev, E. L. Portnoi, and M. K. Trukan, "AlAs–GaAs heterojunction injection lasers with a low room-temperature threshold," *Fiz. Tekh. Poluprov.*, vol. 3, pp. 1328–1332, Sept. 1969. (*Sov. Phys.—Semicond.*, vol. 3, pp. 1107–1110, Mar. 1970.)

[90] C. A. Burrus and B. I. Miller, "Small-area double heterostructure aluminum-gallium arsenide electroluminescent diode sources for optical-fiber transmission lines," *Opt. Commun.*, vol. 4, pp. 307–309, Dec. 1971.

[91] R. W. Dawson and C. A. Burrus, "Pulse behavior of high-radiance small-area electroluminescent diodes," *Appl. Opt.*, vol. 10, pp. 2367–2369, Oct. 1971.

[92] C. A. Burrus, "Radiance of small-area high-current-density electroluminescent diodes," *Proc. IEEE* (Lett.), vol. 60, pp. 231–232, Feb. 1972.

[93] T. P. Lee and C. A. Burrus, "Noise in the detected output of small-area light-emitting diodes," *IEEE J. Quantum Electron.*, vol. QE-8, pp. 370–373, Mar. 1972.

[94] C. J. Nuese and R. E. Enstrom, "Efficient 1.06 -μm emission from $In_xGa_{1-x}As$ electroluminescent diodes," *IEEE Trans. Electron Devices* (Corresp.), vol. ED-19, pp. 1067–1069, Sept. 1972.

[95] A. A. Bergh and P. J. Dean, "Light-emitting diodes," *Proc. IEEE*, vol. 60, pp. 156–223, Feb. 1972.

[96] L. N. Kurbatov, S. S. Shakhidzhanov, L. V. Bystrova, V. V. Krapukhin, and S. I. Kolonenkova, "Investigation of superluminescence emitted by a gallium arsenide diode," *Fiz. Tekh. Poluprov.*, vol. 4, pp. 2025–2031, Nov. 1970. (*Sov. Phys.—Semicond.*, vol. 4, pp. 1739–1744, May 1971.)

[97] T. P. Lee, C. A. Burrus, Jr., and B. I. Miller, "A stripe-geometry double-heterostructure amplified-spontaneous-emission (super-luminescent) diode," *IEEE J. Quantum Electron.*, vol. QE-9, pp. 820–828, Aug. 1973.

[98] R. N. Hall, G. E. Fenner, J. D. Kingsley, T. J. Soltys, and R. O. Carlson, "Coherent light emission from GaAs junctions," *Phys. Rev. Lett.*, vol. 9, pp. 366–378, Nov. 1962.

[99] M. I. Nathan, W. P. Dumke, G. Burns, F. H. Dill, Jr., and G. J. Lasher, "Stimulated emission of radiation from GaAs p-n junctions," *Appl. Phys. Lett.*, vol. 1, pp. 62–64, Nov. 1962.

[100] T. M. Quist *et al.*, "Semiconductor maser of GaAs," *Appl. Phys. Lett.*, vol. 1, pp. 91–92, Dec. 1962.

[101] I. Hayashi, M. B. Panish, P. W. Foy, and S. Sumski, "Junction lasers which operate continuously at room temperature," *Appl. Phys. Lett.*, vol. 17, pp. 109–111, Aug. 1970.

[102] Zh. I. Alferov *et al.*, "Investigation of the influence of the AlAs–GaAs heterostructure parameters on the laser threshold current and the realization of continuous emission at room temperature," *Fiz. Tekh. Poluprov.*, vol. 4, pp. 1826–1829, Sept. 1970. (*Sov. Phys.—Semicond.*, vol. 4, pp. 1573–1575, Mar. 1971.)

[103] I. Hayashi, M. B. Panish, and F. K. Reinhart, "GaAs-$Al_xGa_{1-x}As$ double heterostructure injection lasers," *J. Appl. Phys.*, vol. 42, pp. 1929–1941, Apr. 1971.

[104] a) M. B. Panish, H. C. Casey, Jr., S. Sumski, and P. W. Foy, "Reduction of threshold current density in GaAs-$Al_xGa_{(1-x)}As$ heterostructure lasers by separate optical confinement and carrier confinement," *Appl. Phys. Lett.*, vol. 22, pp. 590–591, June 1973.
b) G. H. B. Thompson and P. A. Kirby, "Low threshold-current density in five-layer-heterostructure (GaAl)As/GaAs localized-gain-region injection lasers," *Electron. Lett.*, vol. 9, pp. 295–296, June 1973.

[105] O. Berolo and J. C. Woolley, "Electroreflectance spectra of $Al_xGa_{1-x}As$ alloys," *Can. J. Phys.*, vol. 49, pp. 1335–1339, Apr.–June 1971.

[106] M. B. Panish and I. Hayashi, "Heterostructure semiconductor lasers," in *Solid-State Science*, vol. 4, R. Wolfe, Ed. New York: Academic Press, to be published.

[107] a) M. A. Afromowitz, private communication.
b) H. C. Casey, Jr., D. D. Sell, and M. B. Panish, "The refracive index of $Al_xGa_{1-x}As$ between 1.2 and 1.8 eV," to be submitted to *Appl. Phys. Lett.*, to be published.

[108] G. H. B. Thompson, "A theory for filamentation in semiconductor lasers including the dependence of dielectric constant on injected carrier density," *Opto-Electron.*, vol. 4, pp. 257–310, Aug. 1972.

[109] J. E. Ripper, J. C. Dyment, L. A. D'Asaro, and T. L. Paoli, "Stripe-geometry double heterostructure junction lasers: Mode structure and CW operation above room temperature," *Appl. Phys. Lett.*, vol. 18, pp. 155–157, Feb. 1971.

[110] J. C. Dyment, L. A. D'Asaro, J. C. North, B. I. Miller, and J. E. Ripper, "Proton-bombardment formation of stripe-geometry heterostructure lasers for 300 K CW operation," *Proc. IEEE* (Lett.), vol. 60, pp. 726–728, June 1972.

[111] T. Tsukada, R. Ito, N. Nakashima, and O. Nakada, "Mesa-stripe-geometry double-heterostructure injection lasers," *IEEE J. Quantum Electron.*, vol. QE-9, pp. 356–361; Feb. 1973. (Earlier preliminary results appeared in *Appl. Phys. Lett.*, vol. 20, pp. 344–345, May 1972.)

[112] A. R. Goodwin, D. H. Lovelace, and P. R. Selway, "Near- and far-field emission distributions of mesa stripe geometry double heterostructure lasers," *Opto-Electron.*, vol. 4, pp. 311–321, Aug. 1972.

[113] L. A. D'Asaro, "Advances in GaAs junction lasers with stripe geometry," *J. Luminesc.*, vol. 7, pp. 310–337, 1973.

[114] a) L. G. Cohen, "Power coupling from GaAs injection lasers into optical fibers," *Bell Syst. Tech. J.*, vol. 51, pp. 573–594, Mar. 1972.
b) D. Kato, "Light coupling from a stripe-geometry GaAs diode laser into an optical fiber with spherical end," *J. Appl. Phys.*, vol. 44, pp. 2756–2758, June 1973.

[115] L. G. Cohen and M. V. Schneider, "Microlenses for coupling junction lasers to optica fibers," *Appl. Opt.*, vol. 13, Jan. 1974.

[116] H. Kressel, H. F. Lockwood, and F. Z. Hawrylo, "Large-optical-cavity (AlGa)As–GaAs heterojunction laser diode: Threshold and efficiency," *J. Appl. Phys.*, vol. 43, pp. 561–567, Feb. 1972.

[117] G. H. B. Thompson and P. A. Kirby, "(GaAl)As lasers with a heterostructure for optical confinement and additional heterostructure for extreme carrier confinement," *IEEE J. Quantum Electron.*, vol. QE-9, pp. 311–318, Feb. 1973.

[118] T. L. Paoli, B. W. Hakki, and B. I. Miller, "Zero-order transverse mode operation of GaAs double-heterostructure lasers with thick waveguides," *J. Appl. Phys.*, vol. 44, pp. 1276–1280, Mar. 1973.

[119] T. L. Paoli and J. E. Ripper, "Direct modulation of semiconductor lasers," *Proc. IEEE*, vol. 58, pp. 1457–1465, Oct. 1970.

[120] H. Kressel and H. P. Mierop, "Catastrophic degradation in GaAs injection lasers," *J. Appl. Phys.*, vol. 38, pp. 5419–5420, Dec. 1967.

[121] D. A. Shaw and P. R. Thornton, "Catastrophic degradation in GaAs laser diodes," *Solid-State Electron.*, vol. 13, pp. 919–924, July 1970.

[122] H. Kressel and N. E. Byer, "Physical basis of noncatastrophic degradation in GaAs injection lasers," *Proc. IEEE*, vol. 57, pp. 25–33, Jan. 1969.

[123] D. H. Newman and S. Ritchie, "Gradual degradation of GaAs double-heterostructure lasers," *IEEE J. Quantum Electron.*, vol. QE-9, pp. 300–305, Feb. 1973.

[124] R. L. Hartman and A. R. Hartman, "Strain induced degradation of GaAs injection lasers," *Appl. Phys. Lett.*, vol. 23, pp. 147–149, Aug. 1973.

[125] B. C. DeLoach, Jr., B. W. Hakki, R. L. Hartman, and L. A. D'Asaro, "Degradation of CW GaAs double heterojunction lasers at 300 K," *Proc. IEEE* (Lett.), vol. 61, pp. 1042–1044, July 1973.

[126] T. L. Paoli and B. W. Hakki, "CW degradation at 300°K of GaAs double heterostructure junction lasers: I. Emission spectra," and B. W. Hakki and T. L. Paoli, "II. Electronic gain," *J. Appl. Phys.*, vol. 44, Sept. 1973.

[127] a) B. C. DeLoach, Jr., "Reliability of GaAs injection lasers," presented at the 1973 Conf. on Laser Engineering and Applications, Washington, D. C. (abstract in *IEEE J. Quantum Electron.*, vol. QE-9, p. 688, June 1973).
b) I. Hayashi, "Progress of semiconductor lasers in Japan," presented at the 1973 Conf. on Laser Engineering and Applications, Washington, D. C. (abstract in *IEEE J. Quantum Electron.*, vol. QE-9, pp. 687–688, June 1973).
[128] R. L. Hartman, J. C. Dyment, C. J. Hwang, and M. Kuhn, "Continuous operation of GaAs–$Ga_{1-x}Al_xAs$ double heterostructure lasers with 30°C half lives exceeding 1000 hours," *Appl. Phys. Lett.*, vol. 23, pp. 181–183, Aug. 1973.
[129] J. E. Geusic, H. M. Marcos, and L. G. Van Uitert, "Laser oscillation in Nd-doped yttrium aluminum, yttrium gallium and gadolinium garnets," *Appl. Phys. Lett.*, vol. 4, pp. 182–184, May 1964.
[130] J. E. Geusic, W. B. Bridges, and J. I. Pankove, "Coherent optical sources for communications," *Proc. IEEE*, vol. 58, pp. 1419–1439, Oct. 1970.
[131] R. B. Allen and S. J. Scalise, "Continuous operation of YAlG:Nd laser by injection luminescent pumping," *Appl. Phys. Lett.*, vol. 14, pp. 188–190, Mar. 1969.
[132] F. W. Ostermayer, "$GaAs_{1-x}P_x$ diode pumped YAG:Nd lasers," *Appl. Phys. Lett.*, vol. 18, pp. 93–96, Feb. 1971.
[133] F. W. Ostermayer, R. B. Allen, and E. G. Dierschke, "Room-temperature CW operation of a $GaAs_{1-x}P_x$ diode-pumped YAG:Nd laser," *Appl. Phys. Lett.*, vol. 19, pp. 289–292, Oct. 1971.
[134] N. P. Barnes, "Diode-pumped solid-state lasers," *J. Appl. Phys.*, vol. 44, pp. 230–237, Jan. 1973.
[135] L. J. Rosenkrantz, "GaAs diode-pumped Nd:YAG laser," *J. Appl. Phys.*, vol. 43, pp. 4603–4605, Nov. 1972.
[136] R. B. Chesler and D. A. Draegert, "Miniature diode-pumped Nd:YAlG lasers," *Appl. Phys. Lett.*, vol. 23, pp. 235–236, Sept. 1973.
[137] J. Stone and C. A. Burrus, "End-pumped-fused-silica neodymium-doped fiber-geometry lasers operating at 1.06 and 1.08 μm wavelength," *Appl. Phys. Lett.*, vol. 23, pp. 388–389, Oct. 1973.
[138] T. Kushida, H. Marcos, and J. E. Geusic, "Laser transition cross section and fluorescence branching ratio for Nd^{3+} in yttrium aluminum garnet," *Phys. Rev.*, vol. 167, pp. 289–291, Mar. 1968.
[139] J. E. Goell, "A repeater with high input impedance for optical-fiber transmission," presented at the 1973 Conf. on Laser Engineering and Applications, Washington, D. C. (abstract in *IEEE J. Quantum Electron.*, vol. QE-9, pp. 641–642, June 1973).
[140] B. G. King and H. J. Schulte, "An experimental large capacity multiplexer for an optical transmission system," presented at the Int. Communications Conf., Seattle, Wash., June 1973 (abstract in the *Proceedings* of the conference).
[141] G. White and C. A. Burrus, "Efficient 100 Mb/s driver for electroluminescent diodes," *Int. J. Electron.*, vol. 34, to be published.
[142] G. White and G. M. Chinn, "A 100-Mb·s^{-1} fiber-optic communication channel," *Proc. IEEE* (Lett), vol. 61, pp. 683–684, May 1973.
[143] R. W. Dawson, private communication.
[144] W. M. Hubbard, "Efficient utilization of optical-frequency carriers for low and moderate bandwidth channels," *Bell Syst. Tech. J.*, vol. 52, pp. 731–765, May–June 1973.
[145] J. C. Dyment, J. E. Ripper, and T. P. Lee, "Measurement and interpretation of long spontaneous lifetimes in double heterostructure lasers," *J. Appl. Phys.*, vol. 43, pp. 452–457, Feb. 1972.
[146] T. Ikegami and Y. Suematsu, "Direct modulation of semiconductor junction lasers," *Electron. Commun.* (Japan), vol. 51-B, pp. 51–57, Feb. 1968.
[147] N. G. Basov, V. V. Nikitin, and A. S. Semenov, "Dynamics of semiconductor injection lasers," *Usp. Fiz. Nauk*, vol. 97, pp. 561–600, Apr. 1969. (*Sov. Phys.—Uspekhi*, vol. 12, pp. 219–240, Sept.–Oct. 1969.)
[148] M. J. Adams, "Rate equations and transient phenomena in semiconductor lasers," *Opto-Electron.*, vol. 5, pp. 201–215, Mar. 1973.
[149] C. J. Hwang, "Properties of spontaneous and stimulated emission in GaAs junction lasers. II," *Phys. Rev.*, vol. B2, pp. 4126–4134, Nov. 1970.
[150] E. Pinkas, B. I. Miller, I. Hayashi, and P. W. Foy, "GaAs-$Al_xGa_{1-x}As$ double heterostructure lasers—Effect of doping on lasing characteristics of GaAs," *J. Appl. Phys.*, vol. 43, pp. 2827–2835, June 1972.
[151] T. Ozeki and T. Ito, "Pulse modulation of DH-(GaAl)As lasers," *IEEE J. Quantum. Electron.* vol. QE-9, pp. 388–391, Feb. 1973.
[152] T. Ozeki and T. Ito, "A 200-Mb/s PCM DH-GaAlAs laser communication experiment," presented at the 1973 Conf. on Laser Engineering and Applications, Washington, D. C. (abstract in *IEEE J. Quantum Electron.*, vol. QE-9, p. 692, June 1973).
[153] W. O. Schlosser, K. Kurokawa, private communication.
[154] M. Chown, A. R. Goodwin, D. F. Lovelace, G. H. B. Thompson, and P. R. Selway, "Direct modulation of double-heterostructure lasers at rates up to 1 Gb/s," *Electron. Lett.*, vol. 9, pp. 34–36, Jan. 1973.
[155] H. W. Thim *et al.*, "Subnanosecond PCM of GaAs lasers by Gunn-effect switches," presented at the 1973 IEEE Int. Solid-State Circuits Conf., Philadelphia, Pa., Feb. 1973 (abstract in *Dig. of Tech. Papers*).
[156] J. E. Carroll and J. G. Farrington, "Short-pulse modulation of gallium-arsenide lasers with Trapatt diodes," *Electron. Lett.*, vol. 9, pp. 166–167, Apr. 1973.
[157] P. Russer and S. Schulz, "Direkte Modulation eines Doppelheterostrukturlasers mit einer Bitrate von 2–3 Gbit/s," *Arch. Elek. Übertragung*, vol. 27, pp. 193–195, Apr. 1973.
[158] J. E. Goell, "A 274-Mb/s optical-repeater experiment employing a GaAs laser," paper presented at the 1973 Conf. on Laser Engineering and Applications, Washington, D. C.; also in *Proc. IEEE* (Lett.), vol. 61, pp. 1504–1505, Oct. 1973.
[159] S. D. Personick, "Receiver design for digital fiber optic communication systems, Parts I and II," *Bell Syst. Tech. J.*, vol. 52, pp. 843–886, July–Aug. 1973.
[160] N. G. Basov, V. N. Morozov, V. V. Nikitin, and A. S. Semenov, "Investigation of GaAs laser radiation pulsations," *Fiz. Tekh. Poluprov.*, vol. 1, pp. 1570–1574, Oct. 1967. (*Sov. Phys.—Semicond.*, vol. 1, pp. 1305–1307, Apr. 1968.)
[161] T. P. Lee and R. H. R. Roldan, "Repetitively *Q*-switched light pulses from GaAs injection lasers with tandem double-section stripe geometry," *IEEE J. Quantum Electron.*, vol. QE-6, pp. 339–352, June 1970.
[162] D. Gloge and T. P. Lee, "Signal structure of continuously self-pulsing GaAs lasers," *IEEE J. Quantum Electron.* (Corresp.), vol. QE-7, pp. 43–45, Jan. 1971.
[163] a) T. L. Paoli and J. E. Ripper, "Optical pulses from CW GaAs injection lasers," *Appl. Phys. Lett.*, vol. 15, pp. 105–107, Aug. 1, 1969.
b) J. E. Ripper and T. L. Paoli, "Optical self-pulsing of junction lasers operating continuously at room temperature," *Appl. Phys. Lett.*, vol. 18, pp. 466–468, May 15, 1971.
c) ——, "Frequency pulling and pulse position modulation of pulsing CW GaAs injection lasers," *Appl. Phys. Lett.*, vol. 15, pp. 203–205, 1969.
[164] B. S. S. Rao, A. Subrahmanyam, and P. Swarup, "A technique of modulating pulsed semiconductor lasers," *IEEE Trans. Commun.*, vol. COM-21, pp. 284–289, Apr. 1973.
[165] S. E. Miller, "Integrated optics: An introduction," *Bell Syst. Tech. J.*, vol. 48, pp. 2059–2069, Sept. 1969.
[166] F. S. Chen, "Modulators for optical communications," *Proc. IEEE*, vol. 58, pp. 1440–1457, Oct. 1970.
[167] I. P. Kaminow and E. H. Turner, "Electrooptic light modulators," *Proc. IEEE*, vol. 54, pp. 1374–1390, Oct. 1966.
[168] E. I. Gordon, "A review of acoustooptical deflection and modulation devices," *Proc. IEEE*, vol. 54, pp. 1391–1401, Oct. 1966.
[169] S. H. Wemple, J. F. Dillon, Jr., L. G. Van Uitert, and W. H. Grodkiewicz, "Iron garnet crystals for magnetooptic light modulators at 1.064 μm," *Appl. Phys. Lett.*, vol. 22, pp. 331–333, Apr. 1973.
[170] F. K. Reinhart, "Electroabsorption in $Al_yGa_{1-y}As$–$Al_xGa_{1-x}As$ double heterostructures," *Appl. Phys. Lett.*, vol. 22, pp. 372–374, Apr. 1973.
[171] F. S. Chen, "Effects of modulator impairments on optical digital communications," to be published.
[172] R. T. Denton, F. S. Chen, and A. A. Ballman, "Lithium tantalate light modulators," *J. Appl. Phys.*, vol. 38, pp. 1611–1617, Mar. 1967.
[173] K. K. Chow and W. B. Leonard, "Efficient octave-bandwidth microwave light modulators," *IEEE J. Quantum Electron.*, vol. QE-6, pp. 789–793, Dec. 1970.
[174] F. S. Chen and W. W. Benson, "A lithium niobate light modulator for fiber optical communications," *Proc. IEEE* (Lett.), to be published.
[175] G. White and G. M. Chin, "Travelling wave electro-optic modulators," *Opt. Commun.*, vol. 5, pp. 374–379, Aug. 1972.
[176] J. F. S. Ledger and E. A. Ash, "Laser-beam modulation using grating effects," *Electron. Lett.*, vol. 4, pp. 99–100, Mar. 1968.
[177] J. M. Hammer, "Digital electro-optic grating deflector and modulator," *Appl. Phys. Lett.*, vol. 18, pp. 147–149, Feb. 1971.
[178] M. A. R. P. deBarros and M. G. F. Wilson, "High-speed electro-optic diffraction modulator for baseband operation," *Proc. Inst. Elec. Eng.*, vol. 119, pp. 807–814, July 1972.
[179] a) S. Wright and M. G. F. Wilson, "New form of electro-optic deflector," *Electron. Lett.*, vol. 9, pp. 169–170, May 1973.
b) J. M. Hammer, D. J. Channin, and M. T. Duffy, "Fast electro-optic waveguide deflector modulator," *Appl. Phys. Lett.*, vol. 23, pp. 176–177, Aug. 1973.
[180] I. P. Kaminow, J. R. Carruthers, E. H. Turner, and L. W. Stulz, "Thin-film $LiNbO_3$ electro-optic light modulator," *Appl. Phys. Lett.*, vol. 22, pp. 540–542, May 1973.
[181] D. J. Channin, "Voltage-induced optical waveguide," *Appl. Phys. Lett.*, vol. 19, pp. 128–130, Sept. 1971.
[182] F. K. Reinhart, "Reverse-biased gallium phosphide diodes as high-frequency light modulators," *J. Appl. Phys.*, vol. 39, pp. 3426–3434, June 1968.
[183] D. Hall, A. Yariv, and E. Garmire, "Optical guiding and electro-

optic modulation in GaAs epitaxial layers," *Opt. Commun.*, vol. 1, pp. 403–405, Apr. 1970.

[184] F. K. Reinhart and B. I. Miller, "Efficient GaAs–$Al_xGa_{1-x}As$ double-heterostructure light modulators," *Appl. Phys. Lett.*, vol. 20, pp. 36–38, Jan. 1972.

[185] W. R. Klein and B. D. Cook, "Unified approach to ultrasonic light diffraction," *IEEE Trans. Sonics Ultrason.*, vol. SU-14, pp. 123–134, July 1967.

[186] A. W. Warner and A. H. Meitzler, "Performance of bonded, single-crystal $LiNbO_3$ and $LiGaO_3$ as ultrasonic transducers operating above 100 MHz," *Proc. IEEE* (Lett.), vol. 56, pp. 1376–1377, Aug. 1968.

[187] N. Uchida, S. Fukunishi, and S. Saito, "High-frequency acousto-optic device using ultrasonically bonded and sputter-machined $LiNbO_3$ transducer," presented at the 1973 Conf. on Laser Engineering and Applications, Washington, D. C. (abstract in *IEEE J. Quantum Electron*, vol. QE-9, p. 660, June 1973).

[188] D. A. Pinnow, "Guide lines for the selection of acoustooptic materials," *IEEE J. Quantum Electron.*, vol. QE-6, pp. 223–238, Apr. 1970.

[189] N. Uchida and N. Niizeki, "Acoustooptic deflection materials and techniques," *Proc. IEEE*, vol. 61, pp. 1073–1092, Aug. 1973.

[190] D. Maydan, "Acoustooptical pulse modulators," *IEEE J. Quantum Electron*, vol. QE-6, pp. 15–24, Jan. 1970.

[191] A. W. Warner and D. A. Pinnow, "Miniature acoustooptic modulators for infrared optical communications," presented at the 1973 Conf. on Laser Engineering and Applications, Washington, D. C. (abstract in *IEEE J. Quantum Electron.*, vol. QE-9, pp. 659–660, June 1973).

[192] a) L. Kuhn, M. L. Dakss, P. F. Heidrich, and B. A. Scott, "Deflection of an optical guided wave by a surface acoustic wave," *Appl. Phys. Lett.*, vol. 17, pp. 265–267, Sept. 1970.
b) R. V. Schmidt, I. P. Kaminow, and J. R. Carruthers, "Acousto-optic diffraction of guided optical waves in $LiNbO_3$," *Appl. Phys. Lett.*, to be published.

[193] P. K. Tien, R. J. Martin, R. Wolfe, R. C. LeCraw, and S. L. Blank, "Switching and modulation of light in magnetooptic waveguides of garnet films," *Appl. Phys. Lett.*, vol. 21, pp. 394–396, Oct. 1972.

[194] L. K. Anderson and B. J. McMurtry, "High-speed photodetectors," *Proc. IEEE*, vol. 54, pp. 1335–1349, Oct. 1966.

[195] L. K. Anderson, M. DiDomenico, and M. B. Fisher, "High-speed photodetectors for microwave demodulation of light," in *Advances in Microwaves*, vol. 5, L. Young, Ed. New York: Academic Press, 1970.

[196] H. Melchior, M. B. Fisher, and F. Arams, "Photodetectors for optical communication systems," *Proc. IEEE*, vol. 58, pp. 1466–1486, Oct. 1970.

[197] H. Melchior, "Demodulation and photodetection techniques," in *Laser Handbook*, F. T. Arecchi and E. D. Schulz-Dubois, Eds. Amsterdam, The Netherlands: Elsevier, North-Holland, pp. 628–739.

[198] ——, "Sensitive high speed photodetectors for the demodulation of visible and near infrared light," *J. Luminesc.*, vol. 7, pp. 390–414, 1973.

[199] R. J. McIntyre, "Multiplication noise in uniform avalanche diodes," *IEEE Trans. Electron Devices*, vol. ED-13, pp. 164–168, Jan. 1966.

[200] S. D. Personick, "New results on avalanche multiplication statistics with applications to optical detection," *Bell Syst. Tech. J.*, vol. 50, pp. 167–189, Jan. 1971.
——, "Statistics of a general class of avalanche detectors with applications to optical communication," *Bell Syst. Tech. J.*, vol. 50, pp. 3075–3095, Dec. 1971.

[201] R. J. McIntyre, "The distribution of gains in uniformly multiplying avalanche photodiodes: Theory," *IEEE Trans. Electron Devices*, vol. ED-19, pp. 703–713, June 1972.
J. Conradi, "The distribution of gains in uniformly multiplying avalanche photodiodes: Experimental," *IEEE Trans. Electron. Devices*, vol. ED-19, pp. 713–718, June 1972.

[202] A. B. Gillespie, *Signal, Noise and Resolution in Nuclear Counter Amplifiers*. London, England: Pergamon, 1953.

[203] H. Melchior, "Semiconductor detectors for optical communications," presented at the 1973 Conf. on Laser Engineering and Applications, Washington, D. C. (abstract in *IEEE J. Quantum Electron.*, vol. QE-9, p. 659, June 1973).

[204] H. C. Spriging and R. J. McIntyre, "Improved multi-element silicon photodiodes for detection of 1.06 μm," presented at the Int. Electron. Devices Meet., Washington, D. C., Oct. 1968.

[205] O. Krumpholz and S. Maslowski, "Schnelle Photodioden mit Wellenlängen-unabhängigen Demodulationseigenschaften," *Z. Angew. Phys.*, vol. 25, pp. 156–160, Mar. 1968.

[206] H. Melchior and W. T. Lynch, "Signal and noise response of high speed germanium avalanche photodiodes," *IEEE Trans. Electron Devices*, vol. ED-13, pp. 829–838, Dec. 1966.

[207] D. P. Mathur, R. J. McIntyre, and P. P. Webb, "A new germanium photodiode with extended long-wavelength response," *Appl. Opt.*, vol. 9, pp. 1842–1847, Aug. 1970.

[208] H. W. Ruegg, "An optimized avalanche photodiode," *IEEE Trans. Electron Devices*, vol. ED-14, pp. 239–251, May 1967.

[209] P. Webb and R. J. McIntyre, "An efficient low noise avalanche photodiode for visible to 1.06 μm," in *Proc. Electro-Optical Systems Design Conf.* (East), pp. 51–56, 1971.

[210] O. Krumpholz and S. Maslowski, "Avalanche Mesaphotodioden mit Quereinstrahlung," *Wiss. Ber. AEG-Telefunken*, vol. 44, pp. 73–79, 1971.

[211] S. Maslowski, "New kind of detector for use in communication systems with glass fiber waveguides," presented at OSA Topical Meet. on Integrated Optics, Las Vegas, Feb. 1972 (summary in *Dig. Tech. Papers*).

[212] S. E. Miller, unpublished work.

[213] E. A. J. Marcatili, unpublished work.

[214] D. Gloge, "Fiber delay equalization by carrier drift in the detector," *Opto-Electron.*, vol. 5, pp. 345–350, July 1973.

[215] W. T. Lindley, R. J. Phelan, Jr., C. M. Wolfe, and A. G. Foyt, "GaAs Schottky barrier avalanche photodiodes," *Appl. Phys. Lett.*, vol. 14, pp. 197–199, Mar. 1969.

[216] G. E. Stilman *et al.*, "High speed InGaAs avalanche photodiodes," *Solid State Res.* (MIT Lincoln Labs.), 1972.

[217] R. C. Eden, J. S. Harris, K. Nakano, and J. C. Chu, unpublished. Reference to this work is made by A. M. Andrews, J. A. Higgins, J. T. Longo, E. R. Gertner, and J. G. Pasko, "High-speed $Pb_{1-x}Sn_xTe$ photodiodes," *Appl. Phys. Lett.*, vol. 21, pp. 285–287, Sept. 1972.

[218] J. L. Shay, K. J. Bachmann, E. Buehler, and J. H. Wernick, "$CdSnP_2$–InP heterodiodes for near infrared light emitting diodes and photovoltaic detectors," to be published in *Appl. Phys. Lett.*, 1973.

[219] J. E. Goell and R. D. Standley, "Integrated optical circuits," *Proc. IEEE*, vol. 58, pp. 1504–1512, Oct. 1970.

[220] P. K. Tien, "Light waves in thin films and integrated optics," *Appl. Opt.*, vol. 10, pp. 2395–2413, Nov. 1971.

[221] S. E. Miller, "A survey of integrated optics," *IEEE J. Quantum Electron.*, vol. QE-8, pp. 199–205, Feb. 1972.

[222] R. V. Pole, S. E. Miller, J. H. Harris, and P. K. Tien, "Integrated optics and guided waves—A report of the topical meeting," *Appl. Opt.*, vol. 11, pp. 1675–1685, Aug. 1972.

[223] A. Yariv, "Active integrated optics," in *Proc. 1971 Esfahan Conf. on Pure and Applied Laser Physics*. New York: Wiley, in press.

[224] E. Garmire, H. Stoll, A. Yariv, and R. G. Hunsperger, "Optical waveguiding in proton-implanted GaAs," *Appl. Phys. Lett.*, vol. 21, pp. 87–88, Aug. 1972.

[225] S. Somekh, E. Garmire, and A. Yariv, "Channel optical waveguide directional couplers," *Appl. Phys. Lett.*, vol. 22, pp. 46–47, Jan. 1973.

[226] H. L. Garvin, E. Garmire, S. Somekh, H. Stoll, and A. Yariv, "Ion beam micromachining of integrated optics components," *Appl. Opt.*, vol. 12, pp. 455–459, Mar. 1973.

[227] J. C. Tracy, W. Wiegman, R. A. Logan, and F. K. Reinhart, "Three-dimensional light guides in single-crystal GaAs–$Al_xGa_{1-x}As$," *Appl. Phys. Lett.*, vol. 22, pp. 511–512, May 1973.

[228] R. A. Logan and F. K. Reinhart, "Optical waveguides in GaAs–AlGaAs epitaxial layers," *J. Appl. Phys.*, to be published.

[229] B. I. Miller, R. A. Logan, T. P. Lee, and C. A. Burrus, to be published.

[230] H. G. Danielmeyer and H. P. Weber, "Fluorescence in neodymium ultraphosphate," *IEEE J. Quantum Electron.*, vol. QE-8, pp. 805–808, Oct. 1972.

[231] H. P. Weber, T. C. Damen, H. G. Danielmeyer, and B. C. Tofield, "Nd-ultraphosphate laser," *Appl. Phys. Lett.*, vol. 22, pp. 534–536, May 1973.

[232] T. C. Damen, H. P. Weber, and B. C. Tofield, "NdLa pentaphosphate laser performance," *Appl. Phys. Lett.*, vol. 23, Nov. 1973.

[233] J. P. van der Ziel, W. A. Bonner, L. Kopf, S. Singh, and L. G. Van Uitert, "Laser oscillation from HO^{3+} and Nd^{3+} ions in epitaxially grown thin aluminum garnet films," *Appl. Phys. Lett.*, vol. 22, pp. 656–657, June 1973.

[234] J. G. Grabmaier, B. C. Grabmaier, R. Th. Kersten, R. D. Plättner, and G. J. Zeidler, "Epitaxially grown Nd-laser-films," *Phys. Lett.*, to be published.

[235] a) I. P. Kaminow and J. R. Carruthers, "Optical waveguiding layers in $LiNbO_3$ and $LiTaO_3$," *Appl. Phys. Lett.*, vol. 22, pp. 326–328, Apr. 1973.
b) S. Miyazawa, "Growth of $LiNbO_3$ single-crystal film for optical waveguides," *Appl. Phys. Lett.*, vol. 23, pp. 198–200, Aug. 1973.

[236] a) J. M. Hammer, D. J. Channin, M. T. Duffy, and J. P. Wittke, "Low-loss epitaxial ZnO optical waveguides," *Appl. Phys. Lett.*, vol. 21, pp. 358–360, Oct. 1972.
b) W. E. Martin and D. B. Hall, "Optical waveguides by diffusion in II–VI compounds," *Appl. Phys. Lett.*, vol. 21, pp. 325–327, Oct. 1972.

[237] P. K. Tien, R. J. Martin, S. L. Blank, S. H. Wemple, and L. J.

Varnerin, "Optical waveguides of single-crystal garnet films," *Appl. Phys. Lett.*, vol. 21, pp. 207–209, Sept. 1972.
[238] A. A. Ballman, P. K. Tien, H. Brown, and R. J. Martin, "The growth of single crystalline waveguiding thin films of piezoelectric sillenites," *J. Crystal Growth*, to be published.
[239] D. P. Schinke, R. G. Smith, E. G. Spencer, and M. F. Galvin, "Thin-film distributed-feedback laser fabricated by ion milling," *Appl. Phys. Lett.*, vol. 21, pp. 494–496, Nov. 1972.
[240] C. V. Shank and R. V. Schmidt, "Optical technique for producing 0.1μ periodic surface structures," *Appl. Phys. Lett.*, vol. 23, pp. 154–155, Aug. 1973.
[241] J. E. Goell, "Barium silicate films for integrated optical circuits," *Appl. Opt.*, vol. 12, pp. 737–742, Apr. 1973.
[242] ——, "Electron-resist fabrication of bends and couplers for integrated optical circuits," *Appl. Opt.*, vol. 12, pp. 729–736, Apr. 1973.
[243] D. B. Ostrowsky and J. C. Dubois, "Electron-beam masking for optical-waveguide fabrication," presented at the OSA Topical Meet. on Integrated Optics, Las Vegas, Feb. 1972 (summary in *Dig. Tech. Papers*).
[244] H. Kogelnik and C. V. Shank, "Stimulated emission in a periodic structure," *Appl. Phys. Lett.*, vol. 18, pp. 152–154, Feb. 1971.
[245] ——, "Coupled-wave theory of distributed feedback lasers," *J. Appl. Phys.*, vol. 43, pp. 2327–2335, May 1972.
[246] M. Nakamura *et al.*, "Laser oscillation in epitaxial GaAs waveguides with corrugation feedback," *Appl. Phys. Lett.*, vol. 23, pp. 224–225, Sept. 1973.
[247] H. W. Yen *et al.*, "Optically-pumped GaAs waveguide laser with fundamental 0.12 micron corrugation feedback," *Opt. Commun.*, to be published.
[248] H. Kogelnik, "Coupled wave theory for thick hologram gratings," *Bell Syst. Tech. J.*, vol. 48, pp. 2909–2947, Nov. 1969.
[249] D. Marcuse, *Light Transmission Optics*. New York: Van Nostrand Reinhold, 1972, ch. 9.
[250] F. W. Dabby, M. A. Saifi, and A. Kestenbaum, "High-frequency cutoft periodic dielectric waveguides," *Appl. Phys. Lett.*, vol. 22, pp. 190–191, Feb. 1973.
[251] W. R. Bennett and J. R. Davey, *Data Transmission*, New York: McGraw-Hill, 1965.
M. Schwartz, W. R. Bennett, and S. Stein, *Communication Systems Techniques*. New York: McGraw-Hill, 1966.
[252] S. D. Personick, "Baseband linearity and equalization in fiber optic communication systems," *Bell Syst. Tech. J.*, vol. 52, Sept. 1973.
[253] W. M. Hubbard, "Comparative performance of twin-channel and single-channel optical-frequency receivers," *IEEE Trans. Commun.*, vol. COM-20, pp. 1079–1086, Dec. 1972.
[254] R. W. Dawson, "Effect of ambient temperature on infrared transmission through a glass fiber," *Bell Syst. Tech. J.*, vol. 51, pp. 569–571, Feb. 1972.
[255] O. Krumpholz, "Optical-coupling problems in communication systems with glass fiber waveguides," presented at the OSA Topical Meet. on Integrated Optics, Las Vegas, Nev., Feb. 7–10, 1972; also, *Appl. Opt.*, vol. 11, pp. 1675–1685, Aug. 1972.
[256] M. Borner, "Optical information transmission with glass fiber waveguides," *Wiss. Ber. AEG-Telefunken*, vol. 44, pp. 41–45, 1971.
[257] C. G. Someda, "Simple, low-loss joints between single-mode optical fibers," *Bell Syst. Tech. J.*, vol. 52, pp. 583–596, Apr. 1973.
[258] D. L. Bisbee, "Optical fiber joining technique," *Bell Syst. Tech. J.*, vol. 50, pp. 3153–3158, Dec. 1971.
[259] R. B. Dyott, J. R. Stern, and J. H. Stewart, "Fusion junctions for glass-fibre waveguides," *Electron. Lett.*, vol. 8, pp. 290–292, June 1, 1972.
[260] *Opto-Electron.*, vol. 5, pp. 261–366, July 1973.

Part II
Fiber Preparation

Glass Fibers for Optical Communications

ROBERT D. MAURER

Invited Paper

***Abstract*—Glass optical waveguides with attenuations below 20 dB/km have made possible a new approach to optical communications. These glass fibers satisfy requirements for transmission over kilometer lengths with experimental systems utilizing existing devices for sources and detectors. The realization of material and fabrication advances necessary for this accomplishment are the topic of this paper. Basic theoretical principles are introduced in a review fashion. The application of these principles in choice of materials and fabrication is described. Results in fiber performance following this framework are given in a section on evaluation, which includes information capacity, attenuation, and some environmental requirements. Preliminary experiments in bundling and cabling are discussed, followed by concluding remarks.**

I. Introduction

THE ATTAINMENT of glass fibers with attenuations below 20 dB/km has radically changed the outlook for optical communications [1]. Formerly, the available transmission devices were marked as complicated and prospectively expensive [2]. Most of them involved the installation of pipes filled with lensing media of various types which had to be controlled by interacting servosystems. Therefore, the prospect of flexible trouble-free transmission lines greatly enhances the practicality of future optical communication systems. Advances in solid-state detectors and light sources, both coherent and incoherent, offer additional support. With these source–detector combinations, losses of 30 to 50 dB can be tolerated with reasonable signal-to-noise ratios. Therefore, a loss of 20 dB/km permits transmission over more than one kilometer without amplification. This attenuation factor is the approximate break-over point for widespread application and, therefore, this article will emphasize techniques for achieving such low losses. The incorporation of fibers in communication systems will be covered in a forthcoming invited paper.[1]

Signal transmission with glass waveguides presents several advantages in addition to the frequently mentioned high bandpass. Among these are low susceptibility to electromagnetic interference, small size, low weight, and dielectric isolation. These advantages may well open up specialized applications prior to widespread communication use. However, the specialized advantages are peripheral to the potential of transmission lines that are economically competitive in both high- and low-bandpass systems, thereby making optical communications superior throughout communications technology.

This review covers the progress in techniques for attaining low-loss (below 20 dB/km) glass waveguides. Cylindrical fibers with three types of radial refractive-index variation are discussed: discontinuous (step) with single-mode propagation, discontinuous with multimode propagation, and continuous (gradient) with multimode propagation. Emphasis will be placed upon demonstrations of feasibility rather than theory. The next section on theory covers only the results necessary to interpret experimental results mentioned later. There follows the bulk of the article with principles found necessary for

This invited paper is one of a series planned on topics of general interest—The Editor.

Manuscript received November 20, 1972; revised January 10, 1973.

The author is with Corning Glass Works, Research and Development Laboratories, Corning, N.Y. 14830.

[1] T. Li, E. A. J. Marcatilli, and S. E. Miller, to be published in the *Proc. IEEE*.

Reprinted from *Proc. IEEE*, vol. 61, pp. 452–462, Apr. 1973.

materials and fabrication and the emergence of these in fiber performance. Finally, some remarks are made on preliminary cable experiments and future directions.

II. Theoretical Background

A. Fiber Form

A few theoretical principles and definitions of waveguide phenomena are necessary for understanding the fabrication results discussed below. Two basically different types of cylindrical fiber offer practical advantages. One type is the step-refractive index variation in which the fiber has a core of one refractive index and a cladding of a lower refractive index. The other type is the gradient refractive index variation in which the fiber has a high refractive index at its axis which decreases continually to a lower refractive index at the surface. For theoretical purposes here, the outer boundary of both types can be neglected. That is, these refractive index variations can be considered to proceed to infinity in the radial direction.

Cylindrical waveguides with step-refractive-index variation are characterized by an important parameter [3]

$$V = \frac{2\sqrt{2}\pi r}{\lambda} (\bar{n}\Delta n)^{1/2} \tag{1}$$

where r is the core radius, λ the free space wavelength, $\bar{n}$ the average refractive index of core and cladding, and Δn the difference in refractive index between core and cladding. Below $V \cong 2.4$, the fibers propagate a single mode, designated HE_{11}. Because of the axial symmetry, this mode is polarization degenerate, although any core ellipticity lifts this degeneracy. The chief advantage of a single-mode waveguide is its high bandpass. Above $V \cong 2.4$, the quantity of propagating modes increases rapidly as V increases, with the approximate number given by [4], [5]

$$N \sim \frac{V^2}{2}. \tag{2}$$

In addition, when all modes are equally excited, an approximate value for the fractional power carried in the cladding can be given as a function of V [4]

$$\frac{\text{cladding power}}{\text{total power}} \equiv p_{cl} = \frac{8}{3V} \tag{3}$$

which may be used with the core power fraction to give the total attenuation [6]

$$\beta = p_{cl}\beta_{cl} + p_c\beta_c \tag{4}$$

where β_{cl} and β_c are the attenuation coefficients of the cladding and core, respectively.

Waveguides with continuous variation of refractive index (gradient guides) are fabricated to have a variation of the form [7], [8]

$$n = n_0 \operatorname{sech} \rho r \tag{5}$$

where n_0 is the refractive index on the fiber axis and ρ the radial variation constant related to the focussing distance. All modes corresponding to meridional rays (HE_{1m} modes) have the same group velocity, neglecting material dispersion (see below). From a macroscopic viewpoint, all meridional rays focus at successive equivalent positions on the axis of the fiber. But skew rays do not focus with the refractive index gradient given by (5) [9]. Therefore, both experimentally and theoretically, deviations from the refractive index variation of (5) are difficult to separate from skew ray effects due to the light source.

B. Materials Properties

A major impediment to the use of glass fibers has been their attenuation. Even today this remains one of the most critical factors for system economics and manufacturing control. The attenuation can be subdivided into absorption, or conversion of light into heat, and scattering, or light escaping the bound modes. Absorption consists of three types: intrinsic, impurity, and atomic "defect" color centers. Scattering also consists of three types: intrinsic, glass inhomogeneity, and aberrations in the radial (cross-sectional) form of the refractive index. All six of these will be discussed in turn.

Intrinsic absorption, by definition, occurs when the material exists in a "perfect" state. Normally, perfect dielectric materials, like glass, are considered perfectly transparent. This is true for most applications but must be examined more closely for optical waveguides where absorption coefficients three orders of magnitude lower are needed and achieved. A knowledge of this absorption is desired not only to assess its contribution to the total absorption measured in fibers but also to assess the fundamental lower limit for any particular material. Glasses transparent in the visible have strong optical absorption bands in the ultraviolet and infrared which extend to some small degree into the visible. It is the "tail" of the ultraviolet that is considered most significant because the infrared bands, which are located beyond 4 μm, are very narrow. As a consequence, their contribution in the 600–1000-nm range is neglected. The ultraviolet absorption is due to atomic transitions involving oxygen and shifts somewhat with wavelength as the glass composition changes [10]. Little is presently known about the shape of these bands and hence their contribution to the loss. This is because fiber absorptions are so much smaller than any bulk ultraviolet absorption studied heretofore that extrapolations from the studied range are meaningless. The only recourse has been to take the smallest absorptive attenuation yet observed and conclude that the intrinsic absorption must be smaller.

Impurity absorption arises predominantly from transition-metal ions such as iron, cobalt, and chromium [11]. The absorption of these ions varies from glass to glass as does their valence state. The absorption peaks are very broad [12] so that it is difficult to identify the offending species from the spectral dependence of fiber absorption. Using iron as an example, both ferrous and ferric ions absorb in the visible range, but the ferrous ion is more troublesome with a strong absorption peak near 1000 nm. Fig. 1 shows the absorption of iron in fused silica. By doping the glass with known amounts of impurity, absorption values can be extrapolated to low concentrations, assuming the absorption is linear in concentration. When the glass has the same preparation history as the fiber, the valence state ambiguity is thus bypassed. Usually this approach shows that concentrations of impurity below a few parts per billion (ppb) are necessary if absorption below 20 dB/km is to be attained.

Another important impurity is "water" which is present as OH^- ions. This contributes sharp, easily identified, absorption bands around 950 and 725 nm [13]. These are, respectively, the third and fourth harmonic of the fundamental

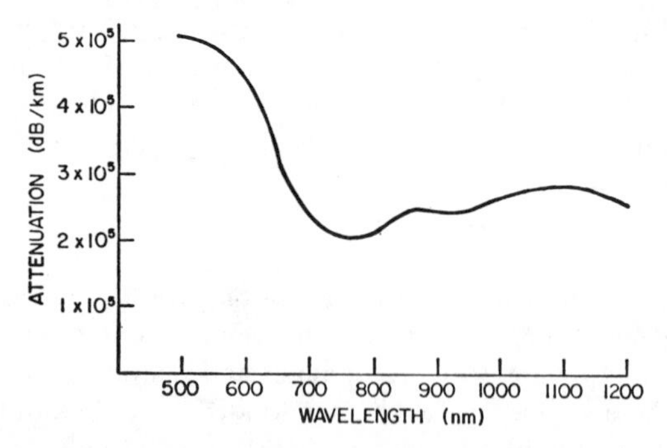

Fig. 1. Absorption spectrum of iron in fused silica for a concentration of 700 ppm atomic. The absorption in the vicinity of 1100 nm is thought due to Fe^{2+} and that at short wavelengths to Fe^{3+}.

vibrational band at about 2.8 μm. By measuring the water removed in a vacuum bakeout concurrently with the strength of the 2.8-μm band, absolute calibration can be obtained. The ratio of the 2.8-μm band to the harmonic bands can then be used for their calibration, even though these other bands are too weak to be measured except in very long paths, such as fibers. In fused silica, the peak of the 950-nm band causes 1.25 dB/km/ppm by weight [14]. The absorption per ion for water does not vary as much from glass to glass as do the transition metal ion absorptions.

The third source of absorption is atomic "defects" in glass structure including, by definition here, unwanted species of elements deliberately added to the glass composition. Oxygen defects in silica glasses have been studied by radiation [15]. These may be related to similar color centers introduced in silicate glasses by the oxidizing conditions of the melt. An example of a deliberately added element yielding an unwanted species might be titanium whose plus-three valance state has a strong absorption in the visible while titanium plus-four does not. These examples should also make it clear that oxidation states can be extremely important in glass fiber attenuation.

Some scattering attenuation arises from the intrinsic nature of glass. All transparent matter scatters light due to thermal fluctuations which, in turn, generate fluctuations in refractive index [16]. Glass differs in that these fluctuations are frozen-in when the material is cooled through the annealing range of temperature [17]. The attenuation coefficient (base e) for scattering alone is called the turbidity and is given by [16]

$$\tau \cong \frac{8\pi^3}{3\lambda^4}(n^2-1)^2 kT\beta \tag{6}$$

for pure liquids. This scattering is due to compressive fluctuations, the only significant type of fluctuation in a pure liquid. Here, k is Boltzmann's constant, T the absolute temperature, and β the compressibility. Because of the frozen-liquid state of glass, high-temperature quantities are used in (6). For pure fused silica the temperature dependence of (6) has been verified and a value for the compressibility derived when the annealing temperature is used for T [18]. If other oxides are added to silica, the compressive-fluctuation scattering changes chiefly through the factors T (annealing temperature) and β. A more important change is a second type of scattering caused by fluctuations in the concentration of these oxides [16]. This additional contribution is given by

$$\tau \cong \frac{16\pi^3 n}{3\lambda^4}\left(\frac{\partial n}{\partial c}\right)^2 \overline{\Delta c^2}\,\delta V \tag{7}$$

where $\overline{\Delta c^2}$ is the mean-squared concentration fluctuation and δV the volume over which it occurs. Generally, if the added oxide tends to raise the bulk refractive index, fluctuations in its concentration will cause greater scattering because these fluctuations represent greater refractive index fluctuations [15]. For most high-refractive-index glasses, concentration fluctuations (7) provide by far the dominant contribution to the total scattering.

Other kinds of glass inhomogeneity can cause scattering. Some compositions tend to segregate into immiscible liquids or to precipitate crystals. Or, improper mixing during melting will result in scattering refractive-index variations. However, these other kinds of glass inhomogeneity are not intrinsic to the material and can be judiciously avoided, in contrast to the thermal fluctuations discussed in the previous paragraph.

The third source of scattering is aberration in the radial form of the refractive index. This causes conversion of light into unbound modes. For continuous refractive index variation waveguides, this scattering imperfection cannot be distinguished from the other inhomogeneities in the material previously discussed. For the discontinuous refractive index variation waveguides, this scattering imperfection is irregularity of form (roughness) in the glass boundary between core and cladding. Such irregularity could cause almost any angular and wavelength dependence of the scattering depending on the form, dimensions, etc., of the irregularity.

Glass dispersion is important through its effect in limiting the bandpass. For optical transitions in the ultraviolet approaching the visible range of wavelengths, the refractive index rises as well as the dispersion. Thus it is generally true that changing a glass composition to increase the refractive index will also increase the dispersion. The detailed results of material dispersion influence on bandpass are beyond the scope of this paper, but to a good approximation material dispersion adds linearly to the other dispersive effects [5], [19]. Often material dispersion is the major factor limiting bandpass, so the general principles influencing bandpass and their magnitude are relevant here. Dispersion causes pulses to broaden as they propagate down the waveguide, and the resultant overlap between adjacent pulses limits pulse rate and, hence, information flow. The broadening is proportional to the waveguide length. For single-mode guides the input pulse frequency breadth broadens the output to a width b,

$$b = \left[a^2 + \left(\frac{L}{a}\frac{\partial^2 h}{\partial \omega^2}\right)^2\right]^{1/2} \tag{8}$$

where a is the Gaussian input pulsewidth, L the guide length, ω the frequency, and h the guide wave vector at center frequency [19]. Material dispersion enters through the derivative factor and is the predominant bandpass limitation for wavelengths less than about 1 μm. Table I gives some results for the material dispersion alone for single-mode waveguides with bandwidth-limited input pulses.

Equation (8) and Table I are also relevant for each mode of a multimode waveguide. However, an important practical difference is that multimode waveguides are to be used with

TABLE I

ILLUSTRATION OF THE EFFECTS OF MATERIAL DISPERSION ON BANDPASS

Glass	Single Mode—Pulse Spectral (Information) Bandwidth Limiting (GHz·km^{-1})	Multimode—Incoherent Source (Carrier) Bandwidth Alone Limiting (MHz·km^{-1})	Multimode Group Velocity Differences Alone Limiting (MHz·km^{-1})
Corning Code 7940 (fused silica)	40	418	39
Schott F-2 (lead silicate)	25	158	43
Typical soda–lime	39	408	40

Note: For the second column, the source was assumed to emit at 800 nm and to be 20 nm broad. The third column was added to compare with the second; numerical aperture was assumed to be 0.15 and the core diameter 100 μm ($V \cong 59$).

incoherent sources of broad spectral width far exceeding the information rate spectral width. Dispersion over this source emission width can influence the bandpass and is the limiting factor in some small-V waveguides, as discussed later.

Pulse broadening in multimode waveguides with step refractive index variation is mainly caused by the behavior of the different modes. When these modes are very weakly coupled, the pulsewidth is due to the different group velocities of each mode. A pulse input into all modes emerges at different times at the receiver [20]. While the mode propagation can be handled in detail [5], estimates are made using ray optics. The transit time difference between the straight through (axial) ray and the ray at the maximum (critical) angle to the axis is taken as the minimum pulsewidth. This yields

$$N_{\max} \cong c/L\,\Delta n. \tag{9}$$

Table I also includes values from (9) for comparison.

The remarks about multimode bandpass leading to (9) are not applicable when the coupling between modes is strong over the distance considered. The bandpass then becomes much higher than the estimate of (9) and material dispersion relatively more important. The most important realization for this discussion is that fabrication affects the coupling, since it is the imperfections beyond the perfect geometry which scatter light from one mode to another and cause the coupling. This is a topic of current interest, but with little information so far experimentally connected to fabrication [59].

Since, as shown, single-mode waveguide bandpass is limited primarily by material dispersion and multimode by the spread in mode group velocities, there is a point of transition. Taking the same source spectral characteristics as used in Table I, this transition can be calculated from (9) and (1). Thus a waveguide with the material dispersion of fused silica and a core of 50-μm radius will have its bandpass limited by the material below $V \cong 20$ (200 modes) and by group velocity dispersion above. As modal group velocity dispersion is decreased by mode coupling or refractive index gradients, the V for transition will increase.

Multimode waveguides with continuous refractive index variation, as previously discussed, can have less spread in group velocities for different modes. Assuming the radial refractive index variation is perfectly constructed, material dispersion can be the only limitation on bandpass. Table I again illustrates the situation. To achieve the highest potential bandpass of such waveguides, a collimated spectrally narrow source must be used with the values of column 2 increased by the proportional decrease in the spectral width over that assumed. Column 1 will then provide an upper limit.

There are many nonlinear optical effects in glass fibers which might be of device use, including Kerr effects, stimulated Brillouin emission, stimulated Raman emission, etc. These devices will not be considered here, but the limitation on power levels imposed by these effects is important. Both stimulated Brillouin emission and stimulated Raman emission dissipate power rapidly and attenuate the signal [21]. The acoustic and optical modes generated in the material absorb this lost energy. The power densities for the onset of these effects are largely independent of the glass compositions [22]. For signals (pump) of bandwidth small compared to the Raman linewidth but comparable to the Brillouin linewidth, the single-mode power is limited to a maximum of about 3×10^{7} W/cm^2, and for broad spectral sources comparable in width to the Raman line, the single-mode power is limited to about 2×10^{9} W/cm^2. For these numbers, a loss at the Stokes line of 4 dB/km was assumed. Therefore, some tradeoff of higher input power level can be made against the information rate limitation due to increased dispersion effects over the broader spectrum.

III. GLASS MATERIALS SYSTEMS

Suitable glass compositions should be capable of manufacture in a homogeneous state, including absence of any traces of phase separation. Since tremendous quantities are potentially needed throughout the world's communication system, no elements should be used which are not abundant in the earth's crust. All three composition systems which have been intensively investigated, the high-silica glasses, soda–lime glasses, and lead silicates, satisfy these requirements. These composition choices will be discussed in light of the previously mentioned theoretical principles.

The important wavelength region in the intrinsic absorption is the "tail," or very low absorption, part of the ultraviolet band about which little is known. Fused silica has the lowest wavelength ultraviolet absorption edge among conventional glasses. Therefore, assuming all glasses have the same shape absorption edge, the intrinsic absorption will be lowest in fused silica and generally higher the higher the refractive index. Since absorptions as low as 3 dB/km at 1.06 μm have been observed in fused silica, the intrinsic absorption is less than this [23]. Fused silica itself is unsuitable for a fiber core because there are no easily prepared glasses of lower refractive index to clad it. However, it has been recognized that doped silica could serve as a core and silica as a cladding [24]. Fibers of this type have shown absorption attenuation as low as 1 or 2 dB/km at 633 nm. In these glasses, at least, intrinsic absorption seems to be negligible. In other glasses, observed absorptions have been over 30 dB/km at 633 nm which probably means that the intrinsic absorption has not even been approached.

The choice among various compositions with respect to transition-metal ion absorption rests on the possibility of obtaining pure starting materials. There is no known way to purify glasses after melting, analogous to zone-refining silicon. The transition-metal ion content of the glass will thus com-

TABLE II

ATTENUATIONS DUE TO REPRESENTATIVE IMPURITIES IN SODA-LIME GLASSES

Ion	Absorption Peak (nm)	Concentration for 20 dB/km	
		Absorption at the peak (ppba)	Absorption at 800 nm (ppba)
Cu^{2+}	800	9	9
Fe^{2+}	1100	8	15
Ni^{2+}	650	4	26
V^{3+}	475	18	36
Cr^{3+}	675	8	83
Mn^{3+}	500	18	1800

prise whatever was in the starting materials plus whatever is acquired in the steps leading to fiber fabrication. These impurities have been most extensively studied in soda-lime glasses [25]. Table II shows the results for several ions taken from this literature. The numbers vary somewhat from glass to glass for a particular element because, among other things, the oxidation state is apt to be different. Changes for a given oxidation state of an element are not great and Table II is a good guide, particularly for alkali-containing glasses. It is apparent that minimizing the number of oxide components in a glass minimizes the chance for impurity pickup. Silicate glasses are preferred for most commercial applications because of their properties, and silica itself, as a constituent, can be obtained in very high purity. This is an indirect result of the work performed for semiconductor improvement, since many of the compounds used to obtain pure silicon can be oxidized to obtain silica. Synthetic quartz also can be made in high purity [26].

Glass atomic "defect" absorption, as previously defined, will depend markedly on the oxidation state of the glass. A discussion of this topic is much too complicated to be included here. Suffice it to say that high temperatures tend to cause reduction because oxygen is evolved as a gas to maximize the entropy. The free energy $U-TS$ always tends toward a minimum. Thus increasing the entropy S offsets the internal energy U. This increase becomes more important the higher the temperature because T appears as a multiplying factor. One particular case is of interest, since it has been utilized to produce low-attenuation waveguides. This is the oxidation of titanium by water in the glass [27]. Reduced titanium has a strong absorption which appears in newly made glass. However, a temperature treatment below the melting temperature apparently reverses the oxidation equilibrium with water in the glass, leading to oxidized titanium and evolved hydrogen gas. Reduced silicon also has been found in glass [28]. Perhaps one of the most fruitful areas for further research is thermal (fiber-forming) generation of light-absorbing atomic defects. Certainly, lower temperature melting glasses will reduce the severity of this problem if it exists. While both soda-lime and lead-silicate glasses are low melting, soda-lime is to be preferred in this regard, since lead is so easily reducible.

Closely allied to absorption caused by oxidation is absorption caused by radiation, since species susceptible to oxidation reduction are also apt to act as trapping sites. Precise data on radiation susceptibility have not been obtained, since observed effects often depend on impurities and data of different workers do not agree. Generally, fused silica seems the

TABLE III

SCATTERING LOSSES IN dB/km FOR SEVERAL REPRESENTATIVE GLASSES AS MEASURED AT 546 nm AND EXTRAPOLATED TO OTHER WAVELENGTHS

	633 nm (dB/km)	800 nm (dB/km)	1060 nm (dB/km)
Corning Code 7940 (fused silica)	4.8	1.9	0.6
Corning Code 8361 (soda-lime type)	8.5	3.3	1.1
Bausch and Lomb 517-645 (borosilicate crown)	7.7	3.0	1.0
Schott F-2 (lead silicate)	47.5	18.6	6.0

Note: These measurements on bulk glass represent only the intrinsic material loss to which must be added any other scattering losses.

most resistant to radiation coloring. Addition of other oxides, especially alkalis or alkaline earths, tends to increase coloration. Thus soda-lime or lead-silicate glasses are much more sensitive. The addition of cerium tends to "protect" against this visible coloration by acting as a competitive trapping center that does not color in the visible range [29], [30]. The magnitude of the coloration effects that have been studied in bulk glasses is much more severe than anything tolerable in an optical waveguide. A new area of investigation will have to be opened here.

The level of intrinsic scattering in glasses is well established, as indicated by Section II. Equations (6) and (7) indicate that the scattering should vary as λ^{-4}. This has been closely verified experimentally [31]. Data on intrinsic scattering are usually measured at one wavelength and extrapolated to others with this law. Table III gives some representative glasses with their attenuation at various wavelengths extrapolated from measurements at 546 nm [17]. Fused silica, soda-lime, and alkali silicates have the lowest scattering levels. Fused silica, of which Corning Code 7940 is an example, usually does not vary more than about ± 10 percent for any method of manufacture. Variations outside this precision can usually be traced to bubbles or some similar imperfection. This makes fused silica a fairly good secondary standard for checking scattering instrumentation. If any low-refractive-index glass is used for a fiber, the bulk scattering should be below ~ 2 dB/km at 900 nm. High-refractive-index glasses, and particularly lead-containing glasses above 1.6, do not appear suitable for low-attenuation waveguides. Glasses similar to Schott F-2 are used for the core of most commercial flexible fiber optics.

The other sources of scattering, inhomogeneity, and radial refractive index variation, offer no clear-cut choices among glasses. Inhomogeneity due to liquid-liquid phase separation is likely to arise in high silica glasses of the alkaline-earth-silica type, or, to a somewhat lesser extent, the alkali-silica type. For lower silica concentrations, borosilicates are a classic example. Such phase separation is difficult to suppress by rapid cooling because it nucleates so easily [32]. A good procedure to avoid immiscibility, which is followed in soda-lime glasses, is to adjust the composition with other oxide additives.

Dispersion choices among low-refractive-index glasses are not significant. Either high-silica or soda-lime glasses are among the best (Table I). As the index is raised, performance sacrifice occurs in some cases.

Diffusion is significant in controlling the radial refractive index profile. Fibers of continuous variation (5) are made by diffusion exchange between ions. A fiber with a discontinuous refractive index change will have the discontinuity smoothed by diffusion when the fiber is drawn. In the former case diffusion should be maximized for manufacturing speed, while in the latter it should be limited. In glass melts, mass transport is governed by convective flow as well as diffusion. Most of the processes of fiber formation occur at high viscosities so that diffusion is dominant. In this range, the mobility of ions is generally inversely proportional to their valence [33], [34]. The diffusion of trivalent and quadrivalent ions is so small that little or no data are available. Unfortunately, this does not mean that it is so small that it is unimportant here, inasmuch as very small distances are involved in single-mode waveguides. However, the valence rule alone is enough to draw some conclusions. Monovalent ions, such as alkali, thallium, and cuprous, are to be avoided if possible in discontinuous-refractive-index waveguides. Conversely, continuous variations made by diffusion should utilize these. Of the monovalent ions, thallium has the largest specific effect on the refractive index. This favors its use, since high-refractive-index gradients increase the effective numerical aperture of the waveguide.

Glass strength will be of continual interest, but it is only a secondary factor in choosing glass-composition systems. The basic primary factors, prehistory (flaw production), and water vapor, are largely independent of glass composition. Since glass is a brittle substance, any crack-like flaw will produce a high stress concentration at its tip and propagate through the specimen. Very little physical contact is necessary to produce such flaws and greatly reduce strength. The fiber surfaçe should be coated to prevent such abuse. Water vapor acts on the tip of such flaws and diminishes the energy required to generate new surface, thereby reducing the strength still further. Fused silica has yielded the highest strengths ever obtained on bulk specimens, even though other glasses show comparable values [35].

IV. Fabrication

For discontinuous refractive index, multimode waveguide fabrication is simpler than single mode, and many of the techniques have been employed for years in the production of commercial flexible fiber optics. The two techniques utilized are referred to as the "rod in tube" and the "double crucible," and descriptions exist in the literature [36]. Fig. 2(a) illustrates the rod-in-tube method which relies on necking down the structure at high temperatures while conserving the cross-section geometry. Conservation of matter shows that

$$\left(\frac{r_b}{r_f}\right)^2 = \frac{L_b}{L_f} \tag{10}$$

where b and f refer to the rod-in-tube blank and the fiber, respectively, r is the radius, and L the length. For the cylindrical geometry, surface tension forces aid rather than distort formation of the desired shape. For more complicated geometries, such as a square cross section, the drawing through the hot zone (low viscosity) must be fast enough to prevent surface tension from greatly distorting the shape. The rounding

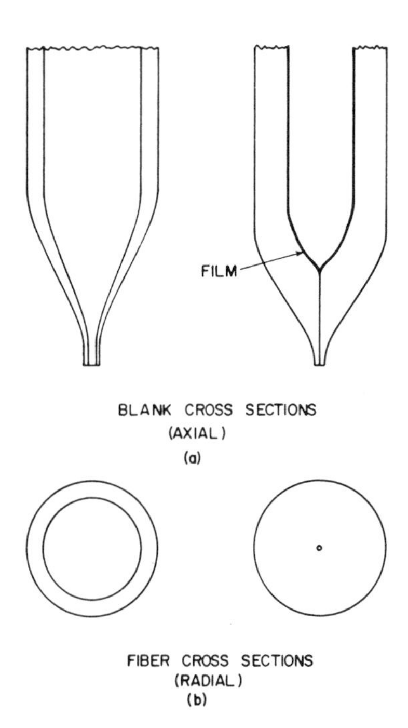

Fig. 2. Fabrication of multimode and single-mode optical waveguides. (a) Typical multimode fabrication blank sometimes called the "rod in tube." (b) A film blank technique used for making single-mode optical waveguides. This film permits simple fabrications of a very small core.

of a corner to the radius of curvature r_c is given by the crude rule of thumb

$$r_c = \frac{\Gamma t}{2\eta} \tag{11}$$

where Γ is the surface tension (about 3×10^{-9} J/m^2), t the time at viscosity, and η the viscosity. The time may be altered to some extent with the traction force. Strictly speaking, it is not possible to draw down a blank with sharp corners without some rounding.

The fabrication of single-mode waveguides is much more difficult because the core is so small. Some numerical calculations can illustrate the difficulty. Examination of (1) shows that Δn must be decreased if r is to be increased, since V must be kept constant at about 2. However, very small Δn is also difficult to control—a value of about 5×10^{-3} seems a reasonable limit. (Control of absolute refractive index in conventional optical glass manufacture is about 1×10^{-3}, but such precision is not yet practical in waveguide technology.) Assuming this lowest practical value for Δn, $\bar{n}\sim1.5$, an operating wavelength of 800 nm and $V=2.0$, leads to $r\cong2$ μm—a very small dimension to fabricate. Satisfactory fiber handling presently requires a radius of more than 50 μm which leads to a cladding: core ratio of 25:1. Equation (10) shows this large ratio also holds for the blank, so that a rod-in-tube blank with the rod 2 mm in diameter requires a 5-cm diameter tube. This would be extremely awkward to fabricate. The difficulty can be circumvented by several successive drawings, each followed by addition of another tube to provide cladding. A more elegant technique is shown in Fig. 2(b) [37]. A film is used, in effect, to obtain a small dimension for the core part of the blank; the hole in the center vanishes on drawing if the viscosity is low enough. Appropriate blank dimensions can be calculated, following the same principle used for calculating (10).

There is another geometrical consideration in making single-mode waveguides as previously discussed. This is the cladding thickness necessary to confine the power; if the wave reaches the cladding surface, losses occur. The radial power distribution in the waveguide depends on V, and a good rule of thumb is that the cladding radius should be about ten times the core radius ($V \sim 2$). This still leaves considerable reduction of the refractive index over that previously assumed without encountering problems of large stiff fibers. There have been no experimental demonstrations of finite cladding loss.

The double-crucible technique employs two concentric crucibles containing the core and cladding glass, respectively. Similarly, there are concentric orifices in the bottoms of the crucibles. This is a low-viscosity process with the fiber being formed directly from the melt. Relative dimensions of core and cladding are achieved by adjusting the viscosity of the glass and the orifice shape. This technique has sufficient versatility to permit the drawing of both single-mode and multimode fibers [38].

Different techniques are used for making continuous-index-variation fibers. One is to draw several concentric tubes of glass with step changes of refractive index between them, relying on diffusion to provide a smooth refractive index profile [39]. A more practical method is ion exchange [40], [41]. A glass fiber or rod is made from homogeneous glass and immersed in a bath of molten salts containing monovalent alkali ions. These ions exchange with ions of higher polarizability (like thallium) within the glass. The process is diffusion controlled. By stopping this process before complete exchange, a radial refractive index profile very similar to (5) is generated by diffusion laws.

These different fabrication methods offer different advantages and problems. One particular area of interest is the quality of interface between core and cladding. In conventional fiber manufacture this becomes contaminated and generates scattering loss [42] as well as, perhaps, an absorptive loss. The double-crucible method insures completely clean interfaces. If the core and cladding glasses are incompatible so that phase separation occurs on mixing at the interface, significant losses could still occur, but this can be handled by composition adjustment. The best answer to this problem is the continuous-refractive-index fiber made by ion exchange which has no interface at all.

A second area of interest is volatility of some glass component which generates inhomogeneity. The double-crucible method is most susceptible to this because of the lower viscosity. Consequently, some homogenization technique must be used within the crucible or a way found to reduce volatility.

Other important points involve the chances for contamination in the process. Whenever crucibles are used to melt glass, extreme care is necessary, because molten glass is highly corrosive and will dissolve any known material to some degree. Thus the crucible itself must be of high purity and made from elements that do not cause optical absorption. The incorporation of platinum from platinum crucibles is a frequent occurrence which causes both absorptive and scattering effects [43]. With ion-exchange processes, similar purity considerations apply to the molten salt.

High-silica glasses can be made without a crucible, using flame hydrolysis techniques. In this approach, a mixture of silicon and other compounds, depending on the oxide additive desired, are burned in a flame. The resultant small glass particles drop to the bottom of a furnace where they build up in a glass slab. This process has the advantage of maintaining high purity through a minimum of hot-glass handling.

Slow molten-salt diffusion processes are an important consideration for manufacturing speed. The approximate exchange time is given by

$$\tau \sim r^2/D \tag{12}$$

where r is the blank radius and D the diffusion constant. In one example, fabrication of a rod blank for subsequent fiber drawing required 432 h [44]. One reason for the lengthy diffusion time in the example above was that r in (12) was artificially large. The diffusion profile best approximates (5) near the fiber axis so that only the center portion was used. As more complicated and accurate methods of heating and diffusing are devised for manufacture of the necessary gradient, ion-exchange times can be expected to drop much below the value above.

Raising the temperature to increase the diffusion constant in (12) is limited by glass deformation and by the temperature at which the molten salt begins to attack and dissolve the glass [45]. Dissolution is due to increase of alkalinity of the salt bath from decomposition, thus molten KNO_3 decomposes above about 500°C with an increase in the K_2O concentration. If the ion exchange is much faster than the dissolution, or etching, the net result can be beneficial in that a fresh clean surface is generated without hampering the primary process. The diffusion rate can also be increased by eliminating divalent ions from the glass, such as the lead which was part of the composition [45]. On the other hand, thallium volatility during melting is high and a low-temperature melting glass, like a lead silicate, is beneficial from this standpoint.

V. Evaluation of Fiber Characteristics

A. *Optical*

The information set forth in the preceding sections can be used to evaluate fiber characteristics in relation to system performance. This section will illustrate the evolution of materials and fabrication principles alone; many other fiber characteristics, such as backscattering, bandpass, etc., which are important for system performance, are beyond the scope of the present article.

Single-mode fibers represent the most critical test of form fabrication and control. Two points are especially important —dimensional control throughout the fabrication and diffusion during fiber drawing. Fig. 3 shows titania concentration from an electron-beam microprobe scan across the core of a single-mode high-silica glass waveguide in which only the core contained titania. The resolution of the microprobe is about 2 μm so the rounding may be due to instrumental effects rather than diffusion. Additional evidence against significant diffusion is the agreement between this observed core size and that calculated from the film thickness of the blank [Fig. 2(b)]. To a first approximation, diffusion does not change the V value of a waveguide even though the core radius changes. The decrease in refractive index compensates for the increase in core radius. The refractive index change is usually linear in concentration for the important element used to increase it (titanium here). Thus $\Delta n = Kc = (KN)/(\pi r^2)$, where N is the number of ions per unit length of fiber core. From (1),

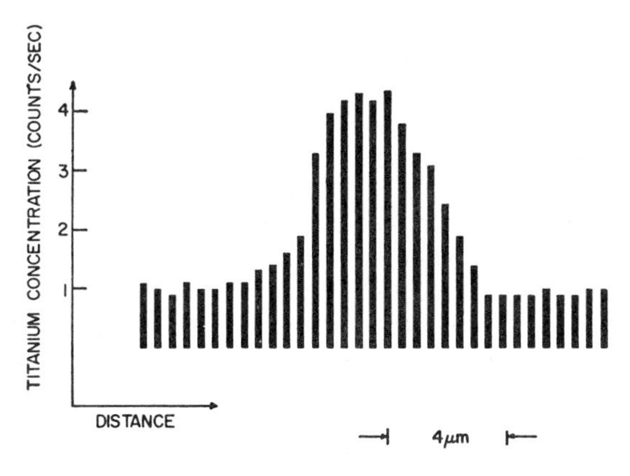

Fig. 3. Experimental measurement of core size in a single-mode optical waveguide. Electron-beam microprobe analysis of titanium concentration perpendicular to the fiber axis with a resolution of about 2 μm. The counts far from the core center are background, since the cladding is pure silica.

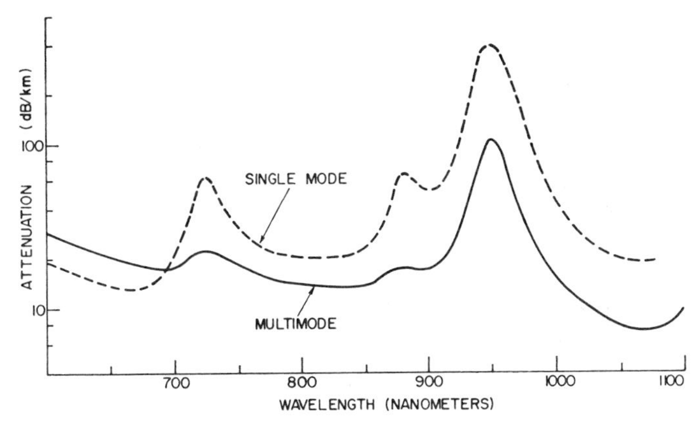

Fig. 4. Attenuation of typical low-loss optical waveguides through the spectral region of interest for solid-state light sources. The cores are titania-doped silica and the cladding pure silica.

TABLE IV

CALCULATED MODE CUTOFF WAVELENGTH (WHERE $V=2.4$) COMPARED WITH OBSERVED

Waveguide	Observed (nm)	Calculated (nm)
A	480	460
B	520	530

Note: These high-silica glass fibers were fabricated by the method of Fig. 2(b).

$V_1/V_2 \cong 1$. More exact numerical calculations show that the propagation vector does not greatly change for profiles as rounded as Fig. 2 when these profiles are due to diffusion [46]. On the other hand, deliberate attempts to change the core by diffusion have succeeded [47]. Table IV shows V values calculated from measurements of core size and bulk core refractive index compared with V values measured by mode cutoff patterns [48]. These experimental waveguides of Table IV thus show sufficient control for research purposes.

The single-mode designation of cylindrical waveguides is somewhat of a misnomer, since the symmetry assures two equivalent orthogonal polarizations and the mode is doubly degenerate. Any asymmetry, such as a radial stress or ellipticity in the core cross section, will destroy this degeneracy. The result is two modes of slightly different propagation characteristics. An idea of this asymmetry can be obtained from studying the relation between the state of polarization of entrance and exit beams [49]. Generally, a linearly polarized input beam will emerge circularly polarized for propagation distances greater than about a meter. On the other hand, input polarization orientations can be found which yield linearly polarized outputs. The results are thus consistent with slight asymmetry, but the origin is not yet known.

Evaluation of fabrication precision for continuous refractive index variations is extremely difficult. Equation (5) is usually handled through the expansion

$$n/n_0 = 1 - \left(\frac{\rho^2}{2}\right)r^2 + \left(\frac{5\rho^4}{24}\right)r^4 - \cdots.$$

Often only the first term is used and a "parabolic refractive index" discussed. Precision requires comparison with all the higher order terms and progress has been made utilizing the fourth order [50].

The most extensive evaluation has been on attenuation, because this was the main impediment to application. Measurement is straightforward with single-mode waveguides, because they are insensitive to the method of input as long as the input stays constant during the measurements. Unwanted light other than the single mode is easily stripped out with index fluid baths around the fiber. Multimode waveguides are entirely different. The attenuation depends markedly on the input conditions, because different modes with different attenuations are excited. Presumably, after some distance down the waveguide, coupling between the modes will lead to a uniform power distribution. However, some fibers have shown traces of the input conditions after propagation over 600 m. At least it is possible to say that the materials and fiber fabrication are consistent with the lowest attenuation observed for the modal distribution that was utilized. Physical intuition suggests that high-order modes will be more highly attenuated if the geometrical ray optic analysis is used. Rays at large angles to the fiber axis (higher order modes) travel farther and hence have higher absorption loss. However, this effect is small for small V. These large angle rays also probably suffer larger scattering loss because most scattering distributions peak in the forward direction, thus less of the scattered light is recaptured by the waveguide. So far, these expectations have been borne out [51].

Fig. 4 shows data for attenuation of two high-silica glass waveguides with titania-doped cores. One is a single mode and one a multimode in order to show that the materials principles apply equally to both. The most notable feature is the presence of sharp absorption peaks due to water, described in Section II [52]. In addition to the 950- and 725-nm bands there is another smaller band at 875 nm which is attributed to a combination of a Si–O vibration with the 950-nm harmonic. The multimode fiber was fabricated with dryer atmospheres and shows that all these bands decrease proportionately. The OH concentrations are about 320 and 80 ppm atomic, respectively.

A breakdown of the attenuation at 633 nm in the single-mode waveguide illustrates the relative importance of the factors discussed in Section II-B. A total loss measurement gives 16 dB/km and a total scattering measurement 8 dB/km. The difference of 8 dB/km is attributed to absorption. No evidence has been found for water absorption at this wavelength; the tails of the bands previously identified do not seem to contribute. This leaves intrinsic, impurity, and defect

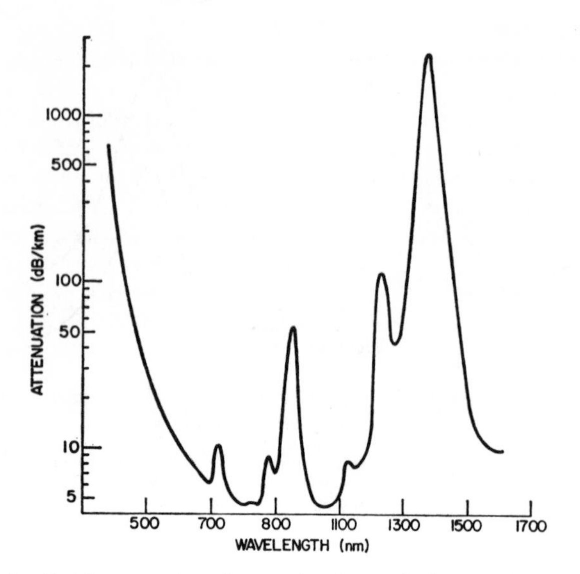

Fig. 5. Total attenuation of a very low-loss optical waveguide through a broad spectral region showing the influence of water impurity. All of the peaks are connected with light absorption due to OH^- present at about 100 ppm atomic.

absorption. Intrinsic absorption is probably below 1 or 2 dB/km, because this is the lowest absorption ever observed in this region, as will be mentioned later. This leaves about 6–7 dB/km due to transition-ion impurities and plus-three titanium. Plus-three titanium probably accounts for a large fraction, since it is a strong absorber (10 dB/km/ppb atomic) and would be difficult to completely oxidize by the method mentioned in Section III.

The scattering attenuation contains about 5-dB/km attenuation which can be attributed to bulk glass scattering, since measurements of doped silica indicate a slight increase of the scattering over the value for pure silica in Table III. The single-mode fiber thus contains about 3-dB/km scattering loss above the material contribution. This extra scattering has been further investigated and found to be predominantly in the forward direction [53]. It may be attributed to radial variation of the core refractive index along the guide and to glass inhomogeneity, with the relative magnitudes of each unknown.

The multimode waveguide of Fig. 4 has measurably different characteristics aside from the lower OH content already mentioned. This is notably a higher scattering attenuation of 11–12 dB/km—some 4 dB/km higher than the single-mode waveguide of the same glass composition. This leads to a higher total attenuation of about the same amount in the 633-nm region. This relative relation between single-mode and multimode scattering losses seems to be general and is perhaps connected with the fabrication. The approximate equality of the absorption between the two types suggests that it is attributable to the basic material, such as titanium plus-three.

Fig. 5 more clearly shows the importance of water in high-silica waveguides with very low loss. The series of water peaks shows decreasing absorption toward shorter wavelengths with contribution throughout the wavelength region of interest for solid-state light sources (750 to 1060 nm). This waveguide had an absorption attenuation of 1–2 dB/km at 633 nm, showing that transition-metal impurities were of small consequence. The largest part of the attenuation at this wavelength, or 9 dB/km, was scattering. The scattering drops rapidly as the wavelength increases and the total loss reaches values as low as 4 dB/km. However, as the scattering loss drops, the curve minima show an increase from other absorption so that the overall shape is parabolic. This strongly suggests that water is the primary source of absorption attenuation, raising the loss beyond about 600 nm.

Low attenuation requires that the cladding glass also be of relatively low attenuation. From (4) it is clear that $\beta_{cl} < (\beta/p_{cl})$, so that the 4-dB/km waveguide ($V \sim 55$) from (3) had a cladding glass loss of less than 80 dB/km, assuming all modes equally excited.

Lead–silicate glass has been reduced in bulk attenuation to 50 ± 30 dB/km at 850 nm [54]. This glass has a scattering loss about 70-percent of Schott F-2 (see Table III) and hence probably is of lower refractive index. Its scattering loss at 850 nm is thus about 12 dB/km with the remainder absorption. Unfortunately, fabrication into a fiber increased the attenuation to about 400 dB/km, due to unknown causes.

Lead–silicate glass has also been used to fabricate gradient index fibers of low attenuation [55]. The glass used was similar to the Schott glasses above with the exception of thallium addition. Judging from the composition, scattering losses should be comparable or slightly higher [17]. Values of attenuation around 600 and 1000 nm were of the order of 100 dB/km. Thus at 600 nm, the absorptive loss has been reduced to the vicinity of 80 dB/km. In gradient index fibers, some types of aberration can generate loss because light reaches the fiber surface and is scattered or absorbed. Therefore, these quoted total losses are an upper limit to the material loss, which might be much lower.

A complete treatment of bandpass evaluation cannot be undertaken here. However, spectral variations in delay time illustrate the properties determining the bandpass [56]. These show that the glass dispersion is the predominant factor in single-mode fibers (Section I).

B. Environmental

Effects of environment on glass waveguide physical integrity and optical performance have not been extensively investigated because of the early stage of development.

Data on thermal changes of attenuation for commercial fibers (~1000 dB/km) have shown little or no transmission change over the range −196°C to +200°C [57]. However, the large room-temperature attenuation indicates that impurity effects were being observed. It thus seems that the impurity contribution to low-loss fibers will not introduce a temperature dependence. Titania-doped silica fibers with a room-temperature attenuation of about 40 ± 1 dB/km showed no change in attenuation when heated to 200°C. Very low attenuations have not yet been studied, but it presently seems that temperature effects will not be important.

Some preliminary observations on radiation effects in low-loss fibers are given in Table V. Aside from military applications, radiation is a unique hazard for optical fibers in conventional uses. Cosmic radiation will generate an exposure of about 1 R per year. Communication installations may thus accumulate 20–40 R over their lifetime. While this is trivial for conventional cable and glass applications, it cannot be ignored for optical waveguides which depend so critically upon

TABLE V

PRELIMINARY DATA ON THE EFFECT OF RADIATION ON ATTENUATION

Dose	633 nm (dB/km)	800 nm (dB/km)	1060 nm (dB/km)
	Neutrons		
0 n/cm²	10	5	7.5
5.5×10^{10} n/cm²	14	7.5	(9)
5.6×10^{10} n/cm²	40	9	
	Gamma Rays		
0 rad	9	4.5	4
300 rad	16	(4.5)	(4)
2300 rad	54	10	

Note: Changes at long wavelengths were too small to measure with precision.

the absence of any color centers. Assuming that a 10-percent change in attenuation is tolerable, the table indicates that the fibers tested would be satisfactory for optical frequencies above about 650 nm but not below. This neglects any beneficial bleaching effects which might occur in a long-time installation. More testing is needed in order to go beyond this very early study.

Fiber strength will undoubtedly be an area of extensive study although, again, experimentation is only beginning. Two types of strength are important—short-term strength and long-term strength, or fatigue. The former is important for the process of installation while the latter is important for installed situations, such as bends, which leave the fiber under continual stress. Requirements for these situations are not yet known and will depend upon installation design.

Present high-silica fibers have a mean strength of over 703 kg/m² (100 000 psi). Since one break in a communications link the order of a kilometer will disrupt the signal, the probability of breakage in this length fiber is needed. The data for the fibers above were obtained on samples only 20 cm long. Therefore, this type of testing will require accurate values for the distribution of strength down to very low probability of breakage in order to predict what may happen over kilometer lengths.

VI. Bundling and Cabling

Bundling is a term applied to grouping many fibers together in some form of substructure of a cable. The bundle might be a group of fibers that could later be encapsulated in a plastic jacket or it might consist of some cable substructure with the individual fibers plastic coated. Both types of encapsulation have been experimentally performed with several plastics such as PVC and FEP Teflon®. At this stage the fibers could be handled and shipped if necessary.

Cabling refers to incorporating these substructures into a component including power wires, fillers, armoring, sheathing, etc., which is suitable for installation. Both near- and distant-term trends are visible. The distant trend is toward single fibers carrying an information channel analogous to wire today. This will require very low probability of fiber breakage plus advances in light sources, splicing, etc. A near-term trend is the use of a fiber bundle to carry an information channel. This gives the advantages of a large cross section to raise the capture efficiency from available light sources, retention of flexibility for installation, and redundancy to guard against fiber breakage.

The evaluation of fiber breakage is a common concern for all envisaged applications. If, under manufacture and installation, the majority of the breakage occurs in a few of the weaker fibers, one approach can be adopted. If, on the other hand, breakage of one fiber interacts in the region of breakage with other fibers in a way that breaks them as well, the situation is more difficult. For the former, more fibers remain unbroken (for a given number of breaks per unit bundle length) than if the breakage were random. In the latter, less are unbroken. For the former case, the random-breakage theory thus gives a conservative result, while for the latter, empiricism may be necessary.

The advantage of using random breakage for engineering is that it provides a single attenuation parameter characterizing broken fiber count for any length bundle when all the fibers in the bundle carry the same information. If b is the number of fibers broken per unit length, then the number of previously unbroken fibers that are broken per unit length is proportional to the unbroken fraction (random probability)

$$\frac{dN_b}{dx} = b\left(\frac{N_u}{N_0}\right)N_0$$

or

$$N_u = N_0 e^{-bx}.$$

For parallel transmission in a bundle, this has the form of an effective "attenuation factor" b. For example, the first bundle of low-loss fibers jacketed had an average attenuation of about 60 dB/km. In 305 m, the breakage represented 39 dB/km ($b=8.9$ km^{-1}), for a total "attenuation" of 99 dB/km. Recently, breakage below 10 dB/km has been attained. This is satisfactory, without further development, for a restricted class of applications, but improvement is necessary for long distances. Protective film coating and advances in handling techniques are part of the answer. The factor b might also be used to express an additional attenuation factor due to breakage following various testing procedures. A wide variety of tests have been carried out for flexible fiber bundles to be used in the automotive industry [48]. This experience has already benefited the optical waveguide work because of the many similar aspects of the technology.

Cabling has only been carried out on an early experimental basis and breakage appeared small but the quantitative analysis has not yet been performed.

VII. Conclusions

Remarkable progress has been made in obtaining glass fibers suitable for optical communication on an experimental basis. Much work remains in developing these so that they are suitable for practical cables. Unexpected beneficial discoveries and inventions may entirely change the character of this development; but even without such assistance, the general course and technical steps toward a standard product can already be envisaged. Fiber development has spurred development of the necessary compatible components to interface these waveguides to other equipment. This includes optical couplers, splicers, etc., in addition to electronic devices. As these components take form, the final requirements for constructing system architecture will be met.

Acknowledgment

The author wishes to thank his colleagues at the Corning Glass Works who contributed suggestions to all parts of this paper, and to those closely involved in the technical work; namely, Dr. D. B. Keck, who assisted in guiding all phases, Dr. F. P. Kapron for theory, Dr. P. C. Schultz for glass research, and Dr. F. Zimar for fiber fabrication.

References

[1] F. P. Kapron, D. B. Keck, and R. D. Maurer, "Radiation losses in glass optical waveguides," *Appl. Phys. Lett.*, vol. 17, pp. 423–425, Nov. 15, 1970.

[2] D. Gloge, "Optical waveguide transmission," *Proc. IEEE*, vol. 58, pp. 1513–1522, Oct. 1970.

[3] E. Snitzer, "Cylindrical dielectric waveguide modes," *J. Opt. Soc. Amer.*, vol. 51, pp. 491–498, May 1961.

[4] D. Gloge, "Weakly guiding fibers," *Appl. Opt.*, vol. 10, pp. 2252–2258, Oct. 1971.

[5] ——, "Dispersion in weakly guiding fibers," *Appl. Opt.*, vol. 10, pp. 2442–2445, Nov. 1971.

[6] A. W. Snyder, "Power loss in optical fibers," *Proc. IEEE* (Lett.), vol. 60, pp. 757–758, June 1972.

[7] A. Fletcher, T. Murphy, and A. Young, "Solution of two optical problems," *Proc. Roy. Soc.*, vol. 223, pp. 216–225, Apr. 22, 1954.

[8] F. P. Kapron, "Geometrical optics of parabolic index-gradient cylindrical lenses," *J. Opt. Soc. Amer.*, vol. 60, pp. 1433–1436, Nov. 1970.

[9] S. Kawakami and J. Nishizawa, "An optical waveguide with the optimum distribution of the refractive index with reference to waveform distortion," *IEEE Trans. Microwave Theory Tech.*, vol. MTT-16, pp. 814–818, Oct. 1968.

[10] G. H. Sigel, Jr., "Vacuum ultraviolet absorption in alkali doped fused silica and silicate glasses," *J. Phys. Chem. Solids*, vol. 32, pp. 2373–2383, Oct. 1971.

[11] D. S. McClure, "Electronic spectra of molecules and ions in crystals," in *Solid State Physics*, vol. 9, F. Seitz and D. Turnbull, Eds. New York: Academic Press, 1959, pp. 399–525.

[12] T. Bates, "Ligand field theory and absorption in spectra of transition-metal ions in glasses," in *Modern Aspects of the Vitreous State*, vol. 2, J. D. MacKenzie, Ed. Washington, D.C.: Butterworths, 1962, pp. 195–254.

[13] M. W. Jones and K. C. Kao, "Spectrophotometric studies of ultra low loss optical glasses," *J. Phys. E*, vol. 2, pp. 331–335, Apr. 1969.

[14] D. B. Keck, P. C. Schultz, and F. Zimar, "Attenuation of multimode glass optical waveguides," *Appl. Phys. Lett.*, vol. 21, pp. 215–217, Sept. 1972.

[15] E. Lell, N. J. Kreidl, and J. R. Hensler, "Radiation effects in quartz, silica, and glasses," in *Progress in Ceramic Science*, vol. 4, J. E. Burke Ed. New York: Pergamon, 1966, pp. 1–93.

[16] K. A. Stacey, *Light Scattering in Physical Chemistry*. New York: Academic Press, 1956, pp. 8–21.

[17] R. D. Maurer, "Light scattering by glasses," *J. Chem. Phys.*, vol. 25, pp. 1206–1209, Dec. 1956.

[18] D. L. Weinberg, "X-ray scattering measurements of long range thermal density fluctuations in liquids," *Phys. Lett.*, vol. 7, pp. 324–325, Dec. 15, 1963.

[19] F. P. Kapron and D. B. Keck, "Pulse transmission through a dielectric optical waveguide," *Appl. Opt.*, vol. 10, pp. 1519–1523, July 1971.

[20] M. DiDomenico, Jr., "Material dispersion in optical fiber waveguides," *Appl. Opt.*, vol. 11, pp. 652–654, Mar. 1972.

[21] N. Bloembergen, *Non-Linear Optics*. New York: Benjamin, 1965, pp. 102–120.

[22] R. H. Stolen, E. P. Ippen, and A. R. Tynes, "Raman oscillation in glass optical waveguide," *Appl. Phys. Lett.*, vol. 20, pp. 62–64, Jan. 15, 1972.

[23] T. C. Rich and D. A. Pinnow, "Total optical attenuation in bulk fused silica," *Appl. Phys. Lett.*, vol. 20, pp. 264–266, Apr. 1, 1972.

[24] R. D. Maurer and P. C. Schultz, "Fused silica optical waveguide," U. S. Patent 3 659 915 assigned to Corning Glass Works.

[25] H. L. Smith and A. J. Cohen, "Absorption spectra of cations in alkali-silicate glasses of high ultra-violet transmission," *Phys. Chem. Glasses*, vol. 4, pp. 173–187, Oct. 1963.

[26] E. D. Kolb *et al.*, "Low optical loss synthetic quartz," *Materials Res. Bull.*, vol. 7, pp. 397–406, May 1972.

[27] D. S. Carson and R. D. Maurer, "Optical attenuation in titania-silica glasses," *J. Non-Cryst. Solids*, to be published.

[28] A. J. Cohen, "The role of germanium impurity in the defect structure of silica and germanias," *Glastech. Ber.*, vol. 32K, pp. VI–53–58, 1959.

[29] J. S. Stroud, J. W. H. Schreurs, and R. F. Tucker, "Charge trapping and the electronic structure of glass," in *Seventh International Congress on Glass*. New York: Gordon and Breach, 1965, pp. 42.1–42.18.

[30] J. S. Stroud, "Color centers in cerium-containing silicate glass," *J. Chem. Phys.*, vol. 37, pp. 836–841, Aug. 15, 1962.

[31] H. N. Daglish, "Light scattering in selected optical glasses," *Glass Technol.*, vol. 11, pp. 30–35, Apr. 1970.

[32] R. D. Maurer, "Crystal nucleation in a glass containing titania," *J. Appl. Phys.*, vol. 33, pp. 2132–2139, June 1962.

[33] J. D. MacKenzie, "Semiconducting oxide glasses," in *Modern Aspects of the Vitreous State*, vol. 3, J. D. MacKenzie, Ed. Washington, D. C.: Butterworths, 1964, pp. 126–130.

[34] R. H. Doremus, "Diffusion in non-crystalline silicates," in *Modern Aspects of the Vitreous State*, vol. 2, J. D. MacKenzie, Ed. Washington, D. C.: Butterworths, 1962, pp. 1–71.

[35] W. B. Hillig, "Sources of weakness and the ultimate strength of brittle amorphous solids," in *Modern Aspects of the Vitreous State*, vol. 3, J. D. MacKenzie, Ed. Washington, D. C.: Butterworths, 1964, pp. 152–194.

[36] N. S. Kapany, *Fiber Optics*. New York: Academic Press, 1967, pp. 110–128.

[37] D. B. Keck and P. C. Schultz, "Method of producing optical waveguide fibers," U. S. Patent 3 711 262 assigned to Corning Glass Works.

[38] K. C. Kao, R. B. Dyott, and A. W. Snyder, "Design and analysis of an optical fiber waveguide for communication," in *Trunk Telecommunications by Guided Waves* (IEE Conf. Publ. 71), pp. 211–217, Sept. 1970.

[39] British Patent 1 266 521 issued to Nippon Selfoc Co., Ltd., Mar. 8, 1972.

[40] T. Uchida *et al.*, "Optical characteristics of a light-focusing fiber guide and its application," *IEEE J. Quantum Electron.*, vol. QE 6, pp. 606–612, Oct. 1970.

[41] A. D. Pearson, W. G. French, and E. G. Rawson, "Preparation of a light focussing glass rod by ion-exchange techniques," *Appl. Phys. Lett.*, vol. 15, pp. 76–77, July 15, 1969.

[42] A. R. Tynes, A. D. Pearson, and D. L. Bisbee, "Loss mechanisms and measurements in clad glass fibers and bulk glass," *J. Opt. Soc. Amer.*, vol. 61, pp. 143–153, Feb. 1971.

[43] N. Bloembergen *et al.*, "Fundamentals of damage in laser glass," National Academy of Sciences Publ. NMAB-271, Washington, D. C., pp. 29–33, July 1970.

[44] H. Kita and T. Uchida, "Light focussing fiber and rod," in *Fiber Optics, SPIE Seminar Proc.*, vol. 21, pp. 117–123, Jan. 1970.

[45] H. Garfinkel, "Cation exchange properties of dry silicate membranes," in *Membranes, Microscopic Systems and Models*, vol. 1. New York: Dekker, 1972, pp. 179–247.

[46] K. B. Chan *et al.*, "Propagation characteristics of an optical waveguide with a diffused core boundary," *Electron. Lett.*, vol. 6, pp. 748–749, Nov. 12, 1970.

[47] O. Krumpholz, "Mode propagation in fibers: Discrepancies between theory and experiment," in *Trunk Telecommunications by Guided Waves* (IEE Conf. Publ. 71), pp. 56–61, Sept. 1970.

[48] E. Snitzer and H. Osterberg, "Observed dielectric waveguide modes in the visible spectrum," *J. Opt. Soc. Amer.*, vol. 5, pp. 499–505, May 1961.

[49] F. P. Kapron, N. F. Borrelli, and D. B. Keck, "Birefringence in dielectric optical waveguides," *IEEE J. Quantum Electron.*, vol. QE-8, pp. 222–225, Feb. 1972.

[50] K. B. Paxton and W. Streifer, "Aberrations and design of graded-index rods used as image relays," *Appl. Opt.*, vol. 10, pp. 2090–2096, Sept. 1971.

[51] D. Gloge, A. R. Tynes, M. A. Duguay, and J. W. Hanson, "Picosecond pulse distortion in optical fibers," *IEEE J. Quantum Electron.*, vol. QE-8, pp. 217–221, Feb. 1972.

[52] D. B. Keck and A. R. Tynes, "Spectral response of low-loss optical waveguides," *Appl. Opt.*, vol. 11, pp. 1502–1506, July 1972.

[53] E. G. Rawson, "Measurement of the angular distribution of light scattered from a glass fiber optical waveguide," *Appl. Opt.*, vol. 11, pp. 2477–2481, Nov. 1972.

[54] A. Jacobsen, N. Neuroth, and F. Reitmayer, "Absorption and scattering losses in glasses and fibers for light guidance," *J. Am. Ceramic Soc.*, vol. 54, pp. 186–187, Apr. 1971.

[55] H. Kita, I. Kitano, T. Uchida, and M. Furakawa, "Light focussing glass fibers and rods," *J. Am. Ceramic Soc.*, vol. 54, pp. 321–326, July 1971.

[56] D. Gloge and E. L. Chinnock, "Fiber-dispersion measurements using a mode-locked krypton laser," *IEEE J. Quantum Electron.* (Corresp.), vol. QE-8, pp. 852–854, Nov. 1972.

[57] R. W. Dawson, "The effect of ambient temperature on infrared transmission through a glass fiber," *Bell Syst. Tech. J.*, vol. 51, pp. 569–571, Feb. 1972.

[58] R. S. Rider, "Environmental testing of flexible jacketed fiber optic bundles," in *Fiber Optics, SPIE Seminar Proc.*, vol. 21, pp. 43–48, Jan. 1970.

[59] D. Marcuse, "Higher-order loss processes and the loss penalty of multimode operation," *Bell Syst. Tech. J.*, vol. 51, pp. 1819–1836, Oct. 1972.

Optical Characteristics of a Light-Focusing Fiber Guide and Its Applications

TEIJI UCHIDA, MOATOAKI FURUKAWA, ICHIRO KITANO, KEN KOIZUMI, AND HIROYOSHI MATSUMURA

Abstract—A lenslike glass fiber guide with a parabolic variation of refractive index has been developed. This optical guide named SELFOC® has the following characteristics: simultaneous transmission of laser beams modulated by wide-band signals through narrow space; optical image transmission; realization of a lens with tiny aperture or with extremely short focal length; and the possibility of being bent with a small radius of curvature without spoiling transmission characteristics.

In the case of a typical fiber guide with length 1 meter and diameter 0.3 mm, transmission loss is about 0.2 dB and depolarization is about 20 dB at wavelength 0.63 μ. The mode pattern of a laser beam after passing through the fiber guide is scarcely deformed.

The fiber guide can be used as a transmission line or lens, in optical communication, optical data processing, and optical instruments.

I. Introduction

A LENSLIKE cylindrical medium whose refractive index varies quadratically with the distance from the axis has been investigated by many researchers [1]–[43]. This medium is technologically important because of its usefulness in optical communication, optical data processing, and optical instruments. The gas lens [3]–[6] is expected to be used as a long-distance transmission line because of low loss. However, the gas lens cannot be bent with a small radius of curvature. Several trials [38]–[41] for manufacturing solid-state lenslike media using special glass systems have also been performed.

The light-focusing fiber guide [44] described here seems to be the first practical solid-state lenslike medium. This fiber guide named SELFOC® was prepared by means of an ion-exchange technique [45]. It contains at least two different ions as network modifiers that diffuse through glass above the annealing temperature. Ions with different electronic polarization per unit volume are chosen. Different distribution of concentration takes place for each ion. This causes a continuous variation of refractive index. In this paper, the results of investigations on optical characteristics of the fiber guide and practical applications are presented.

II. Optical Characteristics

A. *Refractive Index Variation*

The refractive index of an ideal lenslike medium is given [2] by

Manuscript received December 31, 1969; revised April 7, 1970. Part of this paper was presented at the IEEE Conference on Laser Engineering and Applications, Washington, D. C., May 26–28, 1969.

T. Uchida and M. Furukawa are with Nippon Electric Company, Ltd., Kawasaki, Japan.

I. Kitano, K. Koizumi, and H. Matsumura are with Nippon Sheet Glass Company, Ltd., Itami, Japan.

$$n(r) = n_a(1 - \tfrac{1}{2}a_2 r^2), \tag{1}$$

where

r = distance from the axis,
n_a = index of refraction on the axis,
a_2 = a positive constant.

In the case of the gas lens, it is not possible to realize a complete parabolic variation because of gravitational effect [25], whereas in the solid-state media, this is possible if suitable materials are chosen. A good parabolic variation is achieved in the case of SELFOC.

A Gaussian beam of a He–Ne laser was used for the investigation of refractive index variation. Mode matching is required to let the laser beam pass through a piece of SELFOC without mode conversion. From Marcatili's wave solution [7], the spot size ω_0 of the lowest mode in a lenslike medium is given by

$$\omega_0 = \left(\frac{\lambda_0}{n_a\pi}\right)^{1/2} \cdot a_2^{-1/4}, \tag{2}$$

where λ_0 is the free space wavelength. The value of n_a is estimated to be about 1.6. The value of a_2 is determined as follows. Pieces of SELFOC are cut into disks. To determine whether the disks have imaging properties an object is examined through them. The lens-equivalent focal lengths of the disks are then measured. The focal length of a lenslike medium is given [7], [12] by

$$f = \frac{1}{n_a\sqrt{a_2}\sin(\sqrt{a_2}\,t)}, \tag{3}$$

where t is the length of the lenslike media. The value of a_2 can be obtained from f. In the piece of SELFOC in which a parabolic variation is achieved, the constant a_2 is also determined by measuring the periodic length of the sinusoidal path of the beam made visible due to scattering (Fig. 1). The constant a_2 is given as $4\pi^2/L^2$, where L is the periodic length [1]. The spot size ω_0 is calculated from (2). The Gaussian beam with a particular spot size ω_0 at the beam waist is injected normally into the section of SELFOC. The extent of the parabolic variation is estimated from the relation between the output beam slope and the input beam displacement. The slope of the ray (or of the axis of a Gaussian beam) coming out of a lenslike medium is given [12] by

$$\alpha = n_a[-r_i\sqrt{a_2}\sin(\sqrt{a_2}\,t) + r_i'\cos(\sqrt{a_2}\,t)], \tag{4}$$

Reprinted from *IEEE J. Quantum Electron.*, vol. QE-6, pp. 606–612, Oct. 1970.

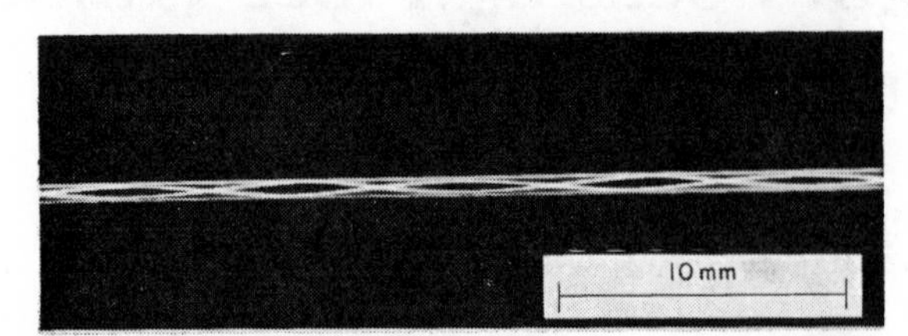

Fig. 1. Simultaneous transmission of laser beams through a piece of SELFOC with a diameter of 0.7 mm. Sinusoidal paths of the beams are made visible due to scattering.

where

r_i = displacement of the ray (or of the axis of a Gaussian beam) at the input of lenslike media,

r_i' = slope of the ray (or of the axis of a Gaussian beam) at the input.

When $r_i' = 0$, α is proportional to r_i. The experimental setup is shown in Fig. 2. A piece of straight SELFOC is fixed on an adjustable mount that can be moved in the direction vertical to the optic axis. Both ends of SELFOC have optically flat surfaces perpendicular to the optic axis. The beam displacement is indicated by the scales of a micrometer that is used for fine movement of the mount. The optical flat set on a goniometer is also used as an adjustable means to cause minute displacement. By tilting the flat, the beam is shifted parallel to the direction of propagation.[1] The displacement is indicated by the scales of the goniometer. The output beam slope is determined from the position of the center of the beam spot on a scaled screen. Examples of the relation between the output beam slope and the input beam displacement are shown in Figs. 3 and 4. The origin is chosen arbitrarily. The mode pattern of the output beam hardly changes from the Gaussian mode in the ranges from -320 to $+360\ \mu$ (Fig. 3) and from -50 to $+50\ \mu$ (Fig. 4) measured at the axis of the beam. From the linearity of the graphs in Figs. 3 and 4, we see that the parabolic variation is achieved in these ranges.

B. Mode Pattern

The mode pattern of the Gaussian beam after passing through a straight or curved piece of SELFOC was examined. Analyses [10], [11] show that in the curved lenslike medium, the center of the Gaussian beam is displaced from the symmetric axis of refractive index, but the beam mode propagates as in a straight lenslike medium. The beam displacement Δ due to a curvature depends on the radius of curvature R and the constant a_2 through the relation $\Delta \simeq (a_2R)^{-1}$. When the beam enters the curvature coaxially ($r = 0$) and paraxially ($r' = 0$), the beam undulates between $r = 0$ and $r = 2/a_2R$ [10]. Therefore, the minimum radius of curvature $R_{\min}$ is given by $R_{\min} = 2/a_2r_{\max}$, where $r_{\max}$ is the range of parabolic variation. The minimum radius of curvature $R_{\min}$ is estimated as 80 mm for a typical SELFOC fiber that has a length of 1 meter, a diameter of 0.3 mm, the coefficient a_2 of 0.5 mm^{-2}, and the parabolic variation range $r_{\max}$ of 50 μ. The mode pattern is not deformed as long as the relation $(\sqrt{a_2}\cdot R_{\min})^{-1} \ll 1$ is maintained [10]. The inequality is satisfied for this fiber. By measuring the horizontal intensity profile of the emerging beam from the fiber, which is curved horizontally, it was confirmed that the fiber could be bent with a radius of curvature of 175 mm

[1] The optical flat is slightly tilted in the vicinity of a position perpendicular to the direction of propagation.

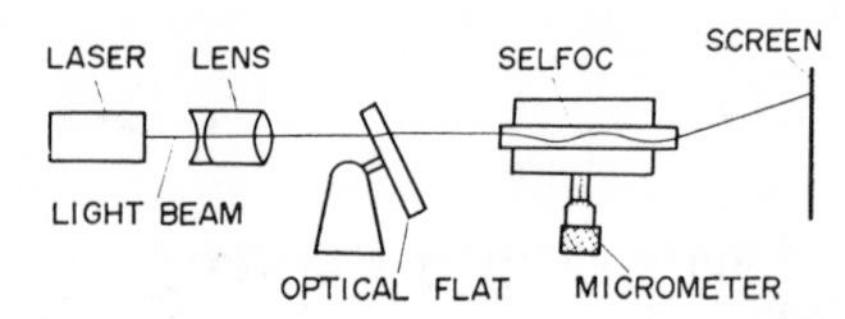

Fig. 2. Experimental setup for measuring the relation between the output beam slope and the input beam displacement.

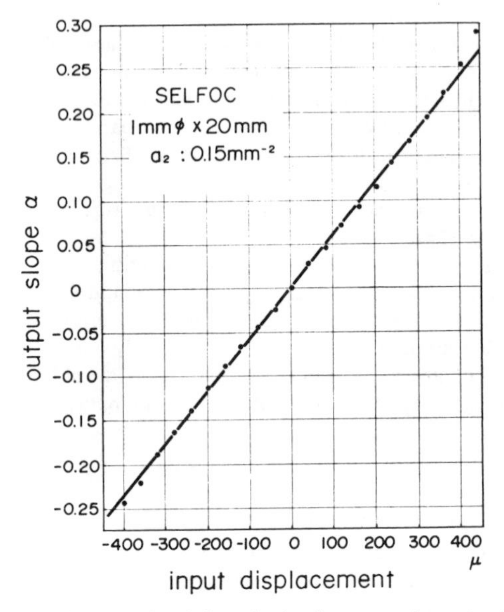

Fig. 3. Typical example of the relation between the output and the input beam displacement; rod: 1 mm $\phi \times 20$ mm.

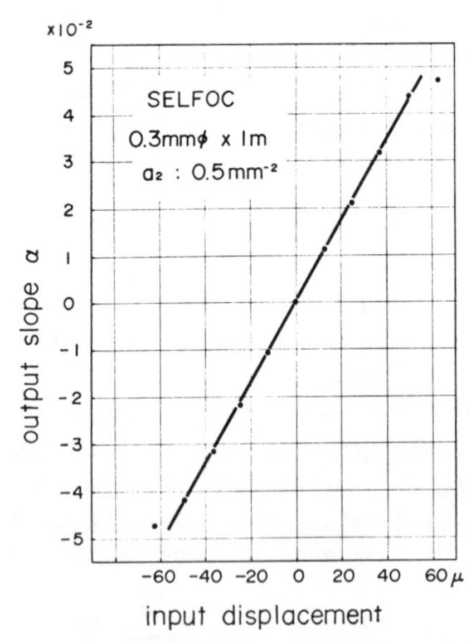

Fig. 4. Typical example of the relation between the output and the input beam displacement; fiber: 0.3 mm $\phi \times 1$ meter.

without spoiling the mode pattern of a laser beam. Fig. 5 shows the intensity profiles of the beams; curve *a* shows the profile of the incident beam on an end of the typical fiber; the profiles of the beams after they pass through the fiber when it is straight and when it is bent with a radius of curvature of 175 mm,[2] are shown in curves *b* and *c*, respectively. There is no significant difference between them.

C. Attenuation

In the development of SELFOC, it is important to minimize the transmission loss. The transmission loss comes from the bulk loss of the material. It is known that the bulk loss of glass is partly due to absorption caused by impurity ions and partly due to scattering resulting from the localized density fluctuation of materials and particle inclusion [46]. Two pieces of SELFOC of different lengths are used to evaluate the bulk loss. The total transmission loss of each piece was first determined from the difference of the input power and the output power measured with a photodetector. The bulk loss was obtained by eliminating the reflective losses at the surfaces of both ends from the total loss. A typical value of the bulk loss for a piece of SELFOC 1 meter long is about 0.2 dB at a wavelength 0.63 μ.[3] It was ascertained that absorption is the main cause of the loss. In special SELFOC fibers 1 meter long that have been made recently, a loss less than 0.1 dB is often obtained. This result lends hope to the use of SELFOC as a long-distance transmission line. Reproducibility of such a low-loss SELFOC fiber is being ameliorated.

D. Depolarization

We need to know the amount of depolarization if we use polarized light in SELFOC. The ratio of field intensities for the crossed and parallel polarization is measured by using two birefringent prisms, one as a polarizer and the other as an analyzer. The measuring limit is about −45 dB. The values obtained are about −20 dB for the piece of SELFOC with length 1 meter and diameter 0.3 mm, and about −42 dB for the one with length 20 mm and diameter 1 mm, respectively. Depolarization in SELFOC seems to arise from the scattering caused by the irregular and inhomogeneous distribution of refractive index and the localized strain in SELFOC.

E. Imaging Property

The resolving power of a SELFOC lens was investigated under the configuration shown in Fig. 6. A resolution test chart[4] and a Kodak high-resolution plate (2000 lines per mm) were used. The pictures of the test chart were taken under the illumination of white light.

The resolving power depends on a field angle θ shown in Fig. 6. The results are as follows in the case of a SELFOC lens with length 3.34 mm, diameter 1.0 mm, and focal length 1.9 mm.

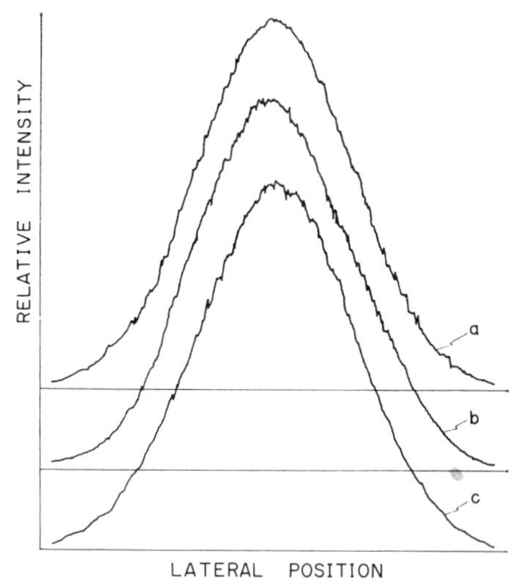

Fig. 5. Horizontal intensity profiles of the beams in relative linear scale. Curve *a* shows the profile of the incident beam on an end of a SELFOC fiber with length 1 meter, diameter 0.3 mm, coefficient a_2 0.5 mm^{-2}, and the parabolic variation range r_{max} 50 μ. Curves *b* and *c* show the profiles of the incident beam after they pass through the fiber when it is straight and when it is bent horizontally with a radius of curvature of 175 mm, respectively.

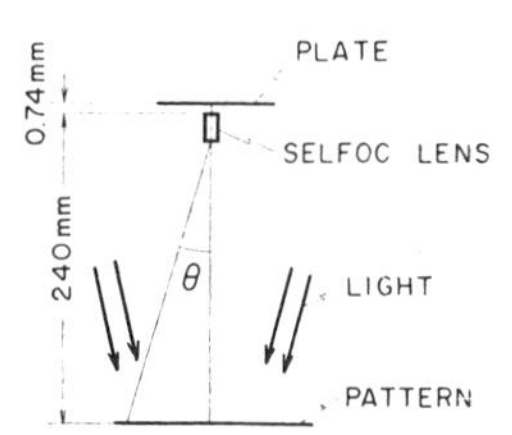

Fig. 6. Configuration for resolving power measurement.

tan θ (Fig. 6)	Resolving Power (lines/mm)
0.012	450
0.058	400
0.104	360
0.217	280

The resolving power is affected by the irregularity of parabolic refractive-index variation, insufficient flatness of terminal surfaces, and chromatic aberration.

III. Applications

The light-focusing fiber guide SELFOC has the following properties. The laser beam modulated by a wide-band signal can be transmitted without distortion of the wave form. Laser beams can be transmitted simultaneously without cross talk. Transmission loss is small. The polarization plane for the laser-beam transmission is conserved. Optical image transmission is possible. A lens with a tiny aperture or with an extremely short focal length can be realized. The fiber guide can be bent with a great curvature without spoiling transmission characteristics.

[2] *b* and *c* are measured at a point about 4 cm away from the emerging end of the fiber.
[3] Similar value of the bulk loss per unit length is obtained for a piece of SELFOC 10 meters long.
[4] Japan Industrial Standard.

SELFOC is used as a transmission line for laser beams. Mode conservation and flexibility make SELFOC useful as a feeder in optical communications systems and also as a guide for optical beam connection between laser devices. A half-meter practical guide was developed. This consists of a SELFOC fiber with a diameter of $0.25 \sim 0.3$ mm contained inside of a stainless-steel pipe with a bore of 1 mm.

As an optical guide for interconnections, the fiber is often cut into a piece with length $NL/2$, where N is an integer and L is the periodic length, i.e., $2\pi/\sqrt{a_2}$. A slight change of position of an incident laser beam on one end of such a SELFOC fiber results only in a slight shift of an output laser beam on the other end of the fiber without changing the slope of the output beam. The change of slope is very troublesome in practical use.

A bundle consisting of a hundred pieces of SELFOC is not very large in diameter, and can be used as a multiplex transmission line by passing each beam through each line of SELFOC.

Many laser beams can pass through a single SELFOC fiber without mutual interactions. Fig. 1 shows sinusoidal paths of two laser beams. Such simultaneous transmission of beams is utilized in space-division multiplexing through a SELFOC line, as indicated in Fig. 7. The cross talk between laser beams in the SELFOC fiber is negligibly small on account of the coherency of each laser beam; scattered light from one laser beam path is scarcely transferred to other laser beam paths.

In the optical communications systems using light pulse trains, time-division multiplexing can be achieved optically by combining the pulsed beams having different phases into one beam. So far, 1) a half-mirror or 2) a combination of a compensator and a polarization separator [47] has been used as a means for pulse combination. However, case 1) a half of the power of each beam is lost at each stage of combination, and case 2) requires a complex apparatus. In both cases only two beams can be combined by each stage.

The time-division multiplexing is easily achieved with a piece of SELFOC, and can be applied to optical communication through the atmosphere. An example is shown in Fig. 8. When modulated optical pulse beams having different phases and angles fall on the optic axis of a SELFOC pulse combiner whose length is odd multiples of a quarter of the periodic length L, both the beams emerge in the same direction. The separation between the axes of the beams at the output end of the combiner is so small compared with the beam spot size at the output side of a transmitting antenna that the beams can be regarded as an approximation of a single beam (see the Appendix). Thus, all the beams can be transmitted and received by common antennas and a common photodetector that is followed by demultiplexing equipment. Fig. 9 shows a block diagram of the experimental setup. In order to obtain easily the phase difference of two beams, the beams emitted from both the output ends of a mode-locked laser were converged into one end of a piece of SELFOC

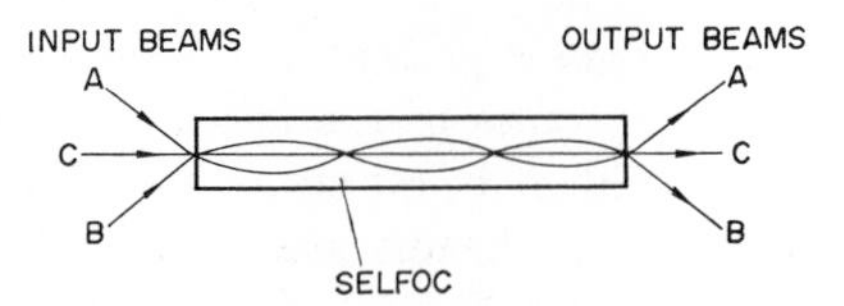

Fig. 7. Space division multiplexing. Beams A, B, and C with different angles of incidence can emerge separately if SELFOC is cut into a suitable length.

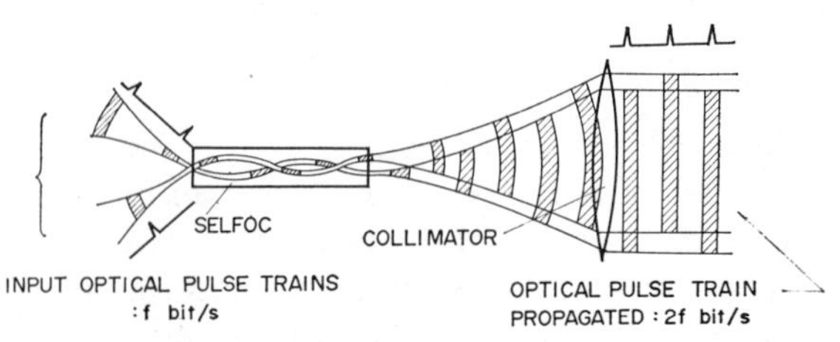

Fig. 8. Time-division multiplexing in the optical communication through atmosphere. Input optical pulse trains have different phases. Emerging from a piece of SELFOC in the same direction in a very narrow separation, they can be transmitted and received by an almost common area of optical antennas and a common photodetector.

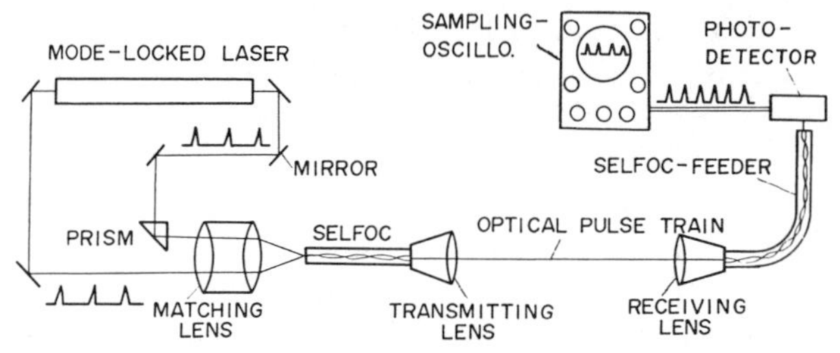

Fig. 9. Experimental setup for doubling of optical pulse-repetition rate using a piece of SELFOC.

with length about 50 mm, diameter 0.3 mm, and a_2 0.5 mm^{-2} placed at an equal distance from both the output ends of the mode-locked laser. A SELFOC fiber 1 meter long with diameter 0.3 mm, and a_2 0.5 mm^{-2} was also used as a feeder to a photodetector. Fig. 10 shows each optical pulse train and the combined one through a piece of SELFOC.

A system of space-division multiplexing with SELFOC is shown in Fig. 11. Suppose that there are m pieces of SELFOC of A and n pieces of SELFOC of B that organize a transmission line in a bundle. If m incident beams are gathered into a bundle through each piece of SELFOC of A, then m bundles of almost parallel beams can be obtained. The m bundles are gathered into one of SELFOC of B. Thus the capacity of multiplexing is m^2n. A capacity of 10 000 beams is possible.

A tapered SELFOC in which the constant a_2 increases gradually along the optic axis toward its thinner end is useful to the connection between SELFOC lines or to the convergence of a laser beam. Long or short pieces of tapered SELFOC were produced. Fig. 12 shows a light beam transmission through a long tapered SELFOC. In addition, a tapered SELFOC 17 mm long was used to enlarge effectively the light-receiving area of an avalanche photodiode, without increase of junction capacity, by attaching the thinner end of the tapered SELFOC at the diode [48].

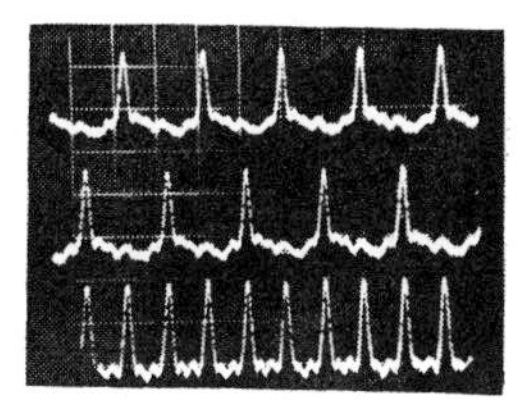

Fig. 10. Repetition-rate doubling. Upper two traces show the incident optical pulse trains combined together through a piece of SELFOC. Lower trace shows the received pulse train; the time scale is 5 ns/div.

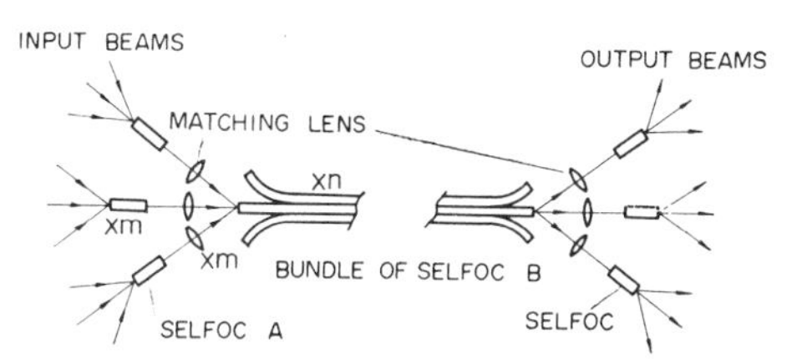

Fig. 11. An optical multiplex communication system. The multiplexing capacity of m^2n can be achieved by using m pieces of SELFOC of A and n pieces of B, and by injecting m beams into each of A.

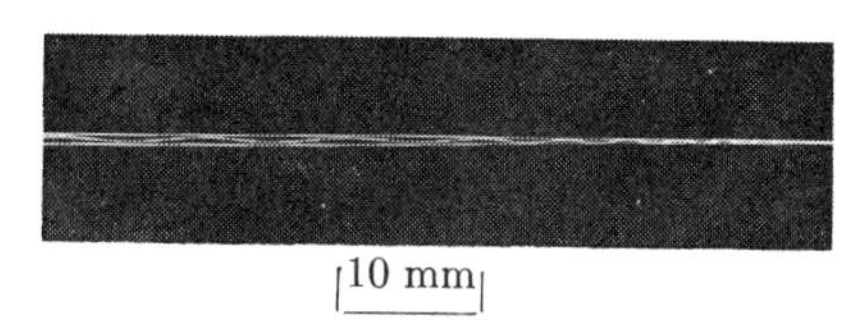

Fig. 12. Light beam transmission through a tapered SELFOC.

SELFOC is useful as an image guide. It was very difficult to produce a lens with an extremely short focal length or a tiny aperture. However, the lens can easily be realized as a disk of SELFOC. From (3), the minimum focal length $f_{\min}$ is given by $1/(n_a \sqrt{a_2})$. With SELFOC having a diameter of 0.25 mm and $a_2 = 2\ \text{mm}^{-2}$, a lens with a focal length of 0.44 mm is obtained. The F number of the lens is approximately 1.8.

Slenderness is a special feature of the SELFOC lens. A needle scope was manufactured for the observation of an affected part in the human body. As shown in Fig. 13, a piece of SELFOC with length 100 mm and diameter 0.75 mm is embedded in a hollow needle. Flexibility is another unique feature of the SELFOC lens. By use of a piece of SELFOC, it is possible to produce a fiber scope with a diameter much smaller than that of the conventional type.

Operation of a bundle of SELFOC lenses as a fly's eye is also important for practical application. One example is shown in Fig. 14, where each SELFOC lens has a diameter of 0.8 mm. For application of a SELFOC lens, the main obstacle is chromatic abberration. Several images of different color were observed when a white letter was viewed through a piece of SELFOC with a length of 300 mm, a diameter of 0.3 mm, and $a_2 = 0.5\ \text{mm}^{-2}$ under the illumination of a mercury lamp. In order to get one image we have to use monochromatic illumination or a narrow-band optical filter.

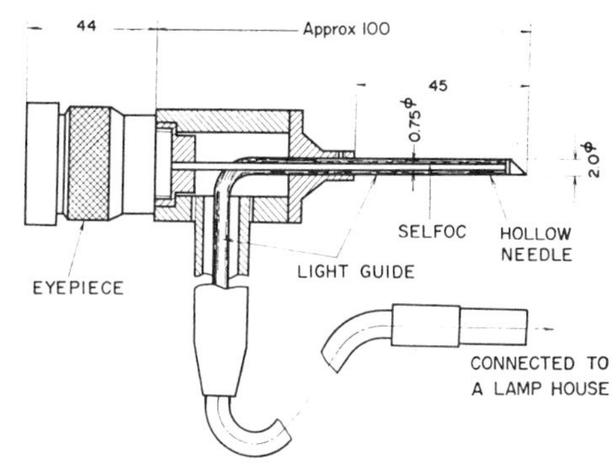

Fig. 13. Structure of the needle scope. A long SELFOC lens embedded in a hollow needle as an image guide.

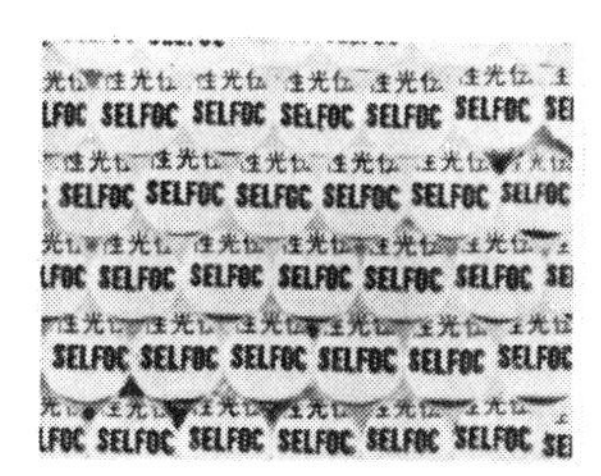

Fig. 14. Multiple images produced by a bundle of SELFOC lenses, whose unit lens has a diameter of 0.8 mm.

IV. Conclusion

A lenslike glass fiber guide with a parabolic distribution of refractive index has been developed. This optical guide named SELFOC has all the following excellent properties:[5] 1) laser beam transmission without band limitation and waveform distortion, 2) simultaneous transmission of a plurality of laser beams without mutual interference, 3) low-loss transmission and conservation of polarization plane for laser beam transmission, 4) optical image transmission, which is impossible over a single glass fiber of conventional type, 5) realization of lenses with tiny aperture and lenses with ultrashort focal length, 6) large flexibility due to its small diameter.

The light-focusing fiber guide is still under development to realize an optical transmission line for intracity communications. Although its typical attenuation loss is about 0.2 dB/m at the wavelength of 0.63 μ, SELFOC guides with losses less than 0.1 dB/m are often produced. Reproducibility of such a low-loss SELFOC is being improved. Moreover another problem is to realize a uniform parabolic distribution of refractive index over a long length.

SELFOC can be widely applicable in the fields of optical communications, optics, and optical data processing. Examples are a feeder, pulse combiner, multiplex transmission line, flexible lens, lens with an extremely short focal length, needle scope, fiber scope, fly's eye lens, and many others.

[5] Any conventional fiber does not have all these properties simultaneously.

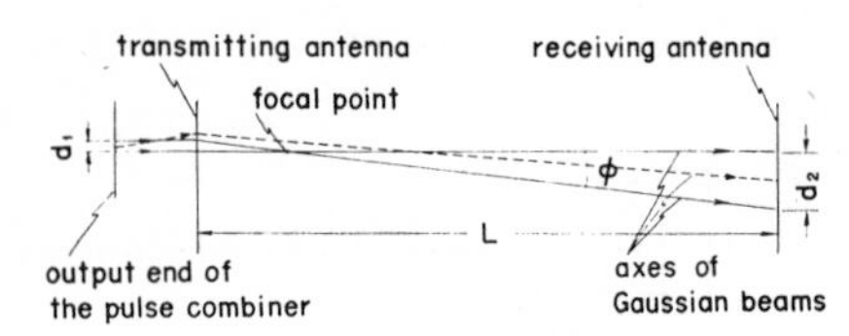

Fig. 15. Beam trajectories in the time-division-multiplex communication through atmosphere.

Appendix

In the far field there are two beams with angle difference ϕ as shown in Fig. 15, and they arrive at different points on the receiving antenna. The separation between these points is given by

$$d_2 = (L - f) \tan \phi \simeq (Ld_1/f) \qquad L \gg f \qquad (5)$$

where L is the distance between the transmitting and receiving antenna, d_1 is the separation between the beams at the output end of the combiner, and f is the focal length of the transmitting antenna. For example, if $L = 1$ km, $f = 1$ meter, and $d_1 = 30\ \mu$, d_2 comes to 3 cm.

On the other hand, the spot sizes of the beams at the receiving antenna usually become much larger than the separation d_2 because of the dancing and blurring effect due to atmospheric turbulence. For example, receiving antennas with apertures of 30 or 45 cm are often employed in our experimental optical communications systems. In addition, the choice of an optical path, indicated by a dotted line in Fig. 15, leads to further decrease of d_2.

Both the transmitted laser beams can be converged by the receiving antenna on a photodetector with a large receiving area such as an avalanche photodiode with a tapered SELFOC [48] and a photomultiplier.

Acknowledgment

The authors would like to thank Y. Degawa, Y. Takeda, I. Someya, S. Nemoto, and M. Uenohara of Nippon Electric Co., Ltd., and M. Miyasaka, H. Kita, and T. Otozai of Nippon Sheet Glass Co., Ltd., for continued guidance and encouragement. Thanks are also due to Y. Ikeda and T. Sumimoto for their help and Y. Matsumura for his cooperation in the early stage of the research.

References

[1] E. T. Kornhauser and G. S. Heller, "A soluble problem in duct propagation," *Proc. Symp. on Electromagnetic Theory and Antennas* (Copenhagen, Denmark), June 1962, pt. 2. Oxford: Pergamon, 1963, pp. 981–990.

[2] D. Marcuse and S. E. Miller, "Analysis of a tubular gas lens," *Bell Sys. Tech. J.*, vol. 43, pp. 1759–1782, June 1964.

[3] D. W. Berreman, "A lens or light guide using convectively distorted thermal gradients in gases," *Bell Sys. Tech. J.*, vol. 43, pp. 1469–1475, July 1964.

[4] ——, "A gas lens using unlike counter flowing gases," *Bell Sys. Tech. J.*, vol. 43, pp. 1476–1479, July 1964.

[5] A. C. Beck, "Thermal gas lens measurements," *Bell Sys. Tech. J.*, vol. 43, pp. 1818–1820, July 1964.

[6] ——, "Gas mixture lens measurements," *Bell Sys. Tech. J.*, vol. 43, pp. 1821–1825, July 1964.

[7] E. A. J. Marcatili, "Modes in a sequence of thick astigmatic lens-like focusers," *Bell Sys. Tech. J.*, vol. 43, pp. 2887–2904, November 1964.

[8] P. K. Tien, J. P. Gordon, and J. R. Whinnery, "Focusing of a light beam of Gaussian field distribution in continuous and periodic lens-like media," *Proc. IEEE*, vol. 53, pp. 129–136, February 1965.

[9] H. Kogelnik, "Imaging of optical modes—Resonators with internal lenses," *Bell Sys. Tech. J.*, vol. 44, pp. 455–494, March 1965.

[10] H. G. Unger, "Light beam propagation in curved Schlieren guide," *Arch. Elek. Übertragung*, vol. 19, pp. 186–198, April 1965.

[11] Y. Suematsu and H. Fukinuki, "Analysis of the idealized light wave-guide using gas lens," *J. Inst. Elec. Commun. Engrs.* (Japan), vol. 48, pp. 58–64, October 1965.

[12] S. E. Miller, "Light propagation in generalized lens-like media," *Bell Sys. Tech. J.*, vol. 44, pp. 2017–2064, November 1965.

[13] D. Marcuse, "Properties of periodic gas lenses," *Bell Sys. Tech. J.*, vol. 44, pp. 2083–2116, November 1965.

[14] ——, "Theory of a thermal gradient gas lens," *IEEE Trans. Microwave Theory and Techniques*, vol. MTT-13, pp. 734–739, November 1965.

[15] W. H. Steier, "Measurements on a thermal gradient gas lens," *IEEE Trans. Microwave Theory and Techniques*, vol. MTT-13, pp. 740–748, November 1965.

[16] H. Kogelnik, "On the propagation of Gaussian beam of light through lenslike media including those with a loss or gain variation," *Appl. Opt.* vol. 4, pp. 1562–1569, December 1965.

[17] S. Kawakami and J. Nishizawa, "Propagation loss in a distributed beam waveguide," *Proc. IEEE* (Correspondence), vol. 53, pp. 2148–2149, December 1965.

[18] J. P. Gordon, "Optics of general guiding media," *Bell Sys. Tech. J.*, vol. 45, pp. 321–332, February 1966.

[19] D. Marcuse, "Comparison between a gas lens and its equivalent thin lens," *Bell Sys. Tech. J.*, vol. 45, pp. 1339–1368, October 1966.

[20] ——, "Deformation of fields propagating through gas lenses," *Bell Sys. Tech. J.*, vol. 45, pp. 1345–1368, October 1966.

[21] Y. Suematsu, K. Iga, and S. Itô, "A light beam waveguide using hyperbolic-type gas lenses," *IEEE Trans. Microwave Theory and Techniques*, vol. MTT-14, pp. 657–665, December 1966.

[22] Y. Aoki, "Light ray in lens-like media," *J. Opt. Soc. Am.*, vol. 56, pp. 1648–1651, December 1966.

[23] E. A. J. Marcatili, "Off axis wave-optics transmission in a lens-like medium with aberration," *Bell Sys. Tech. J.*, vol. 46, pp. 149–166, January 1967.

[24] Y. Aoki and M. Suzuki, "Imaging property of a gas lens," *IEEE Trans. Microwave Theory and Techniques*, vol. MTT-15, pp. 2–8, January 1967.

[25] D. Gloge, "Deformation of gas lenses by gravity," *Bell Sys. Tech. J.*, vol. 46, pp. 357–365, February 1967.

[26] E. T. Kornhauser and A. D. Yaghjian, "Modal solution of a point source in a strongly focusing medium," *Radio Sci.*, vol. 2, pp. 299–310, March 1967.

[27] W. Streifer and C. N. Kurtz, "Scalar analysis of radially inhomogeneous guiding media," *J. Opt. Soc. Am.*, vol. 57, p. 779, June 1967.

[28] S. Kawakami and J. Nishizawa, "Kinetics of an optical wave packet in a lens-like medium," *J. Appl. Phys.*, vol. 38, pp. 4807–4811, November 1967.

[29] P. Kaiser, "Measured beam deformations in a guide made of tubular gas lenses," *Bell Sys. Tech. J.*, vol. 47, pp. 179–194, February 1968.

[30] ——, "The stream guide, a simple, low-loss optical guiding medium," *Bell Sys. Tech. J.*, vol. 47, pp. 761–765, May–June 1968.

[31] S. Kawakami and J. Nishizawa, "An optical waveguide with the optimum distribution of the refractive index with reference to waveform distortion," *IEEE Trans. Microwave Theory and Techniques*, vol. MTT-16, pp. 814–818, October 1968.

[32] W. H. Steier, "Optical shuttle pulse measurements of gas lenses," *Appl. Opt.*, vol. 7, pp. 2295–2300, November 1968.

[33] D. Gloge and D. Weiner, "The capacity of multiple beam waveguides and optical delay lines," *Bell Sys. Tech. J.*, vol. 47, pp. 2095–2108, December 1968.

[34] C. N. Kurtz and W. Streifer, "Guided waves in inhomogeneous focusing media, Pt. I: Formulation, solution for quadratic inhomogeneity," *IEEE Trans. Microwave Theory and Techniques*, vol. MTT-17, pp. 11–15, January 1969.

[35] S. Sawa and N. Kumagai, "General response and stability condition of the periodic beam waveguide consisting of gas lenses," *Trans. Inst. Elec. Commun. Engrs.* (Japan), vol. 52-B, pp. 273–276, May 1969.

[36] S. Sawa and N. Kumagai, "Wave propagation along a serpentine bend of the waveguide consisting of lens-like media," *Trans. Inst. Elec. Commun. Engrs.* (Japan), vol. 52B, pp. 277–283, May 1969.

[37] C. N. Kurtz and W. Streifer, "Guided waves in inhomogeneous

focusing media, Pt. II: Asymptotic solution for general weak inhomogeneity," *IEEE Trans. Microwave Theory and Techniques*, vol. MTT-17, pp. 250–253, May 1969.
[38] S. Kawakami, "Hikari densoro no kangaekata to sono oyo," *Densi Zairyo*, vol. 8, pp. 63–67, May 1969.
[39] A. D. Pearson, W. G. French, and E. G. Rawson, "Preparation of a light focusing glass rod by ion-exchange techniques," *Appl. Opt.*, vol. 15, pp. 76–77, July 1969.
[40] C. Guillement (private communication).
[41] E. G. Rawson and D. R. Herriott, "Analysis of cylindrical graded-index glass rods (GRIN Rods) used as image relays," *Advance Program 1969 Ann. Meeting Opt. Soc. Am.* (Abstract).
[42] M. Imai, M. Suzuki, and T. Matsumoto, "Some considerations on a lens-like medium with aberrations," *Trans. Inst. Elec. Commun. Engrs.* (Japan), vol. 52-B, pp. 491–496, September 1969.
[43] S. Sawa and N. Kumagai, "A new method of analysis of the waveguide consisting of lens-like medium," *Trans. Inst. Elec. Commun. Engrs.* (Japan), vol. 52-B, pp. 624–631, October 1969.
[44] T. Uchida, M. Furukawa, I. Kitano, K. Koizumi, and H. Matsumura, "A light-focusing fiber guide," *IEEE J. Quantum Electronics* (Abstract), vol. QE-5, p. 331, June 1969.
[45] I. Kitano, K. Koizumi, H. Matsumura, T. Uchida, and M. Furukawa, "A light-focusing fiber guide prepared by ion-exchange techniques," *1st Conf. on Solid-State Devices* (Tokyo, Japan), September 1969.
[46] K. C. Kao and G. A. Hockham, "Dielectric-fiber surface waveguides for optical frequencies," *Proc. IEE* (London), vol. 113, pp. 1151–1158, July 1966.
[47] T. S. Kinsel and R. T. Denton, "Terminals for a high-speed optical pulse code modulation communication system: II. Optical multiplexing and demultiplexing," *Proc. IEEE*, vol. 56, pp. 146–154, February 1968.
[48] K. Nishida, Y. Nannichi, T. Uchida, and I. Kitano, "An avalanche photodiode with a tapered light-focusing fiber guide" (to be published).

New Light-Focusing Fibers Made by a Continuous Process

K. Koizumi, Y. Ikeda, I. Kitano, M. Furukawa, and T. Sumimoto

A new continuous manufacturing process for light-focusing glass fibers has been developed using fast ion-exchange in a special double crucible. By use of this new process, new low-loss optical fibers with a transmission loss of 20 dB/km at wavelengths between 0.81 μm and 0.85 μm and a transmission capacity of more than a few Gbit/sec for 1 km were fabricated from low-loss bulk glass prepared by an improved raw powder refinement technique and by clean melting.

I. Introduction

With the advent of optical fibers having a remarkably low transmission loss,[1,2] low-loss fibers are becoming the most promising waveguides in optical communication systems not only for interoffice trunks but also for intercity routes.

During the four years since the advent of light-focusing glass fibers and rods, named SELFOC, in 1968,[3-5] many applications have been developed in the field of optics and optical data processing.[6-8] Their great potential as a waveguide for long-distance optical communications has been widely discussed,[9-17] as well as the possibility of wide bandwidth long SELFOC fibers, because SELFOC fibers have less delay distortion of optical signals than do clad type fibers.[18-20]

A new continuous manufacturing process of light-focusing fibers described here has been developed using fast ion-exchange in a special double crucible. This new process has several merits compared with the conventional manufacturing process of SELFOC fibers.

(1) SELFOC fibers can now be manufactured continuously in a single process, thus facilitating large quantity production.

(2) Group delay resulting from mode conversion is reduced, since the occurrence of irregularities in the refractive index distribution is suppressed, thus ensuring broad bandwidth.

(3) The refractive index gradient is so steep that the laser beam is maintained in the vicinity of the center axis even if the fiber is bent randomly.

The first three authors named are with the Research Laboratory, Nippon Sheet Glass Company, Itami, Japan; the other authors are with the Central Research Laboratories, Nippon Electric Company, Kawasaki, Japan

Received 14 August 1973.

CLEA paper 3.3.

(4) Since the glass is cooled quickly from its molten phase, it is possible to diminish the increase in attenuation loss due to colloid formation or devitrification and to enhance the fiber strength.

In this paper, the manufacturing process and the transmission characteristics of the new light-focusing fibers are reported.

II. Structure and Fabrication

A. Refractive Index Profile

The refractive index profile of the new light-focusing fibers, which we call New SELFOC is different from that of the conventional SELFOC fibers. As shown in Fig. 1, the new SELFOC fibers have a relatively steep change of refractive index only in the core region, while the refractive index of the conventional SELFOC fibers decreases gradually from the center axis to the periphery.

The fibers were manufactured with outer diameters from 150 μm to 250 μm and core diameters below 50 μm; the refractive index difference ΔN between the fiber axis and the core–cladding interface was about 0.02. If we assume that the refractive index distribution in the core region follows the formula:

$$N = N_0 \left(1 - \frac{1}{2}\mathrm{A}r^2\right), \tag{1}$$

where N_0, r, and A are the refractive index on the fiber axis, the distance from the fiber axis, and the refractive index distribution constant, respectively, the constant A of the new SELFOC fibers ranges from 20 mm^{-2} to 300 mm^{-2}. These values are very large compared with those of the conventional SELFOC fibers. By virtue of the large value of the constant A, the laser beam traveling through the new SELFOC fiber is always kept in the vicinity of the fiber axis, even if the fiber is bent randomly.

Reprinted with permission from *Appl. Opt.*, vol. 13, pp. 255–260, Feb. 1974.

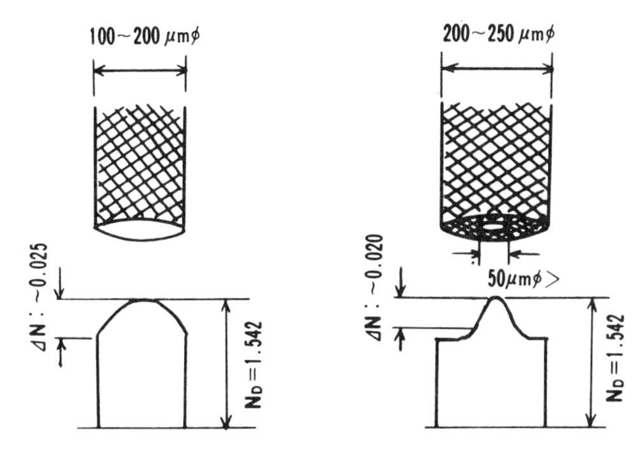

Fig. 1. Refractive index profiles of the conventional SELFOC fibers and the new SELFOC fibers.

B. Fabrication

The new process we describe is similar to the double-crucible method previously used to make clad-type fibers; however, its principle of operation is different. Figure 2 indicates the flow of the two kinds of glass forming the new SELFOC fiber as it is drawn from the double-crucible in the new process. The inner crucible contains the borosilicate glass with Tl ions to form the core, and the outer one contains the borosilicate glass with Na ions to form the cladding of the fiber. The fiber is drawn at a speed of a few tens of meters per minute, while the ion exchange between the Tl and Na ions quickly takes place in the vicinity of the nozzles of the crucibles as a result of the high temperature. As in the conventional process, Tl ions are chosen as the thermally diffusing ions, which contribute to the graded index profile in the fiber. Since Tl ions have the largest electronic polarizability among all monovalent ions in glass combined with a considerably small ionic radius, the ion exchange between the Tl ions in the core glass and the Na ions in the cladding glass easily brings about the value of the refractive index difference ΔN which is desired as a light-focusing medium for optical communication use.

Assume a fiber of core radius a being drawn at a pulling speed v; the flow rate V of the core glass flowing downward in the outer crucible is then given approximately by

$$V = a^2v/R^2, \tag{2}$$

where R is the radius of the core glass flowing downward in the outer crucible. As a measure of the degree of ion-exchange, let us define an ion-exchange parameter K as

$$K = DT/R^2 \tag{3}$$

where D is the diffusion constant of the diffusive ions and T is ion-exchange time. The Ion-exchange length Y, that is, the length along which the core glass contacts the cladding glass is given by

$$Y = TV. \tag{4}$$

Using Eqs. (2), (3), and (4), we can rewrite K as

$$K = YD/a^2v. \tag{5}$$

This expression for K is more convenient, because Y is the main controllable parameter to achieve a desired value of K.

According to the mathematical calculation based on thermal diffusion theory,[21] the profile of the Tl ion concentration distribution near the central axis is well approximated by the parabolic law (1) for the ion-exchange parameter K ranging from 0.01 to 0.1. Since the Tl ions have a far greater electronic polarizability than other diffusive ions, the refractive index distribution in the core region is mainly determined by their concentration profile. Consequently, a paraboliclike distribution of the refractive index is easily obtained by selecting a proper value of K, that is, of Y, D, a, and v.

The manufacturing process of the conventional SELFOC fibers employed an ion-exchange process between glass and molten salt,[4,5] A glass rod containing Tl ions was immersed in a KNO_3 bath at about 500°C. After a long time of immersion for the purpose of ion-exchange, the glass rod was heat-stretched to form a long fiber. This process was not continuous. The new manufacturing process based on the afore-mentioned fast ion-exchange between two kinds of glass in a double crucible at high temperature is continuous, and suitable for mass pro-

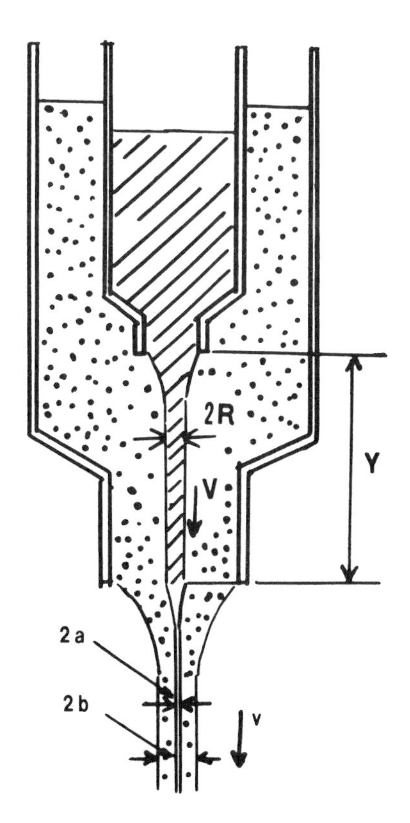

Fig. 2. Ion-exchange process between core and cladding glasses in a double crucible.

duction. In addition, rapid cooling of the glass from the molten phase to the fiber brings about good glass homogeneity and mechanical strength of the fiber. As a result, the new SELFOC fiber can transmit a light beam almost free from mode conversion due to refractive index fluctuation.

III. Transmission Characteristics

A. Low-Loss Transmission

The transmission loss of the new SELFOC fibers in the wavelength range of 0.8–0.9 μm was measured using a stable He–Cd laser with special hollow cathodes, which emits at wavelengths of 0.807 μm, 0.853 μm, and 0.888 μm.[22] This wavelength range coincides with the wavelengths of the diode laser and the light emitting diode sources, which are expected to become the most attractive light sources in optical communication systems. The low-loss values of 20 dB/km, 20 dB/km, and 29 dB/km at the three wavelengths mentioned above have been attained for fibers of various lengths up to 500 m, which had a core diameter of 30 μm. Figure 3 shows the transmission loss measured at the wavelengths of 0.807 μm and 0.853 μm for seven pieces of the new SELFOC fibers of various lengths cut randomly from a long fiber of more than 10 km. As shown in the figure, the loss value of 20 dB/km is confirmed to be independent of fiber length, which indicates that the new process is very stable and produces fibers with constant low-loss values.

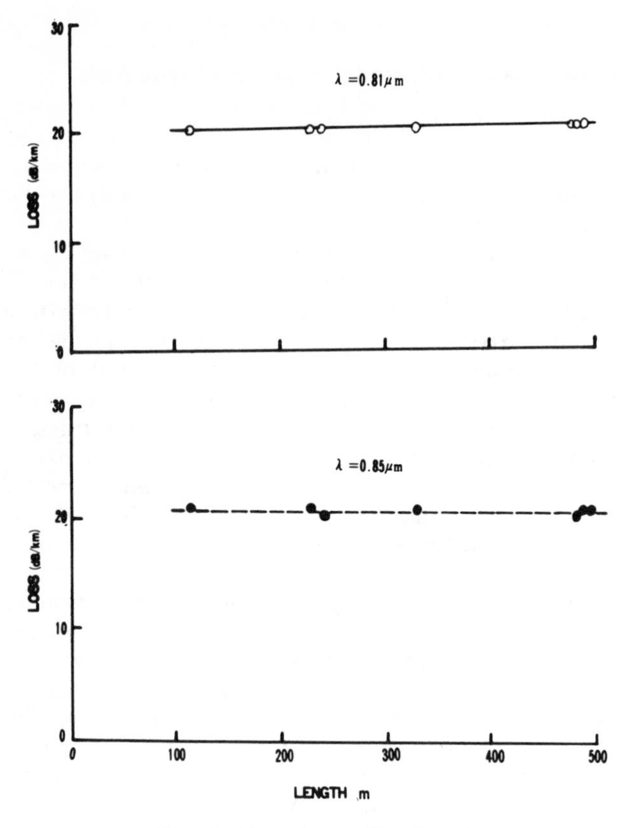

Fig. 3. Total loss vs fiber length.

The low-attenuation area was determined by measuring the output power as a function of the radial displacement of the focused laser beam incident on the fiber front face. Figure 4 indicates a typical result of this measurement for a 500-m long fiber using a He–Cd laser beam at the wavelength of 0.853 μm. The range marked by a loss increase of 3 dB is 25 μm; the range without noticiable increase in loss is more than 10 μm. The considerably wide low-loss range indicates that the new SELFOC fibers have excellent features concerning the launching of a laser beam and the connection of fibers.

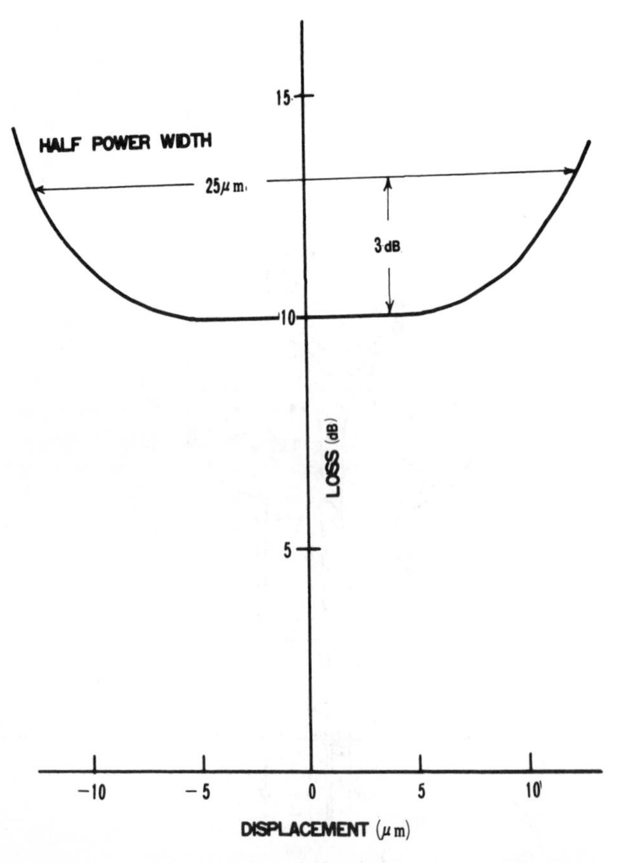

Fig. 4. Total loss vs incident beam displacement (fiber length: 500 m; wavelength: 0.85 μm).

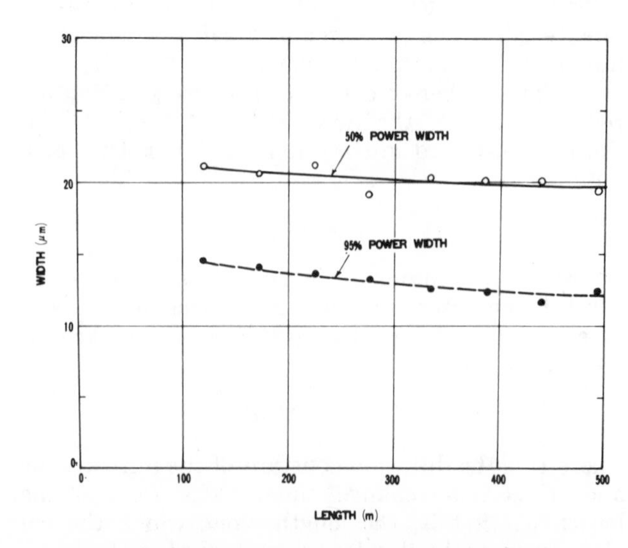

Fig. 5. 50% and 95% power width vs fiber length.

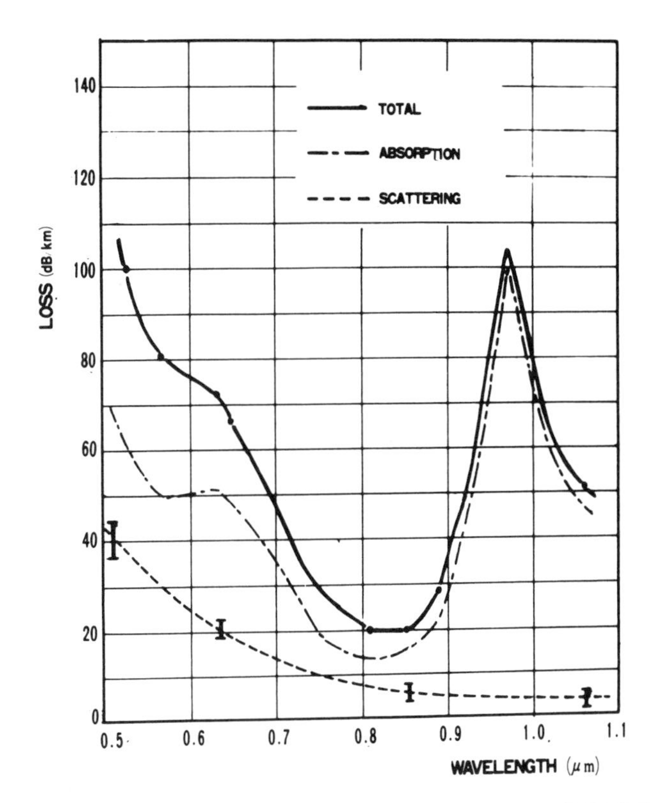

Fig. 6. Total, scattering, and absorption loss of a new SELFOC fiber (fiber length: 500 m; fiber diameter: 230 μm).

Table I. Impurity Concentration Measured by a Mass Spectrometer[a]

Element	Glass (ppb)	Mixed raw powder (ppb)
Iron	600 (7)	320 (7)
Copper	50 (10)	30 (10)
Chromium	20 (10)	10 (10)
Cobalt	N.D. (7)	N.D. (7)
Nickel	10 (10)	N.D. (10)

[a] Numerals in parentheses indicate detection limit.

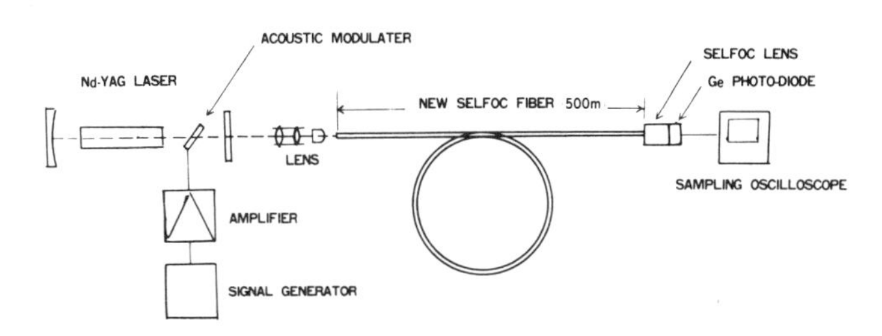

Fig. 7. Pulse-broadening measuring set for the new SELFOC fibers (light pulse width: 0.5 nsec; repetition rate: 300 MHz).

The variation of the low-attenuation area with the increase of fiber length was checked by the same method mentioned above using another sample, 500 m long, having a 25-μm core diameter. Figure 5 indicates that the low-attenuation area, specified as 50% power width or 95% power width, decreases gradually with the increase of fiber length, and tends to approach a constant value. This value is likely to stay constant, even if the measurement is done with longer fibers.

Besides the foregoing total loss measurement, the spectral, total, and scattering losses of the new SELFOC fibers were measured to identify the loss origin. Figure 6 shows a typical example of the spectral total, scattering, and absorption losses of the new SELFOC fibers. In Fig. 6, small circles indicate the total loss values obtained by using a Kr, a He–Ne, a He–Cd, and a Nd^{3+}:YAG laser, while the solid lines show interpolated values based on measurements obtained by using a Xe arc lamp source and interference filters. The scattering loss was measured using a solar cell box as detector and lasers as light sources.[23] The absorption loss was obtained by subtracting the scattering loss from the total loss. As shown in the figure, low transmission loss is realized in the wavelength range 0.80–0.85 μm, while a broad absorption band at shorter wavelengths and a sharp absorption peak at 0.97 μm are recognized. The broad absorption band is identified as a hybrid absorption band of the *d-d* transition of the transition metal ions, especially those of Cr ions and Ni ions, existing as impurities in the glass, and the sharp peak is due to the third harmonic vibration band of OH ions in the glass. By changing the glass composition from flint glass in the conventional process to borosilicate glass and as a result of the adoption of the new process, the absorption loss originating from metal colloid formation and the scattering loss were diminished remarkably compared with those of the conventional SELFOC fibers. The scattering loss was about 6 dB/km and 4 dB/km at the wavelengths of 0.853 μm and 1.06 μm, respectively.

The concentration of impurities was measured by a mass spectrometer. Table I indicates measurements of a sample of bulk core glass and mixed raw powder. As shown in the table, the impurity concentrations of the main transition metal elements have been diminished remarkably because of progress in the raw powder refinement technique and the clean melting method. Although the amount of the Fe impurity is still relatively large compared with those of other impurities, the effect of the Fe ions is not serious because the absorption coefficient for Fe^{2+} ions in the borosilicate glass is comparatively small.

Further improvements in preparing low-loss materials should reduce the loss in the new SELFOC fibers to a level as low as 10 dB/km in the near future.

B. Broad Bandwidth

The pulse transmission characteristics of the new SELFOC fibers were measured using a mode-locked

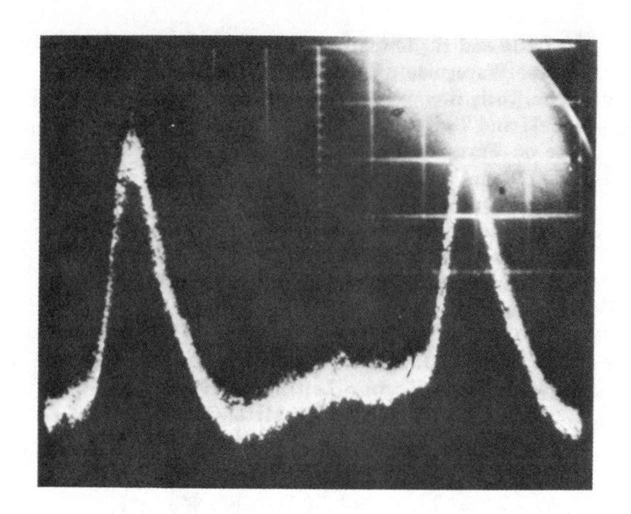

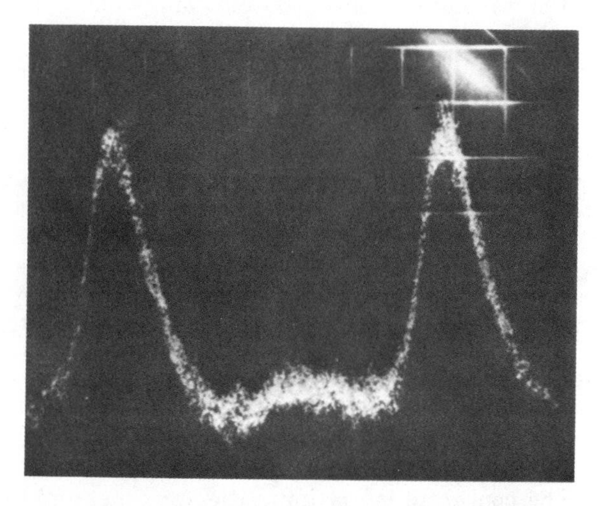

Fig. 8. Input and output pulse shapes (fiber length: 500 m).

Nd^{3+} doped YAG laser, whose pulse-width was 0.5 nsec and whose repetition rate was 300 Mbit/sec. As shown in Fig. 7, the laser beam was launched into the input end of a 500-m fiber in the condition of mode matching. The output light pulses were detected by a Ge photodiode. A piece of SELFOC lens was used to focus the output beam on the Ge photodiode. The output pulse shape was observed by using a sampling oscilloscope. Figure 8 indicates the measured input and output pulse shapes. The half-power width of the input pulse of 0.5 nsec broadened to about 0.65 nsec after traveling through the 500-m long fiber. This seems to indicate that the new SELFOC fibers have a transmission capacity of at least a few Gbit/sec for a 1-km transmission length.

The broad bandwidth of the new SELFOC fibers is due (1) to the parabolic-like distribution properly realized in the core region, and (2) to the greatly reduced refractive index fluctuations as a result of the new continuous process.

IV. Conclusion

A new continuous manufacturing process for light-focusing fibers has been developed using fast ion-exchange in a special double-crucible. By use of this new process, low-loss optical fibers with a transmission loss of 20 dB/km in the wavelength range 0.80–0.85 μm and a transmission capacity of at least a few Gbit/sec for 1 km have been fabricated from low-loss glass prepared by an improved raw powder refinement technique and by clean melting. The broad-band and low-loss characteristics, and the qualification of this new SELFOC fiber for mass production, ensure its potential as a long optical fiber for communication use.

The authors would like to thank T. Uchida of Nippon Electric Co., Ltd. for continued guidance and encouragement. Thanks are also due to A. Ueki, T. Kitano, R. Ishikawa, M. Yoshiyagawa, T. Yamazaki, and Y. Furuse for helpful technical discussions and for joint experimental work.

References

1. F. P. Kapron, D. B. Keck, and R. D. Mauer, Appl. Phys. Lett. **17,** 423 (1970).
2. D. B. Keck, P. C. Schultz, and F. Zimar, Appl. Phys. Lett. **21,** 215 (1972).
3. T. Uchida, M. Furukawa, I. Kitano, K. Koizumi, and H. Matsumura, IEEE J. Quantum Electron. **QE-6,** 606 (1970).
4. I. Kitano, K. Koizumi, H. Matsumura, T. Uchida, and M. Furukawa, "A Light-Focusing Fiber Guide Prepared by Ion-Exchange Techniques," Proc. 1st Conference on Solid State Devices, Tokyo, 26–27 Sept., 1969, Suppl. to J. Japan Soc. Appl. Phys. **39,** 63 (1970).
5. H. Kita, I. Kitano, T. Uchida, and M. Furukawa, J. Am. Ceram. Soc. **54,** 321 (1971).
6. E. G. Rawson, D. R. Herriott, and J. Mckenna, Appl. Opt. **9,** 753 (1970).
7. M. Sakaguchi and N. Nishida, AFIPS Conf. Proc. FJCC, **37,** 653 (1970).

8. K. Matsushita and K. Ikeda, "Newly Developed Glass Devices for Image Transmission," to be published in Proc. SPIE Seminar on Fiber Optics, San Mateo, Calif., 16–18 Oct. (1972).
9. E. A. J. Marcatili, SPIE J. **8,** 101 (1970).
10. P. J. B. Clarricoats and K. B. Chan, Electron Lett. **6,** 694 (1970).
11. Y. Suematsu and K. Furuya, IEEE Trans. Microwave Theory Tech. **MTT-20,** 524 (1972).
12. D. Gloge, Appl. Opt. **11,** 2506 (1972).
13. H. Kirchhoff, Arch. Elektron. Übertragung. **27,** 13 (1973).
14. H. Matsumura and T. Kitano, "Geometrical Analysis in a Lens-like Medium with Fourth-order Abberation," Trans. Inst. Electron. Commun. Eng. Japan, Tech. Group on Quantum Electron. **QE70-37** (1970).
15. T. Kitano, H. Matsumura, M. Furukawa, and I. Kitano, "Experimental Analysis of Fourth Order Abberation in a Lens-like Medium," Trans. Inst. Electron. Commun. Eng. Japan, Tech. Group on Quantum Electron. **QE70-38** (1970).
16. M. Ikeda and H. Yoshikiyo, "Effects of Bends on Multimode Optical Waveguide," Trans. Inst. Electron. Commun. Eng. Japan, Tech. Group on Quantum Electron. **QE72-70** (1970).
17. A Ueki and T. Kitano, "The Effect of Refractive Index Variance on Transmission Characteristics of Long SELFOC Fibers," Trans. Inst. Electron. Commun. Eng. Japan, Tech. Group on Quantum Electron. **QE72-91** (1973).
18. S. Kawakami and T. Nishizawa, IEEE Trans. Microwave Theory Tech. **MTT-16,** 814 (1969).
19. R. Bouillie and J. R. Andrews, Electron. Lett. **8,** 309 (1972).
20. D. Gloge, E. L. Chinnock, and K. Koizumi, Electron. Lett. **8,** 526 (1972).
21. K. B. Chan, P. J. B. Clarricoats, R. B. Dyott, G. R. Newns, and M. A. Savva, Electron. Lett. **6,** 748 (1970).
22. S. Fukuda and M. Miya, "A Metal-Ceramic He-Cd Laser with Sectioned Hollow Cathodes and Output Power Characteristics of Simultaneous Oscillations," to be published.
23. A. D. Pearson and A. R. Tynes, Am. Ceram. Soc. Bull. **49,** 969 (1970).

Optical Transmission in Liquid-Core Quartz Fibers

J. Stone
Bell Telephone Laboratories, Crawford Hill Laboratory, Holmdel, New Jersey 07733
(Received 13 December 1971)

Multimode liquid-core quartz fibers, potentially useful for long-distance optical communication, have been constructed. The optical transmission loss of these fibers has been measured between 6000 and 11 000 Å. Fibers filled with tetrachloroethylene or a mixture of equal parts by volume of tetrachloroethylene and carbon tetrachloride have transmission loss of 20 dB/km or less between 8400 and 8600 Å and between 10 400 and 11 000 Å. These bands are of particular interest because within them operate highly promising oscillators such as the GaAs diode and the Nd:YAG laser. An absorption peak of 80 dB/km at 9600 Å is attributed to a small percentage of OH in the tetrachloroethylene. In the wavelength intervals between 7000 and 7600 Å and between 8200 and 11 000 Å these fibers have losses lower than or about equal to those reported for any other fibers.

We have constructed multimode liquid-core quartz fibers for optical transmission. The fibers have been filled with tetrachloroethylene, C_2Cl_4, or a mixture of tetrachloroethylene and carbon tetrachloride, CCl_4. These fibers have a measured transmission loss of about 20 dB/km or less between 8400 and 8600 Å and between 10 400 and 11 000 Å. It is expected that purification of the liquids will result in even lower losses over a larger range of wavelengths. The search for low-loss optical fibers is currently a very active one.[1] The measured loss values of the liquid-core fibers are sufficiently low to make these fibers candidates for use in long-distance optical communication. In the intervals between 7000 and 7600 Å and 8200 and 11 000 Å the fibers have the lowest transmission losses that have been reported.[2] The only previous work reported on liquid-core optical fibers for optical transmission was reported for glass-cladded fibers by the present author.[3]

The hollow quartz fibers were pulled from quartz tubing, Amersil TO8 commercial grade, 6 mm o.d. and 1-mm wall thickness. An oxygen-hydrogen torch of our own design was used. It consisted of three jets in a triangular arrangement. The jets were made by drilling holes 0.75 mm in diameter in brass escutcheon pins. The fibers were pulled horizontally onto drums 27 cm in diameter in lengths up to about 250 m. Fiber diameters were about 100 μ o.d. and 75 μ i.d. The fibers were slightly out-of-round and the ratio of maximum to minimum diameter was about 1.1.

The liquids used to fill different fibers were Fisher technical grade or Eastman 2418 tetrachloroethylene (index of refraction $n = 1.50$) and Fisher 99% mol pure carbon tetrachloride ($n = 1.458$). The values of n are given at 6328 Å. The Fisher C_2Cl_4 contained a proprietary inhibitor. An attempt to purify the Fisher C_2Cl_4 was unsuccessful and therefore the liquid was used straight from the bottle.

A 130-m length of fiber was filled with Fisher C_2Cl_4 in about 2 h using a fiber-filling cell described earlier.[3] The loss measurements were carried as shown schematically in Fig. 1. Light was launched into the end of the fiber located inside the cell (Fig. 1) using a 5× microscope objective; the thickness of the window in the fiber-filling cell precluded using a higher-powered objective. This objective has a measured numerical aperture of 0.125. The fiber filled with C_2Cl_4 had a numerical aperture given by

$$NA = (2n\Delta n)^{1/2}, \tag{1}$$

where n is the index of refraction of C_2Cl_4 and Δn is the index-of-refraction difference between core and quartz cladding. Taking for quartz $n = 1.457$,[4] we obtain $NA = 0.34$. Therefore, the numerical aperture of the fiber was not completely filled and hence not all the modes of

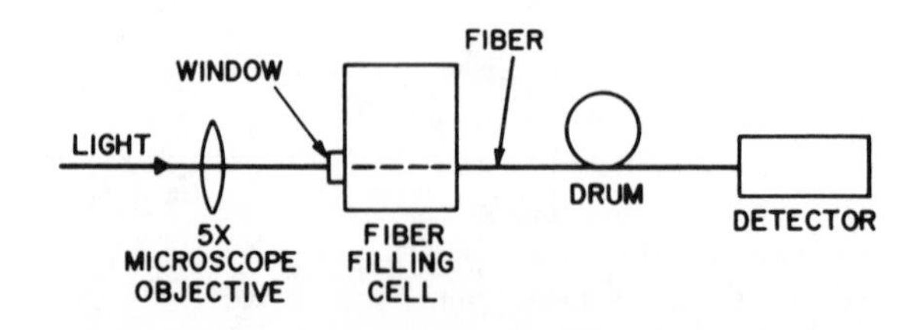

FIG. 1. Schematic arrangement of loss-measurement setup.

Reprinted with permission from *Appl. Phys. Lett.*, vol. 20, pp. 239–240, Apr. 1, 1972.

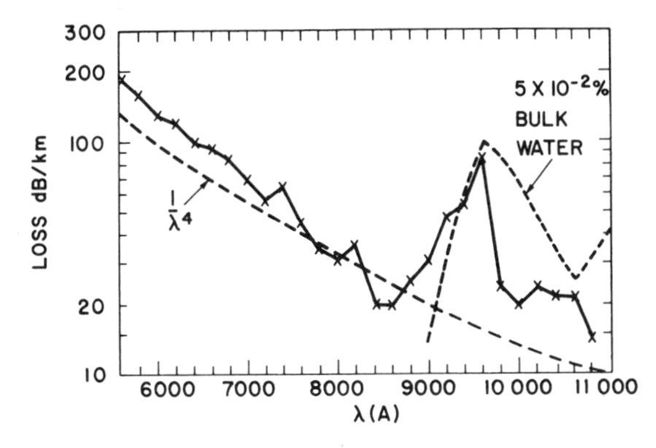

FIG. 2. Measured loss spectrum for a hollow quartz fiber filled with technical grade tetrachloroethylene. Fiber length was 130 m, outside diameter about 100 μ, and inside diameter about 75 μ.

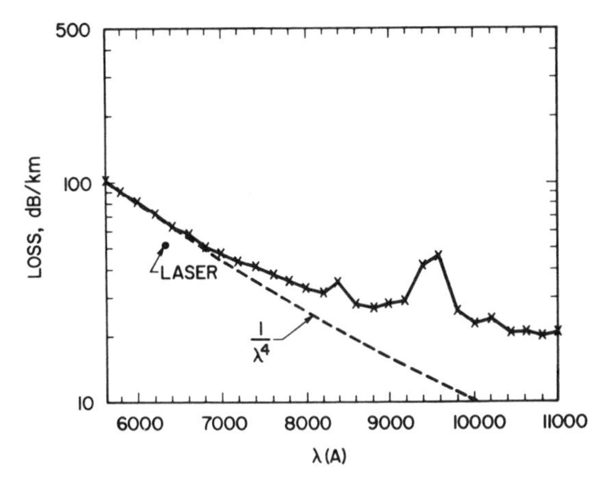

FIG. 3. Measured loss spectrum of a mixture of 50% by volume of tetrachloroethylene Eastman 2418 and 50% carbon tetrachloride Fisher 99 mol% purity. Fiber length was 90 m, fiber cross section similar to that in Fig. 1.

the fiber were excited. However, higher-order modes than expected from the numerical aperture of the objective were probably excited because (i) light was not launched directly down the fiber axis and (ii) mode mixing occurred due to scattering and wall imperfections in the fiber. The loss measurement was made by determining the power out of the fiber end for the full length of the fiber, i.e., 130 m, for intermediate lengths, and for a stub 65 m long. Energy in the cladding was stripped out by immersing the fiber in C_2Cl_4. We used as light sources a He-Ne laser at 6328 Å, TEM_{00} mode, and an Osram XBO 450 high-pressure xenon arc lamp with 100-Å-wide filters every 200 Å between 6000 and 11 000 Å. For the laser the numerical aperture was even smaller than 0.125 since the laser beam did not fill the entire microscope objective. The loss spectrum is shown in Fig. 2. Upon the curve has been superposed a $1/\lambda^4$ curve, where λ is the wavelength, at an arbitrary level to indicate the Rayleigh-scattering dependence and also a curve showing the contribution to the loss to be expected for bulk water at a concentration of $5\times 10^{-2}\%$.[5] The absorption at 9600 Å is due to the third harmonic of the OH vibration line at 2.95 μ[6] and a similar peak can be expected if any alcohol is present. On the supposition that the impurities in C_2Cl_4 included alcohol, we passed the C_2Cl_4 over activated alumina. This did succeed in reducing the absolute height of the peak at 9600 Å. However, fine dust from the alumina remaining in the C_2Cl_4 increased somewhat the level of the loss spectrum which then showed an even closer fit to at $1/\lambda^4$ curve.

Another hollow fiber 90 m long was filled with a mixture of CCl_4 and Eastman 2418 C_2Cl_4. The latter is a higher-purity grade of C_2Cl_4, but it has added to it 0.5% ethanol. The liquids were distilled and then mixed in equal proportions by volume. For the mixture $n = 1.483$, giving $NA = 0.27$ for this fiber. Therefore, the same light-launching arrangement filled a larger proportion of the numerical aperture of the fiber. The loss spectrum is shown in Fig. 3. It is known from absorption measurements in the bulk[7] and scattering data[8] that the loss in CCl_4 is about 20 dB/km at 6328 Å and is mostly Rayleigh scattering. It can be seen from Figs. 2 and 3 that the loss of the mixture at this wavelength is about what would be expected. Also the peak at 9600 Å is still present, indicating that some alcohol or water is still present.

Since CCl_4 has a lower loss than C_2Cl_4, higher proportions of CCl_4 in a mixture should give lower loss. However, for pure CCl_4 the core-cladding index-of-refraction difference is very small so that obtaining guidance of light in the fiber is difficult.[4,9] A fiber filled with a mixture of 20% chlorobenzene ($n = 1.52$) and 80% carbon tetrachloride ($n = 1.470$ for the mixture) had a loss of 45 dB/km at 6328 Å.

Other suitable mixing liquids include deuterated materials such as deuterated benzene. This is because the principal loss mechanism of liquids containing carbon and hydrogen appears to be overtones of the C-H vibration at 3.2 μ.[3] However, for the C-D vibration, these lines occur at 4.4 μ[10] and therefore at a given wavelength, a higher, hence weaker, overtone is observed. This has been verified in fibers filled with normal and deuterated bromobenzene. Also, liquids without any hydrogen atoms, such as $CBrCl_3$ and certain fluorinated liquids of sufficiently high index of refraction, could prove useful.

We wish to acknowledge numerous helpful discussions with E.A.J. Marcatili. H.E. Earl assisted in construction of the equipment. We wish to thank A.R. Tynes for making available his fiber pulling machine.

[1]S.E. Miller, IEEE NEREM, Boston, 1971 (unpublished).
[2]F.P. Kapron, D.B. Keck, and R.D. Maurer, Appl. Phys. Letters **17**, 423 (1970); R.D. Maurer, 1971 IEEE/OSA Conference on Laser Engineering and Applications, Washington, D.C., 1971 (unpublished).
[3]J. Stone, IEEE J. Quantum Electron. (to be published).
[4]Landolt-Bornstein, *Zahlenverte und Funktionen* (Springer-Verlag, Berlin, 1950), Vol. 2, Chap. 8, p. 491.
[5]W.M. Irvine and J.B. Pollack, Icarus **8**, 324 (1968).
[6]N.E. Dorsey, *Properties of Ordinary Water-Substance* (Reinhold, New York, 1940), p. 343.
[7]J. Stone, J. Opt. Soc. Am. (to be published).
[8]I.L. Fabelinskii, *Molecular Scattering of Light* (Plenum, New York, 1968).
[9]Landolt-Bornstein, *Zahlenverte und Funktionen* (Springer-Verlag, Berlin, 1950), Vol. 2, Chap. 8, p. 567.
[10]G. Herzberg, *Infrared and Raman Spectra* (Van Nostrand, New York, 1945), p. 365.

TRANSMISSION LOSS OF TETRACHLOROETHYLENE-FILLED LIQUID-CORE-FIBRE LIGHT GUIDE

Indexing terms: Optical waveguides, Fibre optics, Attenuation measurement

The attenuation of a tetrachloroethylene-filled liquid-core light guide has been measured for wavelengths between 600 and 1600 nm. A number of peaks appear in the curve, and some of these are identified with the effects of residual water. Attenuation troughs appear at 1090, 1205 and 1280 nm where, in each case, the attenuation is less than 8 dB/km.

Introduction: The low-attenuation liquid-core light guide was developed independently during 1971 in Australia[1] and the United States.[2] In December 1971,[3] CSIRO advertised for submissions to be made to it with a view to possible commercial exploitation of this fibre.

More recently, it has been reported that fine silica tubes filled with tetrachloroethylene gave attenuations of less than 20 dB/km in part of the near infrared region of the spectrum,[4] and similar performance has been obtained from hexachlorobutadiene-filled hollow-glass fibres.[5] The range of wavelengths used in these measurements is from about 600 to less than 1200 nm, and attenuation peaks are observed which have been tentatively associated with the presence of water. Such an identification implies that the observed absorption peaks arise from overtones and combinations of fundamental frequencies associated with hydrogen in chemical combination.[6] More reliable identification of the attenuation peaks clearly could be made by extending the measurement range to longer wavelengths, and an example of such extended-range measurements follows.

The results to be presented now were obtained for a tetrachloroethylene-filled silica-case fibre, and these results, while there are detailed differences from fibre to fibre depending on details of fibre manufacture, give a good idea of the nature of the attenuation relationship with wavelength to the long-wave limit of 1600 nm. This study is part of a comprehensive investigation of liquid-cored fibres being made by CSIRO in collaboration with the Australian Post Office Research Laboratories. Work already reported deals with another aspect of the investigation.[7]

Fibre manufacture: The fibre was drawn from 'Heralux' silica tubing, using a precision drawing machine with the stock piece of tubing of 4 mm bore and 12 mm external diameter being heated by an oxygen–propane flame. The fibre was filled at a pressure of 350 bar with 'chromato-quality' tetrachloroethylene, after dehydration by exposure to phosphorus pentoxide.

Measurement technique: The method of measurement is illustrated in Fig. 1. The loss through a relatively long length of fibre was measured by focusing the output from a Bausch and Lomb monochromator, mechanically chopped at a rate of 1 kHz, onto the end of the fibre, and observing the relative output level as a function of wavelength, after detection and amplification in a 1 kHz tuned amplifier. This measurement was carried out on a number of fibres of lengths up to 1 km and having internal diameters between 70 and 110 μm. The differential loss between short and long lengths of one fibre allowed the attenuation per kilometre to be computed.

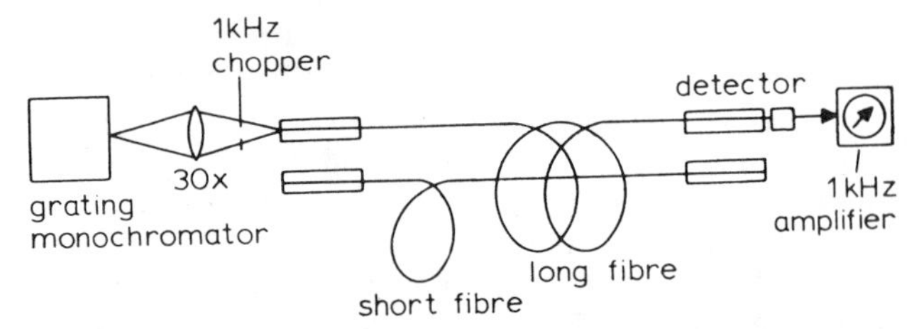

Fig. 1 *Method of measurement*

For wavelengths up to 1100 nm, a silicon *p–i–n* photodiode (HP 4201) was used as the detector, and, for the full band to 1600 nm, a germanium photodiode (OAP 12) was used. To eliminate spurious readings due to 2nd-order grating effects, appropriate filters were needed; nonavailability of a filter having a cutoff wavelength beyond 800 nm limited measurement to 1600 nm.

Assuming homogeneity of the two lengths of fibre (and

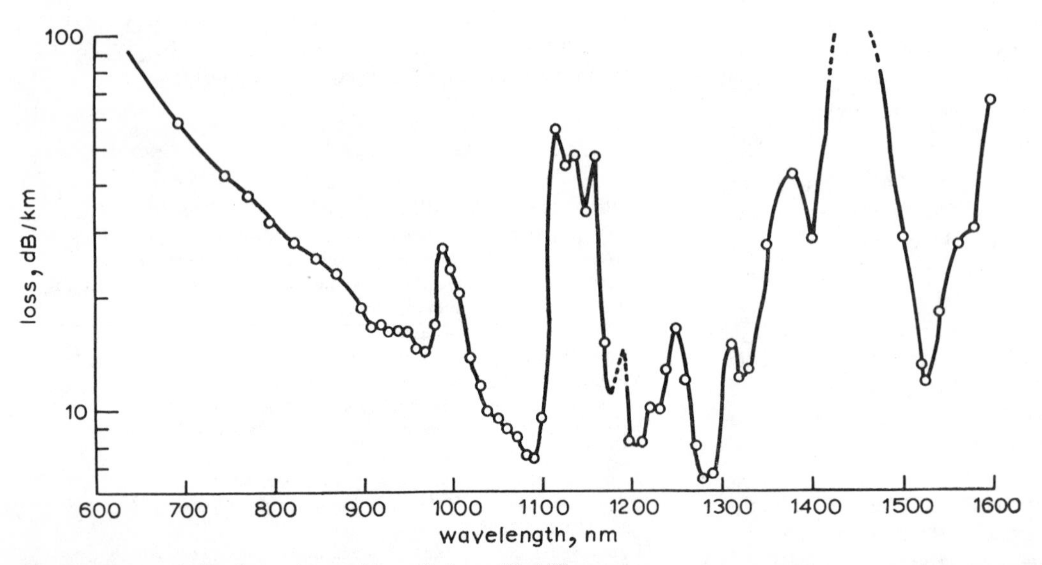

Fig. 2 *Transmission loss of tetrachloroethylene-filled fibre*

Reprinted with permission from *Electron. Lett.*, vol. 8, pp. 533–534, Nov. 2, 1972.

care was taken to ensure this), the method has the advantage that it does not depend on absolute power measurements or calibration of filters, and also that any spurious effects of launching will have been adequately suppressed before detection of the signal. Precision of measurement was of the order of ±0·1 dB, but slight variations with time of the fibre itself increase the overall inaccuracies somewhat.

Results: A length of fibre of 108 μm bore was divided into two lengths, one of 330 and the other of 80 m. The differential transmission loss of these two lengths between 650 and 1600 nm, measured with the germanium photodiode, is shown in Fig. 2. Loss minima occurred at 1090, 1205 and 1280 nm, where the losses were 7·5, 8·0 and 6·5 dB/km, respectively.

In general terms, the losses in these fibres arise from two sources. First, scattering effects at imperfections in the core-cladding interface and inhomogeneities in the core and the cladding cause losses which are dependent on the inverse fourth power of the wavelength. The second loss is that caused by absorption of energy by the material of the core, and to a lesser extent by the cladding.

The results shown in Fig. 2 indicate that the predominant loss in the region between 650 and 900 nm is that due to scattering, while, between 900 and 1600 nm, various absorption peaks are found, which probably arise from overtones and combinations of the fundamental X–H vibrations of impurities in the tetrachloroethylene. This was illustrated by the virtual disappearance of peaks at 910, 950 and 965 nm when the tetrachloroethylene was thoroughly desiccated. Even then, a peak at 950 nm was observed initially; however, over a period of two months, this peak decayed by about 3 dB/km, and this may be due to adsorption of residual traces of water by the surface of the silica. The magnitude of the indicated peak at 1190 nm is uncertain, since the individual fibres peaked at slightly different wavelengths; this may have been due to differential contamination.

Further work on the identification of this and the other peaks in this spectrum is being undertaken.

Acknowledgments: We thank E. Dodge for carrying out the measurements, and acknowledge the permission of the Chief, Division of Tribophysics, CSIRO, and the Australian Post Office to publish this letter.

G. J. OGILVIE
R. J. ESDAILE

26th September 1972

Division of Tribophysics
Commonwealth Scientific & Industrial Research Organisation
University of Melbourne
Parkville, Vict. 3052, Australia

G. P. KIDD

Australian Post Office Research Laboratories
59 Little Collins Street
Melbourne, Vict., Australia

References

1 OGILVIE, G. J.: Australian provisional patent PA 7211/71, 1971
2 'Electronics newsletter', *Electronics*, **44**, 22nd Nov. 1971, p. 25
3 Tender notice: *The age*, 22nd Dec. 1971, p. 26 (and other Australian newspapers)
4 STONE, J.: 'Optical transmission in liquid-core quartz fibers', *Appl. Phys. Lett.*, 1972, **20**, pp. 239–240
5 PAYNE, D. N., and GAMBLING, W. A.: 'New low-loss liquid-core fibre waveguide', *Electron. Lett.*, 1972, **8**, pp. 374–376
6 KAY, W.: 'Near infrared spectroscopy. Pt. 1—Spectral identification and analytical applications', *Spectrochim. Acta*, 1954, **6**, pp. 257–287
7 ROSMAN, G.: 'Variation of delay with launch angle in a liquid-filled fibre', *Electron. Lett.*, 1972, **8**, pp. 455–456

A New Optical Fiber

By P. Kaiser, E. A. J. Marcatili, and S. E. Miller

(Manuscript received November 20, 1972)

Currently there is strong interest in optical fibers for use as a transmission medium, analogous to the use of coaxial or wire pairs in the low-frequency region. Most work is devoted to a fiber structure consisting of a central glass core surrounded by a cylindrical glass cladding having a slightly lower index of refraction. This in turn requires that the chemical composition of the core glass differs from that of the cladding glass, leading to undesired effects at the core-cladding interface and perhaps limiting the minimum fiber losses achievable.

The Nippon Sheet Glass Company and the Nippon Electric Company together have developed a fiber (which they call SELFOC) with an index of refraction decreasing parabolically from the fiber axis to its outer boundary. This fiber requires a continuous variation in chemical composition from the fiber axis outward, with attendant complications in the fabrication process. A related guide requiring a film very thin compared to the wavelength has just been reported.[1]

The unique property of the new fiber is that a viable, handleable transmission medium is created by a structural form that uses only a single low-loss material.

The conception was stimulated by the findings of P. Kaiser, et al., who fabricated unclad round fibers and measured their spectral losses in up to 32-meter unsupported lengths.[2] Recently he found total losses as low as 2.5 dB/km at wavelengths near 1.1 μm using selected samples of low OH content fused silica. Similarly low losses have been measured at 1.06 μm in bulk fused silica by T. C. Rich, et al.[3] It appeared attractive to use material of this kind without the need for modifying the composition to alter the refractive index as is necessary with conventional core-cladding fibers or with graded-index fibers.

Figure 1 shows section views of two possible forms of the single-material (SM) fiber. The usefully guided energy is concentrated pri-

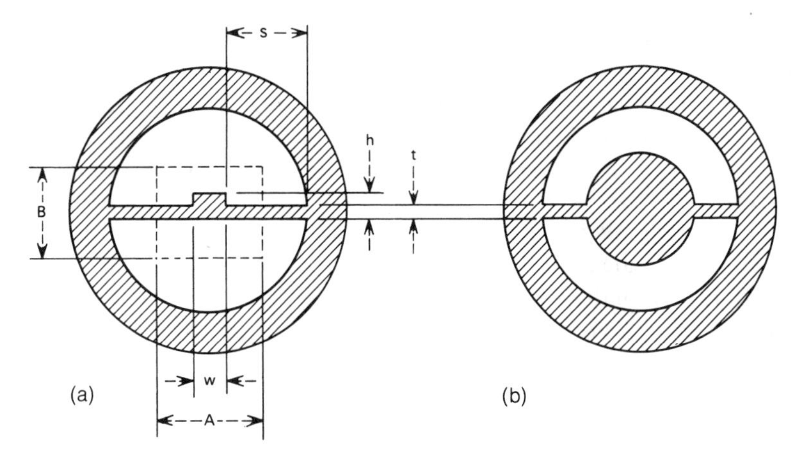

Fig. 1—Cross section of an SM fiber with (a) rectangular and (b) cylindrical core.

marily in the central enlargement, shown rectangular in Fig. 1a and round in Fig. 1b. In Fig. 1a, the central body has dimensions w and h for single-mode operation or dimensions A and B for multimode operation. There is an exponentially decaying field extending outward from the central member in the slab of thickness t; with appropriate spacing between the central enlargement and outer cylinder the guided-wave field at the outside surface can be made negligibly small and the fiber can be handled exactly as can the conventional core-cladding type fiber. Slab modes are possible on the supporting structure, but these are strongly coupled to the outer shell and are readily lost to the surrounding medium. Not shown in the figure is the possibility of adding an absorbing coating on the outer surface for avoidance of crosstalk in a multifiber cable.

The SM fiber structure can have a single propagating mode for any supporting slab thickness t and for any shape of the central enlargement, provided the size of the central enlargement is properly chosen. Practically, though, t must be limited in order to keep the slab dimension s, and consequently the overall size of the guide, reasonably small and still have the exponentially decaying slab field at the outer supporting cylinder small enough.

Analysis has been carried out for both single- and multimode SM fibers of several geometries. A few of the results are abstracted here. For the rectangular-guide case, Fig. 1a, and $t \gg \lambda$, there will be a single propagating mode provided

$$\frac{1}{h^2} + \frac{1}{w^2} \geqq \frac{1}{t^2}\cdot \tag{1}$$

Note that wavelength does not appear in this expression, correct to first order. More exact analysis shows that for $t = 4.89\ \mu m$ and $h = 7.0\ \mu m$, the limiting width w for single-mode operation is $7.07\ \mu m$ at $\lambda = 1.0\ \mu m$ and $6.94\ \mu m$ at $\lambda = 0.6\ \mu m$. In these structures the slab field decays by $1/e$ in 2.80 and 2.75 μm at λ equal to 1.0 and 0.6 μm, respectively.

The wave propagation effects of the slab support can be represented by a uniform-index side support having the same height as the core and an equivalent index $n_e = n_c(1 - \Delta_s)$, where n_c is the index of the w-by-h core. Then, from the equality condition of eq. (1), it can be shown that

$$\Delta_s = \frac{1}{8}\left[\frac{\lambda}{wn_c}\right]^2. \tag{2}$$

Arbitrarily small values of Δ_s can be achieved by making w [and according to (1), also h and t] appropriately large.

For the multimode rectangular guide the number of guided modes may be shown to be

$$N = \frac{\pi}{2}\frac{AB}{t^2}\left\{\frac{1}{1 + \left(\frac{\pi}{2v}\right)^2}\right\} \tag{3}$$

where

$$v = \frac{\pi t}{\lambda}\sqrt{n_c^2 - n^2}, \tag{4}$$

and n is the index of the unshaded region outside the dotted region of Fig. 1a, and A and B are defined in the figure. Note that the number of modes, eq. (3), is (to first order) independent of wavelength–a unique property. For all modes the field decays exponentially along the slab as noted above; for the highest-order mode the field penetration is the largest and decays by $1/e$ in a length l, where

$$l = \frac{\sqrt{2}t}{\pi}\sqrt{1 + \left(\frac{\pi}{2v}\right)^2}. \tag{5}$$

For the multimode SM fiber the equivalent full-height support has an equivalent refractive index $n_c(1 - \Delta_m)$ where

$$\Delta_m = \frac{1}{8}\left(\frac{\lambda}{tn_c}\right)^2. \tag{6}$$

The value of Δ_m can be used to calculate numerical aperture, the tol-

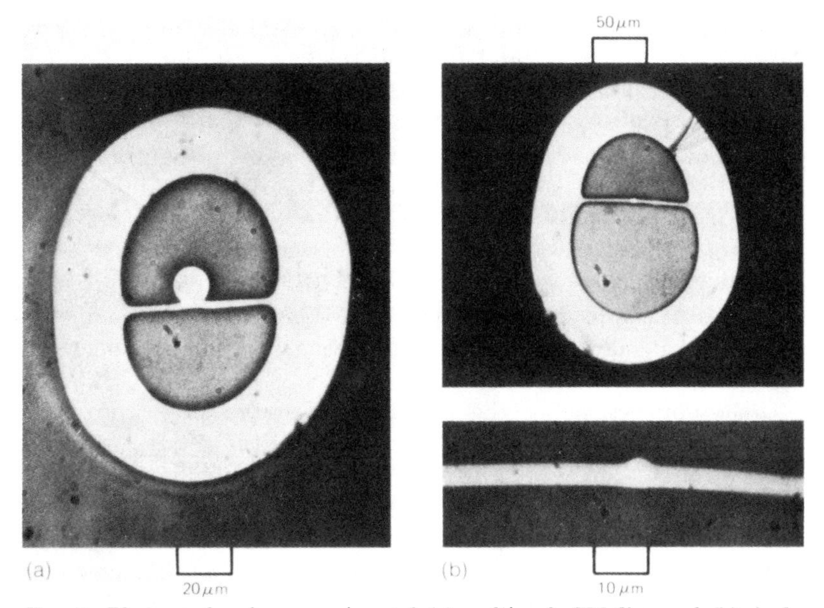

FIG. 2—Photographs of an experimental (a) multimode SM fiber and (b) single-mode SM fiber (top), with magnified core region (bottom).

erable radius of curvature, and modal dispersion. We give here only the numerical aperture,

$$\text{N.A.} = n_c\sqrt{2\Delta_m} = \frac{\lambda}{2t}. \tag{7}$$

SM fibers intended to approximate the geometries shown in Fig. 1 were drawn in an oxygen-hydrogen torch from 6.5 mm i.d., 10 mm o.d. fused quartz tubes containing thin, polished plates and small-diameter rods supported in the center of the tubes. Plates of about 0.2 mm thickness, and core rods of approximately 0.2 mm and 1 mm diameters, resulted in single- and multimode fibers, respectively, whose cross sections are shown in Figs. 2a and b.

For 15-μm-core multimode fibers, a slab thickness t varying between 3 and 4 μm resulted in numerical apertures ranging between 0.11 and 0.08 ($\lambda = 0.6328\ \mu$m), which agree excellently with the predicted values of 0.106 and 0.079, respectively [see eq. (7)]. The h/t ratio of the single-mode guide was about 6.5 μm/4 μm, or 1.625, with the width w amounting to 5 μm.

The spectral losses of the SM fibers were expected to closely approximate those of the unclad fibers drawn from the same material. Whereas this was true for the general shape of the spectral loss curve which was determined between 0.5 and 1.15 μm, the minimum losses were generally higher. For a 300-m-long, Suprasil 2 multimode fiber they amounted to 39, 50, and 28 dB/km at 0.66, 0.80, and 1.06 μm, respectively. Lowest losses of a single-mode fiber having a slightly different geometry than that shown in Fig. 2b were 55 dB/km at 1.06 μm. We believe that residual contamination of the preform elements is the source of the excess losses.

Total scattering losses in the order of 7.5 dB/km at 0.6328 μm demonstrate that the approximately 30 modes (3) of the multimode fiber are well guided and do not lose power into the surrounding cladding to any significant degree.

Other applications of the SM-fiber principle appear promising. Active fiber guides can be created by putting the active material in the central core or by putting it in a liquid surrounding the central member. Integrated optical circuits can utilize the same structure. For example, for a core and slab of index 1.472, slab thickness $t = 0.98\ \mu$m, and a surround index 1 percent less than 1.472, we find the single-mode limit at $h = 1.10\ \mu$m and $w = 6.75\ \mu$m at $\lambda = 0.6328\ \mu$m; the field decays transversely in the slab by $1/e$ in 2.67 μm. Thicker slabs allow larger w and h with single-mode guidance. In early research on optical integrated circuits, J. E. Goell observed wave propagation in a curved guide of the above general form, now understood as another verification of the principle of the SM fibers.

The assistance of H. W. Astle in the fabrication of the SM fibers is gratefully acknowledged.

REFERENCES

1. Nishizawa, J., and Otsuka, A., "Solid-State Self-Focusing Surface Waveguide (Microguide)," Appl. Phys. Lett., *21*, No. 2 (July 15, 1972).
2. Kaiser, P., Tynes, A. R., Cherin, A. H., and Pearson, A. D., "Loss Measurements of Unclad Optical Fibers," presented at the Topical Meeting on Integrated Optics–Guided Waves, Materials and Devices, in Las Vegas, Nevada, February 7–10, 1972.
3. Rich, T. C., and Pinnow, D. A., "Total Optical Attenuation in Bulk Fused Silica," Appl. Phys. Lett., *20*, No. 7 (April 1, 1972), pp. 264–266.

A New Technique for the Preparation of Low-Loss and Graded-Index Optical Fibers

J. B. MACCHESNEY, P. B. O'CONNOR, AND H. M. PRESBY

Abstract—**The lowest loss optical waveguides to date are those of high silica composition prepared by vapor deposition. The present article describes a method for producing waveguides having a GeO_2-SiO_2 core and SiO_2 cladding. These combine low loss with relatively large index differences between core and cladding. Large index differences create problems in optical communication systems because of dispersion effects. Means of index grading so as to decrease dispersion are also described.**

The realization of a practical long distance optical communications system requires optical waveguides which simultaneously exhibit low optical loss and low dispersion of the group velocities of the propagating modes. Chemical vapor deposition (CVD) techniques have previously been shown to hold promise as a means of preparing low-loss waveguides, [1], [2]. We wish to report a new method derived from CVD which is capable of producing waveguides of low optical attenuation and mode dispersion. This method has been applied to the preparation of fibers having a germania borosilicate core and a borosilicate–silica cladding.

In our fabrication technique, deposition is accomplished inside a thin-wall fused-quartz tube (Amersil, Inc. TO8 Commercial Grade), with the tube held in a glass-working lathe so that it is supported on both ends. While rotating, it is heated using an oxy-hydrogen burner which traverses along its length. The reactive gases flowing through the tube form a glassy deposit in the hot zone of the burner, and using multiple passes of the burner, uniform depositions are accomplished along the length of the tube. The gas stream consists of chlorides of silicon, germanium, and boron, ($SiCl_4$, $GeCl_4$, and BCl_3), carried in an oxygen stream. The reaction occurs both at the surface of the tube and in the homogeneous gas stream. The latter produces particles which settle as a powder on the tube walls downstream from the hot zone, and this powdery deposit is subsequently fused to a clear film as the burner traverses along the tube.

For the production of fibers with a germania doped core, a thin layer of borosilicate is deposited inside the support tube to minimize the diffusion of transition metal impurities from the quartz tube into the core, where they might produce serious losses. Next, a layer of silica doped with germania, or silica doped with germanium and boron is deposited. After the deposition is complete, the tube is collapsed into a solid rod, so that the germania rich layer forms the core of the optical waveguide preform. Fig. 1 shows the loss spectrum of a fiber made in this manner. The fiber is 723 m long and has a core of approximately 35 μm and a numerical aperture of 0.235. It can be seen that the losses decrease by approximately λ^{-4}, the expected Raleigh scattering dependence, to a minimum at just under 2 dB/km at 1.06 μm. It should be realized that, since an equilibrium mode distribution may not be attained in the length of fiber measured, losses in dB/km cannot accurately be extrapolated to longer lengths.

Hydroxyl ion related absorptions at 0.72, 0.88, and 0.95 μm are low, amounting to less than 10 dB/km at the 0.95-μm wavelength. We believe that the OH impurities causing these absorptions are due to siloxane present in the $SiCl_4$ starting material. This can be removed by fractional distillation, and loss peaks as low as 2 dB/km above the background at 0.95 μm have been observed. Variations of the process conditions have produced fibers with numerical apertures as high as 0.35.

Multimode optical fibers having numerical aperture in this range exhibit undesirably high mode dispersions, [3], [4]. This characteristic can be ameliorated by grading the composition and thus the index of the core. The index profile attained in this manner can be described by the relation [4]:

$$n = n_0 [1 - 2\Delta (r/a)^{\alpha}]^{1/2}.$$

Here, n is the index at any radius segment r, a is the radius of the whole core and n_0 the maximum index, Δ the index difference between the center and the cladding, and α is the parameter which describes the profile. ($\alpha = \infty$ represents a step index while $\alpha = 2$ defines a parabolic profile.)

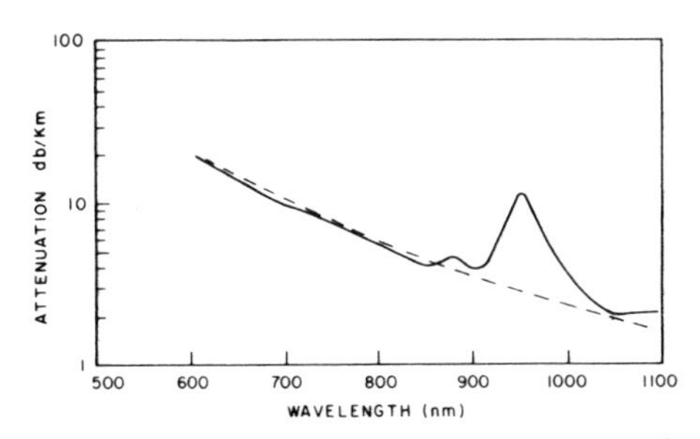

Fig. 1. Spectral loss versus wavelength for step index fiber having B_2O_3-GeO_2-SiO_2 core and B_2O_3-SiO_2, SiO_2 cladding.

Manuscript received May 20, 1974.
J. B. MacChesney and P. B. O'Connor are with the Bell Laboratories, Murray Hill, N.J. 07974.
H. M. Presby is with the Bell Laboratories, Crawford Hill, Holmdel, N.J. 07733.

Reprinted from *Proc. IEEE*, vol. 62, pp. 1280–1281, Sept. 1974.

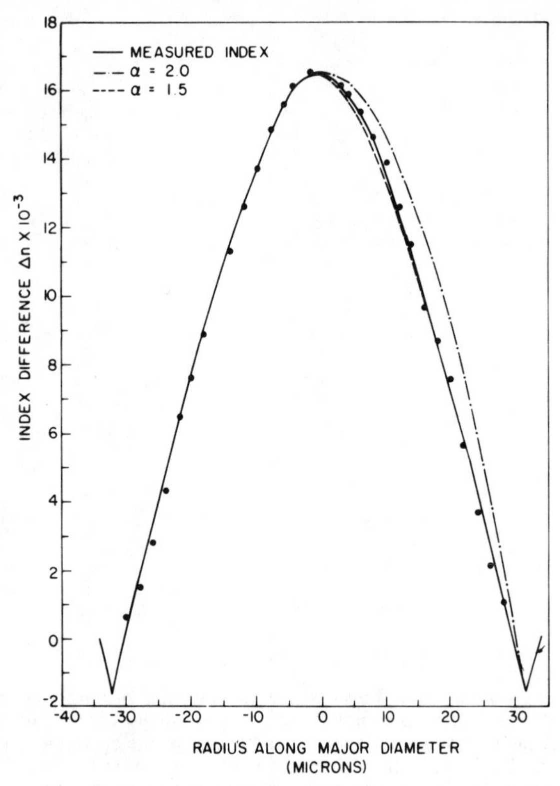

Fig. 2. Index profile for graded index fiber. α values of 2.0 (dash-dot curve) and 1.5 (dashed curve) are provided for comparison to the experimental curve (solid line). The dip of measured curve below $\Delta n = 0$ reflects the use of borosilicate layer to prevent diffusion of impurities into the core from the cladding.

It will be appreciated that index grading can easily be accomplished by gradually varying the flows of reactive gases during the deposition period. As a guide to attaining a desirable profile, the preceding expression was evaluated in terms of radius segments in increments of 0.1 a for $\alpha = 2$. Ten layers of equal thickness (producing ultimately equal increments of r in the fiber) were deposited by proportioning the $GeCl_4$ flow required to produce the maximum index desired into ten increments while keeping the flows of $SiCl_4$ and BCl_3 constant. Deposition times for each layer were proportioned according to the quantity of material required to form the layer. The fiber drawn from a preform prepared in this way was cross sectioned and the profile of its refractive index determined.

The refractive-index profile was obtained using interference microscopy techniques. This profile is plotted in Fig. 2. In terms of the index distribution described by the expression given earlier, the fiber has a maximum index difference equal to 0.0165 and $\alpha = 1.6$. Curves representing α values of 1.5 and 2 are plotted on Fig. 2 for comparison. An α value of 2 is of particular interest since theory predicts greatly reduced pulse dispersion [4] for this distribution.

ACKNOWLEDGMENT

The authors are indebted to F. V. DiMarcello who drew the fibers described and to J. R. Simpson and P. D. Lazay for spectral loss measurements reported here. We also appreciate mass spectrographic data supplied by D. J. Freed, and useful discussions with D. Gloge, W. G. French, G. W. Tasker, and L. G. Van Uitert.

REFERENCES

[1] J. B. MacChesney, R. E. Jaeger, D. A. Pinnow, F. W. Ostermayer, T. C. Rich, and L. G. Van Uitert, *Appl. Phys. Lett.*, vol. 23, p. 340, 1973.
[2] W. G. French, A. D. Pearson, G. W. Tasker, and J. B. MacChesney, *Appl. Phys. Lett.*, vol. 23, p. 338, 1973.
[3] S. E. Miller, E. A. J. Marcatili, and Tingye Li, "Research toward optical-fiber transmission systems Part I: The transmission medium," *Proc. IEEE*, vol. 61, pp. 1703–1726, Dec. 1973.
[4] D. Gloge and E. A. J. Marcatili, *Bell Syst. Tech. J.*, vol. 52, p. 1563, 1973.

Low-Loss Optical Waveguides with Pure Fused SiO_2 Cores

G. WILLIAM TASKER AND WILLIAM G. FRENCH

Abstract—**Optical waveguide fibers have been produced by a chemical-vapor-deposition technique with optical attenuations as low as 1.1 dB/km at 1.02 μm. The application of this technique to the fabrication of graded index fibers with losses below 2 dB/km is also reported.**

Low-loss optical waveguides with pure fused silica cores and borosilicate cladding have previously been reported [1]-[4]. French *et al.* have described a chemical-vapor-deposition (CVD) technique as a means of producing these fibers [1], [3], [4]. We now wish to report that a significant improvement in the CVD method of fabrication has resulted in fibers with attenuations as low as 1.1 dB/km at 1.02 μm. Graded index fibers with losses below 2.0 dB/km have also been achieved.

These advances have been realized by adapting a variation of the CVD fabrication process developed by MacChesney *et al.* [5], [6] for the production of GeO_2 doped SiO_2 core fibers. This technique employs a modified glass-working lathe to support and heat without distortion a thin wall TO8 fused quartz tube (commercial grade Amersil, Inc.) which contains the CVD reactions. By translating the lathe torch along the length of the support tube in multiple passes, uniform depositions of glassy films with a predetermined composition are achieved. Any powder formed by homogeneous reaction in the gas phase is fused as the deposition proceeds.

In our adaptation of this process, a B_2O_3-SiO_2 cladding film is first deposited within the support tube by the reaction of BCl_3 and $SiCl_4$ with O_2 at temperatures between 1600 and 1750°C. When this initial layer has attained a required thickness at the completion of a series of passes, the BCl_3 flow is discontinued and only $SiCl_4$ is allowed to react, resulting in a sharp step in the refractive index profile. For graded index depositions, the BCl_3 flow is automatically stepped down at a programed rate with each torch translation along the support tube until only $SiCl_4$ is reacting during the final pass. All gas flows are then interrupted and the support tube is collapsed into a solid rod preform suitable for drawing into fiber by conventional techniques described elsewhere [7]. This process is inherently free from contamination due to the use of a closed system during the film deposition. Glassy films with extremely low OH concentration are obtained by the direct combination of BCl_3 and $SiCl_4$ with O_2 at the deposition temperature (> 1500°C) characteristic of the technique. Furthermore, the economic feasibility is enhanced by reaction efficiencies > 50 percent.

The optical attenuation of a step index fiber ($\Delta n \sim 0.01$) was measured using light from a tungsten source through a grating monochrometer [8]. This fiber was over 0.5 km in length and was characterized by a 18-μm diameter core, 14.8-μm cladding thickness, a 100-μm overall diameter, and a numerical aperture of 0.17. The spectrum obtained is shown in Fig. 1. Loss minima occurred at 0.86, 0.90, and 1.02 μm. The average losses at these wavelengths were 1.9, 2.4, and 1.1 dB/km, respectively. The exact loss figure obtained is a function of the details of the launching optics, i.e., the number of modes excited in the fiber. For example at 1.06 μm, the loss measured using coherent excitation (low number of modes excited) was 0.90 ± 0.25 dB/km, whereas incoherent excitation of a large number of modes can yield losses as high as 2.0 dB/km. These results represent the lowest losses reported to date in the wavelength regions of interest for optical communication systems. Since a steady-state mode distribution does not hold for our measurements, due to the small numerical aperture of the injected beams and short (0.5-km) fiber lengths, the losses measured may not extrapolate to lengths greater than those employed in the present study.

This CVD method of fabrication is unique in that *both* the core and cladding are produced by chemical vapor deposition. Thus the refractive index profiles of fibers may be accurately tailored to minimize mode dispersion. Fig. 2 illustrates the index profile [9] for fiber fabricated in this manner. Although the index profile of this fiber does not have an exact parabolic variation, the close approximation to a parabola leads to considerable reduction of pulse dispersion when compared to

Manuscript received May 20, 1974.
The authors are with the Bell Laboratories, Murray Hill, N. J. 07974.

Reprinted from *Proc. IEEE*, vol. 62, pp. 1281-1282, Sept. 1974.

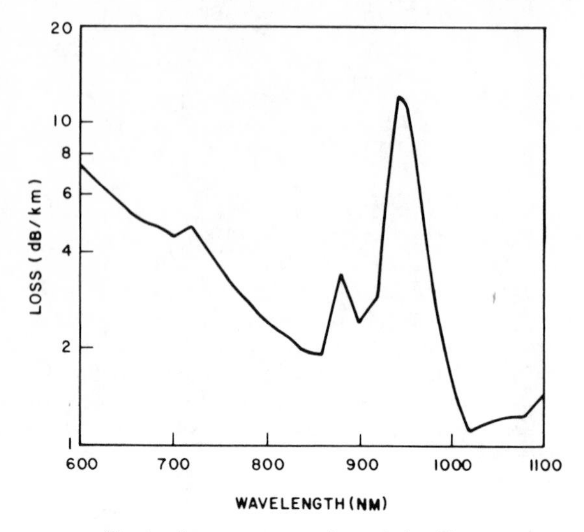

Fig. 1. Loss spectrum of step index fiber.

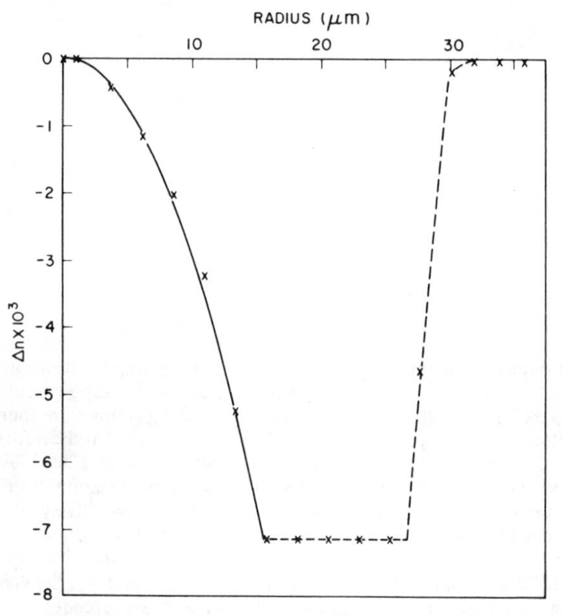

Fig. 2. Refractive index profile of graded index fiber. Solid line is an ideal parabola; points are experimental data determined by interference microscopy.

that which would be expected from a step index fiber with the same total index difference [10]. The loss spectrum for this 0.2-km long graded fiber is presented in Fig. 3. This fiber had a 30-μm core diameter and a 0.17 numerical aperature. Between 0.69 and 1.1 μm the attenua-

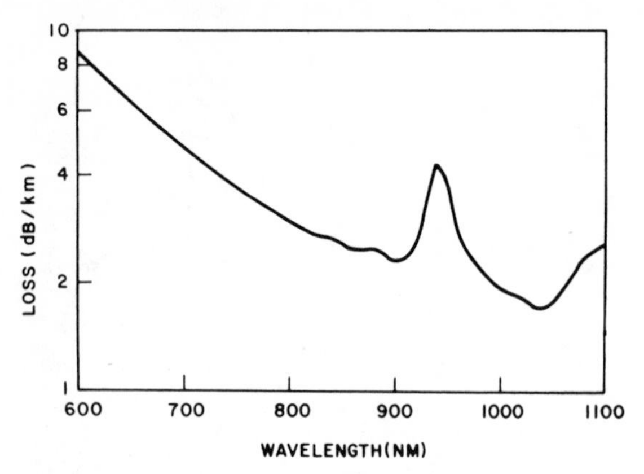

Fig. 3. Loss spectrum of graded index fiber.

tion was less than 5 dB/km. The contribution of the second OH overtone at 0.95 μm to the loss was only 2 dB/km. Minima for this curve occurred at 0.9 and 1.04 μm. The measured losses at these wavelengths were 2.3 and 1.7 dB/km, respectively. This fiber exhibited 4 to 6 dB/km higher losses when large numerical aperature light launching optics was used. This is an indication that the high-order modes are more strongly attenuated in this fiber.

Fiber lengths up to 0.75 km have been drawn from preforms fabricated by our CVD lathe technique. This length was limited only by the capacity of our drawing equipment. Uniformity of the film thickness and composition have been observed to vary only ± 5 percent along this length. Further improvement with regard to length of fiber, dimensional control, and loss characteristics are expected when fabrication parameters have been optimized.

ACKNOWLEDGMENT

The authors are grateful to J. R. Simpson for loss measurements, to J. B. MacChesney and P. B. O'Connor for helpful suggestions, to L. G. Cohen, to C. A. Burrus for the use of unpublished information, and to P. Kaiser and P. Lazay for helpful discussions.

REFERENCES

[1] W. G. French, A. D. Pearson, G. W. Tasker, and J. B. MacChesney, *Appl. Phys. Lett.*, vol. 23, p. 338, 1973..
[2] J. B. MacChesney, R. E. Jaeger, D. A. Pinnow, F. W. Ostermayer, T. C. Rich, and L. G. Van Uitert, *Appl. Phys. Lett.*, vol. 23, p. 340, 1973.
[3] W. G. French and G. W. Tasker, presented at Amer. Ceram. Soc. Glass Div. Meet., Bedford Springs, Pa., 1973.
[4] W. G. French, in *Proc. Tenth Int. Congr. Glass*, Kyoto, Japan, July 1974, to be published.
[5] J. B. MacChesney, P. B. O'Connor, and H. M. Presby, this issue, pp. 1280–1281.
[6] J. B. MacChesney, P. B. O'Connor, F. V. DiMarcello, J. R. Simpson, and P. D. Lazay, in *Proc. Tenth Int. Congr. Glass*, Kyoto, Japan, July 1974, to be published.
[7] A. D. Pearson and W. G. French, *Bell Labs. Rec.*, vol. 50, p. 102, 1972.
[8] P. D. Lazay and J. R. Simpson, to be published.
[9] C. A. Burrus, unpublished information.
[10] D. Gloge and E. A. J. Marcatili, *Bell Syst. Tech. J.*, vol. 52, p. 1563, 1973.

NEW SILICA-BASED LOW-LOSS OPTICAL FIBRE

Indexing terms: Fibre optics, Optical waveguides

A new type of silica-based optical fibre has been made from relatively cheap and abundant materials. The attenuation is very low over the entire range from the near ultraviolet to the gallium-arsenide-laser wavelength. The minimum loss of 2·7 dB/km occurs at 0·83 μm.

Introduction: In many ways, a suitable configuration for an optical-fibre waveguide is a compound glass core surrounded by a compound glass cladding of lower refractive index. Many fibres of this type have been reported, but the best transmissions obtained so far are in the region of 30 to 40 dB/km. The loss is largely due to transition-metal ion impurities, which are difficult to remove completely. A variant of the cladded glass fibre is Selfoc[1] fibre, in which the core has a parabolic variation of refractive index and a minimum loss of less than 20 dB/km has been obtained.

Silica, on the other hand, is produced commercially in a very pure form having a bulk-transmission loss of the order 2 dB/km at a wavelength of 1·06 μm. However, for it to be incorporated as one component of a cladded fibre, a second, compatible material must be found, having a similar softening temperature, expansion coefficient etc., to use as core or cladding, depending on the refractive index. One possibility is to modify the optical properties of silica by the addition of another oxide to form a simple compound glass containing a high proportion of silica. Thus, in 1972, a vapour-deposition technique[2] was used to deposit a titania–silica core glass on the inside of a silica tube, which was subsequently drawn into a fibre. Fibres[3] have also been drawn from rods of silica cladded with a boric-oxide–silica-glass layer, and we have successfully repeated this technique. More recently,[4] low losses have been announced with a modified silica core material, apparently comprising a mixture of silica and germania. Subsequently, there have been further reports of the successful use of the silica–germania mixture[5, 6] and also of the borosilicate-cladded silica fibre.[6]

Although commonly used in the semiconductor industry, and readily available in pure form, germania is expensive and likely to become more so, since germanium is not an abundant element. An alternative, cheaper and more common material would therefore be preferable, providing that it can be combined with silica to form a suitable low-loss glass. We have tried a number of combinations and find that a phosphosilicate glass core in a pure silica cladding provides a very low-loss fibre. Phosphorus is one of the most common elements and is relatively cheap, and the resulting fibre has a number of interesting properties. The most important wavelength range for optical-fibre communication systems is that of the various semiconductor light sources based on gallium arsenide, namely 0·8 to 0·9 μm, and, although not yet fully developed, the fibre has its minimum loss of 2·7 dB/km in this region. Further, the addition of phosphorus pentoxide to silica to form a binary glass does not appear to increase significantly either the intrinsic material absorption or the scattering.

Manufacture: The phosphosilicate glass is made by a controlled chemical-vapour-deposition technique. The starting materials are purified silicon tetrachloride and phosphorus oxychloride, which are vapourised, mixed with oxygen and passed through a tube of silica cladding glass. This tube containing the flowing gas mixture is traversed through a fibre-pulling furnace, which is operated at an appropriate temperature. Simultaneous oxidation and fusion occurs so that a clear phosphosilicate glass is deposited on the inner surface. A suitable thickness is obtained in about 1 h. The composite tube is then simultaneously collapsed and drawn into a fibre, or the operation can be carried out in two separate stages. We use a graphite resistance-heated furnace, which has been developed in these laboratories. The operating temperature, which can be in excess of 2200 °C, is monitored by a thermocouple to allow accurate control and repeatibility.

Fig. 1 shows a typical fibre cross-section, illuminated from the far end. There is a dark spot in the centre, presumably due to some volatilisation of phosphorus pentoxide from the

Fig. 1 *Cross-section of phosphosilicate-core fibre illuminated from far end*

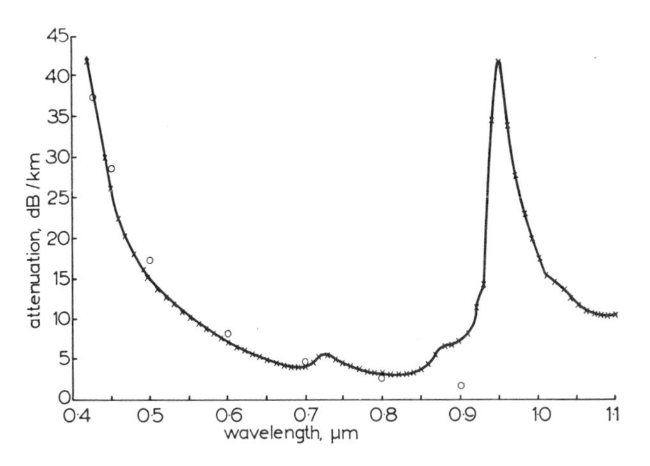

Fig. 2A *Spectral-loss curve for 244 m length of fibre comprising a phosphosilicate core in Suprasil cladding*

The numerical aperture is 0·14. The predicted ultimate loss in pure silica (Reference 7) is shown by the points marked ○

Reprinted with permission from *Electron. Lett.*, vol. 10, pp. 289-290, July 25, 1974.

inner surface of the deposited layer at the temperature (>2000 °C) required for fibre drawings. The fibres typically have a core diameter of 50 μm, an overall diameter of 150 μm and are drawn in lengths of about 1·2 km. The numerical aperture can be varied up to 0·18 or more as desired by control of the relative concentration of phosphorus pentoxide in the core.

Fibre attenuation: The loss in pure bulk silica is mainly by Rayleigh scattering, but with a small absorption component from the intrinsic ultraviolet absorption edge. There may also be impurity bands such as those due to hydroxyl ions. The addition of a second oxide in making a high-silica-content glass for use as core material is expected to introduce an additional component of scattering due to compositional fluctuation. The intrinsic absorption may also be increased, depending on the proximity of the ultraviolet-absorption edge of the additive.

Fig. 2A shows the spectral absorption curve of the SiO_2/P_2O_5 core in Suprasil cladding. The main features are as follows:

(*a*) Most striking, perhaps, is the smoothness of the curve over the short-wavelength portion, where, as indicated by Fig. 2A, the loss is below that of pure silica,[7] showing that the addition of phosphorus pentoxide to silica does not increase either the scattering or the absorption. In addition, there is no evidence of the drawing-induced colour centres at 0·6 μm, which have been observed elsewhere.[8] The minimum attenuation measured at 0·83 μm is 2·7 dB/km. The total loss at 0·633 μm is 5·8 dB/km, compared with the Pinnow *et al.* prediction[7] of 7 dB/km, of which 4·8 dB/km is due to scattering. This implies that our intrinsic absorption is only 1 dB/km and is therefore less than the expected value of ≃ 2 dB/km. Also, at 0·45 μm, again assuming the scatter loss to be that of silica, i.e. 19 dB/km, our intrinsic absorption is 7 dB/km, compared with a predicted value for silica of 10 dB/km. Further, the total scattering in the fibre at 0·633 μm has also been measured directly by an integrating sphere, and the loss of 5·8 dB/km obtained is greater than that of silica, thus implying an even smaller intrinsic absorption. It would appear that the intrinsic absorption is less than that predicted.[7] If this is correct, the ultimate loss at 0·85 μm could be less than 2 dB/km. To determine the true limit of intrinsic absorption, the loss measurements must be extended to wavelengths below 0·4 μm.

(*b*) The effect of the OH impurity can be clearly seen, particularly the peak at 0·95 μm rising to 40 dB/km. We believe that the OH bands are not characteristic of the phosphosilicate core material, but are due entirely to the high hydroxyl content of the Suprasil cladding, which has a bulk loss of 1000 dB/km at this wavelength. The normalised frequency of the fibre represented in Fig. 2A is $V = 20$, and hence of the order of 10% of the power is carried by the cladding, so that the loss contribution of the latter at 0·95 μm will be significant. This has been confirmed by making further fibres with Heralux instead of Suprasil, with the result shown in Fig. 2B. Heralux is a natural fused-quartz product in which the water content is only about 10% of that in Suprasil, and, as expected, the main OH peak has been reduced from 35 to 6 dB/km. Thus it should be possible to eliminate the OH bands* with a further improvement in the manufacturing technique. The minimum loss with Heralux is higher than with Suprasil cladding, because of the higher overall impurity level, but, nevertheless, Heralux is much cheaper, and the loss, even so, is not high.

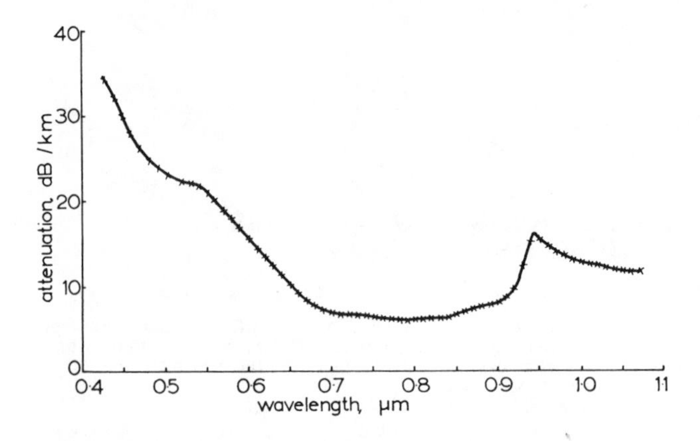

Fig. 2B *Spectral-loss curve for 133 m length of fibre comprising phosphosilicate core in lossy cladding*
Heralux

(*c*) Partially obscured by the hydroxyl absorption peaks at 0·88 and 0·95 μm in Fig. 2 is the third main feature of the spectral-loss curve, namely the broad ferrous-iron absorption band centred at about 1 μm, giving rise to an attenuation of 10 dB/km at 1·06 μm. Iron can produce appreciable absorption even when present in very small concentrations, and we believe the effect to be accentuated in phosphosilicate glasses, which tend to reduce this particular impurity mainly to the ferrous, rather than the ferric, state. Supporting evidence is produced by the very low loss at the blue end of the spectrum, indicating the absence of ferric ions. Nevertheless, although the effect of iron can be serious it should be possible to eliminate it from the core and to obtain a lower loss* at 1·06 μm with the Suprasil cladding by further material purification.

Conclusion: A new type of silica-based fibre has been made from relatively cheap and abundant materials and having an attenuation which is

(*a*) exceptionally low over the full range from the gallium-arsenide-laser wavelength to the near ultraviolet

(*b*) somewhat lower than has been predicted for pure silica.

The minimum attenuation is 2·7 dB/km and occurs at 0·83 μm. By controlling the phosphorus-pentoxide concentration in the core, the numerical aperture can be varied up to 0·18 or more as desired. The intrinsic loss of phosphosilicate glass appears to be no greater than that of pure silica.

Acknowledgments: Grateful acknowledgment is made to I. C. Goyal for carrying out the scattering measurements and to the UK Science Research Council for financial support. We are also indebted to the Pirelli General Cable Works Ltd. for the award of a research fellowship, and to S. Norman and A. Dyer for experimental assistance.

D. N. PAYNE
W. A. GAMBLING

17th June 1974

Department of Electronics
University of Southampton
Southampton SO9 5NH, England

* After this letter was accepted for publication, the attenuation, as measured in a length of 1·2 km, has been reduced appreciably at all wavelengths above 0·85 μm, and is now 2·4 dB/km at 1·1 μm. The OH bands have also been virtually eliminated, and peaks of less than 1 dB/km at 0·95 μm have been obtained

References

1 KOIZUMI, K., IKEDA, Y., KITANO, I., FURUKAWA, M., and SUMIMOTO, T.: 'New light-focussing fibres made by a continuous process', *Appl. Opt.*, 1974, **13**, pp. 255–260

2 KAPRON, F. P., KECK, D. B., and MAURER, R. D.: ' Radiation losses in glass optical waveguides', *Appl. Phys. Lett.*, 1970, **17**, p.. 423–425

3 FRENCH, W. G., PEARSON, A. D., TASKER, G. W., and MACCHESNEY, J. B.: 'A low-loss fused silica optical waveguide with borosilicate cladding', *ibid.*, 1973, **23**, pp. 338–339

4 SCHULTZ, P. C.: 'Preparation of very low-loss optical waveguides'. Presented at the American Ceramic Society meeting, Cincinnati, Ohio, USA, 1973

5 BLACK, P. W., IRVEN, J., BYRON, K., FEW, I. S., and WORTHINGTON, R.: 'Measurements on waveguide properties of GeO_2–SiO_2–cored optical fibres', *Electron. Lett.*, 1974, **10**, pp. 239–240

6 FRENCH, W. G., MACCHESNEY, J. P., O'CONNOR, P. B., and TASKER, G. W.: 'Optical waveguides with very low losses', *Bell Syst. Tech. J.*, 1974, pp. 951–954

7 PINNOW, D. A., RICH, T. C., OSTERMAYER, F. W., and DIDOMENICO, M.: 'Fundamental optical attenuation limits in the liquid and glassy state with application of fibre optical waveguide materials', *Appl. Phys. Lett.*, 1973, **22**, pp. 527–529

8 KAISER, P.: 'Drawing-induced coloration in vitreous silica fibres', *J. Opt. Soc. Am.*, 1974, **64**, pp. 475–481

Light Guide Systems for the Ultraviolet Region of the Spectrum

By Helmut Dislich and Alfred Jacobsen[*]

Whereas diverse and highly refined fiber optics have long been in technical use for the conduction of visible light, traditional but expensive silica glass lenses and prisms have had to be utilized until quite recently for the transmission of ultraviolet light. The reason for this was the lack of suitable materials. UV light cannot be transported to any technically useful extent either with pure glass-glass (core and sheath materials) light guides or with plastic-plastic light guides. UV light guides for wavelengths down to 200 nm can be produced by combination of a silica glass as the core material with a tetrafluoroethylene-hexafluoropropylene copolymer or methylpolysiloxane as the sheath material. Fiber optics of this type constitute a new aid for medical technology, medicine, and many technical and scientific fields, whose range of applications cannot yet be fully envisaged.

1. The Principle of Light Conduction

If a rod or a fiber made of transparent material having a high refractive index n_1 is coated with a transparent sheath of a material having a lower refractive index n_2, light falling on the end surface is conducted through the rod or fiber as a result of total reflection. The light is reflected without loss at the interface between these media (whereas loss occurs on reflection at a mirror surface), and is transmitted through the system in a zigzag path until it emerges at the other end. Important characteristics of such systems are:

1. the light transmittance
2. the aperture angle.

Both should be as large as possible. To achieve this, the core material in particular should have excellent transmittance, and the sheath layer, which should also be transparent, should have an excellent optical contact with the core. This means that the sheath must be in smooth and uniform contact at all points to allow loss-free total reflection. The difference between the refractive indices $n_1 - n_2$ should be as large as possible, since this determines the aperture angle $2\alpha_0$ at which light can be conducted (see Fig. 1) in accordance with

$$n_0 \sin \alpha_0 = \sqrt{n_1^2 - n_2^2}$$

(n_0 is the refractive index of the surrounding medium).

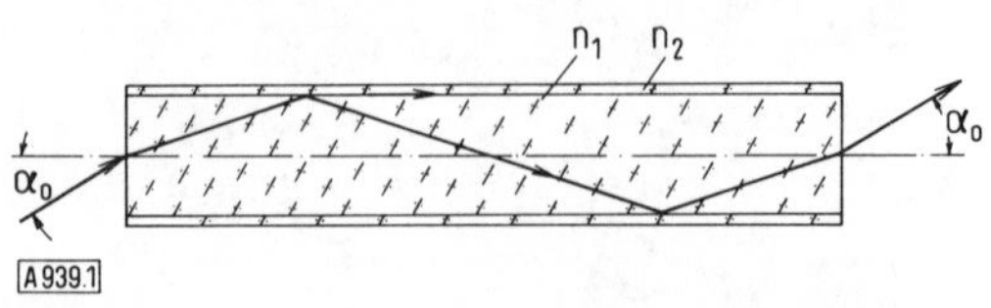

Fig. 1. Principle of light conduction.

1.1. Known Systems

The simplest case of light conduction is when light is transmitted along a straight or only slightly curved (radius

[*] Dr. H. Dislich and Dipl.-Phys. A. Jacobsen
Laboratorien des Jenaer Glaswerk Schott & Gen.
65 Mainz, Postfach 2480 (Germany)

Reprinted with permission from *Angew. Chem.* (*Int. Ed. in English*), vol. 12, pp. 439–444, June 1973.

of curvature >20 diameters) path through an unsheathed rod of glass or plastic. The surrounding air then acts as the low-refractivity sheath. Systems of this type have serious disadvantages since the optically effective sheath surface is exposed, and scratches and dust deposits lead to light scattering and hence to losses. The light guides used in practice are therefore predominantly of the sheathed variety.

The sheathed systems correspond to Figure 1, and those consisting of massive rigid rods require no further description. However, flexible light guides consisting of bundles of fibers have become much better known, if only because of the fact that they can be used to "guide light round corners", and in the case of "coherent bundles of fibers" one can actually see round corners, a possibility that fascinates even the optical layman. The light in this case is led through optically isolated flexible fibers as illustrated in Figure 1. A large number of these fibers are combined into a bundle, the fibers being cemented together at the ends and polished. The bundles are enclosed in a protective metal or PVC tube, and can be widely used in medicine (endoscopy) and in industry. If the arrangement of the fibers in relation to one another at the point where the light enters corresponds to that at the point where the light emerges, transmission of images is possible. If the fibers are unordered, they can be used only for the transmission of light.

Finally, mention should be made of fiber rods; these also consist of a large number of optically isolated fibers, which are fused together over their entire length and are therefore rigid. These are coherent, *i.e.* capable of transmitting images, and can be subsequently bent round relatively sharp corners.

Arising from these basic types there are a large number of interesting variations, which greatly extend the existing possibilities in optics. However, it is not the aim of this report to present a comprehensive description of the current position in this respect. Such an account has appeared elsewhere[1].

2. Materials

To lead up to the subject proper, let us first consider the question of the materials that can be used. As was mentioned above, the decisive factor is the light transmittance. The traditional material of optics is glass. The entire development of fiber optics in the Jenaer Glaswerk Schott & Gen. was carried out with high-quality optical glasses, so that both the cores and the sheaths of the commercially available products[2] consist of glass.

The possibility of using pure plastics systems has been under discussion for a long time. *Kapany*[3] showed that the fiber-forming polymers in the drawn state have disadvantages in comparison with glass fibers because of the anisotropy resulting from their partial crystallinity. He also noted disadvantages with regard to surface quality. However, the flexible systems developed by Du Pont[4,5] are produced from plastics. These consist of a polymethyl methacrylate core and a low-refractivity sheath of partly fluorinated plastics.

We were interested in plastics only as sheath materials, as we wished to take advantage of the outstanding properties of glass as a core material, in particular its good optical transmittance, surface accuracy, and polishing qualities of the end faces where the light enters and emerges.

2.1. The Core Material of the UV Guides

All the possibilities discussed so far are concerned with the conduction of visible light, in some cases extending into the near infrared region. Ultraviolet light with shorter wavelengths could not be conducted, since the transmittance of the core materials was insufficient. For light with wavelengths of down to 200 nm, there is no choice as far as core materials are concerned; practically the only possibility is high-purity silica glass, which must be of outstanding optical quality (in particular free from bubbles), and which can be drawn to fibers. We used Suprasil® (manufacturer: Heraeus-Schott, Quarzschmelze GmbH). For light having a wavelength of 365 nm, it is occasionally possible, within limits, to use special UV-transparent glasses if the optical path length is not excessive. However, this possibility is mentioned only in passing, as it is of no fundamental importance.

2.2. The Sheath Material of UV Fiber Optics

Flint glass has the following refractive indices at various wavelengths: $n_{546}=1.460$; $n_{365}=1.475$; $n_{155}=1.506$.

An optically isolating material for silica glass must therefore consist of a material that has a sufficiently low refractive index at all wavelengths and that is transparent to light of the wavelength in question.

One's first thought is naturally of the wide range of optical glasses, which is shown in Figure 2 in the form of the usual $n_d-\nu_d$ diagram. n_d is the refractive index for the d line (587.6 nm), and ν_d is the Abbé number, a measure of the dispersion.

Though the optical catalog of the Jenaer Glaswerk Schott & Gen. lists several hundred types of optical glasses, it is immediately clear that these do not include one with a sufficiently low refractive index to make it suitable for use as a sheath material. Here we reach a limit to the current optical possibilities of glasses. The present limit of the glass range in the direction of low refractive indices is shown by a broken line in Figure 2. The continuous lines are the limits of the individual glass types. The dot-dash line surrounds the region of the transparent plastics. In the direction of higher refractive indices, the glass region extends far beyond the top of the diagram.

Since glasses offer no further help with regard to a suitable sheath material, the next step is to consider other inorganic, UV-transparent materials with extremely low refractive indices. One such material is magnesium fluoride, which is commonly used as a transparent coating deposited from the vapor phase, *e.g.* as an antireflective coating on optical lenses.

While the properties of the material are promising, the technical difficulties of the vapor deposition of coatings

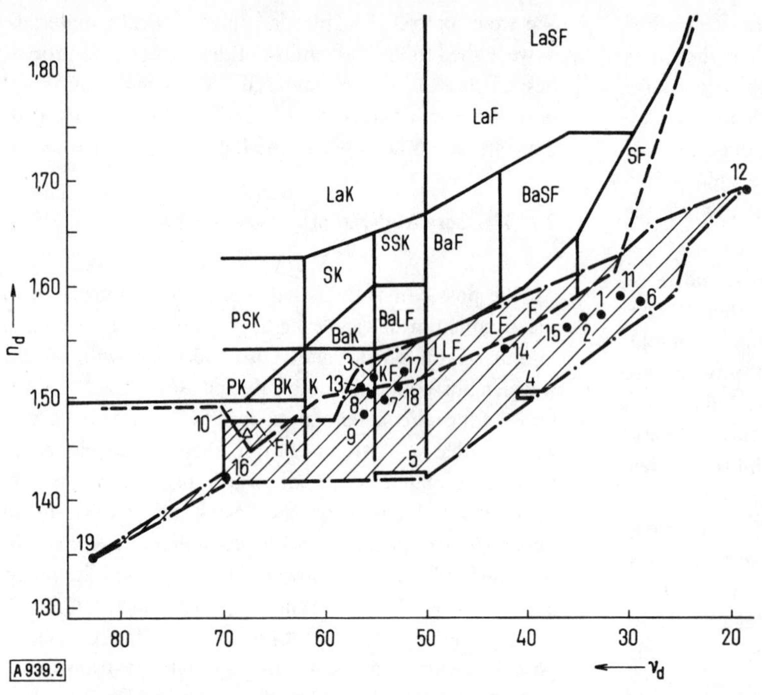

Fig. 2. Glass region and plastic region in the $n_d-\nu_d$ diagram. $\nu_d=(n_d-1)/(n_F-n_C)$ (Abbé number). n_F, n_C: refractive indices for the F and C lines. – – –: present limit of the glass region, –·–·: present limit of the region of transparent plastics having known $n_d-\nu_d$ values. Δ: silica glass; •: plastics 1—19. 1, styrene-acrylonitrile copolymer; 2, poly-*o*-diallyl phthalate; 3, polymethyl α-chloroacrylate; 4, methylphenylpolysiloxane; 5, methylpolysiloxane; 6, poly-α,α-dimethyl *p*-(*p*-hydroxyphenylmethyl)phenyl hydrogen carbonate; 7, polymethyl methacrylate; 8, polydiethylene glycol diallyl carbonate; 9, cellulose acetate; 10, cellulose propionate; 11, polystyrene; 12, poly-*N*-vinylcarbazole; 13, polycyclohexyl methacrylate; 14, polyvinyl chloride; 15, epoxy resin (Araldite CY 206); 16, poly-2,2,2-trifluoroisopropyl methacrylate; 17, methyl methacrylate-α-methylstyrene copolymer; 18, methyl methacrylate-acrylonitrile copolymer; 19, tetrafluoroethylene-hexafluoropropylene copolymer (Teflon FEP).

several μm thick on fibers impose such severe limits that technically useful products are unobtainable[6].

2.3. Optical Properties of Plastics

Another alternative to the classical materials of optical glasses is offered by the transparent plastics, whose region is indicated by the shaded area in the $n_d-\nu_d$ diagram (Fig. 2). It is interesting to note that the region of the transparent plastics extends considerably beyond that of the glasses, and coincides only in part with the latter[7].

A general comparison of the positions in the $n_d-\nu_d$ diagram (optical positions) with the chemical structures of the polymers reveals simple principles, which have been fundamentally known for a long time[8], and which are simply confirmed or expanded by newly developed plastics. Highly aromatic plastics such as polystyrene, polyaryl carbonates, or even poly-*N*-vinylcarbazole have high refractive indices and low Abbé numbers; fluorinated and particularly perfluorinated polymers are characterized by low refractive indices and high Abbé numbers.

With the aid of the Lorentz-Lorenz equation

$$R_M=\frac{n^2-1}{n^2+2}\,\frac{M}{d}$$

R_M = molar refraction, M = molecular weight, d = density, n = refractive index

it is possible to estimate (and even to calculate fairly satisfactorily with the aid of atomic, group, and bond refractions if the density is known) the hypothetical limits for the optical position of the plastics[8]. If hypothetical polymers are speculatively included in this calculation, one arrives at the result that as far as one can see at present, there will presumably be no polymers with refractive indices appreciably above $n_d=1.73$ or below $n_d=1.33$, and that the center of the glass region cannot be reached with plastics.

It may be noted in passing that this places limits on the competition that is sometimes referred to between optical glasses and plastics. The competition arises in the region of overlap between glasses and plastics. In general, however, a more interesting area in optics is that of glass-plastic combinations, particularly here in the case of UV light guides.

Plastics have recently become available whose chemical structure, according to the relations discussed here, suggests that they should have optical properties such as are required of a sheath material for UV light guides. These materials are methylpolysiloxane and tetrafluoroethylene-hexafluoropropylene copolymer (points 5 and 19 respectively in Fig. 2).

2.4. Methylpolysiloxane

Where can we find a material for coating rods or fibers of silica glass which has a refractive index lower than that of silica glass and a good UV transparency? Such properties are embodied in a methylpolysiloxane that became known in an optical quality in the middle 1960's[9], in which the Si—O skeleton of silica glass carries one methyl group per silicon atom, the lowest possible degree of substitution.

```
        CH3       CH3
        |         |
••••O—Si—O—Si—O••••
        |         |
        O         O
        :         |
        :  ••O—Si—O••••
                  |
                  CH3
```

It is obtainable in principle by hydrolysis and polycondensation of a methyltrialkoxysilane CH_3—$Si(OR)_3$[9], and is trifunctionally crosslinked in the cured state. The optically perfectly clear material, while very transparent to UV light, has sufficiently low refractive indices as a result of the incorporation of CH_3 groups. The refractive index has fallen from 1.460 to 1.418 at 546 nm and from 1.475 to 1.453 at 365 nm.

A silica glass rod is immersed in a 50% solution of partly condensed methyltriethoxysilane in alcohol, withdrawn at a constant rate (in addition to other parameters, this mainly determines the coating thickness), and heated to a maximum of 135 °C. The silica glass rod is thus provided with a sheath of very good optical quality, and exhibits an aperture angle of $2\alpha_0 = 29°$ at 365 nm.

While some of the problems in the transport of UV light were solved with these rigid light guides[10,11], flexible UV fiber optics still remained as the principal aim. A silica glass fiber can in principle be coated continuously, as required here. However, the curing times for the above polycondensation are too long for production, particularly when good optical quality is essential. Further investigations were therefore carried out with the object of finding a plastic with a still lower refractive index and a shorter setting time.

2.5. Tetrafluoroethylene-Hexafluoropropylene Copolymer

Point 19 in Figure 2 shows the extreme optical position of the material known by the trade name Teflon FEP®. With $n_{546} = 1.347$, $n_{365} = 1.356$, and $n_{255} = 1.376$, it has a lower refractive index than any other plastic known at present. Because of its branched structure as a copolymer

```
    F   F   CF3 F
    |   |   |   |
•••—C—C—C—C—•••
    |   |   |   |
    F   F   F   F
```

of tetrafluoroethylene and hexafluoropropylene, it does not exhibit the strong crystallization tendency of polytetrafluoroethylene, and so remains sufficiently transparent. Coatings of very fine particles can be applied from aqueous dispersions, and after removal of the water and the dispersing aids by heating, these particles can be fused to form a continuous, optically satisfactory film. This is again a property of the copolymer that it does not share with polytetrafluoroethylene.

We shall not go into details of the technically interesting as well as troublesome development of the continuous coating of a silica glass fiber 100 μm in diameter[12]. The first requirement was to draw a bubble-free, stria-free, and scratch-free fiber of uniform thickness continuously from silica glass. This fiber was passed through the aqueous dispersion of Teflon FEP and then through a long furnace. The processes of drying, evaporation of the dispersing aids, and fusion of the remaining Teflon FEP particles to form a continuous film required accurate adjustment of temperature gradient and of time, mainly because of the need to avoid thermal decomposition of the copolymer. Special care was necessary to lead the relatively fragile fiber over many guiding rollers without damaging its surface[*].

The essential point seems to be that the problem of the conduction of UV light by any desired route to any desired place has been solved[12] by combination of the outstanding UV transmittance of silica glass and its ability to be drawn into fibers with the good UV transmittance and the extremely low refractive index of a polyperfluoroalkane[13], and that this is the only solution found so far.

Figure 3 shows the spectral transmittance of flexible UV light guides in the form of a bundle of fibers.

3. Construction and Uses of UV Light Guides

The coated silica glass fibers have recently been used, mainly during the past year, for the development of many UV-conducting components, which have been included by the Industrial Research Council in the USA (Chicago) among the hundred most significant new technical developments of 1972. The product should be regarded primarily as a new tool for medical technology, medicine, physics, materials testing, photochemistry, and many other fields extending as far as genetics.

3.1. Properties

The system is optically transparent between 200 and 2200 nm; this range is interrupted only by an absorption band at 1400 nm and some weak harmonics of the OH absorptions. The optical transparency was measured with light entering the light guide at an aperture angle of 5°. The significant feature is the relatively high, uniform transmittance in the visible and near UV regions (Fig. 3). Light guides 1800 mm long reach values of more than 40% in the region between 280 and 1300 nm.

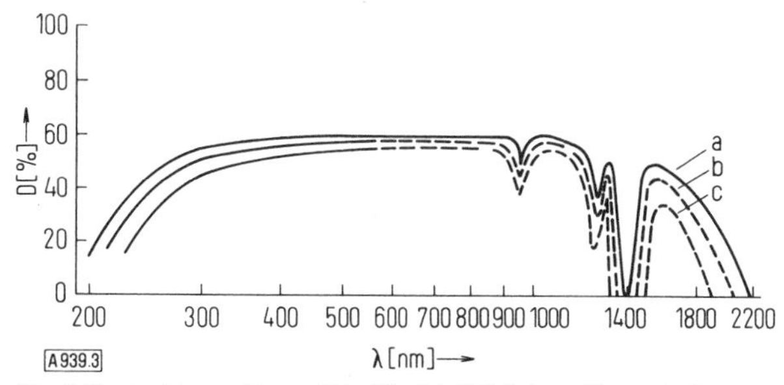

Fig. 3. Spectral transmittance (D) of flexible UV light guides made from silica glass coated with Teflon FEP. Length of light guide: a, 500 mm; b, 1000 mm; c, 1800 mm.

[*] Thanks are due in particular to Dr. *Buyken* for development to the production stage.

The second important property of the UV light guides is their aperture angle. The better the optical contact between the fiber core and the sheath, the smaller are the losses in the numerous total reflections.

Light rays substantially parallel to the optical axis accordingly suffer little loss. Light rays entering the fiber at angles close to the critical angle of incidence α_0 suffer losses that increase with the length of the fiber. Since the angle relationship between the incident light rays and the optical axis remains substantially unchanged, it is found in practice that the aperture angle varies with the length of the light guide. The results of measurements are shown in Figure 4.

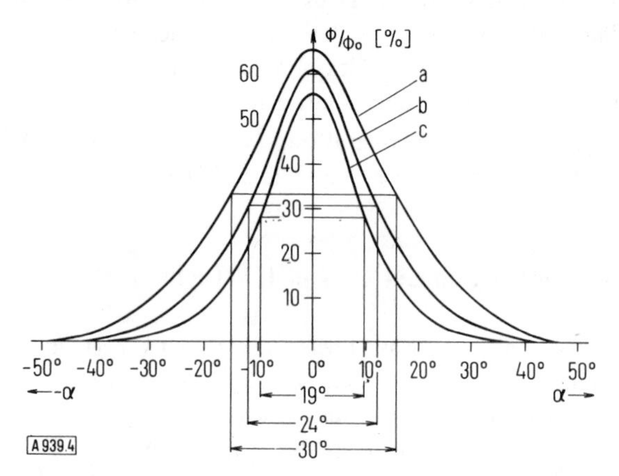

Fig. 4. Angular distribution of the emerging radiation characteristic at 546 nm for silica glass coated with Teflon FEP. Length of light guide: a, 500 mm; b, 1000 mm; c, 1800 mm. The values in the center between the arrows are 2α (50%).

Flexible UV light guides were illuminated uniformly from all directions at the entry end. The intensity distribution of the light emerging from the light guide as a function of the angle with respect to the optical axis shows different opening angles for different lengths if one considers the half widths of the bell-shaped curve. For glass-glass light guides, this value corresponds to the theoretically calculated aperture angle $2\alpha_0$, and is independent of the length of the light guide. This is an indication of the excellent optical contact between fused glasses. The junction between silica glass and the plastic used here is not so perfect, hence the dependence of the aperture angle on the length. One can theoretically expect $2\alpha_0 \approx 68°$; however, the measured aperture given in Fig. 4 is adequate for practical purposes.

Flexible UV light guides possess good stability to UV radiation. The optical transmittance was unchanged by continuous irradiation for a period of several thousand hours with light of wavelength 254 nm at a light density of 1 mW per 4 mm^2 cross section of bundle.

Much greater light intensities can be transmitted at long wavelengths, *e. g.* 320 nm, since solarization effects decrease exponentially with increasing wavelength. X-Radiation in a dose of about 10^5 r causes a decrease in transmittance only in the short-wave region, whereas no adverse effects are detectable in the visible region of the spectrum.

The heat stability of flexible UV light guides depends essentially on their protective tube and on the cement used to bond the fibers at the ends of the bundle. Whereas the light-guide fibers themselves can withstand continuous exposure to temperatures of 250 °C without loss of transmittance, the maximum temperatures to which the ends of the bundles may be exposed are 150—180 °C. On the other hand, no change in transmittance was observed on cooling to −200 °C.

Flexible UV light guides are resistant to water even at temperatures of up to 90 °C. A slight decrease in the optical transmittance occurs only in boiling water.

3.2. Applications of UV Fiber Optics

The uses of UV fiber optics are numerous, and their full extent cannot yet be seen. They considerably facilitate working with UV light. Commercially available UV light sources and the necessary mains equipment are relatively bulky at present. UV light guides enable UV light to be transported to places that are inaccessible to such bulky equipment, and UV light can therefore be used even inside complicated instruments (Fig. 5). Flexible light guides can

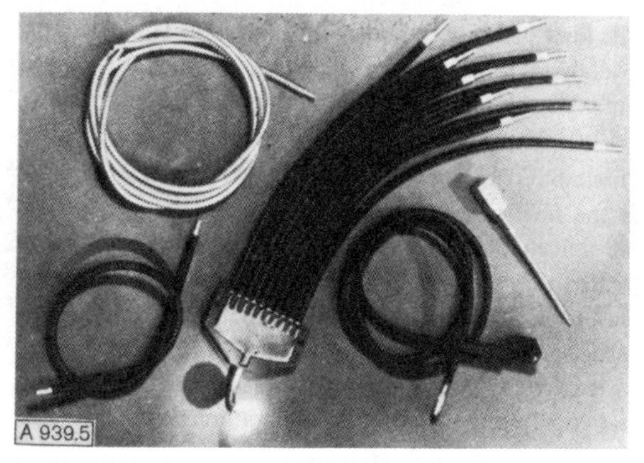

Fig. 5. Different types of light guide.

be led around many corners, like electrical wires, without loss of light. Where relative movement occurs between the UV light source and the receiver, UV light guides form a flexible connection. They serve as secondary light sources in explosion-proof areas or Faraday cages, leading the radiation into these structures easily and with low losses from UV light sources situated outside.

Flexible UV light guides help in many ways in the development of rational automatic analytical equipment for medicine, chemistry, and biochemistry. The transport path of the liquids to be investigated can be considerably shortened, since the light is brought to the point of investigation and the geometry of the UV light sources is of no importance, whereas it was formerly necessary to bring the liquid to the light source.

Wide fields of application arise in medicine, biochemistry, microscopy, and physiology through defined illumination or irradiation of objects with UV light. Certain drugs that accumulate in animal organs and tissues fluoresce on excitation with UV light (Fig. 6). The fluorescence in the visible region of the spectrum can then be observed with other light guides or endoscopes or under the microscope.

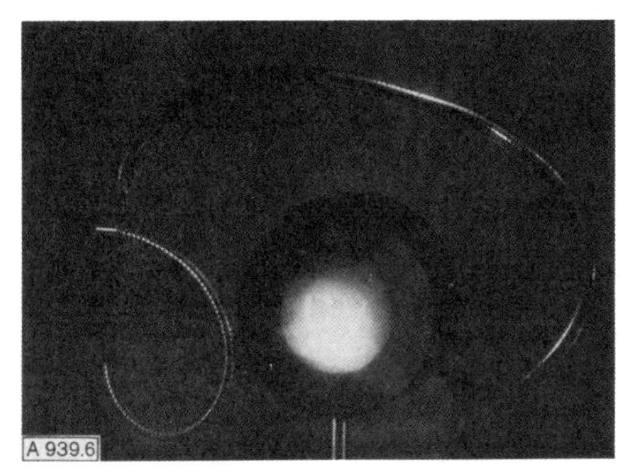

Fig. 6. A flexible light guide transmits UV light and excites fluorescence.

It should be noted that as early as 1964—1966[14] a specific absorption of certain antibiotics (tetracyclines) on carcinomas was observed, and (a decisive point here) the adsorbed antibiotics fluoresced on excitation with UV light. The fiber optics with limited UV transmittance that were available at that time were adequate for these experiments. Following this work, the recognition of cancer with the aid of UV fiber optics has at least become conceivable.

Possibilities for the use of UV light guides also occur in the field of plastics that polymerize and cure under the influence of UV light. Paints and adhesives can be treated very easily with flexible fiber optics. In dentistry, plastics have been developed for application to teeth as varnishes to provide protection lasting several years against caries. With UV light guides, the varnish can be irradiated and cured on the teeth in any dental practice.

Flexible UV light guides combined with light guides for the visible region of the spectrum as two-arm systems form ideal fluorescence light barriers. UV light is led to the point of investigation through one arm. The visible fluorescence light reaches a light-sensitive receiver through the other arm, which cannot transmit reflected UV stray light because of its spectral transmittance. Use is made of this possibility in labeling machines, the designation being printed in code with fluorescent ink and being read automatically. Two-arm UV light guides also give excellent results in automatic letter-sorting machines and stamp-cancelling machines. The actual UV light source is situated far from the reading point, and is accommodated where sufficient space is available in the machine. Numerous UV-conducting components are used in spectrographs, in which the light entering through a slit is led by a cross-section converter to a detector. The cross sections of the light guide can be matched to the geometry of the other optical and electro-optical components (cf. Fig. 6).

The possibilities of UV fiber optics have not yet been exhausted. The applications of UV light can still be greatly extended by means of this new tool for industry and science. Numerous light-guide systems with special dimensions and shapes are available for this purpose.

Received: December 21, 1972 [A 939 IE]
German version: Angew. Chem. *85*, 468 (1973)
Translated by Express Translation Service, London

[1] *A. Jacobsen* and *W. Rimkus*, Feinwerktechnik *71*, 111 (1967).

[2] Information Bulletin No. 7000, Jenaer Glaswerk Schott & Gen.

[3] *N. S. Kapany:* Fiber Optics. Academic Press, New York 1967.

[4] *T. Hager*, *R. G. Brown*, and *B. N. Berick*, SPE (Soc. Plast. Eng.), Sept. *1967*, p. 36.

[5] *A. L. Breen* and *J. R. Breen*, DOS 1494721 (1965), Du Pont.

[6] *P. C. Li*, *D. A. Pontarelli*, *O. H. Olson*, and *M. A. Schwartz*, Amer. Ceram. Soc. Bull. *48*, 214 (1969).

[7] *H. Dislich*, Lecture, Macromolecular Symposium, Helsinki 1972.

[8] *H. Dislich* and *W. Grimm*, unpublished.

[9] *A. J. Burzynski* and *M. R. Eliot*, DOS 1595062 (1965), Owens-Illinois Inc.

[10] *H. Dislich* and *A. Jacobsen*, DBP 1494872 (1965), Jenaer Glaswerk Schott & Gen.

[11] *H. Dislich* and *A. Jacobsen*, Glastech. Ber. *39*, 164 (1966).

[12] *H. Dislich* and *A. Jacobsen*, US-Pat. 3623903 (1966), Jenaer Glaswerk Schott & Gen.

[13] Properties of UV-Fibers, SPIE-16th Annual Technical Meeting, Oct. 1972, San Francisco.

[14] *I. M. Busch* and *W. F. Whitmore jr.*, J. Urol. *97*, 156, 201 (1967).

Low-Loss FEP-Clad Silica Fibers

P. Kaiser, A. C. Hart, Jr., and L. L. Blyler, Jr.

Easy-to-fabricate, low-loss optical fibers are described that consist of a pure fused silica core surrounded with a loosely fitting, extruded, FEP cladding tube. Because of their large numerical aperture (NA), these fibers are particularly well suited for the transmission of the incoherent light emitted by light-emitting diodes. Whereas the losses approached those of the core material for small angle excitation (i.e., 7.6 dB/km at 0.8 μm for a Suprasil 2 fiber of 230-m length), they increased to 14 dB/km for a steady-state NA of 0.3. The measured pulse dispersion of up to 30 nsec agrees well with the expected 24 nsec for a step-index fiber with an NA of 0.3.

Introduction

The development of optical fibers with 2–3 dB/km transmission losses[1–5] has not only enhanced the possibility for an early realization of long distance, high capacity light communication systems, but also has generated interest in utilizing optical fibers for a wide variety of potential short distance applications. Because of their small size, light weight, ability to tolerate small bending radii, and freedom from interference, to name but a few of the many attractive features, optical fibers may eventually be used advantageously in many areas where wire pairs and coaxial cables are used at present.[6] Beyond that, there will be areas of application where only optical fibers hold promise for successful solutions.[7] Consequently, we anticipate a need for many different types of fibers that satisfy particular requirements of specialized applications.

We present here a low-loss fiber that is easy to fabricate and to handle and, because of its large core size and numerical aperture, is well suited for the low capacity, short-to-medium haul transmission of the incoherent light emitted by light-emitting diodes (LED's).[8] The core of the fiber consists of pure fused silica that is the lowest-loss optical-fiber material presently available.[9–11] Because of its low refractive index n of 1.458, there are few glass compositions that qualify as lower index cladding materials.[12] Single material fibers have been developed that circumvent this problem.[13] In an alternate approach, we surrounded the silica core with a perfluoronated-ethylene-propylene (Teflon-FEP 100) cladding tube, whose low refractive index (n_{FEP} = 1.338) results in fibers with potentially large numerical apertures. The choice of FEP as the cladding polymer was made primarily on the basis of its low index—the lowest of any commercial polymer—and its ease of processing in the molten state. Although the losses of FEP were measured to be on the order of 500,000 dB/km as a result of its semicrystalline structure, we found that unclad fibers wound on drums wrapped with FEP film have losses very close to those of freely suspended fibers. This indicated that, at least for low-order mode excitation, the influence of the lossy cladding was negligible. These findings corroborate earlier proposals for the achievement of low-loss optical fibers by means of a reduction of the contact area between a low-loss core and a high-loss cladding by employing, for example, a loosely fitting cladding tube.[14,15]

The concept of using a polymer cladding for silica core fibers is not new. Work is presently underway at various other laboratories to utilize polymer-clad fibers in the optical communication area, and losses in the 70 dB/km range have been obtained for silica fibers coated from a solution with a copolymer of hexafluoropropylene and vinyl fluoride (n = 1.415) (Ref. 16). Our own preliminary work with other solution applied polymer claddings produced losses of similar magnitude, primarily due to interfacial scattering.

FEP-clad silica fibers are also being produced as light guides for the uv spectral region.[17] Losses of 360 dB/km at 0.546 μm have been achieved using uv-grade fused silica. These fibers are fabricated by coating with an aqueous dispersion of FEP particles, which are subsequently fused together through application of heat. This process leads to an intimate contact between the lossy cladding and silica core.

All authors are with Bell Laboratories, New Jersey: P. Kaiser is at Crawford Hill, Holmdel, 07733; and the other authors are at Murray Hill, 07974.

Received 14 June 1974.

Reprinted with permission from *Appl. Opt.*, vol. 14, pp. 156–162, Jan. 1975.

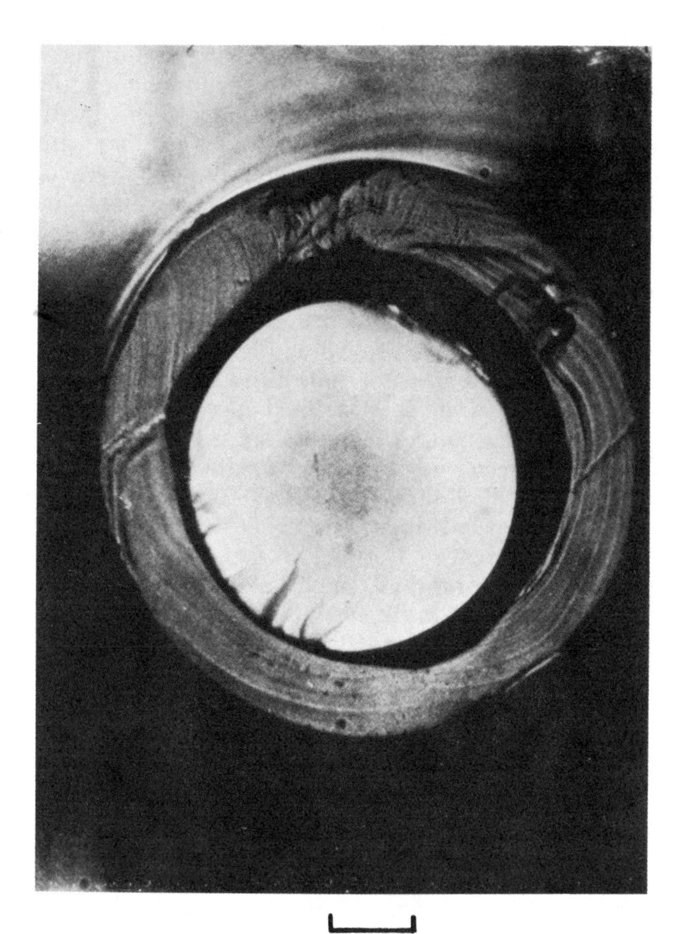

50 μm

Fig. 1. Cross sectional picture of FEP-clad silica fiber.

Furthermore, it is difficult to eliminate voids and other inhomogeneities near the polymer–silica interface, which leads to scattering losses. The approach taken in this work of loosely surrounding the fiber with an FEP cladding tube by a continuous extrusion process avoids these problems and results in substantially lower losses.

Fiber Fabrication and Evaluation

The silica fibers were drawn with an oxy-hydrogen torch from about 7-mm diam preform rods. The pristine fibers were immediately passed through the crosshead of a simple ram extruder mounted on the fiber drawing machine, where they were surrounded with a loosely fitting FEP tube (Fig. 1). A tubing die similar to those commonly used for jacketing wire and cable was employed.

In the tubing technique, the fiber to be coated passes through a core tube that extends to the die exit as shown in Fig. 2. Sufficient clearance is allowed to preclude contact between the fiber and core tube. The polymer is extruded completely independently as a large diameter tube about the fiber and is drawn down, while still molten, to its final size outside the die. As the molten tube is drawn in the ambient air, it begins to solidify as its temperature falls below the melting point (265°C for FEP). By adjusting such design and processing variables as die dimensions, melt temperature, extrusion rate, and line speed, the final dimensions of the polymer tube may be controlled. Thus it is possible, in principle, to form the polymer tube loosely around the fiber without any contact whatsoever until after solidification occurs, thereby eliminating the chance of damaging the polymer surface. In practice, lateral vibrations of the fiber produce contact at the end of the drawdown zone, typically 1–3 cm from the die. Interfacial damage is minimized by employing extrusion conditions that insure a sufficiently solid tube wall at that point.

To produce the fibers reported on here, die dimensions were chosen to result in a drawdown ratio of approximately 100 to 1. Melt temperatures ranging from 310°C to 330°C were found most suitable. Lower temperatures produced a rough tube surface due to viscoelastic melt response, and higher temperatures resulted in breaks of the molten tube during drawdown. Typical fibers produced for study had wall thicknesses of about 50 μm, core diameters of 175 μm, and over-all fiber diameters ranging from 250 μm to 300 μm. The process is not however, limited to these dimensions. For example, we have fabricated FEP-clad silica fibers with core diameters ranging from 75 μm to 500 μm.

Because the silica fibers were immediately protected with the plastic tubes, they maintained a high

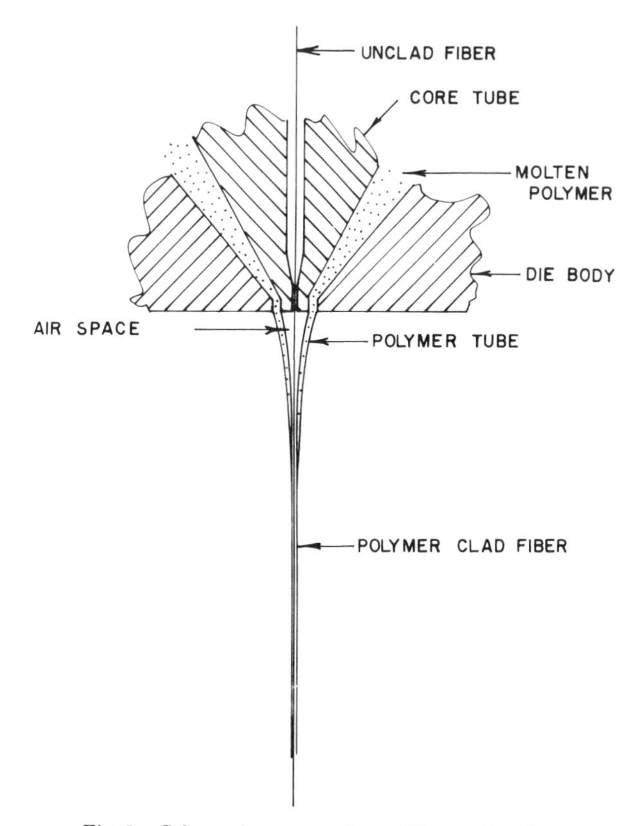

Fig. 2. Schematic cross section of fiber tubing die.

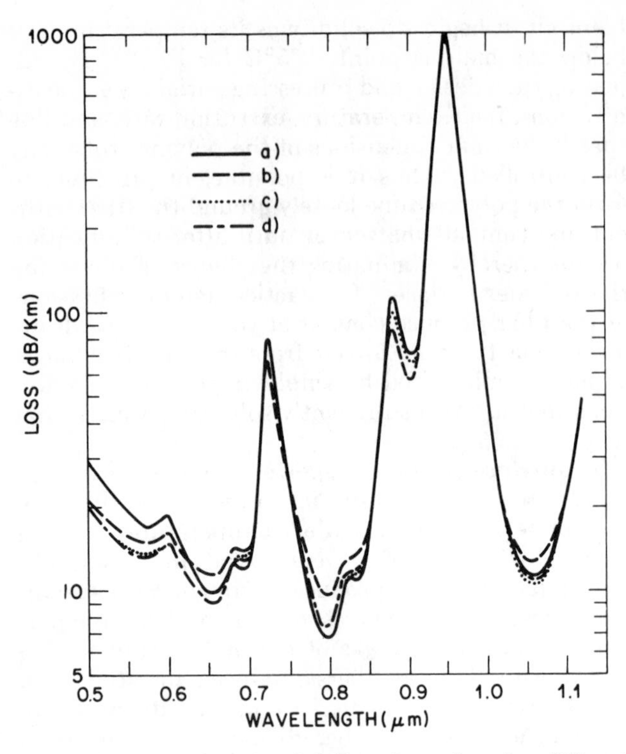

Fig. 3. Spectral losses of unclad fibers drawn from four different batches of Suprasil 2. (a) L = 17 m; (b) L = 57 m; (c) L = 60.5 m; (d) L = 66 m with fiber remaining on FEP-backed drum. Samples (a) to (c) were freely suspended.

tensile strength and were easy to handle. Breaking strengths between 70 kg/mm^2 and 140 kg/mm^2 (100,000 psi and 200,000 psi) were measured with several 1-m long fiber sections.

Because of its low transmission losses on the order of 10 dB/km and less at several wavelengths of interest for optical communication purposes,[18] and because it is readily available at reasonable cost, we chose Suprasil 2, a high water content synthetic vitreous silica, as core material for most of the work. The material losses of four different batches of Suprasil 2 as obtained through unclad-fiber measurements are shown in Fig. 3. As in the case of Suprasil 1 (Ref. 10) or any other synthetic vitreous silica made by the same process, the intense absorption bands visible in the loss spectrum of Suprasil 2 are due to harmonics and combinational bands of the 2.7-μm OH vibration. Practically identical minimum losses of about 9.5 ± 0.5 dB/km at 0.66 μm, 7.0 ± 0.5 dB/km at 0.8 μm, and 10.8 ± 0.5 dB/km at 1.06 μm were obtained with three samples. The slightly higher losses of the fourth sample are believed to be caused by surface contamination and associated scattering losses.

It should be emphasized that the loss curve of one of the samples [curve (d) in Fig. 3] was obtained with a fiber that was wound on a drum covered with a thin film of FEP, whereas in all other cases the fibers were freely suspended. We were thus able to demonstrate that, particularly for low-order mode excitation of the unclad fiber, the FEP backing may not introduce additional losses. These findings are substantially facilitating the unclad fiber measuring technique[19]: it is now no longer necessary to suspend long lengths of fiber, and the measurement can be performed with a protective atmosphere surrounding the fiber with an ensuing reduction in dust accumulation. Second, higher sensitivities and accuracies can now be achieved with this technique since the length of the unclad fiber is no longer dictated by limited laboratory space.

The spectral losses of a 230-m long FEP-clad Suprasil 2 (SS2) fiber are shown in Fig. 4. They, as well as the unclad fiber losses shown in Fig. 3, were measured with a previously described apparatus that employs a Xe arc lamp in conjunction with 35 interference filters covering the spectrum between 0.5 μm and 1.12 μm.[19] For small angle excitation, the losses of the plastic-clad fiber approached the unclad fiber losses of the core material. In fact, when a 300-m long fiber (the original length of the above 230-m fiber) was excited with the collimated light of a He–Ne laser, the resulting 9.3 dB/km loss measured at 0.6328 μm was lower than any SS2 unclad fiber loss measured previously at this wavelength.

As can be seen in Fig. 4, progressively higher losses are obtained when the cone angle of the injected beam is increased. The modal power distribution propagating in the fiber was measured by scanning the far-field at 230-m (far-end) and 1.5-m (near-end) points. (Fig. 5). The numerical apertures listed as parameters in Fig. 4 were computed from the 10-dB widths of the near-end patterns. Whereas initially

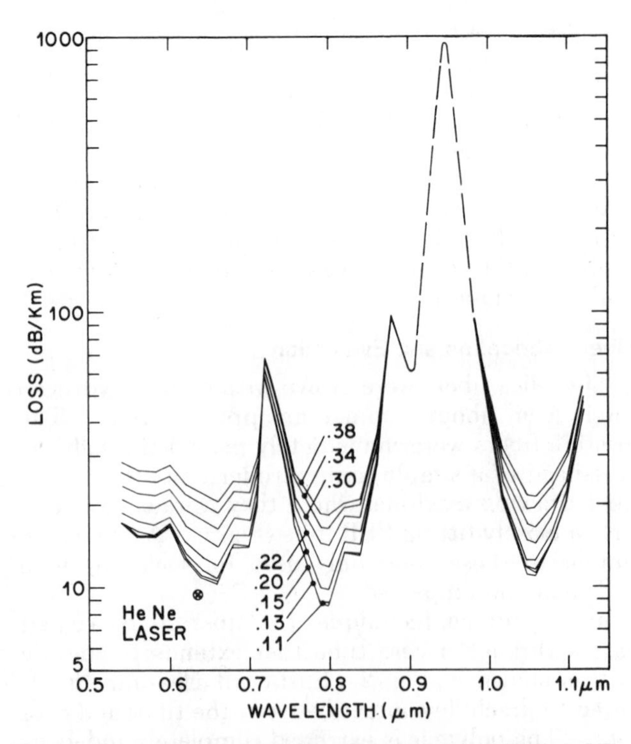

Fig. 4. Spectral transmission losses of a 230-m long FEP-clad Suprasil 2 fiber as a function of the numerical aperture of the injected beam. Also shown is the 9.3-dB/km loss obtained for the original 300-m length of this fiber for He–Ne laser excitation.

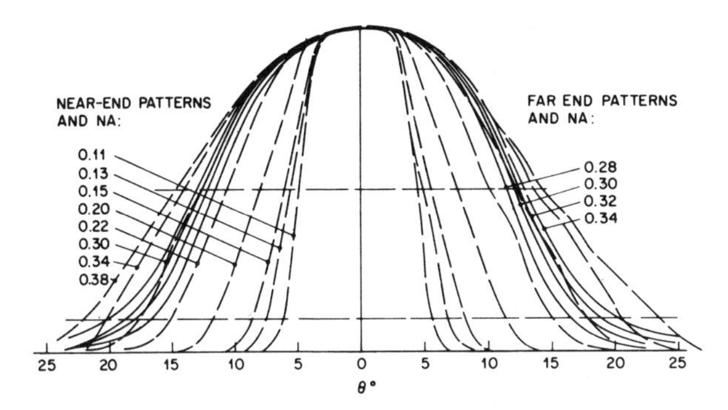

Fig. 5. Near- and far-end radiation patterns of the 230-m long Suprasil 2 fiber for the different launch numerical apertures and spectral losses shown in Fig. 4 (compare with Table I).

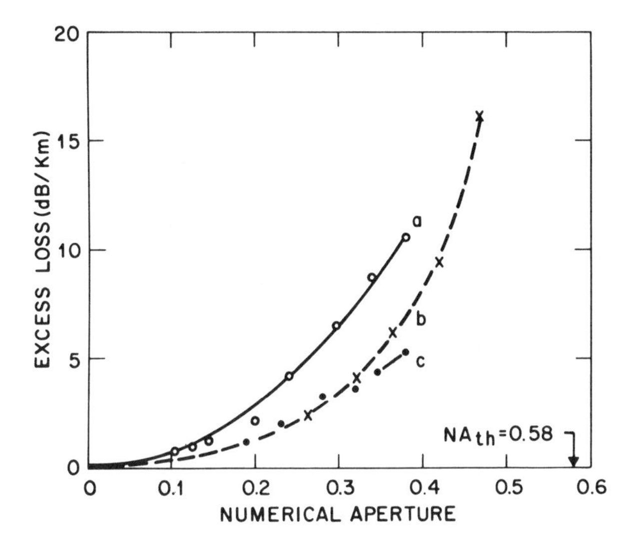

Fig. 6. Increase of the interface- and cladding-related excess losses as function of the 1/10-power-point equivalent launch numerical aperture for FEP-clad silica fibers with (a) Suprasil 2 core, 230-m long (at 0.8 μm); (b) Vitreosil core, 190-m long (at 0.7 μm); and (c) TO8 core, 94-m long (at 0.7 μm).

narrow beams widened due to mode coupling caused by an imperfect interface between the silica core and the FEP tube, initially large angle beams narrowed along the fiber due to preferential attenuation of higher modes because of a lossy cladding and interface.[20,21] The mode pattern that changes least along the fiber represents the steady-state distribution where both mechanisms are in equilibrium. The effective numerical aperture associated with the steady-state condition of the 230-m long FEP-clad SS2 fiber was about 0.30 (= NA_{eff}), and the associated minimum steady-state losses were 14 dB/km at 0.8 μm. This value is about 6.5 dB/km higher than the unclad fiber losses of the SS2 core. It appears that random diameter variations of the core, induced by fluctuating thermal gradients in the flame of the drawing torch, contributed a portion of this excess loss. This aspect is presently under study.

Table I summarizes the near-end and associated far-end NA's, together with the transmission losses at 0.8 μm. The excess losses α as a function of the near-end NA are shown in Fig. 6, curve (a), and can be approximated by

$$\alpha - \alpha_0 = \alpha_1 \, \mathrm{NA}^2, \tag{1}$$

where α_0 represents the loss coefficient of the core material (= 1.75 km^{-1}, corresponding to and extrapolated core material loss of 7.6 dB/km), and α_1 amounts to 16.6 km^{-1}. Since the power that can be coupled from an LED into the fiber increases quadratically with the NA,

$$P_i = C \, \mathrm{NA}^2, \tag{2}$$

where C depends on the radiance of the LED and the area common to the fiber and the diode, the output power for a fiber of length L is given by

$$P_o = C \, \mathrm{NA}^2 \exp[-(\alpha_0 + \alpha_1 \mathrm{NA}^2)L]. \tag{3}$$

Maximum power transmission occurs at an optimum numerical aperture,

$$\mathrm{NA}_{\mathrm{opt}} = (\alpha_1 L)^{-1/2}, \tag{4}$$

which becomes smaller with increasing fiber length. Although one has to take into account that α_1 itself may be length-dependent, it will be generally advantageous for fibers with such transmission characteristics to match the width of the radiation pattern emitted by an LED to the parameter α_1 of a fiber of length L unless the fiber end can essentially be brought in contact with the light-emitting area of the diode. This can, for example, be accomplished by mounting a tiny lens or glass bead in front of the light-emitting area of the LED.[22,23]

Dispersion measurements were performed with 2-nsec wide pulses from a GaAs injection laser.[24] Oscilloscope traces of the input and output pulses obtained for the 230-m long fiber for two different excitation conditions are shown in Fig. 7. Whereas the

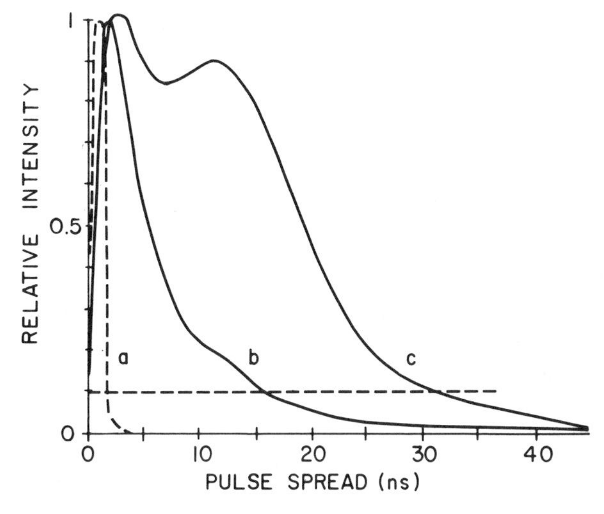

Fig. 7. Pulse dispersion of the 230-m long FEP-clad Suprasil 2 fiber measured with a GaAs laser at 0.9 μm: (a) input pulse; (b) and (c) output pulses for different launch conditions.

Table I. Near- and Far-End Numerical Apertures and Total Losses of a 230-m Long FEP-Clad Suprasil 2 Fiber

NA_{NE}	NA_{FE}	Loss (dB at 0.8 μm)
0.11	0.28	1.93
0.13	0.28	1.98
0.15	0.28	2.02
0.20	0.28	2.23
0.22	0.30	2.71
0.30	0.32	3.24
0.34	0.34	3.75
0.38	0.34	4.16

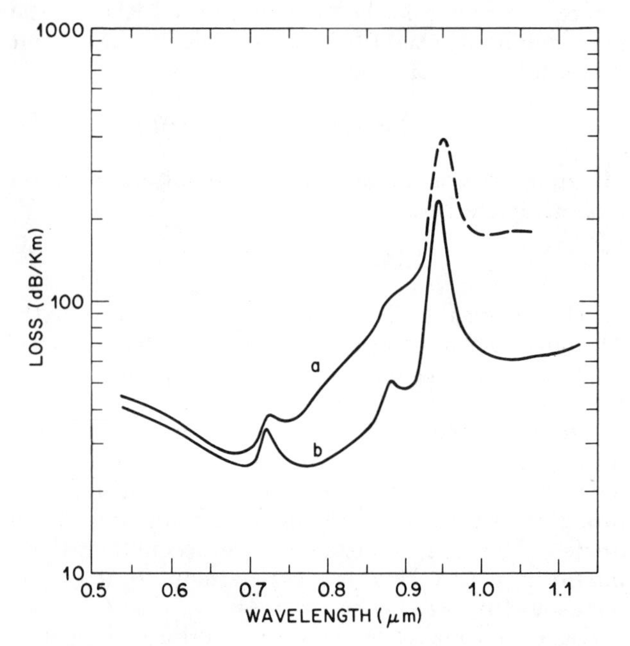

Fig. 8. Spectral loss curves of FEP-clad silica fibers with (a) Vitreosil core, 194-m long; and (b) TO8 core, 94 m long for lower-order mode excitation.

10-dB pulse width for coaxial alignment amounted to 16 nsec, it increased to 30 nsec when the fiber end was intentionally tilted to emphasize the launching of higher order modes. If one neglects mode coupling, the expected pulse spread of a step-index fiber is

$$\tau = NA^2 L/2nc, \tag{5}$$

(c = velocity of light in free space) and, for an NA of 0.3 and with L = 230 m, is 24 nsec. The experimental data are in good agreement with this prediction. The radiation pattern launched into the fiber for GaAs laser excitation was not measured but was estimated to be relatively large due to the large angle of divergence of the laser's radiation in the plane perpendicular to the junction.

As becomes apparent from Fig. 5 and Table I, the output patterns of the 230-m long Suprasil 2 fiber were relatively insensitive to changes of the input beam divergence. This is typical if mode coupling is present and, in turn, tends to reduce the dispersion.[25] However, possibly as a result of the fiber length and coupling length[21] being comparable, no significant reduction in the dispersion was noticed.

The maximum permissible data rate corresponding to a pulse spread τ of 24 nsec for the above 230-m long fiber is roughly given by

$$B_{max} = 1/\tau \cong 0.87/NA^2 L, \text{ in Mb/sec for } L \text{ in km} \tag{6}$$

and amounts to about 40 Mb/sec.

For certain short distance applications it may be economically attractive to use commercial grade fused quartz as the core material.[26] TO8 is particularly well-suited for this purpose, since it has reproducibly low losses on the order of 25 dB/km around 0.7–0.8 μm, and 60–70 dB/km in the 1.06-μm region.[10] The losses of Vitreosil range between 30 dB/km and 100 dB/km at 0.7 μm, and between 200 dB/km and 600 dB/km around 1.06 μm.[27] The small angle excitation loss spectra of FEP-clad fibers made from these materials are shown in Fig. 8. The excess losses, plotted as function of the near-end NA in Fig. 6, are nearly identical for both fibers and are only half those of the SS2 fiber discussed earlier. This improvement is attributed to a superior core–cladding interface, the reasons for which are not yet known. The lower interface losses are associated with a reduction in mode coupling as can be seen in Fig. 9, where the normalized far- and near-end radiation patterns of the 190-m long Vitreosil fiber are shown for different launch NA's. Whereas mode patterns corresponding to near-end NA's of up to 0.35 are essentially transmitted without change, beams associated with larger NA's gradually narrow due to the selective attenuation of higher order modes.

The $1/10$-power-point near- and far-end NA's, together with the corresponding total transmission losses of the FEP-clad Vitreosil and TO8 fibers are listed in Tables II and III. The losses were generally measured with the fibers remaining on the drums on which they were originally wound. Since this tended to increase the contact area between core and cladding, lower transmission losses and reduced mode coupling were typically observed when the fibers were removed from the drums.

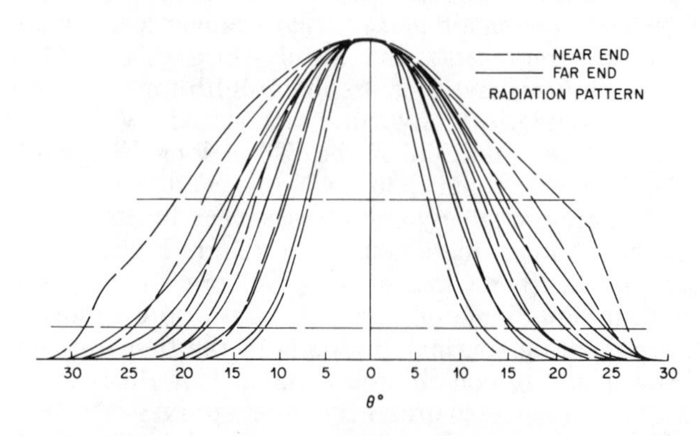

Fig. 9. Near- and far-end radiation patterns of the 190-m long FEP-clad Vitreosil fiber for different launch numerical apertures (compare with Table II).

Table II. Near- and Far-End Numerical Apertures and Total Losses of a 190-m Long FEP-Clad Vitreosil Fiber

NA_{NE}	NA_{FE}	Loss (dB at 0.7 μm)
0.20	0.21	5.43
0.26	0.27	5.64
0.32	0.31	5.97
0.36	0.35	6.35
0.42	0.39	6.95
0.47	0.43	8.25

Table III. Near- and Far-End Numerical Apertures and Total Losses of a 94-m Long FEP-Clad TO8 Fiber

NA_{NE}	NA_{FE}	Loss (dB at 0.7 μm)
0.19	0.16	2.28
0.23	0.19	2.37
0.28	0.23	2.48
0.32	0.26	2.51
0.38	0.31	2.67

Summary and Discussion

Low-loss optical fibers can be made by an in-line extrusion of a low refractive index FEP cladding tube around a pure fused silica fiber. For small NA excitation, the transmission losses of the clad fibers were identical to the unclad fiber losses of the core material. For higher-order mode excitation, the excess losses increased quadratically with the NA of the mode spectrum launched, and about 5 dB/km excess losses were measured for NA's on the order of 0.35. Maximum NA's achieved approached the theoretical value of 0.58 of the FEP–silica interface.

As a consequence of these properties, and since coupling between modes was found to be small in some of the fibers evaluated, FEP-clad silica fibers may be employed in several ways.

(1) *Short distance applications:* because of their large NA, FEP-clad silica fibers are ideal for the transmission of light emitted by LED's. Assuming that the fiber can be filled to its modal capacity (NA = 0.58), the maximum data rate is 2.6 Mb/sec km. This allows, for example, the transmission of 26 Mb/sec over 100-m long fiber lengths encountered in on-premises interconnections. Since the associated cladding-related excess losses are on the order of 20 dB/km (see Fig. 6), the range of the fibers is limited to 1-km distances at most, particularly if commercial grade fused quartz with an intrinsic loss of about 25 dB/km is used as core material. Because TO8 has its highest transparency around 0.7 μm and 0.8 μm, systems employing this material should preferably operate in this wavelength regions.

(2) *Short distance, high capacity applications:* by using LED's with focusing optics (which is typically associated with a loss of power coupled into the fiber) or preferably laser sources, one can limit the NA of the transmitted beam to values below, say, 0.15, and thus achieve data rates of up to 40 Mb/sec km. Because of the reduced excess losses due to smaller NA mode spectra, the range of this fiber under these operating conditions is somewhat larger than that under (1). Now, data rates of about 400 Mb/sec can be transmitted over a 100-m long fiber. Because of the relatively low cost of commercial grade fused quartz and because of the simplicity of the fabrication process, FEP-clad TO8 or Vitreosil fibers are very attractive for the above types of applications.

(3) *Medium range applications:* the range of the fibers can be substantially increased if either high-OH content synthetic silicas like Suprasil 2, or, preferably, low-OH content synthetic silicas like Suprasil W1, W2 or Spectrosil WF are used as the core material. For small angle excitation, transmission losses in the 5 dB/km range are expected to be achievable with Spectrosil WF-cored fibers, whose material losses are about 2 dB/km.[9] The excess losses introduced by the FEP cladding for a 0.15 NA should be about 1–2 dB/km. The corresponding data rate is again 40 Mb/sec km. The preferred operating wavelength coinciding with the region of highest transparency for low-OH content silicas exists in the 1.0–1.2-μm range. On the other hand, only slightly higher losses of 3–4 dB/km have been measured around 0.8 μm and 0.9 μm.

The experimental assistance of H. W. Astle was very much appreciated. We would like to thank S. Gottfried for performing some of the spectral loss measurements; and we are indebted to L. G. Cohen, who permitted us to use unpublished dispersion data. We are grateful to A. D. Pearson and F. V. DiMarcello for the use of their oxy-hydrogen torches at various times.

References

1. D. B. Keck, R. D. Maurer, and P. C. Schultz, Appl. Phys. Lett. **22,** 307 (1973). The fabrication of a 2-dB/km fiber was announced by P. C. Schultz at the Annual Meeting of the American Ceramic Society, Cincinnati, Ohio, 30 April–2 May 1973.
2. P. Kaiser and H. W. Astle, Bell Syst. Tech. J. **53,** 1021 (1974).
3. W. G. French, J. B. MacChesney, P. B. O'Connor, and G. W. Tasker, Bell Syst. Tech. J. **53,** 951 (1974).
4. J. B. MacChesney, P. B. O'Connor, F. V. Dimarcello, J. R. Simpson, P. D. Lazay, and W. G. French, Proc. 10th Internat. Congress on Glass, 8–13 July 1974, Kyoto, Japan.

5. F. W. Dabby, D. A. Pinnow, L. G. Van Uitert, F. W. Ostermayer, M. Savi, and I. Gamlibel, to be published.
6. S. E. Miller, E. A. J. Marcatili, and T. Li, Proc. IEEE **61,** 1703 (1973).
7. R. A. Andrews, A. F. Milton, and T. G. Giallorenzi, IEEE Trans. Microwave Theory Tech. **MTT 21,** 763 (1973).
8. P. Kaiser, A. C. Hart, L. L. Blyler, and H. W. Astle, Spring Meeting Optical Society of America, Washington, D. C., 22–24 April 1974.
9. P. Kaiser, Appl. Phys. Lett. **23,** 45 (1973).
10. P. Kaiser, A. R. Tynes, H. W. Astle, A. D. Pearson, W. G. French, R. E. Jaeger, and A. H. Cherin, J. Opt. Soc. Am. **63,** 1141 (1973).
11. D. A. Pinnow, T. C. Rich, F. W. Ostermayer, and M. DiDomenico Jr., Appl. Phys. Lett. **22,** 527 (1973).
12. L. G. Van Uitert, D. A. Pinnow, J. C. Williams, T. C. Rich, R. E. Jaeger, and W. H. Grodkiewicz, Mater. Res. Bull. **8,** 469 (1973).
13. P. Kaiser, E. A. J. Marcatili, and S. E. Miller, Bell Syst. Tech. J. **52,** 265 (1973).
14. E. A. J. Marcatili, BTL; patent applied for.
15. S. E. Miller, BTL; unpublished work.
16. Y. Suzuki and H. Kashiwagi, Appl. Opt. **13,** 1 (1974).
17. H. Dislich and A. Jacobsen, Angew. Chem. **12,** 439 (1973).
18. P. Kaiser, J. Opt. Soc. Am. **64,** 475 (1974).
19. P. Kaiser and H. W. Astle, J. Opt. Soc. Am. **64,** 469 (1974).
20. D. Marcuse, *Theory of Dielectric Optical Waveguides* (Academic Press, New York, 1974), p. 173.
21. D. Gloge, Bell Syst. Tech. J. **52,** 801 (1973).
22. C. A. Burrus, BTL; unpublished work.
23. W. S. Holden, BTL; unpublished work.
24. L. G. Cohen, BTL; unpublished work.
25. E. L. Chinnock, L. G. Cohen, W. S. Holden, R. D. Standley, and D. B. Keck, Proc. IEEE **61,** 1499 (1973).
26. J. E. Goell, BTL; unpublished work.
27. P. Kaiser, BTL; unpublished work.

Laser Drawing of Optical Fibers

U. C. Paek

A system consisting of a cw CO_2 laser and an ellipsoidal reflector was developed in order to draw fused silica fibers from bulk material. The system was used to obtain unclad fused silica fibers (Suprasil 2) having a total transmission loss of less than 100 dB/km at 6328 Å. Preforms (rod in tube) have also been drawn to fibers of long lengths with diameters as small as 10 μm and variations within 5%. Analysis of the fiber drawing process established relations that allow determination of the design parameters for fiber drawing systems. Experimental results are given that support these relationships.

Introduction

Low loss glass fibers with less than 10 dB/km loss have been considered attractive for use in a variety of optical transmission systems. Fibers with loss as low as 4 dB/km at 1.06-μm wavelength have been reported.[1] There appears to be an intense effort to reduce fiber loss further and develope processes for the economical and reliable manufacture of long fibers. Such efforts include the development of low loss materials and preform structures and the techniques for drawing the preforms into long fibers without contaminating the fiber or otherwise increasing its loss.

We describe in this report the initial work on the development of a versatile fiber drawing process with a CO_2 laser. It is expected that the process will be suitable for a variety of preform configurations and for fused silica as well as lower softening point glasses. The drawing process has been used to draw unclad fused silica fiber (Suprasil 2) with a loss of less than 100 dB/km at 6328 Å. This loss value was the limit of our detection for the length of fiber measured and lower loss is probable. For these measurements a mode-stripping technique was used.[2,3] The fiber diameter was 100 μm.

Many heating methods have been used for fiber drawing, including the oxyhydrogen torch, induction heating, resistance heating, and plasma heating. A CO_2 laser was chosen as the heat source to produce fiber,[4] since it has unique capabilities. The main advantage in using the laser is that its energy can be easily controlled to quickly raise the material's temperature to the required point while not introducing impurities to the workpiece or fiber during processing. This may be of major importance, since suitable fiber core materials[5,6] such as Suprasil W-1 (a product of Amersil, Inc.) must have metallic impurities less than 1 ppm and an OH content of approximately 5 ppm. The rapid energy control available with CO_2 lasers allows for fiber diameter feedback control systems.

A laser system capable of drawing fibers from numerous low loss materials with softening points ranging to above 1580°C was designed and built. It consists of a CO_2 laser (50 W), an ellipsoidal reflector, and a mechanism capable of feeding material and taking up a fiber.

The ratio of these two parameters (feeding and takeup speed) determines the size of a fiber in drawing operation. As yet there is no agreement on the fiber dimension to be used in communication systems.[7,8]

Apparatus

As shown in Fig. 1, a beam from the 50-W CO_2 laser (Coherent Radiation Model 42) is collimated by a beam expander and focused with a germanium lens. Between the lens and reflector, a jet nozzle blows air into the cavity of the reflector along the beam axis. This stream not only prevents any silica vapor generated at the focused area from depositing on the reflecting surface but also provides a controlled atmosphere.

An ellipsoidal reflector, typically having a major axis of 13.6 cm and a minor axis of 5 cm, is cut perpendicular to the major axis at one focal point F_1. Two 0.65-cm-diameter holes at the top and bottom are positioned so that a line drawn through their centers is perpendicular to the major axis and passes through focal point F_2. The reflector is mounted and adjusted so that its major axis is aligned in the

The author is with Western Company, Inc., Engineering Research Center, P.O. Box 900, Princeton, New Jersey 08540.

Received 7 December 1973.

Reprinted with permission from *Appl. Opt.*, vol. 13, pp. 1383–1386, June 1974.

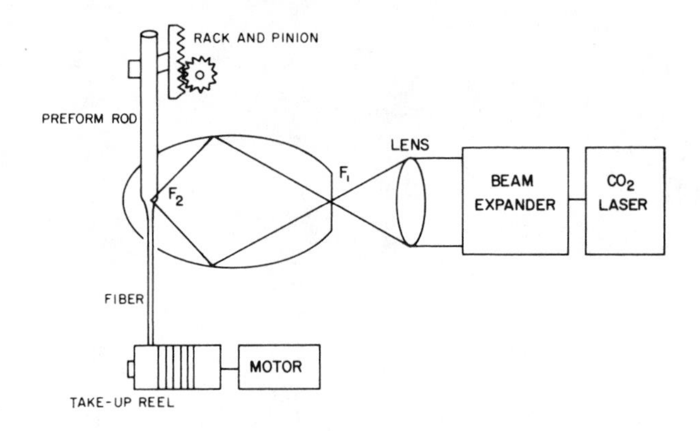

Fig. 1. Schematic fiber drawing system.

beam axis. Focal point F_1 is easily adjusted to coincide with the focused spot of the lens. Accordingly, the laser energy is recollected at focal point F_2, where the workpiece is introduced. The reflecting surface is made of electroplated and vacuum-deposited gold.[9]

The fiber drawing operation can be briefly described by the following procedures (Fig. 1). A fused silica or glass rod (or rod-and-tube combination) is fed into the top port of the reflector (with rack and pinion) and through the bottom port until it extends to about 2.5 cm below the port. The laser is then turned on and a portion of the rod at point F_2 is heated to the softening point. The end of the rod that extends out of the reflector is pulled down and secured on the takeup reel. The fiber is then continuously drawn by rotation of the takeup reel. A photograph of the setup is shown in Fig. 2, and includes the CO_2 laser, reflector, drawing machine, and display system.

Laser Intensity Distribution Around The Workpiece

As shown in Fig. 3, one end of the ellipsoidal reflector is cut at the focal plane located at $x = -c$ perpendicular to the major axis to introduce the laser beam into the reflector cavity with a lens. The beam that is refocused through reflection will contribute the most heat to the workpiece. On the other hand, the beam that strikes the rod directly is substantially defocused and will not appreciably affect the material, and it can therefore be neglected when calculating the intensity distribution around the rod.

In order to derive the intensity distribution, one can assume that a Gaussian beam $I_0 \exp(-w^2/h^2)$ is focused at the focal point F_1 by a lens (with focal length f) where h is a beam radius and w is the radial variable that denotes distance from the propagating axis. The beam arriving at the differential arc ds from the point F_1 will be reflected with an amount of R (reflectivity in percent) and reach the point F_2. It can be seen in Fig. 3 that the arc ds forms a viewing angle $d\phi$ with an origin of F_2. From Fig. 3 we can relate I_1 to the beam reflected from ds towards F_2, where I_1 is defined as intensity distribution over $(d_1/2)\cdot d\phi$, the differential arc of the rod.

$$RfI_0(r_2/r_1)d\phi\sec^2\theta\exp\{-(f/h)^2[y/(x+c)]^2\} = I_1(d_1/2)d\phi,$$

where $r_1 = [(c+x)^2 + y^2]^{1/2}$, $r_2 = [(c-x)^2 + y^2]^{1/2}$ and $w/h = (f/h)[y/(x+c)]$. The ellipse is described by $(x^2/a^2) + (y^2/b^2) = 1$, where a and b represent semimajor and minor axes. Therefore,

$$(I_1/I_0) = (2Rf/d_1)\{[(1-x_1)^2 + y_1^2]^{1/2}[(1+x_1)^2 + y_1^2]^{1/2}/(1+x_1)^2\}\exp\{-(f/h)^2[y_1^2/(1+x_1)^2]\}, \quad (1)$$

where $x_1 = x/c$ and $y_1 = y/c$. Equation (1) indicates that a Gaussian beam with its amplitude I_0 is refocused on the surface of the rod through reflection.

Figure 4(a) shows the upper half portion of the intensity distribution over the rod, indicated by the shaded area in the center of the polar coordinate, since Eq. (1) is symmetric about the x axis. In other words, when the rod is placed right at focal point F_2, it will be heated with the intensity distribution as indicated in Fig. 4(a). Thus, the portion between 35° and 165° is expected to be much more softened

Fig. 2. Fiber drawing setup.

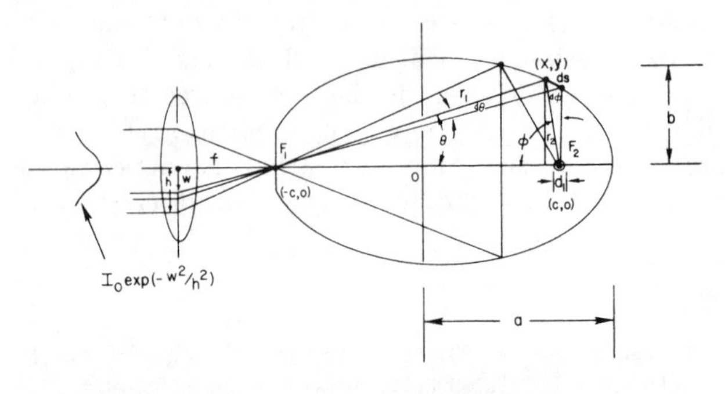

Fig. 3. Description of a beam from a lens, refocused by a reflector to a rod.

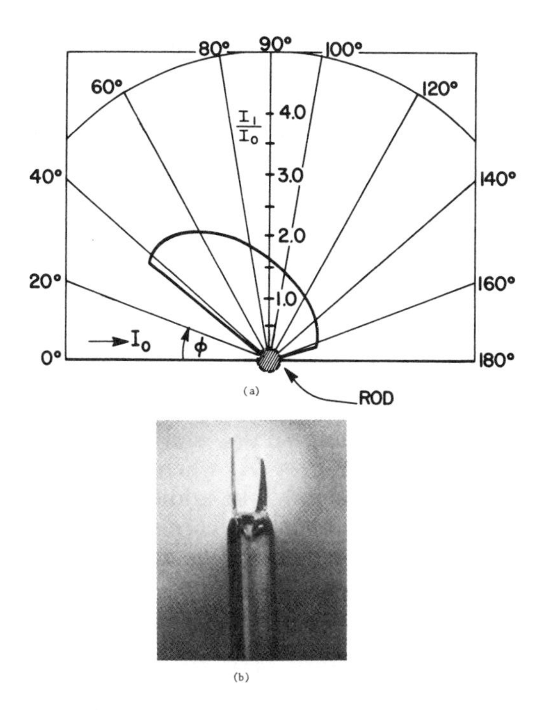

Fig. 4. (a) Angular distribution of laser intensity around the surface of a rod by a reflector (all data used for calculations are $h = 0.5$ cm, $f = 2.5$ cm, $a = 6.8$ cm, $b = 2.5$ cm, and $R = 0.95$). (b) Effect on the rod due to angular distribution (a).

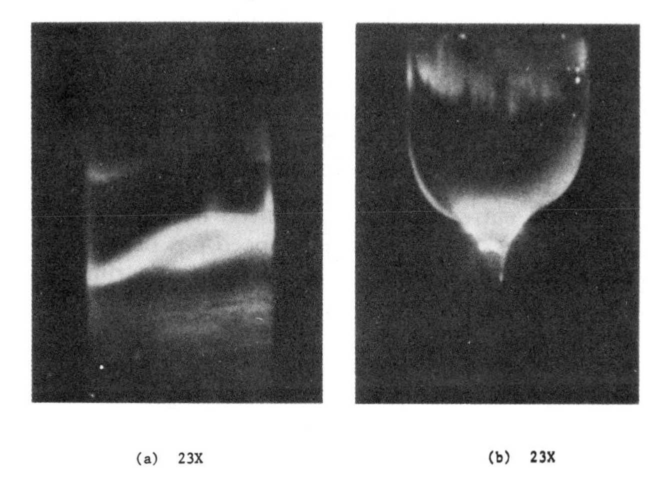

Fig. 5. (a) Rod is heated around its periphery. (b) Heating zone during drawing processes.

than the front and back portion (between 0°–35° and 165°–180°). This nonuniform melting has been observed, as shown in Fig. 4(b). However, the heat distribution can be made essentially uniform around the periphery by a slight defocusing adjustment of the position of focal point F_2 relative to the rod while the heated portion of the rod is observed through the 0.32-cm inspection hole in the reflector surface or a TV monitor display. One finally obtains the result shown in Fig. 5.

When uniform heating is achieved, as in Fig. 5, a fiber is continuously drawn from a preform with its size dependent on the ratio between the feeding and takeup speeds. By applying the mass conservation law (Fig. 6) while assuming identical densities for preform and fiber, we can establish the following relation:

$$V_f d_1^2 = V_t d_2^2 \quad \text{or} \quad V_t = V_f (d_1/d_2)^2, \tag{2}$$

where d_1 is the diameter of the rod, d_2 is the desired fiber diameter, V_f is the feeding speed, and V_t the takeup speed. From Eq. (2), one can easily compute the pulling rate corresponding to a feeding speed for a given ratio of fiber to preform.

From 0.1-cm or 0.2-cm rods, fibers were drawn down to less than 10 μm in diameter. It was found that the diameter variation was within 5%. Larger diameter fibers have also been drawn, with diameters exceeding 250 μm.

Power Requirements

When the rod at the focused area F_2 reaches the softening point, the fiber is drawn with an appropriate feeding speed V_f for a given laser power P. To find correlation between V_f and P, it is necessary to write a governing equation for the model shown in Fig. 7. In this case, the heat leaves the shaded area due to convection and thermal radiation while the laser beam continuously irradiates the area to maintain the softening temperature T_s. Then the heat conduction equation[10] for steady state can be written as

$$-V_f(\partial T/\partial S) = \kappa(\partial^2 T/\partial S^2) - (4HT/\rho C_p d_1), \tag{3}$$

where $\kappa = k/\rho Cp$, k is thermal conductivity, ρ is density, Cp is specific heat, d_1 is diameter of a rod, and S the distance defined in Fig. 7. The total heat transfer coefficient is defined as

$$H = h_c + h_r. \tag{4}$$

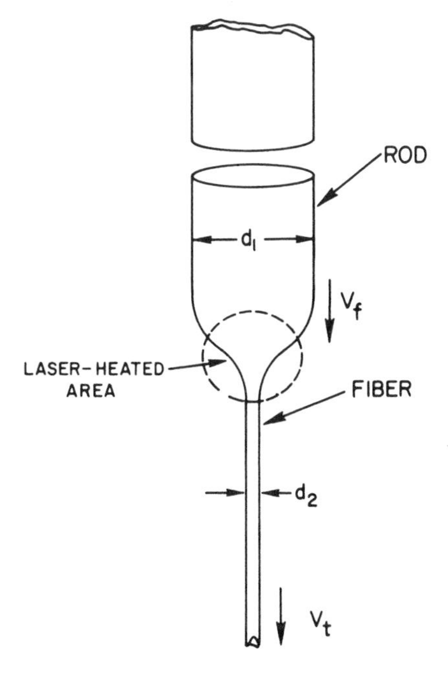

Fig. 6. Necking-down region of a rod.

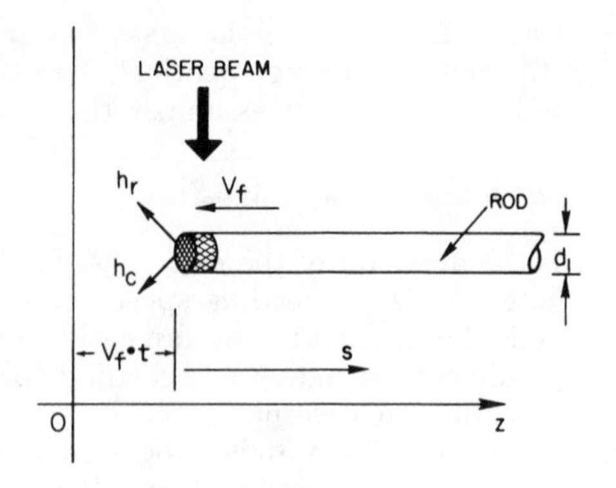

Fig. 7. A model of fiber drawing.

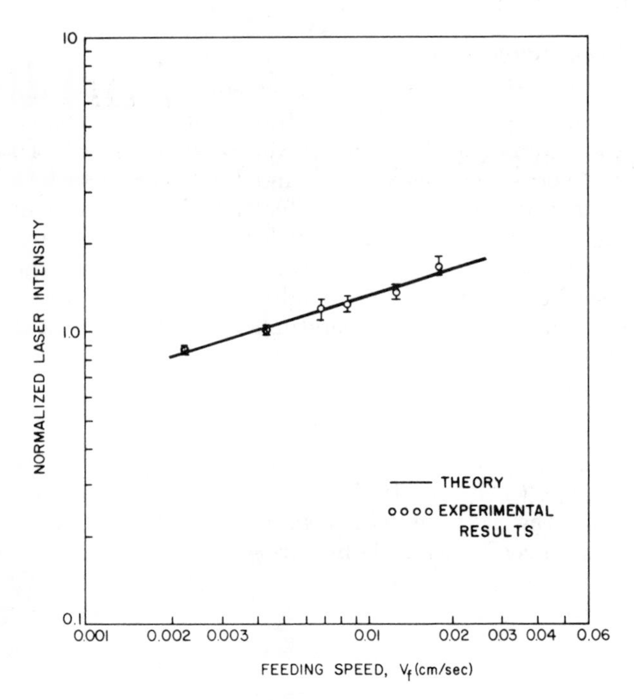

Fig. 8. Analytical and experimental results for relation between laser intensity vs feeding speed V_f (d_1 = 0.2 cm, $H = 10^{-2}$ cal/cm² sec °C, $\eta = 0.5$, and fused quartz used).

where h_c and h_r represent convective and radiative heat transfer coefficients. The boundary conditions are

$$T = T_s, S = 0, \quad T = 0, \quad S \to \infty. \tag{5}$$

In addition, an equation is needed to describe the variable V_f in Eq (3) that can be obtained by taking the energy balance at $S = 0$, neglecting conduction loss through the fiber:

$$-k(\partial T/\partial S) = \eta I - HT_s - V_f \rho C_p T_s, \tag{6}$$

where η is the loss in the optical system (in percentage) and I represents the laser intensity (W/cm²).

In the case of no air jet, an average value of free convection is calculated to be 0.35×10^{-3} cal/sec cm² K and will be higher when the air jet is activated. The coefficient h_r is defined as $\epsilon\sigma T_s^3$, where ϵ is the emissivity of the rod, which is normally equal to unity, and σ is the Stefan-Boltzmann constant, 5.6696×10^{-5} erg/sec cm² K⁴.

The solution to Eq. (3) with boundary condition (5) is

$$T = T_s \exp\left(-\frac{1}{2}\left\{\frac{V_f}{\kappa} + \left[\left(\frac{V_f}{\kappa}\right)^2 + \frac{16H}{kd_1}\right]^{1/2}\right\} \cdot S\right). \tag{7}$$

Finally, we established the relation of V_f and I by substituting Eq. (7) into (6). Then it becomes

$$\eta I = \frac{3}{2}\rho C p V_f T_s + HT_s + (T_s/2)[(\rho C p V_f)^2 + (16Hk/d_1)]^{1/2}. \tag{8}$$

Equation (8) indicates the relationship among the required power, feeding speed, and diameter of the preform. The power $P = IA$, where A is the cross-sectional area of the rod. In addition, it explains physically that sensible heat corresponding to the feeding speed of the rod should be added to the heat loss due to conduction, convection, and radiation to maintain drawing action. Otherwise, the fiber will break. Theoretical [Eq. (8)] and experimental results are given in Fig. 8. with good agreement between the two.

Conclusions

It has been demonstrated that a CO_2 laser (50 W) with an ellipsoidal reflector system is capable of drawing clad and unclad fibers of long lengths, with diameters down to 10 μm, and diameter variations within 5%. For unclad fiber, losses below 100 dB/km have been measured at the wavelength of 6328 Å. We have observed that laser power from 20 to 35 W was required to draw fiber from a 2-mm-diameter rod. The power levels were determined by the feeding speeds, which were in the range of 0.002–0.02 cm/sec. The correlation between the required laser power and feeding speed has been established, and good agreement between experimental results and theoretical calculations was obtained.

The author would like to thank A. L. Weaver for his assistance in the experiments. He would also like to acknowledge many helpful discussions with P. Kaiser.

References

1. D. B. Keck, R. D. Maurer, and P. C. Schultz, Appl. Phys. Lett., **14**, 307 (1973).
2. P. Kaiser, Appl. Phys. Lett. **23**, 45 (1973).
3. P. Kaiser, BTL, Crawford Hill; private communication.
4. R. E. Jaeger, Metallurgical Society of AIME meeting on Preparation and Properties of Electronic Materials, Las Vegas, August, 1973.
5. Optical Fused Quartz and Fused Silica, a catalog published by Amersil, Inc.
6. D. A. Pinnow and T. C. Rich, OSA Topical Meeting Digest on Integrated Optics, TUA-4, Las Vegas, February, 1972.
7. C. A. Burrus, OSA Topical Meeting Digest on Integrated Optics, WB-3, Las Vegas, February, 1972.
8. S. Saito, IEEE Trans. **COM-20**, 725 (1972).
9. W. M. Toscano, ERC-Western Electric Co., Princeton, N. J.; private communications.
10. H. S. Carslow and J. C. Jaeger, *Conduction of Heat in Solids* (Oxford University Press, New York, 1971), p. 387.

Part III
Materials and Loss Evaluation

Spectrophotometric studies of ultra low loss optical glasses II: double beam method

M W Jones and K C Kao
Standard Telecommunication Laboratories Ltd, Harlow, Essex

MS *received 25 November 1968*

Abstract A double beam spectrophotometer is described, suitable for the measurements of very low attenuation coefficients in glass in the wavelength range 500–1000 nm. The instrument is shown to have a balance stability of $\pm 1\times10^{-5}$. The systematic errors involved in the use of the long samples required for these measurements are studied. The instrument is shown to be satisfactory in making measurements of attenuation coefficients of 0·0001 cm^{-1} with an accuracy to $\pm$0·000 01 cm^{-1} when sample pairs of 20 cm length difference are used. Results of measurements on very low loss silica, which illustrate the performance of the instrument, are given.

1 Introduction

The use of cladded glass fibres as a possible transmission medium for optical signals has stimulated this study of attenuation in glasses. For this purpose glasses possessing attenuation coefficients of less than 0·0001 cm^{-1} are required, and the measurement of such low attenuation coefficients is beyond the range of normal spectrophotometers. A single beam instrument for such measurements has been described (Kao and Davies 1968) and forms part 1 of this study. This was shown to be suitable in making measurements to a limit of $\pm$0·000 05 cm^{-1} when sample pairs of 20 cm length difference are used.

A double beam instrument capable of more accurate measurements is described in this paper. The basic balance stability of the instrument is examined, and the problems of systematic errors on measurements with long samples considered. Results are given of measurements on samples of silica having very low attenuation coefficients and these illustrate the performance of the instrument.

2 Description of the instrument

A schematic diagram of the double beam spectrophotometer is shown in figure 1. The instrument differs essentially from conventional types in providing a large light path ($\sim$70 cm) for the insertion of samples and attenuators. This is necessary to accommodate the long glass specimens, up to 25 cm, which are required in accurate measurements of attenuation coefficients of less than 0·0001 cm^{-1}. In designing the system an arrangement with a small number of optical components was sought, and similarity of the paths of the two beams was required. Particular care was taken to obtain a mechanically stable set-up which was free from vibrations. For this reason a fairly massive base plate is used in the chopper assembly, and high quality bearings are employed in the chopper mounting. Flexibility was maintained in the positioning of the lenses and mirrors, and individual components were designed to be adjusted in their mountings and locked in position.

The light source is a 12 v, 50 w tungsten–iodine filament lamp. Good electrical contacts to the tungsten pins of the lamp are necessary for a stable output; spot-welding of nickel tape to these was found convenient and satisfactory. The current for the lamp is supplied by a transistorized power pack

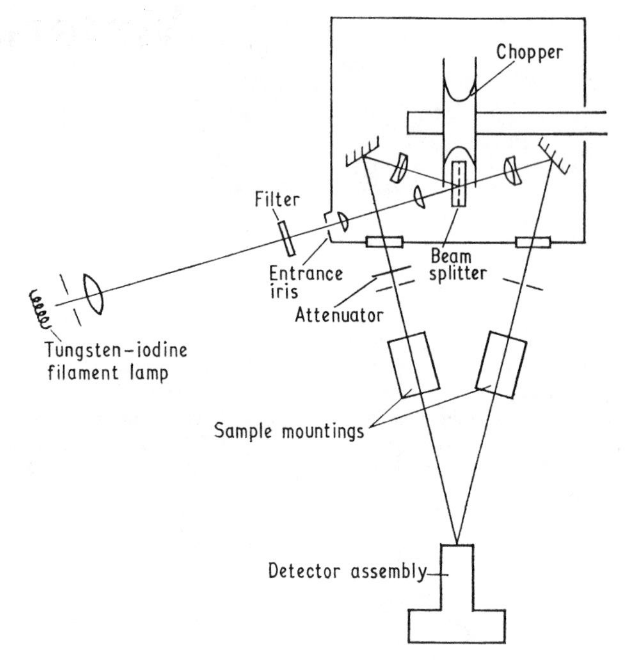

Figure 1 Schematic diagram of double beam spectrophotometer

in parallel with a large capacity 12 v lead–acid accumulator. A resistor in series with the lamp reduces the voltage across its terminals to 11 v, increasing the lamp life without seriously affecting its output. The potential difference across the lamp terminals was measured with a digital voltmeter and showed no fluctuation, indicating a voltage stability of better than 1×10^{-4}.

The lamp filament is imaged on an iris diaphragm by a condenser lens. Adjustment of this diaphragm and that adjacent to the condenser lens allows variation of beam diameter from 3 to 10 mm. These are adjusted together with the other lenses to give a well-collimated beam to the detector.

The selection of light of the required wavelength is done by the use of filters placed in the beam after the condensing lens. A set of filters for the wavelength range 500–1000 nm having a half-bandwidth of 10 nm is used for this purpose. Since absorption bands in glasses are broad at these wavelengths this coverage is adequate.

The beam splitter consists of a partially reflecting aluminium film sandwiched between optical flats 6 mm thick. Although the flats are bloomed to reduce reflected images, significant secondary beams exist. These are removed by iris diaphragms in the light path. The relative intensities of the reflected and transmitted beams change with wavelength; this means that there is a slightly different spectral distribution of light in each of the beams for a particular band of light. The difference in reflection loss due to this effect is negligible with glass samples. Care must be taken, however, in interpreting measurements of loss at wavelengths near absorption edges.

The beams transmitted and reflected by the beam splitter are cut sequentially by a precisely constructed double chopper. This is driven by a $\frac{1}{2}$ h.p. motor run from a Servomex M.C.47 controller. The speed stability of the motor was measured by a counter to record the frequency of a harmonic of the reference signal. Variations of $\pm 2 \times 10^{-4}$ were noted in a period of several minutes, after the motor had been run for some time. This is within the value specified by the manufacturer. A small light source is placed between the blades of the chopper and photodiodes fixed on opposite sides. The positions of the photodiodes are adjusted to provide reference signals in phase and 90° out of phase with the chopped beams.

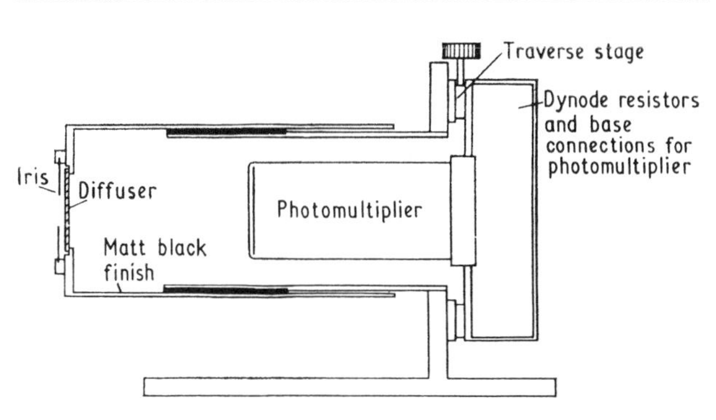

Figure 2 Detector assembly

The samples are clamped in V-blocks which are placed on adjustable mountings. Location points are provided to minimize the change in V-block position with repeated replacement. Evidence from measurements showed that this is satisfactory. Attenuators are required to balance the two beams. A simple system of neutral density filters was found to be satisfactory in practice. These are placed in the beam at a slight angle so that reflected light from the filter is directed away from the detector. Alteration of this angle gives some small variation in the transmitted light and this provides a means for fine adjustment. These filters are very thin so that the beam displacement caused can be ignored.

Reflected light can easily give rise to measurement errors in the instrument. In particular care must be taken to ensure that the introduction of samples into the two beams does not direct unwanted light back to the detector. For this reason, for example, the detector is not symmetrically located with respect to the two beams, but angled slightly so that light from one beam reflected off the front face of the diffuser is not reflected back when a sample is put in the other beam.

The mounting for the photomultiplier used as a detector is shown in figure 2. This allows movement of the photomultiplier in the detector head of ± 10 mm in two directions perpendicular to its axis. Adjustment of the photomultiplier is made for a maximum signal with one beam incident on the diffuser. The beam spot is therefore on the 'effective' axis of the photomultiplier, and systematic errors due to beam diameter change and displacement are minimized. In addition the diffuser–photomultiplier distance is variable from 5 to 15 cm. The mounting accommodates an EMI 9558B photomultiplier (S-20 response) or a Mullard 150 CVP photomultiplier (S-1 response).

The out-of-balance signal is detected by a phase-sensitive system (Aim Electronics system 5-1). In practice some difficulty is caused by the 'spikes' present in the balance signal at crossover point. These give rise to inconvenient overloading of the phase-sensitive detector, and limit the balance stability. A spike suppression circuit is used to overcome this problem, in which the spike is replaced by a constant level signal bounded by constant amplitude spikes of very narrow width. The output from the photomultiplier load is then led successively from the spike suppression circuit through the phase-sensitive detection system and to a pen recorder.

In order to reduce the possibility of sample contamination during measurements a clean-air tent is mounted over the spectrophotometer on the steel supporting table.

3 Balance stability

A fundamental limit to the balance stability is imposed by the discrete nature of the emission of electrons from the photocathode, the so-called shot noise. In order to examine other factors influencing the balance, these fluctuations may be reduced to 1×10^{-5} by use of a high light intensity and suitable choice of circuit time constant in the phase-sensitive detector output. A recorder trace of the balance stability is then taken, sufficient time having been allowed for the various components of the instrument to be stabilized. The figures of balance stability quoted here are estimated r.m.s. fluctuations, taken as one quarter of the deviation containing most of the trace, when drift was absent. The detailed observations were made with an EMI 9558B photomultiplier with an S-20 response. Similar results were obtained for a Mullard 150 CVP photomultiplier with an S-1 response.

Experiments showed that in the absence of a spike-suppression circuit a factor limiting the balance stability is the speed variation of the motor driving the chopper. Fluctuations in the motor speed cause changes in the reference frequency. The phase-shift amplifier is frequency dependent, and thus variations in the phase of the reference signal are produced. The 'spike' signal at the crossover point is 90° out of phase with the reference signal. Variations of the phase of the reference signal cause varying components of this to be seen at balance. It was shown that a motor speed change of $\pm 2 \times 10^{-4}$ (the short term speed stability of the motor) would give balance changes of $\pm 6 \times 10^{-5}$.

The balance stability was measured with different beam intensities. An estimate of the expected shot noise fluctuation was made from the signal strength and the known photomultiplier response. The two values agreed within the experimental error for stabilities up to $\pm 1 \times 10^{-5}$ (time constant $T = 1$ s) over a period of several minutes. However, at the highest stabilities the balance was found to be best characterized by two parameters: a rapid fluctuation due to shot noise, and a slow fluctuation describing the balance change over longer periods of up to one hour. For example, in two experiments over periods of 15 minutes, the r.m.s. rapid fluctuations were estimated to be $\pm 1 \times 10^{-5}$ ($T = 1$ s) and less than $\pm 1 \times 10^{-5}$ ($T = 5$ s). The balance changed slowly in these experiments up to $\pm 5 \times 10^{-5}$ ($T = 1$ s) and $\pm 4 \times 10^{-5}$ ($T = 5$ s), the range indicating the maximum deviations from balance.

The balance stability of interest so far as absorption measurements are concerned is that over the time required for sample insertion and measurements, i.e. several minutes. The results above therefore, show that the intrinsic balance stability of the instrument is satisfactory for measurements to $\pm 0{\cdot}000\,01\ \text{cm}^{-1}$ with sample pairs of 20 cm length difference. The cause of the long-term slow fluctuations in balance, observed to be of the order of $\pm 5 \times 10^{-5}$ in 1 hour, has not been found. The fact that they are slow makes experimental measurements difficult, but the results suggest that the fluctuation is random about the mean. A possible cause is slow variation in the light source, since only a small part of the filament is imaged on the entrance iris diaphragm.

4 Systematic errors

The introduction of samples into the light beams causes signal changes due to systematic errors in addition to those due to loss. These errors have been studied in some detail with the object of reducing their magnitude to less than 2×10^{-4} in a typical measurement.

The two important systematic errors are recognized: (i) beam displacement and (ii) beam diameter change. The former arises, in the case of samples with parallel end faces, because of alignment errors in the beam. The latter is a consequence of a change in focus position of $b(n-1)/n$ for paraxial rays, when a sample of length b and refractive index n is placed in the beam. It is obviously more important in measurements with long samples. In order to minimize the change due to these two effects, a detector arrangement is required so that small changes in beam spot position cause little change in signal. The detector assembly (figure 2) has a positional response which depends on the distance d between the diffuser and the photocathode. If it is assumed that the diffuser is Lambertian, the positional response may be calculated for the arrangement shown. The signal I_0 will change by $-2I_0r^2/d^2$ for a displacement r from the 'axis' position (for small values of r/d).

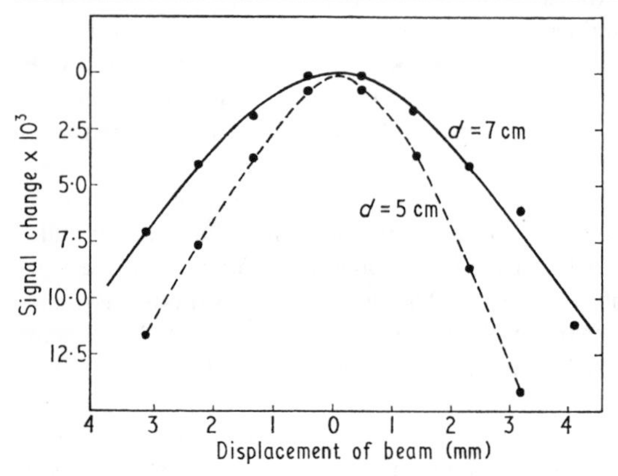

Figure 3 Detector response with displacement for plane diffuser. Photocathode–diffuser distance d, beam diameter 3 mm

The response of the detector with respect to position of beam spot on the diffuser can be examined by displacing the photomultiplier in the detector mounting and noting the change in signal. It is difficult to measure small changes in this way, however, since only a single beam is used. A more satisfactory method made use of the adjustment of a good quality long glass sample in one of the beams. The iris diaphragms were adjusted to give a beam spot of 3 mm diameter on the diffuser, and a displacement up to $\pm 4{\cdot}5$ mm obtained by sample adjustment. In this manner use is made of the better stability of the double beam system. The results obtained are shown in figure 3 for distances between the photocathode and diffuser of 5 and 7 cm. These show a response of the parabolic form anticipated by theory, but signal changes are larger than predicted by the formula above. The difference is due in part to the finite diameter of the beam which gives a larger 'effective' displacement, and in part to the approximation made in the simple theory.

The beam displacement caused by the insertion of a sample with parallel end faces in the beam at an angle α is $\alpha b(n-1)/n$ (notation as above). This means that the alignment for long samples (~ 20 cm) must be very accurate, to better than 3 minutes of arc, if displacements of less than 0·1 mm are to be caused. Also, the sample has to be removed from the beam and replaced without significant changes in alignment. The beam spot was observed on a graduated screen placed in front of the detector. Repeated insertion of samples and realignment suggested that the beam displacement produced was not more than 0·2 mm. This method has the advantage that the parameter of interest is measured directly, but accuracy is limited by the blurred edges of the beam, and also the beam diameter changes for long samples. In the case of good samples with parallel end faces it was found more convenient to align by means of the light reflected from the end faces. The apertures in the beams are stopped down to 1 mm and the samples adjusted for coincidence of the reflected beams and the apertures. This gives an estimated alignment accuracy to a few minutes, since the distance between aperture and farther reflecting sample face is about 30 cm. All these observations taken in conjunction with the response of the detector with position of beam spot (figure 3) indicate that the systematic error due to sample alignment should be less than 1×10^{-4}. Experiments carried out with repeated insertion and re-alignment of samples confirmed this, showing changes of not more than 2×10^{-4}.

In order to minimize the change in diameter of the beam spot on the diffuser when a sample is placed in the beam, the beam collimation needs to be optimum in the region near the detector assembly. Adjustment of the optical system is made and a long sample placed in the beam. The change in beam spot diameter on a screen in front of the detector is readily seen, but blurring of the edges makes exact measurement of the change difficult. This does, however, provide a satisfactory means of optimizing the beam collimation. In order to examine the beam intensity distribution in more detail further experiments were carried out. A 1 mm diameter aperture was traversed across the beam diameter close to the detector and the change in signal with position noted. A long sample was placed in the beam and the process repeated.

Results for a sample of length 23·9 cm ($n = 1{\cdot}67$) are shown in figure 4. Correction has been made here for attenuation and reflection losses, so the curves are normalized. Measurements of this kind suggested that a simple observation of the change in beam intensity in the central region provided a quick and simple method of estimating the beam spot diameter change on sample insertion.

A knowledge of the positional response of the detector and the beam diameter change enables a calculation of the systematic error to be made. Figure 5 shows how this varies with beam diameter for distances between the photocathode and diffuser of 5 and 7 cm and a beam diameter change of 10% on sample insertion. In addition curves are drawn showing the shot noise balance stability (time constant 1 s) at 545 and 825 nm. These are extrapolated in the form expected from single measurements at a beam diameter of 5 mm and a diffuser–photocathode distance of 7 cm. A photomultiplier with an S-1 response is used here. It can thus be seen how the compromise between balance stability and systematic errors is achieved.

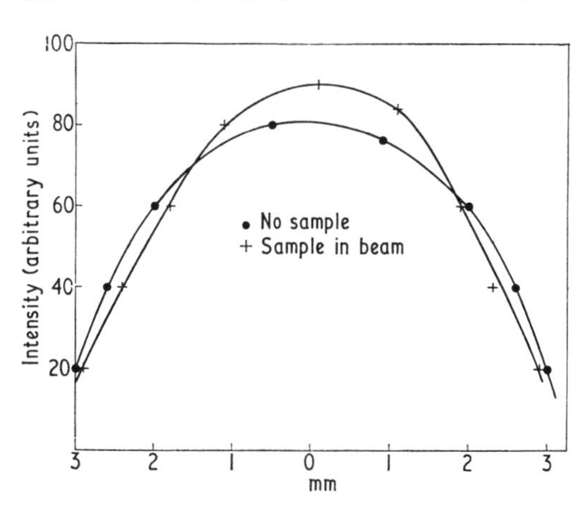

Figure 4 Change in intensity distribution across beam on insertion of sample of length 23·9 cm ($n = 1{\cdot}67$)

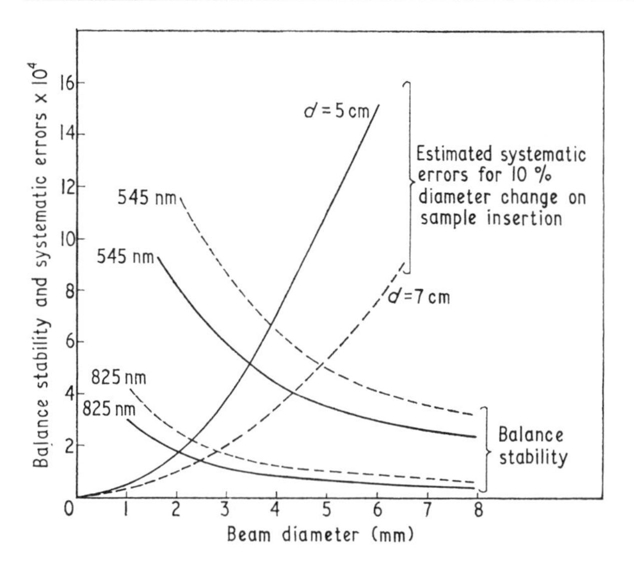

Figure 5 Variation of balance stability (time constant 1 s) with beam diameter (at 545 nm and 825 nm) and estimated systematic errors for different photomultiplier distances d (5 cm, 7 cm)

Some further experimental confirmation was sought to verify estimates of the systematic error due to beam diameter change. Measurements were made using some very low loss silica samples, and the systematic error was altered by changing beam diameter and divergence. It was considered important to try to make observations with the normally used detector arrangement where the systematic error is expected to be very small. The results obtained are shown in figure 6 which shows reasonable correlation between theoretical and experimental changes in spite of their small size.

5 Sample requirements

In order to make measurements of attenuation in glass at the resolution of the double beam spectrophotometer, stringent conditions are imposed upon the samples. A pair of samples must have end faces parallel, preferably within one minute of arc. These samples are cut from a rod of the specimen glass and polished under identical conditions. This is important since the measurements assume equality of reflection losses at the sample surfaces. Confirmation of the similarity of the surfaces is obtained by ellipsometry studies (Wright and Kao 1969). This technique is very sensitive to changes in the

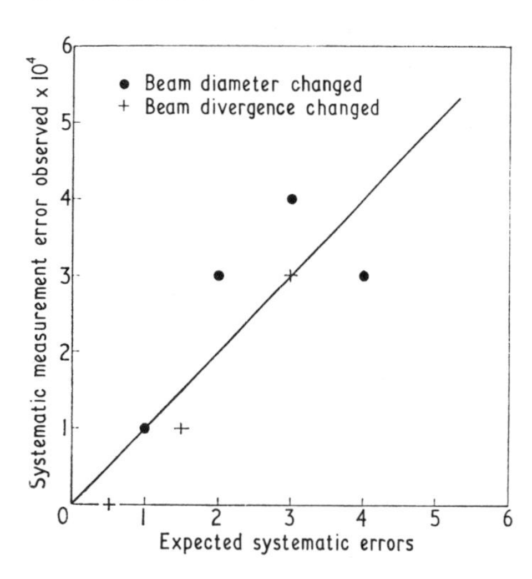

Figure 6 Correlation between observed measurement changes and expected systematic errors

surface and shows that a surface film of lower refractive index than the bulk is characteristic of glasses. An exception is the case of fused silica for which a film of higher refractive index is observed. Measurements on well-polished samples generally gave no difference within the experimental error of the instrument.

Typically a pair of samples of lengths 5 and 20 cm with a cross section of 2 cm (square or circular) has been found convenient for attenuation measurements. The standard preparation procedure adopted is to clean the sample with Teepol and water, scrubbing the end faces gently with a tissue, rinse in deionized water and vapour-degrease for an hour in acetone. The samples are stored in a desiccator prior to measurements.

Care in sampling handling is important, to avoid damage to the end surfaces. Any slight scratches will cause measurement inaccuracies. Contamination of the sample by dust and fragments of cleaning tissue is also a problem. In spite of these difficulties it appeared that visual observation of the surface was a reliable guide to the cleanliness. Changes of up to 1×10^{-3} in measurements were seen with marked dust contamination. The surface after thorough cleaning showed no tendency to collect dust provided it was untouched after the vapour-degreasing. Dust particles can be removed by a fine camel-hair brush, but recontamination is then rapid.

6 Measurement procedure

The samples are cleaned and placed in the V-block holders. Alignment of the samples in the beams is then carried out. The samples are removed, a filter is selected and the attenuator adjusted to balance the two beams. The large variation in the beam signal with wavelength makes it necessary to alter the voltage across the dynode resistor chain of the photomultiplier. This is done to give a signal level of 0·2 v for one of the beams. Fine adjustment of the amplifier gain allows the output signal (i.e. 100%) to be set to a conventional level. The amplifier gain steps are used in conjunction with a pen recorder to give a sensitivity suitable for the loss difference to be measured. The time constant of the output circuit in the phase-sensitive detector is adjusted as necessary.

The balanced beam output is recorded for about 1 minute when steady, the recorder is then switched off and the input to the phase-sensitive detection system shorted. The two samples are placed in the beams, the circuit is reconnected, and the change in signal level recorded for 1 minute. The samples are then removed and the balance level again recorded. This process, which takes about 5 minutes, is repeated if there is any significant balance drift. The loss measurement is accurate to 5%. The whole procedure is repeated for a series of wavelengths over the range 500–1000 nm, taking up to 2 hours. This procedure is satisfactory in measurement of attenuation differences down to 2×10^{-3}, but with smaller changes some further precautions are taken.

A set of measurements is made in the manner described above, care being taken, by repeating some measurements, to ensure that no changes in the sample occur in the process. It is assumed that although this set of results may suffer from a systematic error, differences in attenuation at different wavelengths are correct. Detailed study is then made of the loss at the wavelength of minimum loss. Sample re-cleaning and re-alignment are carried out until reproducible results are obtained. The systematic errors are estimated and correction made for them. The end faces of the samples are examined by ellipsometry to check similarity of surfaces.

7 Performance of instrument

Some measurements of attenuation which illustrate the resolution of the instrument are shown in figure 7. These are for samples of fused silica of different types (obtained from the manufacturers, Schott and Corning). The difference in length between the long and short samples was 15 cm.

It is of interest to compare the curves 1 and 2 as the silica here has a high transmission in the ultra-violet region. This type of silica usually has a high 'water' content and gives characteristic absorption bands in the infra-red (Adams and Douglas 1959).

Attenuation peaks are shown on the curves at 930 nm and 730 nm. These, it is suggested, are due to the third and fourth 'overtones' of the fundamental absorption at 2730 nm due to OH bonds in the silica.

Curve 3 shows the attenuation in Infrasil (Schott) which is designed to have a high transmission in the infra-red, and hence has a low 'water' content. In spite of this, attenuation peaks which we ascribe to OH bonds are indicated. The extremely low attenuation measured in the region of 950 nm was the subject of some detailed study. The loss difference

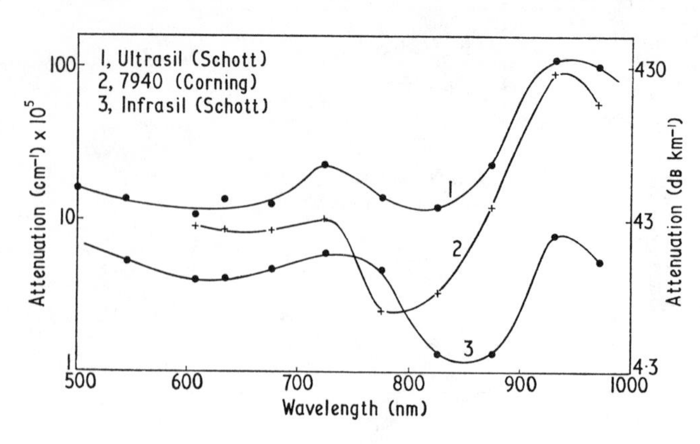

Figure 7 Attenuation in silica samples

observed on insertion of samples was less than 1×10^{-4}. A significant result was obtained only after the correction for systematic errors was applied, giving a positive loss of 2×10^{-4}.

The above measurements show that the instrument can be used to determine attenuation coefficients of 0·0001 cm^{-1} with an accuracy to $\pm$0·000 01 cm^{-1}, providing suitable samples having a 20 cm length difference are available.

Acknowledgments

This work was carried out under a Post Office Research and Development contract. The paper is published by permission of Standard Telecommunication Laboratories Ltd. and of the Senior Director of Development of the Post Office.

References

Adams R V and Douglas R W 1959 *J. Soc. Glass Tech.* **43** 147T–158T

Kao K C and Davies T W 1968 *J. Sci. Instrum.* (*J. Phys.* E) **1** (Ser. 2) 1063–8

Wright C R and Kao K C 1969 *J. Sci. Instrum.* (*J. Phys.* E) **2** (Ser. 2) in the press

Loss Mechanisms and Measurements in Clad Glass Fibers and Bulk Glass

A. R. Tynes,* A. David Pearson,† and D. L. Bisbee*
Bell Telephone Laboratories, Incorporated
(Received 6 April 1970)

Optical-loss mechanisms and transmission losses were investigated in clad glass fibers and in the bulk glass from which they were drawn. Measurements of the total losses, absorption losses, and scattering losses in both the fibers and in the bulk materials are presented and compared. Photomicrographs of scattered light were made at 90° with 0.6328-μm laser light propagating in the fiber core. These revealed the presence of an extremely large number of discrete light-scattering centers, both in the core and in the cladding. Similar photomicrographs of the bulk materials also show the presence of discrete scattering centers. Scattering centers have also been observed by means of an electron microscope; they have been shown to be crystalline, by both electron diffraction and x-ray diffraction.

Index Headings: Glass; Fibers; Scattering; Absorption.

Most present-day applications of fiber optics involve transmission of light over distances ranging from a few centimeters in the case of field flatteners to several meters in applications involving remote sensing. For such applications, the loss of power is of only secondary importance. In communications applications, however, in which information must be transmitted over much greater distances, the loss of power becomes of primary importance. Because typical present-day fibers have extinction coefficients of about 2×10^{-3} cm^{-1}, such losses must be reduced by about two orders of magnitude before the fibers can become useful in long-distance communications.

Before describing the measurements and results of this investigation, it will be instructive to consider the sources of the losses in the fibers and in the bulk glass from which they were drawn. The losses in fibers and bulk glass arise because of absorption and scattering. In general, the scattering loss in the fiber will be greater than in the bulk glass because of the geometrical imperfections in the waveguide structure. The scattering in the bulk material is probably Rayleigh scattering, although the measured Rayleigh ratios[1] in most glasses are generally somewhat greater than predicted.[2] At any rate, the loss owing to Rayleigh scattering in the fiber would be expected to be about the same as in the bulk glass, unless phase separation, or devitrification, occurred during fiber drawing. Absorption loss in the fiber would be expected to be the same as in the bulk glass, although an exception to this has been found and will be discussed.

An estimate of the lowest loss that might be expected, in fibers drawn from optical-quality commercial glasses presently available, can be obtained by assuming that the fiber losses are the same as those in the bulk glass. Because the open literature does not contain much information concerning loss measurements in low-loss optical materials, we have developed a capability for such measurements in our laboratory.[3] Among the optical materials that we have found to have lowest losses are Schott F-2 glass, fused silica, and especially prepared crystalline quartz[4] with total extinction coefficients[5] α_t of 1.3×10^{-3}, 7.3×10^{-4}, and 2.0×10^{-4} cm^{-1}, respectively. If a fiber could be drawn to have the same loss as crystalline quartz its loss would still be about an order of magnitude too great. The extinction coefficients reported include the extinction coefficients for both absorption and scattering, $\alpha_t=\alpha_a+\alpha_s$. Maurer[6] reports Rayleigh ratios as low as 1.6×10^{-6} cm^{-1}; if a fiber could be drawn with a total loss[1,7] corresponding to this Rayleigh ratio, its extinction coefficient would be 2.68×10^{-5} cm^{-1}, which is down to an acceptable level. It appears that if the fiber loss were determined by the loss in the bulk material, the loss would be due largely to absorption.

If low-loss fibers are to be obtained, we must make

* Crawford Hill Laboratory, Holmdel, N. J. 07733.
† Murray Hill, N. J. 07974.

Reprinted with permission from *J. Opt. Soc. Amer.*, vol. 61, pp. 143–153, Feb. 1971.

glass with total extinction coefficients approaching 10^{-5} cm^{-1}. The absorption losses in optical-quality glasses are believed to be due to the presence of ionic impurities, principally those of the transition-series elements. The molar-extinction coefficient for a given ionic impurity depends in some cases rather critically on the type of glass in which it resides, as has been demonstrated by the work of Steele and Douglas.[8] On the basis of their work, Kao[9] has estimated that 1 ppm of Fe^{2+} can result in an extinction coefficient of about 4.6×10^{-5} cm^{-1} at a wavelength of 1 μm. Probably, the levels of most ionic impurities will have to be reduced to not more than 1 ppm and possibly less.

The purpose of the present investigation is not to evaluate the properties of optical materials, but rather to relate optical-loss measurements and loss mechanisms in the fibers to those in the bulk glass from which they were drawn. Thus, the absorption and scattering losses in the bulk glass will be measured and compared to losses measured in the fiber. Any excess scattering loss in the fiber can then be attributed to either waveguide effects, or to phase separation or devitrification during the fiber-drawing process. The amount of scattering loss due to waveguide imperfections will be estimated by measuring the scattering loss for several fibers of different diameters and subtracting from each the expected Rayleigh-scattering loss, as measured in the bulk core glass.[10] Photographic evidence will be presented that shows the presence of waveguide defects as well as discrete scattering centers in both the fiber and in the bulk glass.

FIBER CONFIGURATIONS

The fibers investigated were of two types, as shown in Fig. 1. Figure 1 (left) shows a conventional clad fiber drawn from a rod-tube perform whose dimensions are given. The core rod is Schott SSK-1 glass and the cladding tube is Schott SK-14 glass, with refractive indices of 1.6171 and 1.6038, respectively. For these fibers, the fiber characteristic term[11] $R=(\pi d/\lambda)(n_2^2-n_1^2)^{\frac{1}{2}}$ is (at $\lambda=0.6328$ μm) $1.03\times10^4 d$, where d is the fiber diameter in centimeters and n_2 and n_1 represent the refractive indices of the core and cladding, respectively. Table I lists the fiber-core diameters, characteristic terms R, and some of the waveguide modes[12] each of these fibers is capable of supporting.

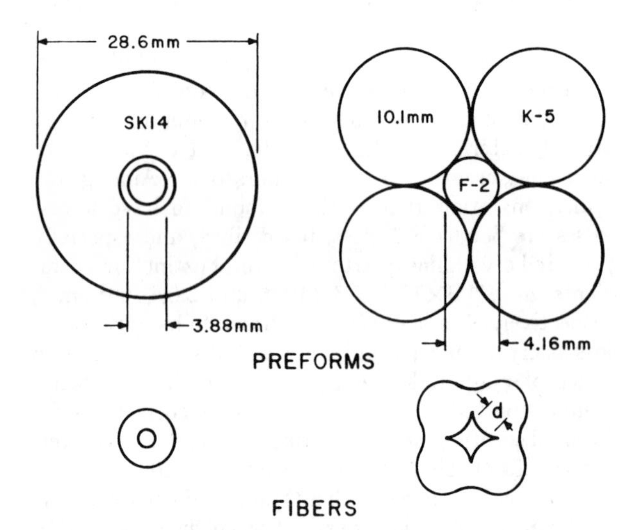

FIG. 1. Fiber configurations. Left, conventional clad fiber. Right, clover-leaf fiber.

The other fiber configuration is shown in Fig. 1 (right). The four cladding rods are Schott K-5 glass and the core rod is Schott F-2 glass with refractive indices of 1.5224 and 1.6200, respectively. This fiber configuration was proposed[13] for the purpose of avoiding the somewhat difficult task of grinding and polishing the inside surface of the tube which constitutes part of the preform in the conventional clad fiber. Because the cores of these fibers have an unusual shape, we do not specify fiber-characteristic terms for them. The diameter specified represents what is probably the minimum effective core diameter. It is important to notice that the voids that exist in the preform are not present in the drawn fiber.

The glasses that were selected for these fibers were chosen primarily because of the close match between the thermal-expansion coefficients of the core and cladding materials. Both fiber configurations have heavy claddings surrounding the cores, which serve to isolate the cores from the outside environment and at the same time make the fibers strong enough to handle with ease. The preform components were fabricated by A. D. Jones Optical Works, Inc., and Laboratory Optical Co.[14] and the fibers were drawn to our specifications by R. A. Humphrey of DeBell and Richardson, Inc.[15]

MEASUREMENTS

The fiber-loss measurements can best be explained with reference to Fig. 2, which shows the scattering detector as well as the detector used to measure the total loss in the fiber. The latter detector was immersed in index-matching oil. The scattering detector[16] consisted of six 1-cm-square silicon solar cells, each sensitive surface of which comprised one interior surface of a cube 1 cm on edge. In this manner, we had an

TABLE I. Description of the conventional clad fibers in terms of diameter, fiber-characteristic term, and some of the modes each is capable of supporting.

Diameter (μm)	R	Modes
3.2	3.3	HE_{11}–TE_{01}–TM_{01}–HE_{21}
5.1	5.25	HE_{12}–EH_{11}–HE_{31}–EH_{21} He_{41}
10.8	11.1	HE_{11}–TE_{01}–TM_{01}–HE_{21} HE_{12}–EH_{11}–HE_{31}–EH_{21} HE_{11}–TE_{02}–TM_{02}–HE_{22} HE_{51}–EH_{31}–HE_{13}–EH_{12} HE_{32}
20	20.6	as above, etc.

equivalent of an integrating sphere, except that all the scattered light fell on a photosensitive surface. Each solar cell was connected in series with a variable potentiometer and all solar-cell-potentiometer combinations were connected in parallel. The reason for the potentioneters was that the sensitivities of the various solar cells were not equal; varying the resistance of the potentiometers permitted adjustment of the output of each combination to the same value, when measuring the power level of the same light source. In order to pass the fiber through this integrating-cube detector, 0.25-mm-diam holes were ground through the centers of two opposite faces of the cube detector.

The main advantage of using an integrating detector of such small dimensions is that it permitted us to probe the scattered light along the length of the fiber. This was very desirable because the scattered light varied markedly (see Fig. 3) and measurements made on short sections that scatter very little encourage us to believe that the scattering losses in fibers can eventually be reduced to an acceptable level.

The total fiber-loss measurements were made by utilizing the detector immersed in a liquid whose refractive index was close to that of the core of the fiber being measured. Thus, it was not necessary to grind and polish the output end of the fiber after each section was removed. We obtained the response of the immersed detector (a silicon solar cell) relative to the integrating-cube detector by measuring the signal output of each when it was exposed to the light from the end of the fiber. The response of the immersion detector is about 3% greater when this fiber–detector combination is immersed in the index-matching liquid than when the combination is in air. For this reason, the relative response of the scattering detector is about 3% greater than the measured value.

To perform a set of loss measurements (i.e., scattering loss and total loss), a fiber about 7 m long was placed on the optical table that holds the laser source and measuring equipment. The fiber was wound in an elongated spiral on the table top so that its minimum bending radius was about 20 cm. The input end of the fiber was prepared by scoring the fiber gently with a razor blade and then bending the fiber sharply until it broke, in a manner similar to breaking a large rod after scoring it with a file.

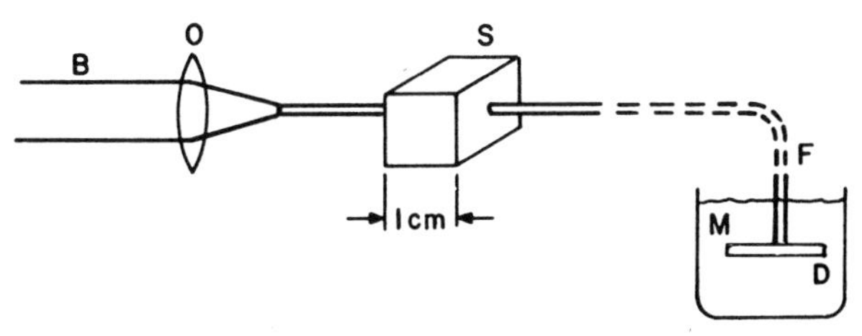

FIG. 2. Schematic diagram of fiber-loss-measuring apparatus: B, chopped beam from laser; S, scattering detector; F, fiber; D, detector; M, index-matching oil.

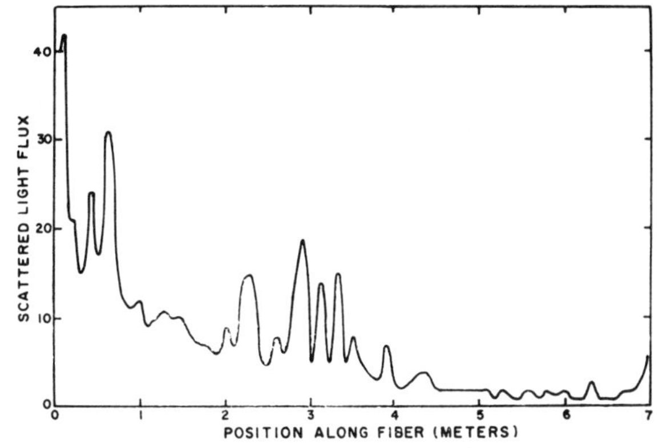

FIG. 3. Typical scattering loss along the fiber. 10.8-μm core, conventional clad fiber.

The 0.6328-μm beam from an He–Ne laser used for exciting the fiber was first split for the purpose of monitoring the laser power level and then chopped so that narrow-band phase-sensitive ac detection could be used. After the fiber was passed through the small holes in the opposite faces of the scattering detector, the laser light was focused onto the input end of the fiber core by means of a 10× microscope objective mounted on a micromanipulator. The output end of the fiber was then viewed end on (using adequate optical attenuation for eye protection) in a medium-power microscope, to make sure that the laser light was propagating in the fiber core.

The fiber-scattering-loss measurements were made by measuring the scattered light at 5-cm intervals over the entire length of the fiber, beginning at the input end. The scattering detector collected the scattered light from a length of 1 cm; the readings were weighted over the full 5-cm interval. Repeated measurements on long fibers agree very well with each other. Figure 3 shows how the scattered light varies along the length of a typical fiber. Figure 4 shows that although the scatter-

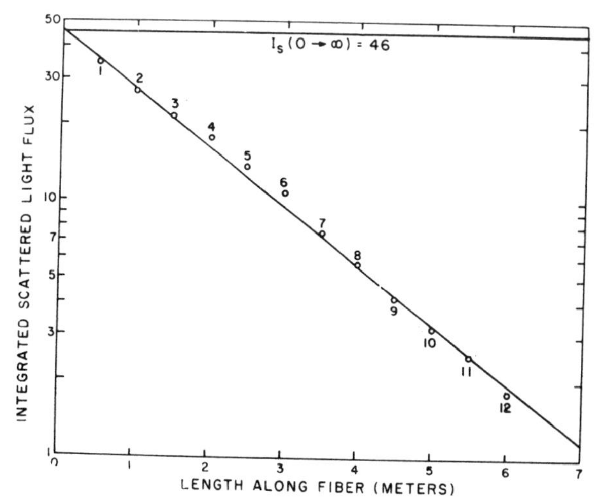

FIG. 4. Measured integrated fiber-scattering loss. 10.8-μm core, conventional clad fiber.

ing shown in Fig. 3 varies markedly along its length, the light lost due to scattering only decays exponentially. This point must be discussed further because understanding of it is central to calculation of the extinction coefficients for scattering and for absorption.

If a quantity of light I_0 is launched into the propagating modes of a fiber, then the light still propagating at a distance x from the input end is given by

$$I(x)=I_0\exp(-\alpha_t x), \tag{1}$$

where $\alpha_t=\alpha_s+\alpha_a$ and α_s and α_a are the extinction coefficients for scattering and absorption, respectively. The amount of light scattered from a length dx of the fiber at the distance x from the input end is given by $dI_s(x)=-I(x)\alpha_s dx=-I_0\exp(-\alpha_t x)\alpha_s dx$. Integration gives

$$I_s(x)=(I_0\alpha_s/\alpha_t)\exp(-\alpha_t x). \tag{2}$$

Similarly for the absorbed light,

$$I_a(x)=(I_0\alpha_a/\alpha_t)\exp(-\alpha_t x). \tag{3}$$

The amount of light I_0 launched into a very long fiber is eventually either scattered out of the fiber or is absorbed by the fiber. The quantities $I_0\alpha_s/\alpha_t$ [Eq. (2)] and $I_0\alpha_a/\alpha_t$ [Eq. (3)] represent the fractions of I_0 that are eventually scattered or absorbed, respectively. Thus, $I_s(x)$ and $I_a(x)$ represent the fraction of $I_0\alpha_s/\alpha_t$ and $I_0\alpha_a/\alpha_t$, respectively, still propagating at a distance x from the input end.

When a set of fiber-loss measurements is made, the scattering measurements are made first and then the total-loss measurements are made by successively removing short lengths of the fiber and measuring the output $I(x)$ of Eq. (1) directly. I_0 is determined from the data obtained from these total-loss measurements by plotting $\log I(x)$ vs x and extrapolating to $x=0$; α_t is determined from the slope of this plot. Now, having determined I_0 and α_t, we can utilize the scattering measurements to determine α_s.

We observe that the scattering detector does not measure $I_s(x)$, but rather the amount of light scattered from a length of the fiber equal to the length of the scattering detector. Thus, the response of the scattering detector of length δx at x is given [from Eq. (2)] by

$$\begin{aligned}\delta I_s(x,\delta x)&=I_s(x-\delta x)-I_s(x)\\&=[I_0\alpha_s/\alpha_t]\exp[-\alpha_t(x-\delta x)]\\&\qquad-[I_0\alpha_s/\alpha_t]\exp[-\alpha_t x],\end{aligned}$$

from which

$$\delta I_s(x,\delta x)=[I_0\alpha_s/\alpha_t]\exp[-\alpha_t x][\exp(+\alpha_t\delta x)-1]. \tag{4}$$

Now, if all the scattered light from $x=0$ to $x=x_0$ is integrated, this would be equivalent to setting $x=\delta x=x_0$ in Eq. (4) and we have

$$\begin{aligned}\delta I_s(0\rightarrow x_0)&=I_s(x=0)-I_s(x=x_0)\\&=(I_0\alpha_s/\alpha_t)[1-\exp(-\alpha_t x_0)]. \end{aligned}\tag{5}$$

Because $\delta I_s(0\rightarrow x_0)$ has been determined experimentally, as has I_0 and α_t, we determine that

$$\alpha_s=[\delta I_s(0\rightarrow x_0)\alpha_t/I_0]/[1-\exp(-\alpha_t x_0)]. \tag{6}$$

We have now determined α_t and α_s directly from experimental measurements and α_a indirectly from the relation $\alpha_t=\alpha_s+\alpha_a$. These values of α_t and α_s, determined from Eq. (1) and Eq. (6), are valid only if the total light and the scattered light $I_s(x)$ decay exponentially. That this is the case is shown in Figs. 4 and 5. Figure 4 was obtained from the scattering measurements by integrating the scattered light from $x=0$ to $x=50$ cm, from $x=0$ to $x=100$ cm, etc., to the end of the fiber. The fraction of the light that is eventually scattered [i.e., $I_0\alpha_s/\alpha_t=I_s(0\rightarrow\infty)$] is shown as the horizontal line labeled $I_s(0\rightarrow\infty)$ in Fig. 4. The downward-sloping line of Fig. 4 is now obtained by subtracting from $I_s(0\rightarrow\infty)$ the integrated scattered light from $x=0$ to $x=50$ cm (point 1), that from $x=0$ to $x=100$ cm (point 2), etc. These points then represent $I_s(x)$ as given by Eq. (2); Fig. 4 shows that $I_s(x)$ decays exponentially. This demonstrates the validity of the theory that is based on an exponential decay of both $I(x)$ and $I_s(x)$.

Figure 5 shows the results of a complete set of fiber-loss measurements made on a clover-leaf fiber whose core size was about 8 μm. The points ○ in the lower curve represent the measurements obtained for the total loss of the fiber; the slope of this curve determines α_t. The extrapolation of this curve to $x=0$ determines I_0. The vertical distance from I_0 down to the upper curve with points △ is taken directly from the corresponding distance in Fig. 4 and represents the measured integrated fiber-scattering loss. Similarly, the vertical distance from I_0 down to the center curve with points ▽ represents the integrated absorption loss obtained by integrating Eq. (3) from zero to x,

$$\delta I_a(0\rightarrow x)=(I_0\alpha_a/\alpha_t)[1-\exp(-\alpha_t x)], \tag{7}$$

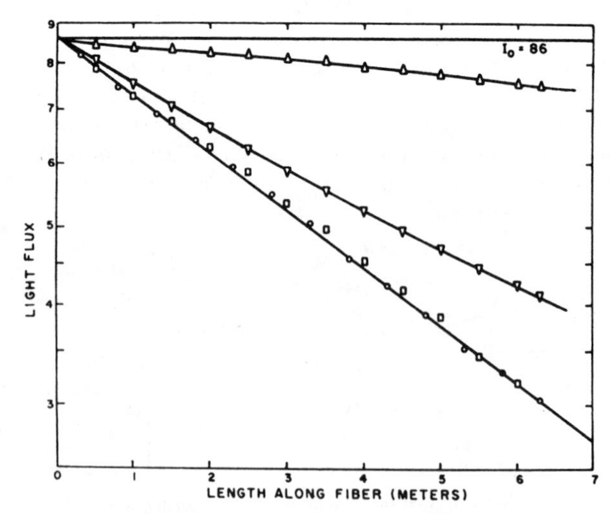

FIG. 5. Complete set of fiber-loss measurements made on a clover-leaf fiber whose core was about 8 μm.

but with the important difference that here the value of α_a is that obtained by direct measurements on the bulk core glass.

If the extinction coefficient for absorption does not change in the fiber-drawing process, then the sum of the losses represented by the difference between I_0 (Fig. 5) and the upper curve and that to the center curve should be equal to the total fiber loss represented by the difference between I_0 and the lower curve. That this is the case can be seen from Fig. 5 where the points □ represent the combined effect of fiber-scattering loss and bulk-absorption loss.

Of course we could, and do, measure the value of α_a in the fiber by measuring both α_t and α_s and determining its value through the relation $\alpha_a=\alpha_t-\alpha_s$. From Fig. 5, we see that this value of α_a agrees with that obtained for the bulk core glass.

When the fiber-loss measurements do not lead to exponentially decreasing light, then the preceding theory cannot be used to determine the extinction coefficients. Such nonexponential behavior is shown in Fig. 6, which also shows that the nonlinearity increases as the fiber-core diameter decreases. Such data can be used, however, because the decay of light becomes exponential about 1.5 m from the input end. To use such data, simply shift the vertical axis ($x=0$) to the right 1.5 m, and pretend that a new amount of light I_0 is launched into the fiber at this point. This is shown in Fig. 7. The determination of the extinction coefficients is now the same as before, except that all the measurements from $x=0$ to $x=150$ cm are ignored.

The results of the various loss measurements are summarized in Table II, in which each set of measurements was obtained from a single length of fiber; consequently, they do not represent average values. In

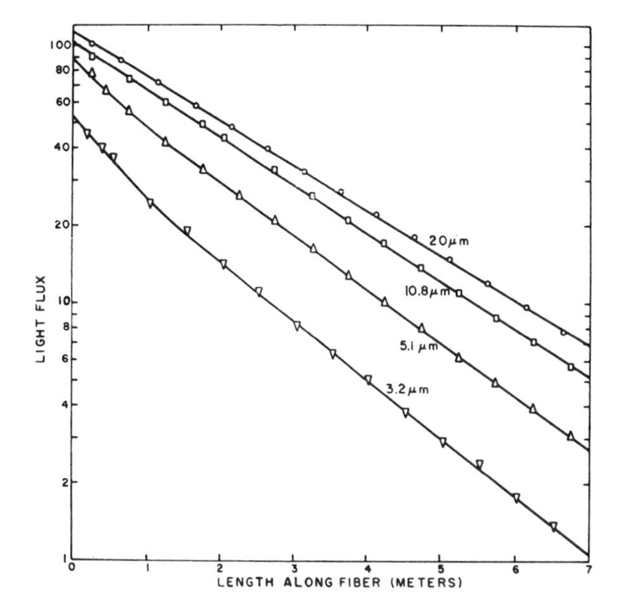

FIG. 6. Nonexponential behavior of total fiber-loss measurements.

the case of the conventional clad fibers, there are systematic changes of the losses for the fibers of different diameters, whereas this is not observed for the clover-leaf fibers. In both cases, the effects of Rayleigh scattering and of waveguide losses are shown, as well as the sources and magnitudes of the losses in the bulk glass from which the fibers were drawn.

For the conventional clad fibers, both α_t and α_s increase as the diameter decreases and additionally the increase in α_t is equal to the increase in α_s. The extinction coefficient for absorption α_a in the fiber should be given by the difference $\alpha_t-\alpha_s$, but here a complication arises, which affects not only the value

TABLE II. Results of loss measurements.

Conventional fiber

Extinction coefficient (10^{-3} cm^{-1})	Core diameter (μm)			
α	20	10.8	5.1	3.2
α_t	4.10	4.20	4.37	4.60
α_s	0.09	0.19	0.38	0.42
α_s'	0.12	0.26	0.52	0.58
$\alpha_a=\alpha_t-\alpha_s'$	3.98	3.94	3.85	4.02
α_R	0.089	0.089	0.089	0.089
$\alpha_w=\alpha_s'-\alpha_R$	0.03	0.17	0.43	0.49

α_a (measured in bulk SSK-1 core glass) = 3.1

Clover-leaf fiber

α	8	5
α_t	1.65	1.63
α_s	0.32	0.27
α_s'	0.43	0.36
$\alpha_a=\alpha_t-\alpha_s'$	1.22	1.27
α_R	0.124	0.124
$\alpha_w=\alpha_s'-\alpha_R$	0.31	0.24

α_a (measured in bulk F-2 core glass) = 1.26

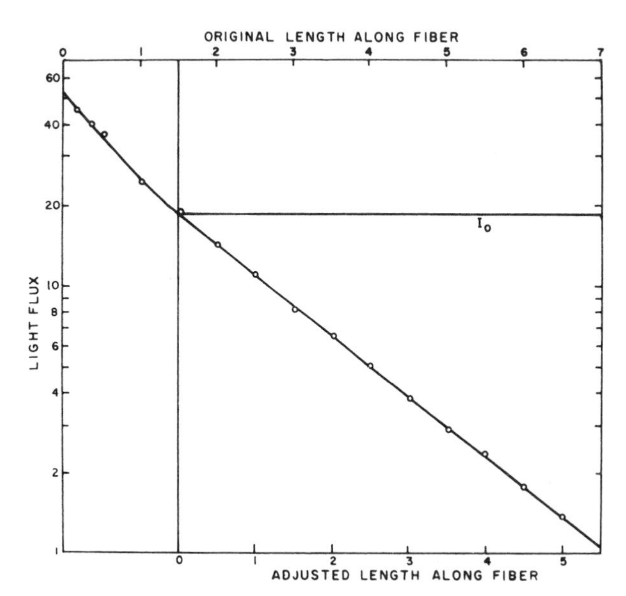

FIG. 7. Correction to permit use of nonexponential-loss measurements.

of α_a, but also the value of α_w that relates to the losses that arise because of waveguide imperfections. This complication arises because a significant fraction of the Rayleigh-scattered light suffers total internal reflection at the cladding–air interface (as well as a smaller amount at the core–cladding interface, which we shall ignore) and hence is propagated in the cladding and consequently is not measured by the scattering detector.

Because of this complication, the measured value of α_s is too small by an amount that we shall now determine. Figure 8 shows a Rayleigh-scattering center located at the center of a fiber. If $\pi/2-\theta_0$ is the critical angle for total internal reflection, then all the light in the forward- and backward-pointing cones will be trapped in the fiber. The solid angle ω generated by the angle θ_0 is given by

$$\omega=\frac{1}{R^2}\int_0^{\theta_0} 2\pi R^2 \sin\theta d\theta = 2\pi(1-\cos\theta_0) \tag{8}$$

and from the definition of the fiber numerical aperture (N.A.), N.A. $=n \sin\theta = n_2 \sin\theta_0 = (n_2{}^2 - n_1{}^2)^{\frac{1}{2}}$, we determine that $\cos\theta_0 = (1-\sin^2\theta_0)^{\frac{1}{2}} = n_1/n_2$ and

$$\omega = 2\pi(n_2-n_1)/n_2. \tag{9}$$

The fraction F of the Rayleigh-scattered light that is trapped is the sum of the solid angles subtended by the forward and backward cones, divided by 4π,

$$F=(n_2-n_1)/n_2. \tag{10}$$

For the fibers used here, we find that the value of F that applies to the scattered light trapped in the fiber cores is 0.82% and 6.2% for the conventional clad fibers and the clover-leaf fibers, respectively. On the other hand, the fraction of the scattered light that is trapped because of total internal reflection at the cladding–air interface amounts to 38% and 34%. A better measure of the fiber scattering loss can be obtained by multiplying the measured value of α_s by $(1+F)$. This may not properly compensate for the scattering that arises because of waveguide-surface imperfections, because Eq. (10) is derived on the assumption that the scattering is isotropic. However, unless the angular scattering spectrum for such imperfections is highly anisotropic, it seems reasonable to use Eq. (10) to determine the correction factor.

In Table II, the quantity α_s' represents the corrected extinction coefficient for scattering. We can now determine the extinction coefficient for absorption by use of the relation $\alpha_a=\alpha_t-\alpha_s'$. We see that the value of α_a is very nearly constant for all the diameters listed. The value of α_a in the conventional clad fibers is about 30% greater than that measured in the bulk SSK-1 core glass. The value of α_a in the fiber is close to that measured in the bulk SK-14 glass used for the cladding; our problem could be solved if we assumed that the light was propagating mainly in the fiber cladding. We feel certain that this is not the case, because we went to great lengths to make sure, before each measurement, that the light was propagating in the fiber core. It appears that the absorption was somehow increased during the fiber-drawing process. This could happen if, for instance, the atmosphere of the drawing furnace was different from that of the furnace in which the glass was originally melted.

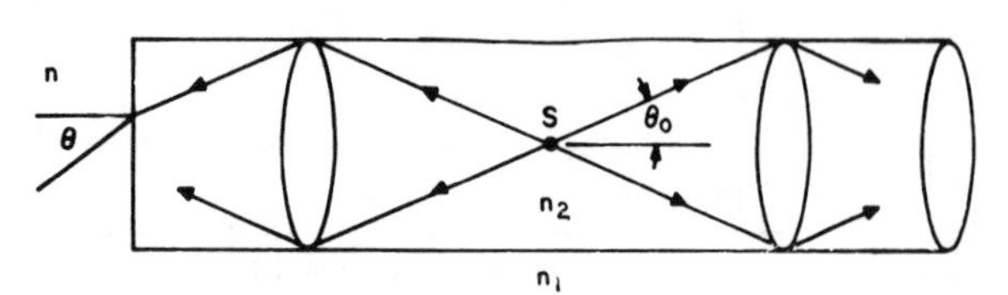

FIG. 8. Fiber geometry for determining the fraction F of trapped Rayleigh-scattered light. S, fiber scattering center.

The scattering losses observed in fibers can be divided into two classes depending on their origin. The Rayleigh-scattering loss in the bulk glass would be expected also in the fiber. In addition to the Rayleigh-scattering loss in the fiber, we should expect additional scattering losses due to waveguide-type losses that arise because of geometrical imperfections in the waveguide structure. Such imperfections may arise because of variations of fiber diameter, ellipticity of the core, and surface roughness at the core–cladding interface. At any rate, it seems useful to separate the scattering due to Rayleigh scattering and waveguide scattering. To do this, we assume that the loss due to Rayleigh scattering is the same in the fiber as that measured in the bulk core glass. This quantity is designated by α_R. The waveguide scattering loss α_w is given by the difference $\alpha_s'-\alpha_R$. For the conventional clad fibers, the value of α_w increases rapidly with decreasing fiber diameter (almost as 1/diameter^2) except for the smallest of these fibers.

In the case of the clover-leaf fibers, there are no anomalies and no variations of significance for the two diameters considered. For these fibers, the value of α_a is for all practical purposes identical to that measured in the bulk Schott F-2 core glass. The somewhat greater scattering loss in these fibers is believed to be due to entrapment of many small air bubbles at the core–cladding interface because many very bright scattering centers can be seen. There are regions between these bright spots where the scattering loss is one half to one quarter of the average scattering loss. In the case of the conventional clad fibers, such scattering centers are seen only occasionally. It appears that, if these small bubbles could be eliminated, it would be possible to obtain scattering losses comparable to the measured Rayleigh scattering in the bulk F-2 core glass.

The measured scattering loss for the 20-μm-diam conventional clad fiber leads to $\alpha_s=9\times10^{-5}$ cm^{-1}, whereas the measured Rayleigh-scattering loss in the bulk Schott SSK-1 core glass results in an $\alpha_R=8.9\times10^{-5}$

cm^{-1}. Because of the factor F in Eq. (10) it would not be too surprising to measure an α_s in a fiber that is less than α_R for the bulk core glass. What is important here is that a fiber has been produced containing sections in which the measured scattering loss has been reduced essentially to that expected due to Rayleigh scattering. The value of 9×10^{-5} cm^{-1} for α_s represents a lower limit for total losses of fibers drawn from currently available Schott SSK-1 core glass.[17] Since Mauer[6] reports measurements of Rayleigh ratios in bulk glass about one quarter of that in our core glass, it appears that, by a judicious selection of glasses, it should be possible to produce fibers with extinction coefficients for scattering (at $\lambda=0.6328\,\mu$m) of the order of 2×10^{-5} cm^{-1}, and even less if light of longer wavelength can be used. The attainment of fibers with total-extinction coefficients of the order of 2×10^{-5} cm^{-1} appears to be a possibility, provided that absorption losses can be reduced to the order of 10^{-5} cm^{-1} which is about two orders of magnitude less than that of most commercial glasses available at this time.

Having presented the results of the various loss measurements and compared the sources of the losses in the fiber and in the bulk glass, we now present qualitative evidence relating to the nature of the scattering losses.

When visible laser light is propagating in a glass fiber it can be seen by the unaided eye, even in a well-lighted room. The reason why it can be seen so easily, in spite of the fact that not much light is scattered, is that the light is concentrated in a very small cross section in the fiber and hence has a rather high power density, which results in a high contrast between the light-scattering region and the background. When a fiber is examined in this manner, the scattering is immediately seen to have two distinct characteristics. The first and most obvious is that the fiber has many rather bright scattering centers that appear as tiny point sources of light. The other is that the fiber glows with a more or less uniformly decreasing luminance along its entire length. Two rather distinct scattering mechanisms appear to be responsible for the observed scattering.

Figure 9 shows two types of point scattering centers, which appear to be typical and can be described in terms of definite types of waveguide defects. Each photograph shows two exposures of the fiber; the lower exposures were taken with conventional illumination; the upper exposures were taken with laser light propagating in the fiber core, with a small amount of conventional illumination so that the entire fiber can be seen. The two dark lines in the central light region of the fiber show the core–cladding interface. The fiber-core diameter is $20\,\mu$m. That these two types of defects are located at the core–cladding interface was demonstrated by rotating the fiber about its axis and observing the position of the defect.

The defect shown in Fig. 9 (top) is believed to be an air bubble entrapped at the core–cladding interface. There are two reasons for believing this. First, careful visual examination of the defect in the microscope shows the defect to be symmetrical and to have no sharp edges. Second, we can see, at B of Fig. 9 with laser light propagating in the fiber core, that there is considerable scattering on both sides of the defect. This is consistent with what would be expected if an air bubble was trapped at the core–cladding interface of the fiber preform. Thus, as fiber is drawn continuously from the preform tip, the air bubble comes ever closer to the tip of the preform. Eventually, the bubble accelerates rapidly into the fiber-forming region of the preform, at which point it is elongated in the direction of the drawing and finally breaks up into a long sequence of much smaller bubbles. Those smaller bubbles are not visible when the fiber is observed in conventional light.

A close examination of the preform from which the conventional clad fibers were drawn supports the belief that the defect shown in Fig. 9 (top) resulted from an entrapped air bubble. When the narrow conical region of the preform remnant is examined in a medium-power microscope, many air bubbles are seen entrapped at the

FIG. 9. Large-point scattering centers. Top, bubble at interface. Bottom, fissure at interface. A: Secondary scattering from bubble, B; C: bubble with conventional illumination; D: scattering from fissure at interface; E: fissure with conventional illumination. The dark lines in the central bright region show the core–cladding interface. Core diameter is 20 μm.

core–cladding interface. This suggests that it might be useful to evacuate the air from the region between the core and cladding in the preform, before the fibers are drawn, although the effects of additional outgassing under these conditions would have to be studied carefully.

Figure 9 (top) also shows another source of fiber scattering, which was made evident by the presence of the air bubble, but which, we believe, is unrelated to it. At A, opposite the bright scattering point on the lower interface, can be seen a diffuse scattering resembling the scattering of city lights from overhead clouds. We believe that this is due to secondary-scattered light from the core–cladding interface, the primary source of which is the light scattered from the bubble at the lower interface. Furthermore, we believe that the secondary-scattered light is observed here because of the smallness of the distances involved and the large quantity of light actually scattered from the bubble.[18]

The other type of point scattering center is shown in Fig. 9 (bottom) where the scattering (upper exposure) is confined to a very small region surrounding the defect and is not elongated as above in the case of the bubble. For this reason, we believe that this defect is a small fissure, or fracture, located at, or very near, the core–cladding interface.

The scattering loss that results from either the air bubbles or the fissures does not constitute a fundamental lower limit to the scattering from fibers. Elimination of these sources of scattering loss appears to be a technological problem for which there is hope that a solution will be found. We have measured scattering losses in short sections (1 or 2 cm) of fibers where the losses are down to the fundamental limit set by Rayleigh scattering.

We must now examine the secondary scattering shown in Fig. 9 (top) more closely because it might constitute a scattering-loss source much more difficult to eliminate. The secondary-scattered light could be the result of light scattered from the bubble being wave-guided circumferentially around the core–cladding interface and then being forward scattered into the microscope. If the secondary scattering is due to interfacial roughness, we should be able to see its effect by taking long time exposures, provided that we are well removed from regions containing bright scattering points, which might otherwise mask the effect we are searching for.

Figure 10 (bottom) shows the results of a 10-h exposure at 100× with laser light propagating in the fiber core. Figure 10 (top) shows the same fiber photographed with conventional illumination for comparison. In Fig. 10 (bottom), many scattering centers are visible and it is difficult to tell whether they are on the core–cladding interface or are distributed more or less uniformly throughout the fiber core. It should be noted that the bright streak that makes up the scattering region is not uniform along the fiber axis and appears to wander around somewhat, but always within the confines of the core.

FIG. 10. Uniformly distributed microscopic scattering centers. Top: in conventional illumination. Bottom: with laser light propagating in fiber core. 10-h exposure, original magnification 100×; shown here 92×.

Figure 11 shows a sequence of photographs similar to that shown in Fig. 10 (bottom), but at a magnification of 500×. Because at a magnification of 500× the depth of field of the microscope is only about 2 μm, we can sample the nature of the scattering centers in separate slices each about 2 μm thick. In order to know the precise location of these slices, however, we must consider the complicating effects of the optical-lens effect produced by the fiber itself. If we consider the actual core–cladding interface (solid inner circle in Fig. 11) to be an object viewed through the fiber from above, we find that the imaging properties of the outer surface of the fiber image this object at the location shown by the dotted circle in Fig. 11. Once the position of this image has been accurately calculated, we can determine the location of the slice under observation by focusing on the top of the outer surface of the fiber and then moving the fiber up toward the microscope objective by a known amount by means of the calibrated focusing stage of the microscope. We estimate that we can determine the location of the slice under observation to about ±1 μm.

Figure 11 shows a sequence of such photographs. The primary motivation for taking these photographs was to determine the relative importance of scattering losses due to interface roughness and that due to Rayleigh scattering throughout the core of the fiber. This question was posed by the scattering shown in Fig. 10 as well as by the secondary scattering shown in Fig. 9, at A.

The results shown in Fig. 11 give a very positive answer to this question. We conclude that the scattering shown in Figs. 10 and 11 is due largely to Rayleigh scattering from discrete scattering centers distributed more or less uniformly throughout the fiber core and that the scattering due to interfacial roughness is of only minor importance, at least for the 20-μm-diam fiber. Furthermore, we conclude that the uniform dull glow of the fiber, which can be seen visually with the unaided eye, is due to Rayleigh scattering from the core of the fiber.

The reasons for our conclusions can be explained by reference to the exposures at levels 4, 5, and 6 in Fig. 11. If both interfacial and volume effects were equally important scattering mechanisms, then levels 4, 5, and 6 would show scattering centers throughout the core, outlined by two well-separated rows of scattering centers corresponding to the interfacial regions lying within the 2-μm depth of field of the microscope. If interfacial scattering were negligible, we would see only the uniform volume scatter. Although level 4 does show a small increase of radiance near the interface, the pictures show essentially uniform volume scattering.

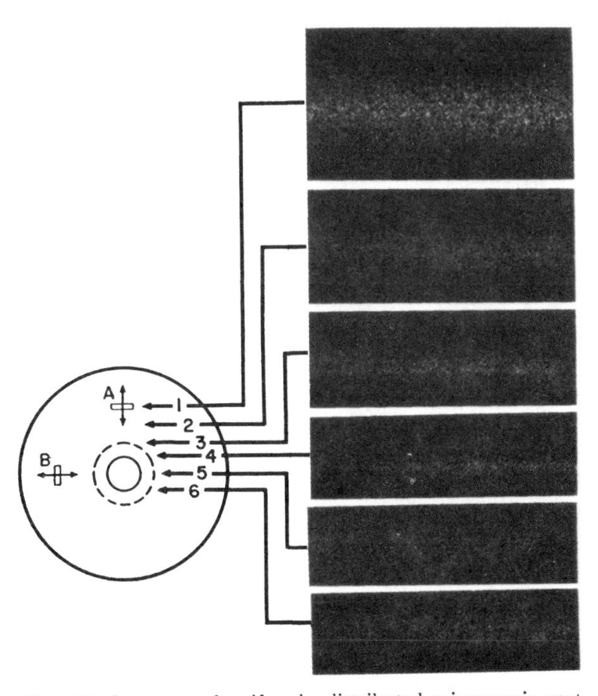

Fig. 11. Sequence of uniformly distributed microscopic scattering centers. These scattering centers are believed to be those that produce Rayleigh scattering in glass. 10-h exposure at 500× (original magnification) (shown here 215×) except at level 1 which is 57 h at 500× (215×). The exposure below the fiber cross section shows the width of the core as observed in the microscope.

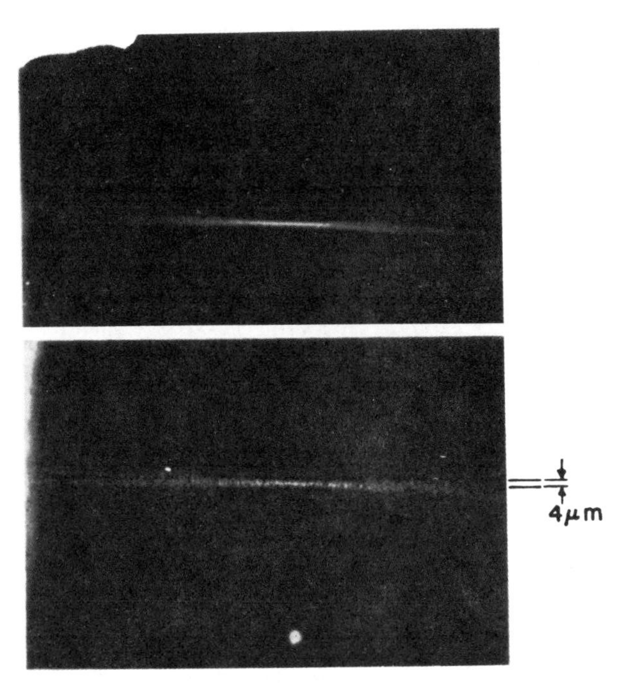

Fig. 12. Top, scattering in liquid compared to scattering in bulk glass, bottom. Original magnification 200×; shown here 154×.

For this reason, we conclude that the scattering due to interfacial roughness is not the primary source of scattering loss in the case of the 20-μm-diam fiber.

So far, we have not demonstrated that the uniformly distributed scattering centers shown in Fig. 11 have any connection with the expected Rayleigh scattering in the core glass. The photograph shown in Fig. 12 (bottom) was taken to determine if the scattering centers that are responsible for Rayleigh scattering in bulk glass could be observed and, if so, whether they look like those shown in Fig. 11.

The photograph of Fig. 12 (bottom) was taken by first focusing incident laser light down to a narrow waist inside a piece of bulk core glass and then photographing the beam waist in 90° scattered light by use of a microscope. The waist in the photograph has a diameter of about 4 μm. After an exposure of about 1 h, we obtained the photograph shown, which shows the presence of many scattering centers that appear to be identical to those observed in the fiber. We believe these scattering centers to be the frozen-in inhomogeneities that are responsible for the Rayleigh scattering in glass. We conclude that the scattering centers observed in the fibers are simply what would be expected because of Rayleigh scattering. Figure 12 (top) shows, for comparison, the type of scattering observed when a glass cell filled with a scattering liquid was substituted for the bulk glass. As expected, no discrete scattering centers are visible, because the molecules are in continuous motion in the liquid. This technique could be used to measure the diameters of focused laser beams.

Figure 11 shows an interesting phenomenon, which we shall now attempt to explain. Scattering centers appear in the photographs taken at all six levels shown. Scattering centers are clearly present at levels 1 and 2.[19] We must now ask why it is that scattering centers are visible at level 1 but are not visible at point B at level 5. After all, point A at level 1 is the same distance from the center of the fiber as point B at level 5 and there is no reason to expect scattering centers at point A but not at point B. When the fiber was rotated 90° about its axis, another exposure at level 1 showed very clearly that not only are there scattering centers at point B, but now they are observed whereas before they were not.

This complication can be explained by assuming that the scattering centers are preferentially aligned circumferentially about the fiber axis because of the fluid-flow characteristics of the fiber as it was being drawn. If the scattering centers are aligned and elongated as shown at positions A and B, the scattering would be highly directional. It is well known that the scattering from a rectangular slit, for example, is much greater in a direction normal to the longer dimension of the slit than in the other direction. Thus, the elongated scattering center at position A would scatter preferentially upward into the microscope and would be observable. The scattering center at point B, on the other hand, would scatter mostly in the horizontal direction and would scatter very little light into the microscope and therefore would not be observed as readily.

One more remark should be made in connection with Fig. 11. The radial distribution of radiance within a fiber can be obtained from a sequence of photographs such as these. Such a technique would be especially useful as a light probe when the light is propagating in a single-mode fiber and the electromagnetic field of the guided wave penetrates outward into the cladding.

DISCUSSION

We have designed, constructed, and calibrated optical-loss-measuring systems that permit us to measure absorption and scattering losses in bulk glasses and in optical fibers. Our optical-loss-measuring set[3] permits us to measure values of α_t as low as 10^{-4} cm^{-1} on either bulk glass or liquid samples. Kao[20] discusses loss-measuring techniques with similar capabilities. The contribution of Rayleigh scattering to the value of α_t in the bulk glass or liquid is determined by utilizing a metallurgical microscope that has been modified for the photoelectric measurement of the scattered light. For this scattering measurement, 10^{-3} W of incident 0.6328-μm laser light is more than sufficient and glass samples as small as 1-mm cube can be utilized. By measuring both α_t and α_s, we determine the bulk extinction coefficient for absorption α_a through the relation $\alpha_t=\alpha_s+\alpha_a$.

In the case of the fibers, the scattering loss is measured by means of our specially designed integrating-cube detector[16] whose dimension is 1 cm on an edge. Such a small scattering detector permits us to probe the scattered light very carefully along the fiber, in search for short sections where the scattered light might be as low as the limit determined by the Rayleigh scattering measured in the bulk glass. Such sections were found for the 20-μm-diam fiber and encourages us to believe that fiber-scattering losses can be reduced to an acceptable level.

Total fiber loss is measured by immersing the output end of the fiber in a vessel containing the detector (a silicon solar cell) and an index-matching oil. By successively removing short lengths of the fiber and measuring the output light for each length, we determine the value of α_t for the fiber. The value of α_a in the fiber is obtained by subtracting the corrected value of α_s, namely α_s', from the value of α_t. Fiber waveguide-type losses are determined by subtracting the value of α_R (i.e., that corresponding to the measured Rayleigh ratio in the bulk core glass) from α_s'.

We have presented the results of our loss measurements in both the fibers and in the bulk glass from which they were drawn. We have shown that the fibers are somewhat more lossy than the bulk glass and that the scattering loss increases with decreasing fiber diameter. For the 20-μm-diam fiber (Table II), the measured scattering loss is only about one third greater than the expected Rayleigh scattering and there are short sections where the scattering loss is down to the Rayleigh scattering limit. Because there are glasses whose Rayleigh ratios are about one quarter of that we have measured, it would appear that fibers with an extinction coefficient for scattering (at $\lambda=0.6328$ μm) of the order of 2×10^{-5} cm^{-1} could be drawn. The main obstacle to obtaining total extinction coefficients approaching 2×10^{-5} cm^{-1} is the high absorption losses in presently available glasses.

The fiber-scattering centers shown in Fig. 11 appear identical to those observed in the bulk glass as shown in Fig. 12. Discrete particles in the bulk Schott SSK-1 glass have been observed by means of an electron microscope and have been identified as being crystalline, by use of both electron-diffraction and x-ray techniques,[21] although the crystalline species has not been identified. We have no evidence to suggest that these crystallites can be identified with the scattering centers we observe by means of the scattered light, although it would not be unreasonable to suggest such an identification.

REFERENCES

[1] The Rayleigh ratio is defined by the equation $R_{90}=IL^2/I_0V$ cm^{-1}, where I is the amount of light scattered at 90° from the incident, unpolarized light beam of radiance I_0. L is the distance from the scattering volume V to the point where I is measured. The relation between the Rayleigh ratio and the extinction coefficient for scattering is given by $\alpha_s=(16\pi/3)R_{90}$ cm^{-1}. See I. L.

Fabelinskii, *Molecular Scattering of Light* (Plenum, New York, 1968), pp. 37–39.

[2] See Ref. 1, Chap. 7.

[3] Unpublished. Our optical-loss-measuring set uses a 0.6328-μm He-Ne laser source for a balanced two-beam bridge. The insertion loss is obtained by placing the sample at the Brewster angle in one arm of the measuring set and then determining the resulting imbalance electronically.

[4] A. A. Ballman, D. M. Dodd, N. A. Kuebler, R. A. Laudise, D. W. Rudd, and D. L. Wood, Appl. Opt. **7**, 1387 (1968).

[5] α_t is defined by the relation $I = I_0 \exp[-\alpha_t x]$.

[6] R. D. Maurer, J. Chem. Phys. **25**, 1206 (1956).

[7] The value of α expressed in cm^{-1} is converted to dB/km by multiplying the value in cm^{-1} by 4.33×10^5.

[8] F. N. Steele and R. W. Douglas, Phys. Chem. Glasses **6**, 246 (1965).

[9] K. C. Kao and G. A. Hackham, Proc. IEE **113**, 1151 (1966).

[10] Unpublished. We have modified a metallurgical microscope for photoelectric detection of scattered light. In this measuring technique, an image of a focused laser beam (as seen in our Fig. 12) is formed on the cathode of a photomultiplier that has been masked off to expose an equivalent area (at sample) measuring 10×20 μm. We can quickly (5 sec) replace the glass sample by a glass cell containing benzene for relative Rayleigh-ratio determinations, or can use the device for absolute measurements. Measurements can be easily made with 10^{-3} W of 0.6328-μm laser light incident on the sample, and samples as small as a 1-mm cube can be utilized.

[11] N. S. Kapany, *Fiber Optics-Principles and Applications* (Academic, New York, 1967), p. 55, Eq. (3.20).

[12] A. Werts, L'Onde Électrique **46**, 967 (1966).

[13] Private communication, S. E. Miller, Bell Telephone Laboratories, Inc., Crawford Hill Laboratory, Holmdel, N. J. 07733.

[14] A. D. Jones Optical Works, Inc., Burlington, Mass. 01803, and Laboratory Optical Co., P. O. Box 387, Plainfield, N. J. 07060.

[15] DeBell and Richardson, Inc., Hazardville, Conn. 06036.

[16] A. R. Tynes, Appl. Opt. **9**, 2706 (1970).

[17] It is important to note that SSK-1 is not the highest-transmission Schott glass available. It was chosen for this early study because of its thermal properties.

[18] By use of the scattering detector, we determined that the bubble scatters about 0.1% of the light in the fiber at that point.

[19] A 10-h exposure at level 1 was the same as that shown at level 2 except slightly less intense. The 57-h exposure at level 1 was for the purpose of emphasizing the presence of the scattering centers.

[20] K. C. Kao and T. W. Davies, J. Sci. Instr. (J. Phys. E) **1** (Ser. 2), 1063 (1968); M. W. Jones and K. C. Kao, J. Sci. Instr. (J. Phys. E) **2** (Ser. 2), 331 (1969); C. R. Wright and K. C. Kao, J. Sci. Instr. (J. Phys. E) **2** (Ser. 2), 579 (1969).

[21] Private communication, G. W. Kammlott, Bell Telephone Laboratories, Inc., Murray Hill, N. J. 07974.

Fundamental optical attenuation limits in the liquid and glassy state with application to fiber optical waveguide materials

D.A. Pinnow, T.C. Rich, F.W. Ostermayer, Jr., and M. DiDomenico, Jr.

Bell Laboratories, Murray Hill, New Jersey 07974

(Received 26 January 1973; in final form 20 March 1973)

Fundamental optical scattering and absorption mechanisms have been identified which limit light transmission in fiber optical waveguide materials. These mechanisms, which are intimately associated with the random structure in the liquid and glassy state, are described and then used as a basis for comparing fiber optical waveguide materials. It is concluded that pure fused silica is a preferred waveguide material, having ultimate total losses of 1.2 dB/km at the Nd:YAG laser wavelength of 1.06 μ, 3.0 dB/km at the $Ga_xAl_{1-x}As$ emission wavelength of approximately 0.8 μ, and 4.8 dB/km at the GaP:Zn,O emission wavelength centered at 0.7 μ.

It is well known that the optical attenuation in fiber optical waveguides is due to the sum of the bulk material attenuation and attenuation due to imperfections in the waveguide structure. It is further known that bulk attenuation is comprised of two parts, absorption and scattering, both of which may be influenced by impurities in the glass. However, it has not been broadly appreciated that fundamental mechanisms in chemically pure liquids and glasses cause intrinsic optical absorption and scattering loss. Both of these mechanisms are intimately associated with the random structure of the liquid and vitreous state. The purpose of this letter is to compare potential fiber optical waveguide materials on the basis of their ultimate intrinsic loss characteristics. In doing this we describe the nature and magnitude of these losses.

Scattering loss in liquids and glasses is known to be due to microscopic variations in the local dielectric constant associated with the random molecular structure of these materials.[1] For pure liquids, such as carbon tetrachloride (CCl_4), the variation in local dielectric constant is due principally to thermally driven fluctuations in the number of molecules within a region having dimensions substantially less than an optical wavelength. The magnitude of the optical scattering coefficient due to this effect can be shown, by using classical electromagnetic theory and thermodynamics, to be

$$\alpha_{\text{scat},\rho}=\frac{8}{3}\frac{\pi^3}{\lambda^4}(n^8p^2)(kT)\beta_T, \tag{1}$$

where λ is the optical wavelength, n is the index of refraction, p is the photoelastic coefficient, k is Boltzmann's constant, T is the absolute temperature, and β_T is the isothermal compressibility of the material. Qualitative insight into this expression can be gained by noting that kT is the driving force for the density fluctuations, β_T is a measure of the compliancy of the medium to the driving force, and n^8p^2 is a term which converts density fluctuations into dielectric constant fluctuations. It should also be noted that this expression has the explicit λ^{-4} Rayleigh dependence, indicating that the scatter loss substantially decreases at longer wavelengths. Since the parameters in Eq. (1) are known for simple liquids, it is possible to immediately calculate their scattering loss. Results are shown in Fig. 1 for CCl_4, which has been considered as a waveguide material.[2]

The expression given in Eq. (1) is based on the assumption of thermodynamic equilibrium, which is valid for liquids but *not* for glasses. The random structure of glasses is not determined by the ambient temperature, but by its fictive temperature, which is closely related to the softening point or more precisely the temperature at which the glass, if heated, would come into thermodynamic equilibrium. By incorporating the fictive temperature concept, Pinnow *et al.*[3] were able to account for light scattering in single-component glasses such as fused silica. Since glasses are known to be substantially less compliant than liquids, one would qualitatively infer from Eq. (1) that the scattering loss would tend to be less in glasses than liquids. However, this effect is partially mitigated by the higher effective temperature of the glass structure, which leads to increased scattering. Results are shown in Fig. 1 for fused silica, which has a fictive temperature of approximately 1700 °K. These results were obtained by the Brillouin spectroscopic method recently described by Rich and Pinnow.[4]

The scatter loss due to density fluctuations is predicted[3] to be less in lower-softening-point glasses than in fused silica. An interesting example is soda-lime-silicate glass (20% Na_2O, 10% CaO, 70% SiO_2) which has a fictive temperature of approximately 800 °K. High-purity samples of this glass have been prepared since it is regarded as a potential waveguide material.[5] The scatter loss in these pure samples has been found experimentally to be greater than that in fused silica. The resolution of this apparent discrepancy is that scattering due to density fluctuations is indeed low; however, there is an additional mechanism which contributes to the scatter loss in multicomponent glasses. This mechanism is attributable to the statistically random distribution of the polarizable components, which leads to an additional contribution to the local variations in the dielectric constant. Ostermayer and Pinnow[6] have recently developed a quantitative model to account for this effect. The resulting scattering coefficient due only to concentration fluctuations in a host glass modified with m constituents is[6]

$$\alpha_{\text{scat},c}=\frac{32\pi^3n^2}{3\lambda^4\rho N_A}\sum_{j=1}^{m}\left[\left(\frac{\partial n}{\partial x_j}\right)_{\rho,T,x_{i\neq j}}+\left(\frac{\partial n}{\partial\rho}\right)_{T,x_i}\left(\frac{\partial\rho}{\partial x_j}\right)_{P,T,x_{i\neq j}}\right]^2 M_jx_j, \tag{2}$$

where n is the refractive index, λ is the wavelength, ρ

Reprinted with permission from *Appl. Phys. Lett.*, vol. 22, pp. 527–529, May 15, 1973.

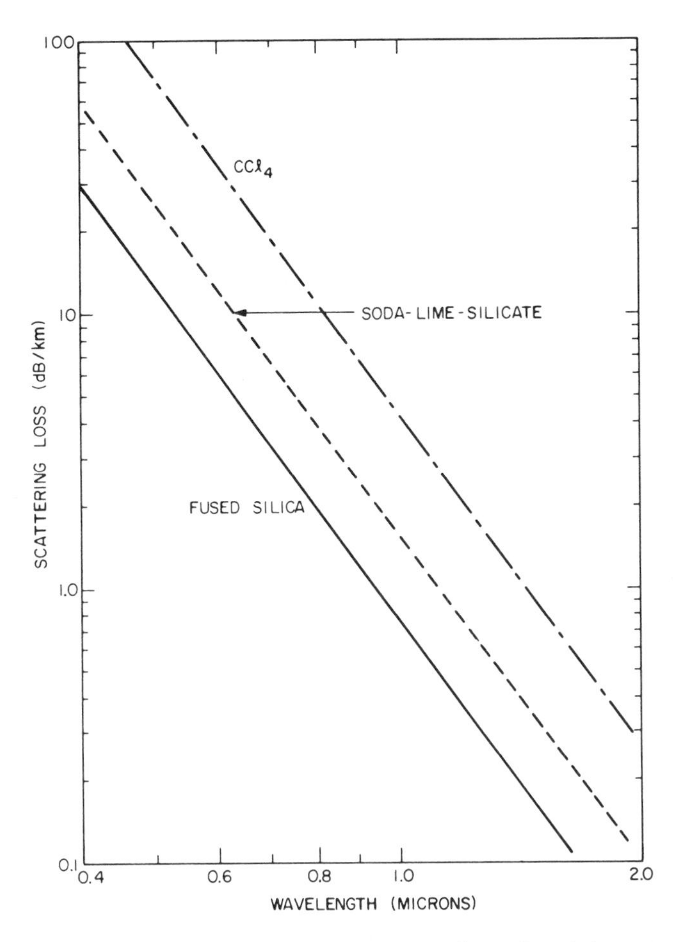

FIG. 1. Intrinsic scattering loss vs optical wavelength for liquid CCl_4 and vitreous fused silica and soda-lime-silicate.

is the density, N_A is Avogadro's number, and M_j and x_j are the molecular weight and the weight fraction, respectively, of the jth modifier. The partial derivates of n can be evaluated from the Gladstone-Dale relation[7] and the partial derivatives of ρ from the Huggins-Sun formula.[8] By using published data on the soda-lime-silicate system,[9] the scattering coefficient calculated from Eq. (2), when added to the scattering coefficient due to density fluctuations calculated from Eq. (1), results in a total scattering loss in agreement with the experimentally determined value shown in Fig. 1.

It is known that the random molecular structure in liquids and glasses gives rise to varying local electric fields on a microscopic scale. Recent theoretical developments by Dow and Redfield,[10] as well as experimental and theoretical work by Wood and Tauc[11] on chalcogenide glasses, provide convincing evidence that such local microfields cause *intrinsic* absorption loss in chemically pure materials in what is normally the transparent region below the fundamental interband absorption edge. Similar experimental results have been observed in the amorphous III-V semiconductor compounds.[12] The mechanism for the loss is due basically to local field-induced broadening of the excitonic levels which are created in optical absorption for energies close to but below the interband edge. The broadening of the exciton levels can be shown[10] to produce a tail on the ultraviolet interband absorption edge which varies exponentially with energy as

$$\alpha_{abs} \approx \exp\left(\frac{E - E_g}{\Delta E}\right). \tag{3}$$

Here E is the photon energy, E_g is the effective energy gap of the material, and ΔE is a parameter which is constant for each material. At present the theoretical model is not sufficiently sophisticated to establish the magnitude of α_{abs} or the value of ΔE. We note however that the exponential form given in Eq. (3) is insensitive to the detailed nature of the microfields and is therefore useful in fitting experimental data as well as gaining insight into the effect. For example, it can be observed that the larger-energy-band-gap materials such as fused silica should have lower intrinsic absorption than the narrower-gap soda-lime-silicate glasses, provided both glasses have comparable ΔE values.

Although experimental absorption data are still sketchy and encumbered by extrinsic impurity effects, we have been able to extract what we believe to be the intrinsic absorption tail in fused silica and soda-lime-silicate glass after studying many samples of extremely-high-purity materials.[13] The results are exhibited in Fig. 2.

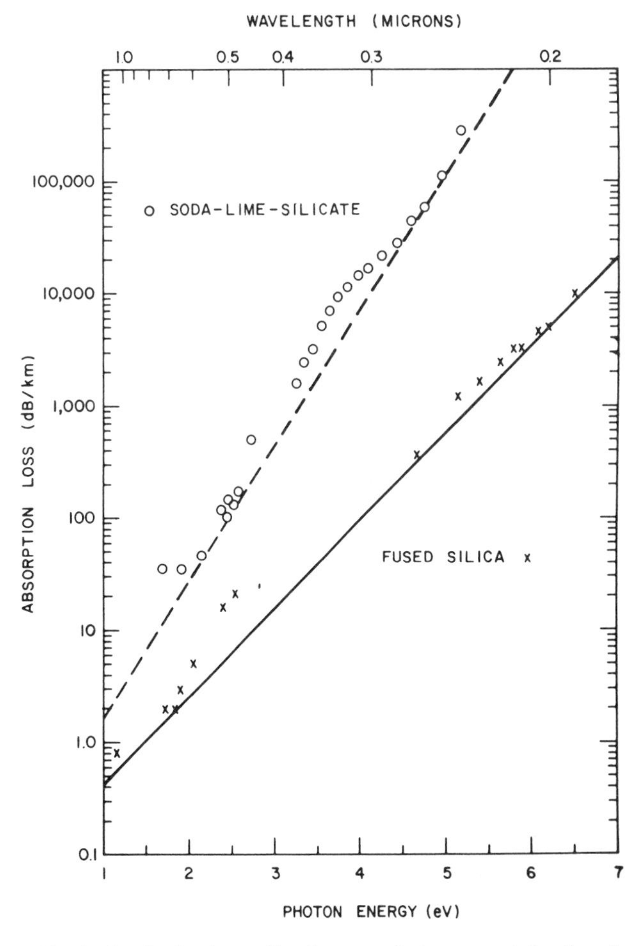

FIG. 2. Intrinsic absorption loss vs photon energy for fused silica and soda-lime-silicate. These curves represent lower bounds for all experimental absorption data. Shown are some of the experimental data points used to develop these curves. The rather broad gap in data for fused silica in the range of 2.5—4.5 eV is due to the limitations in sensitivity and wavelength of the measurement techniques described in the text.

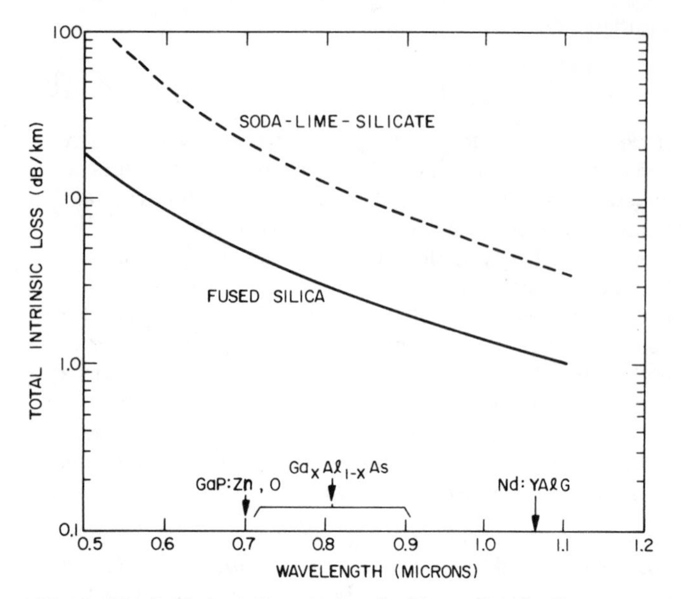

FIG. 3. Total intrinsic loss in fused silica and soda-lime-silicate. The emission wavelengths corresponding to the most promising optical sources are marked on the abscissa.

These data extend from the near-infrared through to the near-ultraviolet portion of the spectrum. The ultraviolet data points shown in this figure were obtained by using a commercial spectrophotometer (Cary 14) operated to the limit of its sensitivity. The lower-loss data points in the visible and near infrared were obtained by the calorimetric technique of Rich and Pinnow[4] using various emission wavelengths from argon, krypton, and Nd:YAG lasers. These results convincingly fit the exponential form given in Eq. (3), although it is apparent that impurity effects, such as the noticeable excess absorption band in the soda-lime-silicate centered at 3300 Å, have not yet been completely eliminated. Nevertheless, the exponential trend in the data, which continues over approximately six decades in attenuation and over 5 eV in photon energy, appears to be intrinsic and not governed by impurity ions which exhibit known characteristic spectral absorption bands[14] rather than the observed exponential dependence. The ΔE values determined from the slopes in Fig. 2 are 0.3 and 0.5 eV for soda-lime-silicate and fused silica, respectively. These values are quite similar to the 0.4 eV observed for amorphous As_2S_3[11] and GaP.[15] Such similarities in the absorption characteristics of greatly different glass types is remarkable. Preliminary data taken on a number of liquids and glasses indicate that ΔE is directly proportional to kT_f, where T_f is the fictive temperature. This implies that in random molecular structures the fictive temperature gives a direct measure of the magnitude of the average local microelectric field term in the Dow-Redfield model for the exponential absorption edge.

In appraising the merits of prospective fiber optical waveguide materials it is useful to consider their fundamental loss limitations. In addition, since these losses are dependent on optical wavelength, it is also possible to establish not only the preferred material but also the preferred optical sources for communications applications.

To unify the data presented above, we have plotted in Fig. 3 the total intrinsic loss (scattering plus absorption) of pure fused silica and pure soda-lime-silicate, two of the most promising waveguide materials. The data are plotted directly as a function of wavelength out to 1.1 μ, which corresponds to the cutoff wavelength of silicon photodetectors. Although the optical loss continues to decrease beyond this wavelength, the prospect of using narrower-band-gap detector materials is not encouraging if one considers that they are less sensitive and may require cooling. Carbon tetrachloride was not included in Fig. 3 because its absorption loss is not known. However, it should be pointed out that in the red and near infrared the scattering loss alone in CCl_4 exceeds the total loss of fused silica. Of the two glasses shown in Fig. 3, it is apparent that fused silica is the preferred material by a factor of 3–4 in loss. Further, it is noted that the total loss in this material decreases with increasing optical wavelength. As a practical consequence, the repeater spacing in a fiber optical communications system limited by intrinsic waveguide loss could be over two times greater at a wavelength of 1.06 μ (Nd:YAG laser) than at 0.8 μ ($Ga_xAl_{1-x}As$ source). The ultimate loss at 1.06 μ in fused silica is 1.2 dB/km.

The authors wish to thank S.H. Wemple and D. Aspnes for helpful discussions, and A.D. Pearson and W.G. French for preparing the high-purity soda-lime-silicate test samples. The authors also wish to thank P. Kaiser for providing some complementary loss data on the soda-lime silicate materials. The experimental assistance of D.D. Badding is gratefully acknowledged.

[1]I. L. Fabelinskii, *Molecular Scattering of Light* (Plenum, New York, 1968), pp. 1–47.
[2]J. Stone, Appl. Phys. Lett. **20**, 239 (1972).
[3]D.A. Pinnow, S.J. Candau, J.T. LaMacchia, and T.A. Litovitz, J. Acoust. Soc. Am. **43**, 131 (1968). An earlier but qualitative discussion of the relationship between fictive temperature and light scattering was given by R.D. Maurer, J. Chem. Phys. **25**, 1206 (1956).
[4]T.C. Rich and D.A. Pinnow, Appl. Phys. Lett. **20**, 264 (1972).
[5]A.D. Pearson, AIME Technical Conference on Recent Advances in Electronic, Optical and Magnetic Materials, San Francisco, Calif. 1971 (unpublished).
[6]F.W. Ostermayer and D.A. Pinnow, Annual Meeting of the American Ceramic Society, Cincinnati, Ohio, 1973 (unpublished).
[7]J.H. Gladstone and T.P. Dale, Proc. Roy. Soc. Lond. **153**, 317 (1863); Proc. Roy. Soc. Lond. **153**, 337 (1863).
[8]M.L. Huggins and K.-H. Sun, J. Am. Ceram. Soc. **26**, 4 (1943).
[9]G.W. Morey, *The Properties of Glass*, 2nd ed. (Reinhold, New York, 1954), Chap. X and XVI.
[10]J.D. Dow and D. Redfield, Phys. Rev. Lett. **26**, 762 (1971); Phys. Rev. B **5**, 594 (1972).
[11]D.L. Wood and J. Tauc, Phys. Rev. B **5**, 3144 (1972).
[12]J. Stuke and G. Zimmer, Phys. Status Solidi B **49**, 513 (1972).
[13]The highest-purity fused silica materials have been Suprasil W1, supplied by Amersil, Inc. The highest-purity soda-lime-silicate materials have been prepared by A.D. Pearson and W.G. French of Bell Laboratories.
[14]*Modern Aspects of the Vitreous State*, edited by J.D. MacKenzie (Butterworth, Washington, D.C., 1952), pp. 195–254.
[15]S.H. Wemple (private communication).

Rayleigh and Brillouin Scattering in K_2O-SiO_2 Glasses

J. SCHROEDER, R. MOHR, P. B. MACEDO, and C. J. MONTROSE

Vitreous State Laboratory, The Catholic University of America, Washington, DC 20017

The intensity and spectral distribution of light scattered by K_2O-SiO_2 glasses (K_2O content up to 40 mol%) were measured. The transverse and longitudinal sound-wave velocities and the photoelastic constants were evaluated from the results. The total intensity of the scattering (and therefore the attenuation caused by it) exhibited a minimum at a concentration of $\approx$25 mol% K_2O. For this composition the attenuation is $\approx$2/3 of that in pure SiO_2. This behavior results from the existence of anomalously small concentration fluctuations in the melt of $K_2O \cdot 3SiO_2$ glass. A qualitative explanation of this result, involving low-temperature immiscibility regions, is presented.

I. Introduction

THE optical attenuation caused by scattering from density fluctuations in glasses in the K_2O-SiO_2 system has been evaluated from ultrasonic data.[1] In the present work, the problem of the total light-scattering in the K_2O-SiO_2 system for glasses containing up to 40 mol% K_2O is treated.

It has been shown[1] that, for this system, the scattering resulting from density fluctuations is minimum at $\approx$20 mol% K_2O; the magnitude of these fluctuations, i.e. $\langle \Delta\rho_\kappa^2 \rangle$, is $\approx$½ that in pure SiO_2. The total scattering in a multicomponent material, however, includes a contribution from local composition fluctuations in addition to the density fluctuations. The present study was therefore undertaken to determine the magnitude of the composition fluctuations in the K_2O-SiO_2 glass system in an attempt to discover if a glass composition exists for which the sum of the density and composition fluctuation scattering is less than the scattering (resulting only from density fluctuations) in pure SiO_2, the lowest-attenuating glass yet reported.

II. Theoretical Background

The spectrum and intensity of the light scattered through an angle ϑ when a glass is illuminated with monochromatic light (frequency$=\nu_0$, wavelength$=\lambda_0$, intensity$=I_0$) polarized normal to the scattering plane are considered. The intensity/unit solid angle scattered by a volume V of the material is given by[2]

$$I = I_0(\pi^2/\lambda_0^4)V^2(1+\cos^2\vartheta)\langle \Delta\varepsilon_\kappa^2 \rangle \tag{1}$$

where $\Delta\varepsilon_\kappa$ is the κth spatial Fourier component of the fluctuation in dielectric susceptibility:

$$\Delta\varepsilon_\kappa = \frac{1}{V}\int_V d\mathbf{r} \exp(-i\kappa\cdot\mathbf{r})\Delta\varepsilon(\mathbf{r}) \tag{2}$$

where $\Delta\varepsilon(\mathbf{r})$ is the deviation at the point $\mathbf{r}$ of the susceptibility from its equilibrium value. The symbols $\langle \ldots \rangle$ in Eq. (1) denote an ensemble average.

The principal features of the scattered light spectrum are the Rayleigh line centered at the frequency ν_0 and a pair of Brillouin lines shifted symmetrically up and down in frequency

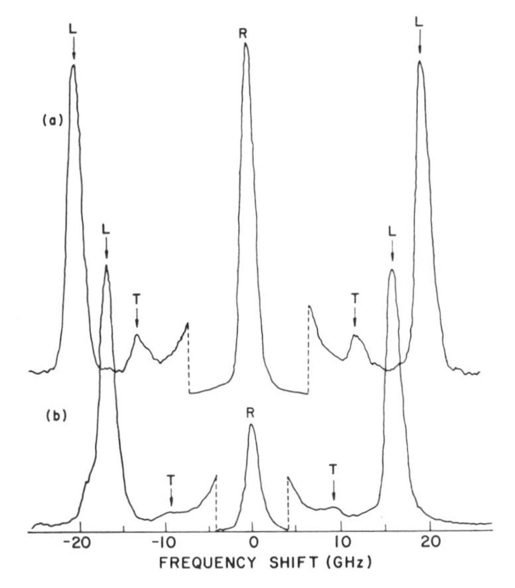

Fig. 1. Spectrum of scattered light from (*a*) fused SiO_2 and (*b*) $K_2O \cdot 3SiO_2$. Rayleigh lines are marked by *R*, Brillouin lines by *L*, and lines associated with transverse waves by *T*. *Vertical dashed lines* indicate scale changes of a factor of 50.

by an amount ν_B, the Brillouin frequency. Typical experimental spectra, in which these lines are labeled R and L, respectively, are shown in Fig. 1. In addition, in a material which can support shear stresses, such as a glass, the scattered spectrum also contains a weaker doublet (the lines labeled T in Fig. 1) shifted from the Rayleigh line by $\pm\nu_S$ which arises from the presence of transverse phonons in the material. The shifts ν_B and ν_S are related to the longitudinal and transverse sound-wave velocities in the material, v and v_S, respectively, by

$$\nu_B = v\kappa/2\pi$$

$$\nu_S = v_S\kappa/2\pi \tag{3}$$

where $\kappa = (4\pi n/\lambda_0)\sin(\vartheta/2)$ is the wave vector associated with the fluctuations responsible for the scattering[2] and n is the refractive index of the medium.

In a single-component substance (e.g. pure SiO_2), the mean-square fluctuation in susceptibility, $\langle \Delta\varepsilon_\kappa^2 \rangle$, can be written in terms of the fluctuations in density, $\langle \Delta\rho_\kappa^2 \rangle$, and temperature, $\langle \Delta T_\kappa^2 \rangle$:

$$\langle \Delta\varepsilon_\kappa^2 \rangle = (\partial\varepsilon/\partial\rho)_T^2 \langle \Delta\rho_\kappa^2 \rangle + (\partial\varepsilon/\partial T)_\rho^2 \langle \Delta T_\kappa^2 \rangle \tag{4}$$

where use has been made of the statistical independence of $\Delta\rho_\kappa$ and ΔT_κ, i.e. $\langle \Delta\rho_\kappa \Delta T_\kappa \rangle = 0$.[3] In general, the second term on the right side of Eq. (4) is negligible, and thus

$$\langle \Delta\varepsilon_\kappa^2 \rangle \approx (\partial\varepsilon/\partial\rho)_T^2 \langle \Delta\rho_\kappa^2 \rangle \tag{5}$$

The mean-square density fluctuations can be calculated from the thermodynamic properties of the medium. For a glass,[1]

$$\langle \Delta\rho_\kappa^2 \rangle = \frac{\rho_0^2}{V}\{kT_f(\beta_T - \beta_S) + kT_f[\beta_S - (\rho_0 v_\infty^2)^{-1}] + kT(\rho_0 v_\infty^2)^{-1}\} \tag{6}$$

Presented at the 74th Annual Meeting, The American Ceramic Society, Washington, DC, May 10, 1972. Received September 2, 1972; revised copy received April 6, 1973.
Supported by the Office of Naval Research.

where ρ_0 is the density, T_f the fictive temperature, i.e. that temperature at which the thermodynamic density fluctuations in the melt are kinetically "frozen into" the glass,[4,5] β_T and β_S are, respectively, the equilibrium isothermal and adiabatic compressibilities of the melt at the fictive temperature, and v_∞ is the high-frequency sound velocity in the glass. For a glass $v_\infty = v$ and is given in terms of the scattered light spectrum by Eq. (3). By combining Eqs. (1), (5), and (6) an expression for the scattered intensity, I, is obtained, i.e.

$$I = I_0(\pi^2/\lambda_0^4)V(1+\cos^2\vartheta)\left(\rho_0\frac{\partial\varepsilon}{\partial\rho}\right)_T^2 \{kT_f(\beta_T-\beta_S) + kT_f[\beta_S-(\rho_0 v_\infty^2)^{-1}] + kT(\rho_0 v_\infty^2)^{-1}\} \quad (7)$$

The first two terms represent intensity that appears in the Rayleigh line of the spectrum; the last describes the intensity of the Brillouin lines.[6,7] The Rayleigh intensity due to density fluctuations, I_R^ρ, is defined by

$$I_R^\rho = I_0\frac{\pi^2}{\lambda_0^4}V(1+\cos^2\vartheta)\left(\rho_0\frac{\partial\varepsilon}{\partial\rho}\right)_T^2 kT_f[\beta_T-(\rho_0 v_\infty^2)^{-1}] \quad (8)$$

and the Brillouin intensity, I_B, by

$$I_B = I_0\frac{\pi^2}{\lambda_0^4}V(1+\cos^2\vartheta)\left(\rho_0\frac{\partial\varepsilon}{\partial\rho}\right)_T^2 kT(\rho_0 v_\infty^2)^{-1} \quad (9)$$

The Landau-Placzek ratio in a single-component glass is given by

$$R_\rho = I_c^\rho/I_B = (T_f/T)(\rho_0 v_\infty^2\beta_T - 1) \quad (10)$$

It is conventional to rewrite the expression for the Brillouin intensity in terms of the photoelastic (Pockels) coefficient, p_{12}, for an isotropic solid[8,9] by using the relation

$$(\rho_0\partial\varepsilon/\partial\rho)_T = n^4 p_{12} \quad (11)$$

Similarly, the intensity of the lines associated with transverse phonons, I_S, is given by

$$I_S = I_0\frac{\pi^2}{\lambda_0^4}V(1+\cos^2\vartheta)(n^4 p_{44})^2 kT(\rho_0 v_S^2)^{-1} \quad (12)$$

where p_{44} is a second photoelastic coefficient and v_S is given by Eq. (3).

In a multicomponent glass system, Eqs. (4) and (5) must be modified to account for local fluctuations in composition. In a binary system, for example, a term of the form[3]

$$(\partial\varepsilon/\partial C)_{PT}^2 \langle\Delta C_\kappa^2\rangle = \left(\frac{\partial\varepsilon}{\partial C}\right)_{PT}^2 \frac{kT'_f}{N'}(\partial\mu/\partial C)_{PT}^{-1} \quad (13)$$

must be added to these equations, where C is the concentration defined as the ratio of the number of moles of solute N to the number of moles of solvent N', μ the chemical potential, and T'_f the fictive temperature associated with thermally arresting composition fluctuations of wave vector κ. In general $T'_f \neq T_f$. The scattering from composition fluctuations appears in the Rayleigh line in the spectrum; the intensity is given by

$$I_R^C = I_0\frac{\pi^2}{\lambda_0^4}V^2(1+\cos^2\vartheta)\left(\frac{\partial\varepsilon}{\partial C}\right)_{PT}^2 \frac{kT'_f}{N'}(\partial\mu/\partial C)_{PT}^{-1} \quad (14)$$

where R_C is defined as the ratio of this intensity to the Brillouin intensity:

$$R_C = \frac{I_R^C}{I_B} = \frac{T'_f}{T}\frac{(\partial\varepsilon/\partial C)_{PT}^2}{(\rho_0\partial\varepsilon/\partial\rho)_{TC}^2}(V/N')\rho v_\infty^2(\partial\mu/\partial C)_{PT}^{-1} \quad (15)$$

The Landau-Placzek ratio is then

$$R_{LP} = I_R/I_B = \frac{I_R^C + I_R^\rho}{I_B} = R_C + R_\rho \quad (16)$$

To evaluate the Pockels coefficient, p_{44}, it is also useful to define the intensity ratio, R_s, as

$$R_s = I_S/I_B = {}^1/_2(p_{44}/p_{12})^2(v_\infty^2/v_S^2) \quad (17)$$

where I_S is the intensity of the transverse phonon lines and v_S is given in Eq. (3).

III. Experimental Procedure

Light-scattering spectra and intensities were measured at a scattering angle of 90° for SiO_2 and for seven K_2O-SiO_2 glasses containing from 8 to 40 mol% K_2O. The major components of the experiment are a 50-mW He-Ne laser (λ_0=632.8 nm) used as the light source, a pressure-swept Fabry-Perot interferometer used to analyze the scattered light spectrum, a photomultiplier tube* as the detector, and standard photon-counting electronics to record the spectrum.

The He-Ne laser was operated multimode with a spectral width of ≈1.5 GHz. The beam has an initial divergence of 0.7 mrad and a diameter of 2 mm. To reduce the beam diameter in the sample without increasing its divergence substantially, a lens with a long focal length was used so that the resulting beam diameter in the sampling region was ≈0.5 mm and the divergence <12′.

The sample was contained in a cubical cell 7.5 cm on a side. The sides of the cell are flat, parallel, and optically polished, and it is filled with an index-matching oil to reduce parasitic scattering from the cell walls and sample surfaces. The cell rests on an *xyz* micropositioner which allows any region of the sample to be placed in the beam, so that seeds and other inhomogeneities in the sample, which do not represent the intrinsic properties of the material, can be avoided.

The scattering angle was established at 90° using a spectrometer.† To increase the light-gathering power of the Fabry-Perot interferometer, a collimating lens was placed so that one focal plane coincided with the beam in the sample. This arrangement effectively increases the acceptance angle of the interferometer and allows a small range of scattering angles about 90° to be accepted by the instrument. To limit the range of scattering angles acceptably, two vertical slits were provided; one, 2 cm wide, was mounted on the collimating lens and the other, 0.5 cm wide, was mounted immediately in front of the cell wall which faced the interferometer. The range of scattering angles thus accepted by the system was ≈90°±1°.

The pressure-swept flat-plate Fabry-Perot interferometer used as the spectrum analyzer had a free spectral range of 50 GHz and a resolution of <2 GHz. The instrumental width determined by the full width at half-maximum of the Rayleigh line observed when scattering from a diffuse scatterer such as ground glass was ≈1.5 GHz and represents the convolution of the interferometer response with the laser profile. The characteristic ring pattern of the Fabry-Perot transmission was imaged by a lens with a focal length of 620 mm so that the ring pattern was centered on a 2.8-mm pinhole. The full acceptance angle of the lens-pinhole system was ≈4.5 mrad.

The light passing through the pinhole was focused on the photocathode of the cooled photomultiplier. The resulting photoelectron pulses were amplified and discriminated by a pulse amplifier and discriminator (PAD).‡ The uniform pulses output by the PAD were fed into a linear ratemeter.§ The analog output of the ratemeter was registered as a function of time by a chart recorder.¶ The effective background counting rate for this system was ≈1 count/s. The counting system was capable of linear operation up to a rate of ≈5×10⁴ counts/s, which was greater than the rates observed in the present experiment.

This apparatus can also easily be used to measure the scattering in a sample relative to a well-characterized standard. The standard is placed directly above or below the sample in the scattering cell. By using the micropositioner, the scattering from either material can be examined without changing the collection geometry. The Rayleigh and/or Brillouin intensities can thus be compared directly.

---

*FW–130, ITT Electron Tube Div., Easton, PA.
†Gaertner Scientific Corp., Chicago, IL.
‡Model 813, Canberra Industries, Meriden, CT.
§Model 1480, Canberra Industries.
¶Model 194, Honeywell Inc., Minneapolis, MN.

Because the K_2O-SiO_2 glasses are rather soft and very hygroscopic, the faces of the samples (parellelepipeds 1 by 1 by 2 cm) could not be polished to anything resembling optical quality. To overcome this difficulty the samples were immersed in mixtures of toluene and standard oils* whose proportions were adjusted to just match the refractive indices of the samples. Thus, unwanted surface scattering was eliminated, and, in addition, the oils protected the samples from attack by water vapor.

The preparation of the samples was described in Ref. 1. Each melt was stirred for at least 5 h to achieve homogeneity and was then allowed to stand for an additional 5 h at high temperature to eliminate any small bubbles. The samples were then poured and annealed at a temperature corresponding to a viscosity of 10^{13} P for 6 h before they were cooled to room temperature.

No attempt was made to control the specimen temperature during the measurements. The temperature of the oil bath in which the samples were immersed, which was monitored using a Cu-constantan thermocouple, varied $<\pm 0.5°C$ during a run. The temperature of measurement was from 21° to 24°C for all runs.

Indices of refraction were measured at room temperature using a modified Abbe refractometer in connection with a 1-mW He-Ne laser (6328 Å). The details of the measurement are described by Boesch *et al.*[10] The accuracy of the refractive indices is ±0.01%. The densities of the glass samples were measured by the buoyancy principle and are estimated to be accurate to ±0.1%.

IV. Results

Typical Brillouin scattering spectra for fused SiO_2 and $K_2O \cdot 3SiO_2$ glasses are shown in Fig. 1. From such spectra the transverse and longitudinal wave speeds, v_S and v_∞, were evaluated at the frequencies ν_S and ν_B, respectively (Table I). The accuracy of the ν_B and v_∞ values is estimated at ±0.1%, for ν_S and v_S at ±0.5%. The Landau-Placzek ratio and the intensity ratio, R_s, were also measured (Table II). The Landau-Placzek ratios are estimated to be accurate to ±3%; the values of R_s are accurate to ±5%. The optical attenuation coefficient due to scattering, α_s (expressed in dB/km), can be calculated from the measured Landau-Placzek ratio and the absolute intensity of the Brillouin lines. In principle, the latter calculation requires that the Pockels coefficients p_{12} be known. To avoid the need for absolute intensity measurements the Brillouin intensities of all the glasses were measured relative to pure SiO_2. The absolute intensity of the Brillouin lines in SiO_2 was obtained by measuring p_{12} for SiO_2 by comparing I_B for SiO_2 to I_B for toluene at 25°C; the p_{12} value obtained was 0.286. This result compares rather well with the values, 0.270 (Ref. 11) and 0.285 (Ref. 12), reported previously. Consequently, α_s could be evaluated from the ratio of I_B for the specimen to that in SiO_2. This relative intensity was measured by mounting the sample glass on the standard SiO_2 sample† and then scanning the SiO_2 sample immediately after the measurement in the K_2O-SiO_2 glass without changing the geometry of the collection optics. The collection solid angle was corrected for the differences in refractive indices of the glasses. The values of α_s obtained in this way are shown in Table II; the accuracy of these data is ±7%. Also shown in Table II is the intensity ratio, R_ρ. The parameters used to compute this ratio, i.e. n, ρ_0, β_T, and T_f, are given in Table III.

The Pockels coefficients, p_{12}, were evaluated for the K_2O-SiO_2 glasses by comparing the Brillouin intensities for these glasses with that of fused silica. The appropriate formula obtained from Eq. (9) is

*R. P. Cargille Laboratories, Inc., Cedar Grove, NJ.
†Homosil, Amersil, Inc., Hillside, NJ.

Table I. Brillouin and Transverse Frequency Shifts and Longitudinal and Transverse Wave Velocities for K_2O-SiO_2 Glasses

Glass	ν_B (GHz)	v_∞ (10^5 cm/s)	ν_S (GHz)	v_S (10^5 cm/s)
SiO_2	19.30	5.920	12.4	3.79
$8K_2O \cdot 92SiO_2$	17.60	5.352	11.0	3.345
$10K_2O \cdot 90SiO_2$	17.23	5.228	10.7	3.24
$15K_2O \cdot 85SiO_2$	16.80	5.053	10.2	3.07
$20K_2O \cdot 80SiO_2$	16.49	4.943	9.9	2.97
$25K_2O \cdot 75SiO_2$	16.30	4.870	9.2	2.75
$33K_2O \cdot 67SiO_2$	16.18	4.806	9.0	2.67
$40K_2O \cdot 60SiO_2$	15.88	4.702	8.49	2.51

Table II. Intensity and Attenuation Data for K_2O-SiO_2 Glasses

Glass	R_{LP}	R_ρ	R_s	α_s (dB/km)
SiO_2	23.2	24.0	0.104	3.88
$8K_2O \cdot 92SiO_2$	45.5	11.9	.051	9.4
$10K_2O \cdot 90SiO_2$	34.8	11.0	.045	7.5
$15K_2O \cdot 85SiO_2$	17.9	10.2	.050	4.13
$20K_2O \cdot 80SiO_2$	13.9	9.8	.026	4.7
$25K_2O \cdot 75SiO_2$	9.5	10.0	.020	2.35
$33K_2O \cdot 67SiO_2$	9.8	10.1	.032	2.6
$40K_2O \cdot 60SiO_2$	10.2	10.2	.036	2.6

Table III. Thermodynamic Parameters and Pockels Coefficients for K_2O-SiO_2 Glasses

Glass	n	ρ_0 (g/cm³)	β_T* (10^{-12} cm²/dyne)	T_f† (°K)	p_{12}	p_{44}
SiO_2	1.4580	2.211	7.0	1600	0.286	0.078
$8K_2O \cdot 92SiO_2$	1.4715	2.271	8.1	820	.278	.055
$10K_2O \cdot 90SiO_2$	1.4748	2.301	8.2	781	.262	.049
$15K_2O \cdot 85SiO_2$	1.4877	2.356	8.4	758	.245	.047
$20K_2O \cdot 80SiO_2$	1.4930	2.386	8.6	743	.240	.039
$25K_2O \cdot 75SiO_2$	1.4978	2.425	8.9	732	.241	.028
$33K_2O \cdot 67SiO_2$	1.5067	2.461	9.4	700	.226	.032
$40K_2O \cdot 60SiO_2$	1.5111	2.491	10.0	679	.132	.019

*Ref. 1.
†Ref. 13.

$$p_{12}^{(gl)} = \left(\frac{I_B^{(gl)}}{I_B^{(o)}}\right)^{1/2} \left(\frac{n^{(o)}}{n^{(gl)}}\right)^{5} \left(\frac{\rho_0^{(gl)}}{\rho_0^{(o)}}\right)^{1/2} \left(\frac{\nu_B^{(gl)}}{\nu_B^{(o)}}\right) p_{12}^{(o)} \tag{18}$$

where the designations (gl) and (o) refer to the glass specimen and to silica, respectively. The coefficients p_{44} were then computed using Eq. (17), and data are given in Table III. The uncertainty in these values is approximately ±10%.

V. Discussion

The experimental Landau-Placzek ratios are shown as a function of composition in Fig. 2; Fig. 3 shows R_C, the ratio of the scattering intensity from concentration fluctuations to the Brillouin intensity. It is apparent that the rather small values of the Landau-Placzek ratio for concentrations above 25% K_2O, and hence in the attenuation coefficient, α_s, are associated with the small intensity of composition fluctuations above the 25% concentration. This rather surprising behavior can be understood, at least qualitatively, from the following.

It is assumed that in the K_2O-SiO_2 system a miscibility gap exists in the concentration range 0 to 25% K_2O; the critical temperature, T_c, and, indeed, the entire "dome" characterizing the two-phase region, are assumed to exist at temperatures below the fictive temperature T'_f. The assumption of the existence of the immiscibility region is plausible in that similar behavior has been observed in the Li_2O-SiO_2 and Na_2O-SiO_2 systems[14]; the low-temperature character of the immiscibility is also not unreasonable if one recalls that the immiscibility

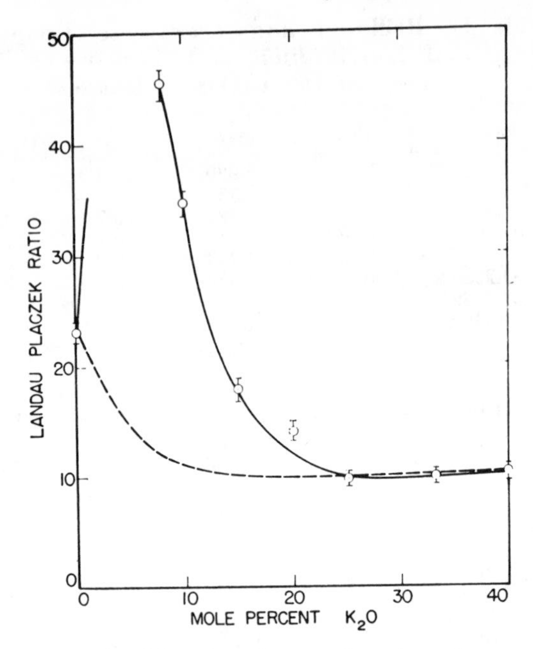

Fig. 2. Landau-Placzek ratio as a function of concentration. *Dashed line* shows contribution from density fluctuations computed from Eq. (10).

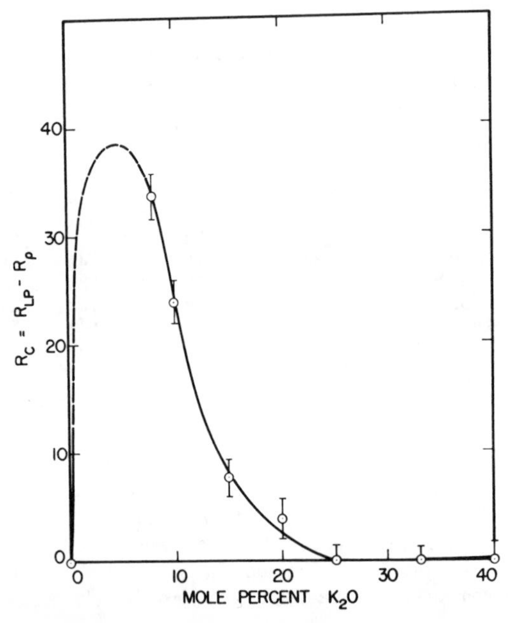

Fig. 3. Contribution from concentration fluctuations to Landau-Placzek ratio, R_C, as a function of concentration.

in the Li_2O-SiO_2 system lies at higher temperatures than that in the Na_2O-SiO_2 system; presumably then the immiscibility in the K_2O-SiO_2 system (if it exists) would lie at lower temperatures than that in Na_2O-SiO_2. In an equilibrium binary mixture, the classical theory of critical fluctuations[3] predicts that

$$(\partial\mu/\partial C)_{PT} \propto T - T_s \tag{19}$$

where T_s is the so-called spinodal temperature.[15] The fluctuations in composition are therefore given by

$$\langle \Delta C_\kappa^2 \rangle \propto [1-(T_s/T)]^{-1} \tag{20}$$

In a glass below its fictive temperature these fluctuations are frozen in at T'_f so that Eq. (20) is replaced by

$$\langle \Delta C_\kappa^2 \rangle \propto [1-(T_s/T'_f)]^{-1} \tag{21}$$

Based on this kind of thinking, it follows that there must exist a minimum in the curve of spinodal temperature versus con-

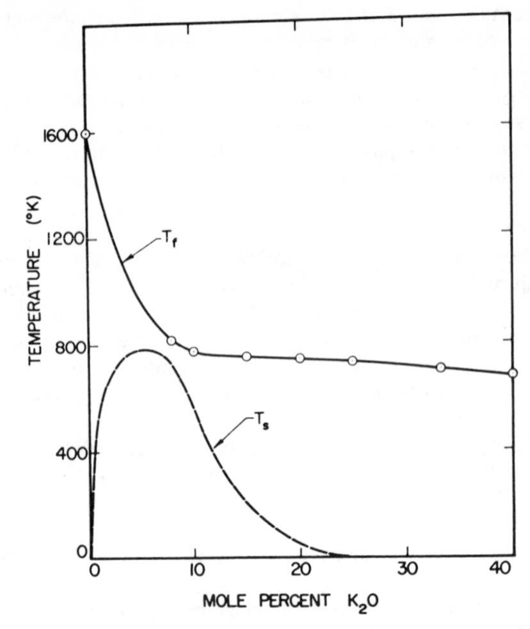

Fig. 4. Possible plot of spinodal temperature, T_s, vs composition consistent with observed data. Fictive temperature, T_f, is also plotted.

centration at 25 mol% K_2O. The type of behavior necessary to account for the data is sketched in Fig. 4. (The curve shown is intended to be illustrative only; a rather wide variety of possible T_s curves is consistent with the results of this work.) There is no obvious reason for, nor simple conclusion to be drawn from, the fact that the region of small concentration fluctuations occurs at and above 25% K_2O content. One could speculate about the existence of complexes of the type $K_2O(SiO_2)_3$ that are relatively stable in the melt and are frozen into the glass as it is cooled; however, such a discussion is certainly premature based on the data available.

VI. Summary

Measurements of the spectrum of light scattered from a set of K_2O-SiO_2 glasses have made possible the evaluation of the longitudinal and transverse sound wave velocities in these materials. Measurements of the intensities of the phonon peaks (shifted lines) in the spectrum have been used to evaluate the Pockels coefficients, p_{12} and p_{44}, for the glasses. In addition, the total scattered intensity exhibits a marked decrease with concentration above 25 mol% K_2O content, due primarily to an anomalous reduction in the concentration fluctuations in this range. This behavior can be rationalized by assuming that miscibility gaps exist in the K_2O-SiO_2 system lying on either side of the composition $K_2O \cdot 3SiO_2$.

The small scattered intensity for the $K_2O \cdot 3SiO_2$ composition and for more K_2O-rich systems means that there is also a decrease in the optical attenuation coefficient due to scattering at this concentration. Indeed the attenuation is roughly $^2/_3$ of that in pure SiO_2 so that this glass offers great potential for use in optical fiber transmission systems. Because of its rather low glass transition temperature ($\approx$450°C) it is also rather easy to fabricate. We are presently experimenting with other alkali silicate glasses and are examining the effects of introducing a third component in the potassium silicate glasses to see if substantial improvements in the chemical durability of the glass can be effected without adversely affecting the optical properties.

References

[1] N. L. Laberge, V. V. Vasilescu, C. J. Montrose, and P. B. Macedo, "Equilibrium Compressibilities and Density Fluctuations in K_2O-SiO_2 Glasses"; this issue, pp. 506–509.

[2] I. L. Fabelinskii, Molecular Scattering of Light; Chapter I. Plenum Press, New York, 1968.
[3] L. D. Landau and E. M. Lifshitz, Statistical Physics, 2d ed.; Chapter 12. Addison-Wesley Publishing Co., Reading, MA, 1969.
[4] H. Mueller, "Theory of Scattering of Light," *Proc. Roy. Soc., Ser. A.*, **166** [2] 425–49 (1938).
[5] R. D. Maurer, "Light Scattering by Glasses," *J. Chem. Phys.*, **25** [6] 1206–1209 (1956).
[6] S. Y. Hsich, R. W. Gammon, P. B. Macedo, and C. J. Montrose, "Light Scattering in Aqueous LiCl Solutions; Evidence for a Low Temperature Immiscibility," *ibid.*, **56** [4] 1663–69 (1972).
[7] D. A. Pinnow, S. J. Candau, J. T. LaMacchia, and T. A. Litovitz, "Brillouin Scattering: Viscoelastic Measurements in Liquids," *J. Acoust. Soc. Amer.*, **43** [1] 131–42 (1968).
[8] J. F. Nye, Physical Properties of Crystals; pp. 254f. Clarendon Press, Oxford, 1960.
[9] (*a*) F. Pockels, "Effect of Elastic Deformation from Uniaxial Stresses on the Optical Behavior of Crystalline Bodies," *Ann. Phys. Chem.*, **37** [1] 144–72 (1889).
(*b*) F. Pockels, "Alteration of the Behavior of Various Glasses by Elastic Deformation," *Ann. Phys. (Leipzig)*, **7** [5] 745–71 (1902).
[10] L. Boesch, A. Napolitano, and P. B. Macedo, "Spectrum of Volume Relaxation Times in B_2O_3," *J. Amer. Ceram. Soc.*, **53** [3] 148–53 (1970).
[11] William Primak and Daniel Post, "Photoelastic Constants of Vitreous Silica and Its Elastic Coefficient of Refractive Index," *J. Appl. Phys.*, **30** [5] 779–88 (1959).
[12] K. Vedam, "Elastic and Photoelastic Constants of Fused Quartz," *Phys. Rev.*, **78** [4] 472–73 (1950).
[13] J. P. Poole, "Low-Temperature Viscosity of Alkali Silicate Glasses," *J. Amer. Ceram. Soc.*, **32** [7] 230–33 (1949).
[14] E. A. Porai-Koshits and V. I. Averjanov, "Primary and Secondary Phase Separation of Sodium Silicate Glasses," *J. Non-Cryst. Solids*, **1** [1] 29–38 (1968).
[15] J. W. Cahn, "Spinodal Decomposition," *Acta Met.*, **9** [7] 795–801 (1961).

Spectral losses of unclad vitreous silica and soda-lime-silicate fibers

P. Kaiser, A. R. Tynes, and H. W. Astle

Bell Telephone Laboratories, Incorporated, Crawford Hill Laboratory, Holmdel, New Jersey 07733

A. D. Pearson, W. G. French, and R. E. Jaeger

Bell Telephone Laboratories, Incorporated, Murray Hill, New Jersey 07974

A. H. Cherin

Bell Telephone Laboratories, Incorporated, Norcross, Georgia 30071

(Received 26 January 1973)

Spectral transmission losses of unclad optical fibers drawn from various types of commercial vitreous silica and of soda-lime-silicate glasses were measured in the wavelength range between 0.5 and 1.12 μm. The automated technique employed was both convenient to use and sensitive enough to measure losses of 6 dB/km, i.e., 75% loss per km, with an estimated accuracy of ± 2 dB/km. Bulk-loss measurements performed with calorimetric and bridge-type techniques were in good agreement with the fiber-loss measurements.

Index Headings: Absorption, Fibers, Glass.

The successful development of low-loss optical fibers for future optical communication systems requires accurate knowledge of the spectral losses of prospective bulk materials. Owing to the availability of only short bulk-sample lengths, the sensitivity of previously used spectrophotometric methods has been limited.[1] Recently, a nondestructive technique for measuring the spectral losses of unclad optical fibers has been introduced.[2] Losses due to surface imperfections and contamination can be kept sufficiently low so that the unclad-fiber losses closely approximate those of the bulk material.

In this paper we describe the spectral transmittance losses of unclad fibers drawn from various grades of commercially available vitreous silica,[3] and of soda-lime-silicate glasses produced at Bell Laboratories.[4] The spectral losses were determined between 0.5 and 1.15 μm by selecting 10-nm portions from the spectrum of a Xe arc lamp at usually 20-nm intervals. The length of the fibers varied from 10 to 27 m; the easier-to-clean shorter sections were employed for the higher-loss materials. Aside from influencing the accuracy, the losses, expressed in dB/km, were independent of fiber length, provided that the surface quality was sufficiently high.

FIBER-DRAWING MACHINE

The schematic representation of the fiber-drawing machine in Fig. 1 shows a variable-speed vertical-feed mechanism for lowering the preform and a 27-cm-diam take-up drum with axial movement for the systematic winding of the fibers. Whereas vitreous-silica fibers were drawn with an oxy-hydrogen torch, the soda-lime-silicate fibers were made by use of an electric furnace. The unclad fibers were drawn both from commercially available 7-mm drawn rods and from ground and polished rods. Provided that the polish was of sufficiently high quality, no differences in the loss curves of the fibers produced from either preform were discernible.

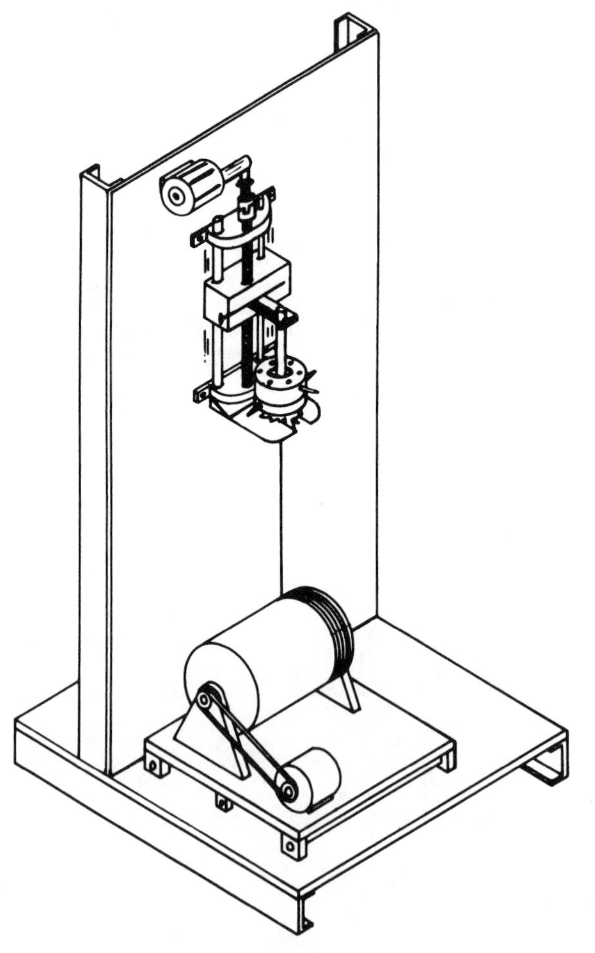

FIG. 1. Fiber-drawing apparatus.

Reprinted with permission from *J. Opt. Soc. Amer.*, vol. 63, pp. 1141–1148, Sept. 1973.

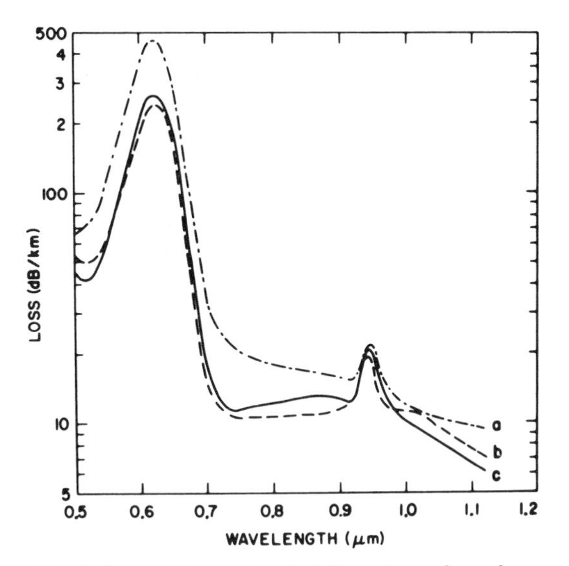

FIG. 2. Spectral losses of unclad fibers drawn from three Suprasil W1 samples purchased at different times.

Immediately preceding the drawing process, both the preform and the plastic-coated drum were carefully cleaned with Windex and distilled water. Electrostatic charges accumulated as a result of the cleaning operation were eliminated with a ^{210}Po α-particle source.[5]

The drawing temperature was controlled by the flow rates of the oxygen and hydrogen fed to the torch. The preform-feed speed and temperature were set to allow for convenient control of the fiber diameter by means of the variable-speed take-up drum. Fibers of about 200-μm diameter were drawn at an approximate rate of 15 m per minute. The drums were placed in a plastic box to prevent dust accumulation on the surface of the freshly drawn fibers.

RESULTS OF SPECTRAL-LOSS MEASUREMENTS

Suprasil W

The spectral losses of three samples of Suprasil W1 (SSW1), a low-water-content vitreous silica of high purity, are shown in Fig. 2. Suprasil W2 (SSW2) is optically slightly less homogeneous than SSW1; but the two materials are considered equally pure chemically, as evidenced by virtually identical spectral losses (Fig. 3). Therefore, the following discussion refers to both materials as Suprasil W (SSW), except where stated otherwise.

Starting at lower wavelengths, the spectral losses of SSW show a high loss peak of about 250 dB/km at 0.625 μm. A uniform low-loss region of typically less than 13 dB/km, i.e., 95% per km, follows between 0.76 and 0.92 μm. The 6–12-dB/km loss peak at 0.945 μm is due to the OH-ion content of SSW.[6] Beyond 0.945 μm, the losses continuously decrease; the smallest loss, 6 dB/km, i.e., 75% per km, was obtained with one SSW1 sample at 1.12 μm, which happens to coincide with the edge of the spectral region investigated. Based on the downward trend of the loss curve at 1.12 μm, minimum losses less than 6 dB/km are expected somewhere between 1.12 and 1.2 μm, because another loss peak of about 17–34 dB/km is expected at 1.24 μm, as explained later.

In contrast to the high optical quality in the near infrared, the transmittance of the SSW fibers is very low in the visible spectrum, owing to the 250-dB/km loss peak mentioned previously. We have not succeeded in assigning this loss band to any particular impurity. Even though we might be tempted to attribute this loss band to the presence of transition-metal ions,[7] many of those that show absorption bands near 0.625 μm also have substantial losses at 1.06 μm, where SSW has its highest transmissivity.

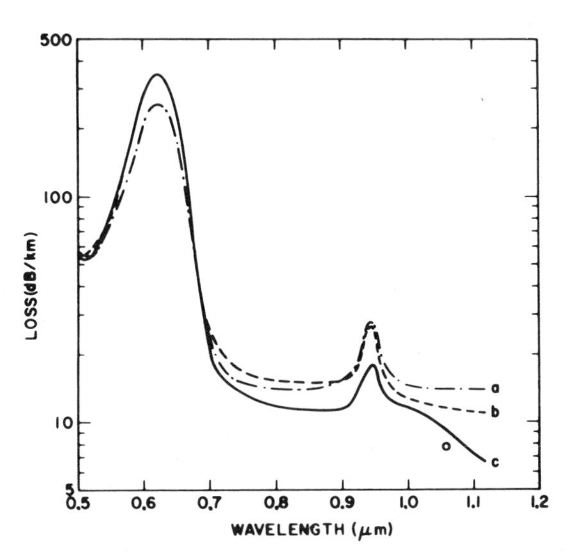

FIG. 3. Spectral losses of unclad fibers drawn from three Suprasil W2 samples purchased at different times. Comparison of bulk (circle) with unclad-fiber losses of sample b at 1.06 μm.

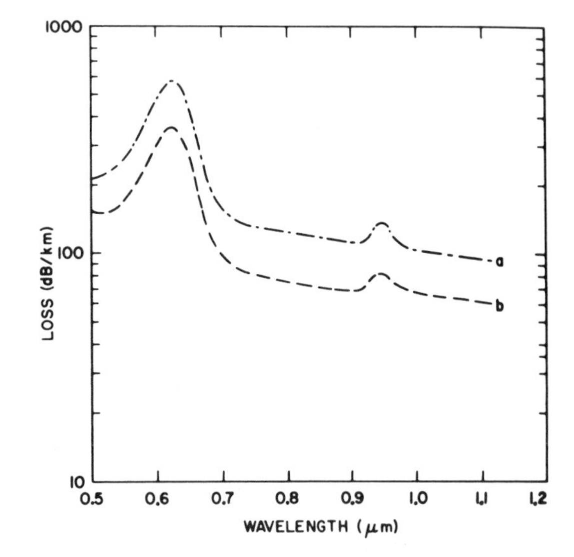

FIG. 4. Spectral losses of unclad fibers drawn from (a) a high bubble-content Spectrosil WF rod and (b) an insufficiently polished Suprasil W2 preform.

Spectrosil WF

The 0.625-μm loss band was also measured in a low-OH-content synthetic silica, Spectrosil WF (SPWF), produced by a different manufacturer (Fig. 4). Unfortunately, the preform rod contained a large number of elongated bubbles, which were drawn down to needles in the fiber and caused scattering losses of the order of 100 dB/km. The excess absorption loss of about 26 dB/km at 0.945 μm is higher than that in SSW. The general wavelength dependence of the SPWF curve is similar to that of a SSW2 fiber drawn from an insufficiently polished preform (Fig. 4). As a result, the surface of this fiber, too, showed imperfections with ensuing high scattering losses.

Suprasil 1

The spectral losses of Suprasil 1 (SS1), a synthetic vitreous silica of very high purity, are displayed in Fig. 5. In contrast to the low OH content of less than 5 ppm of SSW, SS1 contains approximately 1200 ppm OH, which manifests itself in numerous absorption bands throughout the spectral region investigated. The more-pronounced peaks at 0.72, 0.88, and 0.945 μm have already been observed in other fibers and have been assigned to overtones of the fundamental ν_3^1 stretching vibration of the OH ion at 2.72 μm, and to combinational vibrations between the OH overtones and the fundamental SiO_4 tetrahedron vibration at 12.5 μm.[8] Owing to the high sensitivity attained through long sample lengths, we were able to tentatively identify, in addition, the fourth overtone $4\nu_3^1$ of the OH vibration near 0.6 μm, to assign the secondary peak at 0.68 μm to the combination vibration of the third OH overtone and the fundamental SiO_4 vibration, and to observe a combinational vibration of the second OH with the first SiO_4 overtone at 0.82 μm. A small hump in some of the loss curves at 0.64 μm is probably caused by the combinational vibration of the third overtone of the OH and the first overtone of the SiO_4 tetrahedron vibration. An identification of the exact wavelengths of the various absorption peaks was not possible, because of the limited resolution of our wavelength scan.

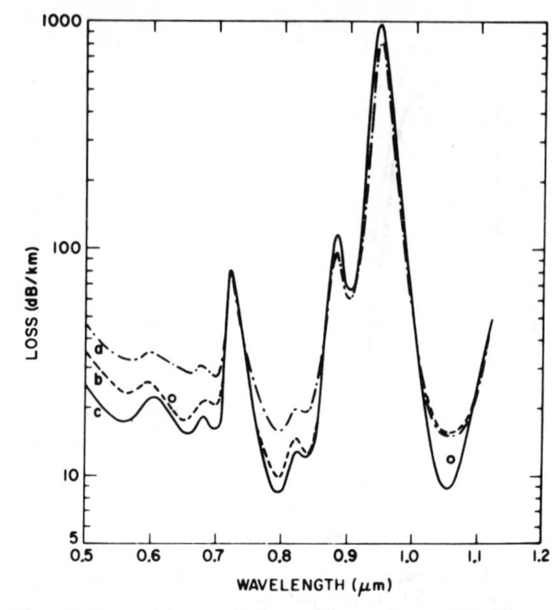

Fig. 5. Spectral losses of three different Suprasil 1 fibers and comparison with bulk losses. The bulk loss at 1.06 μm, shown by the circle, was obtained with the same material as the fiber that gave curve c; the bulk-loss data point at 0.6328 μm (circle) was obtained with a different sample.

Table I. OH overtones and combinational vibrations in Suprasil 1 vitreous silica, and their peak intensities.

Wavelength (μm)	Frequency	Loss (dB/km)	Remarks
0.60	$5\nu_3^1$	6	a,b
0.64	$2\nu_1+4\nu_3^1$	1	a,b
0.68	$\nu_1+4\nu_3^1$	4	a,b
0.72	$4\nu_3^1$	70	b
0.82	$2\nu_1+3\nu_3^1$	4	a,b
0.88	$\nu_1+3\nu_3^1$	90	b
0.945	$3\nu_3^1$	1 000	b
1.13	$2\nu_1+2\nu_3^1$	110	c
1.24	$\nu_1+2\nu_3^1$	2 800	d
1.38	$2\nu_3^1$	65 000	d
1.90	$2\nu_1+\nu_3^1$	10 300	d
2.22	$\nu_1+\nu_3^1$	260 000	d
2.72	ν_3^1	10 000 000	d

[a] Observed for the first time.
[b] Loss values from a Suprasil 1 unclad fiber.
[c] Estimated loss.
[d] Loss from Amersil Catalog.

The peak losses of the various OH absorption bands identified in Fig. 5 are tabulated in Table I. Also included in this listing are other OH overtones and combinational vibrations known to exist between 1.0 and 2.72 μm. Aside from the loss band near 1.13 μm, which, by extrapolation of the loss curve in Fig. 5, was estimated to amount to 110 dB/km, approximate values for these peaks were obtained from optical transmittance curves supplied by Amersil, Inc.[3] When plotted against wavelength, the peak losses due to OH overtones, and their combinational vibrations with the fundamental and first overtone of the SiO_4 vibration, follow smooth curves (Fig. 6). The curves are separated by an approximate factor of 10. However, these data cannot be used for estimation of the losses in intermediate-wavelength regions. Their approximate values can be obtained from Fig. 7, where we used the lowest spectral-loss curve from Fig. 5 and the data from Table I to estimate the spectral losses of SS1 in the whole wavelength range between 0.5 and 2.0 μm.

Under the assumption that the ratios of the various peak losses are constant and amount to 0.11 and 2.8 for the 1.13- and 1.24-μm peaks, relative to the 0.945-μm peak, we can estimate the expected losses of other types of vitreous silica at those wavelengths from a knowledge of their absorption peaks at 0.945 μm. For example, for losses of 6–12 dB/km at 0.945 μm in the case of SSW,

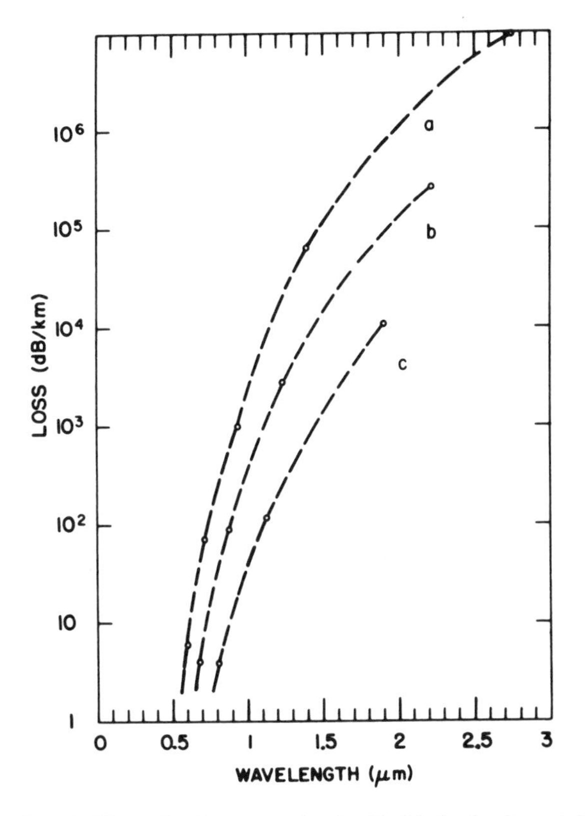

FIG. 6. Absorption losses associated with (a) the fundamental OH vibration and its overtones, (b) the combinational vibrations with the fundamental SiO_4, and (c) first overtone of the SiO_4 vibration in Suprasil 1 vitreous silica.

the expected loss due to OH absorption at 1.13 μm lies between 0.65 and 1.3 dB/km, and at 1.24 μm between 17 and 34 dB/km.

In contrast to the high OH absorption peak of the order of 1000 dB/km at 0.945 μm, a total attenuation of only 8.5 dB/km was measured immediately adjacent to it at 0.795 and 1.05 μm with one of the three SS1 samples tested (Fig. 5, curve c). This demonstrates that the OH absorption lines are sufficiently narrow and do not appreciably affect the losses in intermediate-wavelength regions.

As a result of scattering-loss measurements, we estimate that 50% of the 8.5-dB/km total loss at 0.8 μm is due to absorption, about 25% due to interface scattering, and 25% due to Rayleigh-scattering losses.

The second SS1 sample tested (Fig. 5, curve b) exhibited an almost equal minimum loss of 10 dB/km at 0.8 μm, but exhibited a 6-dB/km higher loss of 15 dB/km at 1.06 μm. Since excess losses due to surface contamination could be shown to decrease with increasing wavelength, the higher loss at 1.06 μm must be attributed to absorption losses caused by impurities. A likely contaminant is the divalent ferrous ion, Fe^{2+}, which has a broad absorption band centered in this spectral region.

The higher losses of the third SS1 sample (Fig. 5, curve a) were caused by imperfections of the fiber surface with consequent greater scattering losses. Without this deficiency, this fiber probably would have exhibited losses at 0.8 and 1.06 μm as low as the first sample. The lower OH absorption peaks may be due to the lower OH concentration of this sample.

On the basis of an experimental absorption peak of SS1 of the order of 1000 dB/km and a 1200-ppm OH content as quoted by the manufacturer, we obtain a calibrated loss of 0.83-dB/km/ppm OH at 0.945 μm. This value is 33% less than the 1.24-dB/km/ppm OH suggested previously.[9] Accordingly, the OH content of the six SSW samples tested is estimated to vary between 7 and 14 ppm, corresponding to the experimental 6–12-dB/km absorption losses. The discrepancy between the experimental data and the manufacturer's quoted value of 5 ppm may be due to an actual OH content exceeding that of the specification, or as a result from drawing in an oxy-hydrogen flame.

Spectrosil

The spectral losses of two Spectrosil (SP) samples are shown in Fig. 8. Similarity in the manufacturing process of SP and SS1 is suggested by nearly identical loss curves. Whereas the second SP sample showed equally low losses at shorter wavelengths, the loss minima at 0.8 and 1.06 μm increased with λ, indicating again a possible contamination with ferrous ions.

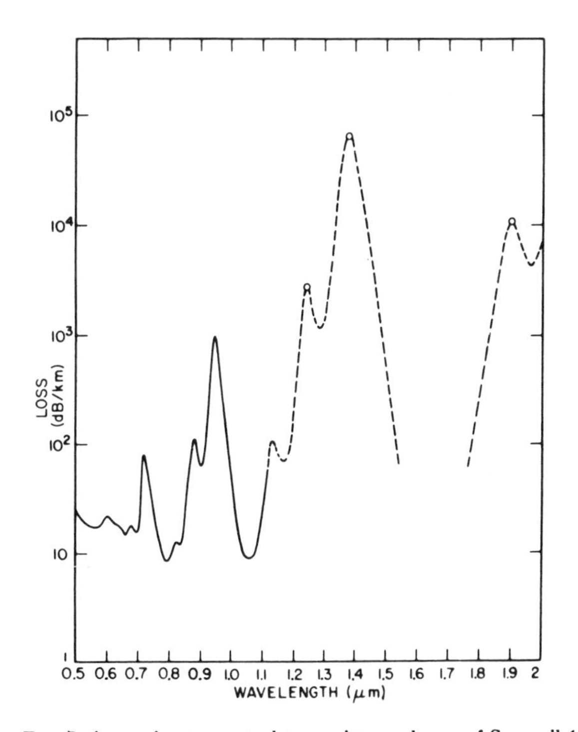

FIG. 7. Approximate spectral transmittance losses of Suprasil 1 vitreous silica in the wavelength range between 0.5 and 2 μm: solid lines, measured with unclad fiber; circle, obtained from Amersil Catalog; broken line, estimated.

Corning 7940

A similar increase of the loss minima with wavelength was observed with Corning 7940,[10] another high-OH-content synthetic vitreous silica (Fig. 9). The lowest losses for this material were measured at 0.54 μm and amounted to 44 dB/km. The approximately 750-dB/km OH peak at 0.945 μm is somewhat less than that of SS1 and SP, resulting in correspondingly smaller but identifiable OH and combinational overtone peaks throughout the spectrum. The calculated OH content of about 900 ppm seems to be typical and has been reported for several 7940 samples.[11]

Infrasil

Spectral losses of unclad fibers drawn from three different grades of commercial fused quartz made from Brazilian rock crystal are discussed next. Approximately one-order-of-magnitude-greater metallic-impurity concentrations result in correspondingly greater losses throughout the spectral range, as shown, for example, by the spectral losses of Infrasil (Fig. 10). The broad absorption band centered around 0.85 μm is probably due to Cu^{2+} ions. Its very low OH content manifests itself in the smallest loss peak observed at 0.945 μm with any vitreous silica, namely, ~4 dB/km. From this, the OH content is calculated to be 4.8 ppm, in agreement with the manufacturer's specification of less than 5 ppm. The lowest total losses of the Infrasil unclad fiber were measured at 1.12 μm and amounted to 38 dB/km. Near 0.625 μm, we again notice the strong loss band found in Suprasil W and Spectrosil WF unclad fibers.

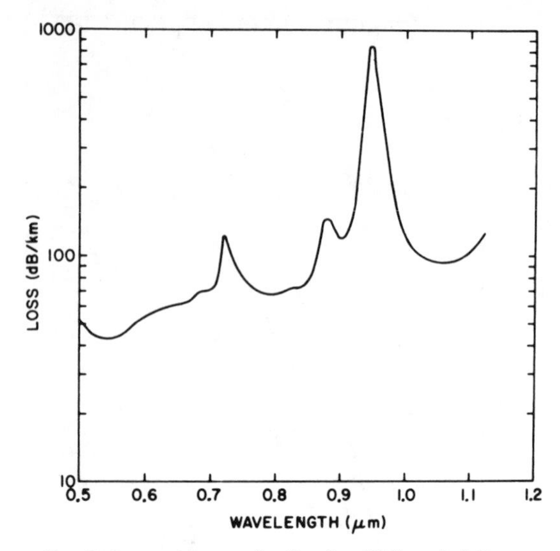

Fig. 9. Spectral losses of a Corning 7940 unclad fiber.

Ultrasil

Although with somewhat less peak loss, the 0.625-μm loss band was also observed in an Ultrasil unclad fiber (Fig. 11). A constant loss of 100 dB/km was measured between 0.69 and 0.92 μm, and some evidence of the OH peaks at 0.72 and 0.88 μm is visible. Based on the measured OH absorption peak of about 108 dB/km at 0.945 μm, the calculated OH content of 130 ppm agrees with the 130 ppm quoted in the literature. Similar to the case of Infrasil, the losses continuously decreased beyond 0.945 μm, with the lowest value of 60 dB/km appearing at 1.12 μm.

TO8 Commercial

Spectral losses of two samples of TO8 commercial-grade fused quartz are presented in Fig. 12. Somewhat

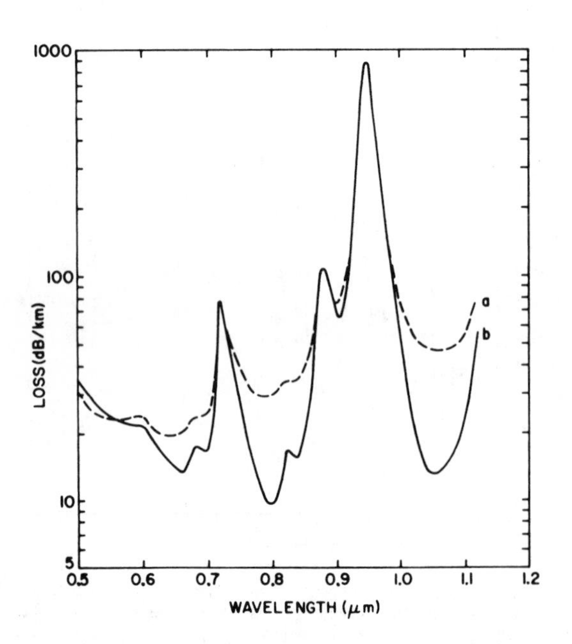

Fig. 8. Spectral losses of two different Spectrosil unclad fibers.

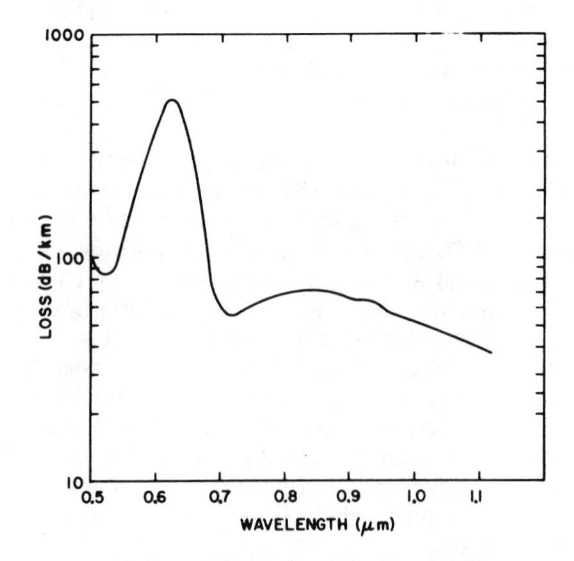

Fig. 10. Spectral losses of an Infrasil unclad fiber.

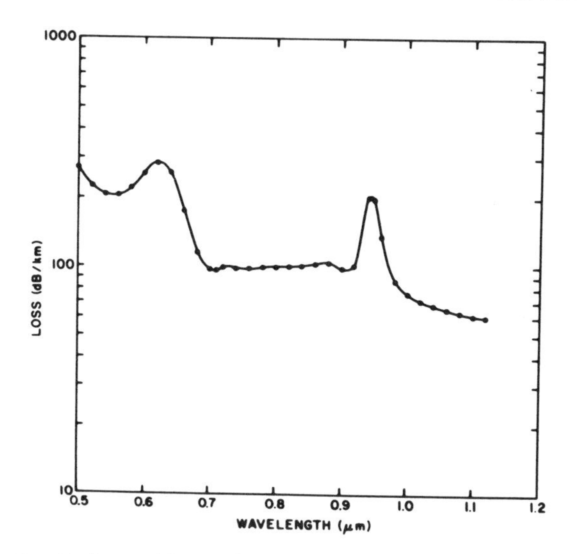

FIG. 11. Spectral losses of an Ultrasil unclad fiber. The circles identify the sample wavelengths used in all measurements.

surprisingly for a commercial-grade material, the two loss curves are very similar; losses as low as 26 dB/km were obtained at 0.70 and 0.77 μm. The OH loss peak of the order of 220 dB/km at 0.945 μm is higher than would be expected on the basis of a 130-ppm OH concentration.[3] The actual OH content might therefore be greater than quoted. One of the TO8 samples shows a small absorption band at 0.625 μm, which is possibly related to the loss band observed with the other materials. The comparatively high losses at greater wavelengths may be attributed to the presence of Fe^{2+} ions.

DISCUSSION OF SPECTRAL LOSSES OF VITREOUS SILICA UNCLAD FIBERS

To facilitate a comparison, the spectral losses of the various types of vitreous-silica fibers are combined in Fig. 13. It should be emphasized that it is still unknown to what degree the samples tested are representative of the various products, particularly in those cases for which only one or two samples were evaluated. The high purity of the synthetic vitreous silicas, Suprasil W and Suprasil 1, manifests itself in transmission losses that are, in general, substantially less than those of fused quartz. OH peaks varying from 4 dB/km for Infrasil to 1000 dB/km for Suprasil 1 and Spectrosil result from widely varying OH-ion concentrations and appear at 0.945 μm in all materials investigated. The OH absorption at shorter wavelengths decreased in proportion to that at 0.945 μm and were not measurable in the lowest-OH-content samples.

The pronounced loss band at 0.625 μm was observed in fibers drawn from vitreous silica that was made synthetically as well as from natural quartz. No assignment of this loss band could be made. An apparent relationship with the OH content is suggested by its absence in the high-OH-content Suprasil 1 and Spectrosil, and its most intense presence in the lowest-OH-content materials: Infrasil, Suprasil W, and Spectrosil WF. With regard to medium-OH-content materials, no definite statement can be made because Ultrasil exhibited a considerable loss peak at this wavelength, whereas one TO8 sample showed a small loss peak and the second sample showed no such peak. So far as we know, this loss band has not been identified in the bulk material; it is likely that it was introduced in the fiber-pulling process. Additional experiments are planned to resolve this question.

In contrast to its high transmission losses in the visible region, the losses of the Suprasil W fibers in the near infrared are sufficiently low to make this material

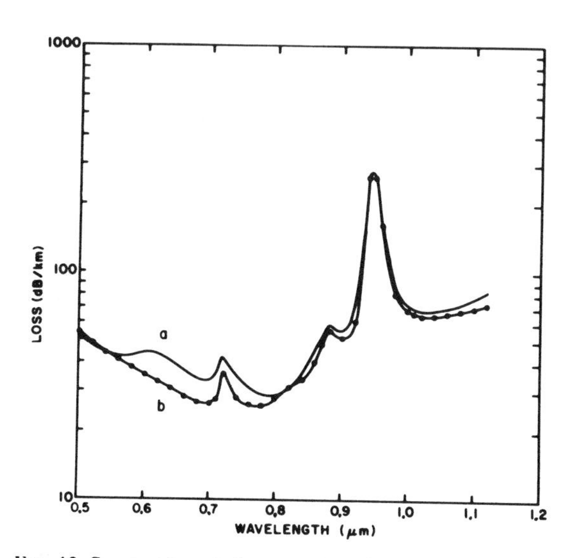

FIG. 12. Spectral losses of two commercial-grade TO8 unclad fibers drawn from different preforms.

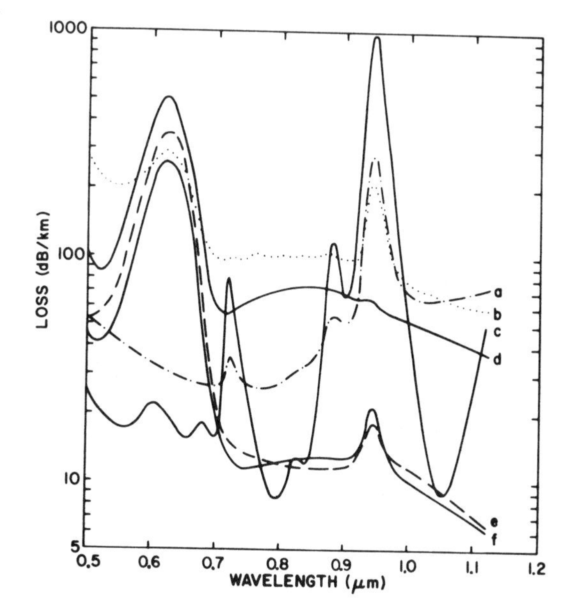

FIG. 13. Spectral losses of unclad fibers made from various grades of vitreous silica. (a) Commercial-grade TO8, (b) Ultrasil, (c) Suprasil 1, (d) Infrasil, (e) Suprasil W2, (f) Suprasil W1.

most significant for fiber-optics communication systems, because of the availability of light-emitting diodes and of GaAs and Nd-YAG laser sources, which emit in the near infrared.

COMPARISON WITH BULK-LOSS MEASUREMENTS

Using a highly sensitive calorimetric technique, Pinnow[10] measured the bulk absorption losses of one Suprasil W2 and one Suprasil 1 preform from which unclad fibers were drawn (see Fig. 3, curve b; Fig. 5, curve c). In Table II, we added 0.6 dB/km for the expected Rayleigh-scattering losses at the 1.06-μm wavelength of the Nd-YAG laser, at which the measurements were performed. The agreement is good for the SSW2 data, where the 4-dB/km-greater loss could be attributed to additional waveguide losses due to imperfections and dust accumulation on the unclad-fiber surface. The reason for the 3-dB/km-higher bulk loss for the SS1 sample is not clear to us. Because different sections of the 1-m-long preform rod were used for the two experiments, we might blame variations of the impurity level along the 7-mm preform rod for this discrepancy. Although this cannot be ruled out, we found very good reproducibility of spectral-loss curves of fibers drawn from different sections of the same Suprasil 1 preform rod. Nevertheless, the argeement between the bulk-material- and fiber-loss measurements can be considered to be reasonably good. Although samples from different batches were used for another such comparison, the 22-dB/km bulk losses of SS1, measured with a bridge-type technique at the He–Ne laser wavelength, 0.6328 μm,[12] compare very well with the approximate 18 dB/km obtained from unclad-fiber measurements (see Fig. 5).

SPECTRAL LOSSES OF UNCLAD SODA-LIME-SILICATE FIBERS

High-purity soda-lime-silicate (SLS) glasses have been produced at Bell Laboratories with the ultimate goal of fabricating low-loss optical fibers. Until now, effects of changes of the composition and impurity level have been evaluated by means of bulk-loss measurements performed at discrete wavelengths, namely, 0.6328 and 1.06 μm.[12] Spectral-loss measurement of unclad fibers drawn from the bulk confirmed those measurements and permitted a comparison throughout the spectral range of interest (Figs. 14 and 15).

TABLE II. Comparison between bulk and unclad-fiber losses measured at 1.06 μm. Samples were taken from the same batch.

	Bulk loss (dB/km)	Unclad-fiber loss (dB/km)
Suprasil W2	7.6	11.6
Suprasil 1	12.0	9.0

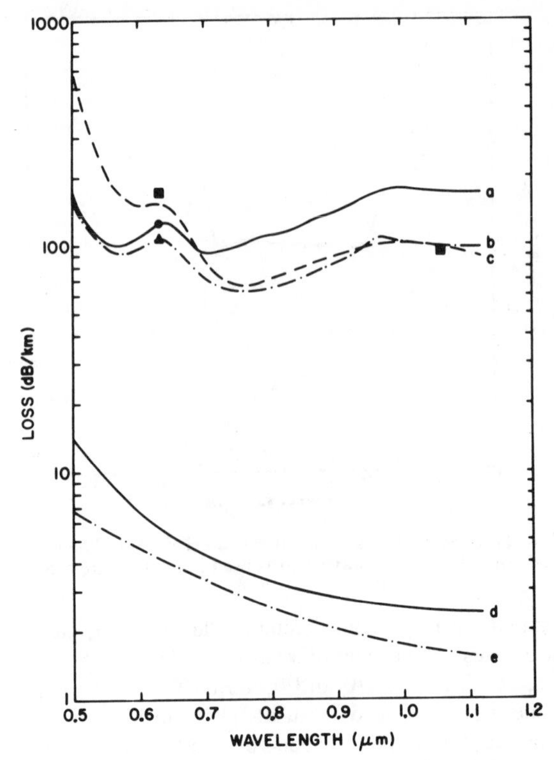

FIG. 14. Spectral losses of unclad fibers drawn from three different soda-lime-silicate glasses, and comparison with bulk-loss measurements performed at 0.6328 μm: ● is associated with sample a, ▲ with sample b, and ■ with sample c. The approximate minimum scattering losses d and e were measured with the same unclad fibers as curves c and b, respectively.

All five compositions tested had relatively high losses at 0.5 μm, which after first decreasing with wavelength, turned into a small absorption band centered around 0.64 μm. Minimum losses occurred between 0.7 and

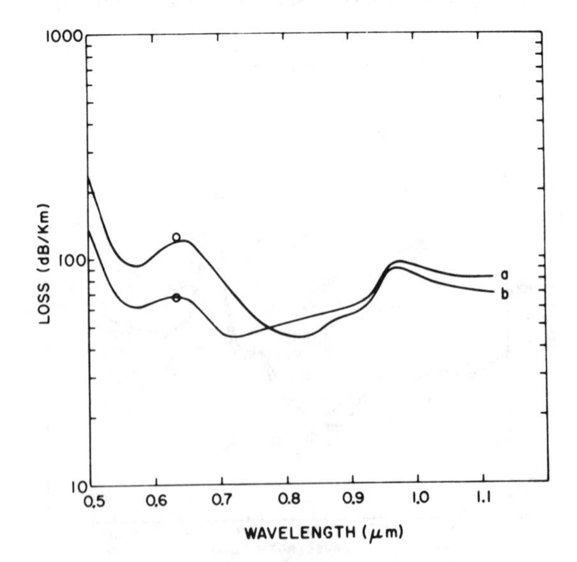

FIG. 15. Spectral losses of unclad fibers drawn from two soda-lime-silicate glasses of higher purity than those in Fig. 14. The circles represent the corresponding bulk-loss measurements.

0.85 μm, depending in detail on the constituents and impurities of the glass. Lowest losses achieved with two of the samples amounted to 45 dB/km. For longer wavelengths, the losses again increased and assumed another maximum at 0.98 μm.

The approximate minimum scattering losses are also shown in Fig. 14 for two of the samples. They illustrate the comparatively small contribution of the scattering to the total losses, particularly at longer wavelengths. Based on this result and on the good agreement between bulk and unclad-fiber measurements at selected wavelengths mentioned earlier, the spectral-loss measurements of unclad soda-lime-silicate fibers represent a fast and accurate method for obtaining the bulk spectral-loss curve.

Two comments concerning unclad-glass-fiber measurements should be made. First, the higher refractive index of SLS glasses of the order of 1.52 necessitated the use of oil instead of glycerol in the slotted measuring cell.[2] Owing to its lower viscosity and surface tension, the oil leaked through the slots and had to be resupplied during the measurement. Fortunately, the response of the detector cell did not critically depend on the liquid level, and accurate measurements could be performed.

The second problem was more complicated. Immediately after fabrication, the unclad SLS fibers developed a high-scattering surface film that was apparently caused by chemical reaction when the fibers were exposed to the atmosphere. Even though this film could be removed with Windex-wetted lens paper, it rapidly redeveloped, forcing us to make the measurement rapidly. For this reason, shorter fiber lengths were utilized and we had to exercise judgment as to the effect of the rapidly deteriorating surface. However, by making repeated measurements on the same fiber, we found that this problem was not serious and that reliable measurements could be obtained.

CONCLUSIONS

Spectral losses of unclad fibers drawn from various types of vitreous silica and soda-lime-silicate glasses were measured between 0.5 and 1.12 μm. The automated technique employed was both convenient to use and sensitive enough to measure fiber losses of the order of 6 dB/km with a reproducibility of ±2 dB/km.

Unclad-fiber losses compared well with bulk losses obtained at discrete wavelengths with other techniques. The question whether this close agreement extends throughout the spectral range is still unresolved. Specifically, we refer to the strong loss band at 0.625 μm observed in low-OH-content vitreous-silica fibers because, in spite of its high absorption, about 250 dB/km, no mention of its occurrence in the bulk material has been found.

The occurrence of fiber losses well below 10 dB/km, i.e., 90% per km, in commercially available vitreous silica presents a challenge to find ways to utilize these high-grade materials for optical-waveguide design.

ACKNOWLEDGMENTS

We thank D. A. Pinnow, who measured the vitreous-silica bulk losses at 1.06 μm with a calorimetric technique. The cooperation of H. E. Earl, who drew two of the vitreous-silica fibers, was greatly appreciated.

REFERENCES

[1]P. J. R. Laybourn, W. A. Gambling, and D. T. Jones, Opto-Electron. **3**, 137 (1971).
[2]P. Kaiser, A. R. Tynes, A. H. Cherin, and A. D. Pearson, Digest of Technical Papers, Topical Meeting on Integrated Optics, Materials and Devices, 7–11 February 1972, Las Vegas, Nev.
[3]Suprasil W1, Suprasil W2, Suprasil 1, Infrasil, Ultrasil, and T08 Commercial Grade were purchased from Amersil, Inc. (see Catalog EM 9227). Spectrosil and Spectrosil WF were purchased from the Thermal American Fused Quartz Company (see Catalog 5M-10-70).
[4]A. D. Pearson and W. G. French, Bell Lab. Rec. **50**, 102 (1972).
[5]Supplied by Minnesota Mining and Manufacturing Company.
[6]M. W. Jones and K. C. Kao, J. Phys. E **2**, 331 (1969).
[7]J. D. Mackenzie, *Modern Aspects of the Vitreous State, Vol. 2* (Butterworths, London, 1962), Ch. 5, p. 195.
[8]D. B. Keck and A. R. Tynes, Appl. Opt. **11**, 1502 (1972).
[9]D. B. Keck, P. C. Schultz, and F. Zimar, Appl. Phys. Lett. **21**, 215 (1972).
[10]This sample was obtained from D. A. Pinnow and was cut from the same block of glass for which absorption losses between 53 and 124 dB/km were measured at 1.06 μm. [See T. C. Rich and D. A. Pinnow, Appl. Phys. Lett. **20**, 264 (1972).]
[11]D. B. Fraser, J. Appl. Phys. **39**, 5868 (1968).
[12]A. R. Tynes, A. D. Pearson, and D. L. Bisbee, J. Opt. Soc. Am. **61**, 143 (1971).

BOROSILICATE GLASSES FOR FIBER OPTICAL WAVEGUIDES

L. G. Van Uitert, D. A. Pinnow, J. C. Williams
T. C. Rich, R. E. Jaeger and W. H. Grodkiewicz
Bell Laboratories
Murray Hill, New Jersey 07974

(Received February 2, 1973; Communicated by N. B. Hannay)

ABSTRACT

Of the existing optical glasses, pure fused silica is known to have the lowest optical attenuation in the red and near infrared portion of the spectrum where optical communications appears most promising. However, to approach the low attenuations afforded by pure fused silica in a waveguide structure requires that a core of fused silica be clad with a glass of slightly lower index refraction. This paper describes an investigation of the binary borosilicate glass system which has led to the realization of a promising cladding material for pure fused silica core fibers.

Introduction

Based on a study of the optical attenuation characteristics of glasses and liquids it has been concluded on theoretical and experimental grounds that pure fused silica has the lowest known optical attenuation in the red and near infrared portion of the spectrum where optical communications appears most promising (1). For example, bulk fused silica of high purity and low water content (Suprasil W-1) has been measured to have a total attenuation at the 1.06 micron emission wavelength of the Nd:YAlG laser of 1.5 ± 0.2 dB/km (comprised of 0.7 dB/km scatter loss and 0.8 dB/km absorption loss) (2). However, to approach such low attenuations in a waveguide structure would require that a core of pure fused silica be clad in some way with a second material of slightly lower index refraction. The lower index of

Reprinted with permission from *Mater. Res. Bull.*, vol. 8, pp. 469-476, Apr. 1973. Published by Pergamon Press Ltd.

refraction cladding is necessary to assure light guidance by the physical process of total internal reflection. The only commercially available glass with a refractive index even slightly less than pure silica is Vycor (3), a fused silica containing small amounts of B_2O_3 and Na_2O. Concurrent work by Kato (4) has shown that Vycor can be used as a cladding material even though the refractive index is only .001 less than that for pure SiO_2. Previous efforts have been largely directed towards "doping" a fused silica core material with any one of a number of additives which are known to increase its index of refraction (5). In these cases pure fused silica is used as a cladding. Although results to date have been very good with these fibers, it is known that even better results could potentially be achieved if the core of the fiber waveguide, where most of the optical energy is confined, were made of undoped silica. This is so because even small additions to silica cause appreciable increases in scattering and absorption loss.

The purpose of this paper is to report the successful formulation of cladding glasses which have many of the desired characteristics necessary to realize the full benefits of a pure fused silica core fiber. Bulk samples of the new glasses, which are relatively high boron containing borosilicate binary mixtures, have been prepared and their properties have been determined. In addition, samples have been used to draw silica core fibers. These fibers have demonstrated satisfactory waveguidance and mechanical strength. It should be possible to achieve low loss in clad fibers with purification of the borosilicate glass and with refinements in the drawing process.

The requirements for a suitable cladding for fused silica are quite stringent. Specifically, it must have characteristics similar to those of fused silica, including reasonably low optical loss, a reasonably well matched thermal expansion coefficient, chemical stability, a high softening temperature, high viscosity in the molten state, etc. These requirements tend to limit the choice of a cladding to an inorganic refractory oxidic type of glass, most probably one high in silica content. However, one finds that of all the inorganic oxides, pure silica has the lowest index of refraction with a single exception, B_2O_3. Pure B_2O_3 has a reported index of refraction of 1.4582 at the sodium D line of 0.589 microns, which is just slightly less than the index of pure SiO_2, 1.4585 (6). Generally, when two pure glasses are mixed, the index of refraction of the composite in the annealed condition varies monotonically between the limiting values of the pure constituents. If this were the case for the B_2O_3 - SiO_2 system, the index of the binary composition would not be sufficiently low to make it useful as a cladding. However, we have found that there is an unusual variation of the index of refraction in the borosilicate system as displayed in Fig. 1 for furnace cooled samples with a definite minimum occurring near the composition $6SiO_2:1B_2O_3$. At this minimum the index of the binary composition is about 0.3% less than that of pure fused silica. This index differential is sufficiently large to make satisfactory waveguides.

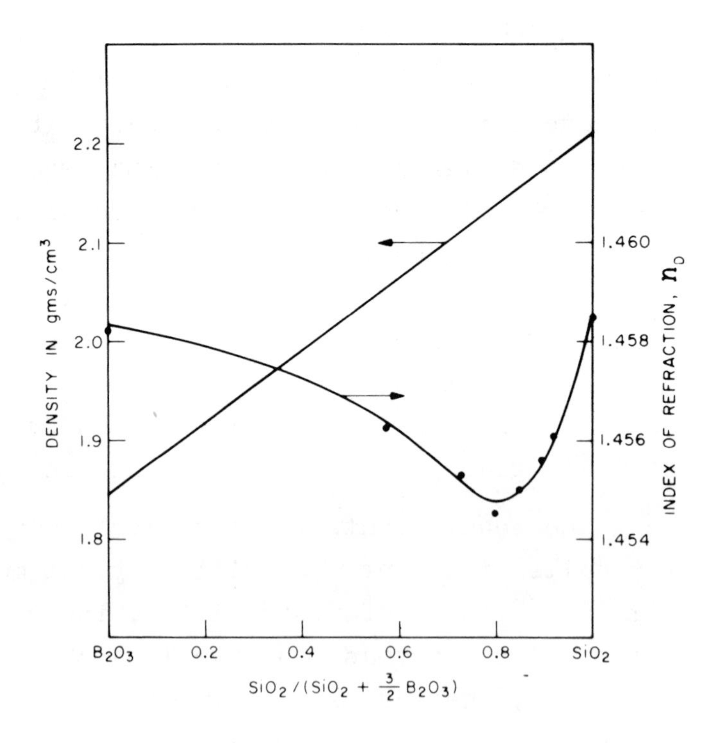

FIG. 1

Variation of density and index of refraction, n_D, as a function of composition in the binary B_2O_3 - SiO_2 system. The composition coordinate $SiO_2/(SiO_2+3/2B_2O_3)$ was selected so that the density variation is linear as explained in the text.

It may be surprising that the simple binary borosilicate system has not been fully investigated many times over by previous workers. Although many papers and even books have been written about the borate glasses, almost all of the known research has been directed towards those compositions with reasonably low working temperatures, substantially less than the high temperatures associated with the silica rich borate mixtures of present interest. The reason why these glasses have been all but avoided is probably due to the degree of difficulty in obtaining homogeneous materials at temperatures in the range of 2,000°C. Preparation details are described in the following section.

Preparation

The starting materials for all of the test samples consisted of Fisher Reagent Grade SiO_2 and B_2O_3. These materials were in powder form (240 mesh) and were combined in varying proportions having mole ratios of SiO_2 to B_2O_3 ranging from 2:1 to 30:1. Mixing was accomplished by ball milling the powders together overnight in a plastic jar. The milled material was loaded into a 100mℓ platinum crucible and sintered at approximately 1350°C for twelve hours in an electric furnace. This produces a compacted and hard milky white substance. After cooling the sintered material was transferred from the platinum to an iridium crucible for a high temperature soak which was found necessary to produce a clear bubble free glass. This soaking, which required from 8 to 30 hours, was performed in an RF induction furnace with an inert atmosphere consisting of a 1 to 1 mixture of argon and nitrogen. The temperature ranged from 1,500°C for the $1SiO_2:1B_2O_3$ mixture to 2,000°C for the higher silica compositions. The glass samples were slowly cooled from the soaking temperature and subsequently removed from the crucible by core drilling. Final preparation of the samples for optical studies was accomplished by sawing, grinding, and polishing procedures identical to those used for fused silica and other standard optical glasses.

Properties

The physical characteristics of pure B_2O_3 are vastly

different from those of SiO_2. B_2O_3 has quite a low softening point, approximately 450°C, and it is known to hydrate rapidly on exposure to the atmosphere. On the other hand, SiO_2 has a very high softening point, approximately 1710°C, and is exceedingly stable chemically in the atmosphere at ambient temperature. One common feature of both glasses is their high viscosity at temperatures well above their respective softening points.

A. Density

The low softening point boron oxide rich compositions up to 1:1 mixtures have been studied in considerable detail by Cousen and Turner (7). Their density data varies linearly between the pure components as shown in Fig. 1 when arbitrarily plotted as a function of the contribution to the relative oxygen content in the glass by SiO_2, as indicated by the mole ratio of SiO_2 (two oxygens) to that of SiO_2 plus 1.5 times that of B_2O_3 (three oxygens). This type of dependence is typical in most compound glasses and is based on the empirical fact that the specific volume of a glass approximates quite well to the sum of the specific volumes of its constituent oxides (8). We have found that the density data of our high silica samples fits very well to this same linear relation if we assume that the ratio of SiO_2 to B_2O_3 in the final glass is identical to the ratio of the starting powders. This is rather strong evidence that essentially no B_2O_3 is lost from the melt during the high temperature sintering and soaking.

B. Chemical Stability

It is well known that the B_2O_3 rich mixtures rapidly hydrate (9). However, it was not previously established at what SiO_2 concentration the glass becomes essentially inert to water. Our tests indicate that the $2SiO_2:1B_2O_3$ mixture is slightly attacked by the atmosphere and is soluble in water. The next higher silica mixture we prepared, a 4:1 blend, was found to be unchanged even after immersion into boiling water for 8 hours. There was no detectable weight change nor any visible signs of hydrate formation on the polished surfaces. It is concluded that glass compositions of 4:1 or higher in silica are stable against attack by water.

The B_2O_3 - SiO_2 phase diagram has been studied in considerable detail by Rockett and Foster (9). They noted that because of their inability to devitrify any of the borosilicate glasses once prepared, they had to use non-standard methods for their studies. This strong resistance of devitrification is, of course, a desirable attribute for a glass to be used in the construction of an optical waveguide. Rockett and Foster concluded that no liquid immiscibility or phase separation occurs in the B_2O_3 - SiO_2 system within the limits of detection of a polarizing microscope or X-ray diffractometer.

C. Optical Properties

The general optical quality of the glasses prepared by the methods discussed in the previous section are quite good. With the exception of occasional bubbles, the glasses are visibly homogeneous and transparent without coloration.

The index of refraction was measured with an Abbe refractometer at an effective wavelength of 0.589 microns. The results are shown in Fig. 1 for the compositions which we studied. There is a definite minimum in the index for the samples studied which occurs at the value of approximately 0.8 on the abscissa, corresponding to a 6:1 molar composition. At the minimum the differential index of refraction between the compound glass and pure fused silica is approximately 0.3%. We have confirmed by infrared absorption studies that the OH content in these samples is sufficiently low (approximately 150 ppm) that the unusual variation in index can not be due to the presence of such an impurity. It is, however, related to the inverse glass nature of pure SiO_2 (10) as will be discussed elsewhere (11).

We have also used the refractometer to measure the dispersion of the index of refraction in the mixtures. The results indicate that, to within the accuracy of the measurements, the dispersions are all equal to that of pure fused silica.

D. Thermal Expansion

Cousen and Turner (7) have reported that the room temperature thermal expansion coefficient of pure B_2O_3 is 15×10^{-6}/°C. As silica is added they found that the thermal expansion coefficient decreases. For the 1:1 molar composition the expansion coefficient is 5×10^{-6}/°C. We find that this trend continues. The low temperature expansion coefficient of the 8:1 molar composition was found to be 1.3×10^{-6}C. At higher temperatures in the range of 550 to 650°C the expansion coefficient of this compound glass becomes equal to that of pure SiO_2 (0.5×10^{-6}/°C) and then increases once again. Measurements were made up to approximately 900°C, well above the glass transition temperature which was found to occur at approximately 725°C. Based on these results we were optimistic that the thermal expansion mismatch between the silica rich glass and the pure fused silica is sufficiently small to allow fabrication of sound waveguide structures. Initial work in fabricating waveguides has verified this point, although changes had to be made in the structure and the fiber drawing process.

Conclusion

Silica rich borosilicate glasses have been formulated that should be suitable for cladding pure silica core fibers to make improved optical waveguides. The motivation for this work was to take full advantage of the exceptionally low optical loss afforded by using pure fused silica in the core of a waveguide structure. Efforts are currently underway to fabricate low loss waveguides incorporating this new cladding material.

Acknowledgment

The authors wish to acknowledge with thanks the experimental assistance of D. D. Badding.

References

1. T. C. Rich and D. A. Pinnow, Appl. Phys. Letters 20, 264 (1972) and also D. A. Pinnow, T. C. Rich, F. W. Ostermayer and M. DiDomenico, Jr., submitted for publication.

2. T. C. Rich and D. A. Pinnow, unpublished results.

3. Corning Glass Works Inc., Catalog No. 7912.

4. D. Kato, Appl. Phys. Letters 22, 3 (1973).

5. R. D. Maurer and P. C. Schultz, U. S. Patent 3,659,915, May 2, 1972.

6. G. W. Morey. The Properties of Glass, p. 370. Reinhold Publishing Corp., N. Y. (1954).

7. A. Cousen and W. E. S. Turner, J. Soc. of Glass Tech. 12, 169 (1928).

8. G. W. Morey. The Properties of Glass, p. 221. Reinhold Publishing Corp., N. Y. (1954).

9. T. J. Rockett and W. R. Foster, J. Amer. Ceramic Soc. 48, 75 (1965).

10. H. Solmang and K. Von Stoesser, Glastechische Berichte 8, 463 (1930).

11. S. H. Wemple, D. Pinnow, T. C. Rich, R. E. Jaeger and L. G. Van Uitert, to be submitted for publication.

Part IV
Propagation Theory

Propagation Effects in Optical Fibers

DETLEF GLOGE, MEMBER, IEEE

(*Invited Paper*)

Abstract—**The round dielectric waveguide exhibits a surprising variety of characteristics that are not accurately inferable from the slab model. The forceful effort of recent years has grately extended the knowledge of these structures and added new and exciting modifications. An attempt to unify these results in a simplified picture is made. Specific phenomena relevant to optical fiber design and fabrication are then brought into focus. Some of the problems discussed are cross sectional loss variations, various core index profiles and the tolerances required in their preparation, the necessary cladding thickness, directional changes, and sources of mode coupling affecting signal distortion and loss.**

I. INTRODUCTION

THE LAST few years have seen a rapid increase both in technological know-how and theoretical understanding of optical fibers and, along with it, a new variety of fiber structures. At the same time, the issue of material loss, which had barred fibers from the communications field longer than necessary, was so convincingly solved that other aspects are now becoming a prime concern of optical communications research. This then seems to be a good time to take stock, to organize the knowledge gained, and to assess the available options. Accordingly, a great number of review articles have appeared in a rapid sequence all over the world. To name only the most recent in the order of their appearance, there is an article by Maurer [1] addressed mainly to the technology of fiber preparation. *Opto-Electronics* devoted its July and September issues of 1973 to the subject of fiber optics featuring reviews of the state of the art in Britain [2] and Germany [3]. An article by Ohnesorge [4] advanced some of the less conventional ideas of communication systems application for optical fibers. Miller *et al.* [5] have prepared a very comprehensive review of the current knowledge relating it to potential applications in the conventional communications network. The quite different though equally immediate potential of fibers for military applications becomes apparent in an article [6] which appeared in this TRANSACTIONS in December, 1973. The conventional technology of fiber bundles [7] seems to be more

Manuscript received March 27, 1974; revised August 15, 1974.

The author is with the Crawford Hill Laboratory, Bell Laboratories, Holmdel, N. J.

Reprinted from *IEEE Trans. Microwave Theory Tech.*, vol. MTT-23, pp. 106–120, Jan. 1975.

readily applicable in this case, where multiterminal information transfer over short distances on board aircraft or ship is the objective. To complete our "review of the reviews," we should mention a 6-page summary of the state of the art [8], interesting for those looking for a concise overview, and, of course, some of the textbooks available on the theory of dielectric wave guidance [9]–[11]. It is quite likely that review articles on optical fibers will continue to appear within similar or even shorter intervals. An English-language paper on the very extensive and successful work in Japan [12], for example, would be quite welcome. On the other hand, those interested in the field are by now quite well informed about the history and breakthrough of communication-oriented fiber optics and can find more than 300 references on the details in the reviews cited earlier.

Rather than adding to this list, I wish to narrow the objectives of this paper in an attempt to present a concise and unified picture of the propagation characteristics of the round dielectric waveguide both in its ideal and its imperfect state, in order to provide some simple results relevant to design and systems questions in optical transmission applications. Many of these results deviate sufficiently from the well-known characteristics of the two-dimensional dielectric-film model to arouse some caution as to the validity of the usual extrapolations of film concepts to the fiber description. One of the objectives of this article is, therefore, a closer look at the limits of our knowledge of the round dielectric structure and the identification of areas where further investigations are necessary. In line with these objectives, references are cited merely to direct the reader to additional material on a given subject matter and are not necessarily the first or only publications on the subject.

We begin in Section II with a discussion of the guidance concepts governing the round dielectric waveguide. Aside from the classical core-cladding structure, this includes cylindrical graded-index profiles and slab-supported guides. Section III takes a look at realistic fibers in a practical environment and attempts to incorporate the influence of cross sectional loss variations, finite cladding thickness, directional changes, and random imperfections in the laws of propagation. The emphasis is on measurable variables and measured results. The dispersive effect of the material and the waveguide, delay between fiber modes, and the resulting signal distortion are the subject of Section IV.

II. GUIDANCE PRINCIPLES

Trapped or lossless propagation of light in dielectric waveguides relates to idealized structures which are straight and unperturbed in propagation direction, have unperturbed surroundings (infinitely thick cladding) and are made from lossless materials. This is the structure we discuss in this section, confident that slight modifications will later suffice to adapt the results to practical conditions; perturbation methods to achieve this are discussed in the next section. The classical clad glass fiber [7] traps light by total internal reflection at the core boundaries where the index drops to the slightly lower cladding value. Modern structures like the graded-index fiber whose index decreases gradually outward from the axis employ a continuous focussing process to achieve trapping, which can be considered as a kind of distributed internal reflection [13]. Propagation in the single-material fiber [14] is not characterized by an index change at all, but solely by the cross sectional configuration. Whatever the underlying principle of guidance, all structures can be designed to support one or many trapped modes of propagation. Beginning with the classical clad structure, we shall discuss its single-mode configuration in greater detail, hoping that this will aid in the understanding of the other two.

The round clad fiber is one of the few dielectric waveguide structures, for which Maxwell's equations have rigorous (though fairly unwieldy) solutions and these have been discussed in many places [9], [10]. The very existence of such solutions has probably kept theorists from searching for simplifying approximations even after such approximations were found, and recognized as useful, in the case of the parabolically graded index profile (see, for example [15]). The approximations recognize the fact that, ultimately for technological reasons, the change from the core index n_1 to the cladding index n_2 is typically small. In other words,

$$n_2 = n_1(1 - \Delta) \tag{1}$$

and $\Delta \ll 1$. Under these conditions, field solutions can be found both for the flat and the graded core index profile, which are essentially transverse electromagnetic and linearly polarized [15]–[17]. Strictly speaking, all these solutions (except for the fundamental) are superpositions of more complicated field solutions that appear degenerate in these approximations, but in reality break apart upon propagation over long distances as a result of small differences in the propagation constants. The approximations are completely satisfactory, however, in predicting the modal power distributions and group velocities; this is essentially what is needed in fiber-optical communication systems, which, for the time being, are restricted to direct (power) modulation and detection for economical reasons.

In the case of the classical fiber structure, which has an abrupt index step at the core-cladding boundary, the approximation ignores essentially the slight difference (of order Δ^2) in the matching conditions for the electric and magnetic field components tangential to that boundary. As a result, the boundary conditions simply require continuity of all transverse field amplitudes and their radial derivatives through the boundary. In a cylindrical structure, these conditions can be met by the well-known trial solution

$$\boldsymbol{E}(r,\phi) = E(a)\begin{Bmatrix} J_l(ur/a)/J_l(u) & \quad r < a \\ K_l(wr/a)/K_l(w) & \quad r > a \end{Bmatrix}\cos l\phi, \tag{2}$$

where $k = 2\pi/\lambda$ is the free-space wave number, a the core radius, and $E(a)$ the maximum field amplitude at the interface; u and w are a pair of parameters, whose mutual interrelation is determined by the matching conditions

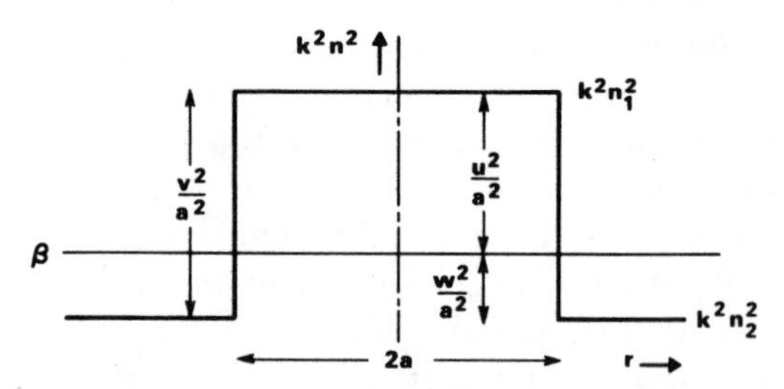

Fig. 1. Dielectric profile of the single-mode fiber with uniform core index.

[16], [17]. How u and w relate to the propagation constant β and the structural parameter v (usually called the v-value) is evident from the dielectric profile sketched in Fig. 1. We find

$$v = (u^2 + w^2)^{1/2} = ak(n_1^2 - n_2^2)^{1/2} \approx akn_1(2\Delta)^{1/2} \quad (3)$$

and

$$\beta^2/k^2 = n_2^2 + (n_1^2 - n_2^2)w^2/v^2. \quad (4)$$

In general, the problem of characterizing the modes of a certain fiber can be approached in the following way: the measurement of a, Δ, and the wavelength of operation determines v. The parameters u and w then result from (3) and the (transcendental) "characteristic equation" derived from the matching conditions [16], [17]. Equation (4) finally permits the computation of β. For many technical problems, u, w, or β are of little direct use, except as a basis for calculating the more important mode power and group velocity relations. For example, the question of how bulk loss differences in core and cladding affect the mode loss requires some knowledge of the power distribution in core and cladding. In the classical fiber structure, the mode power and group velocity relations can be derived from an auxiliary parameter [17], [11]

$$j_l = -\frac{J_l^2(u)}{J_{l+1}(u)J_{l-1}(u)}. \quad (5)$$

If P is the total power in a certain mode and P_1 and P_2 are the power fractions in core and cladding, we find [17]

$$\frac{P_1}{P} = 1 - \frac{P_2}{P} = (1 + j_l)w^2/v^2. \quad (6)$$

The power density at the core-cladding interface, averaged over the circumference, is

$$p(a) = \frac{P}{\pi a^2} j_l w^2/v^2. \quad (7)$$

The density $p(a)$ determines the amount of mode loss caused by imperfections in the interface. The group velocity relations of interest are discussed in Section IV.

This outline would be of little help to the reader without some knowledge of how to obtain u or w. Before this is discussed, we must take a closer look at situations of practical interest and the numbers of modes involved in these cases. The approximate theory, on which our discussions are based, stipulates that the number of independent field solutions (degenerate mode groups) is equal to the number of zeros of all Bessel functions J_l which are smaller than v, plus the fundamental solution [16], [17]. If $v < 2.4$, only the fundamental mode propagates. This mode has the transverse field distribution $J_0(ur/a)$ in the core. The cladding field decays monotonically with the distance from the interface; it reaches farther and farther into the cladding as v decreases. A theory which considers the cladding thickness as unlimited finds this mode to be trapped even as v approaches zero.

There are two distinctly different operating conditions for fibers used in transmission systems: single-mode and multimode operation. The former avoids the signal-impairing effect of mode delay differences and therefore provides the ultimate in transmission bandwidth. For this reason, the single-mode fiber has a definite potential for wide-band optical communication; it is likely to gain in importance once sources become available [18] that have sufficient spatial coherence to excite the one mode of the single-mode fiber efficiently. On the other hand, single-mode operation limits the core diameter and the index difference as well as the tolerances acceptable for these two parameters, a fact which may complicate large-scale fabrication and field splicing of such fibers. Secondly, and maybe more importantly, since the single-mode fiber accepts only the equivalent of one radiation mode from any source, it is practically useless in combination with incoherent sources (luminescent diodes) and deficient, if the source is multimode. For all these reasons and because recent fiber art has devised schemes of alleviating the signal impairing effect of mode delay differences in multimode fibers, multimode operation is at least of equal importance. Typically, such fibers are designed to support a great number of modes in order to fully bring their advantages to bear.

Coming back now to the computation of the parameters u or w from v, we distinguish the two cases of single-mode and multimode operation. As we shall see, the second case allows a rather summary treatment of all modes yielding closed-form approximations of satisfactory accuracy. Rather than enumerating similar approximations valid in specific regions of the single-mode regime, we simply plot here all important parameters characterizing the fundamental mode in a way that allows an accurate reading everywhere. For this reason, the ordinate of Fig. 2 is chosen linear around unity and logarithmic below. Plotted along the abscissa is the v-value in the region between $v = 0.6$ and 2.4, which covers the single-mode regime of interest. In addition to the parameters u and w, we have plotted w^2/v^2, which determines β according to (4), $j_0 w^2/v^2$, which is proportional to $p(a)$ of (7), and $(1 + j_0)w^2/v^2$, which determines the power distribution (6). Accordingly, P_1/P can be read off the left side and P_2/P off the right side of Fig. 2.

Although, theoretically, the fundamental has no cutoff as v decreases and hence propagates even at lowest frequencies, it is obvious from Fig. 2 that the power contained by the core is only 0.1 percent at $v = 0.6$ and decreases rapidly for lower v, making effective guidance

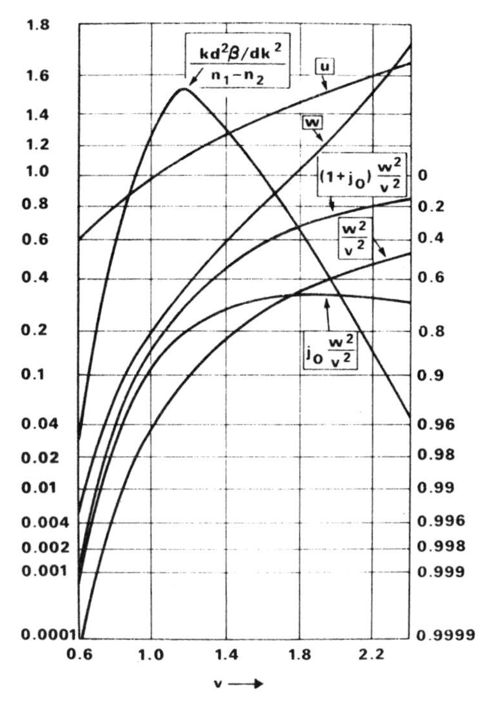

Fig. 2. Various characteristics of the fundamental mode in the step-index fiber plotted versus the v-value.

below $v = 0.6$ virtually impossible. More interesting than the lower bound is the region towards large v, because, at optical wavelengths, the objective is in making the core radius a and hence v as large as possible, to alleviate fabrication and splicing tolerances. A safe operating point may be $v = 2$. Any remaining effort in maximizing a is limited to a reduction of Δ. The parameter Δ, on the other hand, must at least be of the order of 0.3 percent if excessive bending loss is to be avoided; this condition is explained by (41) of Section III. As a result, we find with the help of (3) that the core diameter of the single-mode fiber can measure 5 to 6 wavelengths at the most.

Plots similar to Fig. 2 can be produced for all modes of the classical fiber structure when the region of v is extended to larger values [17]. Such detailed knowledge of every one of the hundreds of modes propagating in typical multimode fibers can be rather confusing. On the other hand, one arrives at very satisfactory closed-form expressions of general validity for all but the very lowest order modes, if one makes use of the Debye approximations for the Bessel functions [19]. Although these approximations are rather simple and straightforward as far as the mathematical side of the problem is concerned, a direct derivation from physical principles has several side benefits. One is an effortless association of modes with rays [20], the other is an easy extension of these principles to fibers with a nonuniform core index, which will be of use later on. The direct derivation uses the method of Wentzel, Kramers, Brillouin, and Jeffreys (WKBJ) modified to apply to cylindrical structures [21], [22]. This approach is outlined in the Appendix. The following more casual argument is based on the same ideas but requires less mathematics [23].

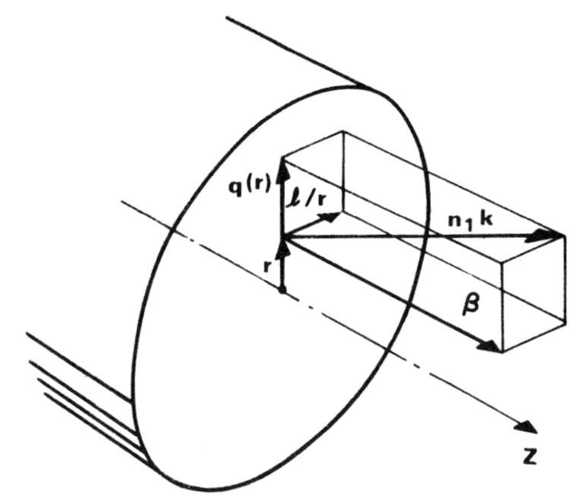

Fig. 3. Wave vector diagram in the propagation region of a multimode fiber.

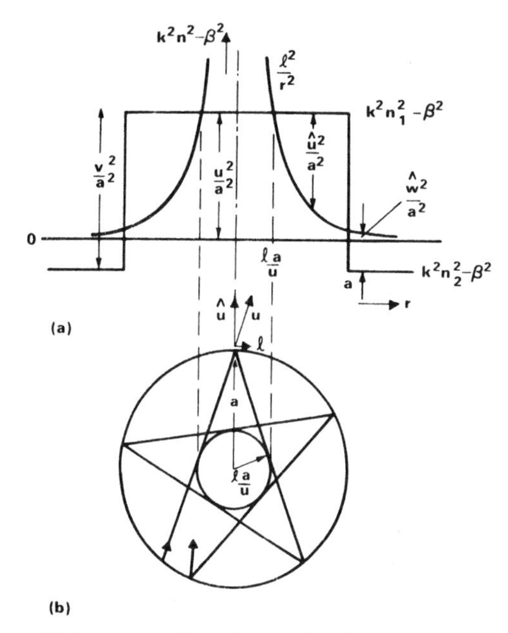

Fig. 4. (a) Dielectric profile of the classical fiber structure (uniform core index) showing squared magnitudes of vector components for azimuthal mode order l.

We assume the existence of locally plane waves in the core guidance region. Fig. 3 shows a decomposition of the local wave vector (pertaining to a given mode) into its components in the cylindrical coordinate system (r, ϕ, z). The axial component is the propagation constant β of the mode. The azimuthal periodicity indicated by $\cos l\phi$ in (2) results in a ϕ-component l/r at r. Since the magnitude of the wave vector is nk, the radial component becomes

$$q(r) = (k^2n^2 - \beta^2 - l^2/r^2)^{1/2} \tag{8}$$

with $n = n_1$ in the core and $n = n_2$ in the cladding. Fig. 4 is a plot of a dielectric profile similar to Fig. 1, showing the square magnitudes of the various components as a function of the radius. Within the region $la/u < r < a$, in which q is real, we introduce a parameter

$$\hat{u} = qa = (u^2 - l^2a^2/r^2)^{1/2}. \tag{9}$$

This region represents the core area, in which a periodic field solution exists. Outside of this region, both in the core and in the cladding, the radial component becomes imaginary, causing the mode field to decay monotonically with the distance from the boundaries. The decay parameter in the cladding is

$$\hat{w} = (w^2 + l^2a^2/r^2)^{1/2}. \tag{10}$$

The cylindrical surface of radius la/u in the core represents a caustic [20] for rays which travel within the region $la/u < r < a$ and have the direction of the local wave vector. A cross sectional projection of one of these rays is shown in Fig. 3(b). The entire group represents the "congruence" of rays [20] associated with the mode under discussion. Before we define this mode more fully, let us digress briefly to take a closer look at the ray picture. The angle which all rays of a congruence form with the axis obeys the relation

$$\sin\theta = u/akn_1. \tag{11}$$

When the rays leave the fiber end face into air, $\sin\theta$ increases to

$$\sin\Theta = u/ak \tag{12}$$

as a result of refraction. For small Θ, if we were to set up a screen perpendicular to the axis at a distance ak mm away from the end face, all these rays would impinge on the screen approximately at a distance u mm from the axis. In typical multimode fibers, ak may be of the order of 100 and u varies between 0 and v, the latter being of the order of 30. In accordance with lens optics definitions the product of the refractive index and the sine of the maximum ray angle, that is,

$$n_1 \sin\theta_{\max} = v/ak = (n_1^2 - n_2^2)^{1/2} \approx n_1(2\Delta)^{1/2} \tag{13}$$

is called the numerical aperture of the fiber.

A complete identification of the modes of the cylindrical structure requires two mode numbers: the azimuthal order number discussed earlier and a meridional order number m, which identifies the number of field maxima of the radial field solution (this count includes the maximum at $r = 0$ for $l = 0$). The WKBJ approximation for m is obtained from a count of the number of half periods comprised within the radial phase change between the caustic and the interface. This results in the relation

$$m\pi = \int_{l/u}^{1} \hat{u}\, d\rho \tag{14}$$

with $\rho = r/a$. A rigorous and more accurate derivation of this relationship can be found in the Appendix. However, (14) gives usually a satisfactory description of the modes of a typical multimode fiber. In the case of the classical index profile, we can solve (14) after inserting (9) and obtain

$$m = (1/\pi)[(u^2 - l^2)^{1/2} - l \arccos(l/u)]. \tag{15}$$

Fig. 5 illustrates this relation for different ratios u/v. Modes are marked in this plot by a uniform raster of spots of density $1/v^2$ (see upper right of Fig. 5). Each spot to the left of the line $u/v = 1$ represents a degenerate quadruplet of trapped modes of different polarization or orientation. There are no spots along the ordinate since $m = 1,2,\cdots$; spots falling on the abscissa ($l = 0$) represent doublets [11]. If $v \gg 1$, as for typical multimode fibers, we can count the total number of modes simply by integrating (15) with respect to l. The result is the mode volume

$$M = 4\int_0^v m(l)\, dl = v^2/2. \tag{16}$$

Given a certain mode (m,l), Fig. 5 provides the u-value of that mode. The characteristic ray (or group of rays)

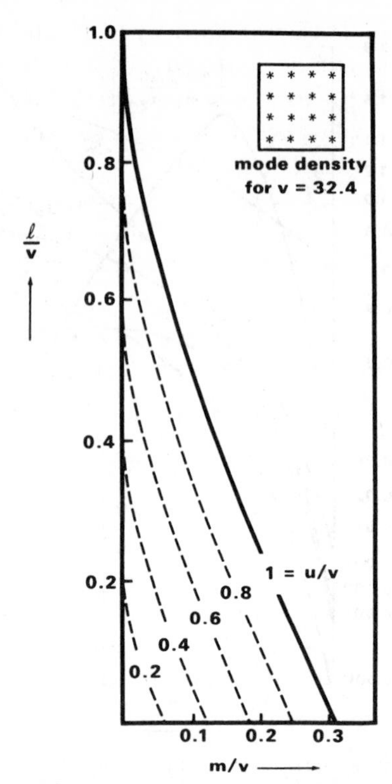

Fig. 5. Plot of mode number l versus m for different parameters u, all normalized with respect to the v-value; density of modes is $1/v^2$ (see upper right); all spots to left of $u/v = 1$ designate degenerate groups of trapped modes.

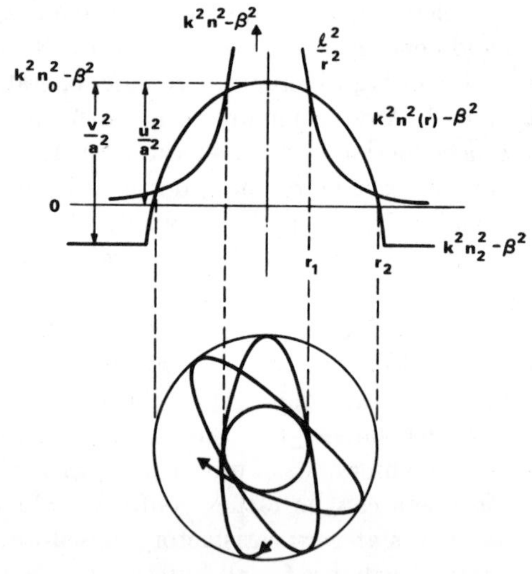

Fig. 6. (a) Dielectric profile of a graded index fiber showing caustics for mode characterized by u, l. (b) Cross sectional projection of a ray characterizing above mode.

of that mode is then defined by (11) and the caustic at $r = al/u$. Equation (4) and the relation $w^2 = v^2 - u^2$ yield the propagation constant β. As for the power distribution, a more accurate match of the field solutions (A9) at the interface is necessary to obtain the approximation

$$j_l \approx \frac{u^2}{w^2}\left[1 - \frac{1}{h + (w^2 + l^2)^{1/2}}\right] \quad (17)$$

with $h = 1$ for $l = 0$ and $h = 0$ for $l \neq 0$. By using this relation in (6), the cladding power becomes

$$P_2 = \frac{Pu^2/v^2}{h + (w^2 + l^2)^{1/2}}. \quad (18)$$

Note, that $P_2 = P/(h + l)$ at cutoff, where $u = v$ and $w = 0$. Thus, contrary to the mode behavior in the slab (or for $l = 0, 1$ in the fiber), a good fraction of most fiber modes is concentrated in the core even at cutoff. This is a result of the fact that the cladding field solution (A9) decreases as r^{-l} at cutoff, where $\hat{w} = l/r$.

The convenience of detection and measurement in the optical far-field renders the modal far-field distribution often more important than the mode field itself. We recall that the exit angle Θ of a ray characterized by m and l obeys the relation (12) and is only a function of u and not explicitely of m and l. We can therefore infer that, as $ka \rightarrow \infty$, the far-field radiation of all modes characterized by u converges on a cone with half-apex angle Θ. A different, though somewhat more general, interpretation of the same relationship stipulates that the power $dP(u)$ of all modes falling between two lines characterized by u and $u + du$ in Fig. 5 can be collected in a ring of radius u and width du on a screen a distance ak away from the fiber end (measure u and ak in millimeters, say).

Fig. 6 depicts the dielectric profile and a ray projection in the case of a graded core index profile. The area of periodic field solutions is now limited by two caustics. Rays characterized by u exit at the angle Θ only, if they leave the guide exactly from the center (only possible for meriodional rays). All other exit positions of rays characterized by u lead to exit angles smaller than $\Theta(u)$. Accordingly, the far-field relations for $ka \rightarrow \infty$ in the graded-index case are not as simple as in the case of the uniform core. A straightforward modification of (14) results in the characteristic integral

$$m = \frac{1}{\pi}\int_{r_1}^{r_2} [k^2n^2(r) - \beta^2 - l^2/r^2]^{1/2}\, dr \quad (19)$$

where we have again neglected a small term additive to m. A class of profiles of particular interest has the form

$$n(r) = n_0 \begin{Bmatrix} (1 - 2\Delta\rho^g)^{1/2}, & \rho < 1 \\ (1 - 2\Delta)^{1/2}, & \rho > 1 \end{Bmatrix} g > 1. \quad (20)$$

It includes the classical fiber with uniform core discussed so far ($g \rightarrow \infty$) as well as the parabolic index distribution ($g = 2$); the latter is important because it provides an effective equalization of the modal group velocities [13]. We return to the discussion of this effect in Section IV.

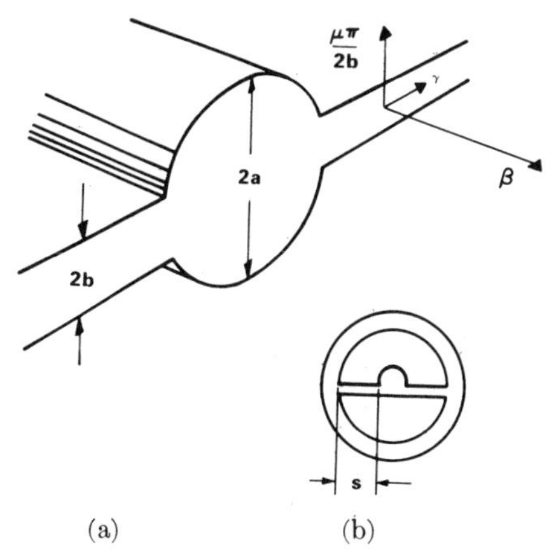

(a) (b)

Fig. 7. (a) Sketch of single-material fiber showing vector diagram in membrane. (b) Typical cross section showing membrane length s.

An exact solution of (19) is not known, but I have found the approximation

$$l + 4\left(\frac{2}{2+g}\right)^{2/g} m = \left(\frac{g}{g+2}\right)^{1/2}\left(\frac{2}{g+2}\right)^{1/g} u\left(\frac{u}{v}\right)^{2/g} \quad (21)$$

very satisfactory. It is exact for $g = 2$ and very good as long as g is not too large. The maximum error results for $g \rightarrow \infty$ and can be inferred from a comparison with (15). An integration of (21) similar to (16) yields the mode volume

$$M_g = \frac{1}{2}\frac{gv^2}{g+2} \quad (22)$$

which can be shown [23] to be exact for all g within the limits of the approximations that apply to (19). As (22) shows, the mode volume of the graded profile is typically less than that of the uniform core when both have the same total index difference. Parabolic grading, for example yields $M_2 = M_\infty/2$.

The fiber configuration of Fig. 7 permits the trapped propagation of a desired number of modes in a core which is suspended between two membranes made of the same material as the core [14]. Although there is no index change between the core and the membranes, the latter have a function similar to that of the cladding in previous configurations. In line with the idealizations applied to those, we assume for the time being that the membranes extend to infinity to both sides of the core. To understand the function of the membrane, consider the wave vector diagram of an arbitrary (plane) wave field in the membrane (see Fig. 7) [24]. The large index change between the membrane and the surrounding air enforces a transverse wavenumber of a magnitude close to $\mu\pi/2b$, where $\mu = 1,2,\cdots$ and b is the half width of the membrane. The vector summation yields

$$k^2n_1^2 = \beta^2 + \left(\frac{\mu\pi}{2b}\right)^2 + \gamma^2. \quad (23)$$

If the membrane field is part of a core mode, β is the propagation constant of that mode and hence, with the notation of Fig. 1,

$$\frac{u^2}{a^2} = \left(\frac{\mu\pi}{2b}\right)^2 + \gamma^2. \tag{24}$$

Trapping of core modes requires that all field solutions are of the evanescent type along the membranes and hence γ must be imaginary even for the smallest μ, that is, $\mu = 1$. Consequently, $u/a \leq \pi/2b$, where the equality denotes cutoff for the mode characterized by u. Since per definition $u = v$ at cutoff, we can define an effective v-value of this fiber configuration of the form

$$v_e = \pi a/2b. \tag{25}$$

By using v_e instead of v in (3) and all subsequent equations as well as in Figs. 2 and 5 we obtain essentially all mode characteristics of this fiber structure, including the conditions for single- or multimode operation.

III. PROPAGATION LOSSES

Mode attenuation results first of all from the dissipative and scattering loss of the core and cladding materials and the interface between the two. Secondly, loss can be caused by the finite cladding width and the (sometimes intentionally) lossy jacket around the cladding. Thirdly, a radiation loss is suffered by modes which are not fully trapped. All three sources of loss usually affect different modes differently. Even small loss differences can cause virtual extinction of some modes in comparison to others, if the fiber is sufficiently long. In general, coupling of modes as a result of perturbations along the fiber balances the effect of loss differences by continuously transferring power into modes otherwise lost. Ultimately, this transfer causes a loss in all modes.

Let us first ignore the complicating influence of coupling. This assumption seems quite relevant to potential communications applications, since careful preparation and handling of fibers has been shown to reduce coupling to negligible amounts even in long fiber lengths [25]. Under these conditions a loss of, say, α_1 dB/km in the core and α_2 dB/km in the cladding produces a mode loss

$$\alpha_I(m,l) = \alpha_1 P_1/P + \alpha_2 P_2/P. \tag{26}$$

with P_1 and P_2 from (6). If we ignore the case $l = 1$, we have $h = 0$ in (17) and

$$\alpha_I = \alpha_1 + (\alpha_2 - \alpha_1)\frac{u^2}{v^2}(w^2 + l^2)^{-1/2}. \tag{27}$$

for the classical step-index profile. Fig. 8 is a plot similar to Fig. 5 showing lines of constant α_I. In the case of large v, the loss of few modes exceeds α_1 even if α_2 is large. Similar modal loss coefficients can be derived with the help of (7), if a source of loss exists in the interface between core and cladding.

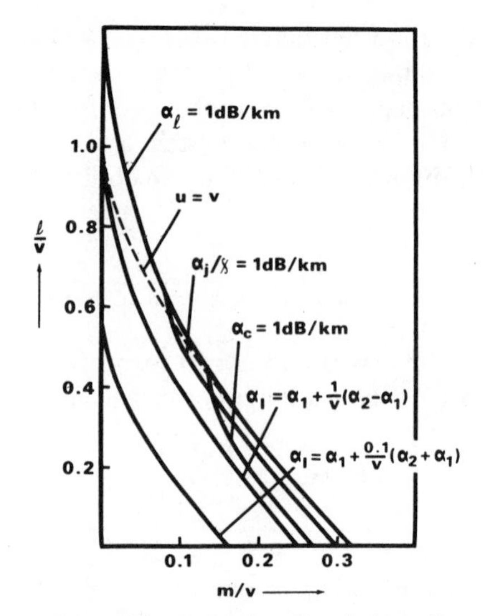

Fig. 8. Same plot as Fig. 5 (broken line is boundary for trapped modes); solid lines are lines of constant loss for various loss phenomena: α_l—leakage loss, α_j/χ—jacket loss, α_c—curvature loss, α_I—loss as a result of cross sectional loss variation of the form (26).

A graded profile is generally obtained by adding or exchanging one or more components in the host glass. It is likely that the additive causes a certain excess loss or, in the case of an exchange, that the loss of one component is different from that of the other. In either case, the loss variation is likely to follow the index variation so that the loss at a distance r from the axis becomes

$$\alpha(r) = \alpha_0 + (\alpha_2 - \alpha_0)\frac{n_0^2 - n^2(r)}{n_0^2 - n_2^2}. \tag{28}$$

The loss suffered by a given mode is in this case

$$\alpha_{II} = \int_0^\infty \alpha(r)p(r)r\,dr \Big/ \int_0^\infty p(r)r\,dr \tag{29}$$

where $p(r)$ is the power density of that mode at r. To solve (29) we use the relation [26]

$$\frac{\beta}{k}\frac{d\beta}{dk} = \int_0^\infty n^2(r)p(r)r\,dr \Big/ \int_0^\infty p(r)r\,dr \tag{30}$$

and $d\beta/dk = ct(u)$ from (60) of Section IV. We obtain

$$\alpha_{II} = \alpha_0 + (\alpha_2 - \alpha_0)\frac{2}{g+2}\frac{u^2}{v^2}. \tag{31}$$

Note that (31) does not converge into (27) for $g \to \infty$, but yields $\alpha_{II} = \alpha_0$. This is so because (31) is based on the assumption of negligible evanescent fields, which was introduced to obtain (19). This approximating assumption is valid and useful in cases when both index and loss vary within the regions of propagating field solutions (between the two caustics), because the different extent of these regions for different modes is then likely to be the overriding influence in causing loss differences among the modes. This effect disappears of course for $g \to \infty$, and the second-order effects expressed by (27) are then the predominant source of loss differences.

Although assumed infinitely thick in Section II, the fiber cladding measures typically only tens of micrometers in thickness and is covered by a lossy jacket to avoid cross-talk to other fibers. The resulting perturbation of the mode fields of Section II is nevertheless small because of the rapid cladding field decay of almost all modes. For this reason, the power loss can be obtained in good approximation from the unperturbed mode field with the help of the matching conditions at the jacket interface [27]. If the (complex) jacket index is n_j and its real part is not too different from the cladding index, the ratio between the radial and the axial power flow at the interface becomes [27]

$$\chi = \mathrm{Re}\,[(kn_j/\beta)^2 - 1]^{1/2}. \tag{32}$$

Assuming a cladding thickness s, we can relate the power loss α_j to the averaged power density $p(a+s)$ at the jacket interface by integrating around the circumference and dividing by the total mode power P. The result is

$$\alpha_j = 4.34\pi(a+s)\chi p(a+s)/P \text{ in dB}. \tag{33}$$

After inserting (7) and (A9), we have

$$\alpha_j = \frac{4.34}{a}\chi\frac{\hat{w}(a)}{\hat{w}(a+s)}j_l\frac{w^2}{v^2}\exp\left(-2\int_a^{a+s}\hat{w}\,dr/a\right) \tag{34}$$

with $\hat{w}$ from (10) in the case of the step index profile. If $l = 0$,

$$s = \frac{a}{2w}\ln\,(4.34\chi j_l w^2/v^2 a\alpha_j). \tag{35}$$

In the case of the single-mode fiber operating at $v = 2$, we find most parameters appearing in (35) from Fig. 2 and obtain $s = 8a$ for $\alpha_j/\chi = 1$ dB/km, in good agreement with results of [27]. The great variety of potential jackets makes it difficult to find a representative value for χ. It may be of the order of unity or smaller. On the other hand, it is evident from (35) that s is not very sensitive to the ratio α_j/χ.

The fairly complicated rigorous solution of (34) for $l \neq 0$ is omitted here. The line $\alpha_j/\chi = 1\ dB/km$ in Fig. 8 depicts the result for the specific example:

core radius a	=	25 μm
cladding thickness s	=	20 μm
free wavelength λ	=	1 μm
relative index difference Δ	=	1 percent
core refractive index n_1	=	1.46.

Trapped modes to the right of this line have a loss ratio α_j/χ larger than 1 dB/km. Their number represents approximately a fraction

$$\eta_j \approx \frac{1}{2\pi}\frac{a^2}{s^2 v^3}\frac{\ln^3\,(1/a\alpha_j)}{\ln\,(1+s/a)} \tag{36}$$

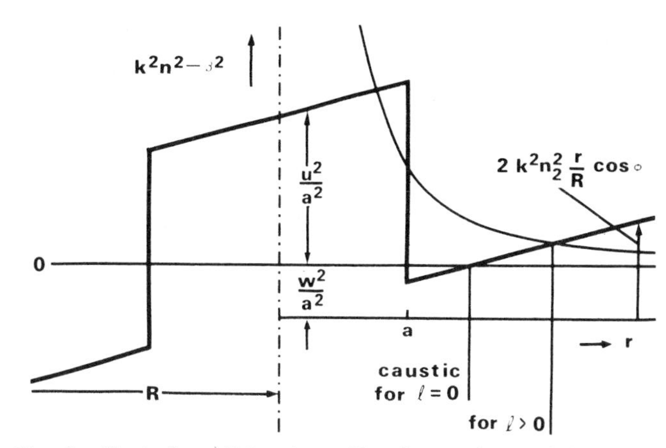

Fig. 9. Equivalent dielectric profile of curved step index fiber; center of curvature is far left (outside field of view) at distance R from guide axis.

of the total mode volume. For the parameters listed above, η_j amounts to about 6 percent. As a rule of thumb, the cladding thickness should be

$$s \approx 36a\eta_j^{-1/2}v^{-3/2} \tag{37}$$

for $\alpha_j/\chi = 1$ dB/km to be restricted to a fraction η_j of all modes. The same relationship holds for the membrane length in the case of the single-material fiber (Fig. 7(b)). The core-membrane structure is typically surrounded by a tube of the same material as core and membranes. This tube provides strength and stiffness and the necessary shield against contamination of the core surface. Naturally, in this case, χ obeys a relationship different from (32).

The bending of a dielectric waveguide produces a source of radiation loss in the cladding; Marcatili and Miller [29] explain this phenomenon as follows: to maintain a guided mode field with equiphase fronts on radial planes, a fraction of the mode field on the outside of the bend would have to exceed the plane wave velocity in the cladding medium. Since this is impossible, the energy associated with this part of the mode field is lost to radiation. For the purpose of evaluating this effect, we reduce the problem to that of a straight guide by conformal mapping [20]; the result is the index distribution shown in Fig. 9. It is easy to convince oneself that the index slope of Fig. 9 causes very nearly the same phase velocity differences as a bend with its center at a large distance R to the left of the profile center. It is also evident that the incessant increase of the index towards the right eventually creates a real radial wave vector component in the cladding, and, as a result, periodic field solutions and a radiative power flow extending to infinity.

Because of the absence of circular symmetry, the field solutions with the profile of Fig. 9 are not of the form (2) or (A9). This fact greatly complicates an estimation of the radiative loss and has limited most published work to the lowest mode orders. Reference [29] is an excellent survey of the literature. The following remarks add some recent results applicable to aribtrary mode orders [30].

As Fig. 9 indicates, the extent of the evanescent field (between the interface and the caustic at which the field turns radiative) increases with increasing azimuthal mode order l. As a result and because of the rapid decay of the evasescent field, meridional mode orders suffer more loss in fiber bends than corresponding azimuthal orders. Furthermore, modes that are degenerate in the straight fiber exhibit differences in loss if they differ in orientation with respect to the plane of curvature: fields that are even or symmetric with respect to this plane behave differently from fields that are odd. The following results are based on WKBJ field solutions for the profile of Fig. 9 expressed in parabolic cylinder coordinates [30]. The power loss per unit length in decibels is

$$\alpha_c = \frac{4.34}{a^2 n_2 k} j_l \frac{w^2}{v^2} (4w^2 + \zeta^{4/3} B^{-1/3,1/2}) \cdot \exp^{-1} (w^2 B^{2/3} + \zeta^{2/3} - \tfrac{4}{3} B^{-1/3})^{3/2} \quad (38)$$

with

$$B = \frac{2}{3} \frac{R}{n_2^2 k^2 a^3} \quad (39)$$

and

$$\zeta = \pi \begin{cases} l + \frac{1}{2}, & \text{for even modes} \\ l, & \text{for odd modes.} \end{cases} \quad (40)$$

In the case of the single-mode fiber operated at $v = 2$, the first term in the exponential of (38) dominates. For $\alpha_c = 1$ dB/km, we obtain with the help of the parameters of Fig. 2

$$R = (5 + 0.2 \ln \Delta) \lambda \Delta^{-3/2}. \quad (41)$$

A relative index difference of 0.3 percent corresponds to a core diameter of 5.6 μm and permits a bending radius of 23 mm, which is in the vicinity of the mechanically safe bending limit.

The line $\alpha_c = 1$ dB/km in Fig. 8 represents a computation of the bending loss on the basis of (38) for the fiber characterized earlier and bent to a radius of 15 mm. In general, the fraction of modes having a loss α_c larger than 1 dB/km is approximately

$$\eta_c = 0.1 \frac{[\frac{4}{3} B^{-1/3} + \ln^{2/3} (\Delta^{1/2} a \alpha_c)]^{5/2}}{v \Delta (R k n_2)^{2/3}}. \quad (42)$$

For the example above, $\eta_c = 8$ percent.

So far, we have ignored modes outside the cutoff line denoted by $u = v$ in Fig. 8. For these modes, the zero level in Fig. 4(a) falls below the cladding level $k^2 n_2^2 - \beta^2$. This implies radiative field solutions throughout the cladding and intolerable loss if $l = 0$. For $l \neq 0$, however, the radiative field solutions exist only beyond the caustic at $r = al(u^2 - v^2)^{-1/2}$, where the function l^2/r^2 in Fig. 4 intersects the level $k^2 n_2^2 - \beta^2$. For large l, leakage through the evanescent field region between the interface and the caustic is small [31], and hence these modes can propagate long distances even though $u > v$. To obtain a simple estimate, we make the assumption that $u^2 - v^2 \ll l^2$, because this condition permits the largest extent of the evanescent field region and thus promises the least leakage. For typical multimode fibers for long-distance transmission which have $v < 50$, this is a valid approximation, since the loss of modes not fulfilling this condition is so high that they are of no further interest. Similar power flow considerations as for the jacket problem lead to the loss coefficient [31]

$$\alpha_l = 4.34 \frac{l - 1}{a^2 \beta} \left(1.85 \frac{u^2 - v^2}{l^2 - 1}\right)^l. \quad (43)$$

The line $\alpha_l = 1$ dB/km in Fig. 8 depicts the result for the parameters listed earlier. Modes to the left of this line must be considered as propagating even though they are theoretically "cut off." The relative increment in mode volume as a result of these modes is approximately

$$\eta_l = 0.1 (a \alpha_l / \Delta^{1/2})^{1/v}. \quad (44)$$

This result holds for $v < 50$, a condition that seems to be fulfilled for typical multimode fibers envisaged in optical communication applications. The fiber characterized earlier has $\eta_l = 6$ percent.

It is interesting to note that the last three loss processes (α_j, α_c, and α_l), which all originate from some form of leakage through the evanescent field region, show a sharp rise of loss at a certain mode order. This permitted us to define this effect in terms of a reduction of the total mode volume rather than as a loss per unit length. If the leaky modes are not excited and coupling is absent, a loss based on these effects should essentially be avoidable. As we shall see, it is the presence of coupling which turns these effects into an actual loss per unit length.

Mode coupling is caused predominantly by perturbations which have a periodicity in propagation direction equal to the beat wavelength Λ pertaining to the two modes that are coupled [32]. This wavelength is the distance within which the phase of one of the modes lags a total of 2π behind the other. The phase lag per unit length is

$$\kappa = \frac{2\pi}{\Lambda} = \left| \frac{\partial \beta}{\partial m} dm + \frac{\partial \beta}{\partial l} dl \right| \quad (45)$$

where $dm = 0, \pm 1, \pm 2, \cdots$ and $dl = 0, \pm 1, \pm 2, \cdots$ are the differences in the order numbers between the two modes. With the help of (15), we obtain, for example,

$$\kappa = \frac{\tan \theta}{a} \frac{dm\pi + dl \arccos (l/u)}{(1 - l^2/u^2)^{1/2}} \quad (46)$$

for the classical uniform core index. Most perturbations are random and of a kind which strongly favors coupling between neighboring modes having a long beat wavelength or small κ. The combinations $dm = 0, \pm 1$ and $dl = 0, \pm 1$ are therefore of particular importance. Moreover, the nature of the perturbation excludes certain transitions; directional changes of the guide axis, for example, permit only $dl = \pm 1$. Even with these restrictions, the κ-values of neighboring modes of the cylindrical step-index, as obtained from (46), are functions of m and l that are too complicated for a rigorous evaluation of mode coupling in a multimode fiber. All studies so far have therefore used approximations for the κ-values similar to the one obtainable by applying (45) to (21). In that case, if $g \to \infty$, the minimum κ-value between neighboring modes is

$$\bar{\kappa} = \frac{u}{\beta a^2} = \frac{\tan \theta}{a} \tag{47}$$

for the uniform core.

As an important source of mode coupling, let us consider the effect of random directional changes of the axis of a multimode fiber of the classical type (uniform core). We assume that we know the "power spectrum" $\Phi(\kappa)$ of the curvature. The power coupling coefficient pertaining to two modes with phase lag $\bar{\kappa}$ is then

$$C = \tfrac{1}{4}\Phi(\kappa)(akn_1)^2 = \Phi v^2/8\Delta. \tag{48}$$

Because of our approximation (47), C is a function of u only and not explicitely of m and l. To simplify our problem even further, we assume that also the loss distribution, which may be caused by a combination of the loss phenomena discussed earlier, is only a function of u. We therefore write it as $\alpha(u)$. In that case, the transition to a mode continuum permits us to reduce the coupling among all M modes to some form of diffusion phenomenon governed by a partial differential equation of the form [34]

$$\frac{\partial}{\partial u}\left(C\frac{\partial Q}{\partial u}\right) = \alpha(u)\frac{\partial Q}{\partial z}. \tag{49}$$

where $Q(u)$ is the power distribution in the mode groups characterized by u.

No matter what power distribution is excited at the fiber input, coupling and the loss processes involved eventually establish a dynamic equilibrium which transforms $Q(u)$ into a distribution $P(u)$, such that $Q = P\exp(-Az)$, where P is the lowest eigenfunction and A the lowest eigenvalue of (49). In other words, the power distribution assumes a function which minimizes the loss A. Equation (49) then becomes

$$\frac{\partial}{\partial u}\left(C\frac{\partial P}{\partial u}\right) = [\alpha(u) - A]P. \tag{50}$$

Note that $P(u)$ is the far-field power density discussed earlier in the limit that $ka \to \infty$.

A good phenomenological description of measured results [35], [36, fig. 5] which leads to the Poschl–Teller differential equation [22], [37] is provided by

$$\alpha(u) = \epsilon \tan^2 \frac{\pi}{2}\frac{u}{v}. \tag{51}$$

For this loss distribution and if we assume $\Phi(\bar{\kappa}) = \Phi_c$ to be independent of $\bar{\kappa}$, the lowest eigenvalue A of (50) obeys the relation

$$\Phi_c = \frac{2\Delta A}{1 + \epsilon/A}. \tag{52}$$

One of the best fibers made to date had for example [35], [36], $\Delta = 1$ percent and $\epsilon \approx A \approx 1$ dB/km; hence $\Phi_c = 0.0023\ \text{km}^{-1}$. The results do not change significantly if Φ, rather than being independent of $\bar{\kappa}$, is a slowly decreasing function of κ with $\Phi_c = \Phi(\bar{\kappa}_{max})$. To understand the physical significance of the value obtained for Φ_c, let us assume that Φ results from a number of minor, but relatively abrupt, directional changes distributed randomly over 1 km of fiber length. Let the directional change be 0.1 degree of angle occurring within 0.1 mm of fiber length (radius of curvature 57 mm). In that case, 76 directional changes per kilometer are sufficient to cause the value Φ_c obtained above. Note that it is the change of curvature, not curvature itself, which produces coupling and coupling loss. As noted earlier, a much stronger, but constant curvature of 15 mm radius produces an elimination of some high-order modes, but essentially no loss, if these modes are not excited and coupling is absent.

A study of the parabolic profile for the case $\epsilon = 0$ (abrupt loss increase at $u = v$) can be found in [38]. In that case, $\bar{\kappa} = \bar{\kappa}_{max}$ for all modes so that only $\Phi(\bar{\kappa}_{max})$ must be considered. It leads essentially to the relation (52) with $\epsilon = 0$.

IV. Delay Distortion

A number of promising applications of fibers are in communication systems which utilize some form of digital envelope modulation of the optical signal [5]. Accordingly, fiber performance is usually characterized in terms of the degradation of an optical pulse propagating through the fiber. We shall follow this practice; alternative descriptions like the baseband frequency characteristic of the fiber can, at least in principle, be obtained from the above results by a simple Fourier transformation [39]. The delay per unit length of a light pulse at a given carrier frequency f_0 is

$$t = \frac{1}{c}\frac{d\beta}{dk}\bigg|_{f=f_0}. \tag{53}$$

If the carrier has a spectral width B which is broad compared to that of the detected pulse envelope, the pulse spread per unit fiber length as a result of the change of $d\beta/dk$ with f is approximately [40]

$$\tau = \frac{1}{c}\frac{B}{f_0}k\frac{d^2\beta}{dk^2}\bigg|_{f=f_0}. \tag{54}$$

The propagation constant β is a function of k, not only because the index changes with frequency (material dispersion), but, in addition, because β in (5) is a function of the v-value which in turn is proportional to frequency. This effect is here called "waveguide dispersion." A third pulse impairment is a consequence of the fact that (53) is a function of the mode number, so that a pulse spread arises in multimode fibers even if the frequency dependence is neglected.

As far as their effect on the signal is concerned, material and waveguide dispersion are interrelated in a complicated way; however, by computing one in the absence of the other, we can show that the material effect usually dominates and the waveguide effect can be neglected. We assume first that the carrier is a plane wave propagating in a dielectric of index $n(f)$. We have $\beta = nk$ and, since $dk/k = df/f = -d\lambda/\lambda$, we obtain from (54)

$$\tau = \frac{1}{c}\frac{B}{f_0}\lambda^2 \frac{d^2n}{d\lambda^2}\bigg|_{\lambda=c/f_0}. \tag{55}$$

The coefficient $\lambda^2 d^2n/d\lambda^2$ computed from index data of a silica-rich core material [41] is plotted in Fig. 10. Also shown is the result of a direct measurement of the effect [42] at a wavelength of 0.8 μm. Typical luminescent diodes made from Al–Ga–As have a spectral width (between $1/e$ points) of 4 percent and hence produce a τ of 4 ns/km when operated at 0.8 μm [43]. The effect could be substantially reduced, if such sources could be operated at longer wavelengths, possibly by using In–Ga–As instead of Al–Ga–As [44].

Next consider a classical (step-index) fiber made from a dispersionless material. To calculate $d\beta/dk$, we write (30) in the form

$$\frac{\beta}{k}\frac{d\beta}{dk} = n_1^2\frac{P_1}{P} + n_2^2\frac{P_2}{P} \tag{56}$$

and obtain with the help of (4) and (6)

$$\frac{d\beta}{dk} = \frac{1}{\beta k}\left[\beta^2 + \frac{w^2}{a^2}j_l\right] \approx n_2 + (n_1 - n_2)\frac{w^2}{v^2}(1 + 2j_l). \tag{57}$$

In order to compare this with the coefficient in Fig. 10, we have plotted

$$(n_1 - n_2)^{-1}kd^2\beta/dk^2 \tag{58}$$

as obtained from (57) for the fundamental mode versus v in Fig. 2. The coefficient reaches a maximum at $v = 1.2$, but decreases to about 0.28 at a typical operating point of $v = 2$. Thus $kd^2\beta/dk^2 = 0.4$ to 0.04 percent for $\Delta = 1$ to 0.1 percent, as compared to $\lambda^2 d^2n/d\lambda^2 = 3.1$ percent at 0.8 μm wavelength.

Waveguide dispersion coefficients as high as those indicated in Fig. 2 occur in multimode fibers only for those

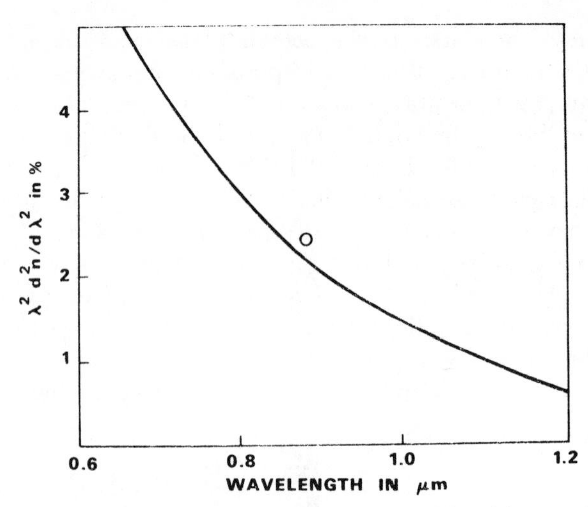

Fig. 10. Material dispersion coefficient for silica-rich core [41] plotted versus wavelength. Dot indicates measured value [42].

modes which are operated relatively close to cutoff. Even if these modes are fully transmitted, they constitute a sufficiently small number to have little influence on the pulse distortion as far as their waveguide dispersion is concerned. In these fibers, it is the delay difference between individual modes which distorts the pulse.

To study this effect, we introduce (17) into (57) and write

$$\frac{d\beta}{dk} = \frac{n_1^2 k}{\beta}\left[1 - \frac{u^2}{v^2}\frac{2\Delta}{h + (w^2 + l^2)^{1/2}}\right]. \tag{59}$$

The term n_1^2k/β is easily identified as the ray-optics approximation for $d\beta/dk$. As a mode approaches cutoff, w and l can be small enough to reduce the ray optics delay n_1^2k/β by a significant amount, producing what is known in the slab structure as the Goos–Haenchen shift [9], [45]. In fact, for $l = 0$, this shift coincides exactly with that of the TE slab modes [45]. Most modes in multimode fibers propagate sufficiently far from cutoff that n_1^2k/β is a satisfactory approximation for all. Let us now estimate the magnitude of these delays directly for the general class of graded profiles (20), of which the classic step-index profile is a special member. We use (21) and (53) to find [23]

$$t(u) = \frac{n_0^2 k}{c\beta}\left(1 - \frac{4\Delta}{g+2}\frac{u^2}{v^2}\right) \tag{60}$$

which reduces to $n_0^2k/c\beta$ for $g \to \infty$. For arbitrary g, the delay of the mode of lowest order is $t(0) = n_0/c$. The highest orders have $u = v$ and $t(v) = (n_0/c)(1 - 2\Delta)^{1/2}$. The maximum difference is therefore $t_{max} - t_{min} \approx n_0\Delta/c$ for $g \to \infty$. If $\Delta = 1$ percent this amounts to about 50 ns/km.

Optimal equalization occurs for [23]

$$g = 2 - 2\Delta \tag{61}$$

which characterizes a profile very close to the parabolic. In that case $t(0) = t(v) = n_0/c$, but all other modes have

$t(u) < n_0/c$, the fastest arriving at $t(v/\sqrt{2}) = n_0\Delta^2/8c$. For arbitrary g, the delay difference between the slowest and the fastest mode is

$$t_{max} - t_{min} = \frac{n_0}{c} \begin{cases} \Delta(g - 2 + 2\Delta)/(g + 2), & 2 < g \\ (g - 2 + 4\Delta)^2/32, & 2 - 2\Delta < g < 2 \\ (g - 2)^2/32, & 2 - 4\Delta < g < 2 - 2\Delta \\ \Delta(2 - 2\Delta - g)/(g + 2), & 1 < g < 2 - 4\Delta. \end{cases} \quad (62)$$

Evidently, good equalization occurs in a very narrow region of g values and requires accurate control of the grading process during the fiber or preform preparation. That these requirements can be met very closely was demonstrated by the early Selfoc fibers [46], whose profile had a g-value of very nearly $2 - \Delta$ in a large part of the cross section [47]; experiments proved that these fibers showed indeed an amazingly good mode equalization [48]. For those profiles, which belong to the class (20), but whose g-values deviate from the optimal, we can calculate the maximum delay spread as a function of the maximum deviation dn of the index from the optimal anywhere between $r = 0$ and $r = a$. The ratio $dn/n_0\Delta$ is called the profile error; we use it as a parameter in Fig. 11. Plotted in Fig. 11 is the index difference or the numerical aperture which would lead to a given delay spread per kilometer for various profile errors. Also shown is the spread caused by material dispersion in silica fibers, when the carrier source is an Al–Ga–As luminescent diode operated at 0.8 μm.

Equation (62) must be considered as an upper bound for the pulse broadening possible as a result of mode delay differences. The actual broadening is usually much smaller; two effects are responsible for this. One is the selective loss of certain modes or mode groups as a result of the loss effects discussed in Section III. The other is mode coupling which tends to average the delay by "switching the light around" among the various modes. To study the first effect, consider the example of a graded-index fiber whose material loss varies in the fiber cross section according to the relation (31). The reason for such a variation was explained earlier. Let us exclude the singularity in the vicinity of the parabolic distribution for the time being and assume that the g-value of our fiber deviates substantially from 2. A profile error of 10 percent, for example, corresponds approximately to $g = 3$. If the material loss were the same everywhere and all modes were excited equally, the arrival of all modes at the end of a transmission path L would fall into the interval

$$Ln_0/c < T < (Ln_0/c)\Delta(g - 2)/(g + 2),$$

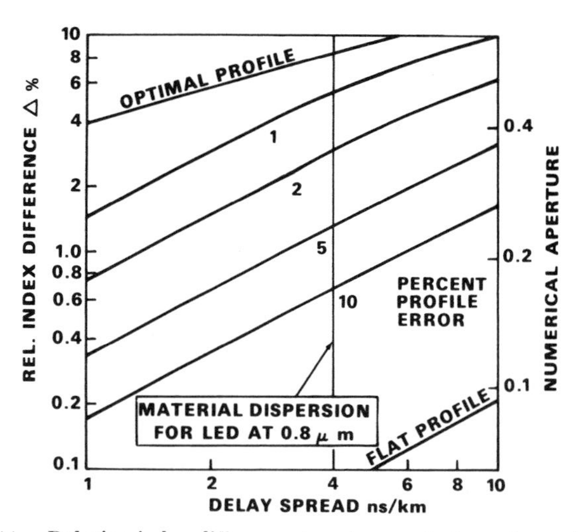

Fig. 11. Relative index difference Δ and numerical aperture which produce delay spread plotted along absissa for various profile errors $dn/n_0\Delta$.

where T is related to the mode parameter u by [23]

$$T(u) = \frac{n_0 L}{c}\Delta\frac{g - 2}{g + 2}\frac{u^2}{v^2}. \quad (63)$$

The power density per unit time interval would be proportional [23] to $T^{2/g}$. If the two loss values α_0 and α_2 of (28) are different, we find the loss as a function of u from (31), and using (63), can write the power distribution as proportional to

$$T^{2/g} \exp\left[-0.46(\alpha_2 - \alpha_0)cT/n_0\Delta(g - 2)\right]. \quad (64)$$

with α_0 and α_2 in dB/km. The rms value of this distribution is

$$\sigma = \frac{n_0\Delta}{c}\frac{(1 - 2/g)(1 + 2/g)^{1/2}}{0.86g(\alpha_2 - \alpha_0)} \quad (65)$$

as long as $\sigma \ll T(v)$ of (63). The rms value is a good measure of the expected pulse broadening and of the limits of the information rate of transmission [49]. Note that (65) is independent of the transmission distance L. As an example, let us assume $\alpha_0 = 20$ dB/km, $\alpha_2 = 40$ dB/km, $\Delta = 1$ percent, and $g = 3$. As long as $L > 2$ km, the rms width can approximately be computed from (63) using $T^{2/g}$ as the power distribution. One finds an rms value of 2.8 ns/km. For $L > 2$ km, (65) applies and the rms width asymptotically approaches a value of 3.5 ns. This obviously desirable limitation of the pulse broadening is achieved by extinguishing some of the high-order modes. If these modes represent a necessary and important part of the carrier as in the case of an incoherent source, the overall loss resulting from this extinction may represent an intolerable penalty payed for the improvement in signal distortion. For the example discussed earlier, this penalty is 19 dB after 4 km. At that point, the rms value of 3.5 ns is about 3 times shorter than that expected without mode-dependent loss.

Mode coupling produces a similar signal improvement [50]. In fact, all by itself and if limited to trapped modes, it achieves this improvement without a loss penalty. In practice, the influence of mode coupling is difficult to separate from the loss effects. In the presence of both, the signal improvement as well as the loss penalty are complicated functions of the interdependence of coupling and loss in the various modes [51]. So far only the simplest models have been considered. A convenient treatment of the problem begins with (49) considering the time dependence of $Q(u,T,z)$ by an additional term $t(u)\partial Q/\partial T$ with t from (60) and $\partial Q/\partial T$ being the partial derivative with respect to time. Closed-form solutions of the resulting partial differential equation have been given only for the step-index profile, large ϵ in (51) and $C(u)$ = constant [52]. However, it can be shown [50]–[52] that, for a transmission length $L \gg \Delta/C(v)$, when the dynamic equilibrium distribution is established, the power output becomes a Gaussian in time, whose rms value σ increases as $L^{1/2}$. This relation obtains under a wide variety of conditions independent of specific fiber characteristics. The loss penalty is then equal to 4.34 AL in dB with A being the equilibrium loss coefficient obtained from (50). Let σ_0 be the rms width of the output power distribution in the absence of coupling and (mode-dependent) loss. Since σ_0 is proportional to L, the product

$$G = 4.34 \left(\frac{\sigma}{\sigma_0}\right)^2 AL \text{ in dB} \tag{66}$$

is independent of L and has come to be used as the figure of merit of a given (or artifically introduced) combination of coupling and loss.

The most desirable loss distribution would be described by a small coefficient ϵ in (51), leading to a sharp increase of $\alpha(u)$ at $u = v$, which accounts for the transition from trapped to leaky modes. Most discussions of the problem therefore consider a first approximation with $\epsilon = 0$. A variety of coupling functions $C(u)$ have been considered, among them the class [53]

$$C(u) = C(v)(v/u)^{\nu}. \tag{67}$$

To summarize the results, Fig. 12 presents a plot of $G(z)$ for $\epsilon = 0$ and a uniform core index. The figure of merit of the parabolically graded fiber [38] with $\epsilon = 0$ is $G = 0.27$.

V. CONCLUSIONS

We have tried to give a consistent picture of the theory of the optical fiber, as far as it is most relevant to design and systems questions in optical transmission applications. In most cases, we have opted for clarity and simplicity rather than utmost accuracy and hope that those interested in better accuracy can find it in the references cited. The main approximation underlying all problems discussed here is the assumption of essentially forward directed propagation and, following from that, transverse electromagnetic field solutions. We have used the WKBJ approach for all multimode fibers, even in the case of a uniform core index, because it provides a clear mode picture, a simple correspondence between modes and rays and an effortless transition to a mode continuum. In addition, it is easily extendable to graded-index profiles. Higher order approximations extending beyond the paraxial results are obtained where necessary, as, for example, in the computation of the group delay for near-parabolic profiles. Emphasis was placed on those characteristics of fibers, which deviate significantly from those of slab or film guides; this is particularly important for multimode fibers which transmit a large number of modes with high azimuthal orders. Practical aspects of fiber design, as, for example, the influence of a finite cladding width, curvature, cross sectional loss variations, material dispersion, and the effect of index profile tolerances were assessed. The large variety of potential applications made it unpractical to consider specific designs; we hope that the results are presented in a sufficiently simple way so that the reader can use them to solve his specific problems.

APPENDIX

WKBJ APPROXIMATIONS FOR CYLINDRICAL STRUCTURES

The general wave equation becomes separable in a cylindrical coordinate system (r,ϕ,z), if the refractive index n is a function of r only. In that case, the differential equation for the radial field dependence $E(r)$ assumes the form

$$\frac{\partial^2 E}{\partial r^2} + \frac{1}{r}\frac{\partial E}{\partial r} + \left[k^2 n^2(r) - \beta^2 - \frac{l^2}{r^2}\right] E = 0. \tag{A1}$$

We set

$$E = F \exp\left[ikS(r)\right] \tag{A2}$$

where F is a coefficient independent of r. Upon substitution into (A1), we have

$$ikS'' + (ikS')^2 + ikS'/r + (k^2n^2 - \beta^2 - l^2/r^2) = 0 \tag{A3}$$

where the primes denote differentiation with respect to r. We now assume that n changes slowly within a distance comparable to the wavelength λ, so that an expansion of $S(r)$ in powers of λ converges rapidly. (For the classical structure, we exclude the area around the index step; the step can later be accounted for by suitable matching conditions.) After substituting

$$S(r) = S_0 + \frac{1}{k} S_1 + \cdots \tag{A4}$$

into (A3) and equating equal powers of λ, one obtains the following equations for the first two terms of the expansion (A4):

$$-(kS_0')^2 + (k^2n^2 - \beta^2 - l^2/r^2) = 0 \qquad \text{(A5)}$$

$$ikS_0'' - 2kS_0'S_1' + ikS_0'/r = 0. \qquad \text{(A6)}$$

Integration of these equations yields

$$S_0 = \pm \int^r (n^2 - \beta^2/k^2 - l^2/k^2r^2)^{1/2}\, dr \qquad \text{(A7)}$$

and

$$S_1 = (i/4) \ln (r^2n^2 - \beta^2r^2/k^2 - l^2/k^2) \qquad \text{(A8)}$$

plus constants of integration which are omitted for clarity. In order to construct the complete solutions, we have to distinguish between three regions (see Fig. 4 or 6): the tube in which propagating field conditions obtain (S_0 real), and the two regions inside and outside of that tube in which S_0 is imaginary.

Let us consider the classical index distribution of Fig. 4 as an example. In the case of lossless propagation, standing-wave conditions obtain for the cross sectional field distribution in the propagation region. Hence, in order to obtain a solution of the form $\cos(kS_0 + \psi)$, we must consider both signs of (A7) in this region. The phase term ψ is determined by the matching conditions at the inner caustic. In the case of the classical structure, we have [20] $\psi = -\pi/4$. The field outside the propagation region vanishes for $r \to \infty$. We therefore choose the sign in (A7) to produce a decaying exponential for increasing r. To obtain the field solutions in the propgation region ($n = n_1$) and in the cladding region ($n = n_2$), we use the abbreviations (9) and (10) in (A8) and (A9), insert the latter equations into (A4) and finally write (A2) with the help of (A4) in the form

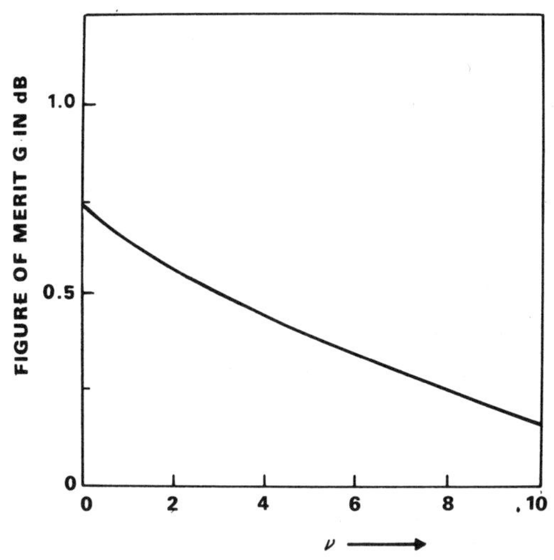

Fig. 12. Figure of merit $G(\nu)$ for coupling coefficients proportional to $(u/v)^\nu$, for $\epsilon = 0$ (abrupt loss increase at $u = 0$), and a step-index profile.

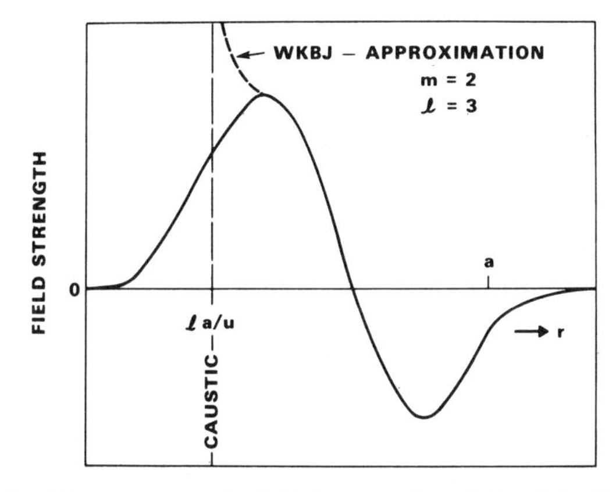

Fig. 13. Transverse mode field for $m = 2$ and $l = 3$ (solid line). WKBJ approximation coincides with exact solution everywhere except as shown by dashed line.

$$E(r) = \begin{cases} F_1(2a/\pi\hat{u}r)^{1/2} \cos\left[-\dfrac{\pi}{4} + \displaystyle\int_{al/u}^{r} \hat{u}\, dr/a\right] & \dfrac{al}{u} < r < a \\[2ex] F_2(\pi a/2\hat{w}r)^{1/2} \exp\left[-\displaystyle\int_a^r \hat{w}\, dr/a\right] & r > a \end{cases} \qquad \text{(A9)}$$

where the constants F_1 and F_2 have been chosen such that the solutions coincide with the Debye approximations of the Bessel functions $J_l(ur/a)$ and $K_l(wr/a)$, respectively [19]. They describe the Bessel functions with surprising accuracy even to the lowest order numbers. Fig. 13 illustrates the example $m = 2$, $l = 3$.

Just as in the case of the more accurate solution (2), the relations between u and w and between F_1 and F_2 are obtained from the match of the field solutions (A9) and their radial derivatives at $r = a$. However, a satisfactory solution for the majority of the modes of a multimode fiber is based on the assumption $F_2 = 0$ which ignores the evanescent fields altogether. In this case, propagating modes exist [20], [22] when

$$\int_{la/u}^{a} \hat{u}\, dr/a = (m - \tfrac{1}{4})\pi \qquad \text{(A10)}$$

where $m = 1,2,\cdots$ is the meridional order number. As the $\pi/4$-term on the right is significant only for a few low-order modes, it is neglected in the text.

ACKNOWLEDGMENT

The author wishes to thank D. Marcuse for helpful discussions.

REFERENCES

[1] R. D. Maurer, "Glass fibers for optical communications," *Proc. IEEE*, vol. 61, pp. 452–462, Apr. 1973.
[2] M. M. Ramsay, "Fiber optical communications within the United Kingdom," *Opto-Electron.*, vol. 5, pp. 261–274, July 1973.
[3] S. Maslowski, "Activities in fiber-optical communications in Germany," *Opto-Electron.*, vol. 5, pp. 275–284, July 1973.
[4] H. Ohnesorge, "Neue Möglichkeiten fur Nachrichtensysteme auf der Basis des Laser-Glasfaserkanals," Nachrichtentech. Rep. 17, Berlin, 1973.
[5] S. E. Miller, E. A. J. Marcatili, and T. Li, "Research toward optical-fiber transmission systems," *Proc. IEEE*, vol. 61, pp. 1703–1725, Dec. 1973.
[6] R. A. Andrews, A. F. Milton, and T. G. Giallorenzi, "Military applications of fiber optics and integrated optics," *IEEE Trans. Microwave Theory Tech.* (*1973 Symposium Issue*), vol. MTT-21, pp. 763–769, Dec. 1973.
[7] N. S. Kapany, *Fiber Optics.* New York: Academic, 1967.
[8] D. Gloge, "Optical fibers for communication," *Appl. Opt.*, vol. 13, pp. 249–254, Feb. 1974.
[9] N. J. Kapany and J. J. Burke, *Optical Waveguides.* New York: Academic, 1972.
[10] D. Marcuse, *Light Transmission Optics.* New York: Van Nostrand, 1972.
[11] ——, *Theory of Dielectric Optical Waveguides.* New York: Academic, 1974.
[12] T. Uchida, "Progress of optical fibers and optical integrated circuits in Japan," presented at the 1973 Conf. Laser Engineering and Application, Washington, D. C. (Abstract in *IEEE J. Quantum Electron.*, vol. QE-9, p. 687, June 1973.)
[13] S. E. Miller, "Light propagation in generalized lens-like media," *Bell Syst. Tech. J.*, vol. 44, pp. 2017–2064, Nov. 1965.
[14] P. Kaiser, E. A. J. Marcatili, and S. E. Miller, "A new optical fiber," *Bell Syst. Tech. J.*, vol. 52, pp. 265–269, Feb. 1973.
[15] H. Kogelnik and T. Li, "Laser beams and resonators," *Appl. Opt.*, vol. 5, pp. 1550–1567, Oct. 1966.
[16] A. W. Snyder, "Asymptotic expressions for eigenfunctions and eigenvalues of a dielectric or optical waveguide," *IEEE Trans. Microwave Theory Tech.* (*1969 Symposium Issue*), vol. MTT-17, pp. 1130–1138, Dec. 1969.
[17] D. Gloge, "Weakly guiding fibers," *Appl. Opt.*, vol. 10, pp. 2252–2258, Oct. 1971.
[18] M. B. Panish, "Heterostructure injection lasers," this issue, pp. 20–30.
[19] M. Abramowitz and I. A. Stegun, *Handbook of Mathematical Functions.* New York: Dover, 1964, p. 366.
[20] S. J. Maurer and L. B. Felsen, "Ray methods for trapped and slightly leaky modes in multilayered or multiwave regions," *IEEE Trans. Microwave Theory Tech.*, vol. MTT-18, pp. 584–595, Sept. 1970.
[21] C. N. Kurtz and W. Streifer, "Guided waves in inhomogeneous focusing media," *IEEE Trans. Microwave Theory Tech.*, vol. MTT-17, pp. 250–263, May 1969.
[22] J. P. Gordon, "Optics of general guiding media," *Bell Syst. Tech. J.*, vol. 45, pp. 321–332, Feb. 1966.
[23] D. Gloge and E. A. J. Marcatili, "Multimode theory of graded-core fibers," *Bell Syst. Tech. J.*, vol. 52, pp. 1563–1578, Nov. 1973
[24] E. A. J. Marcatili, "Slab-coupled waveguides," *Bell Syst. Tech. J.*, vol. 4, pp. 645–674, Apr. 1974.
[25] W. A. Gambling, D. N. Payne, and Y. Matsumura, "Gigahertz bandwidths in multimode, liquid-core, optical fiber waveguide," *Opt. Commun.*, vol. 6, pp. 317–322, Dec. 1972.
[26] K. M. Case, "On wave propagation in inhomogeneous media," *J. Math. Phys.*, vol. 13, p. 360, 1972.
[27] D. Marcuse, "The coupling of degenerate modes in two parallel dielectric waveguides," *Bell Syst. Tech. J.*, vol. 50, pp. 1791–1816, July–Aug. 1971.
[28] E. A. J. Marcatili and S. E. Miller, "Improved relations describing directional control in electromagnetic wave guidance," *Bell Syst. Tech. J.*, vol. 48, pp. 2161–2187, Sept. 1969.
[29] E. G. Neumann and H. D. Rudolph, "Radiation from bends in dielectric rod transmission lines," this issue, pp. 142–149.
[30] D. Gloge, "Mode fields and loss in curved fibers," unpublished work.
[31] A. W. Snyder and D. J. Mitchell, "Leaky rays cause failure of geometric optics on optical fibers," *Electron. Lett.*, vol. 9, pp. 437–438, Sept. 1973.
[32] S. E. Miller, "Some theory and applications of periodically coupled waves," *Bell Syst. Tech. J.*, vol. 48, pp. 2189–2219, Sept. 1969.
[33] D. Marcuse, "Mode conversion caused by surface imperfections of a dielectric slab waveguide," *Bell Syst. Tech. J.*, vol. 48, pp. 3187–3215, Dec. 1969.
[34] D. Gloge, "Optical power flow in multimode fibers," *Bell Syst. Tech. J.*, vol. 51, pp. 1767–1783, Oct. 1972.
[35] D. B. Keck, R. D. Maurer, and P. C. Schultz, "On the ultimate lower limit of attenuation in glass optical waveguides," *Appl. Phys. Lett.*, vol. 22, pp. 307–309, Apr. 1973.
[36] D. B. Keck, "Spatial and temporal power transfer measurements on a low-loss optical waveguide," *Appl. Opt.*, vol. 13, pp. 1882–1888, Aug. 1974.
[37] G. Pöschl and E. Teller, "Bemerkungen zur Quantenmechanik des Anharmonischen Oszillators," *Z. Phys.*, vol. 83, pp. 143–151, 1933.
[38] D. Marcuse, "Losses and impulse response of a parabolic-index fiber with random bends," *Bell Syst. Tech. J.*, vol. 52, pp. 1423–1437, Oct. 1973.
[39] D. Gloge, E. L. Chinnock, and D. H. Ring, "Direct measurement of the (baseband) frequency response of multimode fibers," *Appl. Opt.*, vol. 11, pp. 1534–1538, July 1972.
[40] R. B. Dyott and J. R. Stern, "Group delay in glass fiber waveguide," *Electron. Lett.*, vol. 7, pp. 82–84, 1971.
[41] R. P. Kapron and D. B. Keck, "Pulse transmission through a dielectric optical waveguide," *Appl. Opt.*, vol. 10, pp. 1519–1523, July 1971.
[42] D. Gloge, E. L. Chinnock, and T. P. Lee, "GaAs twin-laser setup to measure mode and material dispersion in optical fibers," *Appl. Opt.*, vol. 13, pp. 261–263, Feb. 1974.
[43] R. W. Dawson, "Pulse widening in a multimode optical fiber-excited by a pulsed GaAs LED," *Appl. Opt.*, vol. 13, pp. 264–265, Feb. 1974.
[44] C. J. Nuese, R. E. Enstrom, and M. Ettenberg, "Room-temperature laser operation of InGaAs junctions," *Appl. Phys. Lett.*, vol. 24, pp. 83–85, Jan. 1974.
[45] K. Artmann, "Berechnung der Seitenverzerrung des total-reflektierten Strahles," *Ann. Physik*, vol. 2, pp. 87–102, 1948.
[46] T. Uchida *et al.*, "Optical characteristics of a light-focusing fiber guide and its applications," *IEEE J. Quantum Electron.*, vol. QE-6, pp. 606–612, Oct. 1970.
[47] T. Kitano, H. Matsumura, M. Furukawa, and I. Kitano, "Measurements of the fourth-order aberration in a lens-like medium," *IEEE J. Quantum Electron.*, vol. QE-9, pp. 967–971, Oct. 1973.
[48] D. Gloge, E. L. Chinnock, and K. Koizumi, "Study of pulse distortion in Selfoc fibers," *Electron. Lett.*, vol. 8, pp. 526–527, Oct. 1973.
[49] S. D. Personick, "Receiver design for digital fiber optic communication systems," *Bell Syst. Tech. J.*, vol. 52, pp. 843–886, July–Aug. 1973.
[50] ——, "Time dispersion in dielectric waveguide," *Bell Syst. Tech. J.*, vol. 50, pp. 843–859, Mar. 1971.
[51] D. Marcuse, "Pulse propagation in multimode dielectric waveguides," *Bell Syst. Tech. J.*, vol. 51, pp. 1199–1232, July–Aug. 1972.
[52] D. Gloge, "Impulse response of clad optical multimode fibers," *Bell Syst. Tech. J.*, vol. 52, pp. 801–816, July–Aug. 1973.

Asymptotic Expressions for Eigenfunctions and Eigenvalues of a Dielectric or Optical Waveguide

ALLAN W. SNYDER

Abstract—**An asymptotic technique is presented, resulting in an analytically simple self-consistent description of the modes of a circular dielectric structure. When the dielectric difference between the rod and surrounding medium is small, the asymptotic expressions are valid for all frequencies. Even when the inside dielectric constant is twice the outside, less than a 10 percent error is usually involved. A simple functional expression for the eigenvalues of both the circular rod and the dielectric slab results from the analysis, thus eliminating the need for numerical or graphical methods.**

I. Introduction

Although the exact form of the modes for a circular dielectric cylinder are well known [1]–[3], they are in general complicated, cumbersome, and require numerical solution for the eigenvalues. However, very simple and highly accurate asymptotic expressions for the eigenfunctions and eigenvalues have been derived. These expressions are shown to be valid for all frequencies when the dielectric difference between the rod and the surrounding medium is small. It is common, particularly at optical frequencies,[1] to have guiding structures with δ as small as 0.01 where

$$\delta = 1 - (\epsilon_2/\epsilon_1) \tag{1a}$$

$$= 1 - (n_2/n_1)^2 \tag{1b}$$

where ϵ_1, ϵ_2 are the dielectric constants of the inside and outside media, and n_1, n_2 are the index of refraction of the inside and outside media, respectively. In fact, the human retinal receptors are also a fiber optic bundle of small δ [4], [5]. Even when the inside dielectric is twice the outside ($\delta=0.5$), less than a 10 percent error results at cutoff in the asymptotic expressions. Since most formulations for excitation, radiation, and scattering are based on approximations identical to those used to derive the asymptotic expressions, it is not necessary and in fact conveys no further information to manipulate the exact equations. Furthermore, with the asymptotic analysis considerably more physical insight is possible.

Other authors have noted simplifications, principally determined numerically, for the situation of $\delta \ll 1$ [3], [6]. However, there appears to be no unified analysis exploiting the functional simplicity under asymptotic conditions. Here an asymptotic analysis providing an analytically simple self-consistent description of dielectric waveguides is presented. Application of the theory to scattering and excitation is presented in [7], to multimode propagation along cylindrical tapered dielectric rods in [8].

II. Formulation of the Asymptotic Solution

In source-free homogeneous media, electromagnetic Cartesian field components satisfy the scalar wave equation

$$(\nabla^2 + K^2)\alpha = 0 \tag{2}$$

subject to boundary conditions at each discontinuity. An asymptotic solution to ϕ is found by assuming [9], [10]

$$\phi \sim \sum_n (\phi^{(n)}/K^n) = \phi^{(0)} + O(1/K) \tag{3}$$

in regions (x, y, z) of $K \gg 1$. $K=2\pi/\lambda$ is the wavenumber; λ is the local wavelength, and $\phi^{(0)}$ is locally a plane wave [9]. For guiding structures of arbitrary geometry, the asymptotic description is not feasible unless K is large for all (x, y, z).

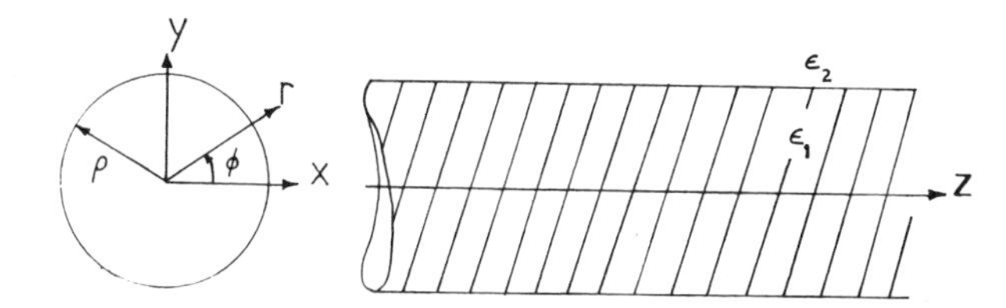

Fig. 1. Uniform circular guiding structure of dielectric constant ϵ_1 inside and ϵ_2 outside the rod. $\epsilon_1 > \epsilon_2$.

Consider the special but important case of the circular dielectric cylinder as illustrated in Fig. 1. Suppose $\rho K \gg 1$ within the rod but arbitrary outside. Then the asymptotic representation of the fields for $r<\rho$ must be matched at $r=\rho$ to the exact solution for $r>\rho$. ρ is the radius of the cylinder. However, since the eigenfunctions and the eigenvalue equation are known, they can be directly expanded asymptotically, thus preventing the necessity of having to reconstruct Bessel and Hankel functions from a series representation. Since it is the objective of this paper to obtain the simplest description for the circular dielectric rod only, the last procedure is followed. An understanding of the uniform dielectric cylinder is constructive when extending the analysis to more general open bounded guiding structures. Asymptotic solutions, including ray optical interpretations, of bounded two-dimensional waveguides are given in [11].

Manuscript received February 5, 1969; revised July 7, 1969. This study was partially supported by the Post Office of Great Britain. (*Editor's note:* Circumstances prevented the presentation of this paper at the International Microwave Symposium, Dallas, Tex., May 5–7, 1969.)

The author is with the Department of Electronic and Electrical Engineering, University College London, London W.C. 1, England.

[1] The condition $\delta \ll 1$ is mandatory for desired bandwidth, since distortion (derivative of group velocity with respect to frequency) is proportional to δ.

Reprinted from *IEEE Trans. Microwave Theory Tech.*, vol. MTT-17, pp. 1130–1138, Dec. 1969.

III. Exact Expressions for the Modes and Eigenvalues of a Circular Dielectric Cylinder

In this section a normalization procedure is introduced which is essential for the asymptotic analysis of Section IV. Since propagation is a function of ρ, ω, and δ it is convenient to define a normalized frequency V,

$$V^2 = (\rho\omega)^2\mu\epsilon_1\delta = u_p^2 + w_p^2. \tag{4}$$

ω is the angular frequency, μ is the permeability of the media, and the subscript p is used to indicate the pth mode. The longitudinal field components are solutions of the normalized wave equations

$$\left[\rho^2\nabla_t^2 + \begin{pmatrix} u_p^2 \\ -w_p^2 \end{pmatrix}\right]\alpha_p = 0 \qquad \begin{matrix} R < 1 \\ R > 1. \end{matrix} \tag{5}$$

$R = r/\rho$ is the normalized radius and $\nabla_t = \nabla^2 - \partial^2/\partial z^2$. Due to cylindrical symmetry, the pth mode has vector fields

$$\bar{E}_p'(x, y, z) = \{\bar{e}_p'(x, y) + \bar{E}_{pz}'(x, y)\} e^{j(\omega t - \beta_{pz}/\rho} \tag{6a}$$

$$\bar{H}_p'(x, y, z) = \{\bar{h}_p'(x, y) + \bar{H}_{pz}'(x, y)\} e^{j(\omega t - \beta_{pz}/\rho)} \tag{6b}$$

where $\bar{e}_p'$, $\bar{h}_p'$ are the transverse fields and $\bar{E}_{pz}'$, $\bar{H}_{pz}'$ are the longitudinal fields. β_p, the normalized modal propagation constant, is defined as

$$\beta_p^2 = k^2(1 - \theta_p^2) \tag{7}$$

where

$$\theta_p = \sqrt{\delta}u_p/V = u_p/k. \tag{8}$$

k is the normalized propagation constant of a z-directed plane wave in medium ϵ_1.

$$k = (v/\sqrt{\delta}) = 2\pi\rho/\lambda. \tag{9}$$

λ is the wavelength in medium ϵ_1; δ is defined by (1), and u_p is the eigenvalue of (5).

When ϵ_1, ϵ_2, and μ are functions of the transverse coordinates only, $\bar{e}_p'$ and $\bar{h}_p'$ are derivable from the scalar components E_{pz}' and H_{pz}' [1]:

$$\begin{pmatrix} u_p^2 \\ -w_p^2 \end{pmatrix}\bar{h}_p' = j\rho\beta_p\left\{\nabla_t H_{pz}' + (k/\beta_p) \cdot\sqrt{\epsilon_1/\mu}\begin{pmatrix} 1 \\ 1-\delta \end{pmatrix}\hat{z}\times\nabla_t E_{pz}'\right\} \qquad \begin{matrix} R < 1 \\ R > 1 \end{matrix} \tag{10a}$$

$$\begin{pmatrix} u_p \\ -w_p \end{pmatrix}\bar{e}_p' = j\rho\beta_p\left\{\nabla_t E_{pz}' - (k/\beta_p) \cdot\sqrt{\mu/\epsilon_1}\,\hat{z}\times\nabla_t H_{pz}'\right\} \qquad \begin{matrix} R < 1 \\ R > 1. \end{matrix} \tag{10b}$$

$\hat{z}$ is the unit vector in the z direction.

The eigenfunctions found from (5) and (10) are arranged in a form ideally suited to an asymptotic analysis [3], [6], [12]. For the hybrid (HE_{lM}, EH_{lM}) modes,[2]

$$E_z' = \begin{pmatrix} J_l(uR) \\ \eta_3 K_l(wR) \end{pmatrix} g_1(\phi) \qquad \begin{matrix} R < 1 \\ R > 1 \end{matrix} \tag{11a}$$

$$H_z' = \sqrt{\epsilon_1/\mu}\,(F_2\beta/k)\{g_1(\phi)/g_2(\phi)\}E_z' \tag{11b}$$

$$e_R' = j\beta\begin{bmatrix} (1/u)\{\gamma_1 J_{l-1}(uR) + \gamma_2 J_{l+1}(uR)\} \\ (\eta_3/w)\{\gamma_1 K_{l-1}(wR) - \gamma_2 K_{l+1}(wR)\} \end{bmatrix} g_1(\phi) \qquad \begin{matrix} R < 1 \\ R > 1 \end{matrix} \tag{11c}$$

$$e_\phi' = j\beta\begin{bmatrix} (1/u)\{\gamma_1 J_{l-1}(uR) - \gamma_2 J_{l+1}(uR)\} \\ (\eta_3/w)\{\gamma_1 K_{l-1}(wR) + \gamma_2 K_{l+1}(wR)\} \end{bmatrix} g_2(\phi) \qquad \begin{matrix} R < 1 \\ R > 1 \end{matrix} \tag{11d}$$

$$h_R' = jk\sqrt{\epsilon_1/\mu}\begin{bmatrix} (1/u)\{-\gamma_3 J_{l-1}(uR) + \gamma_4 J_{l+1}(uR)\} \\ -(\eta_3/w)\{\gamma_5 K_{l-1}(wR) + \gamma_6 K_{l+1}(wR)\} \end{bmatrix} g_2(\phi) \qquad \begin{matrix} R < 1 \\ R > 1 \end{matrix} \tag{11e}$$

$$h_\phi' = jk\sqrt{\epsilon_1/\mu}\begin{bmatrix} (1/u)\{\gamma_3 J_{l-1}(uR) + \gamma_4 J_{l+1}(uR)\} \\ (\eta_3/w)\{\gamma_5 K_{l-1}(wR) - \gamma_6 K_{l+1}(wR)\} \end{bmatrix} g_1(\phi) \qquad \begin{matrix} R < 1 \\ R > 1. \end{matrix} \tag{11f}$$

When $l = 0$, the fields degenerate to the TM_{0M} and TE_{0M} set

TM_{0M} Modes

$$E_z' = \begin{cases} J_0(uR) \\ \eta_3 K_0(wR) \end{cases} \qquad \begin{matrix} R < 1 \\ R > 1 \end{matrix} \tag{12a}$$

$$e_R' = j\beta\begin{bmatrix} J_1(uR)/u \\ -\eta_3 K_1(wR)/w \end{bmatrix} \qquad \begin{matrix} R < 1 \\ R > 1 \end{matrix} \tag{12b}$$

$$\bar{h}' = (k/\beta)\sqrt{\epsilon_1/\mu}\,\hat{z}\times\bar{e}\begin{pmatrix} 1 \\ 1-\delta \end{pmatrix} \qquad \begin{matrix} R < 1 \\ R > 1 \end{matrix} \tag{12c}$$

TE_{0M} Modes

$$H_z' = (\beta/k)\sqrt{\epsilon_1/\mu}\begin{pmatrix} J_0(uR) \\ \eta_3 K_0(wR) \end{pmatrix} \qquad \begin{matrix} R < 1 \\ R > 1 \end{matrix} \tag{13a}$$

$$e_\phi' = j\beta\begin{bmatrix} -J_1(uR)/u \\ \eta_3 K_1(wR)/w \end{bmatrix} \qquad \begin{matrix} R < 1 \\ R > 1 \end{matrix} \tag{13b}$$

$$\bar{h}' = (\beta/k)\sqrt{\epsilon_1/\mu}\;\times\bar{e}' \tag{13c}$$

[2] The subscript l is for azimuthal variations and M is for radial variations [3].

where the g functions are

$$g_1(\phi) = \begin{Bmatrix} \sin l\phi \\ \text{or} \\ \cos l\phi \end{Bmatrix} \tag{14a}$$

$$g_2(\phi) = \begin{Bmatrix} \cos l\phi \\ \text{or} \\ -\sin l\phi \end{Bmatrix}, \tag{14b}$$

ϕ is the azimuthal angle as illustrated in Fig. 1,

$$\eta_3 = J_l(u)/K_l(w), \tag{15}$$

and the γ's are given as

$$2\gamma_1 = F_2 - 1 \tag{16a}$$

$$2\gamma_2 = F_2 + 1 \tag{16b}$$

$$2\gamma_3 = F_1 - 1 \tag{16c}$$

$$2\gamma_4 = F_1 + 1 \tag{16d}$$

$$2\gamma_5 = 2\gamma_3 + \delta \tag{16e}$$

$$2\gamma_6 = 2\gamma_4 - \delta \tag{16f}$$

where

$$F_1 = (uw/V)^2[\{\eta_1 + (1-\delta)\eta_2\}/l \tag{17a}$$

$$= (1/F_2) - \theta_p^2 w^2 \eta_2 / l \tag{17b}$$

$$F_2 = (V/uw)^2[l/(\eta_1 + \eta_2)] \tag{18}$$

$$\eta_1 = J_l'(u)/uJ_l(u) = \pm \{J_{l\mp 1}(u)/uJ_l(u)\} \mp l/u^2 \tag{19a}$$

$$\eta_2 = K_l'(w)/wK_l(w) = -\{K_{l\mp 1}(w)/wK_l(w)\} \mp l/w^2. \tag{19b}$$

The prime notation on the J_l and K_l functions in (19) is used for derivatives with respect to the argument. The subscript p is dropped throughout the paper unless emphasis is desired. J_l is the Bessel function of order l, and K_l is the modified Hankel function of order l.

The eigenfunctions are algebraically complex, and except for $l=0$ the z-directed modal wave impedance $|\bar{e}_p'|/|\bar{h}_p'|$ is a complicated function of R and ϕ.

Subjecting the fields to the boundary conditions provides the eigenvalue equation

$$F_2 - F_1 = \theta_p^2 F_2 \tag{20}$$

where θ_p is given by (8) and the F functions by (17) and (18). Substituting (17b) into (20) leads to another form of the eigenvalue equation which is useful for the asymptotic analysis:

$$(1/F_2)^2 - 1 = \theta_p^2\{(w_p^2\eta_2/lF_2) - 1\}. \tag{21}$$

w_p is related to u_p via (4). Both (20) and (21) are complicated transcendental equations involving Bessel and modified Hankel functions; therefore u_p can only be solved numerically.

IV. Derivation of Asymptotic Forms

A. Eigenvalue Equations

Inspection of the eigenvalue equation (20) or (21) reveals that when $\theta_p \ll 1$ considerable simplification results. It is therefore reasonable to anticipate a representation of u_p by a series in powers of θ_p, i.e.,

$$u_p(V_1\theta_p) \sim \sum_{i=0}^{\infty} \theta_p^i u_p^i(V) = U + \theta_p u^{(1)} + \theta_p^2 u^{(2)} + \cdots \tag{22}$$

where $U = u_p^{(0)}(V)$ is the zero-order eigenvalue and V is treated as an independent variable in the expression. From the definition of θ_p given by (8), (22) is observed to be a series in $(1/k)$ similar to (3). The normalization or scaling procedure, i.e., the introduction of U and V, is useful when applying perturbation methods [13]. All quantities that are functions of θ_p are expanded as in (22); for example,

$$F_2(\theta_p) \sim F_2^{(0)} + \theta_p F_2^{(1)} + \theta_p^2 F_2^{(2)} + \cdots \tag{23}$$

$$\eta_2(\theta_p) \sim \eta_2^{(0)} + \theta_p \eta_2^{(1)} + \theta_p^2 \eta_2^{(2)} + \cdots \tag{24}$$

$$J_l(u_p) \sim J_l(U) + \theta_p u^{(1)} J_l'(U) + \cdots \tag{25}$$

$$w_p(V, \theta_p) \sim W + \theta_p w^{(1)} + \theta_p^2 w^{(2)} + \cdots \tag{26}$$

where $W = w_p^{(0)}(V)$.

Substitution of the series representation for the relevant functions into the eigenvalue equation (20) or (21) and into the equation for V given by (4) leads to the following forms when like powers of θ_p are equated.

Zero Order in $\theta_p - U_p$:

$$V^2 = U^2 + W^2 = (\rho\omega)^2 \mu\epsilon_1\delta \tag{27}$$

$$F_1^{(0)} = F_2^{(0)} = \mp 1. \tag{28}$$

From (18) and (28)

$$(UW/V)^2(\eta_1^{(0)} + \eta_2^{(0)}) = \mp l. \tag{29}$$

With the definition of η_1 and η_2 given by (19), the eigenvalue equation for all modes is[3]

$$\boxed{(UJ_l/J_{l\mp 1}) = \pm (WK_l/K_{l\mp 1}).} \tag{30}$$

In (30) and throughout the paper if the arguments of J_l and K_l are not shown they are understood to be U and W, respectively. The hybrid modes are catalogued from (28) as HE_{lM} for the upper sign and EH_{lM} for the lower sign [3]. Therefore, in (28), (30), and throughout the paper when the double sign notation is used the upper sign is taken for the HE and the lower sign for the EH modes. This identification follows for the $l\mp 1$ subscripts on the J and K functions as well; e.g., the $l=1$ HE_{1M} eigenvalue equation is

$$(UJ_1/J_0) = (WK_1/K_0). \tag{31}$$

A useful alternative eigenvalue equation is found when recurrence relations (99) and (106) from the Appendix are applied to (30):

[3] Except for the $\text{HE}_{lM}(l>2)$ and $\text{EH}_{lM}(l>1)$ modes (30) provides an exact u_p both at cutoff (where $\theta_p = \sqrt{\delta}$) and when $k\to\infty (\theta_p \to 0)$. Therefore, (30) is correct at $k=0$ for the HE_{11} mode. This result is partially a consequence of the u_p/k expansion (22) instead of the usual $1/k$ given by (3).

$$(UJ_{l\mp 2}/J_{l\mp 1}) = \mp (WK_{l\mp 2}/K_{l\mp 1}). \tag{32}$$

Equations (30) and (32) are frequently implemented to simplify equations for various dielectric waveguide formulations.

First Order in $\theta_p - u_p^{(1)}$: From (20) and (21),

$$F_1^{(1)} = F_2^{(1)} = 0. \tag{33}$$

From (4)

$$w_p^{(1)} = -(U_p/W_p)u_p^{(1)}. \tag{34}$$

Equating to zero the first-order expression for F_2 found from (18) leads to

$$u_p^{(1)} = w_p^{(1)} = 0 \tag{35}$$

after some algebra. In other words, $u_p \sim U_p$ to order θ_p^2. Although it is possible to continue the above process to obtain higher order terms, for the present purposes only the zero-order term, U_p, is of interest. However, since the field expressions are directly proportional to F_1 and F_2 through the γ's as given by (16), the next nonzero term of the F's is of interest.

Second Order in $\theta_p - F_1^{(2)}, F_2^{(2)}$: From (20),

$$F_1^{(2)} = F_2^{(2)} - F_2^{(0)}. \tag{36}$$

From (21),

$$F_2^{(2)} = -(F_2^{(0)}/2)\{(W^2\eta_2^{(0)}/lF_2^{(0)}) - 1\}. \tag{37}$$

Using (19b) and (28), (37) becomes

$$F_2^{(2)} = WK_{l\mp 1}/2lK_l. \tag{38}$$

Using (36), (38), and recurrence relation (105) from the Appendix,

$$F_1^{(2)} = WK_{l\pm 1}/2lK_l. \tag{39}$$

Derivative of Eigenvalue (Zeroth Order): A functional representation for $dU/dV = U'$ is frequently required. With the zero-order eigenvalue equation, it is possible (although somewhat algebraically tedious) to derive a simple expression for U'. From the zero-order eigenvalue equation (30)

$$\pm U' = (1/J_lK_{l\mp 1})\left[\frac{d}{dV}(WK_lJ_{l\mp 1}) \mp U\frac{d}{dV}(K_{l\mp 1}J_l)\right]. \tag{40}$$

By using (27)

$$\frac{dW}{dV} = (V - UU')/W. \tag{41}$$

The derivative quantities of (40) then become

$$\frac{d}{dV}(WK_lJ_{l\mp 1}) = N_1U' + N_2 \tag{42}$$

$$\frac{d}{dV}(J_lK_{l\mp 1}) = N_3U' + N_4 \tag{43}$$

where

$$N_1 = WK_lJ'_{l\mp 1} - U(WK'_l + K_l)/W \tag{44a}$$

$$N_2 = V(WK'_l + K_l)J_{l\mp 1}/W \tag{44b}$$

$$N_3 = K_{l\mp 1}J'_l - UJ_lK'_{l\mp 1}/W \tag{44c}$$

$$N_4 = VJ_lK'_{l\mp 1}/W. \tag{44d}$$

U' is then cast as

$$U' = N_t/N_b \tag{45}$$

where

$$N_t = N_2 \mp UN_4 \tag{46}$$

$$N_b = -N_1 \pm UN_3 \pm J_lK_{l\mp 1}. \tag{47}$$

After considerable algebraic manipulation and use of the Appendix,

$$N_b = -[(l \mp 1)V^2J_lK_{l\mp 1}/W^2] \pm (V^2J_lK_l/W) \mp (l \mp 1)V^2K_lJ_{l\mp 1}/UW \tag{48}$$

$$N_t = (V/WJ_l)[\mp UK_l\{[2(l \mp 1)J_{l\mp 1}/UJ_l] - 1\} - WK_{l\mp 1}J_{l\mp 1}/J_l]. \tag{49}$$

If (30) is alplied to the $K_lJ_{l\mp 1}$ term of (48) and to the $J_{l\mp 1}/J_l$ terms of (49), there results

$$N_b = \pm V^2J_lK_{l\mp 2}/W \tag{50}$$

$$N_t = \pm (UV/W)[K_{l\mp 2} - K_{l\mp 1}^2/K_l]. \tag{51}$$

Division of N_t by N_b results in

$$U' = (U/V)[1 - (1/\xi)]. \tag{52}$$

ξ, a parameter which occurs frequently with circular dielectric rod studies, is defined as

$$\xi = K_lK_{l\mp 2}/K_{l\mp 1}^2. \tag{53}$$

Approximate Forms for Near and Far from Cutoff: Near cutoff is given by the condition

$$W_p \to 0. \tag{54}$$

Then from (27)

$$U_p = V + O(W_p/V)^2. \tag{55}$$

Far from cutoff is given by the condition

$$W_p \gg 1(V \gg U_p). \tag{56}$$

From (27),

$$W_p = V + O(U_p/V)^2. \tag{57}$$

With these expressions and the approximate forms for the J_l and K_l functions in the Appendix (for large and small arguments), very simple equations can be derived for most cases of interest.

Discussion of Zero-Order Eigenvalue Equations: Since $u^{(1)}$ is zero, $u_p = U_p$ to order θ_p^2. Therefore, when θ_p^2 can be neglected, U_p is highly accurate. From the definition of θ_p (8), it is confirmed that far from cutoff (56) $\theta_p^2 \ll 1$ for all δ; however, at cutoff (55) $\theta_p^2 = \delta$. Since θ_p is largest at cutoff, if $\delta \ll 1$ the asymptotic zero-order solution U_p is valid for *all* frequencies,[3] i.e., for all V. In fact, U_p obtained from (30) is shown [6] numerically to have an error for all frequencies of less than 1 percent when $\delta \leq 0.2$ and less than 10 percent when $\delta = 0.5$ ($\epsilon_1 = 2\epsilon_2$). Although (30) requires numerical solution, it has a very useful functional representation. From the asymptotic forms of K_l (see Appendix), ξ defined by (53) is shown to be

$$\xi = 1 + 1/V \qquad (V \gg U_p). \tag{58}$$

Then, the derivative expression for U_p (52) becomes

$$\frac{dU_p}{dV} = U_p/V^2 \tag{59}$$

which has the solution

$$\boxed{U_p(V) = U_p(\infty)e^{-1/V}} \tag{60}$$

where $U_p(\infty)$ is found from (30) to be the roots of $J_{l\mp 1}$, i.e.,

$$\begin{aligned} U_p(\infty) &= \text{roots of } J_{l\mp 1} \\ &= 2.405 \ \text{HE}_{11} \\ &= 3.832 \ \text{TM}_{01}, \text{TE}_{01}, \text{HE}_{21} \\ &= 5.135 \ \text{EH}_{11}, \text{HE}_{31} \\ &= 5.520 \ \text{HE}_{12} \\ &= 6.370 \ \text{EH}_{21}, \text{HE}_{41} \\ &= 7.016 \ \text{TM}_{02}, \text{TE}_{02}, \text{HE}_{22}. \end{aligned} \tag{61}$$

Equation (60) has been shown by the author to be valid for modes on dielectric slabs or films of width 2ρ. $U_p(\infty)$ is then found from the appropriate eigenvalue equation, e.g., $U_p(\infty) = \pi/2$ for the TE_{01} mode. A comparison of the approximate expression for U given by (60) with that of the numerical solution of (30) is displayed in Fig. 2. The dashed lines represent the approximate solution. Except near cutoff they are observed to be in excellent agreement. Since guides are operated above cutoff, U_p as given by (60) is a perfectly adequate representation when θ_p^2 is small.

Physical Significance of θ_p: θ_p has an interesting physical identification. From Fourier analysis a mode can be thought of as a superposition of plane waves each at a characteristic angle α_p with the z axis. Therefore, β_p is identified as the component of the plane wave propagation constant in the z direction, i.e.,

$$\beta_p^2 = k^2\begin{pmatrix} 1 \\ 1-\delta \end{pmatrix} \cos^2 \alpha_p \qquad \begin{matrix} R < 1 \\ R > 1 \end{matrix} \tag{62a}$$

$$= k^2(1 - \theta_p^2). \tag{62b}$$

Equation (62b) follows from (7). Equating (62a) and (62b) for $\theta_p^2 \ll 1$ leads to

$$\theta_p^2 = \begin{cases} \alpha_p^2 & R < 1 \\ \alpha_p^2 + \delta(1 - \alpha_p) & R > 1 \end{cases} \tag{63}$$

where R is the normalized radius r/ρ. In other words, θ_p is the characteristic angle that a cone of plane waves makes with the dielectric boundaries ($R \leq 1$) to form the pth mode. As discussed previously θ_p is small far from cutoff and equal to $\sqrt{\delta}$ at cutoff. Therefore, when $\delta \ll 1$, the modes at any frequency are formed by nearly z-directed plane waves. To order θ_p^2 the normalized modal propagation constant (7) for all modes is that of a z-directed plane wave in medium ϵ_1, i.e.,

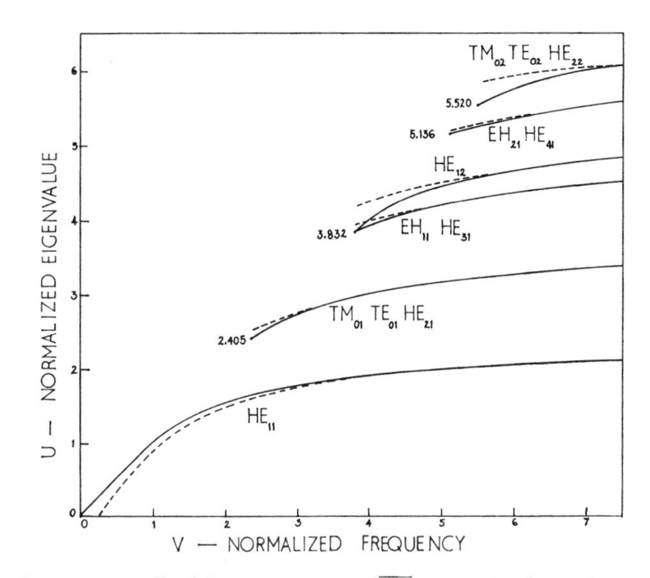

Fig. 2. Normalized frequency $V = \rho\omega\sqrt{\mu\epsilon_1\delta}$ versus the eigenvalue U_p. The solid lines are determined numerically from (30); the dashed lines are the approximate solution given by (60).

$$\beta_p = k + O(\theta_p^2). \tag{64}$$

B. Eigenfunctions

A direct consequence of $\beta_p = k + O(\theta_p^2)$ is the simplification of (10):

$$\bar{h}_p' = \sqrt{\epsilon_1/\mu}\begin{pmatrix} 1 \\ 1-\delta \end{pmatrix} \hat{z} \times \bar{e}_p' \qquad \begin{matrix} R < 1 \\ R > 1 \end{matrix} \quad + O(\theta_p^2). \tag{65}$$

However, since θ_p^2 is neglected, either $\delta \ll 1$ or $V^2 \gg U_p^2$ (i.e., far from cutoff condition). If $V^2 \gg U_p^2$, the fields can be shown by using (57) and the asymptotic forms of the K_l function to decay rapidly for $R > 1$. It can be shown that to order U^2/V^3, 100 percent of the power propagates within the rod. In other words, far from cutoff, $\bar{e}_p'$ for $R > 1$ is extremely small. Therefore, when θ_p^2 is neglected, (65) is equivalent to

$$\bar{h}_p' = \sqrt{\epsilon_1/\mu}\, \hat{z} \times \bar{e}_p' \qquad \text{for all } R. \tag{66}$$

The z-directed modal impedance $|\bar{e}_p'|/|\bar{h}_p'|$ is $\sqrt{\mu/\epsilon_1}$, i.e., that of a z-directed plane wave in medium ϵ_1, and knowledge of $\bar{e}_p'$ is sufficient to specify the transverse fields. Expanding fields (11), (12), and (13) in an asymptotic series as in (22), and equating like powers of θ_p leads to an approximate expression for the field.

Zero Order in θ_p: From the definition of the γ's given by (16), the knowledge of F_1 and F_2 given by (28) is required. Therefore

$$\gamma_1^{(0)} = \gamma_3^{(0)} = \begin{cases} 1 & \text{HE} \quad (67\text{a}) \\ 0 & \text{EH} \quad (67\text{b}) \end{cases}$$

$$\gamma_2^{(0)} = \gamma_4^{(0)} = \begin{cases} 0 & \text{HE} \quad (67\text{c}) \\ 1 & \text{EH}. \quad (67\text{d}) \end{cases}$$

With (64) and (67) the fields are shown to be

$$E_z' = \begin{pmatrix} J_l(UR) \\ \eta_3 K_l(WR) \end{pmatrix} g_1(\phi) \qquad \begin{matrix} R \le 1 \\ R \ge 1 \end{matrix} \tag{68a}$$

$$= \begin{cases} J_0(UR) & R \le 1 \\ (J_0/K_0)K_0(WR) & R \ge 1 \end{cases} \qquad \text{(TM modes)} \tag{68b}$$

$$H_z' = \mp \sqrt{\epsilon_1/\mu}\,[g_2(\phi)/g_1(\phi)]E_z' \tag{69a}$$

$$= \sqrt{\epsilon_1/\mu}\begin{bmatrix} J_0(UR) \\ (J_0/K_0)K_0(WR) \end{bmatrix} \qquad \begin{matrix} R \le 1 \\ R \ge 1 \end{matrix} \qquad \text{(TE modes)} \tag{96b}$$

$$e_R' = \mp jk\begin{pmatrix} J_{l\mp 1}(UR)/U \\ \pm\, \eta_3 K_{l\mp 1}(WR)/W \end{pmatrix} g_1(\phi) \qquad \begin{matrix} R \le 1 \\ R \ge 1 \end{matrix} \tag{70a}$$

$$= jk\begin{pmatrix} J_1(UR)/U \\ -\,(J_0/K_0)K_1(WR)/W \end{pmatrix} \text{TM modes} \qquad \begin{matrix} R \le 1 \\ R \ge 1 \end{matrix} \tag{70b}$$

$$e_\phi' = -\,jk\begin{pmatrix} J_{l\mp 1}(UR)/U \\ \pm\, \eta_3 K_{l\mp 1}(WR)/W \end{pmatrix} g_2(\phi) \qquad \begin{matrix} R \le 1 \\ R \ge 1 \end{matrix} \tag{71a}$$

$$= jk\begin{pmatrix} -\,J_1(UR)/U \\ (J_0/K_0)K_1(WR)/W \end{pmatrix} \text{TE modes} \qquad \begin{matrix} R \le 1 \\ R \ge 1 \end{matrix} \tag{71b}$$

where the $g(\phi)$ functions are defined by (14) and η_3 by (15).

These fields can be considerably simplified when multiplied by a common factor, i.e., define $\overline{E}_q$ as

$$\begin{aligned} \overline{E}_q &= (jU/kJ_{l\mp 1})\overline{E}_q' && \text{(hybrid modes)} \\ &= -\,(jU/kJ_1)\overline{E}_q' && (l = 0 \text{ modes}). \end{aligned} \tag{72}$$

Application of (30) to the $R>1$ portion of the resulting equations leads, for the longitudinal components, to

$$E_{pz} = \mp \sqrt{\mu/\epsilon_1}\begin{Bmatrix} \tan l\phi \\ \text{or} \\ -\text{ctn}\, l\phi \end{Bmatrix} H_{pz} \tag{73a}$$

$$= j\theta_p\{J_l/J_{l\mp 1}\}\begin{Bmatrix} \sin l\phi \\ \text{or} \\ \cos l\phi \end{Bmatrix} f_{l\pm 1}(R)$$

$$= -\,j\theta_p(J_0/J_1)f_0(R) \qquad \text{(TM modes)} \tag{73b}$$

$$H_{pz} = -\,j\sqrt{\epsilon_1/\mu}\,\theta_p(J_0/J_1)f_0(R) \qquad \text{(TE modes)} \tag{74}$$

and for the transverse components, to

$$e_R = \pm\, f_l(R)\begin{Bmatrix} \sin l\phi \\ \text{or} \\ \cos l\phi \end{Bmatrix} \tag{75a}$$

$$= -\,f_0(R) \qquad \text{(TM modes)} \tag{75b}$$

$$e_\phi = f_l(R)\begin{Bmatrix} \cos l\phi \\ \text{or} \\ -\sin l\phi \end{Bmatrix} \tag{75c}$$

$$= f_0(R) \qquad \text{(TE modes)} \tag{75d}$$

where the radial dependence is given as

$$f_l(R) = \begin{cases} J_{l\mp 1}(UR)/J_{l\mp 1} & R \le 1 \\ K_{l\mp 1}(WR)/K_{l\mp 1} & R \ge 1. \end{cases} \tag{76}$$

$\overline{h}_p$ is found from (66). Note that the z-field components are of order θ_p and at the boundary of the cylinder $(R=1)f_l(R)=1$.

A coordinate transformation on (75) leads to

$$\overline{e}_p = \left[\pm\,\hat{x}\begin{Bmatrix} \sin(l\mp 1)\phi \\ \text{or} \\ \cos(l\mp 1)\phi \end{Bmatrix} + \hat{y}\begin{Bmatrix} \cos(l\mp 1)\phi \\ \text{or} \\ -\sin(l\mp 1)\phi \end{Bmatrix}\right] f_l(R) \tag{77a}$$

$$= \begin{Bmatrix} \hat{x} \\ \text{or} \\ \hat{y} \end{Bmatrix} f_1(R) \qquad \text{HE}_{1M} \tag{77b}$$

$$= [-\hat{x}\cos\phi + \hat{y}\sin\phi]f_0(R) \qquad \text{TM}_{0M} \tag{77c}$$

$$= [-\hat{x}\sin\phi + \hat{y}\cos\phi]f_0(R) \qquad \text{TE}_{0M} \tag{77d}$$

where $\hat{x}$ and $\hat{y}$ are unit vectors. The modes are plane polarized, and except for the HE_{1M} modes, the polarization is ϕ dependent.

At cutoff ($W\to 0$) and

$$f_l(R) \to 1 \qquad \begin{matrix} R \ge 1 & \text{HE}_{1M} \text{ modes} \\ R \le 1 & \text{HE}_{11} \text{ mode only.} \end{matrix} \tag{78}$$

Therefore the transverse fields of the HE_{11} mode at cutoff for all R and the HE_{1M} modes for $R\ge 1$, are exactly those of a plane wave. Furthermore, $E_z\to 0$ for HE_{1M} modes as $W\to 0$.

In order to relate modal coefficients directly to power, it is convenient to define orthonormal eigenfunctions $\overline{e}_p$ and $\overline{h}_p$ such that

$$\int \overline{e}_p \times \overline{h}_q^* \cdot \hat{z}\,da = \sqrt{\epsilon_1/\mu}\int \overline{e}_p\cdot\overline{e}_q da = \begin{cases} 1 & p = q \\ 0 & p \ne q. \end{cases} \tag{79}$$

Integration is performed over the entire infinite cross sectional area ($da = r\,dr\,d\phi$). The dot product form of $\mathbf{e}_p\times\mathbf{h}_q^*\cdot\hat{z}$ follows from (66). The complex conjugated form of orthogonality is valid for lossless media only [14] (* is the complex conjugate). This demands that $\bar{e}_p$ have the form

$$\bar{e}_p = \bar{e}_p/\sqrt{\psi_p} \tag{80}$$

where the normalization constant ψ_p is defined as

$$\psi_p = \sqrt{\epsilon_1/\mu}\int |\bar{e}_p|^2 da \tag{81a}$$

$$= 2\pi\rho^2\sqrt{\epsilon_1/\mu}\int_0^\infty Rf^2{}_l(R)dR. \tag{81b}$$

The definition of $f_l(R)$ provided by (76) and the integral relations (103b), (110b) of the Appendix applied to (81) lead to

$$2\int_0^\infty Rf_l^2(R)dR = (K_{l\mp2}K_l/K_{l\mp1}^2) - (J_{l\mp2}J_l/J_{l\mp1}^2). \tag{82}$$

From (30) and (32),

$$J_{l\mp2}J_l/J^2{}_{l\mp1} = -(W/U)^2K_{l\mp2}K_l/K^2{}_{l\mp1}. \tag{83}$$

Substituting (83) into (82) using (27) yields a simple expression for ψ_p:

$$\psi_p = \sqrt{\epsilon_1/\mu}\,\pi\rho^2(V/U_p)^2\xi \tag{84}$$

where ξ is defined by (53).

The normalized power density S_z is independent of ϕ when only one mode is present:

$$S_z = \sqrt{\epsilon_1/\mu}\,|\bar{e}|^2 = \sqrt{\epsilon_1/\mu}\,f^2{}_l(R)/\psi_p. \tag{85}$$

Since energy propagates throughout the entire transverse plane, it is useful to define a parameter η which characterizes the modal power within the rod,

$$\eta = \frac{\text{power of mode } p \quad R < 1}{\text{total power of mode } p} = 2\pi\sqrt{\epsilon_1/\mu}\int_0^1 |e_p|^2 RdR. \tag{86}$$

By following the procedure used to obtain (84), η is given as

$$\eta = (U/V)^2\{(W/U)^2 + 1/\xi\}. \tag{87}$$

The group velocity v_g is of interest particularly for optical communication systems:

$$v_g = \frac{d}{d\beta}(\rho\omega) = (1/\sqrt{\mu\epsilon_1})v_{Ng} \tag{88}$$

where v_{Ng} is the normalized group velocity defined as

$$v_{Ng} = (1/\sqrt{\delta})\frac{dV}{d\beta}. \tag{89}$$

Using the derivative expression dU/dV from (52) and β from (7), it is shown after some algebra that

$$\frac{d\beta}{dV} = \sqrt{\delta}\{1 - (\theta_p{}^2/2)(1 - 2/\xi)\}. \tag{90}$$

Then the normalized group velocity is

$$v_{Ng} = 1 + (\theta_p{}^2/2)(1 - 2/\xi) \tag{91}$$

where ξ is given by (53). Distortion is proportional to the derivative of v_{Ng} with respect to V, and is therefore directly related to δ. This is one of the reasons why optical communication fibers are constructed with $\delta \ll 1$.

V. Conclusions

The asymptotic forms of the eigenfunctions and eigenvalues are presented and shown to be highly accurate when $\theta_p{}^2$ defined by (8) is neglected. One has only to compare these results with the exact equations of Section III to appreciate their utility. Since formulations for excitation and scattering are usually based on approximations similar to those used to derive the asymptotic forms, it conveys no further information to use the exact equations [7].

The most useful relations are summarized as follows.

A circular dielectric rod of ϵ_1 inside and ϵ_2 outside is considered. When terms of order $\theta_p{}^2 = \delta(U_p/V)^2$ can be neglected ($\delta = 1 - \epsilon_2/\epsilon_1$), the following results are valid. The transverse orthonormal fields $\bar{e}_p$, $\bar{h}_p$ are normalized such that

$$\sqrt{\epsilon_1/\mu}\int \bar{e}_p\cdot\bar{e}_q da = \begin{cases}1 & p = q\\ 0 & p \neq q\end{cases} \tag{92}$$

where the integration is over the infinite cross section.

$$\sqrt{\psi_p}\,\bar{e}_p = \sqrt{\psi_p}\sqrt{\mu/\epsilon_1}\,\bar{h}_p \times \hat{Z}$$

$$= \left\{\pm\hat{R}\begin{Bmatrix}\sin l\phi\\ \text{or}\\ \cos l\phi\end{Bmatrix} + \hat{\phi}\begin{Bmatrix}\cos l\phi\\ \text{or}\\ -\sin l\phi\end{Bmatrix}\right\} f_l(R) \tag{93a}$$

$$= f_0(R) \qquad \text{TE}_{0M} \text{ modes} \tag{93b}$$

$$= -f_0(R) \qquad \text{TM}_{0M} \text{ modes} \tag{93c}$$

where

$$f_l(R) = \begin{cases}J_{l\mp1}(UR)/J_{l\mp1} & R \leq 1 \quad (94a)\\ K_{l\mp1}(WR)/K_{l\mp1} & R \geq 1 \quad (94b)\end{cases}$$

$$\psi_p = \rho^2\pi\sqrt{\epsilon_1/\mu}(V/U)^2\xi \quad \text{(the normalization factor)} \tag{95}$$

$$\xi = K_lK_{l\mp2}/K_{l\mp1}^2 \tag{96}$$

and $\hat{z}$, $\hat{R}$, and $\hat{\phi}$ are unit vectors. When the double sign notation is used the upper sign is for HE_{lM}, the lower for EH_{lM} modes. $J_l = J_l(U)$ and $K_l = K_l(W)$. The longitudinal fields are of order θ_p. Except very close to cutoff, U_p is given as

$$U_p(V) = U_p(\infty)e^{-1/V} \tag{97}$$

where $U_p(\infty)$ are the roots of $J_{l\mp1}$ given by (61) and V is defined by (4). Table I presents a summary of the important functions both near and far from cutoff conditions.

Appendix

Bessel and Modified Hankel Function Relationships

This Appendix presents relations of use for dielectric waveguide studies. All of the equations are arranged in the $l\mp1$ form of the text, thereby simplying the calculations.

TABLE I

USEFUL EXPRESSIONS FOR CIRCULAR DIELECTRIC WAVEGUIDES WITH NEAR AND FAR FROM CUTOFF FORMS

Function	Near Cutoff ($W \ll 1$)	Far Above Cutoff ($W \gg 1, V \gg U$)
Normalized frequency $V^2 = (\rho\omega)^2\mu\epsilon_1\delta = U^2 + W^2$	$U \cong V$	$W \cong V$
$\theta_p = \sqrt{\delta}\,U/V = U/k$	$\sqrt{\delta}$	No special form
Eigenvalue equation $\pm\{UJ_l/J_{l\mp1}\} = WK_l/K_{l\mp1}$ or alternatively $\mp\{UJ_{l\mp2}/J_{l\mp1}\} = WK_{l\mp2}/K_{l\mp1}$	$\{UJ_l/J_{l\pm1}\} = $ $2(l-1)$ HE_{lM} $l \geq 2$ 0 all other modes $J_l = 0$ (except $\text{HE}_{lM} \geq 2$) $J_{l-2} = 0$ $\text{HE}_{lM} l \geq 2$	$\pm\{UJ_l/J_{l\mp1}\} = V$ $J_{l\mp1} \to 0$ as $V \to \infty$ $U(V) = U(\infty)e^{-1/V}$ $U(\infty) =$ roots of $J_{l\mp1}$
$\xi = K_lK_{l\mp2}/K^2_{l\mp1}$	$2\ln(2/1.78W)$ $l = 0, l = 2$ HE_{lM} $1/\{W\ln(2/1.78W)\}^2$ $l = 1$ HE_{lM} $(l-1)/(l-2)$ $l > 2$ HE_{lM} $(l+1)/l$ $l \geq 1$ EH_{lM}	$1 + 1/V$
$\eta = \dfrac{\text{modal power } R < 1}{\text{total modal power}} = \left(\dfrac{U}{V}\right)^2\left\{\left(\dfrac{W}{V}\right)^2 + \dfrac{1}{\xi}\right\}$	$(l-2)/(l-1)$ $l > 2$ HE_{lM} $l/(l+1)$ $l \geq 1$ EH_{lM} 0 all other modes	$1 - U^2/V^2$
$\dfrac{dU}{dV} = (U/V)(1 - 1/\xi)$	$1/(l-1)$ $l > 2$ HE_{lM} $1/(l+1)$ $l \geq 1$ EH_{lM} 1 all other modes	U/V^2
Normalized group velocity $v_{Ng} = 1 + (\theta_p^2/2)(1 - 2/\xi)$	$1 + (\theta_p^2/2)(3-l)/(l-1)$ $l > 2$ HE_{lM} $1 + (\theta_p^2/2)(1-l)/(l+1)$ $l > 1$ EH_{lM} $1 + \theta_p^2/2$ all other modes	$1 + (\theta_p^2/2)\{(2/V) - 1\}$

Note: $U = U_p(V)$; $W = W_p(V)$; $K_l = K_l(W)$; $J_l = J_l(U)$. When the double sign notation is used, the upper sign is taken for HE_{lM} modes, the lower for EH_{lM} modes.

Bessel Function Formulas

$$J_{l\mp1} = (2lJ_l/U) - J_{l\pm1} \tag{98}$$

$$J_{l\mp2} = \{2(l\mp1)J_{l\mp1}/U\} - J_l \tag{99}$$

$$J_l' = \mp(lJ_l/U) \pm J_{l\mp1} = (J_{l-1} - J_{l+1})/2 \tag{100}$$

$$J_{l\mp1}' = \mp J_l \pm (l\mp1)J_{l\mp1}/U \tag{101}$$

$$J_{-l} = (-1)^lJ_l \tag{102}$$

$$\int_0^Z ZJ_{l\mp1}(UZ)J_{l\mp1}(\Delta Z)dZ$$

$$= \begin{cases} \mp\{Z/(U^2-\Delta^2)\}[\Delta J_{l\mp1}(UZ)J_l(\Delta Z) \\ \quad - UJ_{l\mp1}(\Delta Z)J_l(UZ)] \qquad \Delta \neq U & (103a) \\ (Z^2/2)[J^2_{l\mp1}(UZ) - J_{l\mp2}(UZ)J_l(UZ)] \\ \qquad \Delta = U & (103b) \end{cases}$$

$$\int^Z Z^lJ_{l-1}(Z)dZ = Z^lJ_l. \tag{104}$$

Modified Hankel Formulas

$$K_{l\mp1} = \mp(2lK_l/W) + K_{l\pm1} \tag{105}$$

$$K_{l\mp2} = \{\mp2(l\mp1)K_{l\mp1}/W\} + K_l \tag{106}$$

$$K_l' = \mp(lK_l/W) - K_{l\mp1} = -(K_{l-1} + K_{l+1})/2 \tag{107}$$

$$K_{l\mp1}' = -K_l \pm (l\mp1)K_{l\mp1}/W \tag{108}$$

$$K_l = K_{-l} \tag{109}$$

$$\int_Z^\infty ZK_{l\mp1}(WZ)K_{l\mp1}(\Delta Z)dZ$$

$$= \begin{cases} \{Z/(W^2-\Delta^2)\}[WK_{l\mp1}(\Delta Z)K_l(WZ) \\ \quad - \Delta K_{l\mp1}(WZ)K_l(\Delta Z)] \qquad \Delta \neq W & (110a) \\ (Z^2/2)[K_{l\mp1}(\Delta Z)K_l(\Delta Z) - K_{l\mp2}(\Delta Z)] \\ \qquad \Delta = W & (110b) \end{cases}$$

$$\int^Z Z^lK_{l-1}dZ = -Z^lK_l \tag{111}$$

$$\int ZK_{l\mp1}(WZ)J_{l\mp1}(\Delta Z)dZ$$

$$= \{Z/(W^2+\Delta^2)\}[\pm\Delta K_{l\mp1}(WZ)J_l(\Delta Z) - WK_l(WZ)J_{l\mp1}(\Delta Z)] + (\Delta/W)^{l\mp1}/(\Delta^2 + W^2) \tag{112}$$

Approximate Forms

For $W \ll 1$ neglecting terms $O(W^2)$,

$$K_0 = \ln 2/1.781W \tag{113}$$

$$K_l = (l-1)!2^{l-1}W^{-l} \qquad l \geq 1 \tag{114}$$

$$(K_0/K_1) = W \ln (2/1.781W) \tag{115}$$

$$(K_{l-1}/K_l) = W/2(l-1) \qquad l \geq 2 \tag{116}$$

$$(K_{l+1}/K_l) = 2l/W \qquad l \geq 1. \tag{117}$$

For $W \gg 1$ neglecting terms $O(1/W^2)$,

$$K_l \sim \sqrt{\pi/2W}\, e^{-W}[1 + (4l^2 - 1)/8W] \tag{118}$$

$$(K_{l\mp 1}/K_l) = 1 + (1 \mp 2l)/2W. \tag{119}$$

For $U \ll 1$,

$$J_l = U^l/l!2^l. \tag{120}$$

For $U \gg 1$,

$$J_l \sim \sqrt{2/\pi U} \cos \{U - \pi(2l+1)/4\}. \tag{121}$$

Acknowledgment

The author is grateful to Prof. A. L. Cullen for providing many valuable discussions and a stimulating research atmosphere and would like to thank Prof. P. J. B. Clarricoats, R. B. Dyott, and Dr. K. C. Kao.

References

[1] J. A. Stratton, *Electromagnetic Theory.* New York: McGraw-Hill, 1941.

[2] P. J. B. Clarricoats, "Propagation along bounded and unbounded dielectric rods," IEE Monograph 409E, October 1960.

[3] E. Snitzer, "Cylindrical dielectric waveguide modes," *J. Opt. Soc. Am.*, vol. 51, pp. 491–498, May 1961.

[4] J. M. Enoch, "Nature of the transmission of energy in the retinal receptors," *J. Opt. Soc. Am.*, vol. 51, pp. 1122–1126, October 1961.

[5] G. Biernson and A. W. Snyder, "A model of vision employing optical mode patterns for colour discrimination," *IEEE Trans. Systems Science and Cybernetics*," vol. SSC-4, pp. 173–181, July 1968.

[6] G. Biernson and D. J. Kinsley, "Generalized plots of mode patterns in a cylindrical dielectric waveguide applied to the retinal cones," *IEEE Trans. Microwave Theory and Techniques*, vol. MTT-13, pp. 345–356, May 1965.

[7] A. W. Snyder, "Excitation and scattering of modes on a dielectric or optical fiber," *IEEE Trans. Microwave Theory and Techniques*, this issue, pp. 1138–1144.

[8] ——, "Coupling of modes on a cylindrical, tapered, dielectric rod," *Proc. IEEE* (Letters), vol. 57, pp. 737–739, April 1969.

[9] M. Kline, "An asymptotic solution of Maxwell's equations," *Commun. Appl. Math.*, vol. 4, pp. 225–262, 1954.

[10] M. Kline and I. Kay, *Electromagnetic Theory and Geometric Optics.* New York: Interscience, 1965.

[11] S. J. Maurer and L. B. Felsen, "Ray optical techniques for guided waves," *Proc. IEEE*, vol. 51, pp. 1718–1729, October 1967.

[12] A. W. Snyder, "Surface waveguide modes along a semi-infinite dielectric fiber excited by a plane wave," *J. Opt. Soc. Am.*, vol. 56, pp. 601–610, May 1966.

[13] J. D. Cole, *Perturbation Methods in Applied Mathematics.* Waltham, Mass.: Blaisdell, 1968.

[14] R. B. Adler, "Waves on inhomogeneous cylindrical structures," *Proc. IRE* (Correspondence), vol. 40, pp. 338–339, March 1952.

Weakly Guiding Fibers

D. Gloge

Thin glass fibers imbedded into a glass cladding of slightly lower refractive index represent a promising medium for optical communication. This article presents simple formulas and functions for the fiber parameters as a help for practical design work. It considers the propagation constant, mode delay, the cladding field depth, and the power distribution in the fiber cross section. Plots vs frequency of these parameters are given for 70 modes

I. Introduction

Recently, glass fibers have been produced that permit the transmission of optical signals over several kilometers.[1] In general, these fibers support many modes, which propagate at different velocities.[2] Since this causes signal distortion over long distances, fibers that transmit only a limited number of modes are of special interest.[3] A fiber waveguide consists of a thin central glass core surrounded by a glass cladding of slightly lower refractive index. Most modes can be suppressed by making the core thin and the index different between core and cladding small.[3] Typically, a difference of a few parts in a thousand is feasible. This avoids propagation of most modes. The modes that do propagate are weakly guided, but in general the guidance is sufficient to negotiate bends with radii of tens of centimeters.[4]

Maxwell's equations have exact solutions for the dielectric cylinder,[2] but even with the simplifying assumption that the cladding be infinitely thick these solutions are too complicated to be evaluated without computer. Recent efforts in simplifying the theory for weakly guided modes had promising results,[5] but in the region of interest, they did not lead to the kind of simple formulas one would wish to have for fiber design work. The following paper is aimed at such formulas and functions. It is meant as a help for engineering applications directed toward fiber communication systems. Most results are valid for all frequencies and propagation conditions—even at cutoff—with an accuracy of the order of the index difference between core and cladding.

The author is with Bell Telephone Laboratories, Inc., Crawford Hill Laboratory, Holmdel, New Jersey 07733.

Received 28 May 1971.

II. Mode Parameters and Characteristic Equation

Consider the cylindric core of radius a depicted in Fig. 1(a). The refractive index of the core is n_c. Let the cladding material of index n extend to infinity. We shall use both cartesian coordinates (x,y) and cylindrical coordinates (r,ϕ). The propagation constant β of any mode of this fiber is limited within the interval $n_c k \geq \beta \geq nk$, where $k = 2\pi/\lambda$ is the wavenumber in free space. If we define parameters

$$u = a(k^2 n_c^2 - \beta^2)^{\frac{1}{2}} \qquad (1)$$

$$w = a(\beta^2 - k^2 n^2)^{\frac{1}{2}}, \qquad (2)$$

the mode field can be expressed by Bessel function $J(ur/a)$ inside the core and modified Hankel function $K(wr/a)$ outside the core. The quadratic summation

$$v^2 = u^2 + w^2 \qquad (3)$$

leads to a third parameter,

$$v = ak(n_c^2 - n^2)^{\frac{1}{2}}, \qquad (4)$$

which can be considered as a normalized frequency. By matching the fields at the core–cladding interface, we obtain characteristic functions $u(v)$ or $w(v)$ for every mode; the propagation constant and all other parameters of interest can be derived from these functions.

For weak guidance, we have

$$\Delta = (n_c - n)/n \ll 1. \qquad (5)$$

In this case, we can construct modes whose transverse field is essentially polarized in one direction. This can be deduced from the results of Ref. 5, but in order to elucidate the approximation involved, let us try a direct derivation. We postulate transverse field components

$$E_y = H_x \begin{Bmatrix} Z_0/n_c \\ Z_0/n \end{Bmatrix} = E_l \begin{Bmatrix} J_l(ur/a)/J_l(u) \\ K_l(wr/a)/K_l(w) \end{Bmatrix} \cos l\phi. \qquad (6)$$

Reprinted with permission from *Appl. Opt.*, vol. 10, pp. 2252–2258, Oct. 1971.

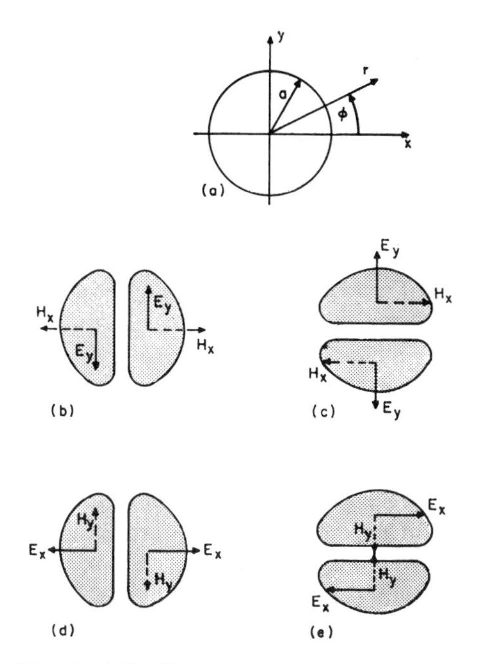

Fig. 1. Sketch of the fiber cross section and the four possible distributions of LP_{11}.

Here, as in the following, the upper line holds for the core and the lower line for the cladding; Z_0 is the plane wave impedance in vacuum, and E_l the electrical field strength at the interface. Figure 1(b)–(e) illustrate the case $l = 1$. Since we have the freedom of choosing $\sin l\phi$ or $\cos l\phi$ in Eq. (6) and two orthogonal states of polarization, we can construct a set of four modes for every l as long as $l > 0$. For $l = 0$, we have only a set of two modes polarized orthogonally with respect to each other.

The longitudinal components can be obtained from the equations[6]

$$E_z = \frac{iZ_0}{k}\begin{Bmatrix} 1/n_c^2 \\ 1/n^2 \end{Bmatrix}\frac{\partial H_x}{\partial y}, \tag{7a}$$

and

$$H_z = (i/kZ_0)(\partial E_y/\partial x). \tag{7b}$$

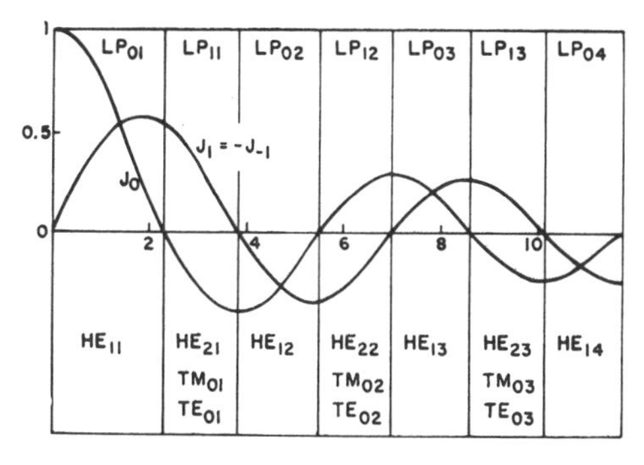

Fig. 2. The regions of the parameter u for modes of order $l = 0,1$.

By introducing Eq. (6), we therefore have

$$E_z = \frac{-iE_l}{2ka}\left\{\begin{array}{l} \dfrac{u}{n_c}\dfrac{J_{l+1}(ur/a)}{J_l(u)}\sin(l+1)\phi + \dfrac{u}{n_c}\dfrac{J_{l-1}(ur/a)}{J_l(u)}\sin(l-1)\phi \\ \dfrac{w}{n}\dfrac{K_{l+1}(wr/a)}{K_l(w)}\sin(l+1)\phi - \dfrac{w}{n}\dfrac{K_{l-1}(wr/a)}{K_l(w)}\sin(l-1)\phi \end{array}\right\}, \tag{8a}$$

$$H_z = \frac{-iE_l}{2kZ_0a}\left\{\begin{array}{l} u\dfrac{J_{l+1}(ur/a)}{J_l(u)}\cos(l+1)\phi - u\dfrac{J_{l-1}(ur/a)}{J_l(u)}\cos(l-1)\phi \\ w\dfrac{K_{l+1}(wr/a)}{K_l(w)}\cos(l+1)\phi + w\dfrac{K_{l-1}(wr/a)}{K_l(w)}\cos(l-1)\phi \end{array}\right\}. \tag{8b}$$

For small Δ, the longitudinal components [Eqs. (8a), (8b)] are small compared to the transverse components. The factors involved are u/ak and w/ka which because of Eqs. (1) and (2) are both of the order $\Delta^{\frac{1}{2}}$. Repeated differentiation of Eqs. (8a) and (8b) leads to transverse components which are not identical with the postulated field [Eq. (6)] but small of order Δ compared to it. We shall neglect these fields in the following. It is this approximation that determines the accuracy of our assumption of linearly polarized modes.

To match the fields at the interface let us write Eq. (6) in terms of cylindrical components. We then have

$$E_\phi = \frac{1}{2}E_l\begin{Bmatrix} J_l(ur/a)/J_l(u) \\ K_l(wr/a)/K_l(w) \end{Bmatrix}[\cos(l+1)\phi + \cos(l-1)\phi], \tag{9a}$$

$$H_\phi = -\frac{1}{2}\frac{E_l}{Z_0}\begin{Bmatrix} n_cJ_l(ur/a)/J_l(u) \\ nK_l(wr/a)/K_l(w) \end{Bmatrix} \times [\sin(l+1)\phi - \sin(l-1)\phi]. \tag{9b}$$

If we set $n_c = n$ in Eqs. (8) and (9) and use the recurrence relations for J_l and K_l, we can match all tangential field components at the interface by the one equation

$$u[J_{l-1}(u)/J_l(u)] = -w[(K_{l-1}(w)/K_l(w)]. \tag{10}$$

This is the characteristic equation for the linearly polarized (LP) modes. Setting $w = 0$ yields the cutoff values $J_{l-1}(u) = 0$. For $l = 0$, this includes the roots of the Bessel function $J_{-1}(u) = -J_1(u)$, which we shall count so as to include $J_1(0) = 0$ as the first root. We then obtain the cutoff values indicated in Fig. 2 for LP_{0m} and LP_{1m}. In the limit of $w \to \infty$, we have $J_l(u) = 0$. Thus, the solutions for u are between the zeros of $J_{l-1}(u)$ and $J_l(u)$. Every solution is associated with one set of modes designated LP_{lm}. For $l \geq 1$, each set comprises four modes.

The accuracy of the characteristic equation can be improved if we retain n and n_c as different in Eqs. (8)

and (9). In this case, however, terms with $(l + 1)\phi$ and $(l - 1)\phi$ satisfy two different characteristic equations:

$$(u/n_c)[J_{l\pm1}(u)/J_l(u)] = \pm(w/n)[K_{l\pm1}(w)/K_l(w)]. \quad (11)$$

By using the recurrence relations for J_l and K_l, one can easily show that these two equations converge into Eq. (10) for $n_c = n$. For $n_c \neq n$, this degeneracy ceases to exist; each mode LP_{lm} breaks up into modes with terms $(l + 1)\phi$, which can be identified as $HE_{l+1,m}$, and modes with terms $(l - 1)\phi$ which form $EH_{l-1,m}$ or TE_m and TM_m.[2,5] This association is indicated in the lower half of Fig. 2 for the cases $l = 0$ or 1. A more rigorous proof of the previous results is given in the Appendix, where Eqs. (10) and (11) are derived directly from the exact characteristic equation. The following calculations are based on Eq. (10), which is found to be sufficiently accurate for most practical applications.

III. Approximate Analytic Solution

By using Eq. (3) and differentiating both sides of Eq. (10) with respect to v, one can write the characteristic Eq. (10) in the form[5]

$$du/dv = (u/v)[1 - \kappa_l(w)], \quad (12)$$

where

$$\kappa_l(w) = K_l^2(w)/K_{l-1}(w)K_{l+1}(w). \quad (13)$$

For large w, we have $\kappa_l \approx 1 - (1/v)$. This can be used to solve Eq. (12) for large v.[5] Unfortunately, parameters of interest like the propagation constant or the field depth in the cladding depend on the difference $v^2 - u^2$. As this difference becomes small in the region of interest, the relative error introduced by the above approximation becomes intolerably large. To be useful, an approximation of u must improve toward smaller v.

With this in mind, we replace Eq. (13) by

$$\kappa_l \approx 1 - (w^2 + l^2 + 1)^{-\frac{1}{2}}, \quad (14)$$

which not only approximates Eq. (13) for large w, but provides a reasonable fit throughout. We now use Eq. (3) and replace w^2 by $v^2 - u^2$. Yet since u stays in the narrow region between successive roots of adjacent Bessel functions, we may write

$$w \approx (v^2 - u_c^2)^{\frac{1}{2}}, \quad (15)$$

replacing u by its cutoff value u_c. For the mode LP_{lm}, u_c is the mth root of $J_{l-1}(u)$. The approximations (14) and (15) are satisfactory for all modes except $LP_{01} = HE_{11}$, whose mode parameters u, v, and w all approach zero simultaneously.

If we exclude HE_{11}, we can now use Eqs. (14), (15), and the boundary value u_c at cutoff to solve Eq. (12). The result is

$$u(v) = u_c \exp[\arcsin(s/u_c) - \arcsin(s/v)]/s, \quad (16)$$

with

$$s = (u_c^2 - l^2 - 1)^{\frac{1}{2}}. \quad (17)$$

In the case of the HE_{11} mode, a more careful approximation is necessary, although the basic approach is similar to the one outlined previously. Without going into detail, we list the result

$$u(v) = (1 + \sqrt{2})v/[1 + (4 + v^4)^{\frac{1}{4}}] \quad \text{for } HE_{11}. \quad (18)$$

We mentioned earlier that u is bound between successive zeros of the Bessel functions J_{l-1} and J_l. We thus know the asymptotic value u_∞ for $v \to \infty$ exactly. Equations (16) and (18) approximate these values within an error of 2%; this is a good indication of the accuracy of Eqs. (16) and (18). For $v \gg s$ (far enough from cutoff), we can reduce Eqs. (16) and (18) to

$$u(v) = u_\infty[1 - (1/v)] \quad (19)$$

for all modes, using the mth root of $J_l(u)$ for u_∞.

IV. Propagation Constant and Mode Delay

With the help of $u(v)$, we can calculate the propagation constant β from Eq. (1). In order to make the results independent of particular fiber configurations, however, we shall not plot β directly but the ratio

$$b(v) = 1 - (u^2/v^2) = [(\beta^2/k^2) - n^2]/(n_c^2 - n^2), \quad (20)$$

which, for small index difference, reduces to

$$b \approx [(\beta/k) - n]/(n_c - n). \quad (21)$$

From this and Eq. (5) we obtain the propagation constant

$$\beta = nk(b\Delta + 1) = nk[1 + \Delta - \Delta(u^2/v^2)]. \quad (22)$$

Since β and b are proportional, the quantity b can be understood as a normalized propagation constant. Figure 3 shows $b(v)$ for 18 LP modes. A comparison with exact computer solutions showed deviations that were too small to be displayed in Fig. 3.

Direct detection of intensity-modulated light signals recognizes dispersion effects only in the envelope of the

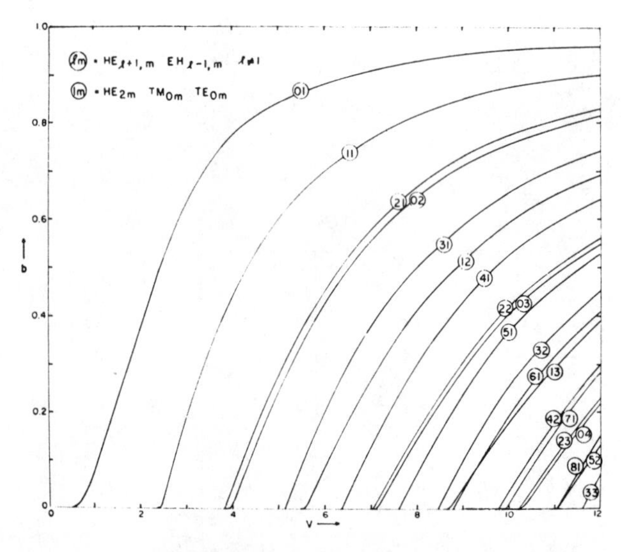

Fig. 3. Normalized propagation parameter $b = (\beta/k - n)/(n_c - n)$ as a function of the normalized frequency v.

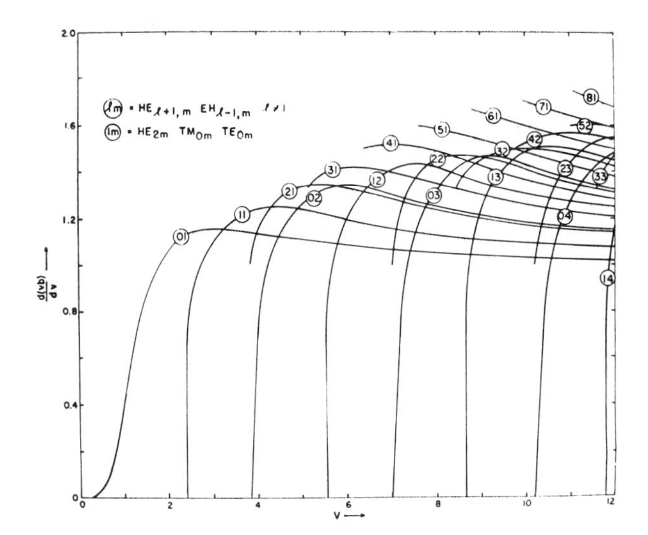

Fig. 4. Normalized group delay $d(vb)/dv$ as a function of v.

light signal. This envelope is influenced by the group delay

$$\tau_{gr} = (L/c)(d\beta/dk). \tag{23}$$

Here c is the vacuum velocity of light and L the length of the fiber. When we differentiate Eq. (22), we must consider the k dependence of n, Δ, and b. Yet, if the dispersion of the core and the cladding glass is approximately the same, Δ is independent of k. Moreover, for all glasses, $kdn/dk \ll n$. If we ignore products of Δ with $(k/n)(dn/dk)$, we obtain

$$\tau_{gr} = \frac{L}{c}\{[d(nk)/dk] + n\Delta[d(vb)/dv]\}. \tag{24}$$

The first part of Eq. (24) characterizes the material dispersion, which is the same for all modes. The second part, which represents the group delay on account of waveguide dispersion, is governed by the derivative $d(vb)/dv$. Because of Eqs. (12) and (20), this derivative can be expressed by

$$d(vb)/dv = 1 - (u/v)^2(1 - 2\kappa), \tag{25}$$

with κ from Eq. (13). This function is plotted in Fig. 4. Far from cutoff, it approaches unity for all modes. At cutoff, $d(vb)/dv = 2\kappa(0)$. This results in cutoff values $d(vb)/dv = 0$ for $l = 0$, 1, and $2[1 - (1/l)]$ for $l \geq 2$. As Fig. 4 shows, the mode of largest order l has the largest group delay. The difference between this and the slowest mode is approximately $1 - (2/l)$. For large v, we have $l_{max} \approx v$. Thus we obtain a group spread $[(1 - (2/v)](n_1 - n_2)L/c$ for a fiber that propagates many modes.

V. Power Flow and Power Density

The Poynting vector in axial direction can be calculated from the cross product of the transverse fields given in Eq. (6). Integration over the cross section of core and cladding leads to tabulated integrals[5,9]; the results are

$$P_{core} = [1 + (w^2/u^2)(1/\kappa)](\pi a^2/2)(Z_0/n_c)E_l^2 \tag{26}$$

and

$$P_{clad} = [(1/\kappa) - 1](\pi a^2/2)(Z_0/n)E_l^2 \tag{27}$$

for the power flow in core and cladding, respectively. If we ignore the small difference between n_c and n, the total power in a certain mode becomes

$$P = P_{core} + P_{clad} = (v^2/u^2)(1/\kappa)(\pi a^2/2)(Z_0/n)E_l^2. \tag{28}$$

Practical fibers have small heat and scattering losses which cause significant attenuation over long distances. In general, these losses are attributable to certain parts of the fiber and proportional to the power propagating in this part. For considerations of this kind, it is convenient to use the power fractions

$$P_{core}/P = 1 - (u^2/v^2)(1 - \kappa) \tag{29}$$

and

$$P_{clad}/P = (u^2/v^2)(1 - \kappa), \tag{30}$$

which are plotted in Fig. 5. As expected, the mode power is concentrated in the core far away from cutoff. As cutoff is approached, the power of low order modes ($l = 0{,}1$) withdraws into the cladding, whereas modes with $l \geq 2$ maintain a fixed ratio of $l - 1$ between the power in core and cladding at cutoff.

The power density is related to the mode power P by

$$p(r) = \kappa\frac{u^2}{v^2}\frac{2P}{\pi a^2}\begin{Bmatrix} J_l^2(ur/a)/J_l^2(u) \\ K_l^2(wr/a)/K_l^2(w)\end{Bmatrix}\cos^2 l\phi.$$

By averaging over ϕ at $r = a$, we obtain the mean density

$$\bar{p}(r) = \kappa(u^2/v^2)\frac{P}{\pi a^2}\begin{Bmatrix} J_l^2(ur/a)/J_l^2(u) \\ K_l^2(wr/a)/K_l^2(w)\end{Bmatrix} \tag{31}$$

At the core–cladding interface, we have $r = a$ and

$$\bar{p}(a) = \kappa(u^2/v^2)(P/\pi a^2). \tag{32}$$

The normalized density $\pi a^2\bar{p}(a)/P$ is plotted in Fig. 6. For modes of order $l = 0{,}1$ this density approaches zero both at cutoff and far away from it, having a maxi-

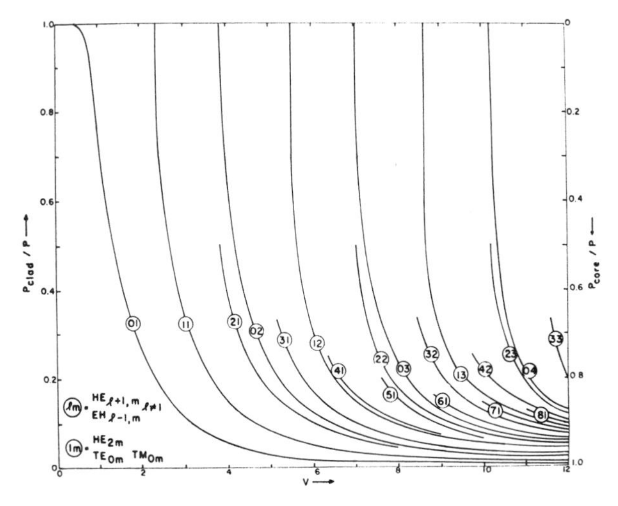

Fig. 5. Portion of the mode power which propagates in the cladding plotted vs v.

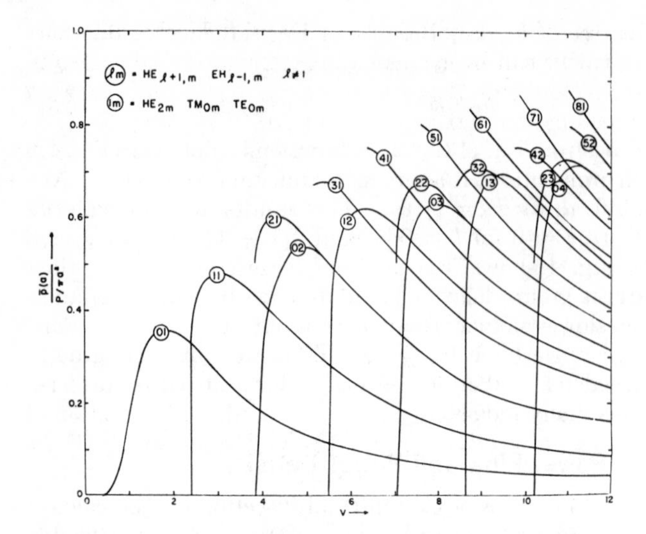

Fig. 6. Normalized power density at the core-cladding interface plotted vs v.

This function describes the cutoff power distribution in the cladding. It decreases with the distance from the axis for all but the lowest azimuthal order, whose cladding field is independent of the radius. These results are of course based on our theoretical model of a core imbedded into an unbounded cladding.

VI. Approximations for Multimode Fibers

Fibers that transmit a large number of modes are of particular interest in connection with incoherent light sources, as, for example, light-emitting diodes. This is because the amount of light that the fiber accepts from this source increases with the number of modes it transmits. Clearly, fibers with a large mode volume neither can nor need be evaluated in as much detail as was done in previous sections. A much simpler mode picture is required. To find this, let us introduce a somewhat simpler way of counting the modes.

Figure 8 illustrates the front end of a fiber and the cone of light that this fiber accepts. The cone is limited by those rays that, after entering the front

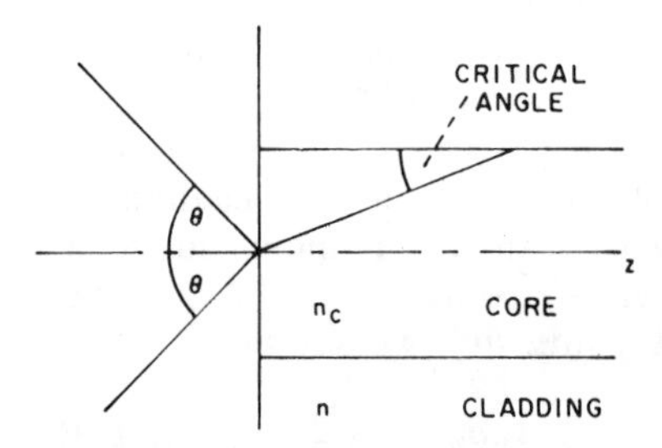

Fig. 8. Sketch of the fiber front face and the cone of light that the fiber accepts.

end of the core, are totally reflected at the interface between core and cladding. These rays form an angle

$$\theta \approx \sin\theta = (n_c^2 - n^2)^{\frac{1}{2}} \tag{33}$$

with the axis. Consider now the free-space modes which can enter the front area πa^2 of the core.[8] There are pairs of modes that are perpendicularly polarized with respect to each other. Each pair occupies a cone of solid angle $\pi\delta^2$, where[8]

$$\delta = \lambda/\pi a. \tag{34}$$

The total number of free-space modes accepted by the fiber is consequently

$$N \approx 2(\theta/\delta)^2 \tag{35}$$

or, because of Eq. (4),

$$N \approx v^2/2. \tag{36}$$

This is also the number of modes transmitted by the fiber. We find this confirmed if we count all cutoff values $u_c < v$, bearing in mind that the lowest value

mum in between. Modes with $l \geq 2$ have $\bar{p}(a) = [1 - (1/l)]P/\pi a^2$ at cutoff.

For $r \gg a/w$, we can replace the K functions in Eq. (31) by their approximation for large argument and obtain

$$\bar{p}(r) \approx \kappa(u^2/v^2)(P/\pi ar)\exp[-2w(r - a)/a] \quad \text{for } r \gg a,$$

as long as w is not too small. The power density decreases exponentially with the distance from the interface. The parameter w is plotted in Fig. 7. It decreases sharply as cutoff is approached and is zero at cutoff. For sufficiently small w we may set $u = v$ and replace the K functions in Eq. (31) by their approximation for small argument, obtaining

$$\bar{p}(r) \approx \kappa_l(P/\pi a^2)(a/r)^l \quad \text{for } r > a,\ w = 0.$$

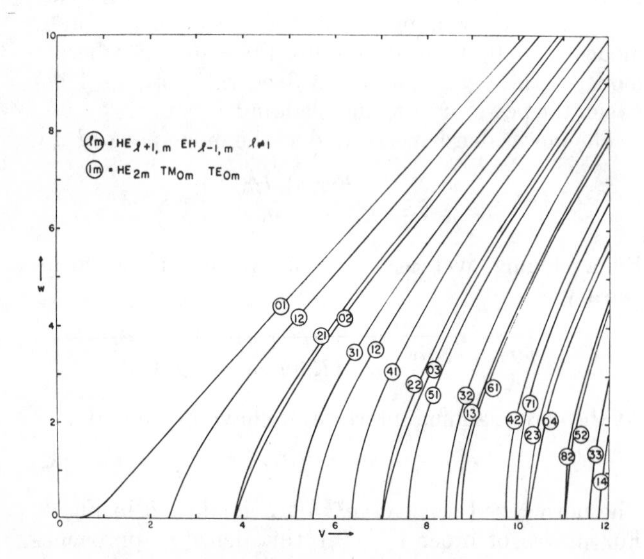

Fig. 7. Cladding parameter w plotted vs v.

represents two modes and all others four. Equation (36) permits us to label the fiber modes in the sequence of their cutoff values. Vice versa, we can predict the cutoff of the νth mode approximately at

$$u_c \approx (2\nu)^{\frac{1}{2}}. \quad (37)$$

This rule, of course, holds only for large ν.

Let us now use this counting method to describe fibers with large mode volume. We restrict ourselves to operation far from cutoff, ignoring, among the large number of modes, those few that are appreciably close to cutoff. We thus set $\kappa = 1$. We replace u by u_c, bearing in mind that the parameter u never departs much from its cutoff value u_c. Because of Eq. (20), the propagation parameter b then becomes

$$b = 1 - (u_c^2/v^2) = 1 - (\nu/N). \quad (38)$$

Equation (25) can be simplified accordingly to yield the group delay parameter

$$d(vb)/dv = 1 + (u_c^2/v^2) = 1 + (\nu/N). \quad (39)$$

We can estimate the power density at the interface by converting Eq. (32). This yields

$$\bar{p}(a) = (P/\pi a^2)(u_c^2/v^2) = (P/\pi a^2)(\nu/N). \quad (40)$$

The power flow in the cladding can be obtained from Eq. (30) by using the approximation $1 - \kappa \approx (v^2 - u_c^2)^{-\frac{1}{2}}$, which is obtained from Eq. (14). Equation (30) then reads

$$P_{\text{clad}} \approx P u_c^2/v^2(v^2 - u_c^2)^{\frac{1}{2}} = P\nu/N(2N - 2\nu)^{\frac{1}{2}}. \quad (41)$$

The incoherent source, in general, excites every fiber mode with the same amount of power. We find the average power distribution in a fiber under these conditions by averaging Eqs. (40) and (41) over all N modes, treating the mode index ν as a continuous variable. We then obtain for the density at the interface

$$[\bar{p}(a)/P]_{\text{tot}} = \frac{1}{\pi a^2 N}\int_0^N \frac{\nu}{N}\,d\nu = \frac{1}{2\pi a^2}. \quad (42)$$

It is interesting to note that this quantity is independent of N or the parameter v. The average cladding power follows from an integration of Eq. (41). The result is

$$(P_{\text{clad}}/P)_{\text{tot}} = \frac{1}{N}\int_0^N \frac{\nu d\nu}{N(2N - 2\nu)^{\frac{1}{2}}} = \frac{4}{3}N^{-\frac{1}{2}}. \quad (43)$$

The power flow in the cladding decreases proportional to $N^{-\frac{1}{2}}$ or, because of Eq. (36), proportional to $1/v$.

As an example, consider a fiber with a core index of 1.5, $\Delta = 0.003$, and a core radius of 25 μ. This results in $v = 20.3$ at 0.9-μ wavelength. The group delay between the fastest and the slowest mode is 13.5 nsec after 1 km. The fiber transmits 206 modes. Roughly 10% of the power propagates in the cladding.

VII. Conclusions

The theory of dielectric waveguides greatly simplifies if the difference between the refractive indices of core and cladding is small. In this case, linearly polarized modes with simple fields and a simple characteristic equation can be defined. Approximate analytic solutions can be derived which are exact at cutoff, showing a maximum relative error of up to 2% for very high frequencies. The mode parameters that follow from these solutions have sufficient accuracy for most practical applications. We consider the propagation constant (Fig. 3), mode delay (Fig. 4), the cladding field depth (Fig. 7), and the power distribution in the fiber cross section (Figs. 5 and 6). Far from cutoff, the group velocity of all modes is smaller than the plane wave velocity in the core and decreases as cutoff is approached. For certain modes, however, this trend reverses shortly before cutoff is reached. The fraction of mode power that propagates in the cladding is the larger the closer the mode to cutoff. For most modes, however, it is just a small fraction of the total power even at cutoff. The modes with the two lowest azimuthal orders are an exception. Their cutoff distributions are characterized by plane wave fields in the cladding which contain practically all the mode power. In multimode fibers with incoherent input, the power is in general equally distributed among all modes. In this case, we find that the power density at the interface is independent of the number of modes, but the power flow in the cladding decreases proportional to the root of that number.

I am grateful to E. A. J. Marcatili for helpful suggestions.

Appendix: Another Derivation of the Simplified Characteristic Equations

As mentioned earlier, the problem of the dielectric cylinder with sharp index step can be solved exactly; using the nomenclature defined in Sec. II, one can write the exact characteristic equation in the form

$$(Q - D - 2\Delta\{[(l \pm 1)/\omega^2] \pm [K_l(\omega)/\omega K_{l\pm 1}(\omega)]\})(Q - D) = Q^2[1 - 2\Delta(u^2/v^2)], \quad (A1)$$

where

$$Q = (l \pm 1)(v^2/u^2w^2), \quad (A2)$$

$$D = [J_l(u)/uJ_{l\pm 1}(u)] \mp [K_l(w)/wK_{l\pm 1}(w)], \quad (A3)$$

and

$$2\Delta = (n_c^2 - n^2)/n_c^2. \quad (A4)$$

The upper sign holds for HE_{l+1} modes and the lower sign for EH_{l-1}, TM, and TE. Equation (A4) agrees with Eq. (5) in the case of small index differences $n_c - n$. If Δ is set to zero in Eq. (A1), we find $D = 0$; Eq. (A3) then becomes the simplified characteristic Eq. (10). For small Δ, D is also small. Let us now simplify Eq. (A1) to the extent that we retain terms linear in Δ or D. This results in

$$D = \Delta\{Q(u^2/v^2) - [(l \pm 1)/w^2] \mp [K_l(w)/wK_{l\pm 1}(w)]\} \quad (A5)$$

and with Eq. (A2)

$$D = \mp\Delta[K_l(w)/wK_{l\pm 1}(w)]. \quad (A6)$$

By introducing this into Eq. (A3) and inserting Eq. (5), we find

$$(u/n_c)[J_{l\pm1}(u)/J_l(u)] = \pm(w/n)[K_{l\pm1}(w)/K_l(w)]. \qquad \text{(A7}$$

This is exactly the characteristic Eq. (11).

References

1. F. P. Kapron, D. B. Keck, and R. D. Maurer, Appl. Phys. Lett. **17,** 423 (1970).
2. E. Snitzer, J. Opt. Soc. Am. **51,** 491 (1961).
3. K. C. Kao and G. A. Hockham, Proc. IEE **113,** 1151 (1966).
4. E. A. J. Marcatili, Bell Syst. Tech. J. **48,** 2103 (1969).
5. A. W. Snyder, IEEE Trans. Microwave Theory Techniques **MTT-17,** 1130 (1969).
6. S. A. Schelkunoff, *Electromagnetic Waves* (Van Nostrand, New York, 1943), p. 94.
7. A. W. Snyder, IEEE Trans. Microwave Theory Techniques **MTT-17,** 1138 (1969).
8. G. Toraldo di Francia, J. Opt. Soc. Am. **59,** 799 (1969).
9. P. Moon and D. E. Spencer, *Field Theory Handbook* (Springer, Berlin, 1961), p. 192.

Radiation Losses of Dielectric Waveguides in Terms of the Power Spectrum of the Wall Distortion Function

By DIETRICH MARCUSE

(Manuscript received July 23, 1969)

In an earlier paper I described a perturbation theory of the radiation losses of a dielectric slab waveguide. The statistical treatment of the radiation losses was based on the correlation function of the wall distortion. This paper discusses the results of the radiation loss theory in terms of the power spectrum of the function describing the thickness of the slab. We found that only those mechanical frequencies θ of the power spectrum contribute to the radiation loss that fall into the range $\beta_0 - \mathrm{k} < \theta < \beta_0 + \mathrm{k}$. ($\beta_0$ = propagation constant of guided mode, k = free space propagation constant.) The mechanical frequencies near both end points of this mechanical frequency range contribute more to the radiation loss than the region well inside of this range.

We also discuss the far-field radiation pattern caused by a strictly sinusoidal wall distortion.

I. INTRODUCTION

In an earlier paper I developed a perturbation theory of the mode conversion effects between guided modes and of the radiation losses of a given guided mode caused by deviations from perfect straightness of the waveguide wall.[1] For simplicity, the discussion had been limited to a waveguide in the form of an infinitely extended dielectric slab.

The statistical discussion had been based on the description of the wall distortion by means of a correlation function. In Ref. 1 an exponential correlation function had been assumed. However, it has been established that the shape of the correlation function has little influence on the radiation losses.

It is possible to base the discussion of radiation losses not on correlation functions, but on the mechanical power spectrum of the wall distortion function. This study provides information as to how the various mechanical frequencies of the wall distortion function contribute to the radiation losses.

The analysis of Ref. 1 was based on the use of radiation modes of the dielectric slab which represent standing waves in directions transverse to the propagation direction of the guided modes. The question naturally arises how a superposition of these standing waves can result in radiation flowing away from the rod. This question is answered by examining the far field radiation pattern caused by a sinusoidal distortion of one wall of the dielectric waveguide. This paper gives the relation between the length of the mechanical period, the wavelength of the guided mode, and the direction of the main lobe of the radiation.

II. RADIATION LOSS AND POWER SPECTRUM

The amplitudes of the modes of the continuous spectrum were derived in Ref. 1, equations (65) and (69). We have

$$g_e(\rho, L) = \frac{Lk^2}{2i(\pi)^{\frac{1}{2}}}(n_g^2 - 1)\frac{\rho(\cos \kappa_0 d \cos \sigma d)[\varphi(\theta) - \psi(\theta)]}{\left[\beta\left(\beta_0 d + \frac{\beta_0}{\gamma_0}\right)(\rho^2 \cos^2 \sigma d + \sigma^2 \sin^2 \sigma d)\right]^{\frac{1}{2}}} \tag{1}$$

for the even modes, and

$$g_0(\rho, L) = \frac{Lk^2}{2i(\pi)^{\frac{1}{2}}}(n_g^2 - 1)\frac{\rho(\cos \kappa_0 d \sin \sigma d)[\varphi(\theta) + \psi(\theta)]}{\left[\beta\left(\beta_0 d + \frac{\beta_0}{\gamma_0}\right)(\rho^2 \sin^2 \sigma d + \sigma^2 \cos^2 \sigma d)\right]^{\frac{1}{2}}} \tag{2}$$

for the odd modes. The functions

$$\varphi(\theta) = \frac{1}{L}\int_0^L [f(z) - d]e^{-i\theta z}\, dz, \tag{3}$$

$$\psi(\theta) = \frac{1}{L}\int_0^L [h(z) + d]e^{-i\theta z}\, dz, \tag{4}$$

with

$$\theta = \beta_0 - \beta \tag{5}$$

are the Fourier transforms of the wall distortion functions $f(z) - d$ and $h(z) + d$. [$x = f(z)$ is the boundary of the dielectric-air interface, $x = d$ describes the wall of the perfect guide, and $x = h(z)$ is the distorted boundary near $x = -d$.]

The meaning of the constants appearing in equations (1) to (5) is:

β_0 = propagation constant of guided mode (propagating in z-direction),

β = component of the propagation constant of the continuum mode in z-direction,

k = propagation constant in free space,

L = length of guide section with wall distortions,

n_g = dielectric constant of slab,

$$\rho = (k^2 - \beta^2)^{\frac{1}{2}} \tag{6}$$

$$\sigma = (n_g^2 k^2 - \beta^2)^{\frac{1}{2}}, \tag{7}$$

$$\kappa_o = (n_g^2 k^2 - \beta_0^2)^{\frac{1}{2}}, \tag{8}$$

$$\gamma_0^2 = (\beta_0^2 - k^2)^{\frac{1}{2}}. \tag{9}$$

The y-component of the electric radiation field caused by the wall distortions is given by

$$E_y = \int_0^\infty [g_e(\rho, L)\mathcal{E}_e(\rho, z) + g_0(\rho, L)\mathcal{E}_0(\rho, z)]\, d\rho. \tag{10}$$

The functions $\mathcal{E}_e$ and $\mathcal{E}_o$ are the even and odd radiation modes. The ratio of scattered power to incident guided mode power is obtained from

$$\frac{\Delta P}{P} = \int_{-k}^{k} (\mid g_e(\rho, L) \mid^2 + \mid g_0(\rho, L) \mid^2) \frac{\beta}{\rho}\, d\beta. \tag{11}$$

For simplicity we assume that one wall of the slab is perfect

$$h(z) = -d, \tag{12}$$

so that

$$\psi(\theta) = 0, \tag{13}$$

the relative scattering loss, follows from equations (1), (2), and (10)

$$\frac{\Delta P}{P} = \int_{-k}^{k} \frac{1}{d^2} L \mid \varphi(\theta) \mid^2 I(\beta)\, d\beta \tag{14a}$$

with

$$I(\beta) = \frac{(kd)^4}{4\pi} (n_g^2 - 1)^2 \frac{\cos^2 \kappa_0 d}{\beta_0 d + \dfrac{\beta_0}{\gamma_0}} (\rho d) \left[\frac{\cos^2 \sigma d}{(\rho d)^2 \cos^2 \sigma d + (\sigma d)^2 \sin^2 \sigma d} + \frac{\sin^2 \sigma d}{(\rho d)^2 \sin^2 \sigma d + (\sigma d)^2 \cos^2 \sigma d} \right]. \tag{14b}$$

Since $\varphi(\theta)$ is the Fourier component of the wall distortion function its absolute square value

$$|\varphi(\theta)|^2 \tag{15}$$

is the "power spectrum" of $f(z) - d$. It is apparent from equation (14) that $\Delta P/P$ depends on the power spectrum of the wall distortion function. Incidentally, equation (14) is not a statistical expression, but holds for a specific dielectric slab waveguide. We entered the power spectrum in the combination $L|\varphi|^2$ in equation (14) since this combination is independent of L for a randomly varying function $f(z) - d$.

Equation (14) allows us immediately to determine the range of mechanical frequencies θ which contribute to the radiation loss. The integral in equation (14) is extended from $-k$ to k, the β range of continuous radiation modes. The range of mechanical frequencies contributing to the scattering loss is therefore given by

$$\beta_0 - k < \theta < \beta_0 + k. \tag{16}$$

This is an important result since it states that those parts of the power spectrum which lie outside of the range, equation (16), do not contribute to radiation loss.

This last statement must not be misconstrued to mean that a waveguide with a sinusoidal wall distortion extending over length L

$$f(z) = d + a \sin \theta' z \qquad 0 \leqq z \leqq L \tag{17}$$

with θ' lying outside the range of equation (16) does not lose power by radiation. The power spectrum of equation (17) is

$$|\varphi(\theta)|^2 = \left[\frac{a}{L} \frac{\sin(\theta' - \theta)\frac{L}{2}}{\theta' - \theta}\right]^2. \tag{18}$$

A term with $\theta' + \theta$ in the denominator has been neglected in equation (18). The accuracy of this approximation improves with increasing values of L.

It is apparent from equation (18) that $|\varphi(\theta)|^2$ has non-vanishing values for $\theta \neq \theta'$ so that there is some small contribution to radiation loss even if θ' lies outside of the range of equation (16).

However, if we consider the limit $L \to \infty$ we can approximate the power spectrum, equation (18), by a δ-function:

$$\lim_{L\to\infty} |\varphi(\theta)|^2 = \frac{\pi a^2}{2L} \delta(\theta - \theta'). \tag{19}$$

In this special case the expression (14a) for the scattered power becomes

$$\frac{\Delta P}{P} = \frac{\pi}{2}\left(\frac{a}{d}\right)^2 I(\beta_0 - \theta'). \tag{20}$$

The scattering from a dielectric waveguide with a wall distortion function whose power spectrum is a δ-function is proportional to $I(\beta_0 - \theta')$.

The function $I(\beta)$ is plotted in Fig. 1 for $n_g = 1.01$, $kd = 8.0$, and $\beta_0 d = 8.041$. The scattering caused by a wall distortion with a δ-function spectrum (a sinusoidal wall distortion of infinite length) is nearly independent of the value of $\beta = \beta_0 - \theta'$ over most of the β-range. There are two sharp peaks at $\beta \approx k$ and $\beta \approx -k$. The physical reasons for the sharp increase in loss at these values is easy to understand if we consider the direction of the radiation pattern as a function of θ'. We show in Section III [equation (35)] that the angle α between the waveguide and the main radiation lobe is given by

$$\cos \alpha \doteq \frac{\beta}{k} = \frac{\beta_0 - \theta'}{k}. \tag{21}$$

The two peaks of the function $I(\beta)$, or correspondingly of the radiation loss, are associated with

$$\alpha \approx 0 \quad \text{and} \quad \alpha \approx \pi. \tag{22}$$

This shows that the radiation loss is high when the radiation pattern is

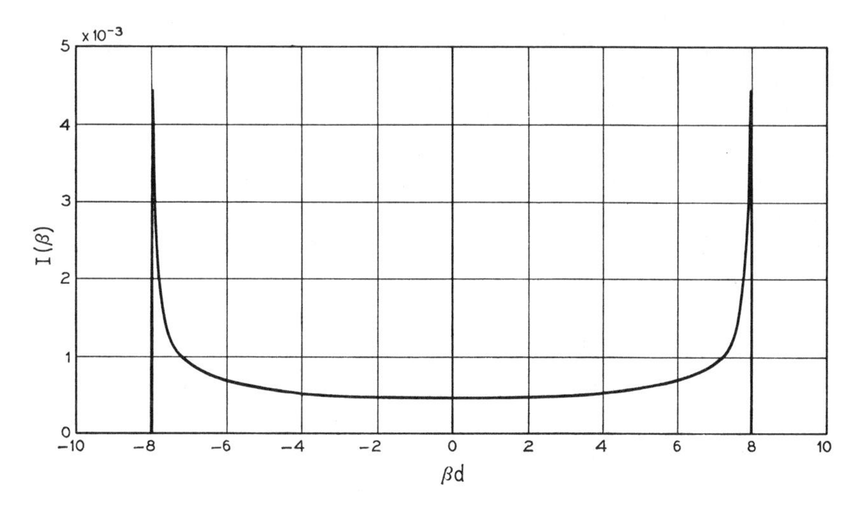

Fig. 1 — Graphical representation of the function $I(\beta)$ [eq. (14b)]. $n_g = 1.01$, $kd = 8.0$, $\beta_0 d = 8.041$.

directed very nearly parallel to the surface of the waveguide. The radiation modes gain more power if the guided mode can interact with them over a longer distance. An observation of this loss peak is reported in Ref. 2.

A power spectrum with sharp peaks much like that of equation (18) or (19) is not likely to occur for dielectric waveguides with random imperfections of the dielectric interface. It is much more reasonable to expect that such waveguides may have spectral distributions which are nearly independent of θ over a certain range of θ values. In the limit of a "white" spectrum,

$$| \varphi(\theta) |^2 = \text{constant}, \tag{23}$$

the scattering loss is proportional to the integral over the function $I(\beta)$ shown in Fig. 1. The two peaks contribute very little to this integral. Numerical integration of $I(\beta)$ of Fig. 1 including and excluding the peaks resulted in the values:

$$\int_{-8}^{8} I(\beta)\, d\beta = 0.011, \quad \int_{-7.8}^{7.8} I(\beta)\, d\beta = 0.0096,$$

$$\text{and} \quad \int_{-7.5}^{7.5} I(\beta)\, d\beta = 0.0087.$$

This result is reassuring for the use of the perturbation theory which was used to derive equation (14). The perturbation theory is based on the assumption that power is converted from the guided mode to the radiation field but that no power is converted back from the radiation field to the guided mode. This approximation is certain to yield better results if the radiation pattern is directed away from the rod. In other words, the perturbation theory will work poorest in the region of the peaks of Fig. 1. However, for spectra that do not particularly favor the regions of these peaks, the contribution of those regions (which at the same time give the least reliable results) to the total radiation loss is only slight.

III. THE FAR FIELD RADIATION PATTERN

The far field pattern of the radiation field (that is excited by the lowest order even guided mode traveling in the dielectric slab with sinusoidal perturbation of one wall) can easily be calculated from equation (10). The even and odd radiation modes were given in Ref. 1 (for $|x| > d$)

$$\mathcal{E}_y^{(e)} = \left[\frac{2\omega\mu P}{\pi\beta(\rho^2\cos^2\sigma d + \sigma^2\sin^2\sigma d)}\right]^{\frac{1}{2}}$$
$$\times\ [\rho\cos\rho(|x| - d)\cos\sigma d - \sigma\sin\rho(|x| - d)\sin\sigma d]e^{i(\omega t - \beta z)} \qquad (24)$$
$$\mathcal{E}_y^{(0)} = \frac{x}{|x|}\left[\frac{2\omega\mu P}{\pi\beta(\rho^2\sin^2\sigma d + \sigma^2\cos^2\sigma d)}\right]^{\frac{1}{2}}$$
$$\times\ [\rho\cos\rho(|x| - d)\sin\sigma d + \sigma\sin\rho(|x| - d)\cos\sigma d]e^{i(\omega t - \beta z)}. \qquad (25)$$

With $\psi(\theta) = 0$ and

$$\varphi(\theta) \approx \frac{a}{iL}\exp\left[i(\theta' - \theta)\frac{L}{2}\right]\frac{\sin(\theta' - \theta)\frac{L}{2}}{\theta' - \theta} \qquad (26)$$

and with the help of equations (1) and (2) we get from equation (10)

$$E_y = -\frac{ak^2}{(2)^{\frac{1}{2}}\pi}(\omega\mu P)^{\frac{1}{2}}(n_g^2 - 1)\frac{\cos\kappa_0 d}{\left(\beta_0 d + \frac{\beta_0}{\gamma_0}\right)^{\frac{1}{2}}}$$
$$\times\int_0^\infty \frac{\rho}{\beta}\left\{\frac{\cos\sigma d[\rho\cos\rho(x - d)\cos\sigma d - \sigma\sin\rho(x - d)\sin\sigma d]}{\rho^2\cos^2\sigma d + \sigma^2\sin^2\sigma d}\right.$$
$$\left.+\frac{\sin\sigma d[\rho\cos\rho(x - d)\sin\sigma d + \sigma\sin\rho(x - d)\cos\sigma d]}{\rho^2\sin^2\sigma d + \sigma^2\cos^2\sigma d}\right\}$$
$$\times\exp\left[i(\theta' - \theta)\frac{L}{2}\right]\frac{\sin(\theta' - \theta)\frac{L}{2}}{\theta' - \theta}\times e^{i(\omega t - \beta z)}\,d\rho. \qquad (27)$$

In the far field with $x \to \infty$ and $z \to \infty$ (but L finite) we can obtain an approximate solution of the integral in equation (27) by the method of stationary phase.[3] The sine and cosine functions of argument $\rho(x - d)$ can be expressed as sums of exponential functions. The most important terms of the integrand of equation (27) are, therefore, of the form

$$\exp[-i(\beta z \pm \rho x)]. \qquad (28)$$

This exponential term is an extremely rapidly varying function of ρ as $x \to \infty$ and $z \to \infty$. All other terms in the integrand vary slowly by comparison. According to the method of stationary phase the contribution to the integral comes predominantly from a region that is determined by

$$\frac{\partial}{\partial\rho}(\beta z \pm \rho x) = 0. \qquad (29)$$

With the help of equation (6), equation (29) leads to the condition

$$\frac{x}{z} = \pm \frac{\rho_0}{\beta} \tag{30}$$

or

$$\rho_0 = k \sin \alpha \tag{31a}$$

$$\beta = k \cos \alpha \tag{31b}$$

with

$$\cos \alpha = \frac{z}{(x^2 + z^2)^{\frac{1}{2}}} = \frac{z}{r}. \tag{32}$$

For $x > 0$ and $z > 0$ only the $+$ sign in equation (30) is possible. This is an important point. It shows that even though the radiation modes, equations (24) and (25), represent standing wave patterns in x-direction only, the outward traveling part of the decomposition of the standing wave into traveling waves makes a contribution to the radiation field, equation (27).

All terms of the integrand with the exception of equation (28) can be taken out of the integral. The remaining integration can be carried out using the expansion

$$\beta z + \rho x = k(x \sin \alpha + z \cos \alpha) - \frac{1}{2} \frac{z}{k \cos^3 \alpha} (\rho - \rho_0)^2 + \cdots$$

$$\cdot \int_0^\infty e^{-i(\beta z + \rho x)} \, d\rho = (1 + i)(\pi)^{\frac{1}{2}} \frac{(k)^{\frac{1}{2}} \cos \alpha}{(r)^{\frac{1}{2}}} e^{-ik(x \sin \alpha + z \cos \alpha)}. \tag{33}$$

The far field is therefore obtained in the form

$$E_y = \frac{1}{(\pi)^{\frac{1}{2}}} \exp\left(i \frac{\pi}{4}\right) a k^{\frac{3}{2}} (\omega \mu P)^{\frac{1}{2}} (n_o^2 - 1) \frac{\cos \kappa_0 d}{\left(\beta_0 d + \frac{\beta_0}{\gamma_0}\right)^{\frac{1}{2}}}$$

$$\cdot \frac{\rho_0^2 \sin 2\sigma_0 d - i\rho_0 \sigma_0 \cos 2\sigma_0 d}{(\rho_0^2 + \sigma_0^2) \sin 2\sigma_0 d - 2i\rho_0 \sigma_0 \cos 2\sigma_0 d} \frac{\sin (\theta' - \theta) \frac{L}{2}}{\theta' - \theta}$$

$$\cdot \exp\left[i(\theta' - \theta) \frac{L}{2}\right] e^{i\rho_0 d} \frac{1}{(r)^{\frac{1}{2}}} e^{i[\omega t - k(x \sin \alpha + z \cos \alpha)]}. \tag{34}$$

The index zero was added to σ to indicate that it must be evaluated from equations (7) and (8) using ρ_0 of equation (31a).

Equation (34) reveals several important features of the far field of radiation. This field is essentially a plane wave traveling in the direction of α ($\tan \alpha = x/z$, and x and z are the coordinates of the point of observation).

The field intensity is inversely proportional to the square root of the distance r from (the sinusoidally distorted) waveguide section. The dependence on distance is inversely proportional to $(r)^{\frac{1}{2}}$ rather than r because the waveguide is infinitely extended in y-direction (see Ref. 1).

The main radiation lobe occurs at the maximum value of $[\sin (\theta' - \theta)L/2]/(\theta' - \theta)$ that is at $\theta = \theta'$ or from equations (5) and (31b) at

$$\cos \alpha_m = \frac{\beta_0 - \theta'}{k} \tag{35}$$

(β_0 = propagation constant of guided mode).

The width of the main lobe depends on the length L of the sinusoidally distorted waveguide section. The difference in angle between the peak of the lobe and the first null determines the half width of the main lobe

$$\Delta\alpha = \frac{2\pi}{Lk \sin \alpha} \text{ for } \alpha \neq 0. \tag{36a}$$

The width of the main radiation lobe is inversely proportional to L. The lobe is narrowest for $\alpha = \pi/2$ and becomes wider as α decreases toward zero. If the peak of the main lobe is at $\alpha = 0$, we obtain

$$\Delta\alpha = \left(\frac{4\pi}{Lk}\right)^{\frac{1}{2}} \text{ for } \alpha = 0. \tag{36b}$$

The peak amplitude of the main radiation lobe is not strongly dependent on α. The increase in radiated power in forward direction ($\alpha = 0$) which is apparent from Fig. 1 is caused by the broadening of the radiation lobe with decreasing angle.

IV. CONCLUSION

The radiation loss of dielectric waveguides caused by deviations from perfect straightness of the waveguide walls depends on the "power spectrum" of the wall deviation function. A sinusoidal wall perturbation gives rise to radiation into a particular direction in space. Each Fourier component of the Fourier expansion of the wall distortion function is responsible for radiation into a particular direction. The width of the radiation lobes is wide for scattering directions parallel to the rod so that those Fourier components responsible for forward and backward scattering contribute more to the radiation loss than those causing scattering in other directions. However, this preferential loss behavior is not very pronounced, so that the Fourier components responsible for forward and backward scattering contribute only a small amount of the total radiation loss caused by a broad power spectrum.

The coupling between two guided modes of the dielectric waveguide is also governed by equation (5). Only one component of the power spectrum of the wall distortion function influences the coupling between two guided modes, while the entire range of mechanical frequencies, equation (16), determines the radiation loss.

The general predictions of this theory have been experimentally verified. Microwave experiments on a periodically corrugated teflon rod have shown that the radiation losses are negligibly small if the period of the corrugation is such that θ lies outside of the interval indicated by equation (16).[2] However, if θ falls inside of the interval, equation (16), considerable radiation losses do occur. The peak of the radiation losses shown in Fig. 1 and the direction and width of the radiation lobes have also been observed in agreement with this theory.

REFERENCES

1. Marcuse, D., "Mode Conversion Caused by Surface Imperfections of a Dielectric Slab Waveguide," B.S.T.J., this issue, pp. 3187–3215.
2. Marcuse, D., and Derosier, R. M., "Mode Conversion Caused by Diameter Changes of a Round Dielectric Waveguide," B.S.T.J., this issue, pp. 3217–3232.
3. Mathews, J., and Walker, R. L., "Mathematical Methods of Physics," New York: W. A. Benjamin, 1965, pp. 85–86.

Time Dispersion in Dielectric Waveguides

By S. D. PERSONICK

(Manuscript received October 16, 1970)

In dielectric waveguides operating at optical frequencies, the primary cause of time dispersion of narrow pulses can be mode conversion. In this paper we argue that under certain assumptions a dielectric waveguide acts as a linear system in intensity. That is, given the intensity input, the intensity output is equal to the input convolved with an intensity impulse response. We show that contrary to intuition, the width of the impulse response gets narrower when coupling between guided modes increases. Using the perturbation results of D. Marcuse, we obtain an interesting model of energy propagation down imperfect guides. We conclude that the intensity response width increases as the square root of the guide length for sufficiently long guides and approaches a gaussian shape for sufficiently long guides.

We conclude from the theory that the dispersion in dielectric waveguides may be orders of magnitude below that which was previously expected in guides of sufficiently long length having properly controlled large amounts of mode conversion. These theoretical results have not yet been verified experimentally.

I. INTRODUCTION

In multimode dielectric waveguides operating at optical frequencies, the primary cause of time dispersion of narrow pulses can be mode conversion. In a geometrically perfect guide with more than a single mode, energy initially launched in a given mode remains in that mode as it propagates down the guide. Physical guides have imperfections from perfect geometric shape (e.g., roughness at the core-cladding interface of a nominally right circular cylindrical guide) which allows energy to couple between modes during propagation down the guide. Since group velocities differ in general amongst the modes, a pulse of energy initially launched in a single mode or combination of modes will be broadened due to the spread of propagation times of different parts of the energy.

In this paper we argue that under certain assumptions, a dielectric waveguide acts as a linear system in intensity as well as in voltage. That is, we show that the relationship between the input intensity and output intensity of the guide is defined in terms of an intensity impulse response. We argue that this intensity impulse response, for sufficiently long guides, has a mean-square width about its mean which increases only linearly with length. Further, in the limit of very long guides, we argue that the response shape is gaussian. We also show that the greater the coupling between modes, the less the time dispersion—a result which at first contradicts intuition. Finally, we obtain quantitative results and an interesting model of an optical guide under certain assumptions.

We conclude from the theory that the dispersion in dielectric waveguides may be orders of magnitude below that which was previously expected in guides of sufficiently long length having properly controlled *large* amounts of mode conversion. These theoretical results have not yet been verified experimentally.

II. AN OUTLINE OF THE ARGUMENTS

We next outline the steps of the derivations to follow, so that the reader can follow the train of thought.

We start with the fact that the optical guide is a linear system in voltage. That is, if we expand the input signal in spatial modes and expand the output signal in the same modes, then the time varying coefficients of the modes at the output are related to the coefficients at the input by a set of voltage impulse responses. We then make an assumption about the associated set of transfer functions (Fourier transforms of the impulse responses) which allows us to argue that the set of average output intensities and the set of input intensities are also related by a set of impulse responses. Thus the guide is also linear in intensity under the assumptions.

We next argue that for sufficiently long guides, these intensity impulse responses coupling a chosen input mode coefficient and a chosen output mode coefficient are indifferent to the modes chosen except perhaps for a magnitude scale factor.

Finally, this allows us to show that for guides longer than the above scale, the intensity impulse response which is now in common for all input-output pairs has a mean-square deviation about its mean which grows linearly in length, and which approaches the gaussian shape in the limit of long guides.

III. THE OPTICAL GUIDE AS A LINEAR SYSTEM IN INTENSITY

3.1 *The Random Channel*

We now argue that under a simple assumption, the expected value of the intensity at the output of a random channel is related to the intensity of the input to the channel by a simple convolution with an intensity impulse response. We start with input baseband signal $a(t)$. We use $a(t)$ to linearly modulate a carrier $m(t)$ which may be coherent or a stationary (wide sense) random process centered at frequency f_0. We assume that the result $x(t) = a(t)m(t)$, has bandwidth B, i.e., its spectrum extends from $-B/2 + f_0$ to $B/2 + f_0$.

We pass $x(t)$ through a time invariant filter with a random impulse response $h(t)$ representing the channel, resulting in the final output $y(t)$. We define a Fourier transform relationship between the function $\Lambda(f)$ and the function $\lambda(t)$

$$\Lambda(f) = \int_{-\infty}^{\infty} \exp\,[i2\pi f(t)]\lambda(t)\,dt, \qquad \Lambda(f) \Leftrightarrow \lambda(t). \tag{1}$$

By simple linear system theory if

$$X(f) \Leftrightarrow x(t), \quad H(f) \Leftrightarrow h(t), \quad Y(f) \Leftrightarrow y(t), \tag{2}$$

then

$$Y(f) = X(f)H(f).$$

Define the envelopes of $x(t)$ and $y(t)$ by

$$x(t) = \sqrt{2}\,\mathrm{Re}\,\{x_e(t)\exp(i2\pi f_0 t)\}, \qquad y(t) = \sqrt{2}\,\mathrm{Re}\,\{y_e(t)\exp(i2\pi f_0 t)\}. \tag{3}$$

The intensity of the input and output signals are defined as

$$I_{\mathrm{in}}(t) = |x_e(t)|^2 = a^2(t)\,|m_e(t)|^2, \qquad I_{\mathrm{out}}(t) = |y_e(t)|^2, \tag{4}$$

where $m_e(t)$ = carrier envelope.

Assumption: Stationarity of channel transfer function

$$\langle H^*(\alpha)H(f+\alpha)\rangle = \Gamma(f) \tag{5}$$

provided $f_0 - B/2 < \alpha, f + \alpha < f_0 + B/2$.

The assumption, while apparently arbitrary, is essential to the results which follow. Perturbation results of Marcuse,[1] to be discussed in Section 4.1, indicate that equation (5) may be satisfied for the input-output temporal transfer function of a given spatial eigenmode of an optical dielectric waveguide with mechanical imperfections, provided the mechanical imperfections satisfy constraints also to be discussed.

Define for any function $U(\alpha)$

$$U_+(\alpha) = U(\alpha) \qquad \alpha \geqq 0, \qquad\qquad (6)$$
$$= 0 \qquad \alpha < 0.$$

Then it has been shown that (See Appendix C)

$$|y_e(t)|^2 \Leftrightarrow 2 \int_{-\infty}^{\infty} Y_+(f + \alpha) Y_+^*(\alpha)\, d\alpha. \qquad (7)$$

Then clearly

$$|y_e(t)|^2 \Leftrightarrow 2 \int_{-\infty}^{\infty} X_+(f + \alpha) H_+(f + \alpha) H_+^*(\alpha) X_+^*(\alpha)\, d\alpha. \qquad (8)$$

Using equation (5) we obtain

$$\langle |y_e(t)|^2 \rangle \Leftrightarrow 2\Gamma(f) \int_{-\infty}^{\infty} \langle X_+(f + \alpha) X_+^*(\alpha) \rangle\, d\alpha, \qquad (9)$$
$$\langle |y_e(t)|^2 \rangle \Leftrightarrow \Gamma(f) I_{\mathrm{in}}(f),$$

where

$$I_{\mathrm{in}}(f) \Leftrightarrow \langle I_{\mathrm{in}}(t) \rangle.$$

Thus*

$$\langle I_{\mathrm{out}}(t) \rangle = \langle I_{\mathrm{in}}(t) \rangle * \gamma(t) \qquad (10)$$

where

$$\gamma(t) \Leftrightarrow \Gamma(f).$$

Thus we have a linear system relationship between the channel input and output intensities.

3.2 *Extension to Vector Channels*

Suppose we have a vector channel (corresponding to multimode guide) consisting of a vector of L input functions

---

* The notation $x(t) * y(t)$ signifies convolution:

$$x(t) * y(t) \triangleq \int_{-\infty}^{\infty} x(t - u) y(u) du.$$

$$\mathbf{X}(t) = \begin{bmatrix} x_1(t) \\ \vdots \\ x_L(t) \end{bmatrix}$$

and L output functions

$$\mathbf{Y}(t) = \begin{bmatrix} y_1(t) \\ \vdots \\ y_L(t) \end{bmatrix}.$$

The input and output functions are related by an $L \times L$ matrix of impulse responses

$$\boldsymbol{\theta}(t) = [h_{ij}(t)] \tag{11}$$

where we have

$$y_l(t) = \sum_{j=1}^{L} h_{lj}(t) * x_j(t), \tag{12}$$

i.e.,

$$\mathbf{Y}(t) = \boldsymbol{\theta}(t) * \mathbf{X}(t).$$

Now consider a cascade of two vector channels having impulse response matrices ${}^1\boldsymbol{\theta}(t)$ and ${}^2\boldsymbol{\theta}(t)$. The input passes first through channel 1 and then through channel 2. The output of channel 2 is given by

$$\mathbf{Y}(t) = {}^2\boldsymbol{\theta}(t) * {}^1\boldsymbol{\theta}(t) * \mathbf{X}(t). \tag{13}$$

Define the envelope of $y_k(t)$, $y_{ke}(t)$: we know that

$$|y_{k_e}(t)|^2 \Leftrightarrow 2 \int_{-\infty}^{\infty} Y_{k+}(\alpha + f) Y_{k+}^*(\alpha)\, d\alpha,$$

$$|y_{k_e}(t)|^2 \Leftrightarrow 2 \int_{-\infty}^{\infty} \sum_j \sum_l \sum_m \sum_n {}^2H_{kl+}(f+\alpha)\, {}^1H_{lj+}(f+\alpha) X_{j+}(f+\alpha) \tag{14}$$

$$\cdot {}^2H_{kn+}^*(\alpha)\, {}^1H_{nm+}^*(\alpha) X_{m+}^*(\alpha)\, d\alpha.$$

Assumption a. Stationarity of Mode Transfer function.
b. Mode Transfer functions uncorrelated.

$$\langle {}^1H_{lj}(f+\alpha)\, {}^2H_{kl}(f+\alpha)\, {}^1H_{nm}^*(\alpha)\, {}^2H_{rn}^*(\alpha) \rangle$$

$$= {}^1\Gamma_{lj}(f)\, {}^2\Gamma_{kl}(f)\; \delta_{l,n}\; \delta_{j,m}\; \delta_{k,r} \tag{15}$$

where

$$\delta_{X,Y} \triangleq \text{Kronecka delta}$$

for $\{f + \alpha, \alpha\} \epsilon$ input signal bands.

We are implying that randomness, especially in phase, erases correlation between the transfer functions of different modes. The validity of this assumption for optical guides with more than one mode will be discussed in Section 4.1.

It then follows that

$$\langle |y_{ke}(t)|^2 \rangle \Leftrightarrow 2 \int_{-\infty}^{\infty} \sum_{j=1}^{L} \sum_{l=1}^{L} {}^1\Gamma_{lj}(f){}^2\Gamma_{kl}(f)\langle X_{j+}(f + \alpha)X^*_{j+}(\alpha)\rangle \, d\alpha,$$
$$\langle |y_{ke}(t)|^2 \rangle \Leftrightarrow \sum_{j=1}^{L} \sum_{l=1}^{L} {}^2\Gamma_{kl}(f){}^1\Gamma_{lj}(f) I_{\text{in}j}(f) \tag{16}$$

where

$$I_{\text{in}j}(f) \Leftrightarrow \langle | x_{je}(t) |^2 \rangle.$$

Thus under assumption (15) we have

$$\langle |y_{ke}(t)|^2 \rangle = \sum_{j=1}^{L} \sum_{l=1}^{L} {}^2\gamma_{kl}(t) * {}^1\gamma_{lj}(t) * \langle |x_{je}(t)|^2 \rangle \tag{17}$$

where

$$\gamma_{kl}(t) \Leftrightarrow \Gamma_{kl}(f).$$

Forming the matrix ${}^1G(t)$ with elements ${}^1\gamma_{kl}(t)$ and similarly ${}^2G(t)$; the vectors of input and output *intensities* are related by

$$| Y_e |^2 = {}^2G(t) * {}^1G(t) * |X_e|^2. \tag{18}$$

Thus the vector channel is a vector linear system in intensity as well as voltage [compare equation (18) to equation (13)].

3.3 *A Limit Theorem for a Cascade of Vector Channels*

Now consider a cascade of a large number M of vector channels, each behaving as described in Sections 3.1 and 3.2, i.e., if $|Y_e(t)|^2$ is the vector of the average intensity responses at the output, $|X_e(t)|^2$ the input intensity vector; we have

$$|Y_e(t)|^2 = \left(\mathop{*}_{j=1}^{M} \mathbf{G}_j(t) \right) * |\mathbf{X}_e(t)|^2 = \mathbf{G}_T * |\mathbf{X}_e(t)|^2$$

where

$$\mathbf{G}_T = \mathop{*}_{j=1}^{M} G_j(t) = \mathbf{G}_M * \mathbf{G}_{M-1} \cdots * \mathbf{G}_1 . \tag{20}$$

We would like to argue now that for sufficiently large M, all the elements of G_T are identical in waveform, differing at most by a constant. In other words, we would like to argue that the shape of the intensity response between any input mode and any output mode is indifferent to the choices of input and output modes, for sufficiently long guides, except perhaps for the magnitude of the responses.

We shall prove our results for a lossless two-mode guide. Define $\gamma_{ij}(t, b)$ as the $j \to i$ intensity impulse response for b sections of guide. We obtain (see Appendix A for derivation)

$$\gamma_{21}(t, b) = \sum_{R=1}^{b} \gamma_{11}(t, b - R) * \gamma_{21}(t) * \gamma_{22}^{*R-1}(t) \tag{21}$$

where

$$\gamma_{22}^{*R-1}(t) \triangleq \gamma_{22}(t) * \gamma_{22}(t) \cdots R - 1 \text{ times}$$

and

$$\gamma_{11}(t, 0) = \delta(t).$$

Now assume that the guide is lossless, i.e.,

$$\int_{-\infty}^{\infty} \{\gamma_{11}(t) + \gamma_{21}(t)\} \, dt = 1. \tag{22}$$

Further define

$$\gamma_{ij}(t) = a_{ij}\rho_{ij}(t),$$

$$\int_{-\infty}^{\infty} \rho_{ij}(t) \, dt = 1, \qquad 0 < a_{ij} < 1. \tag{23}$$

Thus

$$\gamma_{21}(t, b) = \sum_{R=1}^{b} \gamma_{11}(t, b - R) * \rho_{21}(t) * \rho_{22}^{*R-1}(t)(a_{21}a_{22}^{R-1}). \tag{24}$$

For the lossless guide, and b sufficiently large $\gamma_{11}(b - R) \approx \gamma_{11}(t, b)$ for $R \ll b$. Furthermore a convolution of $\gamma_{11}(t, b - R)$ with $\rho_{21}(t) * \rho_{22}^{*R-1}(t)$ is approximately equal to $\gamma_{11}(t, b - R)$ for $R \ll b$ since the response $\gamma_{11}(t, b - R)$, which is a convolution of $b - R$ terms, has a narrow spectrum compared to the other R term convolution for $b \gg R$. Furthermore $a_{21}a_{22}^{R-1} \to 0$ for R large. Thus for b sufficiently large

$$\gamma_{21}(t, b) \approx \gamma_{11}(t, b)(a_{21}/(1 - a_{22})). \tag{25}$$

Similarly we have

$$\gamma_{11}(t, b) = \sum_{R=1}^{b-1} \gamma_{12}(t, b - R) * \rho_{21}(t) * \rho_{11}^{*R-1}(t) a_{21} a_{11}^{R-1},$$

$$\approx \gamma_{12}(t, b) \frac{a_{21}}{1 - a_{11}} = \gamma_{12}(t, b), \tag{26}$$

(since $a_{21} + a_{11} = 1$). Thus

$$\mathbf{G}(t, b) \cong \rho(t, b) \begin{bmatrix} \dfrac{1}{1 + \eta} & \dfrac{1}{1 + \eta} \\ \dfrac{1}{1 + 1/\eta} & \dfrac{1}{1 + 1/\eta} \end{bmatrix} = \rho(t, b)\mathbf{A}, \tag{27}$$

where $\eta = a_{21}/a_{12}$, and $\rho(t, b) \triangleq \gamma_{11}(t, b)/\int_{-\infty}^{\infty} \gamma_{11}(t, b)\, dt$. Finally we obtain the response of a guide of kb sections

$$\mathbf{G}(t, kb) = \rho^{*k}(t, b)\mathbf{A}.$$

Note that $\mathbf{A}$ is idempotent, i.e., $\mathbf{A}^2 = \mathbf{A}$ and $\rho(t, b)$ is a positive unit area function.

3.4 *Application to Long Optical Guide*

For a multimode lossless of guide sufficiently long length, $l = kL$, with finite coupling between all modes, we can generalize equation (27) to conclude that the intensity impulse response between an input and output mode is a constant times some positive unit area function $\rho(t, L)$ convolved with itself l/L times when L is a scale on which equation (27) holds in the generalized case (more than two modes). Since the central limit theorem states that the convolution of a large number of unit area positive functions approaches a gaussian shape,* we conclude that the impulse response should approach a gaussian shape in the limit of long guides. Further, the impulse response's second moment about its mean increases linearly with increasing guide length for guides longer than L.† That is, the second moment about the mean of the response is

$$M_2(l) = M_2(L)l/L. \tag{28}$$

If τ_1 is the "differential delay" (time/meter) of propagation in the

* Provided that $\int_{-\infty}^{\infty} \rho(t, L)t^2 dt < \infty$, when we add similar independent random variables with finite second moments, the probability density of the sum, which is the *convolution* of the individual densities, approaches a gaussian shape.

† The second central moment of a convolution is the sum of the individual second central moments.

slowest modes and τ_2 is the differential delay in the fastest mode, then*

$$M_2(L) \leqq \frac{(\tau_1 - \tau_2)^2 L^2}{4} = \frac{(\Delta\tau L)^2}{4}. \tag{29}$$

Therefore,

$$M_2(l) \leqq \frac{(\Delta\tau)^2}{4} Ll$$

for any L where equation (27) holds [for an N mode case we have an $N \times N$ matrix multiplying $\rho(t, l)$].

IV. QUANTITATIVE RESULTS

4.1 *Perturbation Theory*

We shall now apply the above results to the case of a lossless slab dielectric waveguide previously studied by Marcuse.[1] We expand the input field to the guide as

$$\epsilon(t, x, 0) = \sum \epsilon_k(t, 0)\psi_k(x) \tag{30}$$

where x is the cross-sectional position parameter and the $\psi_k(x)$ are the eigenmodes of the guide. The field a distance l down the guide is written as

$$\epsilon(t, x, l) = \sum \epsilon_k(t, l)\psi_k(x). \tag{31}$$

We have a linear voltage impulse response relationship between the vector of input voltages $[\epsilon_k(t, 0)]$ and the vector of output voltages $[\epsilon_k(t, l)]$. Defining $E_k(\omega, l)$ as the Fourier transform of $\epsilon_k(t, l)$ we have

$$E_k(\omega, l) = \sum_j C_{kj}(\omega, l) E_j(\omega, 0). \tag{32}$$

Marcuse has shown that a *perturbation* theory solution for the $C_{kj}(\omega, l)$ is given by

$$C_{kj}(\omega, l) = \lambda_{kj} \exp [i\beta_j(\omega)l] \int_0^l g(z) \exp \{i[\beta_k(\omega) - \beta_j(\omega)]z\} \, dz \tag{33}$$

where λ_{kj}, is a constant weakly dependent upon ω and $g(z)$ is the wall perturbation from straightness. It is assumed $k \neq j$ [For $k = j$, $C_{kj}(\omega, l) = 1$]. We have therefore

* The right side of equation (29) is the mean-square intensity impulse response width if the response consists of an impulse of area $\frac{1}{2}$ at the shortest delay and an impulse of area $\frac{1}{2}$ at the longest delay.

$$C_{kj}(\omega, l)C^*_{np}(\omega + \sigma, l)$$
$$= \lambda_{kj}\lambda^*_{np} \exp \{i(\beta_j(\omega) - \beta_p(\omega + \sigma))l\} \int_0^l \int g(z)g(z')$$
$$\cdot \exp \{i[(\beta_k(\omega) - \beta_j(\omega))z - (\beta_n(\omega + \sigma) - \beta_p(\omega + \sigma))z']\} \, dz \, dz'. \tag{34}$$

Defining the correlation function

$$\langle g(z)g(z')\rangle = R_g(z - z') \tag{35}$$

(we assume $g(z)$ is a wide sense stationary process) we obtain

$$\langle C_{kj}(\omega, l)C^*_{np}(\omega + \sigma, l)\rangle = \lambda_{k1}\lambda^*_{np} \exp (i \, \Delta\beta_2 l) \int^l \int R_g(z - z')$$
$$\cdot \exp [i(\beta_k(\omega) - \beta_j\omega)(z - z')] \exp [i \, \Delta\beta(\omega, \sigma)z'] \, dz \, dz' \tag{36}$$

where

$$\Delta\beta = (\beta_k(\omega) - \beta_n(\omega + \sigma)) - (\beta_j(\omega) - \beta_p(\omega + \sigma)),$$
$$\Delta\beta_2 = \beta_j(\omega) - \beta_p(\omega + \sigma).$$

If $R_g(z - z')$ drops off quickly for $(z - z')$ in an interval of length l, then we have the approximate result

$$\langle C_{kj}(\omega, l)C^*_{np}(\omega + \sigma, l)\rangle \simeq \lambda_{kj}\lambda^*_{np} l S_g(\beta_k(\omega) - \beta_j(\omega))$$
$$\cdot \exp (i \, \Delta\beta_2 l) \frac{1 - \exp [i \, \Delta\beta(\omega, \sigma)l]}{i \, \Delta\beta(\omega, \sigma)l} \tag{37}$$

where $S_g(\cdot)$ = Fourier Transform of $R_g(\cdot)$, and is assumed to be constant as a function of $\beta_k(\omega) - \beta_i(\omega)$ for ω within the excitation bandwidth. For $k = n$, $j = p$

$$\Delta\beta \simeq \left[\frac{\partial\beta_k(\omega)}{\partial\omega} - \frac{\partial\beta_j(\omega)}{\partial\omega}\right]\sigma \triangleq \Delta\tau_{kj}\sigma,$$
$$\Delta\beta_2 = \left(\frac{\partial\beta_j}{\partial\omega}\right)\sigma. \tag{38}$$

That is we assume σ is small enough so that there is negligible dispersion of energy travelling in a single mode. Thus, the intensity impulse response between input j and output k is (see Fig. 1)

$$\gamma_{kj}(t, l) \simeq \lambda_{mj}\lambda^*_{mj}S_g(\beta_m(\omega) - \beta_j(\omega))f(t - \tau_j l) \tag{39}$$

where

$$f(t) = 1, \qquad t \epsilon [0, \Delta\tau_{mj}l];$$
$$= 0, \qquad \text{otherwise;}$$

for l small enough for the perturbation theory to hold.

From equations (36) through (38), it is clear that for cases where we do not have $k = n, j = p$ the correlation function $C_{kj}(\omega)C_{np}^*(\omega + \sigma)$ will be negligible provided $\Delta\beta(\omega, \sigma)$ is sufficiently large in the band of input frequencies. Thus we can use the perturbation length impulse response to find a long-guide response by means of equation (20).

From equation (37) we see that we increase coupling between modes by making the mechanical perturbation spectrum large at frequencies which correspond to the difference of the inverses of the phase velocities of the modes at excitation frequencies. It should be emphasized that while making the perturbation spectrum high at frequencies that couple guided modes, we will wish to avoid making it too high at frequencies that couple guided to unguided radiation modes since such coupling results in loss.

4.2 *A Hydraulic Model of Dispersion*

We shall now show that the perturbation results imply a model which is an interesting interpretation of the propagation process, and which allows easy computation of the response of a long guide.

Suppose energy traveled down the guide as follows. We start with a large number of indivisible bundles of energy at the guide input. Each bundle begins propagating down the guide randomly jumping from mode to mode. At any point down the guide, a bundle travels at the group velocity associated with the mode it is currently in. At any position, the probability that a bundle will jump to mode k, given that it is in mode j, in the next increment of distance dl is $\lambda_{kj}\lambda_{kj}^* S_g(\beta_k(\omega) - \beta_j(\omega))\, dl$.

Since we have a very large number of bundles, the output response of the guide in intensity should have the same shape as the probability distribution of the arrival time at the output of an individual bundle. For a short guide of length L, the probability that a bundle is in mode k given that it started in mode j is $| \lambda_{kj} |^2 S_g(\beta_k(\omega) - \beta_j(\omega))L$ and its arrival time distribution is given exactly by Fig. 1. Since this distribution is the perturbation solution for the intensity impulse response of a short guide, we see that the hydraulic model gives the same result as the perturbation theory. Further, a little thought will show (see Appendix B) that the extrapolation from a short guide to a long guide in the hydraulic model is analytically the same as equation (20). Thus any technique which can be used to determine the intensity impulse response characteristics using the hydraulic model will be valid for the solution of equation (20) using the perturbation results.

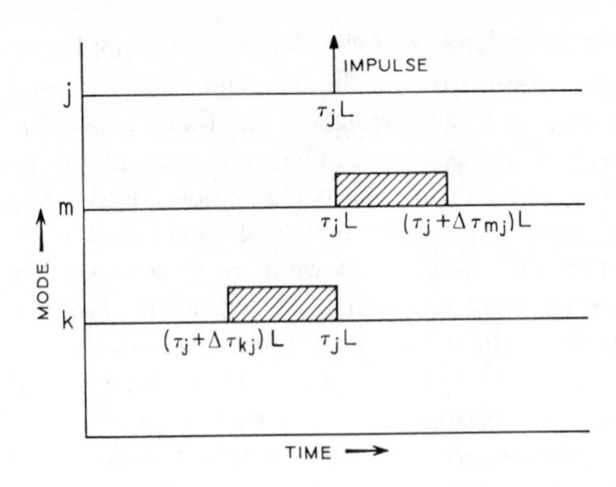

Fig. 1—Output intensity response for modes j, m and k, given mode j is excited.

4.3 *The Solution of Some Hydraulic Model Response*

4.3.1 *Characteristics*

We now wish to determine some probability density moments of the arrival time of a bundle of energy at the output of a long guide recalling that this has the shape of the guide impulse response.

Let $H(l')$ be the mode a bundle is in at distance l' down the guide. In that mode the bundle travels with differential delay (time/meter) $\tau_H = \tau(l')$. The total propagation time down the guide is

$$T = \int_0^l \tau(l')\, dl'. \tag{40}$$

The expected propagation time down the guide is

$$\langle T \rangle_{av} = \int_0^l \langle \tau(l')\, dl' \rangle_{av}\, . \tag{41}$$

The variance about the mean is

$$\begin{aligned} \langle (T - \langle T \rangle_{av})^2 \rangle_{av} &= \left\langle \int_0^l \int_0^l (\tau(l') - \langle \tau \rangle_{av}(l'))(\tau(l'') - \langle \tau \rangle_{av}(l''))\, dl'\, dl'' \right\rangle \\ &= \left[\int_0^l \int R_\tau(l', l'')\, dl'\, dl'' \right] - \langle T \rangle_{av}^2 \end{aligned} \tag{42}$$

where $R_\tau(l', l'') = E(\tau(l')\tau(l''))$. We need the correlation function $R_\tau(l, l')$ and the mean $\langle \tau \rangle_{av}(l')$.

4.4 *Calculation for a Lossless Two-Mode Guide*

For the lossless two-mode guide we have an energy bundle making a Poisson number of mode changes in any length L with mean $|\lambda_{12}|^2 S_g(\beta_1(\omega) - \beta_2(\omega))L$. The correlation function $R_\tau(l, l')$ (assuming we start off randomly in one of the modes) is that of a random telegraph wave[2] and is given by

$$R_\tau(l, l') = \tfrac{1}{4}\,|\Delta\tau_{12}|^2 \exp(-2\,|l - l'|/l_c) + \left(\frac{\tau_1 + \tau_2}{2}\right)^2 (l - l')^2 \qquad (43)$$

where

$$1/l_c = |\lambda_{12}|^2\, S_g(\beta_1(\omega) - \beta_2(\omega)),$$

and

$$\Delta\tau_{12} = \left(\frac{\partial\beta_1}{\partial\omega} - \frac{\partial\beta_2}{\partial\omega}\right) = \tau_1 - \tau_2 .$$

We obtain the mean and second moment about the mean of the intensity response of a guide of length L.

$$\langle T\rangle_{\mathrm{av}} = \left(\frac{\tau_1 + \tau_2}{2}\right)L,$$

$$\langle(T - \langle T\rangle_{\mathrm{av}})^2\rangle_{\mathrm{av}} = \frac{(\Delta\tau_{12})^2 l_c L}{4}\left[1 - \frac{l_c}{2L}(1 - \exp(-2L/l_c))\right], \qquad (44)$$

$$\lim_{[L/l_c]\to\infty} \langle(T - \langle T\rangle_{\mathrm{av}})^2\rangle_{\mathrm{av}} = \frac{(\Delta\tau_{12})^2 L l_c}{4},$$

where

$$l_c = [|\lambda_{12}|^2\, S_g(\beta_1(\omega) - \beta_2(\omega))]^{-1},$$

$$\lim_{[L/l_c]\to 0} \langle(T - \langle T\rangle_{\mathrm{av}})^2\rangle_{\mathrm{av}} = \frac{(\Delta\tau_{12})^2 L^2}{4},$$

(compare equation (44) to equation (29)).

4.5 *Extension to the Two-Mode Guide With Loss*

We can use the hydraulic model to extend the above results to a two-mode guide with loss and differential loss.

Assume that when travelling at distance dl in mode j, a bundle of light has probability $\alpha_j dl$ of being absorbed.

If in travelling down a guide of length L, the bundle spends a distance L_1 in mode 1 and L_2 in mode 2, then the probability that it is not absorbed is

$$P = \exp\,[-(\alpha_1 L_1 + \alpha_2 L_2)] = \exp\,[-((\alpha_1 - \alpha_2)L_1 + \alpha_2 L)]. \quad (45)$$

The number of bundles entering the guide at time zero and arriving at the output end at time t in the presence of loss equals the number that would arrive at time t in the absence of loss times the probability that a bundle with total travel time t is not absorbed. But we have

$$L_1 + L_2 = L,$$

$$t = \tau_1 L_1 + \tau_2 L_2 = (\tau_1 - \tau_2)L_1 + \tau_2 L,$$

$$P = \exp\,[-(\alpha_1 L_1 + \alpha_2 L_2)] = \exp\left\{-\left(\Delta\alpha\,\frac{(t - \tau_2 L)}{\Delta t_{12}} + \alpha_2 L\right)\right\} \quad (46)$$

where

$$\Delta\tau_{12} = \tau_1 - \tau_2 \quad \Delta\alpha = \alpha_1 - \alpha_2.$$

With a little algebra we obtain

$$P = \exp\left[-\left\{\frac{\Delta\alpha}{\Delta\tau_{12}}\,(t - \langle\tau\rangle_{\text{av}}L) + \langle\alpha\rangle_{\text{av}}L\right\}\right] \quad (47)$$

where

$$\langle\alpha\rangle_{\text{av}} = (\alpha_1 + \alpha_2)/2,$$

$$\langle\tau\rangle_{\text{av}} = (\tau_1 + \tau_2)/2.$$

Thus the intensity impulse response for a two-mode guide with loss is equal to the lossless response multiplied by P of equation (47).

Note that the gaussian shape for long guides still holds because the product of a gaussian and an exponential envelope is a shifted gaussian.

V. CONCLUSIONS

We can conclude at least one important result. Long optical fiber waveguides need not have large dispersion due to random imperfections if properly controlled mode coupling exists. From equation (44) we see that a mechanical perturbation spectrum which is peaked at frequencies that couple guided modes will lower dispersion. However, to avoid loss, we must not make the mechanical perturbation spectrum too high at frequencies that couple guided and radiating modes.

The above conclusions have been obtained by D. T. Young and H. E. Rowe,[3] for the two-mode guide by solving the coupled line equations directly under the assumption of white noise coupling.

APPENDIX A

We wish to derive equation (21) from the following relationship

$$G(t, b) = [\gamma_{ij}(t, b)] = [\gamma_{ij}(t)]^*[\gamma_{ij}(t)]^* \cdots (b \text{ Times}),$$
$$i, j = 1, 2. \qquad (48)$$

Equation (48) implies the following model shown in Fig. 2. The transfer function $\gamma_{21}(t, b)$ is the overall transmission response between input 1 and output 2. This can be obtained by adding up the transmission responses over all different paths between input 1 and output 2 using any desired bookkeeping scheme. Every path between input 1 and output 2 must pass through the $\gamma_{21}(t)$ function *for the last time* in some section. If a path passes through the $\gamma_{21}(t)$ function for the last time in the fifth section from the end, then it must pass through four $\gamma_{22}(t)$ functions on its way to output 2. The sum of the path transfer functions between input 1 and the input to the $\gamma_{21}(t)$ function in the fifth section from the end is $\gamma_{11}(t, b - 5)$. Thus the contribution to the overall transfer function between input 1 and output 2 due to all paths which pass through a $\gamma_{21}(t)$ function for the last time in the fifth section from the end is $\gamma_{11}(t, b - 5) * \gamma_{21}(t) * \gamma_{22}^{*4}(t)$. Equation (21) merely expresses the sum of the contributions over all positions of last passage through a γ_{21} (t) function.

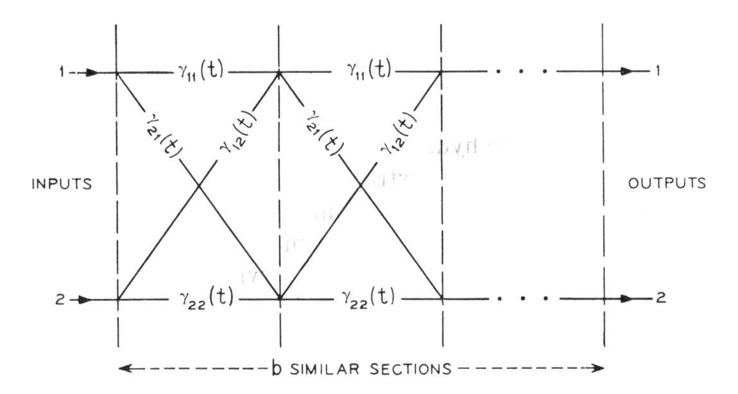

Fig. 2—b-section guide.

$$\gamma_{21}(t, b) = \sum_{R=1}^{b} \gamma_{11}(t, b - R) * \gamma_{21}(t) * \gamma_{22}^{*R-1}(t). \tag{21}$$

APPENDIX B

We wish to show that the equations for obtaining the intensity response of a long guide, given the intensity response of a short guide, for the hydraulic model are identical to the extrapolation equations for the intensity response given by equation (20).

Suppose we have the probability density, for a short guide, that a bundle of energy starting off in mode j of the guide (at time zero) arrives at the output of the guide at time t in mode i. Thus we have the matrix of densities $P(t, L)$ where L is the guide length and the elements $p_{ij}(t, L)$ are the previously described densities. Let $I_j(t, 0)$ be the probability density that a bundle of light arrives at input j at time t. Let $I_j(t, L)$ be the probability density that a bundle arrives at the output position L in mode j at time t. Let $I(t, \cdot)$ be the corresponding vectors. Using the laws of addition of random variables we obtain

$$I_i(t, L) = \sum_s p_{is}(t, L) * I_s(t, 0)$$

of in matrix notation

$$I(t, L) = P(t, L) * I(t, 0)$$

therefore

$$I(t, kL) = \left(\overset{k}{\underset{1}{*}} P(t, L)\right) * I(t, 0).$$

We see that the probability density of the output arrival mode and time of a bundle of energy for a long guide, which corresponds to the intensity response, is extrapolated from the short-guide response exactly as in equation (20). Thus since the perturbation results of Marcuse correspond to the hydraulic model in the limit of short guides and satisfy the conditions for extrapolation using equation (20), it follows that properties of the hydaulic model solution for long guides will correspond to the solution of equation (20) starting with these perturbation results. This is true no matter what techniques we use to find these hydaulic model properties.

APPENDIX C

We wish to establish that for a narrowband high frequency signal

$$y(t) = (y_e(t) \exp(i\omega_0 t) + y_e^*(t) \exp(-i\omega_0 t))/\sqrt{2}$$

of carrier frequency $f_0 = \omega_0/2\pi$ and envelope $y_e(t)$, the intensity is given by

$$|y_e(t)|^2 = 2 \int \left[\int_{-\infty}^{\infty} Y_+^*(f) Y_+(f+\alpha) \, df \right] \exp(-i2\pi\alpha t) \, d\alpha,$$

$$\Leftrightarrow 2 \int_{-\infty}^{\infty} Y_+^*(f) Y_+(f+\alpha) \, df.$$

Define

$$Y_e(f) = \int_{-\infty}^{\infty} y_e(t) \exp(i2\pi f t) \, dt, = \sqrt{2} \, Y_+(f + f_0),$$

[provided $y_e(t)$ is narrowband compared with f_0];

$$|y_e(t)|^2 = \int_{-\infty}^{\infty} Y_e(f) \exp(-2i\pi f t) Y_e^*(f') \exp(i2\pi f' t) \, df \, df',$$

$$= \int_{-\infty}^{\infty} \int \exp[-i2\pi(f - f')t][Y_e(f) Y_e^*(f')] \, df \, df',$$

$$= \int_{-\infty}^{\infty} \int^{\infty} \exp(-i2\pi\gamma t)[Y_e^*(f') Y_e(f' + \gamma)] \, d\gamma \, df',$$

where $\gamma = f - f'$;

$$= 2 \int_{-\infty}^{\infty} \exp(-i2\pi\gamma t)[Y_+^*(g) Y_+(g + \gamma) \, d\gamma \, dg]$$

where $g = (f' + f_0)$.

Q.E.D.

REFERENCES

1. Marcuse, D., "Mode Conversion Caused By Surface Imperfections In A Dielectric Slab Waveguide," B.S.T.J., *48,* No. 10 (December 1969), pp. 3187–3215.
2. Davenport, W. L., and Root, W. B., *Random Signals and Noise,* New York: McGraw-Hill, 1958, pp. 61–62.
3. Young, D. T., and Rowe, H. E., "Impulse Response Of A Two Mode Random Guide," Unpublished work.

Part V
Signal Distortion

GROUP DELAY IN GLASS-FIBRE WAVEGUIDE

Indexing terms: Optical waveguide, Fibre optics

The group delay in a single-mode glass fibre appears to be dominated by the glass dispersion but could, in principle, be reduced considerably by using glass with a certain index/frequency characteristic. In a fibre communication system, with a GaAs laser as transmitter, the large frequency spread of the present-day laser is the most serious factor limiting the bandwidth.

Computer programs have been written which calculate the group velocity, dispersion and fraction of the total power carried in the core for the HE_{11} mode, which is the interesting mode for optical communications. The dispersion, or change in group velocity with frequency, in an optical guide can be separated into two parts: that part which is the property of the guiding structure and is due to the variation of guide wavelength with frequency, and that part which is intrinsic to the propagating medium itself; the optical dispersion of the glass.

Guide dispersion: It is convenient to define a normalised frequency:

$$V = \frac{2\pi fa}{c}(n_1{}^2 - n_2{}^2)^{\frac{1}{2}} \quad \quad (1)$$

where

n_1 = core index

n_2 = cladding index

f = frequency

a = core radius

c = velocity of light in free space

and also to define a normalised index difference:

$$\delta = 1 - \left(\frac{n_2}{n_1}\right)^2 \quad \quad (2)$$

At low V, where most of the energy is carried in the cladding, the group velocity is asymptotic to the velocity of an unbounded wave in that medium. At high V, the energy is mostly carried in the core, and the group velocity accordingly tends to the unbounded velocity of the core material. At some intermediate point between these extremes, the rate of change of group velocity with frequency is zero. Fig. 1 shows dV_g/df against V computed for typical values of core and cladding index. Unfortunately the zero-dispersion point, at $V = 3$, turns out to be in the region above cutoff for the higher-order modes, which is at $V = 2{\cdot}405$. There remains the choice, for keeping the dispersion as small as possible, of working the guide either as near to the higher-mode cutoff as practicable, or else at a much lower V where the dispersion is asymptotically approaching zero. Fig. 1 also shows the fraction of the energy carried in the core plotted against V. Near the higher-mode cutoff, the core carries some 80% of the total energy, while, at the point for comparable dispersion at low V, the fraction of the energy carried in the core is very small, and the waveguide virtually loses its guidance. With small differences in the core and cladding index, the dispersion at any value of V is proportional to the normalised index δ, so that, to minimise the guide dispersion alone, it is necessary to have as small a difference in indices as possible and to choose the operating point to be as near as practicable to the higher-mode cutoff.

Glass dispersion: The intrinsic dispersion of the glass itself depends on the type of glass, being in general greater for those with higher indices. It is relatively large in the visible region of the spectrum compared with its value at infrared frequencies. A good approximation for the index n of a glass can be obtained from a polynomial of the form[1]

$$n^2 = A_0 + A_1 f^{-2} + A_2 f^2 + A_3 f^4 + A_4 f^6 + A_5 f^8 \quad (3)$$

which has been written into the program to modify the guide dispersion to include the intrinsic dispersions of the core and cladding glasses. Fig. 2 shows the computed total rate of change of group velocity with frequency for two soda–lime–silica glasses as core and cladding using the index data from Reference 1. Calculations on a number of glasses show that a typical working value for dV_g/df is in the region of 10^{-8} m/cycle, which, for a practical index difference between core and cladding, swamps the guide dispersion.

Glass characteristics for low dispersion: The group velocity is given by

$$V_g = 2\pi \frac{df}{d\beta} \quad \quad (4)$$

where β is the waveguide propagation constant related to the guide wavelength λ_g by

$$\beta = \frac{2\pi}{\lambda_g}$$

Then

$$\frac{dV_g}{df} = 2\pi \frac{d}{df}\left(\frac{df}{d\beta}\right) = 2\pi \frac{d}{df}\left(\frac{1}{d\beta/df}\right)$$

$$= \frac{-2\pi \dfrac{d^2\beta}{df^2}}{\left(\dfrac{d\beta}{df}\right)^2} \quad \quad (5)$$

For zero dispersion, i.e.

$$\frac{dV_g}{df} = 0, \quad \frac{d^2\beta}{df^2} = 0$$

β is given by

$$\beta = \frac{2\pi f n_1}{c}\left(1 - \frac{u^2\delta}{V^2}\right)^{\frac{1}{2}} \quad \quad (6)$$

u is the dimensionless eigenvalue, which is a solution of the guide characteristic equation. The ratio u^2/V^2 tends to a maximum of unity as V tends to zero, and $(u^2/V^2)\delta$ will therefore be much less than unity for small δ. Then eqn. 6 can be written

$$\beta \simeq \frac{2\pi f n_1}{c} \quad \quad (7)$$

The condition for zero dispersion then becomes

$$\frac{d^2(fn_1)}{df^2} = 0 \quad \quad (8)$$

Reprinted with permission from *Electron. Lett.*, vol. 7, pp. 82–84, Feb. 11, 1971.

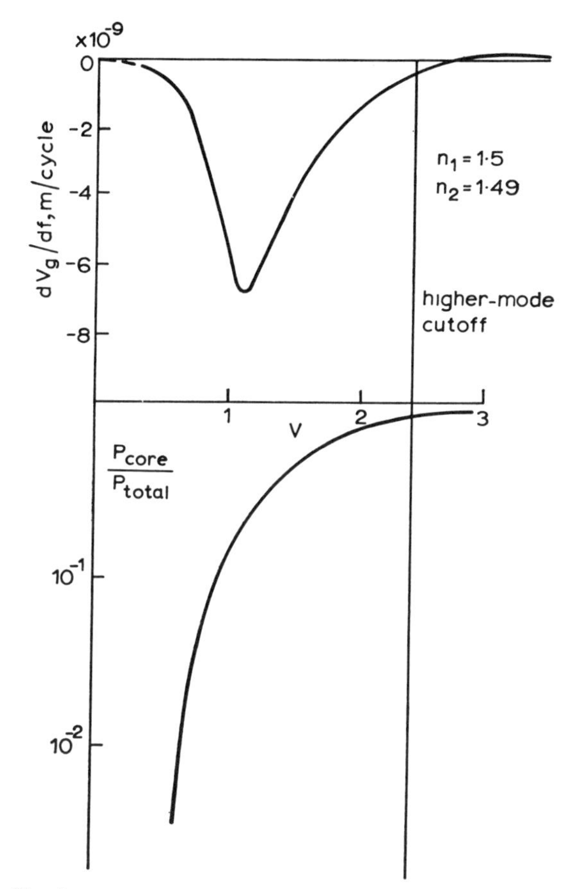

Fig. 1

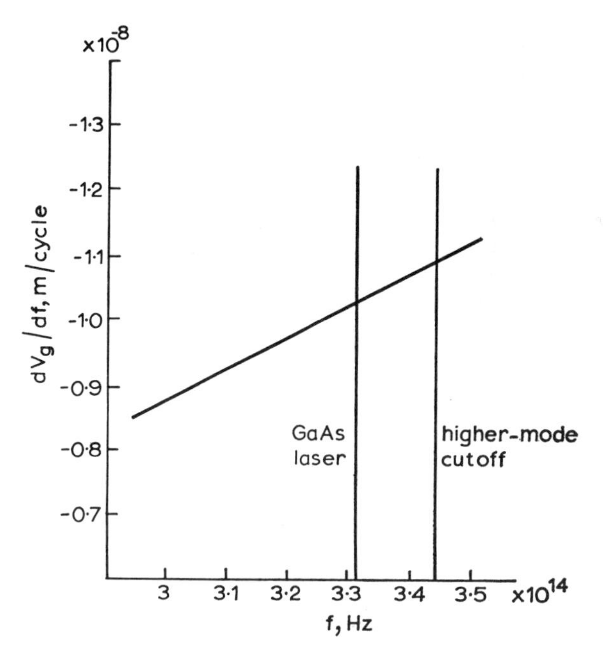

Fig. 2

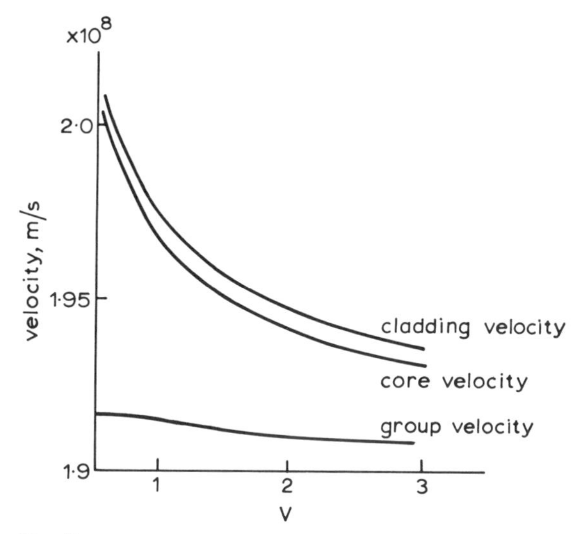

Fig. 3

which gives

$$n_1 = A + B/f \quad . \quad . \quad . \quad . \quad . \quad . \quad . \quad . \quad . \quad . \quad (9)$$

where A and B are arbitrary constants.

Fig. 3 shows the computed group velocity for a guide with a core index which obeys this law and with δ constant. By using eqn. 7 for β in determining the condition for zero dispersion, we are ignoring a term associated with the guiding properties of the fibre, and some residual dispersion must therefore be expected when eqn. 9 is satisfied. Computer calculations have shown that this is of the same order as the guide dispersion with n_1 and n_2 kept constant and with the same δ. For example, for $\delta = 0{\cdot}005$, the residual dispersion is about two orders of magnitude less than the overall dispersion quoted previously for real glasses.

With so many demands already being made on the glasses, it is perhaps optimistic to expect to find a pair of glasses which also have the necessary dispersion properties, but it may be possible to develop such glasses by a study of glass composition and its effect on index and dispersion.

Group delay in a communication system: Since no known complete system exists which uses a single-mode glass-fibre waveguide, some assumptions have to be made about the type of transmitter and the kind of modulation. For economic and practical reasons, the GaAs laser with some form of pulse-code modulation seems the most likely combination at the moment. Present room-temperature GaAs lasers have a minimum line width of about 5 Å representing a frequency spread of $1{\cdot}86 \times 10^{11}$ Hz. Supposing that the pulse-repetition rate is small compared with this frequency spread; then, over a transmission line distance S metres, there will be a spread in transit time of

$$\Delta t = \frac{S}{V_{g1}} - \frac{S}{V_{g2}} \simeq S\left(\frac{n}{c}\right)^2 \Delta V_g \quad . \quad . \quad . \quad . \quad . \quad (10)$$

where n is the average index of the glasses and V_{g1} and V_{g2} are the group velocities at the extremes of the spectrum. But

$$\Delta V_g = \frac{dV_g}{df}\Delta f \quad . \quad . \quad . \quad . \quad . \quad . \quad . \quad . \quad . \quad . \quad (11)$$

so that

$$\Delta t = \left(\frac{n}{c}\right)^2 \Delta f \frac{dV_g}{df} S \quad . \quad . \quad . \quad . \quad . \quad . \quad . \quad (12)$$

Δt can be expressed as a fraction of the pulse-repetition period T:

$$\frac{\Delta t}{T} = \left(\frac{n}{c}\right)^2 \frac{S}{T} \Delta f \frac{dV_g}{df} \quad . \quad . \quad . \quad . \quad . \quad . \quad . \quad (13)$$

Taking a practical case with a line length of 1 km, a pulsed laser linewidth of 5 Å, and a total dispersion of 10^{-8} m/cycle, with a pulse-repetition frequency of 10^9 Hz, the fractional spread in transit time

$$\frac{\Delta t}{T} = 0{\cdot}05$$

Taking $\Delta t/T = 0{\cdot}5$ as an upper limit, this gives a maximum length of transmission line of 10 km. Group delay is not a limiting factor in the sort of system which seems to be feasible in the near future, but any substantial increase in system bandwidth and/or decrease in waveguide loss, with consequent increase in repeater spacing, would make some form of correction necessary.

Effect of a slightly elliptical core: The elliptical dielectric waveguide has been analysed by Yeh,[2] who defines the two modes that correspond to the HE_{11} mode in circular guide as the even ${}_eHE_{11}$ mode and the odd ${}_oHE_{11}$ mode with electric fields in the direction of the minor and major axes, respectively. The ${}_oHE_{11}$ mode has a better 'binding geometry' than the ${}_eHE_{11}$ mode and would therefore be favoured by random scattering effects. An estimate of the difference in group velocity between modes can be made by representing the slightly elliptical guide as two superimposed circular guides whose diameters are the major and minor axes of the ellipse. The rate of change of group velocity with core radius is

$$\frac{dV_g}{da} = \frac{dV_g}{dV}\frac{dV}{da} = \frac{2\pi f}{c} n\delta^{\frac{1}{2}} \frac{dV_g}{dV} \quad . \quad . \quad . \quad . \quad . \quad . \quad (14)$$

Again, expressing the resultant difference in group delay Δt as a fraction of the repetition period T,

$$\frac{\Delta t}{T} = \left(\frac{n}{c}\right)^2 S \frac{2\pi f}{c} \delta^{\frac{1}{2}} \frac{dV_g}{dV} \Delta a \quad . \quad . \quad . \quad . \quad . \quad . \quad (15)$$

Taking a practical example with $S = 10^3$ m, $f = 3 \times 10^{14}$ Hz and $T = 10^{-9}$ s, at $\delta = 0{\cdot}01$ and $V = 2$ the normalised rate of change of group velocity dV_g/dV is computed to have a value of $-2{\cdot}3 \times 10^5$ m/s. For a change Δa in core radius of 10^{-8} m (approximately 1%), the fractional difference in group delay is

$$\frac{\Delta t}{T} = 0{\cdot}024$$

The authors gratefully acknowledge helpful discussions with Prof. P. J. B. Clarricoats of Queen Mary College, London, and with A. W. Snyder of Yale University.

R. B. DYOTT
J. R. STERN

11th January 1971

Telecommunications Headquarters
PO Research Department
Brook Road, Dollis Hill, London NW2, England

References

1 'Schott optical glass catalogue', Jena Glassworks, Mainz, W. Germany
2 YEH, C.: 'Elliptical dielectric waveguides', *J. Appl. Phys.*, 1962, **33**, pp. 3235–3243

Material Dispersion in Optical Fiber Waveguides

M. DiDomenico, Jr.

In weakly guiding dielectric (glass) fiber waveguides the factors governing the envelope delay distortion of optical signals can be separated conveniently into material and waveguide contributions. By making use of the fact that the normal dispersion of the index-of-refraction of glasses accurately fits a single oscillator Sellmeier equation, simple relations of general validity are obtained for the material contribution to the group delay and frequency dependence of the group delay in dielectric fiber waveguides. Using these results expressions are obtained for the material contribution to the information bandwidth of fiber optic waveguides excited by wideband sources (light emitting diodes) and narrow band sources (lasers). It is shown that in single mode guides the bandwidth limitation imposed by material dispersion is negligible, while in multimode guides the difference in group delay of higher order modes imposes the most severe bandwidth limitation.

Introduction

The delay distortion experienced by optical signals in dielectric fiber waveguides is due basically to the effects of material dispersion and to waveguide characteristics. In a recent paper Gloge[1] has shown that for weakly guiding fibers (i.e., fibers having small refractive index differences between core and cladding) material and waveguide dispersion effects can be separated as additive effects. The purpose of this paper is to calculate in a general and simple way the material contribution to the group delay and frequency dependence of the group delay in weakly guiding fibers. By using the results of a previous analysis[2] for the normal dispersion of the electronic dielectric constant of solids, simple relations of general validity are obtained for the delay distortion due to material effects.

Dispersion of the Optical Dielectric Constant

It has been shown[2] that the normal dispersion of the electronic (optical) dielectric constant in materials accurately satisfies a Sellmeier relation of the form

$$n^2 - 1 = E_0E_d/(E_0^2 - E^2), \tag{1}$$

where n is the refractive index, E is the photon energy, and E_0 and E_d are, respectively, oscillator energy and dispersion energy parameters. To a good approximation the oscillator energy E_0 is related to the lowest direct bandgap E_t by $E_0 \approx 1.5\ E_t$. In single crystals the dispersion energy E_d is found empirically[2] to scale with nearest neighbor cation coordination number N_c, anion valency Z_a, and effective number of dispersion electrons N_e (usually $N_e = 8$), i.e., $E_d = \beta N_c Z_a N_e$. The scaling constant β is two valued taking on discrete ionic, β_i, and covalent, β_c, limits ($\beta_i = 0.26$ eV and $\beta_c = 0.37$ eV). Thus crystals with the same chemistry and structure have the same value of E_d. As a result, in the region of transparency ($E \ll E_0$) we obtain from Eq. (1) the relation,

$$n^2 - 1 = E_d/E_0, \tag{2}$$

indicating that to a good approximation the dielectric susceptibility varies inversely with E_0 or the bandgap.

Because glasses are amorphous materials, one might expect E_d to be smaller than in single crystals due to broken bonds, i.e., a smaller N_c. This is certainly borne out when one compares E_d for crystalline quartz (SiO_2) $E_d = 18.3$ eV with E_d for fused silica (SiO_2) $E_d = 14.7$ eV. The oscillator energy for both crystal quartz and fused silica is $E_0 \approx 13.5$ eV. When we compare the oscillator and dispersion energies for a large group of representative glasses we find no systematic trends in E_d as observed in single crystals.

Shown in Fig. 1 are plots of $1/(n^2 - 1)$ vs E^2 (or $1/\lambda^2$) for several glasses. These glasses were selected from the Schott catalog (except SiO_2) and span the range of indices of refraction available from typical crown and flint glasses ($1.45 < n < 2$). The intercept at $E = 0$ in Fig. 1 gives E_0/E_d, and the slope gives $1/E_0E_d$. Note that the index data fit the single oscillator Sellmeier expression very well. A large number of the glasses given in the Schott catalog have been analyzed in terms of Eq. (1). Table I lists values of E_0 and E_d for a representative sample of glasses together with the code designation given in the Schott catalog. The Sellmeier oscillator parameters listed in Table I should be compared with the extensive listing of parameters for single crystals given in Ref. 2.

The author is with Bell Telephone Laboratories, Inc., Murray Hill, New Jersey 07974.

Received 14 October 1971.

Reprinted with permission from *Appl. Opt.*, vol. 11, pp. 652–654, Mar. 1972.

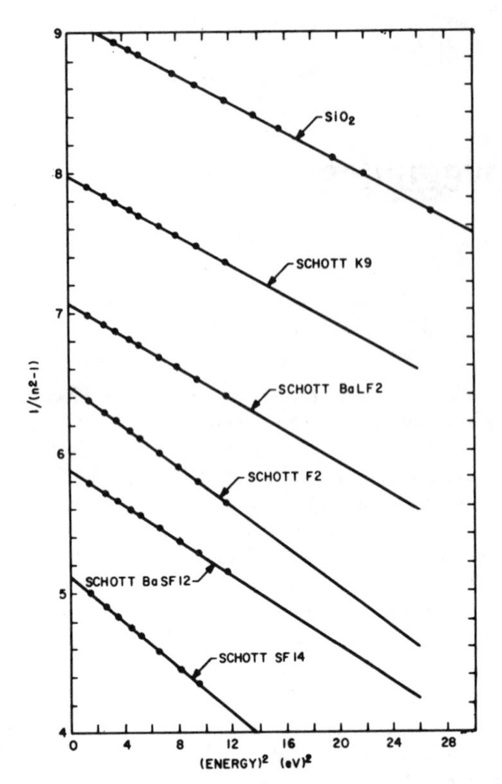

Fig. 1. Plot of refractive-index factor $1/(n^2 - 1)$ vs E^2 for several selected glasses.

Group Delay and Frequency Dependence of Group Delay-Material Contribution to Information Bandwidth of Fibers

The group velocity in a dispersive medium is given by

$$v_g = c/(n + Edn/dE), \quad (3)$$

where E is the photon energy and c is the speed of light. In the region of transparency Eq. (1) accurately represents the dispersion of the optical susceptibility. By differentiating Eq. (1) we find that to an excellent approximation in the region $E \ll E_0$ (i.e., in the visible for glass materials)

$$dn/dE = E_d E/nE_0^3. \quad (4)$$

Thus the group delay per unit length $T = 1/v_g$ is given by

$$T = (n/c)[1 + (E_d E^2/n^2 E_0^3)]. \quad (5)$$

Using the oscillator parameters listed in Table I we conclude that throughout the visible part of the spectrum the material contribution to the group delay is simply n/c, the value in a nondispersive medium.

Also of interest is the frequency dependence of the group delay. In optical bandwidth B (expressed in units of energy) the differential group delay per unit length is given by

$$s = B(d/dE)(1/v_g). \quad (6)$$

It can be shown from Eqs. (1) and (3), after some algebra, that in the region of transparency $E \ll E_0$,

$$s \approx 3BE_d E/cnE_0^3. \quad (7)$$

This is the material contribution to the waveguide time dispersion. Gloge[1] has shown that the total waveguide time dispersion is the superposition of the material contribution and a negative waveguide characteristic. Using values appropriate to SiO_2 we find that in the visible $E \approx 2$ eV,

$$cs/\Delta \approx 0.05, \quad (8)$$

where $\Delta = B/E$ is the relative bandwidth. Referring to the data in Table I we conclude that for most glasses E_0 is smaller than the 13.4 eV value for SiO_2, and, as a consequence, the cs/Δ values will be somewhat higher than indicated by Eq. (8).

If a Gaussian pulse of width τ is transmitted through a dispersive medium of length L the output pulse width τ_0 is

$$\tau_0^2 = \tau^2 + s^2L^2. \quad (9)$$

This equation applies for a pulse whose spectral width is small compared to the emission spectrum of the source[1] or for a bandwidth-limited Gaussian pulse.[3] In the first case B in Eq. (7) is the spectral width of the source, while in the second case B is the reciprocal of the pulse width, i.e., $B = h/\tau$, where h is Planck's constant ($h = 4.14 \times 10^{-15}$ eV-sec). Thus for a broadband source (e.g., an electroluminescent diode) transmitting a wide pulse we have

$$s \approx 3\Delta E_d E^2/cnE_0^3 \approx 170\Delta(\text{nsec/km}), \quad (10)$$

whereas for a bandwidth-limited pulse from, for example, a mode-locked laser we have

Table I. Dispersion Parameters for a Representative Selection of Glasses from the Schott Catalog

Material	n at λ = 0.588 μ	E_0 (eV)	E_d (eV)
SiO_2	1.4586	13.4	14.7
PK2	1.5182	12.96	15.9
PSK2	1.5687	12.51	16.94
BK1	1.5101	12.64	15.72
K9	1.5148	12.2	15.3
BaLF2	1.5710	11.35	16
BaK4	1.5688	11.86	16.8
SK20	1.5596	12.42	17.3
KF6	1.5174	11.5	14.5
BaLF2	1.5710	11.25	15.94
SSK4	1.6177	11.67	18.25
LLF2	1.5407	10.86	14.36
BaF9	1.6433	10.8	17.67
LF1	1.5731	10.26	14.5
F2	1.6200	9.4	14.5
F6	1.6364	9.3	14.8
BaSF12	1.6700	9.74	16.6
SF6	1.8052	7.78	16.23
SF14	1.7618	8.1	15.9
LaK N14	1.6968	11.7	21.2
LaF N2	1.7440	10.37	20.3
LaSF N3	1.8061	9.13	19.6

$$s \approx \frac{3hE_dE}{cnE_0^3\tau} \approx \frac{0.35}{\tau} \times 10^{-12} \frac{\text{nsec}}{\text{km}}, \quad (11)$$

where τ is the pulse width expressed in seconds. Using Eqs. (10) and (11) we can calculate a material contribution to the bandwidth of a dielectric fiber waveguide. When $sL = \tau$ the output pulse width is $\sqrt{2}\tau$. We define the base bandwidth or information bandwidth of the fiber as $b = 1/\tau = 1/sL$. In this base bandwidth negligible envelope delay distortion occurs due to the frequency dependence of the group delay. Accordingly for a broadband source we have from Eq. (10),

$$bL_{\text{km}} \approx cnE_0^3/3\Delta E_dE^2 \approx (5.9/\Delta)\text{MHz-km}, \quad (12)$$

or

$$\tau \approx \frac{3\Delta E_dE^2}{cnE_0^3} L \approx 170\Delta L_{\text{km}}\,\text{nsec}, \quad (13)$$

where L_{km} is the length expressed in kilometers. For a bandwidth-limited pulse Eq. (11) gives

$$b\sqrt{L_{\text{km}}} \approx (cnE_0^3/3hE_dE)^{\frac{1}{2}} \approx 53\ \text{GHz-km}^{\frac{1}{2}}, \quad (14)$$

or

$$\tau \approx [(3hE_dE/cnE_0^3)L]^{\frac{1}{2}} \approx 18\sqrt{L_{\text{km}}}\,\text{psec}. \quad (15)$$

We emphasize that Eqs. (10)–(15) only give the material dispersion contribution to the time dispersion of a dielectric fiber waveguide. To calculate the total time dispersion one must add to s the contribution due to waveguide dispersion as discussed by Gloge.[1] We also emphasize that the numerical examples apply specifically to SiO_2 at $E \approx 2$ eV. Other glasses will have similar behavior as indicated by the data in Table I.

The analysis leading to Eqs. (12)–(15) applies to each individual waveguide mode in a dielectric fiber waveguide. Gloge[1] has shown that in weakly guiding fibers, where the refractive index difference between core and cladding is small, the *total* differential group delay s in single mode, and to a good approximation in multimode fibers, is determined principally by material dispersion. Thus in a single mode waveguide the envelope of a pulse will experience a group delay T and an envelope delay distortion determined by the differential group delay s given by Eqs. (10) or (11). Equation (12) implies for a single mode fiber, a bandwidth length limitation of $\approx$600 MHz-km for a source with 1% relative bandwidth (e.g., a light emitting diode). Equation (14) implies a bandwidth square-root-of-length limitation of $\approx$50 GHz-km$^{\frac{1}{2}}$ for a bandwidth limited laser source. In multimode fibers each mode experiences a different group delay T_m but, to a good approximation, the same differential group delay s. Therefore, in a multimode waveguide the envelope delay distortion of a pulse is determined by the difference in arrival times or group delay of different higher order modes, as well as by the differential group delay s. It can easily be shown that the group delay of higher order modes imposes a more serious bandwidth limitation than the material dispersion contribution. For example, for a multimode fiber having a core-cladding index difference of 10% the bandwidth can be as low as 1 MHz-km.

We may conclude that the bandwidth limitation due to material dispersion in single mode guides is negligible for most glasses of interest, and that the bandwidth limitation due to the difference in group delay of higher order modes is of primary importance in multimode guides with index differences larger than 1%. In a recent study Kapron and Keck[4] have numerically calculated the information bandwidth of a single mode fiber including both material and waveguide dispersion effects. Their numerical calculations and our bandwidth estimates for single mode fibers are in good agreement.

Conclusions

We have shown that the dispersion of the index of refraction of glasses accurately fits a single oscillator Sellmeier equation. Two parameters, an oscillator energy E_0 and a dispersion energy E_d, completely characterize the refractive properties of glasses in the visible portion of the spectrum. This result has been reported previously for single crystals,[2] where it was shown empirically that the dispersion energy is a unique parameter related to crystal structure, chemistry, and ionicity. No simple behavior has been found for the dispersion energy parameter of glasses.

Gloge[1] has shown previously that in weakly guiding fibers, material and waveguide contributions to the group delay and differential group delay can be separated. By combining the results of his analysis together with the Sellmeier description of the index of refraction, simple equations of general validity have been obtained for the material contribution to the group delay and frequency dependence of the group delay in the visible region of the spectrum. In dielectric fiber waveguides the material contribution is the dominant contribution to the frequency dependence of the group delay. Our results indicate that the limitation to the information bandwidth imposed by the material is negligible in single mode and multimode dielectric fiber waveguides. In multimode guides the difference in group delay experienced by higher order modes imposes the most severe bandwidth limitation.

I wish to thank S. Singh for kindly providing the data in Table I and R. G. Smith for his comments on the manuscript. I also wish to thank D. Gloge for a very informative discussion on time dispersion in dielectric fiber waveguides.

References

1. D. Gloge, Appl. Opt. **10**, 2442 (1971).
2. S. H. Wemple and M. DiDomenico, Jr., Phys. Rev. **B3**, 1338 (1971).
3. C. G. B. Garrett and D. E. McCumber, Phys. Rev. **A1**, 305 (1970).
4. F. P. Kapron and D. B. Keck, Appl. Opt. **10**, 1519 (1971).

Multimode Theory of Graded-Core Fibers

By D. GLOGE and E. A. J. MARCATILI

(Manuscript received March 29, 1973)

New technologies of fiber manufacture and a demand for unusual fiber qualities in communication systems have intensified the interest in a comprehensive theory of multimode fibers with nonuniform index distributions. This paper deals with a general class of circular symmetric profiles which comprise the parabolic distribution and the abrupt core-cladding index step as special cases. We obtain general results of useful simplicity for the impulse response, the mode volume, and the near- and far-field power distributions. We suggest a modified parabolic distribution for best equalization of mode delay differences. The effective width of the resulting impulse is more than four times smaller than that produced by the parabolic profile. Of course, practical manufacturing tolerances are likely to influence this distribution. A relation is derived between the maximum index error and the impulse response.

I. INTRODUCTION

Conventional optical fibers consist of a high-index core surrounded by a cladding of lower index. The index step between core and cladding contains the light inside the core and isolates it from the outer fiber surface, whose quality is usually difficult to control. In a more general way, inside guidance can be accomplished by any index profile which decreases from a maximum inside the fiber to a lower (cladding) value. The specific shape of the profile has an effect on the distribution of the guided optical power in the fiber and on the overall loss encountered, but, more importantly, the profile profoundly influences the velocities of the various propagating modes. A good example is the parabolic index distribution which was predicted to nearly equalize the group velocities of the propagating modes.[1,2] The Selfoc fiber which closely approximates these conditions has indeed since exhibited an extremely narrow impulse response.[3,4]

These effects greatly enhance the chances of multimode fibers to be used in optical communication systems. On the other hand, a theory of the interrelations between index profile, impulse response, and power distribution is presently only available for the two special cases of the uniform and the parabolic core index. This paper provides a more general theory and studies a broad class of index profiles potentially useful in communication applications. The uniform and the parabolic profile are special cases within this class.

Our concern with multimode fibers for communication applications allows us to make four simplifying assumptions:

(*i*) The index profile is circular symmetric.

(*ii*) The core diameter measures hundred wavelengths or more and, hence, a great number of modes can propagate.

(*iii*) The total index change within the guiding core region is only a few hundredths, so the propagating modes can be considered essentially as transverse electromagnetic.[5]

(*iv*) Index variations within the distance of a wavelength are negligible, and the conditions of geometrical optics (or the zeroth order of the WKB method) apply.

Except for these four restrictions and the requirement of guidance, the index profile can be of the most general form. It can, for example, have an index depression in the center and one or several ring-shaped index maxima.[6]

For the sake of clarity, this paper is restricted to the simpler type of profile illustrated in Fig. 1. We assume the index profile will decrease monotonically from the center and converge into a flat cladding region which guarantees isolation from the outside surface.

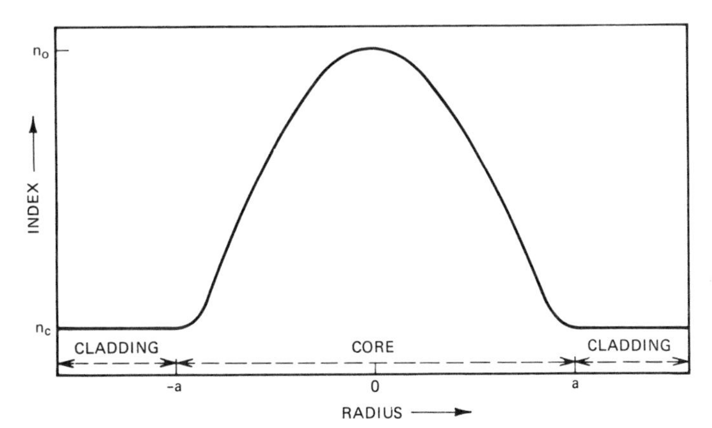

Fig. 1—Cross-sectional sketch of circular symmetric index profile in multimode fiber.

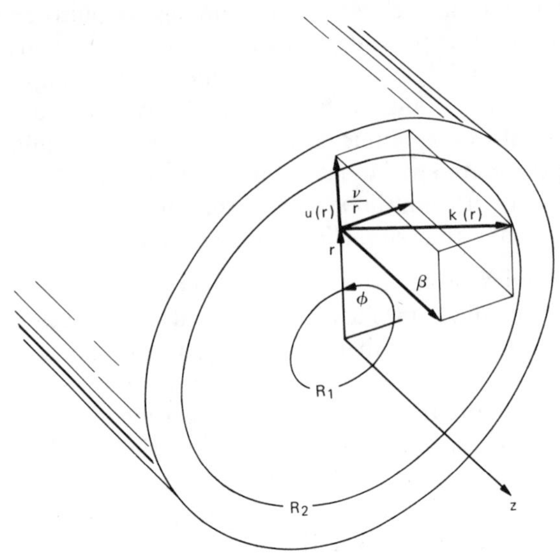

Fig. 2—Wave vector diagram in the propagating region of a multimode fiber.

Apart from the index profile, there are, of course, other influences which affect the impulse response and the optical power distribution inside the fiber. Mode excitation, loss differences in the process of propagation, and coupling among the modes play a part. To isolate the effect of the index profile, we assume here the ideal case of uniform loss, absence of coupling, and equal and simultaneous excitation of all propagating modes at the input. For the computation of the impulse response, the input is assumed to be an infinitely narrow pulse of unit energy.

II. MODE DESIGNATION AND MODE COUNT

All guided modes are essentially transverse electromagnetic and, with some proviso, can be decomposed into linearly polarized pairs.[5,7] Because of the circular symmetry of the index n, the modes have a circular periodicity and can be identified in the conventional way by an azimuthal order number ν. To characterize the radial field distribution, we need an additional mode number μ. The propagation constant β of a particular mode (μ, ν) can then be approximately determined by the WKB method.[6,8] Figures 2 and 3 give a physical description of these relationships. In Fig. 2, the local wave number

$$k(r) = 2\pi n(r)/\lambda \tag{1}$$

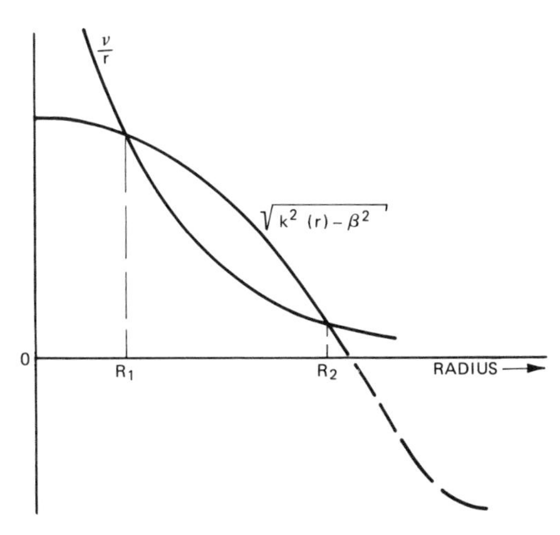

Fig. 3—Sketch defining regions of periodic and aperiodic field characteristics of a mode of azimuthal order ν.

is decomposed into its components in a cylindrical coordinate system (r, ϕ, z). The unknown radial component becomes

$$u(r) = [k^2(r) - \beta^2 - \nu^2/r^2]^{\frac{1}{2}}. \tag{2}$$

Given β and ν, we can find two radii R_1 and R_2, at which $u(r)$ vanishes (see Fig. 3). These radii define a ring-shaped region within which u is real, causing a radial periodicity of the mode field. Outside this region, the field is aperiodic.

Radially decreasing (or evanescent) field conditions obtain outside, when the phase inside (approximately) adds up to an integer number of half periods between R_1 and R_2. Consequently, if μ designates this number of half periods,

$$\mu\pi = \int_{R_1}^{R_2} u(r)dr = \int_{R_1}^{R_2} [k^2(r) - \beta^2 - \nu^2/r^2]^{\frac{1}{2}} dr. \tag{3}$$

We would have obtained the same result by way of the WKB method, with the only difference that μ and ν^2 would be replaced by $\mu + \frac{1}{4}$ and $\nu^2 + \frac{1}{4}$. These corrections are important in the case of small μ or ν, and particularly for the fundamental mode which has $\mu = \nu = 0$. On the other hand, to obtain a general view of the mode structure, we can ignore the $\frac{1}{4}$-terms as long as we refrain from discussing individual low-order modes.

For the purpose of a total mode count, let us consider the limits of μ, ν, and β. The requirement of evanescent field conditions in the cladding (index n_c in Fig. 1) limit β to a minimum value

$$\beta_c = 2\pi n_c/\lambda. \tag{4}$$

Modes with smaller β find propagating conditions in the cladding and are no longer bounded by the core profile. Condition (4) defines mode cutoff. The largest value for ν results for $\beta = \beta_c$ and $\mu = 0$, and alternatively μ is largest for $\beta = \beta_c$ and $\nu = 0$. We obtain the total number of modes M from a summation of (3) over all ν from 0 to $\nu_{\max}$. If $\nu_{\max}$ is a large number, we may consider ν a continuous variable and replace the sum by an integral. In this case,

$$M = \frac{4}{\pi} \int_0^{\nu_{\max}} \int_{R_1(\nu)}^{R_2(\nu)} [k^2(r) - \beta_c^2 - \nu^2/r^2]^{\frac{1}{2}} dr d\nu. \tag{5}$$

The factor 4 in front of the expression allows for the fact that each combination μ, ν designates a (degenerate) group of four modes of different polarization or orientation.[5] Figure 4 illustrates the area of the double integration indicated in (5). A change of order in the integration leads to

$$M = \frac{4}{\pi} \int_0^a \int_0^{r(k^2-\beta_c^2)^{\frac{1}{2}}} (k^2 - \beta_c^2 - \nu^2/r^2)^{\frac{1}{2}} d\nu dr, \tag{6}$$

where a is the radius at which the index $n(r)$ reaches the cladding value n_c. Integrating (6) with respect to ν yields

$$M = \int_0^a [k^2(r) - \beta_c^2] r dr = \left(\frac{2\pi}{\lambda}\right)^2 \int_0^a [n^2(r) - n_c^2] r dr. \tag{7}$$

For small index differences, the integral represents the volume under the (circular symmetric) profile plot. It may be worth noting, though, that the substance of this relation is not limited to circular symmetry.

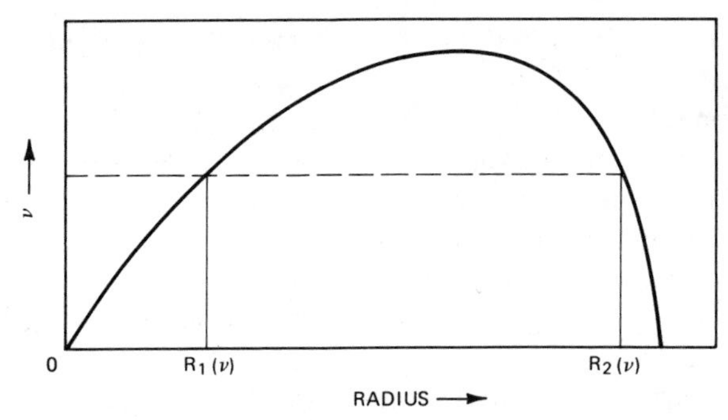

Fig. 4—Region of double integration in eq. 5.

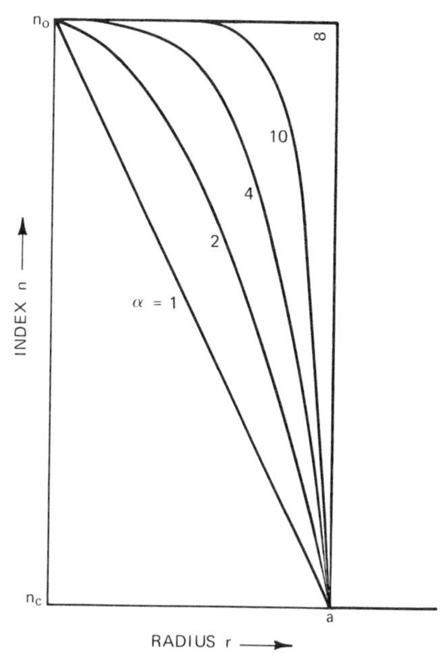

Fig. 5—A few of the index profiles defined by $n = n_0[1 - 2\Delta(r/a)^\alpha]^{\frac{1}{2}}$ for small Δ.

For later use we write (7) in the somewhat different form

$$m(\beta) = \int_0^{R_2(0)} [k^2(r) - \beta^2]^{\frac{1}{2}} r dr\,, \tag{8}$$

where $m(\beta)$ denotes the number of modes having a propagation constant larger than β. The upper limit $R_2(0)$ of the integration is the radius at which $k(r) = \beta$.

Let us now consider a particular class of profiles defined by

$$n(r) = \begin{cases} n_o[1 - 2\Delta(r/a)^\alpha]^{\frac{1}{2}} & \text{for} \quad r < a \\ n_o[1 - 2\Delta]^{\frac{1}{2}} & \text{for} \quad r > a\,, \end{cases} \tag{9}$$

where α is a parameter between 1 and ∞. Figure 5 illustrates the cases $\alpha = 1$, 2, 4, 10, and ∞. All profiles reach a constant cladding value at $r = a$. The core profile has a cone shape for $\alpha = 1$, becomes nearly parabolic for $\alpha = 2$, and converges to the case of the step profile for $\alpha = \infty$. Using (1) we introduce (9) into (8) and obtain

$$m(\beta) = a^2 \Delta k_o^2 \frac{\alpha}{\alpha + 2} \left(\frac{k_o^2 - \beta^2}{2\Delta k_o^2} \right)^{(2/\alpha)+1}, \tag{10}$$

where

$$k_o = 2\pi n_o/\lambda. \tag{11}$$

For $\beta = \beta_c$ from (4), the total mode number becomes

$$M = \frac{\alpha}{\alpha + 2} a^2 k_o^2 \Delta . \tag{12}$$

It is proportional to the index difference and the core cross section. The uniform profile accepts twice as many modes as the parabolic one and three times more than the cone-shaped one.

III. IMPULSE RESPONSE

Consider all modes to be excited by the same narrow pulse at the input. Each mode transports an equal amount of energy to the fiber end. The individual pulses are expected to suffer a certain distortion, depending on the β–ω characteristic of each mode and dispersion in the dielectric. We assume, however, that the resultant broadening is small, or at least not much larger than the group delay differences between adjacent modes. Because of this effect and other limitations in the system response, the pulses from individual modes are likely to fuse into one continuous output pulse called the impulse response. Since all modes carry the same energy, the power profile of the impulse response is equal to the mode density per unit time interval. In the following theory, the continuity of the impulse response results not from the broadening of the individual mode responses, but from the assumption that μ and ν are continuous functions.

The straightforward method of computing the impulse response starts from (3) to find the propagation constant β for each pair, μ, ν. The group delay in a fiber of length L is then

$$\tau(\mu, \nu) = \frac{Ln_o}{c} \frac{d\beta(\mu, \nu)}{dk_o}, \tag{13}$$

where c is the vacuum velocity of light. A simplification of this approach for the purpose of numerical computations is indicated in the appendix. Once $\tau(\mu, \nu)$ is known, the impulse response results from a count of the combinations μ, ν which arrive between τ and $\tau + d\tau$. This number plotted versus τ then constitutes the impulse response.

For the special class of profiles defined by (9), group delay and impulse response can be computed in a much simpler way. First we postulate that, in this case, the relation between τ and β according to (13) is independent of μ and ν. If this holds—and we shall prove it later with the help of eq. (16)—we can replace β by τ in (3) and still perform the same integration over ν which led to (8) and, more specifically, to (10). Solving the result of this integration for τ yields

$$\tau = \frac{Ln_o}{c}\frac{d}{dk_o}\left[k_o^2 - \left(\frac{2m}{a^2}\frac{\alpha+2}{\alpha}\right)^{\alpha/(\alpha+2)}(2\Delta k_o^2)^{2/(\alpha+2)}\right]^{\frac{1}{2}}. \tag{14}$$

This result can easily be verified by solving (10) for β and introducing it into (13). With the help of (10) and the abbreviation

$$\delta = \tfrac{1}{2}(1 - \beta^2/k_o^2), \tag{15}$$

eq. (14) takes the form

$$\tau = \frac{Ln_o}{c}\frac{1 - 4\delta/(\alpha+2)}{(1-2\delta)^{\frac{1}{2}}}. \tag{16}$$

This expression proves indeed to depend on β alone (and not explicitly on m), thus justifying the approach chosen.

To obtain the impulse response, we can now introduce (16) into (10) and differentiate with respect to τ. Although this is not difficult to do, it leads to rather unwieldy expressions. We shall therefore merely consider some special cases of interest. To normalize the impulse response for total unit energy, we divide (10) by (12) and obtain

$$\frac{m}{M} = \left(\frac{\delta}{\Delta}\right)^{(2/\alpha)+1}. \tag{17}$$

Furthermore, since δ can at most assume the value Δ (for $\beta = \beta_c$) and is therefore small compared to unity within the scope of our theory, we develop (16) into a power series in terms of δ and obtain

$$\tau = \frac{Ln_o}{c}\left(1 + \frac{\alpha-2}{\alpha+2}\delta + \frac{3\alpha-2}{\alpha+2}\frac{\delta^2}{2}\right). \tag{18}$$

We relate τ to the total propagation time Ln_o/c and introduce a new time reference, which ignores the delay common to all modes. Hence,

$$t = \frac{\tau c}{Ln_o} - 1 = \frac{\alpha-2}{\alpha+2}\delta + \frac{3\alpha-2}{\alpha+2}\frac{\delta^2}{2}. \tag{19}$$

In this time frame, the fundamental mode arrives at $t = 0$.

As long as α is not too close to 2, the linear term in (19) dominates. Therefore,

$$\delta = \begin{cases} \dfrac{\alpha+2}{\alpha-2}t & \text{except for } \alpha \approx 2 \\ \sqrt{2t} & \text{for } \alpha = 2. \end{cases} \tag{20}$$

Insert this into (17) and differentiate with respect to t to obtain the impulse response

$$\frac{1}{M}\frac{dm}{dt} = \begin{cases} \dfrac{\alpha+2}{\alpha}\left|\dfrac{\alpha+2}{\alpha-2}\dfrac{1}{\Delta}\right|^{(2/\alpha)+1} |t|^{2/\alpha} & \text{except for } \alpha \approx 2 \\ \dfrac{2}{\Delta^2} & \text{for } \alpha = 2. \end{cases} \tag{21}$$

As δ varies from 0 to Δ, the time t changes from 0 to

$$T = \begin{cases} \dfrac{\alpha-2}{\alpha+2}\Delta & \text{except for } \alpha \approx 2 \\ \dfrac{\Delta^2}{2} & \text{for } \alpha = 2. \end{cases} \tag{22}$$

Outside of this time interval, the impulse response is zero. Figure 6 shows plots of (21) for the profiles sketched in Fig. 5. A change from $\alpha = \infty$ to $\alpha = 10$, which implies a relatively small change in the profile, narrows the impulse response by $\frac{1}{3}$. The response becomes

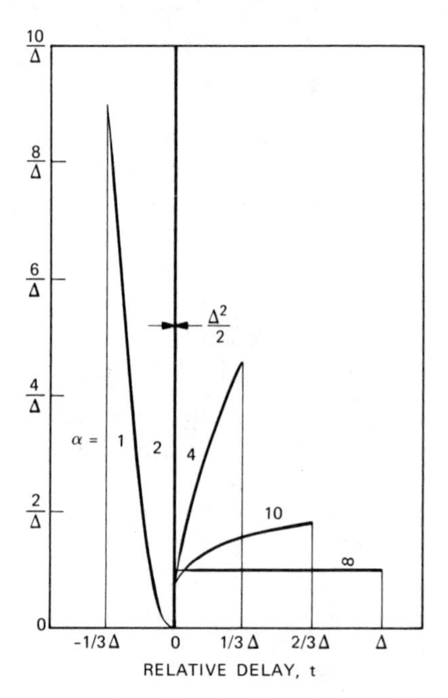

Fig. 6—Impulse response of multimode fibers having the profiles of Fig. 5.

extremely narrow for $\alpha \approx 2$, then broadens again, as α decreases further. For $\alpha < 2$, the high-order modes overtake the fundamental and arrive earlier.

In the vicinity of $\alpha = 2$, where both terms of (19) contribute, the impulse response is a rather complicated function. The most interesting of these cases is the one for which the impulse response has the narrowest possible width. This optimum condition arises for

$$\alpha_{\text{opt}} = 2 - 2\Delta\,, \tag{23}$$

which yields

$$t = \tfrac{1}{2}(\delta^2 - \Delta\delta). \tag{24}$$

In this case, the modes of highest and lowest order both arrive at the same time $t = 0$; all other modes are faster, the fastest one being determined by $\delta = \Delta/2$. It arrives at

$$t = -\frac{\Delta^2}{8}\cdot \tag{25}$$

Equation (24) has two solutions for δ. Hence, (17) yields two values for the same t, indicating that two mode groups, a high and a lower order, contribute to the impulse response at every particular instant in time. By introducing δ into (17), differentiating with respect to t, and then adding the two contributions, we find the impulse response

$$\frac{4}{\Delta^2}\left(1 + \frac{8t}{\Delta^2}\right)^{-\frac{1}{2}}\cdot \tag{26}$$

This function is plotted in Fig. 7. It peaks at $t = -\Delta^2/8$ and decreases towards $t = 0$. Because of the normalization introduced in (19), the absolute temporal width is

$$\frac{Ln_o}{c}\frac{\Delta^2}{8}\cdot \tag{27}$$

The time slot in which a pulse of this kind can be transmitted is narrower than that, because 70 percent of the power is concentrated in the first half of the interval (27).

A practical implementation must, of course, allow for a certain tolerance or error in the profile, as a result of which the total width of the impulse response is likely to exceed (27). To obtain some indication of the pulse broadening as a result of this index deviation, we assume that the erroneous profile is still of the type (9), but has

$$\alpha = \alpha_{\text{opt}} + d\alpha. \tag{28}$$

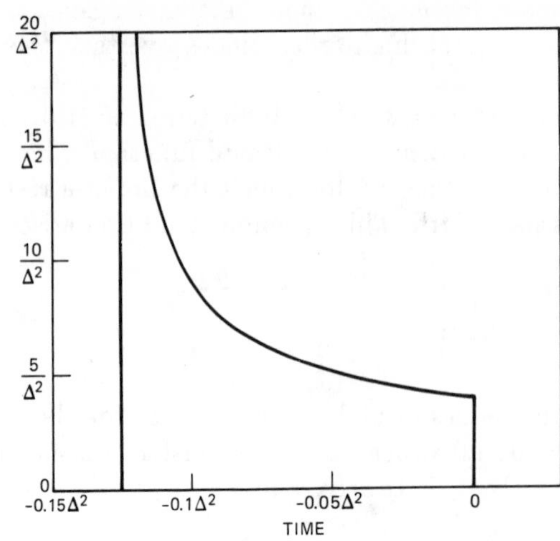

Fig. 7—Impulse response in the case of optimal profile shape.

The maximum index deviation from the optimum profile then appears approximately at

$$r = ae^{-\frac{1}{2}} \tag{29}$$

and has the value

$$dn_{\max} = d\alpha \frac{n_o\Delta}{2e}, \tag{30}$$

where e is the base of the natural logarithm. As a result of this profile error, the normalized width of the impulse response becomes

$$\tfrac{1}{8}(\Delta + \tfrac{1}{2}|d\alpha|)^2 \tag{31}$$

or, in absolute terms,

$$\frac{Ln_o}{8c}\left(\Delta + \frac{e}{n_o\Delta}\,|dn_{\max}|\right)^2. \tag{32}$$

Consider a guide with a maximum index $n_o = 1.5$ and an index variation $\Delta = 2$ percent. If the profile is optimal, mode delay should produce an effective broadening of only 0.25 ns/km. An index deviation of 10^{-4} from the optimal profile increases the broadening to 0.53 ns/km.

IV. NEAR- AND FAR-FIELD POWER DISTRIBUTION

We take again into account the fact that the core cross section measures many wavelengths in diameter. If this cross section is illuminated by an incoherent source (exciting all modes uniformly), the power incident per unit solid angle at any point in the cross section is constant. To compute the power accepted by the fiber, we merely have to know the solid angle of acceptance at any point. We find this angle from the wave vector diagram of Fig. 2, which yields

$$\cos\theta(r) = \beta/k(r). \tag{33}$$

The maximum angle θ_c results for $\beta = \beta_c$; hence,

$$\cos\theta_c(r) = \frac{\beta_c}{k(r)} = \frac{n_c}{n(r)}. \tag{34}$$

Using this relation, we can define a local numerical aperture at the fiber front face

$$A(r) = n(r)\sin\theta_c(r) = [n^2(r) - n_c^2]^{\frac{1}{2}}. \tag{35}$$

The power accepted at r is then

$$p(r) = p(0)\frac{A^2(r)}{A^2(0)} = p(0)\frac{n^2(r) - n_c^2}{n^2(0) - n_c^2}. \tag{36}$$

If all modes propagate equally attenuated and without coupling, the

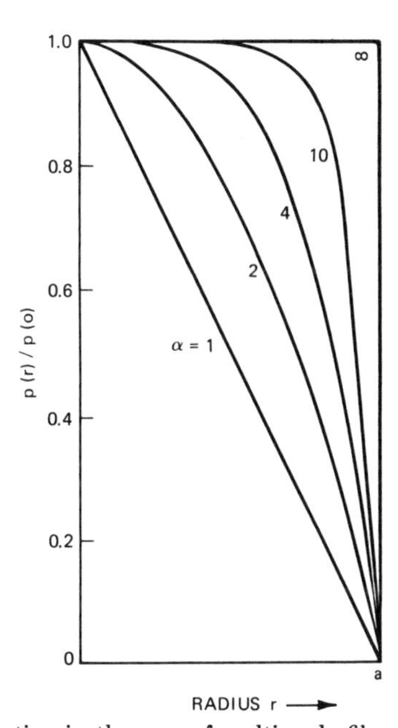

Fig. 8—Power distribution in the core of multimode fibers having the profiles of Fig. 5.

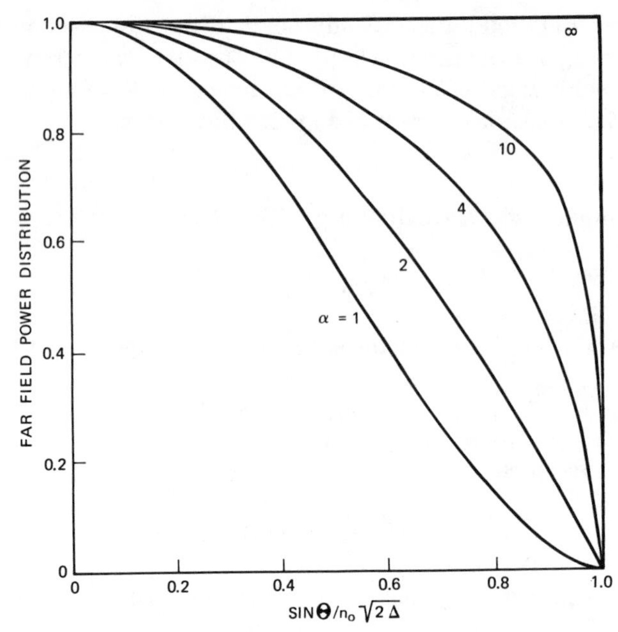

Fig. 9—Power distribution in the far field of multimode fibers having the profiles of Fig. 5.

same power distribution should hold for the fiber end face. The class of profiles described by (9) has

$$A(r) = n_o(2\Delta)^{\frac{1}{2}}[1 - (r/a)^{\alpha}]^{\frac{1}{2}} \tag{37}$$

and

$$p(r) = p(0)[1 - (r/a)^{\alpha}]. \tag{38}$$

The agreement between the profile plots (Fig. 5) and the near-field power plots (Fig. 8) is not a coincidence, but holds in general as long as the total index variation is small.

Under the conditions assumed here, every incremental area of the core cross section at the fiber end uniformly illuminates its cone of acceptance. For this reason, all those areas that have a numerical aperture

$$A(r) \geqq \sin\theta \tag{39}$$

contribute equally to the far-field power at θ. For the class of profiles described by (9), the areas contributing to θ are within a circle whose radius is obtained by solving (37) for r. Consequently,

$$P(\theta) = P(0)\left(1 - \frac{\sin^2\theta}{2n_o^2\Delta}\right)^{2/\alpha} \tag{40}$$

is the far-field power distribution. Figure 9 shows a plot of (40) for the profiles of Fig. 5. The uniform illumination for $\alpha = \infty$ changes to a parabolic distribution for $\alpha = 2$. All plots must be understood as the average power expected under the idealized conditions mentioned earlier. Monochromatic mode excitation results in mode interference phenomena and a local fine structure, which can greatly modify the average distribution considered here.

V. CONCLUSIONS

By assuming somewhat idealized conditions for mode excitation, coupling, and loss in a multimode fiber, we can isolate the influence of the index profile upon mode volume, near- and far-field power distribution, group delay, and impulse response. Surprisingly simple relations exist for a special class of profiles which comprises most multimode fibers of interest. Particular attention is given to a near-parabolic profile which accomplishes optimal delay equalization of all modes. If the (relative) index difference between center and periphery of this profile is Δ, mode delay broadens the impulse response by a fraction $\Delta^2/8$ of the total propagation time. This amounts to about 0.25 ns/km for $\Delta = 2$ percent. On the other hand, an index deviation of 10^{-4} from the optimal profile increases the broadening to 0.53 ns/km.

APPENDIX

Some Further Relations for the Group Velocity

The numerical evaluation of β as a function of μ, ν, and k_o and its subsequent differentiation to obtain τ are usually tedious and time-consuming. A substantial simplification results from a direct computation of τ by applying the operation

$$\tau = \frac{Ln_o}{c} \frac{\partial\mu/\partial k_o}{\partial\mu/\partial\beta} \tag{41}$$

to (3). The result is

$$\tau = \frac{L}{c} \frac{\int_{R_1}^{R_2} k(r)n(r)dr/u(r)}{\int_{R_1}^{R_2} \beta dr/u(r)}. \tag{42}$$

To understand the physical significance of this relation, consider a ray propagating along the fiber core of Fig. 2 in such a way that it has the direction of $k(r)$ at r. A line element along this ray is

$$ds = (dr^2 + r^2 d\phi^2 + dz^2)^{\frac{1}{2}} \tag{43}$$

and therefore

$$\frac{ds}{dr} = \frac{k(r)}{u(r)} \quad \text{and} \quad \frac{dz}{dr} = \frac{\beta}{u(r)}. \tag{44}$$

The condition $u(r) = dr = 0$ at R_1 and R_2 indicates a reflection (turn-around) of the ray. The ray performs periodic undulations between R_1 and R_2, simultaneously moving sideways in a helical fashion. By introducing (44) into (42), we obtain

$$\tau = L \frac{\oint n(r)ds/c}{\oint dz}, \tag{45}$$

where $\oint$ denotes integration over a full period of the ray. The denominator describes the axial length of one ray period, and the numerator the propagation time along the ray within this length. Multiplied by the fiber length, this ratio yields the total group delay. This result emphasizes the equivalence between ray theory and the zeroth-order WKB approach followed in this paper.

Within this order of approximation, the only quantities that depend on the wavelength are the mode numbers. Normalization of these numbers and subsequent transition to continuous variables eliminates the wavelength entirely; group velocity and impulse response are then independent of wavelength. More specifically, if we write

$$\rho = \mu/ak_o \quad \text{and} \quad \sigma = \nu/ak_o \tag{46}$$

and

$$n = n_o[1 - 2d(r)]^{\frac{1}{2}}, \tag{47}$$

eq. (3) assumes the form

$$\rho = \frac{1}{\pi a} \int_{R_1}^{R_2} [2\delta - 2d - (\sigma a/r)^2]^{\frac{1}{2}} dr, \tag{48}$$

and (42) becomes

$$\tau = \frac{Ln_o}{c} (1 - 2\delta)^{-\frac{1}{2}} \frac{\int_{R_1}^{R_2} (1 - 2d)dr/[2\delta - 2d - (\sigma a/r)^2]^{\frac{1}{2}}}{\int_{R_1}^{R_2} dr/[2\delta - 2d - (\sigma a/r)^2]^{\frac{1}{2}}}. \tag{49}$$

These two equations are sufficient to calculate group velocity and impulse response in the case of large mode numbers.

REFERENCES

1. Miller, S. E., "Light Propagation in Generalized Lens-Like Media," B.S.T.J., *44*, No. 9 (November 1965), pp. 2017–2064.
2. Kawakami, S., and Nishizawa, T., "An Optical Waveguide with the Optimum Distribution of the Refractive Index with Reference to Waveform Distortion," IEEE Trans. Microwave Theory and Tech., *MTT-16* (October 1968), pp. 814–818.
3. Uchida, M., Furukawa, M., Kitano, I., Koisumi, K., and Matsumura, H., "A Light-Focusing Fibre Guide," IEEE J. Quan. Elec. (Digest of Technical Papers), *QE-5* (June 1969), p. 331.
4. Gloge, D., Chinnock, E. L., and Koizumi, K., "Study of Pulse Distortion in Selfoc Fibers," Elec. Letters, *8*, 21 (October 19, 1972), pp. 526–627.
5. Gloge, D., "Weakly Guiding Fibers," Appl. Opt., *10*, 10, pp. 2252–2258.
6. Gloge, D., and Marcatili, E. A. J., "Impulse Response of Fibers with Ring-Shaped Parabolic Index Distribution," B.S.T.J., *52*, No. 7 (September 1973), pp. 1161–1168.
7. Matsuhara, M., "Analysis of Electromagnetic-Wave Modes in Lens-Like Media," J. Opt. Soc. Am., *63*, 2 (February 1973), pp. 135–138.
8. Morse, P. M., and Feshbach, H., *Methods of Theoretical Physics*, New York: McGraw-Hill, 1953, p. 1092.

PULSE DISPERSION FOR SINGLE-MODE OPERATION OF MULTIMODE CLADDED OPTICAL FIBRES

Indexing terms: Fibre optics, Optical waveguides

Pulse dispersions as low as 0·4 ns/km have been measured in multimode cladded fibres at a normalised frequency $V = 125$ and for a constant bend radius of 5·5 cm. Particularly when the number of launched modes is small, the pulse dispersion, as well as the polarisation and angular width of the output beam, are strong functions of the degree of mode conversion.

Introduction: An earlier report[1] of pulse dispersion in multimode, liquid-core, optical fibres showed a strong dependence on curvature, and indicated a minimum value of 1·6 ns/km at a bend radius of just under 1 m. However, the cladding capillary from which the fibres were made was rewound from the drawing machine under a constant tension, and, although the supporting drums were of polystyrene, some small distortion nevertheless occurred. An appreciable amount of mode conversion was thereby caused, the presence and dominating effect of which has been convincingly demonstrated by the present experiments, as well as by the low pulse propagation delay obtained for a narrow input beam when the input angle of incidence is varied, compared with that in an ideal fibre.[2] The former measurements have thus been repeated for a range of input conditions, which includes that for single-mode excitation[3] and bending stresses. A considerably smaller dispersion has been obtained, which is comparable with that predicted[4] and measured[5-7] in graded-refractive-index fibres. Both the polarisation and the angular width of the output beam are also strong functions of mode conversion, particularly for small input angular beamwidths, and may be used as a measure of the pulse dispersion.

Experiment: The measurements of dispersion, defined as the increase in the halfpower width of the pulse caused by propagation along the fibre, were carried out as described previously[1] with a mode-locked helium–neon laser operating at 0·633 μm. The input beam was plane polarised, owing to the Brewster windows on the laser tube, and the polarisation of the output beam was measured with an analyser,* and is defined as

$$(I_{max}-I_{min})/(I_{max}+I_{min})$$

where I_{max} and I_{min} are the maximum and minimum intensities, respectively, observed as the analyser is rotated. The output angular-intensity pattern was plotted by an automatic scanning system using a *p–i–n* photodiode to give a display on an oscilloscope. The halfwidth of the beam is defined as the angle at which the beam intensity falls to e^{-2} times that on the axis.

The fibre consisted of hexachlorobuta-1, 3-diene in a cladding[8] of internal diameter 57 μm and having a numerical aperture of 0·46. †The normalised frequency is thus $V = 125$, and roughly 7 800 modes are capable of propagating. The fibre was wound on a drum of 5·5 cm radius, initially at normal tension, but, to reduce the degree of mode conversion without changing the radius of curvature, it was gradually slackened until finally it was quite loose on the drum. Similar results have been obtained for lengths of 61, 150 and 400 m, but only those for the 61 m length are given here, as they are the most complete and the effect of mode filtering[9] due to the lossy cladding is largely avoided.

* HN22
† Chance–Pilkington ME1

Results: The four curves in each of Fig. 1*a* and *b*, which were obtained for different tensions on the drum, show, as before,[1] that, when the angular width of the input beam, and therefore the number of launched modes, is made larger, there is an increase in the output beamwidth and also, because of the greater spread in group velocities, an increase in the dispersion. However, they now indicate, in addition, that, as the fibre is progressively slackened, at constant bend radius, the dispersion and the output beamwidth both fall, thus clearly showing that distortion of the fibre due to bending stress causes mode conversion. When the fibre is only loosely coiled on the drum, and the distortion is a minimum, it can be seen from curve (iv) of Fig. 1*b* that the output width is only marginally greater than that at the input, and the amount of mode conversion occurring is quite small. The output width and the dispersion are thus determined almost entirely by the number of modes launched, and curvature as such has little effect. It is clear, therefore, that the mode conversion evident in curves (i) to (iii) is due almost entirely to distortion, and not bending, of the fibre. This has been confirmed by inducing distortion in other ways; for example, through the application of a moderate transverse pressure.

To minimise the dispersion, the number of propagating modes must be reduced, and this can be done in an unstressed fibre by an appropriate choice of the launching conditions. In fact, the input beamwidth of 0·6° in Fig. 1 corresponds nearly to the launching of only the single HE_{11} mode,[3] and it can be seen that the corresponding pulse dispersion is 0·4 ns/km. This represents a reduction by a factor of 25 compared with that observed previously[1] for the same radius of curvature, and is comparable with the values obtained in the latest Selfoc fibre.[5-7]

For such quasi-single-mode operation, the polarisation of the output beam (0·88) is high, providing further evidence that little mode conversion is occurring. If, during propagation along the fibre, power is transferred to modes of different polarisation, then of course there will be depolarisation of the output radiation. It is interesting to note that, for an input angular width of 0·6°, the polarisation of the output falls from 0·88, for the unstressed fibre, to 0·62, 0·3 and 0·01 for curves (iii), (ii) and (i), respectively. Thus distortion of the fibre not only increases the dispersion and the output beamwidth, but also causes depolarisation. The latter is easily measured and can be used to give an indication of the degree of mode conversion and dispersion. The dependence of dispersion on depolarisation and beamwidth, for a 0·6° launch, is shown in Fig. 2. A direct and simple relationship between these parameters has not previously been reported.

It is interesting to compare the results with an analysis[10] of propagation in a curved cylindrical cladded multimode fibre. The theory shows that each of the HE_{1m} modes launched by a symmetrical Gaussian beam couples only to those modes designated $HE_{2,m}$, $HE_{2,m-1}$, $TE_{0,m}$ and $TE_{0,m-1}$ and there is a periodic exchange of energy between them. In an unstressed fibre, only a small increase in the number of modes, and hence in the output beamwidth, is therefore expected to be caused by curvature, and the experimental results, particularly curve (iv) of Fig. 1*b*, are consistent with this prediction. For single-mode launching (HE_{11}), corresponding to an input angular width of 0·6°, a periodic coupling occurs only to the TE_{01} (or TM_{01}) and HE_{21} modes, so that a form of quasi-single-mode operation results in which the effects observed are a weighted average due to these, and only these, three modes. Again the low values of dispersion, depolarisation and width of the output beam for the unstressed curved fibre are consistent with the theory, which is discussed in detail elsewhere.[10]

Reprinted with permission from *Electron. Lett.*, vol. 10, pp. 148–149, May 2, 1974.

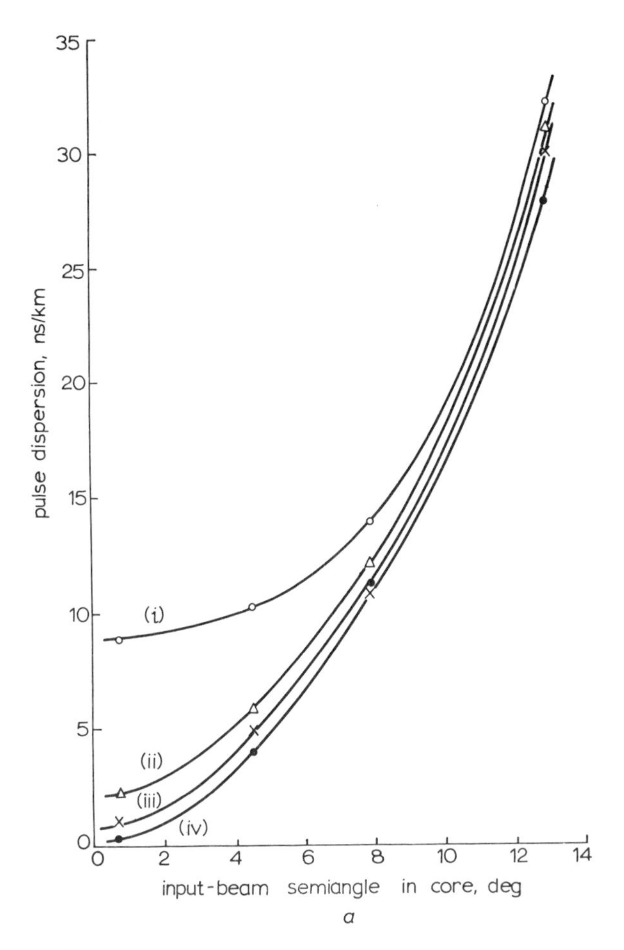

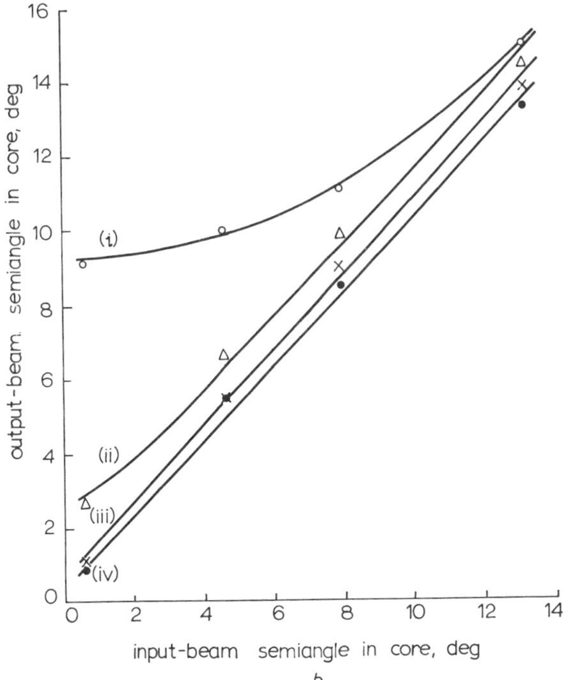

Fig. 1 *Pulse dispersion and output angular beamwidth*

a Pulse dispersion
b Beamwidth
Angular widths refer to core; values in air are approximately 50% greater. Distortion decreases from curve (i) to curve (iv)

Conclusions: In the unstressed Southampton liquid-core fibre, a negligible amount of mode conversion occurs over the lengths measured so far, i.e. up to 400 m. As a result, quasi-single-mode operation has been obtained even when the whole length is coiled, showing that bends with a radius of curvature as small as 5·5 cm do not inherently cause more than a minimal degree of mode conversion. The corresponding dispersion of 0·4 ns/km is smaller than has been reported for any other cladded fibre and is comparable with that of the latest Selfoc[7] fibre. The results are consistent with a theory[10] of propagation in a curved cylindrical fibre. It is much easier to launch a single (fundamental) mode into multimode fibre than into either a single-mode or Selfoc fibre. For example, to obtain a reasonable launching efficiency with single-mode fibre, not only must the input spot size be correctly matched, but also the transverse positioning of the beam must be accurate to a fraction of a micrometre and the tolerance on the angular alignment is small. Similar restrictions apply to Selfoc fibres if multimode operation is to be avoided. On the other hand, with multimode fibres, transverse misalignments of several tens of micrometres can be tolerated. Whether effectively single-mode operation of multimode fibres can be obtained in practical applications will depend on (*a*) the production of geometrically accurate fibres having no intrinsic, 'built-in', stresses and (*b*) the development of methods of armouring and cabling that protect the fibre from external stresses during manufacture, handling, laying etc. Suitable sources would also be necessary.

The results show, in addition, that distortion can induce mode conversion, the degree of which can be easily monitored either through the depolarisation, or the increase in angular width, of the output beam. We have observed the same effect in all the other (solid-core) fibres we have tested, including those with a small numerical aperture. When the numerical aperture (0·46) of our liquid-core fibres is not completely filled, mode conversion increases the dispersion, but has little effect on the transmission loss, because of the small amount of coupling to radiation modes. This is probably not so with fibres of low numerical aperture, where, conversely, mode conversion will affect the loss rather than the dispersion.

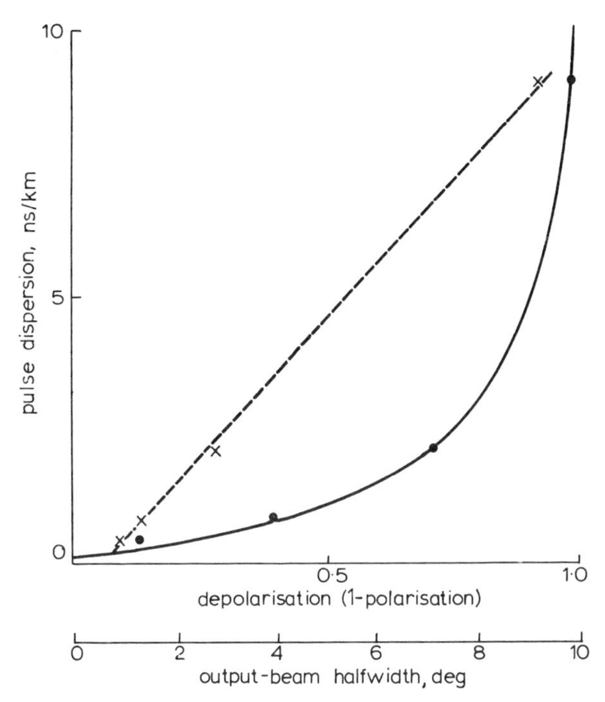

Fig. 2 *Pulse dispersion as function of depolarisation and output angular beamwidth*

Single-mode launching into curved multimode fibre
● function of depolarisation
× function of output angular beamwidth

Acknowledgments: Grateful acknowledgment is made to the UK Science Research Council for supporting the work, and for the award of a grant to one of us (H.M.), as well as to the Pirelli General Co. for a research fellowship (D.N.P.).

W. A. GAMBLING
D. N. PAYNE
H. MATSUMURA

Department of Electronics
University of Southampton
Southampton SO9 5NH, England

27th March 1974

References

1 GAMBLING, W. A., PAYNE, D. N., and MATSUMURA, H.: 'Gigahertz bandwidths in multimode, liquid-core optical fibre waveguide', *Opt. Commun.*, 1972, **6**, pp. 317–322
2 GAMBLING, W. A., PAYNE, D. N., and MATSUMURA, H.: 'Optical fibres and the Goos–Hänchen shift', *Electron. Lett.*, 1974, **10**, pp. 99–101
3 GAMBLING, W. A., PAYNE, D. N., and MATSUMURA, H.: 'Mode excitation in a multimode optical-fibre waveguide', *ibid.*, 1973, **9**, pp. 412–414
4 GAMBLING, W. A., and MATSUMURA, H.: 'Pulse dispersion in a lens-like medium', *Opto-Electron.*, 1973, **5**, pp. 429–437
5 BOUILLIE, R., and ANDREWS, J. R.: 'Measurement of broadening of pulses in glass fibre', *Electron., Lett.*, 1972, **8**, pp. 309–310
6 GLOGE, D., CHINNOCK, E. L., and KOIZUMI, K.: 'Study of pulse distortion in Selfoc fibres', *ibid.*, 1972, **8**, pp. 526–527
7 KOIZUMI, K., IKEDA, Y., KITANO, I., FURUKAWA, M., and SUMIMOTO, T.: 'New light-focussing fibres made by a continuous process', *Appl. Opt.*, 1974, **13**, pp. 255–260
8 PAYNE, D. N., and GAMBLING, W. A.: 'The preparation of multimode glass- and liquid-core optical fibres', *Opto-Electron.*, 1973, **5**, pp. 297–307
9 GAMBLING, W. A., PAYNE, D. N., and MATSUMURA, H.: 'Effect of loss on propagation in multimode fibres', *Radio & Electron. Eng.*, 1973, **43**, pp. 683–688
10 GAMBLING, W. A., PAYNE, D. N., and MATSUMURA, H.: 'Propagation in curved multimode cladded fibres'. Presented at AGARD meeting on electromagnetic wave propagation involving irregular surfaces and inhomogeneous media, The Hague, Netherlands, 1974

Spatial and Temporal Power Transfer Measurements on a Low-Loss Optical Waveguide

Donald B. Keck

Experimental measurements of the spatial and temporal transfer of power of a 225-m length of low-loss optical waveguide have been made. In particular, measurement of the angular attenuation showed substantial loss of the high order modes, which reflected itself in an ~8.2 nsec/km decrease in measured dispersion. Additionally there was a reduction of the effective numerical aperture from 0.15 to 0.12. Negligible mode coupling was observed in this particular waveguide, which allowed a phenomenological calculation of temporal output for an assumed uniform excitation of all modes. This agreed well with experimental measurements. Calculation of this output from knowledge of the index profile is presently not in agreement, and some possible reasons are indicated.

Introduction

With the attenuation problem in glass optical waveguides for communications better understood,[1] and with recent progress such that it now is possible to attain values in the vicinity of 2 dB/km in the near infrared, a greater effort is being spent worldwide on trying to characterize the waveguide for its use in communications systems. Much of this work centers on defining the dispersion or information carrying capacity of these devices, which, for certain applications and types of waveguide, could ultimately provide the transmission distance limitation.

One can separate this dispersion problem into three parts: material dispersion, intramodal dispersion, and intermodal dispersion. It has been shown[3,4] in the case of certain single-mode waveguides that material dispersion dominates intramodal dispersion in the 600–1000-nm spectral region. More generally, these two coupled with source bandwidth will define the limit of information carrying capacity. In this respect fused silica is one of the better glass choices[4] for minimizing dispersion, giving rise to pulse spreading of a few tenths of a nanosecond/kilometer for a GaAs laser input. On the other hand, intermodal dispersion within a multimode guide having a step refractive index profile, can be very much larger than this, having values of several tens of nanoseconds/kilometer for typical low-loss waveguides. There are several methods for reducing the effect of intermodal dispersion with distance. Probably the best known is by use of a nearly parabolic index profile, which tends to equalize the group velocities of the propagating modes.[5,6] An example of this type of waveguide is the SELFOC waveguide, which indeed has exhibited extremely small intermodal dispersion.[7] Another effect that tends to reduce pulse spreading with distance is intermodal coupling. It has been calculated that for random coupling between modes, the spreading should follow a square root of length rather than a linear dependence of length.[8] This effect has also been observed in present optical waveguides.[9] Lastly there are techniques for decreasing dispersion at the expense of increased attenuation, obviously making them less attractive. These include selective mode attenuation, selective mode excitation (in the absence of mode coupling), and selective mode detection.

This work measures the magnitude of some of the above factors governing dispersion for a specific multimode waveguide and correlates them with actual pulse transmission measurements. In particular, the factors included are mode coupling, preferential mode attenuation, and the effect of the refractive index profile. A desirable goal would be to predict the output pulse shape from the known input pulse shape and various waveguide parameters such as radial index profile, for example. This is not possible at the present time, and a phenomenological model was chosen; that is, the measurement of the transfer function for a given piece of waveguide can predict fairly well the waveguide output for an arbitrary source input.

Experimental Technique

The sample that was measured was a high silica, multimode waveguide approximately 220 m long, with an attenuation of 4 dB/km at 1.05 μm. Its

The author is with Corning Glass Works, Corning, New York 14830.

Received 10 January 1974.

Reprinted with permission from *Appl. Opt.*, vol. 13, pp. 1882–1888, Aug. 1974.

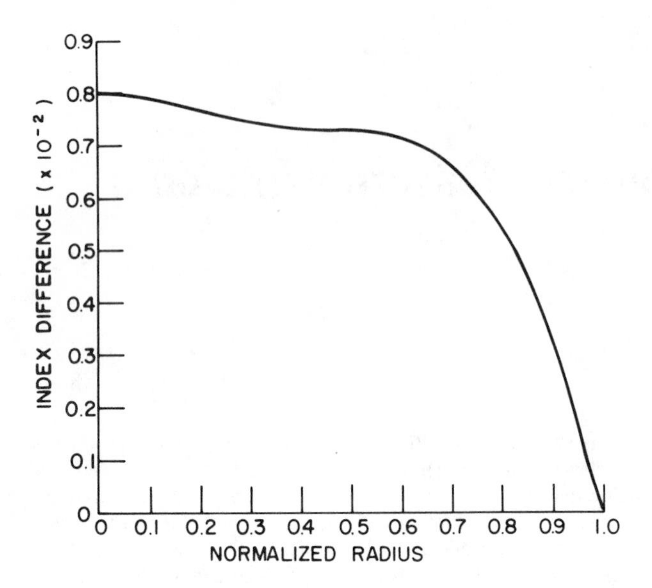

Fig. 1. Plot of radial index difference between core and cladding for experimental waveguide.

spectral response is typical of that previously reported.[1] The radial index distribution for this guide, shown in Fig. 1, was obtained from an electron microprobe scan of the entire end with a subsequent radial averaging to produce the profile shown. Although it is substantially a step profile, the departure from a step will be shown to have a pronounced effect on the dispersion. The equivalent step waveguide numerical aperture is 0.15 with a core diameter of 75 μm, giving rise to approximately 1540 propagating modes. For all measurements the fiber was wrapped on a 16.5-cm radius of curvature drum.

Measurements were made of both the spatial and temporal transfer characteristics of the waveguide utilizing very similar equipment. That used for the temporal transfer is shown schematically in Fig. 2. In both cases a plane wave input from a laser source was made incident on the waveguide as a function of angle. The intent was to have a uniform areal excitation and, through a scan of the input angle in a plane containing the fiber axis, to allow for excitation of all possible fiber modes.

In the case of the spatial measurements a He–Ne laser beam with a diameter of 800 μm and a divergence of 1.1 mrad was used. The intensity distribution over the end face of the fiber was measured to be constant to ±3%. Appropriate mode stripping was employed near the input end to remove cladding light. At the fiber output end, measurements of both the total transmitted power as well as its angular distribution were obtained for the various input angles and for two different fiber lengths. The obvious far-field scan of a pinhole to obtain the angular distribution produced unusable intensity fluctuations due to the coherent input beam and the resulting modal interference pattern. This was avoided in large part by an averaging technique. A lens was used to Fourier transform the angular into a radial distribution. An iris diaphram was then smoothly driven in the lens focal plane and the transmitted intensity measured with a large area silicon solar cell. The data was digitally punched on paper tape, and the computer obtained derivative of this data gave the desired angular distribution.

For the temporal measurement the source was an RCA SG 2001 laser diode emitting at 905 nm driven in a self-pulsing mode by discharging a 15-cm, 50-Ω coaxial line through a Hg reed relay. The technique is very similar to that described by Gloge *et al.*[10] The light was collimated with a 10× microscope objective and allowed to fall on the fiber input. Again uniform areal excitation was achieved with small angular divergence. Detection was accomplished by a TIXL 55 Si avalanche photodiode and displayed on a sampling oscilloscope having a 25-psec rise time. Figure 3(a) shows a typical pulse through zero fiber length for this system. The observed ~400-psec FWHM pulse is presently photodiode limited.

Observations

Spatial Transfer Measurements

Figure 4 shows a semilog plot of the measured relative transmitted power as a function of the input plane wave angle for fiber lengths of 228 m and 2 m.

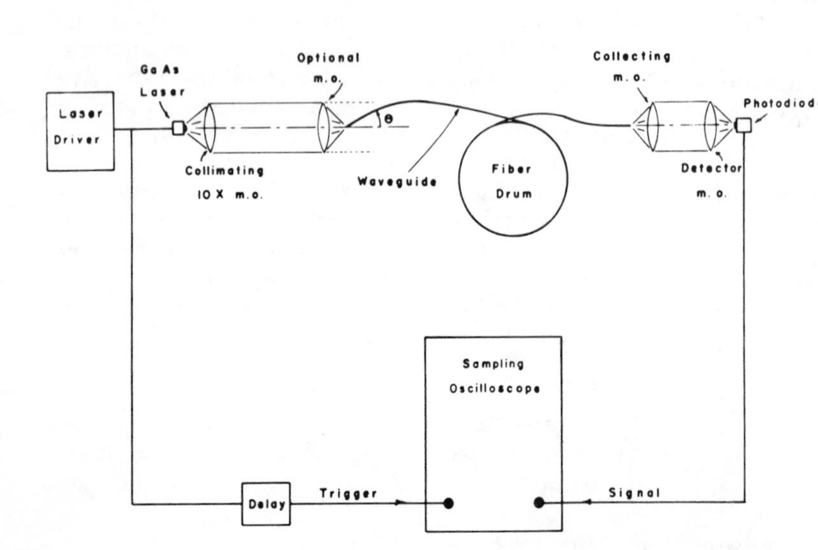

Fig. 2. Schematic of experimental apparatus.

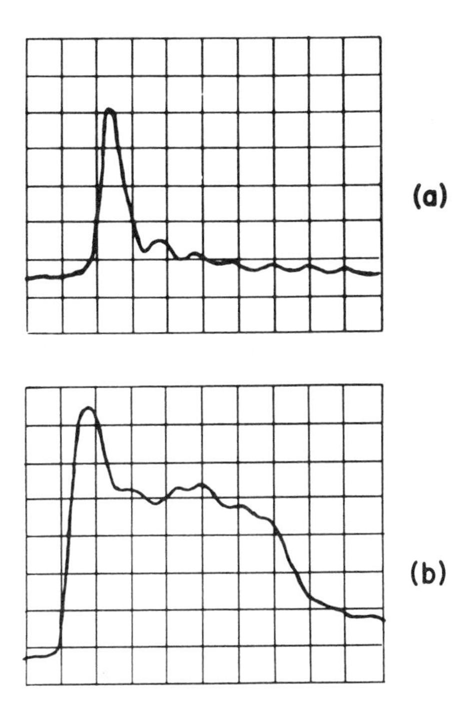

Fig. 3. Observed temporal output for the (a) direct input laser pulse and (b) the transmitted pulse through 225 m of waveguide with 20× input lens. Time scale is 500 psec/div for both measurements.

Several things can be obtained from the data. Fairly obviously, as shown in this plot, the proper manner to specify waveguide transmission is by the input modal coupling efficiency, as given by the short length transmission and the attenuation per unit length of each mode as determined from their ratio of long and short length transmissions. For this particular guide the normalized coupling efficiency, as a function of angle, is essentially unity for input angles of less than ~7°. Beyond ~8° the efficiency drops at a rate of 0.36 dB/deg. This interesting observation is not presently explained by any skew ray analysis of the step index profile and cannot be attributed to the angular attenuation.

The attenuation per unit length as a function of input angle for the guide was calculated from these data and is shown in Fig. 5. The high order modes are strongly attenuated. This plot points up a difficulty in specifying the attenuation of a waveguide merely by the ratio of two transmitted intensities at different lengths. Quite clearly, one could measure two very different attenuation values of predominantly high order modes are excited or if predominantly low order modes are excited. This condition would occur if, for example, an LED were used in one measurement and a weakly focused laser beam in the second measurement. For extremely low attenuation such as 2 dB/km, accurate specification of the attenuation will have very important ramifications regarding applications, and a proper method must be found in light of the above discussion.

Finally, these data may be used to specify an effective numerical aperture for the waveguide. The numerical aperture for this guide taken from the index data of Fig. 1 is about 0.15, giving an acceptance angle of approximately 8.5°. In view of the strong attenuation of the high angle modes, this is an overestimate. A more realistic value might be obtained, for example, from the −3 dB transmission point for the entire waveguide length. Using this definition one obtains an effective numerical aperture for this waveguide of 0.12 and a corresponding acceptance angle of about 7°.

Attention is next turned to the output angular distribution. For a plane wave excitation in a perfect step guide one would expect the output to form a cylindrically symmetric distribution. The output angle should equal the input angle, and the angular width will be governed by the input width, diffraction effects, and mode coupling both at input and during propagation. It is the latter that is of present concern. Figure 6 shows the waveguide far field normalized radiant intensity for three representative input angles, 0°, 4°, 7° and for the two lengths, ~2 m and 228 m. Although not shown in this one-dimensional plot, the output distribution is indeed azimuthally symmetric and falls at the input angle to within experimental accuracy. Mode coupling should manifest itself as a spreading of the angular distribution with length. While there may be some coupling to lower order modes as shown by comparing the 4° and 7° data for the two lengths, the amount of power coupled is a small fraction of that in the excited modes. Whatever mode distribution

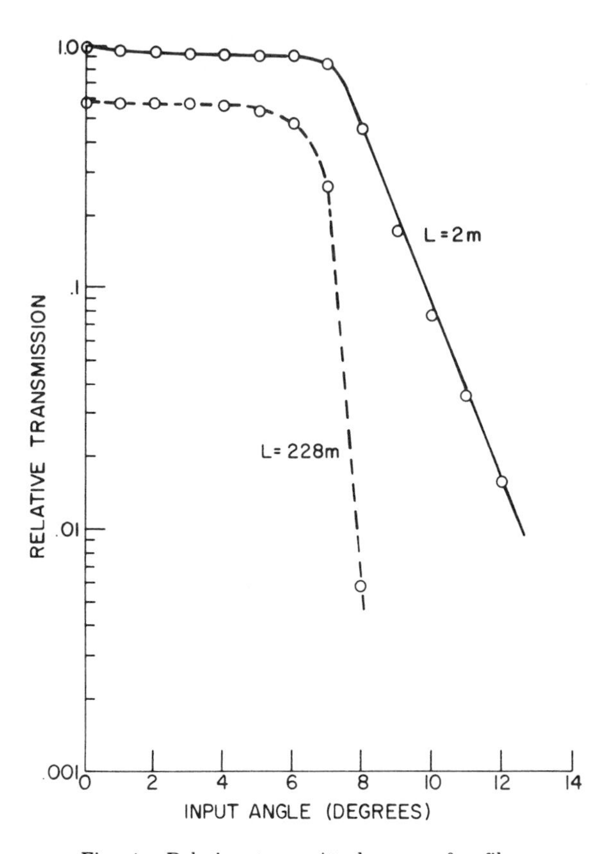

Fig. 4. Relative transmitted power for fiber lengths of 228 m (solid) and 2 m (dashed) as a function of input plane wave angle.

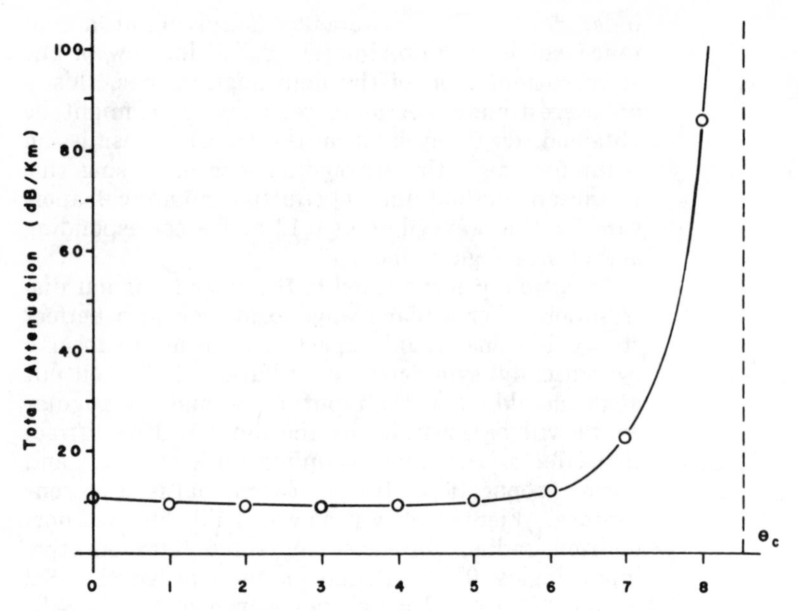

Fig. 5. Measured angular attenuation of waveguide as a function of input plane wave angle at 632.8 nm.

is excited by the plane wave input is essentially preserved over this 228-m length. One concludes from these data that coupling should have a negligible effect in determining the temporal transfer characteristics of this waveguide. This observation is consistant with that on a similar waveguide where a characteristic mode coupling length was found to be approximately 550 m.[9] The FWHM of the output angular distribution is seen to be ~70 mrad, 30 mrad, and 20 mrad for the above input angles, respectively. In all cases this is larger than either the angular width of the input beam (~1 mrad) or would be predicted by diffraction (~10 mrad). The remaining increase must be attributed to excitation of adjacent modes at the waveguide input due to the index profile of the waveguide.

Temporal Transfer Measurement

Two measurements were made on the waveguide. First, the output temporal spread was measured while attempting to excite all modes uniformly by sharply focusing the GaAs laser output onto the fiber with a 20× microscope objective. The oscilloscope trace of this is shown in Fig. 3(b). It is noted that the total pulse width (10% points) is ~4 nsec. Second, the relative arrival time and pulse shape were recorded for plane wave excitation as a function of input angle. This progression of arrivals is shown in Fig. 7. The observed pulse spreading at higher angles must include the excitation of adjacent modes mentioned earlier because the spreading is greater than that expected from input beam divergence. Nothing here would indicate any strong mode coupling, which is in agreement with the spatial measurements. Even without mode coupling, pulses approximately 0.1 nsec broader than the short length pulse are obtained in the vicinity of 0°. This should be due largely to material dispersion. A material dispersion limit of ~0.16 nsec/km is calculated for the present GaAs source, in reasonable agreement with the observation. Using this selective excitation technique, Gambling *et al.*[11] have pointed out that very high bit rates should be achievable in multimode waveguides. Additionally, the observations here show the possibility of angularly multiplexing several beams onto a single multimode waveguide.

For a straight waveguide with a step index profile, the relative arrival time between a pulse making internal angles of θ and 0° with the waveguide axis is given by

$$\Delta t = (N_1 L/c)[\sec\theta - 1], \qquad (1)$$

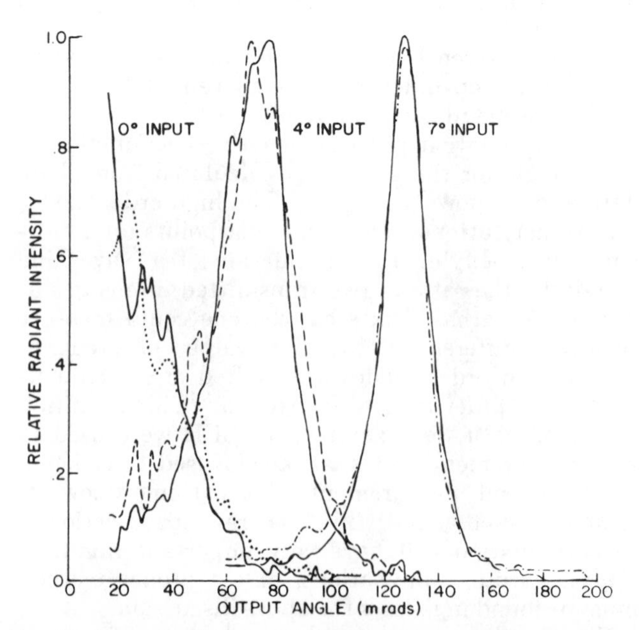

Fig. 6. Normalized transmitted radiant intensity for input angles of 0°, 4°, and 7° for waveguide lengths of 2 m (solid) and 228 m (dashed).

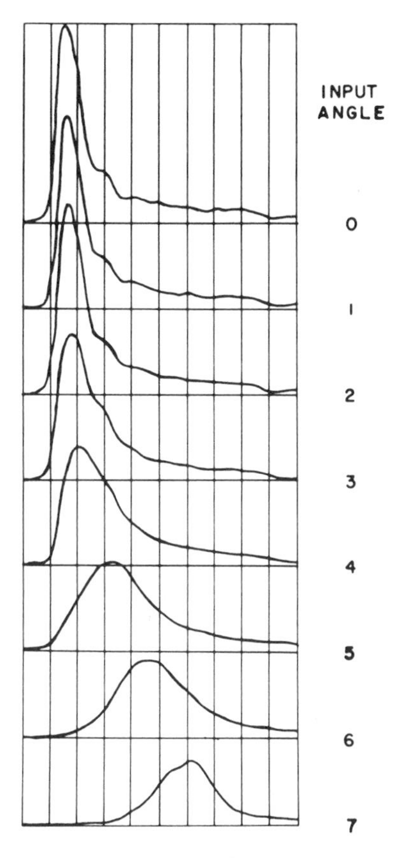

Fig. 7. Progression of pulse arrivals as a function of plane wave input angle. Time scale, 500 psec/div.

where N_1 is the core group index, L the guide length, and c the velocity of light. Figure 8 shows both the measured data points for the waveguide and calculated (dashed curve) relative propagation delay of an assumed step index profile. The measured arrivals are depressed by approximately 0.5–0.7 nsec beyond 4° from those calculated for the step waveguide. The greatest dispersion measured was only 12.7 nsec/km rather than the calculated 25.5 nsec/km. A portion of this difference is easily explained by the angular attenuation data in Fig. 5. The attenuation beyond ~7.5° precludes observation of the high order, late arriving modes, giving rise to a pulse dispersion of 17.3 nsec/km, 8.2 nsec/km less than calculated for a step index profile.

Discussion

With uniform excitation over the entire waveguide input end, the data in Fig. 8 can be used to generate a matrix of transmitted power values as a function of both output arrival time and input angle. With this matrix it should be possible to predict the output pulse for an arbitrary input excitation. This was done for the 20× microscope objective excitation assuming that all angles were equally excited. The focused spot did not encompass the entire end face, but the calculation was made assuming that it did. The convolution integral of the assumed input distribution and the measured fiber function were evaluated for each point in time, and the resulting pulse shape is shown in Fig. 9 along with the measured pulse output from Fig. 3(b) normalized to the peak intensities. The pulse shape agreement gives confidence in this phenomenological model for predicting the temporal output from an arbitrary input distribution.

One would like to perform the above calculation knowing only certain waveguide parameters such as the radial index distribution and the degree of mode coupling. For this waveguide the latter may be neglected. In a recent work, Gloge and Marcatili[12] obtain solutions to the cylindrical waveguide problem for a radially graded index profile of the form:

$$n(r) = n_0[1 - 2\Delta(r/a)^{\alpha}]^{1/2}, \quad (2)$$

where n_0 is the axial index, Δ is the fractional index

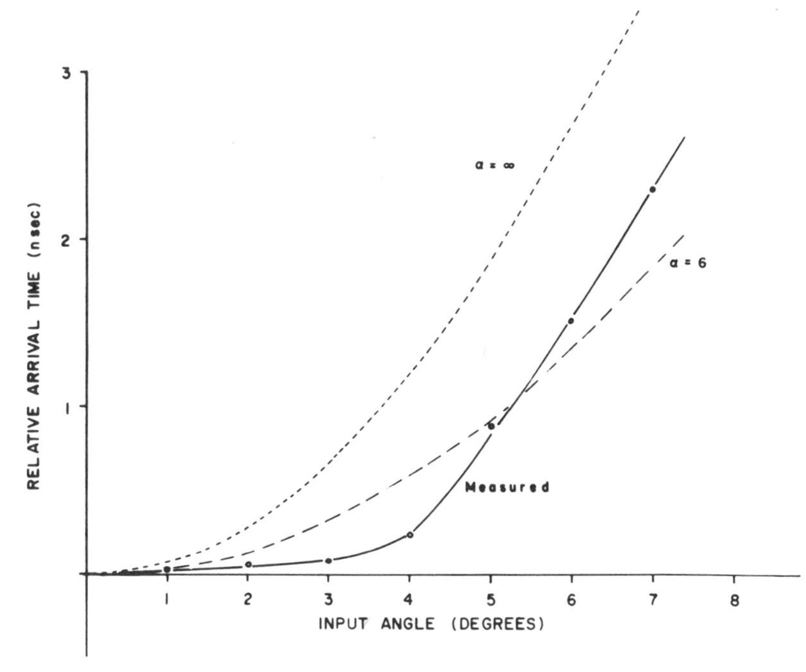

Fig. 8. Relative delay as a function of input angle. Shown here are the measured data points for this waveguide and the calculated delay for two values of the parameter α. The value $\alpha = \infty$ corresponds to a step index profile.

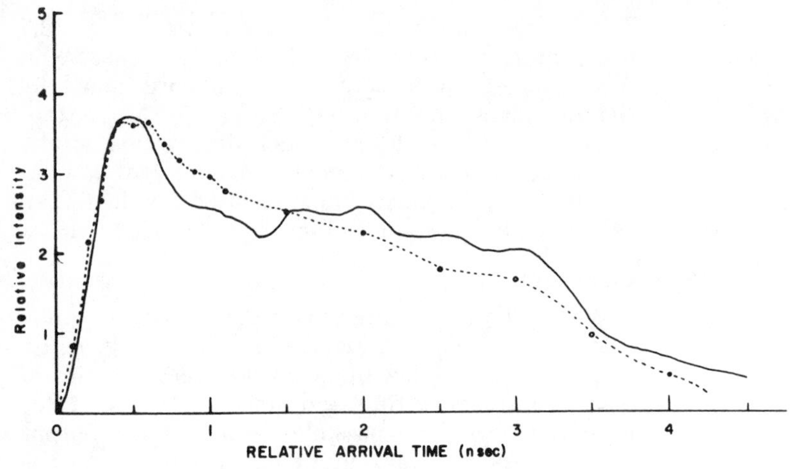

Fig. 9. Comparison of measured (solid) and calculated (dashed) temporal outputs from a 225-m optical waveguide for assumed uniform excitation of all modes.

difference between core and cladding, and a is the core radius. They show that for α not too near 2, the delay for any mode is simply

$$t \approx \frac{n_0 L}{c}\left[1 + \left(\frac{\alpha - 2}{\alpha + 2}\right)\delta + \left(\frac{3\alpha - 2}{\alpha + 2}\right)\delta^2 + \ldots\right], \quad (3)$$

where $\delta = \frac{1}{2}[1-\beta^2/k_0^2]$ and β is the axial propagation constant of the mode, with $k_0 = 2\pi n_0/\lambda$ the axial wavenumber. One sees from this that for small δ ($\delta_{max} = \Delta$) to a first approximation the delay relative to the corresponding step guide ($\alpha = \infty$) mode is reduced by the factor $(\alpha-2)/(\alpha+2)$. In the step index profile waveguide a uniform plane wave incident at a given angle does excite a single mode. In the graded case, however, a spectrum of modes is excited, each of whose relative arrival is diminished by the factor $(\alpha-2)/(\alpha+2)$ relative to that of the step index. Therefore the resulting pulse shape is difficult to predict. If the profile is not too near $\alpha = 2$, each plane wave input angle can be assumed to excite a narrow spectrum of modes in the vicinity of the corresponding step index mode. The relative arrival of this group is then simply assumed to be diminished by the $(\alpha-2)/(\alpha+2)$ factor. A best fit to the data in Fig. 1 gives approximately a value $\alpha = 6$. From the theory this would predict a decrease in the relative step index arrival time as a function of input angle of 0.5. This curve is also shown in Fig. 8. While there is crude agreement, it is not sufficiently good to allow accurate pulse shape predictions. This is probably due to a combination of the assumption that a given angle corresponds to a given mode for the graded case and to uncertainties in the measurement of the refractive index profile of the waveguide. Present techniques for measuring the index profile of a waveguide[13] suffer resolution problems in the regions of rapidly changing index.

Another complication that can enter is the fact that the waveguide was wrapped with a 16.5-cm radius of curvature for all these measurements. Gambling *et al.*[11] have directly measured the resulting arrival time depression with decreasing bend radius for a liquid core waveguide. After accounting for the obvious refractive index differences between our two waveguides, their measurements indicate that bending alone could more than account for the observed data. They also, however, observe a spread in the angular output distribution for a given curvature that is not observed in our measurements. Possible differences in method of excitation could account for their observation of a much larger effect. In the present experiments the input angle was scanned in a plane perpendicular to that of the curved waveguide. Meridional rays would not be affected by curvature in this case, and therefore a reduced effect might be expected. In short the degree to which bending is affecting the measurements is not known, and the fiber transfer function can only be approximately predicted from fundamental waveguide parameters.

Conclusion

Although a complete theoretical explanation of the transmission characteristics of multimode optical waveguides is still not at hand, a number of guide properties have been measured and correlated with their effect on the transfer functions. Attenuation of higher order modes for this low-loss waveguide decreased the specific dispersion by 8.2 nsec/km and reduced the effective numerical aperture. Both spatial and temporal measurements indicate that negligible mode coupling exists in ~228 m of this waveguide, and therefore the possibility of angular multiplexing exists. The departure of the refractive index profile of this guide from the perfect step is believed to be primarily responsible for the decrease in the observed pulse spreading from that of the step profile. Because present theory does not accurately predict the temporal pulse shape, a phenomenological method was used successfully. It should be remembered that extrapolating these measured dispersion results to very long lengths will predict a pulse spread that is always larger than that which occurs if mode coupling begins to play a role.

The author wishes to thank his colleagues at Corning, R. D. Maurer, J. D. Crow, and R. Olshansky, and members of the Bell Laboratories technical staff, E. A. J. Marcatili and D. Gloge, for many illuminating discussions. He also acknowledges the help of T. A. Cook in obtaining the experimental data. The comments of the reviewer concerning the mode-angle correspondence in a graded profile waveguide were appreciated and incorporated into the text.

This work was supported under Contract N00014-73-C-0293 from the U. S. Office of Naval Research.

References

1. D. B. Keck, R. D. Maurer, and P. C. Schultz, App. Phys. Lett. **22,** 307 (1973).
2. P. C. Schultz, Paper 30-6-73, 75th Meeting American Ceramic Society, Apr. 29–May 3, Cincinnati, Ohio.
3. F. P. Kapron and D. B. Keck, Appl. Opt. **10,** 1519 (1971).
4. D. Gloge, Appl. Opt. **10,** 2442 (1971).
5. S. E. Miller, Bell Syst. Tech. J. **44,** 2017 (1965).
6. S. Kawakami and J. Nishizawa, IEEE Trans. Microwave Theory Tech. **MTT-16,** 814 (1968).
7. D. Gloge *et al.*, Electron. Lett. **8,** 526 (1972).
8. S. D. Personick, Bell Syst. Tech. J. **50,** 843 (1971).
9. E. L. Chinnock *et al.*, Proc. IEEE **61,** 1499 (1973).
10. D. Gloge, E. L. Chinnock, and T. P. Lee, IEEE J. Quantum Electron. **QE-8,** 844 (1972).
11. W. A. Gambling *et al.* Electron. Lett. **8,** 568 (1972).
12. D. Gloge and E. A. J. Marcatili, Bell Syst. Tech. J. **52,** 1563 (1973).
13. C. A. Burrus *et al.* Proc IEEE **61,** 1498 (1973).

The Length Dependence of Pulse Spreading in the CGW-Bell-10 Optical Fiber

E. L. CHINNOCK, L. G. COHEN, W. S. HOLDEN, R. D. STANDLEY, AND D. B. KECK

***Abstract*—Previous measurements of pulse broadening in the CGW-Bell-10 optical fiber showed very low dispersion (<2 ns/km). We recently measured pulse spreading as a function of length in this fiber. The data indicate a (length)$^{1/2}$ dependence, at wavelengths of 0.6328, 0.9, and 1.06 μm, for long fiber lengths (>550 m) which are of practical significance.**

I. Introduction

A companion letter [1] describes the optical characteristics of the CGW-Bell-10 optical fiber and gives the pulse spreading observed in this 1-km-long fiber. In this letter we describe additional measurements which establish the length dependence of the pulsewidth at excitation wavelengths of 0.6328, 0.9, and 1.06 μm. We observe pulsewidths which widen with the square root of length at all three wavelengths for long fiber lengths (>550 m) which are of practical importance.

In what follows we briefly describe the experimental apparatus and outline the significant results.

II. Experimental Apparatus

The overall experiment was arranged as shown in Fig. 1. A Nd:YAG laser, intracavity mode-locked by a $LiNbO_3$ crystal driven at 300 MHz, was the 1.06-μm source. For the 0.6328-μm measurement only the source and detectors were changed; in this case, a He–Ne laser, intracavity mode-locked by a KDP crystal driven at 100 MHz, was used. The 0.9-μm source was a GaAs laser diode driven at a 100-Hz repetition rate. Low capacitance small-area silicon diode detectors were employed at 0.6328 and 0.9 μm. A germanium diode detector was used to record 1.06-μm pulses.

The Nd:YAG and GaAs sources provided detector-limited 3-dB pulsewidths of less than 200 ps, while the He–Ne output pulse was about 600 ps wide.

The temporal spread of the input pulse was recorded on a sampling oscilloscope using each of the sources. The angular distribution at the output end of the fiber was also measured by electrically scanning its far field through a small pinhole. Next, the fiber length was reduced by approximately 100 m by carefully scribing the cladding with a tungsten carbide tool and breaking the fiber in tension. We determined the remaining length of fiber by measuring the relative time delay between input and output pulses of 0.9-μm light. This procedure was followed until the length was reduced to 90 m.

III. Results

Fig. 2 shows the half-power output pulsewidth plotted versus fiber length with excitation wavelength as a parameter. The 1.06-μm data point for 2 km of propagation was obtained by series-connecting two 1-km pieces of similar fiber (CGW-Bell-10 and I-350 [1]). An almost identical result was obtained by recording pulses returning from one round trip of propagation along a single fiber immersed in a mercury end reflector.

Results at all three wavelengths indicate a square-root of length pulsewidth dependence for fibers longer than 550 m. This qualitative behavior implies that mode coupling is very significant in this particular fiber [2]. The effects of mode mixing and the graded shape of the core refractive index profile probably both account for the small absolute pulse spread. However, we are not certain which is more important.

Pulsewidths at 1.06 μm remained dependent on (length)$^{1/2}$ for fiber lengths as short as 90 m. On the other hand, pulsewidths at 0.9 μm changed from a linear to a square-root dependence after 550 m, implying that mode mixing was incomplete until that length was reached. The output angular distribution measurements were consistent with these wavelength-dependent discrepancies. At 1.06 μm, this distribution was not significantly influenced by the numerical aperture of the injection lens. However, the output distribution measurement on a 90-m fiber length was smaller by 25 percent when 0.9-μm light was injected directly into the fiber without a lens.

Fig. 1. Experimental arrangement for pulse transmission measurements at 1.06 μm. Only the source and detectors were changed for measurements at 0.6328 and 0.9 μm.

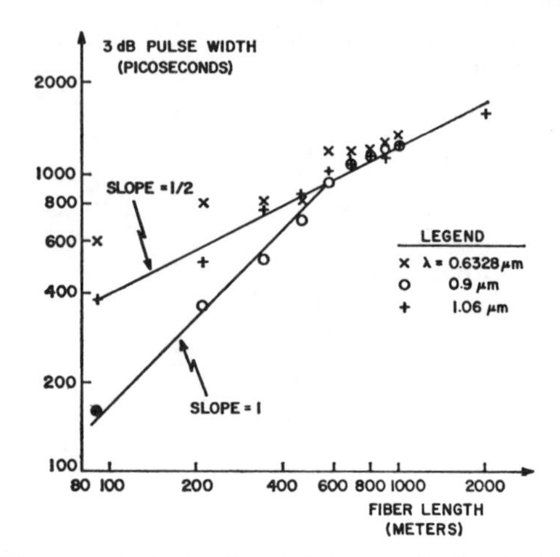

Fig. 2. Half-power output pulsewidth plotted versus fiber length with excitation wavelength λ as the parameter. Input pulses at 0.6328 μm were about 600 ps wide. Data point ● was obtained by deconvolving the detector-limited 0.9-μm source pulsewidth (120 ps) from the measured fiber output pulsewidth (200 ps). Pulsewidths at 1.06 μm were detector-limited at 200 ps.

The cause of the linear length dependence of pulsewidth on short lengths at 0.9 μm is not thoroughly understood. Nevertheless, we can suggest one difference between the latter and the 1.06-μm experiments which may be relevant. The GaAs laser source produces an elliptical angular distribution with an aspect ratio greater than 10 to 1. Excitation of the fiber from such a source should yield an initial modal power distribution significantly different from that obtained with the circularly symmetric Nd:YAG beam. This could have an important effect on output pulsewidth. Additional experiments to further our understanding of these observations have begun.

IV. Conclusions

The length dependence of pulse spreading in the CGW-Bell-10 multimode optical fiber has been determined. We have found that for fiber lengths of practical significance, the pulse spread is proportional to the square root of length in the wavelength range of 0.6328–1.06 μm. This implies that mode coupling effects are very significant. Wavelength-dependent discrepancies at short fiber lengths may have occurred because the various sources excited different distributions of fiber modes. It is also possible that the discrepancies are consistent with the spectral distribution of imperfections which cause mode coupling.

It should be emphasized that these results pertain only to this particular fiber and should not be extended to include all multimode waveguides when designing optical communication systems. Mode mixing effects and the unique core refractive index profile of this particular fiber together determined the fiber dispersion. We are not certain how much each contributed to the absolute values of pulse spreading.

References

[1] C. A. Burrus *et al.*, "Pulse dispersion and refractive-index profiles of some low-noise multimode optical fibers," this issue, pp. 1498–1499.

[2] S. D. Personick, "Time dispersion in dielectric waveguides," *Bell Syst. Tech. J.*, vol. 50, pp. 843–859, Mar. 1971.

Manuscript received May 23, 1973.

E. L. Chinnock, L. G. Cohen, W. S. Holden, and R. D. Standley are with Crawford Hill Laboratory, Bell Laboratories, Holmdel, N. J. 07733.

D. B. Keck is with Research and Development Laboratories, Corning Glass Works, Corning, N. Y. 14830.

Reprinted from *Proc. IEEE*, vol. 61, pp. 1499–1500, Oct. 1973.

STUDY OF PULSE DISTORTION IN SELFOC FIBRES

Indexing terms: Fibre optics, Distortion

A 100 ps gallium-arsenide laser pulse was observed to increase in width by only 100 ps in 70 m of a fibre with an internal refractive-index gradient (Selfoc). This is shown to be in satisfactory agreement with a simple ray analysis which neglects the index discontinuity at the fibre wall.

Incoherent (or multimode) sources, to be used in optical communication systems, require multimode waveguides for efficient transmission. Typically, guides of this kind cause a distortion of the signal because of the velocity differences among the modes. The distortion can theoretically be minimised in a dielectric whose refractive index decreases as the square of the guide radius. The Selfoc fibre best approximates these conditions.[1] Its index profile closely follows the parabolic law throughout much of the inner fibre region. Modes that propagate essentially in this region are expected to show only very small differences among their group velocities.[2] The most serious deviation from the parabolic profile occurs at the edge of the fibre where the index merges into the value of the surrounding medium. The nature of this transition (a steep drop, for example, or a flat continuation) determines the number of 'outer' guided modes whose group velocities differ appreciably from one another and those of the inner modes. How much these outer modes take part in the signal transport is difficult to evaluate.

On the other hand, if properly exploited, the potential equalisation effect inherent in Selfoc fibres may eventually permit the design of much faster optical communication systems than is achievable with homogeneous-core multimode fibres. The latter have shown considerable signal distortion in recent measurements.[3] Since other investigations[5] have demonstrated the equivalence of this and the Selfoc fibre with respect to mode volume and bending losses, Selfoc fibres (or some suitable form of them) could become an important multimode waveguide.

We have tried an experimental evaluation by observing the propagation of very short pulses in Selfoc fibres.[5] Fig. 1 illustrated the setup which was essentially the same as that used in previous similar measurements.[3] A heterojunction gallium-arsenide (GaAs) laser was driven by short 7 A current pulses to excite only the first of a train of self-pulsing spikes. This resulted in a short pulse at 900 nm, which had a peak power of 50 mW and a width of about 100 ps. The repetition rate was 100 Hz. A microscope objective collimated the divergent laser output to form a beam roughly 1 mm in diameter. The beam was split and transmitted, either directly or via the fibre, to two silicon *p–i–n* diodes of the same kind. These were low-capacitance diodes with a particularly small detector area ($50 \times 50\ \mu m$), capable of resolving a 100 ps pulse. The diodes were directly followed by an oscilloscope sampling head with a nominal response time of 25 ps.

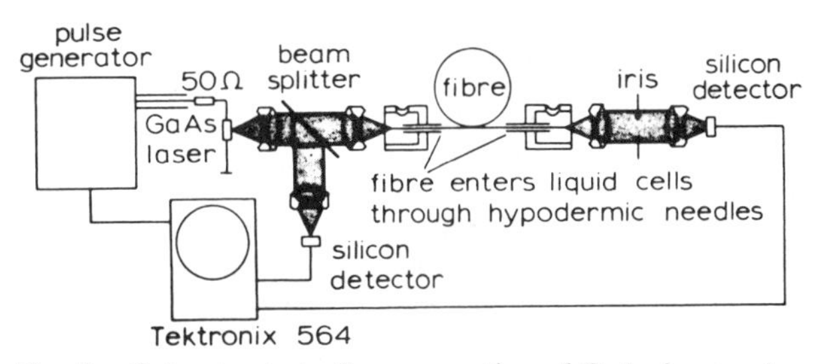

Fig. 1 *Setup to study the propagation of GaAs laser pulses in Selfoc fibres*

Several fibre samples of different lengths, but with essentially the same characteristics, were available for the measurement. The diameter, which was nominally $d = 0{\cdot}1$ mm, varied by roughly 10% along the length. The refractive index, which had a value of $n_0 = 1{\cdot}562$ on axis, decreased to $n_p = 1{\cdot}540$ near the periphery. This index difference corresponded to a numerical aperture of $NA = (n_0^2 - n_p^2)^{\frac{1}{2}} = 0{\cdot}26$ for modes guided by the internal index variation (not by wall reflections). There were $(\pi d NA/2\lambda)^2 = 2000$ of those modes[5] at $\lambda = 900$ nm. The loss at 900 nm was between 60 and 80 dB/km, depending on the fibre sample.

Reprinted with permission from *Electron. Lett.*, vol. 8, pp. 526–527, Oct. 19, 1972.

The fibre ends were prepared by repeated breaking until satisfactory flatness was achieved across the entire cross-section. The end was then inserted (through a hypodermic needle) into a cell filled with index-matching liquid, as shown in Fig. 1. A glass window at one side of the cell admitted the GaAs laser light. Focused on the fibre front face by a microscope objective, the light covered an input cone of light with a half-angle of 0·06 rad. At the fibre end, a similar arrangement was used to collimate the output. An additional microscope objective was then used to focus the output beam on the small junction region of the silicon detector.

Fig. 2 shows sampling-oscilloscope traces of the pulses measured. One horizontal division corresponds to 200 ps. Amplification to equal peak values was applied in all cases except *d*, where the output was not sufficient to permit it. Fig. 2*a* shows a typical GaAs laser pulse before, and Fig. 2*b* after, transmission through 24 m of fibre. A small increase in pulsewidth is noticeable. This increase depended slightly on the injection arrangement. A misalignment of the input beam, for example, caused the output pulsewidth to increase to close to 200 ps. A smaller, hardly noticeable, increase resulted when the input beam was focused more strongly to fill essentially the entire numerical aperture of the fibre.

Possibly to augment these effects in a longer transmission path, another 20 m piece was joined to the first fibre resulting in a total path length of 44 m. The fibre ends at the joints were prepared as previously described. Held by a micropositioner, one junction end was then aligned with the other. A drop of matching liquid was added to minimise the reflection losses. This coupling technique resulted in a loss of typically 1 dB per joint. Fig. 2*c* shows the output pulse at the end of the 44 m path. The pulse in Fig. 2*d* was obtained after adding another 26 m piece, which brought the total length to 70 m. The sequence of Fig. 2 reveals an increase in pulsewidth from 100 to 200 ps. Although insufficient to determine the delay distortion as a function of length, these results establish an upper limit for the pulse broadening caused by this Selfoc fibre.

For lack of a satisfactory theoretical model, an analysis of the results must be speculative. We first consider only those modes which are essentially guided by the index slope and not by the fibre wall. As was demonstrated by a simple ray-optics argument,[2, 6] these 'inner' modes exhibit group-delay differences of the order $T\Delta^2$ during their propagation time T. Here $\Delta = 1 - n_p/n_0$ is the relative difference between the axial and the peripheral index. For a fibre length $L = 70$ m, the propagation time is $T \sim n_0 L/c = 350$ ns and, since $\Delta = 0{\cdot}0143$, we have $T\Delta^2 = 70$ ps. In addition to the multimode effects, we expect the frequency dependence of the

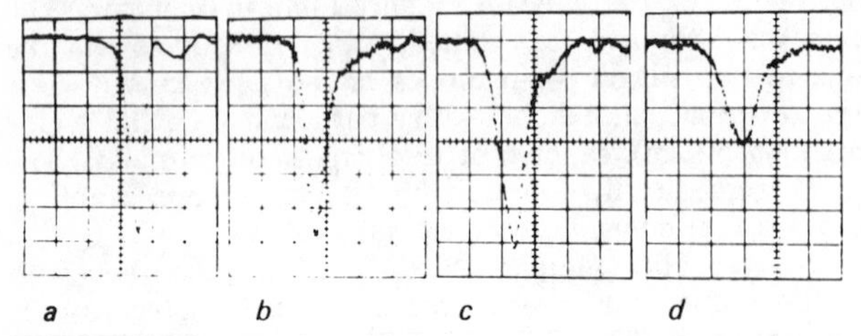

Fig. 2 *Sampling oscilloscope traces of the pulses*

a Measured at the fibre input
b Measured after 24 m
c Measured after 44 m
d Measured after 70 m of Selfoc fibre
Horizontal scale: 200 ps division
Amplitudes were adjusted for equal peak height in all cases except *d*, where the output signal was insufficient

refractive index to cause some pulse distortion. According to Reference 7, the additional increase in width is roughly $0{\cdot}04\,(T/n_0)\,(\Delta\lambda/\lambda)$, where $\Delta\lambda$ indicates the wavelength region covered by the pulse spectrum. The factor 0·04 comprises the (average) index properties of glass at $\lambda = 900$ nm.[7] For the GaAs laser, $\Delta\lambda \sim 1$ nm; the additional increase in width is consequently 10 ps in 70 m. The composite broadening expected from these two effects is indeed close to what was measured. Extrapolating these results linearly with fibre length, as stipulated by this model, leads to a 1 ns output pulse width after 1 km.

It remains to be explained why we detect none of those modes which are guided by the peripheral index step (between n_p and the value n_s of the surrounding medium). These modes exhibit group-delay differences of the order of $T(n_p - n_s)$, which should be significant (of the order of nanoseconds) even in relatively short fibres. It can be argued, of course, that our standard injection method (0·06 rad halfangle) avoided exciting these 'outer' modes. On the other hand, a misalignment or stronger focusing at the input did not show any significant evidence of them. It is possible that the energy in these modes was spread out in a long pulse tail, which disappeared in the noise background of our measurement. If this was the case, the total energy content of the tails was certainly small, for the measured c.w. power was in good agreement with the measured pulse power when averaged over the 10 ms repetition time slot. Very probably, most of these modes are quickly attenuated by imperfections and contamination of the fibre surface (and of the two junctions).

In fact, a selective attenuation may not only affect what we here call the 'outer' modes, but some of the 'inner' modes as well. This, and mixing among the modes, may require a substantial modification of the simple model suggested earlier. In particular, since these effects become most apparent in long fibres, the linear extrapolation of these results with length, tentatively suggested in the previous paragraph, should be considered with caution.

D. GLOGE
E. L. CHINNOCK

Bell Telephone Laboratories
Crawford Hill Laboratory
Holmdel, NJ 07733, USA

18th September 1972

K. KOIZUMI

Nippon Sheet Glass Co.
Itami, Japan

References

1 UCHIDA, T., FURUKAWA, M., KITANO, I., KOIZUMI, K., and MATSUMURA, H.: 'A light-focusing fiber guide'. IEEE conference on laser engineering and applications, Washington, DC, 1969
2 KAWAKAMI, S., and NISHIZAWA, T.: 'An optical waveguide with the optimum distribution of the refractive index with reference to waveform distortion', *IEEE Trans.*, 1969, **MTT-16**, pp. 814–818
3 GLOGE, D., CHINNOCK, E. L., and LEE, T. P.: 'Self-pulsing GaAs laser for fiber dispersion measurements', *IEEE J. Quantum Electron.*, 1972 (to be published)
4 GLOGE, D.: 'Bending loss in multimode fibers with graded and ungraded core index', *Appl. Opt.*, 1972 (to be published)
5 BOUILLIE, R., AND ANDREWS, J. R.: 'Measurement of broadening of pulses in glass fibres', *Electron. Lett.*, 1972, **8**, pp. 309–310
6 GLOGE, D.: 'Optical waveguide transmission', *Proc. Inst. Elec. Electron. Eng.*, 1970, **58**, pp. 1513–1522
7 GLOGE, D.: 'Dispersion in weakly guiding fibers', *Appl. Opt.*, 1971, **10**, pp. 2442–2445

Part VI
Fiber Strength and Stability

Fiber Optic Cable Test Evaluation

ROBERT L. LEBDUSKA

Naval Electronics Laboratory Center, 271 Catalina Boulevard, San Diego, California 92152

(Received May 22, 1973)
(Revised August 6, 1973)

ABSTRACT

The test results presented were obtained over a 9 month period at the Naval Electronics Laboratory Center at San Diego. The test objective was to assess the structural and other physical property characteristics of incoherent, plastic jacketed, glass fiber-optic cable types of current manufacture. Tests were conducted to evaluate the capabilities of these type products to sustain structural integrity and operability after subjection to test exposures simulating Naval shipboard environmental conditions. A total of 29 types of tests were performed utilizing over 200 fiber-optic cables. A major part of this task involved the development of diagnostic test techniques for the evaluation of the test induced cable property modifications.

A final report detailing the fiber-optic cable performance obtained in this test program is available by writing to the Library, Naval Electronics Laboratory Center, San Diego, California 92152. The report is entitled "Fiber-Optic Cable Test Evaluation" and identified as NELC TR-1869.

INTRODUCTION

This test program was instituted to identify the current performance status and/or deficiencies in fiber optic cable products when such items were subjected to test environments and other physical conditions simulating applications of Military/Naval operation. In lieu of more appropriate guidelines, it was expected that the use of fiber optic cabling hardware would be to replace and interface with operational electrical cable systems. Since the cabling environments will be common, military wire cable specifications were used as a guide in the assignment of environmental test profiles. Primary test environments were selected to evaluate the mechanical strength, thermal tolerance, and chemical indifference of the selected fiber optic cables. Tests of cable mechanical strength properties included bending radius, tensile, terminal, twist, mandrel, cyclic flexibility, vibration, mechanical shock, and jacket abrasion. Environmental tests performed consisted of thermal cycle and extended high and low temperature regimes, humidity cycle and steady state profiles and salt fog exposures. Chemical bath immersion tests included salt, mineral, oil, sodium hydroxide and sulphuric acid as test reagents. Results obtained from these tests demonstrated the ability of these type products to withstand a wide range of physical exposure.

TEST METHODS

All physical property and environmental tests were conducted within the physical plant of NELC's Environmental Test Facility. Fiber optic cable degradations may exhibit themselves as physical property modifications and/or optical transmission losses. Also, light transmission changes may be related to various light coupling mechanisms as well as fiber breakage levels. To better evaluate the nature and extent of these property changes, diagnostic tests were devised to discriminately measure the physical and optical property changes induced by the test conditions. It was considered worthwhile to differentiate between discrete fiber breakage levels and cable transmission losses due to other factors. Also determination of discrete fiber break location along the cable length or within the terminals can shed light on the degrading mechanism. Accordingly, diagnostic methods were devised to measure quantitatively the pre-test and post-test cable light transmission and fiber breakage levels.

Reprinted with permission from *Optical Eng.*, vol. 13, pp. 49–55, Jan./Feb. 1974.

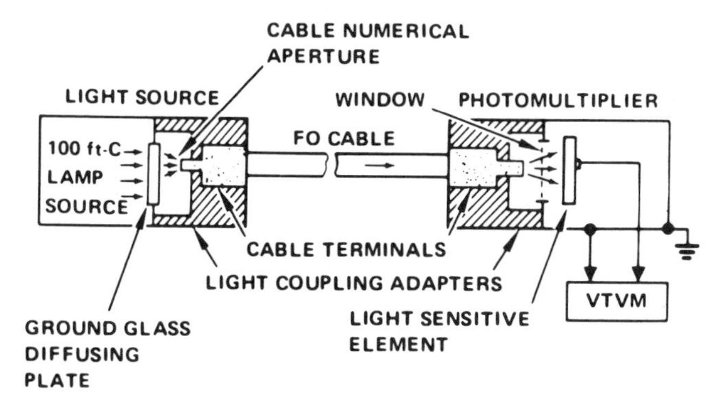

Figure 1. Photometric Measurement Technique.

Photometric Measurement

The Photometric technique employed is shown in Figure 1. The diffusing screen provided uniform input light flux to the cable terminal fiber matrix within the average numerical aperture of the fibers. All fibers will now contribute to the integrated total light transmitted by the fiber bundle and measured by the photomultiplier. Repeatability of data acquisition is provided by the fixed geometry of the light coupling adapters, the input power regulation of the active units, by permitting time for thermal stabilization prior to data collection and through consistency in the mechanics of cable insertion into the adapters. The accuracy of the data taken by this method was calculated to be better than ±5 percent. Total transmission measurements using white light primarily represent a means for corroborating individual fiber breakage levels determined by other methods and attaching quantitative significance to other mechanisms which degrade optical performance.

Fiber Break Detection

Assessment of discrete fiber breakage level within a test cable at any point of time during the test exposure was obtained by microscopic examination of the cable terminal faces. The technique required simultaneous illuminating sources for both cable terminating ends as shown in Figure 2.

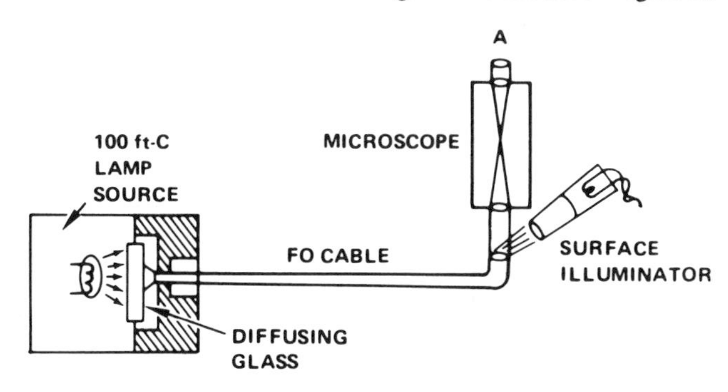

Figure 2. Fiber Break Detection Technique.

A diffusing input light source was provided to assure that all axial fibers will receive light flux within their respective numerical aperture acceptance cones. The surface illuminator provided the means for illumination of all fiber elements within the bundle field. These fibers will appear as dark (black) elements when input source illumination is absent. With both sources operating the broken fibers appear as black circular elements within the illuminated fiber element field. Breakage level measurements taken by reversing the input and output cable ends substantiate the accuracy of the method. This method was believed to be quite accurate and constituted the diagnostic "backbone" for the cable evaluations made under the specific test conditions imposed.

Photometric/Fiber Breakage Calibration

To establish a linearity characteristic between the photometric data and the visual break count information, a fiber optic cable was modified by removing a short, central section of the jacket thereby exposing the fibers. Sequential severing of a measured and increasing number of fibers was done with fiber break counts and transmission measurement made before and after each cutting operation. Good data correlation was obtained between the two measurement methods.

Fiber Break Location

A number of techniques were devised for determination of individual and bulk fiber break locations. It was evident that definition of this parameter would assist in the understanding and evaluation of the optical and physical degrading factors which influence the cables' performances. For some cables, jacket transparency permitted observation of individual fiber breaks. By shining and scanning a high intensity light spot along the cable length, the location of the individual breaks were seen in the microscope to illuminate (from black to bright white) as the break location was reached. Photo-integration methods were used to measure the number of fiber breaks along a given length of transparent or translucent cable. If a beam of light is scanned along the length of cable of interest, those breaks within this length will be observed as brightly lit fiber elements. Using the time exposure property of the camera it is possible to document the integrated summation of those breaks along this path length.

Other Test Methods

Some few continuous fibers along the cable length will provide varying levels of gray coloration at the detector end making true break assessment difficult. Also, some broken fiber cores will partially illuminate their separated end faces and appear gray in the bundle field. A positive identification for fiber breakage involves breakage or interrupted continuity of the clad layer. When breakage occurs, no clad rings are observed around those fibers which are candidates for the broken fiber list by virtue of their gray appearance. Transmission of steep angle incidence light into the source end of a fiber bundle (observed at lengths up to 15 feet) permits observation of the clad rings on all continuous fibers at the detector terminal end. This effect is seen in Figure 3.

Terminal end/surface finish conditions were assessed by visual observation of the surfaces under moderate magnification with very low incident angle light upon the surface. Small irregularities in surface topography are readily seen in relief under these conditions, but the evaluations made are only qualitative in nature.

TEST CABLE SELECTION

Test cables were obtained from three manufacturers. The cable types selected were physically as similar as possible and consistent with normal "production-run" manufacturing technology. No extensive comparison between different manufactured products was made, only those parameters of the cables tested which were considered pertinent to an understanding of the results obtained are noted. All cables used for these tests were of an incoherent light transmitting configuration. Materials employed consisted of multiple light transmitting glass fiber cores, coated with glass clad layers, enclosed in a PVC (Polyvinylchloride) plastic jacket with the fiber bundle suitable lubricated within the jacket. Lubricants

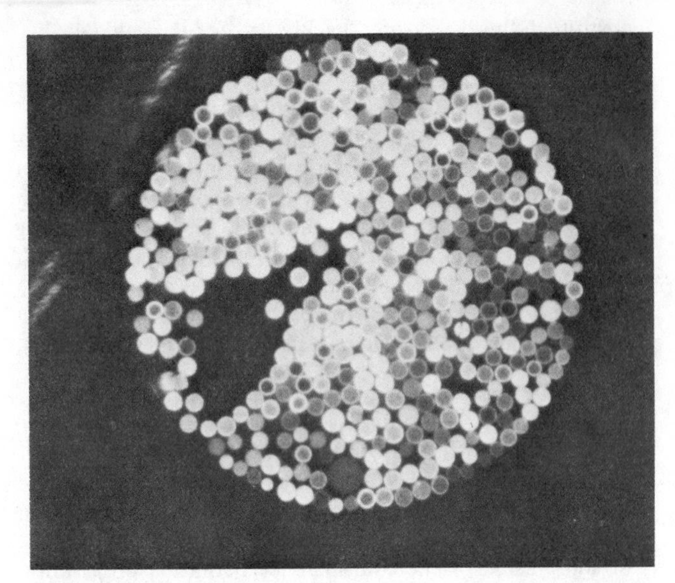

Figure 3. Fiber Clad Appearance.

used by the manufacturers of these cables generally consisted of liquid silicone materials of a proprietary nature. The terminations consisted of metallic fittings with the bundle end faces ground and polished. Most of the cables tested were of the Corning Glass Works types 5010 and 5011. Primary property relationships of the test cables are shown in Figure 4.

TEST PERFORMANCE

Pre-Test Cable Characterization

Those test cables assigned to the first few tests were subjected to an inspection procedure and "handling routine." This qualification method consisted of visual inspection of cable external parts for mechanical integrity; fiber breakage level assessment; terminal face optical quality; and minimal, manually conducted, physical testing of the cables to assure test usability. It became apparent, after a number of cable inspections, that this entire procedure was not required to assure cable qualification. Thereafter, only terminal finish condition and breakage level measurement properties were assessed. Acceptable cables (those with less than 8 percent fiber breakage and reasonably good terminal finish conditions) were tagged (numbered) at the cable end corresponding to the detector terminal. All subsequent photometric measurements and fiber break counts could henceforth be related to a given terminal to improve data accuracy. Photos of before-test fiber breakage level and photometric measurements of cable transmission were taken of all qualified test cables. This library of photo-documentation of the terminal surface(s) throughout the test performance period greatly assisted in the cable evaluation process.

Mechanical Strength Tests

Mechanical strength tests were selected as those which primarily effect the integrity of the cable glass fiber core or bundle as opposed to those tests which may degrade the cable jacketing and/or protective exterior. The military requirement for a high level of flexural strength and flexibility in data transmission type cables is dictated on a per-application basis. Normal shipboard installations will provide inflexible cable runways with bulkhead feed-through coupling over a major portion of the cable run. Therefore, the tests of this section were intended only to relate the durability of the flexural properties of selected F.O. cables to a variety of flexural stimulae. The primary tests of interest involve cable modifications due to bending radius, tensile, cyclic flexibility, mandrel strength and twist. Although a cold temperature (-40°C) was not performed in bend as specified, it is evident that the flexural properties of the test cables will be largely influenced by the plastic jacket cold temperature characteristics.

Bending Radius Test

These tests, performed at ambient temperature conditions, consisted of applying five close-wound cable turns sequentially about a stepped mandrel progressing from the largest diameter step to the smallest. Fiber breakage and photometric data was taken before and after each mandrel step. Figure 5 shows the cable loss as a function of mandrel diameter for the cable types noted. This data suggests that minimum bend properties are related to the diameters of the bundle fibers. Figure 6 is a plot of the minimum mandrel size as a function of fiber diameter for an arbitrarily specified loss

Manufacturer	Type	Fiber Dia. (MILS)	Approx. No. of Fibers	Bundle Dia.	Cable Dia. (MILS)
Corning Glass Works	5010	1.8	450	45	87
Corning Glass Works	5011	1.8	900	62	120
Bausch & Lomb	32-01	0.8	3100	62	125
American Optical	LGM-3	3.0	1500	125	225
Corning Glass Works	Clear Jacket	3.0	220	45	80

Figure 4. Test Cable Properties.

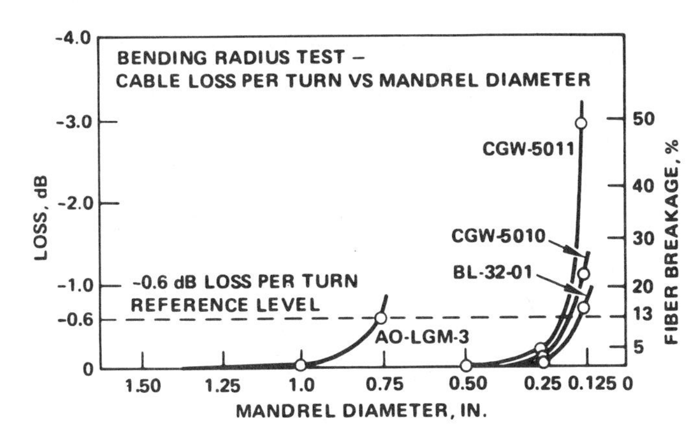

Figure 5. Test Cable Bend Data.

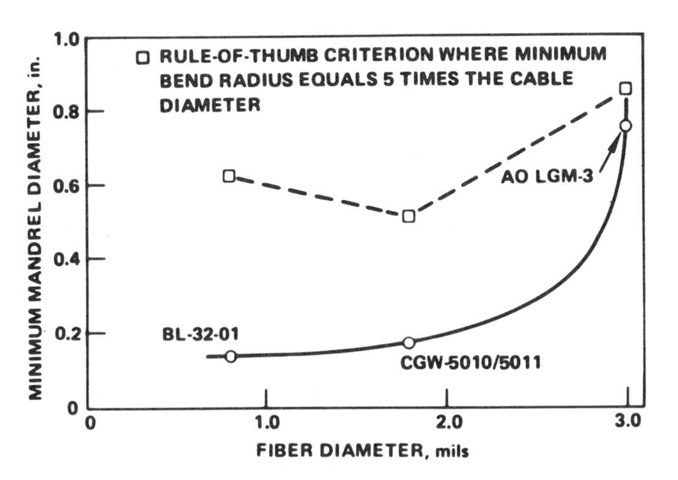

Figure 6. Test Cable Bend Limits.

factor of 3 dB (50% breakage) for 5 turns or 0.6 dB per turn.

Tensile Test

Tensile tests were performed on cables with severed jackets permitting data to be taken on the tensile properties of the fiber bundle to the exclusion of the jacket and also to evaluate the terminal/bundle epoxy bond integrity. The test method consisted of tensile loads applied in suitable steps, a constant time delay between steps, fiber breakage and transmission data taken between steps and with the loading cycle repeated a number of times. An assigned 0.6 dB loss (13% fiber breakage) acceptance level indicated tensile load sustaining capabilities of about 25 and 50 pounds for cables with 1 and 2 square mil bundle areas, respectively. This corresponds to tensile stresses of 25 thousand psi of bundle area. Figure 7 shows a typical fiber breakage pattern for a cable upon application of the tensile load. Figure 8 shows a terminal following bundle recession due to tensile exposure.

Terminal-Tension and Torque Test

These tests applied to the complete, as manufactured, cable types. The method of tensile testing was identical to that described under Tensile Tests. Cable tensile data is shown in Figure 9 for the noted cable types. Negligible cooperative effects appear to be taking place between the bundle and jacket where pure tensile loading is concerned.

Terminal-torque measurements indicate that torque levels of greater than about 3 oz-inches cannot be sustained by

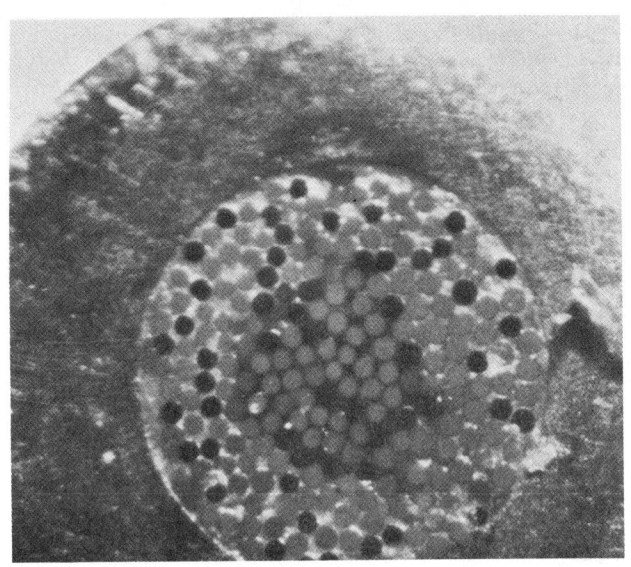

Figure 7. Fiber Breakage Under Tension.

Figure 8. Bundle Recession Due to Tension.

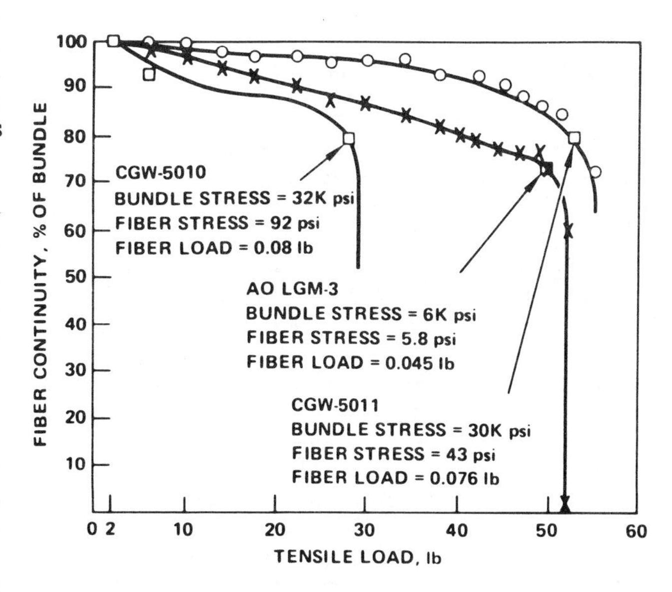

Figure 9. Terminal-Tensile Test Data.

these plastic jacketed cables. The method used to measure torque employed a direct reading gage attached to the rotated terminal. Test results indicated that the ability of the epoxy to sustain bonded integrity under tensile load generally provided a good match to the cable bundle's ability to sustain fiber continuity under comparable loads.

Mandrel Test

These type tests were considered to be fundamental to the evaluation of glass-fiber type cables to endure the stresses normally imposed upon them during installation. The cable was fixtured at its ends and the tensile load applied to one cable end with a ninety degree contact radius between the mandrel and cable. Mandrel diameters used were 0.130, 0.250, 0.50, and 1.0 inches for the cables tested. Figure 10 shows the results of mandrel test data for the CGW-5010 type cables. The fiber optic cables exhibited mean fiber

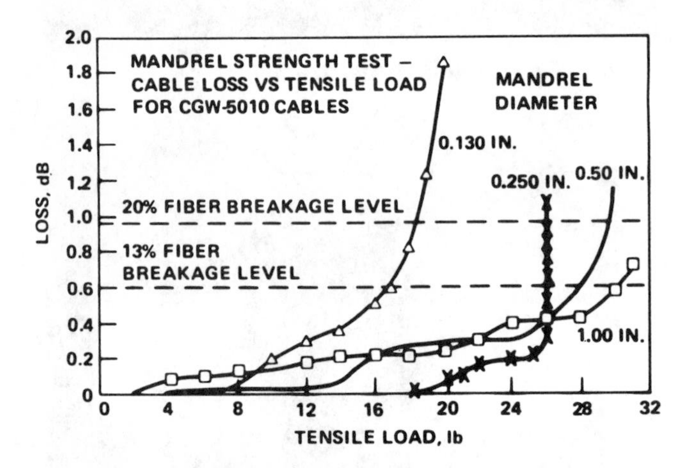

Figure 10. Cable Mandrel Test Data.

breakage levels of about 20 percent at tensile loads of 12 and 18 lb. for the 5011 and 5010 CGW types, respectively. These mandrel strength limitations basically stem from the bending radius property of the glass fibers which admittedly are inferior in bend to metallic fibers. The need for mandrel performance, however, in this small bend radius area does not appear to be great. Practical cable fabrication, installation and routing techniques will obviate this requirement. By comparison, tests using larger mandrel diameters show that for 0.250 inch and 0.50 inch diameter mandrels, essentially the full tensile loads (25 and 50 lb.) can be sustained by the 5010 and 5011 cable types, respectively.

Cyclic Flexibility Test

This test consisted of cable degradation assessment at intervals of about 2000, ±90°, cycles of cable flexure about a one inch diameter mandrel. The flexure rate was controlled at 1 cycle per second. Cyclic exposure goals of 10,000 cycles were selected and acceptance based upon meeting the 13 percent fiber breakage (0.6 dB transmission loss) level after test exposure. Figure 11 shows flexure data for the cable types noted. Catastrophic fiber breakage in the mandrel area was evidenced by "work-hardening" of the jacket with tendency to rupture. Subsequently, the fibers broke rapidly due to the small bend radius provided by the edge of the broken jacket. The test results indicate that the flexural limitations found were jacket dependent and did not reflect deficiencies or limitations within the bundle compositions. Some improvement in jacketing (particularly concerning flexibility at reduced temperatures) should provide cable acceptance to the above stated test condition.

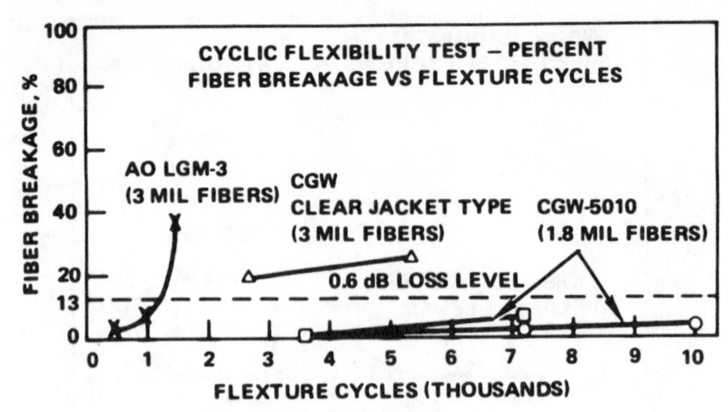

Figure 11. Cable Flexibility Test Data.

Twist Test

Twist test conditions were devised to assess the optic cable's ability to sustain fiber continuity on a rotation per unit cable length basis. The test results indicate that twist levels not exceeding about 3 rotations per cable foot (cables of the CGW 5010 and 5011 types under normal ambient conditions) will not appreciably degrade the cable's performance. The synergistic effect between the fiber bundle and the jacket as regards fiber breakage was clearly evident. Individual fiber breaks were noted to predominate within the region of maximum jacket deformation. Also, the breaks were peripheral and tended to spiral within the jacket.

Vibration Test

The vibration test exposure consisted of 1 hour resonance dwells at the resonant frequency (only one such frequency was found per cable within the 0 to 2000 hertz range of interest) with a vibratory input of 0.100 inch double amplitude for frequencies below 26 hertz and a full 3.5g input for frequencies from 26 to 2000 hertz. All cables were stress-relieved with shrinkable tubing for a length of ±1 inch from any interior cable support and/or terminal support location. Standard shipboard installation practice provides unsupported lengths of cables in cable-ways that range from perhaps 2 to 5 feet. Preliminary cable tests indicated that the maximum transmissibilities (vibratory gains) were obtained at low (1 pound) tensile loads for the 3 foot cables used. Since this represented the potentially most cable-damaging condition, the majority of the vibration tests used this tensile condition. The test cables were subjected to specification levels and durations with negligible evidence of cable degradation. A number of after-test cable terminal faces were seen to produce a specular appearance, many times in the peripheral fiber area, but also on occasion in the general fiber field. It appears that a fiber/epoxy bonding weakness is in evidence. Occasionally, a small amount of uncured epoxy or cable lubricant was seen to well-up from these fissures.

Shock Test

Three mechanical shocks in each direction were applied along two mutually perpendicular axis (lateral and longitudinal) of the test cables (total of 12 shocks). Both 1/2-sine and sawtooth shocks approaching specification levels were applied to the test cables. The test results with shocks up to 100g indicated again negligible light transmission degradation. The vibration and shock tests were performed sequentially upon the same cables, further supporting the conclusion of negligible effect on cable integrity and performance.

ENVIRONMENTAL TESTS

Life Test at Elevated Temperature

This dry air, 85°C, cable exposure for a duration of 250

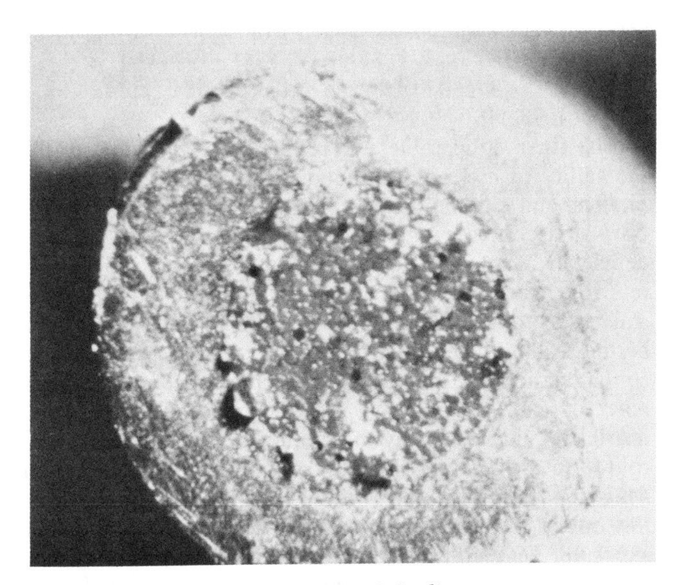

Figure 12. Terminal After Thermal Cycling.

hours indicated essentially no increase in fiber breakage level and/or cable transmission loss.

Thermal Cycle – Short Duration Test

This test, consisting of cable exposure to 5, 1/2-hour, temperature cycles from -55°C to +105°C, indicated some cable transmission loss with negligible fiber breakage. The losses were attributed to terminal degradations due primarily to the high temperature exposure. The measured cable losses were less than 1 dB for all cables tested. The range of temperatures used was quite severe from the terminal and jacket standpoint. Jacketry tends to become very pliable and deforms easily. The epoxy/fiber bonds and bundle terminal bonds are also greatly affected. Figure 12 shows the source terminal of a CGW-5010 cable taken after exposure to all 5 thermal test cycles. Epoxy and/or bundle lubricant is seen to be welling-up, primarily from the contact between the bundle and the terminal inner wall.

Thermal Cycle – Extended Duration, Low Temperature Test

This thermal test exposure, consisting of 3 cycles from ambient to -40°C for a test duration of 24 hours minimum, provided negligible effect upon the test cables' properties and/or performance. This is gratifying in the light that low temperature conditions may cause differential contraction of dissimilar parts, loss of resiliency of packings, congealing of lubricants and cable jacketing/terminating problems.

Thermal Cycle – Extended Duration, High Temperature Test

High temperature conditions may cause parts of complex construction to exhibit dimensional modification due to differential expansion of the dissimilar materials. Epoxies may discolor, crack, bulge, check, or craze. Closure and sealing may be affected. This test, consisted of 3 cycles of hot "soaks" at 49°C (6 hours) and at 68°C (4 hours) for a total 3 cycle duration of 36 hours. Again, negligible test cable fiber breakage or light transmission loss was noted after the test exposure.

Humidity – Steady State Test

These tests are intended to evaluate the influence of absorption and diffusion of moisture and moisture vapor upon material properties. This is an accelerated environmental test, providing a continuous exposure of the complete test cable to high relative humidity at an elevated temperature and consisted of a 240 hours test cable soak at 90 to 95 percent relative humidity and 100°F chamber temperature. No change in cable breakage level or cable attenuation was observed due to this test exposure.

Humidity – Cyclic Test

This test consisted of 10 cycles of temperature from -10°C to +65°C at humidities from 80 to 98 percent and a 15 minute vibration period per test cycle. The total 10 cycle test period was in excess of 240 hours in duration. The combined environmental exposure imposed by this test produced the greatest cable optical degradation observed for any environmental condition. Terminal faces were generally occluded after about 5 exposure cycles and fiber breakage level was difficult to assess but believed to be negligible. Cable transmission losses varied from about 2 to 4 dB after the 10 cycle exposure period and terminal cleaning. These losses are related to terminal surface modifications as seen in Figure 13 showing the detector terminal of a 5010 type cable after 9 test cycles respectively. The combined exposure to thermal, humidity, and vibration conditions can be expected to accelerate the degradation contributed by each test parameter independently.

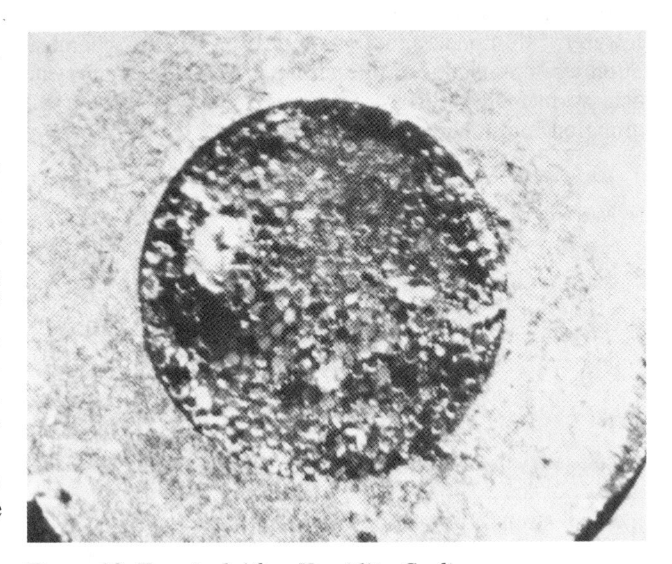

Figure 13. Terminal After Humidity Cycling.

Salt Spray Test

The salt spray test consisted of a 48 hour exposure to an atomized 5 percent salt/water vapor at 35°C. Test results indicated that cable effects were confined to the terminal regions. Corrosion of brass terminals formed chemical compounds which migrated across the terminal faces and created adherent films which resisted cleansing and provided cable losses of the order of 1 dB. It appears that terminal material type and sealing methods are primary considerations for Naval acceptance of these type products.

CHEMICAL TESTS

Salt Bath Test

Negligible cable property or performance degradation was noted after exposure to the 1 hour, 2 thermal cycle, 5 percent salt solution/water immersion test.

Oil Bath Test

This test consisting of cable submersion into mineral oil at 100°C for 5 minutes provided no change in individual fiber continuity level. However, the thermal softening of the PVC jacket and the expansion of trapped air within the jacket interior on occasion created a rupture in the jacket wall with the release of the air pocket. The 5 minute immersion was evidently too short an exposure time to provide terminal face epoxy bonding degradation as noted in the short duration, thermal cycling tests.

Acid Bath Test

This test consisted of a 46 hour cable immersion at 22°C in a 5 percent volume concentrated solution of sulphuric acid in water. Test results again showed that cable transmission losses of about 1 dB resulted from terminal chemical action with the reaction byproducts coating the terminal face.

Alkali Bath Test

This test consisted of a 46 hour immersion in a 1 normal solution of sodium hydroxide in water at 22°C. Again, the cable losses recorded were related to chemical action between the metallic terminals and the chemical reagent. However, additional epoxy modification due to chemical action was observed. Definite terminal face etch phenomena such as increased surface roughness, epoxy void and pit formation and topographic honeycomb effects were observed. The cable losses ranged from 1 to 1.5 dB after the test exposure.

CONCLUSIONS

The fiber optic cable test results presented indicate that the ability of these type products to withstand the gamut of physical and environmental conditions imposed by Military/ Navy usage is very great. In fact, no physical property deficiency was observed which would not or could not be rendered acceptable through normal and proper cable design and fabrication processes. Fundamental in this regard would be the sealing of terminal interfacing regions against possible contamination and perhaps certain chemical reagents. Stress relief of cables at terminal attachment locations and bend restrictive elements may be required along certain cable routes to assure fiber integrity around sharp corners. Condensation due to humidity will likely provide a greater loss in the optical link than will material degradation due to this exposure. Temperatures in excess of 85°C should be suspect for producing terminal degradation effects.

From the above generalized conclusions of the fiber optic cable test results, there appears to be no reason why cables of the type tested here could not be used for limited applications in a shipboard environment with high expectation for component survival and performance continuance.

Optical Power Handling Capacity of Low Loss Optical Fibers as Determined by Stimulated Raman and Brillouin Scattering

R. G. Smith

The effect of stimulated Raman and Brillouin scattering on the power handling capacity of optical fibers is considered and found to be important especially when low loss optical fibers are used. A critical power below which stimulated effects may be neglected is defined for forward and backward Raman scattering and for backward Brillouin scattering. This critical power is determined by the effective core area A, the small signal attenuation constant of the fiber α, and the gain coefficient for the stimulated scattering process γ_0, by the approximate relation $P_{\text{crit}} \approx 20A\alpha/\gamma_0$. For a fiber with 20-dB/km attenuation and an area of 10^{-7} cm^2 $P_{\text{crit}} \approx 35$ mW for stimulated Brillouin scattering. For stimulated Raman scattering P_{crit} is approximately two orders of magnitude higher. It is concluded that these effects must be considered in the design of optical communication systems using low loss fibers.

Introduction

Low loss optical fibers are currently being considered as transmission media for optical communication systems. At low power densities the losses of an optical fiber will be determined by spontaneous Raman, Brillouin, and Rayleigh scattering, absorption losses in the bulk material, scattering at the core-cladding interface, and mode conversion from low-loss trapped modes to high loss cladding modes. On the other hand at high power densities within the fiber, Raman and Brillouin scattering processes can become stimulated, introducing power dependent loss mechanisms.[1-3] In the case of forward stimulated Raman scattering the principal effect is a frequency shift of the radiation transmitted by the fiber to lower frequencies. Sufficiently large frequency shifts could produce amplitude distortion at the receiver if the detector is intrinsically frequency sensitive or if narrow band filters are used. On the other hand backward-wave stimulated scattering processes (either Raman or Brillouin) will result in a severe attenuation of the forward traveling, information carrying wave, due to the transfer of energy to the stimulated backward wave.

In this paper both forward and backward stimulated scattering processes are discussed. Raman scattering may be either forward or backward-wave in character whereas stimulated Brillouin scattering is exclusively a backward-wave process in amorphous media and in the geometry considered. The differential equation governing the stimulated processes is solved under the assumption that the primary radiation—termed the pump—follows the natural exponential attenuation of the fiber, i.e., pump depletion due to the stimulated process is neglected. In this approximation the equation for the forward and backward stimulated radiation fields may be solved exactly. In both cases the criterion that the stimulated wave be everywhere less than the forward traveling injected wave yields a critical input power above which nonlinear effects play an important role.

Using available data for the magnitudes of the nonlinear scattering cross sections it is found that stimulated Brillouin scattering should determine the maximum power handling capacity of a fiber. The maximum power that can be launched is given to a good approximation by $P_{\text{crit.}}^{\text{Brillouin}} \approx 20(\alpha A/\gamma_0)$, where α is the attenuation constant and A the area of the fiber and γ_0 is the gain coefficient of the nonlinear process. Using this relation and the value of γ_0 for Brillouin scattering a fiber with a core area of 10^{-7} cm^2 and an attenuation of 20 dB/km gives $P_{\text{crit.}} = 35$ mW for either cw or long pulse radiation. For short pulses of radiation this critical power is greater due to the backward-wave nature of the interaction and the finite line width of the Brillouin scattering.

Analysis

In this section we evaluate the input power density to an optical fiber for which the nonlinear process, either forward or backward wave stimulated scattering becomes important. The criterion is arbitrarily interpreted to be that set of conditions for which the new

---

The author is with Bell Telephone Laboratories, Inc., Murray Hill, New Jersey 07974.

Received 5 June 1972.

Reprinted with permission from *Appl. Opt.*, vol. 11, pp. 2489–2494, Nov. 1972.

radiation field generated by the nonlinear interaction is comparable to the incident field. For *forward* Raman scattering we require that the magnitude of the Stokes shifted wave be less than the injected wave at all points along the fiber. For *backward* scattering, either Raman or Brillouin, we require that at the entrance face the backward traveling wave be less than the injected wave. All calculations are made assuming single pass amplification of spontaneous emission. For Raman scattering feedback due to scattering could be a problem due to the isotropy of the Raman gain but is not specifically considered here.

Forward Wave Interaction

Consider a uniform transmission medium of effective cross sectional area A and length L with an attenuation constant α. For convenience we label the injected radiation with the subscript p, refer to it as the pump, and denote its frequency by ω_p. Let the pump be injected at $z = 0$, traveling in the $+z$ direction with a power P_p and an effective power density $S_p \approx P_p/A$. In the absence of any nonlinear interaction the pump propagates as

$$P_p(z) = P_p(0) \exp(-\alpha_p z), \tag{1}$$

with an identical dependence for the power density. Assuming that the guiding medium is Raman active, the differential equation for a forward traveling wave at the Stokes frequency is given by

$$[(d/dz) + \alpha_s]P_s(z) = \gamma S_p(z) P_s(z), \tag{2}$$

where the subscript s refers to the Stokes component and P_s is the Stokes power at any point. The gain constant γ depends upon the Raman scattering cross section and is frequency dependent.

A similar equation can be written for the pump wave where the nonlinear interaction is explicitly taken into account. Since we are looking for an upper bound to the input power where nonlinear effects become important we make the simplifying assumption that the pump is not attenuated by the nonlinear interaction. In this case Eq. (2) becomes

$$[(d/dz) + \alpha_s]P_s(z) = \gamma P_s(z) S_p(0) \exp(-\alpha_p z), \tag{3}$$

where $S_p(0)$ is the pump intensity at the input. The solution to Eq. (3) is

$$P_s(z) = P_s(0) \exp\left\{-\alpha_s z + \frac{\gamma S_p(0)}{\alpha_p}[1 - \exp(-\alpha_p z)]\right\}. \tag{4}$$

At the output of the fiber, $z = L$, assuming that $\alpha_p L \gg 1$, Eq. (4) becomes

$$P_s(L) \simeq P_s(0) \exp\left[-\alpha_s L + \frac{\gamma S_p(0)}{\alpha_p}\right]. \tag{5}$$

Thus under the assumption that the pump attenuates according to Eq. (1), Eq. (5) says that the electronic gain at the Stokes frequency is $[\gamma S_p(0)/\alpha_p]$, equivalent to the gain produced by the incident pump intensity for a distance $l_{\text{eff}} = 1/\alpha_p$.

Equation (5) assumes a Stokes wave injected at ($z = 0$). In practice no Stokes wave would be injected; any Stokes power appearing at the output would be due to amplified spontaneous Stokes scattering occurring throughout the length of the fiber. It is shown in Appendix A that the summation over the length of all spontaneous emission weighted by its net gain is equivalent to assuming an input flux at $z = 0$ of one photon per mode (longitudinal and transverse) of the fiber. In this case the total Stokes flux at $z = L$ is given by

$$P_s(L) = \sum_{\substack{\text{transverse}\\ \text{modes}}} \int d\nu(h\nu) \exp\left[-\alpha_s L + \frac{S_p(0)}{\alpha_p}\gamma(\nu),\right], \tag{6}$$

where we now take into account the frequency dependence of the gain coefficient. Assuming a Lorentzian gain profile with full width at half maximum $\Delta\nu_{fwhm}$ and assuming the peak gain to be large, Eq. (6) becomes

$$P_s(L) = \left\{\sum_{\substack{\text{transverse}\\ \text{modes}}} (h\nu_s) \exp\left[-\alpha_s L + \frac{S_p(0)\gamma_0}{\alpha_p}\right]\right\} B_{\text{eff}}, \tag{7}$$

where γ_0 is the peak gain coefficient. The effective bandwidth or number of longitudinal modes is thus

$$B_{\text{eff}} = \frac{\sqrt{\pi}}{2} \frac{\Delta\nu_{fwhm}}{[S_p(0)\gamma_0/\alpha_p]^{\frac{1}{2}}}, \tag{8}$$

and the effective input Stokes power is

$$P_s(0)|_{\text{eff}} = (h\nu_s)(B_{\text{eff}})(\text{number of transverse modes.}) \tag{9}$$

In order that the nonlinear process not be of importance we demand that the Stokes power computed from Eq. (7) be less than the signal power at $z = L$, i.e.,

$$P_s(0)|_{\text{eff}} \exp\left[-\alpha_s L + \frac{S_p(0)\gamma_0}{\alpha_p}\right] < P_p(0) \exp(-\alpha_p L). \tag{10}$$

An absolute upper limit to the input pump power at which point nonlinear effects must be considered would be when Eq. (10) is satisfied as an equality. We define this pump power as the critical power, P_{crit}. For a single transverse mode fiber and assuming $\alpha_s = \alpha_p$ the relation from which the critical power is determined is

$$\frac{\sqrt{\pi}}{2}(h\nu_s)\left(\frac{\gamma_0}{A\alpha_p}\right)\Delta\nu_{fwhm} = \left(\frac{\gamma_0 P_{\text{crit}}}{A\alpha_p}\right)^{3/2} \exp\left(-\frac{\gamma_0 P_{\text{crit}}}{A\alpha_p}\right). \tag{11}$$

As an example consider the case of Raman scattering in crystalline quartz at a pump wavelength of 1 μ. The peak Raman gain coefficient is $\gamma_0 \approx 5 \times 10^{-10}$ cm/W.[4] Assuming a single mode fiber of cross sectional area 10^{-7} cm^2 (diameter $\approx 3\ \mu$), an attenuation coefficient of $\alpha \approx 5 \times 10^{-5}$ cm^{-1} (20 dB/km) for both the Stokes and pump waves, and taking $\Delta\nu_{fwhm} \approx 10$ cm^{-1}, the relation for the critical power becomes

$$\exp(100 P_{\text{crit}}) = 2 \times 10^8 P_{\text{crit}}^{3/2},$$

where P_{crit} is expressed in watts. The critical power determined from this relation is $P_{\text{crit}} = 160$ mW. The value of the critical power is reasonably insensitive to the choice of $\Delta\nu_{fwhm}$. For example changing the line width of a factor of 10 changes the critical power by approximately 15%. On the other hand the critical power is extremely sensitive to α, γ_0, and A. For example with $\alpha_p = 1 \times 10^{-5}$ (4.3 dB/km), $P_{crit} = 30$ mW and for $\alpha_p = 5 \times 10^{-6}$ (2 dB/km), $P_{crit} = 12$ mW, assuming $A = 10^{-7}$ cm² and $\gamma_0 = 5 \times 10^{-10}$ cm/W. For the range of parameters involved here the critical power in watts is given to a very good approximation by the relation

$$P_{crit} \approx 16(A\alpha_p/\gamma_0), \qquad (12)$$

where A is given in cm², α in cm^{-1}, and γ_0 in cm/W.

The gain coefficient, used in the above example, $\gamma_0 = 5 \times 10^{-10}$ cm/W, applies to Raman scattering in α-quartz.[4] Stolen *et al.*[2] report that the line width for Raman scattering in amorphous glasses is of the order of several hundred cm^{-1}, and the peak gain constant is smaller than the value quoted above by roughly an order of magnitude. It is thus concluded that a fiber with an area of 10^{-7} cm² and an attenuation constant of 20 dB/km should be capable of handling an optical power of the order of 1 W without problems from stimulated forward Raman scattering. For lower loss fibers the power handling capacity will be correspondingly reduced.

Backward Wave Interaction

Because of the flat character of the dispersion relation for optical phonons near the center of the Brillouin zone, phase matching of the Raman process may be achieved for both forward and backward scattering; phase matching requirements for Brillouin scattering in isotropic media prohibit forward scattering and so SBS will be a backward wave interaction in the fiber geometry. The analysis for the backward-wave interaction below applies to either Raman or Brillouin scattering, whereas the preceding analysis applies only to Raman scattering.

For the backward-wave interaction we apply the same approach as for the case of the forward-wave interaction except that we consider a Stokes wave, either Brillouin or Raman, traveling in a direction opposite to the pump. Taking $z = 0$ as the point at which the pump is injected and assuming the pump to propagate as $\exp(-\alpha_p z)$ the gain experienced by a Stokes wave injected at some point z_0, defined as the ratio $P_s(0)/P_s(z_0)$ is, Appendix B,

$$G = \exp\{-\alpha_s z_0 + [\gamma S_p(0)/\alpha_p][1 - \exp(-\alpha_p z_0)]\}. \qquad (13)$$

The total backward-traveling, stimulated Stokes power reaching the entrance face is then found by summing all contributions from spontaneous emission along the fiber multiplied by the gain, Eq. (13). In Appendix B it is shown that for *purely spontaneous scattering* this summation is approximately equivalent to the injection of a single Stokes photon per mode at the point along the fiber where the nonlinear gain exactly equals the natural loss of the fiber, i.e., where $\gamma S_p(0)\exp(-\alpha_p z) = \alpha_s$. When $\alpha_s = \alpha_p$ the net effective gain for this single fictitious photon is [see Eq. (B6)]

$$G_{eff}(\nu) = \frac{\exp\{[S_p(0)\gamma(\nu)]/\alpha_p\}}{[S_p(0)\gamma(\nu)/\alpha_p]}. \qquad (14)$$

The total backward Stokes radiation is thus

$$P_s(0) = \sum_{\substack{\text{transverse}\\ \text{modes}}} \int d\nu(h\nu)\cdot G_{eff}(\nu). \qquad (15)$$

Neglecting the frequency dependence of the denominator in Eq. (14) and assuming a Lorentzian line shape, the effective bandwidth is given by Eq. (8) and the effective input Stokes power by Eq. (9). For the case of Brillouin scattering where the phonons are thermally activated the effective input Stokes power must be multiplied by $(kT/h\nu_a + 1) \approx kT/h\nu_a$, where k is Boltzmann's constant, T the temperature, and ν_a the frequency of the acoustic phonon.[5] This factor, equal to the average phonon occupation number, will be typically of the order of 100 to 200 depending upon the pump wavelength and the velocity of sound in the medium.

The critical power for the backward wave process is arbitrarily defined to be that input power for which the backward stimulated Stokes power equals the input pump power at $z = 0$. For a single transverse mode fiber the critical powers for stimulated Raman and Brillouin scattering are found from the following relations:

$$\frac{\sqrt{\pi}}{2}(h\nu_s)\left(\frac{\gamma_0}{\alpha_p A}\right)\Delta\nu_{fwhm}^{\text{Raman}} = \left(\frac{\gamma_0 P_{crit}'}{A\alpha_p}\right)^{5/2}\exp\left(-\frac{\gamma_0 P_{crit}'}{A\alpha_p}\right), \qquad (16)$$

(Raman)

$$\frac{\sqrt{\pi}}{2}\left(\frac{\nu_s}{\nu_a}\right)(kT)\left(\frac{\gamma_0}{\alpha_p A}\right)\Delta\nu_{fwhm}^{\text{Brill.}} = \left(\frac{\gamma_0 P_{crit}''}{A\alpha_p}\right)^{5/2}\exp\left(-\frac{\gamma_0 P_{crit}''}{A\alpha_p}\right).$$

(Brillouin) (17)

The relation for P_{crit}' for backward Raman scattering gives a greater critical power than forward scattering by roughly 25%, the relation being approximately

$$P_{crit}' = 20(A\alpha_p/\gamma_0). \qquad (18)$$

For typical values associated with Brillouin scattering and low loss single-mode fibers the critical power is approximately given by

$$P_{crit}'' \approx 21(A\alpha_p/\gamma_0). \qquad (19)$$

Equations (12), (18), and (19) all give roughly the same relation, the numerical factor in each case being approximately the natural logarithm of the gain required to bring the spontaneous emission to the level of the input pump power.

The expression for the peak gain coefficient for stimulated Brillouin scattering is[5]

$$\gamma_0 = (2\pi^2\nu_s\nu_a M_2)/c^2\alpha_a, \qquad (20)$$

where ν_s and ν_a are the frequencies of the Stokes wave and the acoustic phonon, respectively, α_a is the attenuation constant of the acoustic intensity, c is the velocity of light, and M_2 is the elastooptic figure of merit given by[6]

$$M_2 = n^6 p^2 / \rho V_a^3. \qquad (21)$$

In Eq. (21) n is the index of refraction, p the elastooptic constant, ρ the density, and V_a the velocity of sound. For fused silica $M_2 = 1.51 \times 10^{-18}$ sec^3/g.[6]

Pine[7] has measured the room temperature line width for spontaneous Brillouin scattering in fused silica at $\lambda = 6328$ Å, obtaining a value of approximately 80 MHz. The corresponding attenuation constant is $\alpha_a = 2\pi\Delta\nu_{fwhm}/V_a \approx 800$ cm^{-1}. Walder and Tang[8] estimate an attenuation constant of 560 cm^{-1} at $\lambda = 6943$ Å from stimulated Brillouin measurements in the same material. In estimating the acoustic attenuation constant appropriate for $\lambda_p = 1.064\ \mu$ and a corresponding acoustic frequency of 16.4 GHz, we take Pine's directly measured value and assume a linear dependence on acoustic frequency giving $\alpha_a \simeq 500$ cm^{-1} ($\lambda_p = 1.06\ \mu$). The peak gain coefficient thus becomes $\gamma_0 \approx 3 \times 10^{-9}$ cm/W. Taking $\alpha_p = 5 \times 10^{-5}$ cm^{-1} ($\approx$20 dB/km) and $A = 10^{-7}$ cm^2, the critical power for stimulated Brillouin scattering, Eq. (19), is $P''_{crit} = 35$ mW. This value will be correspondingly lower if fibers of smaller cross sectional areas or lower attenuation constants are used. For example Rich and Pinnow[9] have reported bulk losses at 1.064 μ of less than 3 dB/km.

The critical powers derived above for stimulated forward and backward scattering processes implicitly assume a single frequency pump source. For a pump source of finite line width Eqs. (11), (16), and (17) approximately hold so long as the line width of the pump is less than the line width of the spontaneous scattering, $\Delta\nu_{fwhm}$. For Raman scattering $\Delta\nu_{fwhm}$ is sufficiently large that Eqs. (11) and (16) are expected to be valid for most laser sources. On the other hand for Brillouin scattering at $\lambda_p = 1.06\ \mu$, $\Delta\nu_{fwhm} \approx 50$ MHz. Thus in Eq. (17) P''_{crit} refers to the power within a 50-MHz bandwidth and hence under most circumstances to the power in a single laser mode. In terms of the power handling capacity of a fiber, Eq. (17) is interpreted to imply that stimulated Brillouin scattering will not be of importance provided the power in any individual laser mode, or more specifically the power within any bandwidth of size $\Delta\nu_{fwhm}$, is less than P''_{crit}.

The critical powers have also been derived assuming steady-state or cw operation. Consider the effects of using a pulsed source as would apply when a digital format is used. For the forward-wave interaction Eq. (12) would still apply to the peak pump power unless the pulse width of the pump were so short that either the bandwidth of its envelope exceeded the line width of the spontaneous emission in which case the transient scattering problem must be treated, or the difference in group velocities between the Stokes wave and the pump times the pulse width exceeds $1/\alpha_p$ in which case the interaction length is effectively shortened.

For backward wave interactions where the pump is pulsed the interaction length is limited to a value $l_{eff} \approx v_g\Delta t/2$, where v_g is the group velocity and Δt is the pulse duration, due to the opposite direction of propagation of the waves. For a single pulse the critical pump power will exceed the values determined by Eqs. (18) and (19) by the factor $\approx 1/(l_{eff}\alpha_p)$. In the limit of very short optical pulses where the inverse of the pulse width exceeds the line width of the scattering process the transient nature of the scattering problem must be taken into account. This effect is most pronounced for stimulated Brillouin scattering and will further raise the threshold for stimulated scattering. There remains, however, the possibility of a succession of pump pulses each amplifying the backward traveling Stokes wave. For a consecutive string of ones, each with a duty cycle of D the critical power exceeds the value given by Eqs. (18) or (19) by the factor $1/D$. Such an occurrence represents a worst case and is probably relatively unlikely to occur.

In the above calculations the modal character of the various fields have been neglected. For stimulated Raman scattering the principal modification caused by considering the detailed spatial variation of the fields is to introduce a filling factor or overlap integral into the expressions for the critical powers. For most cases this integral will be of the order of unity, and only a slight modification of the critical powers is expected. The same filling factor arguments apply to the case of stimulated Brillouin scattering in a single mode fiber. On the other hand for multimode fibers the Brillouin scattering threshold will be further modified by the fact that the different transverse modes within the fiber propagate with different phase velocities. In this case a particular backward traveling Stokes mode will interact with a given forward traveling pump wave with a corresponding phonon frequency given by

$$\nu_a = \nu_p(V_a/c)(n_m + n_m'), \qquad (22)$$

where n_m and n_m' are the effective refractive indices for the forward and backward modes. A given Stokes Brillouin wave will effectively interact with only those pump modes that give an acoustic frequency within the acoustic line width. This limitation on the range of modes which can effectively interact further raises the critical power for stimulated Brillouin scattering in multimode guides.

Discussion

The effect of stimulated Raman and Brillouin scattering on the optical power handling capacity of low loss optical fibers has been evaluated. The results are presented for each scattering process in terms of a critical power, defined as the input power level to the fiber for which the particular nonlinear scattering process is clearly of importance. Approximate expressions for the critical powers for forward and backward Raman scattering and backward Brillouin scattering are found in Eqs. (12), (18), and (19), respectively. Because of the exponential dependence of these stimulated processes on pump power level, a reduction of the input power from the critical value by approximately 1 dB reduces the level of the stimulated scattering by approximately 20 dB. Thus the power that may be sent down a fiber without serious interference from nonlinear effects will be typically of the order of 1 dB below P_{crit}, and hence the critical power so defined serves as a useful benchmark.

The ultimate power handling capacity of a fiber will depend upon the fiber parameters A, α, γ_0 as well as the spectral characteristics of the optical pump. When the pump source is a laser operating in a single longitudinal mode and hence with a narrow line width, stimulated Brillouin scattering will have the lowest critical power. The estimated critical power for a fused silica fiber with $A = 10^{-7}$ cm^2 and $\alpha = 5 \times 10^{-5}$ cm^{-1} ($\approx$20 dB/km) is 35 mW for a cw pump. The critical power for stimulated Raman scattering is one to two orders of magnitude greater than this value. When the pump source is multimode, stimulated Brillouin scattering will not be of importance so long as the pump power within any frequency interval $\Delta\nu_{fwhm}$ is less than the critical value. For an extremely broadband pump source stimulated Raman scattering will ultimately set the upper limit to the power handling capabilities of the fiber.

Appendix A: Forward Wave Interaction

The differential equation for the photon occupation number of a given Stokes mode is

$$[(d/dz) + \alpha_s]N_s = \gamma S_p(z)(N_s + 1), \tag{A1}$$

where the gain constant on the right-hand side is chosen to agree with the classical limit, and the additional factor of unity takes into account the spontaneous emission whose rate is

$$(d/dz)N_s^{\text{spontaneous}} = \gamma S_p(z). \tag{A2}$$

Neglecting spontaneous emission, the solution to Eq. (A1) is

$$\frac{N_s(z_2)}{N_s(z_1)} = \exp\left[\alpha_s(z_1 - z_2) + \int_{z_1}^{z_2} \gamma S_p(z)dz\right]. \tag{A3}$$

We next argue huristically that the spontaneous emission from each incremental length—$\gamma S_p(z)dz$—will be amplified according to Eq. (A3). At any point z the photon occupation number of a given mode due to spontaneous emission, in addition to any injected photons $N_s(0)$, is given by

$$N_s(z) = \int_0^z d\xi \gamma S_p(\xi) \exp\left[\alpha_s(\xi - z) + \int_\xi^z \gamma S_p(\eta)d\eta\right] + N_s(0) \exp\left[-\alpha_s z + \int_0^z \gamma S_p(\eta)d\eta\right], \tag{A4}$$

which is easily verified to be the general solution to Eq. (A1). Under the assumption of a pump intensity of the form $S_p(z) = S_p(0) \exp(-\alpha_p z)$, Eq. (A4) becomes

$$N_s(z) = \int_0^z d\xi \gamma S_p(0) \exp\left\{-\alpha_p\xi + \alpha_s(\xi - z) + \frac{\gamma S_p(0)}{\alpha_p} \times [\exp(-\alpha_p\xi) - \exp(-\alpha_p z)]\right\} + N_s(0) \times \exp\left\{-\alpha_s z + \frac{\gamma S_p(0)}{\alpha_p}[1 - \exp(-\alpha_p z)]\right\}. \tag{A5}$$

Consider the case where there is no injected Stokes wave, $N_s(0) = 0$ and for simplicity assume $\alpha_s = \alpha_p$. At the exit plane, $z = L$ Eq. (A5) becomes

$$N_s(L) = \exp(-\alpha_s L) \int_0^L \gamma S_p(0)d\xi \exp\left\{\frac{\gamma S_p(0)}{\alpha_p} \times [\exp(-\alpha_p\xi) - \exp(-\alpha_p L)]\right\}. \tag{A6}$$

For fiber systems of interest $\alpha_p L \gg 1$ and the second exponential factor in braces is approximately zero. When $\gamma S_p(0)/\alpha_p \gg 1$, as will be the case when stimulated emission is important, the major contribution to the integral will occur while $\alpha_p\xi < 1$. Expanding exp $(-\alpha_p\xi) \simeq 1 - \alpha_p\xi$ Eq. (A6) becomes

$$\begin{aligned} N_s(L) &\cong \exp\left[-\alpha_s L + \frac{\gamma S_p(0)}{\alpha_p}\right] \int_0^L \gamma S_p(0)d\xi \exp[-\gamma S_p(0)\xi] \\ &= \exp\left[-\alpha_s L + \frac{\gamma S_p(0)}{\alpha_p}\right][1 - \exp(-\gamma S_p L)] \\ &\approx \left[-\alpha_s L + \frac{\gamma S_p(0)}{\alpha_p}\right], \end{aligned} \tag{A7}$$

where $\gamma S_p L > \gamma S_p/\alpha_p \gg 1$. From Eq. (A5) this is seen to be identical to the result where $N_s(0) = 1$ and spontaneous emission is neglected. Hence the net result of the amplification of all spontaneous emission is equivalent to the fictitious injection of a single photon at the input plane, $z = 0$.

Appendix B: Backward Wave Interaction

Let the incident pump wave travel in the $+z$ direction from $z = 0$ to $z = L$ with the direction of propagation of the Stokes wave in the $-z$ direction. The differential equation for the photon occupation number of the backward wave is

$$[(d/dz) - \alpha_s]N_s = -\gamma S_p(z)(N_s + 1). \tag{B1}$$

Neglecting the spontaneous emission term the solution to Eq. (B1) is

$$\frac{N_s(z_2)}{N_s(z_1)} = \exp\left[\alpha_s(z_2 - z_1) + \int_{z_2}^{z_1} \gamma S_p(z)dz\right]. \tag{B2}$$

The solution to Eq. (B1) including spontaneous emission as well as $N_s(L)$ photons injected at $z = L$ is

$$N_s(z) = \int_z^L d\xi \gamma S_p(\xi) \exp\left[\alpha_s(z - \xi) + \int_z^\xi \gamma S_p(\eta)d\eta\right] + N_s(L) \exp\left[\alpha_s(z - L) + \int_z^L \gamma S_p(\eta)d\eta\right], \tag{B3}$$

which satisfies Eq. (B1). For a pump taking the form $S_p(z) = S_p(0) \exp(-\alpha_p z)$ Eq. (B3) becomes

$$N_s(z) = \int_z^L d\xi \gamma S_p(0) \exp\left\{\alpha_s z - (\alpha_s + \alpha_p)\xi + \frac{\gamma S_p(0)}{\alpha_p}\right.$$

$$\left.\times [\exp(-\alpha_p z) - \exp(-\alpha_p \xi)]\right\} + N_s(L) \exp\left\{\alpha_s(z - L)\right.$$

$$\left.+ \frac{\gamma S_p(0)}{\alpha_p} [\exp(-\alpha_p z) - \exp(-\alpha_p L)]\right\}. \quad \text{(B4)}$$

With no injected Stokes wave the total amplified spontaneous emission at the input plane, $z = 0$ is

$$N_s(0) = \int_0^L d\xi \gamma S_p(0) \exp\left\{-(\alpha_s + \alpha_p)\xi\right.$$

$$\left.+ \frac{\gamma S_p(0)}{\alpha_p} [1 - \exp(-\alpha_p \xi)]\right\}. \quad \text{(B5)}$$

For simplicity assume $\alpha_s = \alpha_p = \alpha$ and consider the limit $\alpha L \gg$, then

$$N_s(0) = \frac{\exp[\gamma S_p(0)/\alpha_p]}{[\gamma S_p(0)/\alpha_p]}, \quad \text{(B6)}$$

the result used in Eq. (15). Consider now Eq. (B1). The net gain for a real photon is $g_{\text{net}} = \gamma S_p(z) - \alpha_s$. For an exponentially decaying pump $g_{\text{net}} = 0$ when

$$\gamma S_p(0) \exp(-\alpha_p z_0) = \alpha_s$$

or

$$\alpha_p z_0 = \ln[\gamma S_p(0)/\alpha_s]. \quad \text{(B7)}$$

Assume that a single photon is injected at $L = z_0$ with no spontaneous emission and take $\alpha_s = \alpha_p$, Eq. (B4) then gives

$$N_s(0) = 1/[\gamma S_p(0)/\alpha_p] \exp\{[\gamma S_p(0)/\alpha_p] - 1\}. \quad \text{(B8)}$$

In the limit $\gamma S_p(0)/\alpha_p \gg 1$, Eq. (B8) reduces to Eq. (B6), and hence the net effect of amplified spontaneous emission from the full interaction length is equivalent to the injection of a single photon per mode at z_0 given by Eq. (B7).

References

1. E. P. Ippen, Appl. Phys. Lett. **16**, 303 (1970).
2. R. H. Stolen, E. P. Ippen, and A. R. Tynes, Appl. Phys. Lett. **20**, 62 (1972).
3. E. P. Ippen and R. H. Stolen, Paper F9, 7th International Quantum Electronics Conference, Montreal, May 1972.
4. W. D. Johnston, Jr., I. P. Kaminow, and J. G. Bergman, Jr., Appl. Phys. Lett. **13**, 190 (1968).
5. C. L. Tang, J. Appl. Phys. **37**, 2945 (1966).
6. D. A. Pinnow, in *Handbook of Lasers*, R. J. Pressley, Ed. (Chemical Rubber Co., Cleveland, 1972).
7. A. S. Pine, Phys. Rev. **185**, 1187 (1969).
8. J. Walder and C. L. Tang, Phys. Rev. Lett. **19**, 623 (1967).
9. T. C. Rich and D. A. Pinnow, Appl. Phys. Lett. **20**, 264 (1972).

Effect of Neutron- and Gamma-Radiation on Glass Optical Waveguides

Robert D. Maurer, Ernst J. Schiel, Stanley Kronenberg, and Robert A. Lux

R. D. Maurer is with Corning Glass Works, Research and Development Laboratories, Corning, New York 14830; the other authors are with U. S. Army Electronics Command, Fort Monmouth, New Jersey 07703.
Received 16 May 1973

Multimode glass optical waveguides with very low attenuation (4 dB/km) have been made by laboratory techniques.[1] Since further research to reduce this already low absorption is no longer of primary interest, efforts have been directed toward other desirable properties of optical waveguides. Among these is the resistance to the environmental hazard of nuclear (neutron and gamma) radiation. Nuclear effects are not only of great importance to military applications, but also to conventional applications. Worldwide background radiation yields an accumulated dose of 10 R over a twenty-year period[2], corresponding to approximately 4×10^{14} ions/cm^3. An increase in attenuation α (base e) can be estimated from $\alpha = N\sigma$, with N the number of absorbing centers per unit volume and σ their optical cross section[3] of $(0.1\ \text{Å})^2$ gives a possible attenuation increase of 170 dB/km ($\alpha = 4 \times 10^{-4}$ cm^{-1}). Radiation levels encountered under certain nuclear environments are orders of magnitude more severe. The attenuation of conventional fiber optics can be expected to increase substantially under these conditions. This Letter, however, shows that it is possible to reduce the increase in attenuation in the near ir spectral region in specially prepared ultrapure glass waveguides.

One set of experiments was conducted with neutron radiation ($E \approx 14$ MeV) from a neutron generator (tritium source) and another set with gamma radiation ($E \approx 1.3$ MeV) from a Co^{60} source. Twenty five- to fifty-meter sections of multimode, high-silica-glass fiber waveguide[4] were measured in their original state at Corning Glass Works and then subjected to irradiation at USAECOM, Fort Monmouth. Then the samples were returned to Corning for transmission measurements[5] with an elapsed time of one to two weeks. In addition to the first irradiations and measurements two other cycles followed. The radiation data are summarized:

	Sample 1 (neutron irradiated)	Sample 2 (gamma irradiated)
First exposure	5.5×10^{10} n/cm^2	300 rads
Second exposure	5×10^{11} n/cm^2	2000 rads
(accumulated dose)	(5.55×10^{11} n/cm^2)	(2300 rads)
Third exposure	8.17×10^{11} n/cm^2	2000 rads
(accumulated dose)	(1.372×10^{12} n/cm^2)	(4300 rads)

The Fort Monmouth Co^{60} irradiation facility is calibrated by Victoreen air ion chambers whose calibration is traceable to National Bureau of Standards calibrations. The accuracy of the gamma ray dose determinations is ± 5%. The uniformity of the dose received by the fibers is also ±5%. The neutron dose was measured with IM-185 tissue-equivalent dosimeters, which were calibrated by aluminum activation foils. Neutron doses are accurate to ±10%. Samples of fibers were mounted on a rotating holder so that uniformity of dose was within ± 5%. In addition to the neutron dose the neutron irradiated fibers received a gamma ray dose of about 7×10^{-10} rad/n/cm^2.

The results are shown in Figs. 1 and 2. In general, radiation induced changes are large at short wavelengths and decrease monotonically to very small effects at longer wavelengths. Beyond about 800 nm, attenuation stays below 20 dB/km for the radiation levels used here. Because of the short length of sample, changes of less than about ±4 dB/km are not reliable. The changes are approximately linear with accumulated radiation at any given wavelength. This indicates that N, as defined above, is still increasing linearly, and the traps, which cause the color centers, have not saturated. On the other hand, the attenuation, which is much lower than estimated above, must be decreased by either extensive recombination or an unusually small cross section.

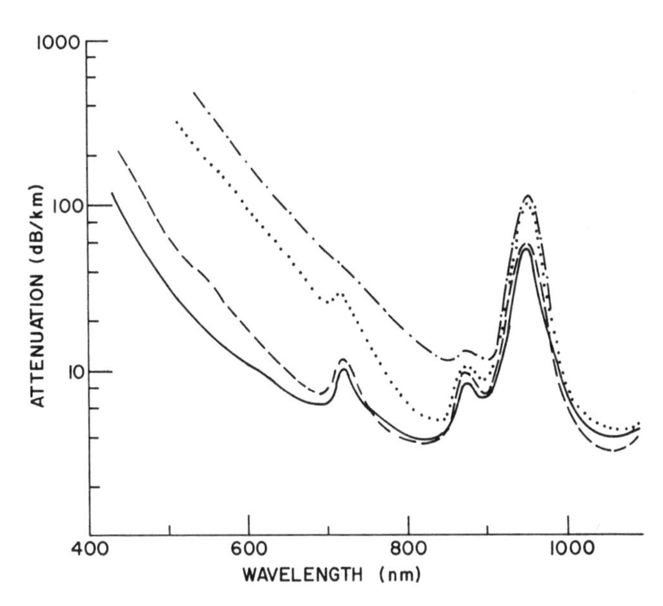

Fig. 1. Change in attenuation of a multimode optical waveguide with cumulative Co^{60} gamma irradiation: —— unirradiated, ---- 300 rads, ... 2300 rads, -.-.- 4300 rads.

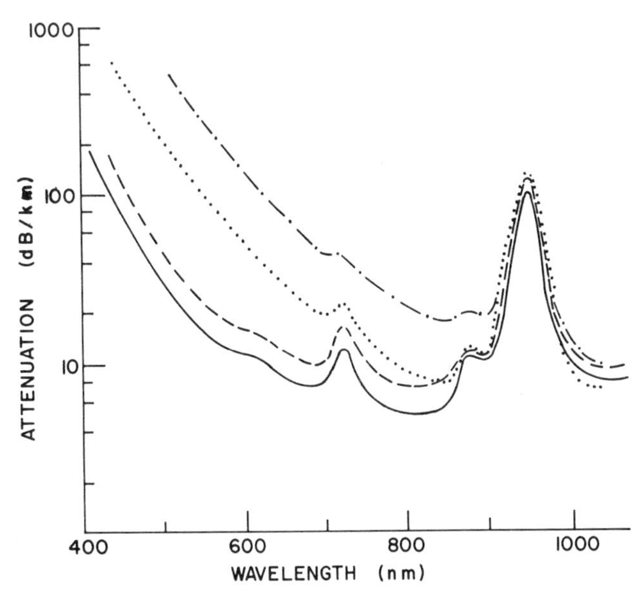

Fig. 2. Change in attenuation of a multimode optical waveguide with cumulative 14 Mev neutron irradiation: —— unirradiated, ---- 5.5×10^{10} n/cm^2, ... 5.6×10^{11} n/cm^2, —.— 1.4×10^{12} n/cm2.

Reprinted with permission from *Appl. Opt.*, vol. 12, pp. 2024–2026, Sept. 1973.

In addition to these experiments, a sample of Corning commercial fiber optic was irradiated with 3000 rads of gamma rays. The attenuation showed similar monotonic wavelength dependence but with an increase of about 20,000 dB/km at 550 nm and 3400 dB/km at 900 nm. Radiation induced color centers often thermally bleach at room temperature. No such effects were noticed during the course of this experiment, which lasted several weeks.

These experiments yield a single important result: it is feasible to fabricate glass optical waveguides with high radiation resistance in the near ir. This result should not be generalized since only one type of fiber was tested. Indeed, the result on the commercial fiber shows that conventional optical waveguides will not perform as these tested glass waveguides of high purity. Nevertheless, the demonstration of feasibility of radiation resistance is an extremely encouraging result for further development and future applications.

At wavelengths longer than 800 nm a communication link designed for a total loss of 20 dB/km would still be operable after exposure to radiation levels similar to those used in these experiments. Fortunately light emitting diode and injection laser signal sources (GaAs and GaAlAs) emit in the near ir (800–900 nm) as well as the Nd doped laser at 1.06 μm. Transmission links subjected to long term terrestrial levels (about 10 R as mentioned above) would change by less than 0.4 dB/km for signal wavelengths beyond about 600 nm. If thermal bleaching occurred during this long period, the change would be even less. It is highly probable that transient color centers exist and would be observable under very high radiation rates. The conductivity from transient charges has been observed during irradiation of fused silica.[6] However, such effects should not be important except in very high radiation environments.

References

1. D. B. Keck, R. D. Maurer, and P. C. Schultz, Appl Phys. Lett. **22,** 307 (1973).
2. A. Hollaender, ed., *Radiation Biology, Vol. 1 High Energy Radiation,* (McGraw Hill, New York, 1954), p. 582.
3. J. S. Stroud, J. Chem. Phys. **35,** 844 (1961).
4. R. D. Maurer, Proc. IEEE, **61,** 452 (1973).
5. D. B. Keck and A. P. Tynes, Appl. Opt., **11,** 1502 (1972).
6. V. E. Culler and H. E. Rexford, *Gamma-Radiation Induced Conductivity in Glass,* Annual Report 1960 of the Conference on Electrical Insulation, NAS-NRC Publication 973, (National Academy of Sciences, Washington, D.C. 1960), pp. 27–30.

Part VII
Connection and Splicing

Optical Fiber Joining Technique

By D. L. BISBEE

(Manuscript received July 15, 1971)

*This paper describes a method of thermally fusing clad glass fibers, end to end, to obtain a good mechanical joint with low transmission loss. Methods of preparing fiber ends and aligning them for joining are discussed. Two sizes of fibers were joined (10.8-μm core and 20-μm core clad fibers with outside diameters of 75 μm and 150 μm respectively).**

I. INTRODUCTION

There is a great deal of interest in using glass fibers as optical waveguides to carry information in much the same way as wires or metallic waveguides do. If glass fibers are to be used in this way, they will need to be joined just as wires and metallic waveguides must be joined.

A method for joining single fibers was developed. Clad glass fibers were joined which had cores of 10.8-μm and 20-μm diameters and overall diameters of 75 μm and 150 μm respectively. The cores were Schott SSK-1 glass and the cladding of Schott SK-14 glass which have glass transition temperatures of 621°C and 649°C respectively. Good mechanical joints which can be made quickly with transmission losses as low as 11.5 percent were obtained, but lower losses should be possible with a little more effort.

II. FIBER END PREPARATION

To get a good joint, good fiber ends are needed. Polishing or etching the fiber ends has been suggested, but we have found that if a fiber is broken properly it will have an end that is suitably flat over most of its surface and perpendicular to the axis of the fiber as seen under a microscope. Figure 1 shows two good ends of 10.8-μm core, 75-μm o.d. fibers magnified 500X. The break can be made by scoring the fiber with a razor blade and breaking it or by laying the fiber across a sharp metallic edge and positioning a Tesla coil so that its discharge is con-

* The fibers were manufactured by DeBell and Richardson, Inc., of Hazardville, Connecticut.

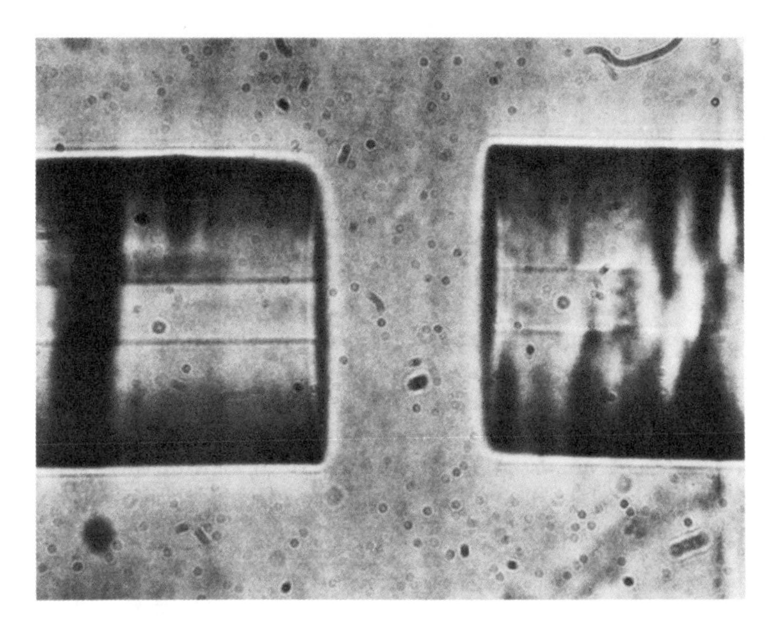

Fig. 1—Two cleanly broken ends of 10.8-μm core, 75-μm o.d. fibers magnified 500X.

centrated at the point where the fiber touches the metal, then breaking it by bending or pulling. The sparking must continue for several seconds with greater time for greater fiber diameter. To prevent the deposition of metal on the fiber because of the electrical discharge, platinum wire was used as the metallic contact. Fibers of 250-μm o.d. or greater can be scored with a file or diamond before breaking.

If the fiber is bent to the breaking point after it is scored or treated with an electrical discharge, one end will have a lip protruding and the other end will have a corresponding absence of material. This can be seen in Fig. 2. This lip is sometimes as long as the diameter of the fiber and would prevent the end from being brought close enough to another fiber end to permit joining them. If the fiber is pulled instead of bent until it breaks, the lip is not produced on most occasions, so this is the recommended procedure.

Very small fibers will sometimes soften and bend from the heat when the discharge from the Tesla coil strikes them, but with care, fibers as small as 25-μm o.d. have been broken with good ends suitable for joining.

III. ALIGNMENT

After the ends of the fiber have been prepared, they must be mounted so that they can be aligned. Teflon-coated tweezers to hold the fiber ends were mounted, one on a general purpose 3-dimensional micromanipulator and one on a precision 3-dimensional micromanipulator with a positioning resolution of 0.127 μm.

The Teflon-coated tweezers are gentle with the fiber and allow it to slip when the fiber contracts after having been heated, as will be discussed later.

From measurements of light output versus fiber end displacement we find that a misalignment of less than 2 μm in the 10.8-μm core fiber gives 10 percent less transmission than when the ends are aligned. Losses due to fiber offsets are covered in detail in Ref. 1.

One can determine when the ends are aligned by viewing them through a microscope and assuming that the core is concentric with the outside of the cladding. To get two perpendicular views of the fiber, a mirror can be mounted so the fiber can be viewed directly from the front and

Fig. 2—A characteristic break in a 10.8-μm core, 75-μm o.d. fiber showing the protruding lip.

through the mirror from the side. Two problems in this method of alignment are, first, the core and cladding may not be concentric, and second, rather sophisticated optics are needed to see an alignment error of 1 μm or less. If the core and cladding are made concentric and if a microscope of 200X or greater is used, one can probably align them well enough by this means.

Another method we have used of determining optimum alignment is to send laser light down the fiber to a detector and adjust the fiber ends for maximum transmission. A problem with this method is that if the end of the fiber is broken at an angle with respect to the normal to the fiber axis, the maximum transmission will be obtained when the fiber ends are misaligned to compensate for the offset in the beam direction caused by the angle of refraction at the nonnormal surface. This offset is y in Fig. 3. This error is small, though, if the angle is small. With a fiber of 80-μm diameter, core index of 1.6, and a surface at the end that is 10 degrees from the normal, the ends of the fibers would have to be offset 0.73 μm to correct for the beam misalignment. This would introduce about 5 percent loss in our 10.8-μm core fiber. An angle of 10 degrees is large, so one should be able to do much better than that.

IV. JOINING THE FIBERS

Several unsuccessful attempts were made to join fibers with epoxy alone and epoxy in a glass sleeve. When using epoxy alone, the resultant joint was too weak to keep the fiber ends aligned when transverse pressure was applied. When using sleeves, the tolerance between sleeve i.d. and fiber o.d. had to be very close, of the order of 1 μm to keep the fibers aligned properly, and this tolerance is hard to obtain. Further, a bubble formed at the fiber junction inside several sleeve joints. Thus, this method was considered limited in practicality.

A method that worked was the fusing of the fiber ends. Number 24 nichrome wire was wound around two metal posts so as to leave an

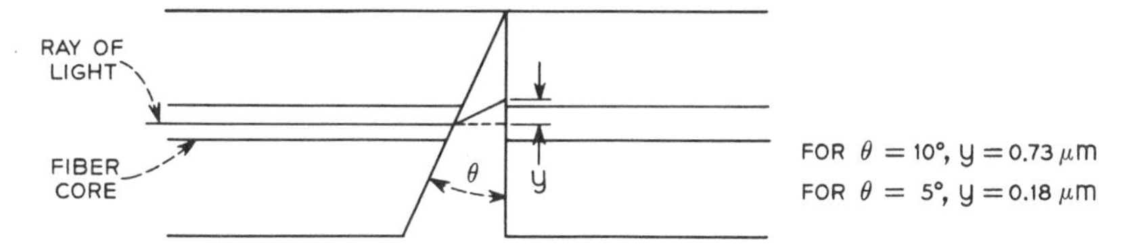

Fig. 3—Beam refraction at a nonnormal fiber end.

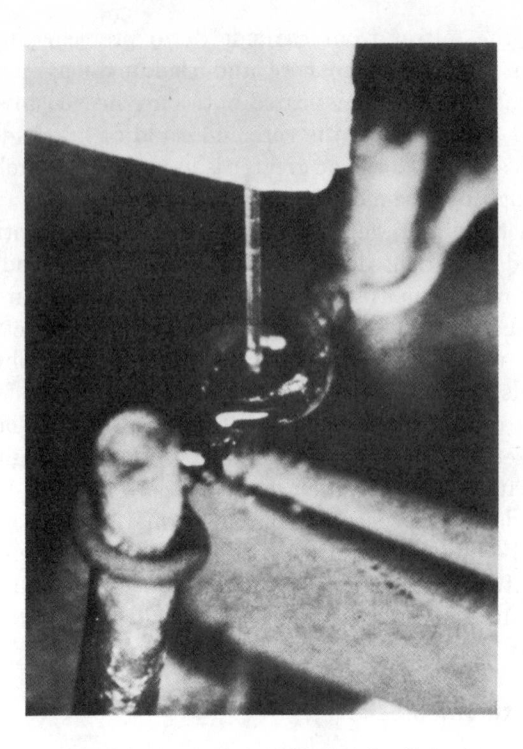

Fig. 4—Heating wire with fiber ends held in place by Teflon coated tweezers.

opening between the wires through which the fiber ends could pass as shown in Fig. 4.

The fiber ends were aligned, leaving a space of about 15 μm between them for thermal expansion. About 14 amperes of current were passed through the nichrome wire which surrounded the fiber ends causing the wire to heat up and fuse the fiber ends together. The longitudinal expansion of the fiber when heated closed the 15-μm gap that was left between the fiber ends. Of course, when the fiber cooled it shrank again, but the fiber could slip in the Teflon-coated tweezers when shrinking. The ends fused together in about 30 seconds after the heat was applied. To tell when the ends were fused, a lamp was placed so that the specular reflection from the fiber ends could be seen in the microscope. Disappearance of the reflection indicated the surface had vanished and the ends were fused. Figure 5 is a microphotograph of a fused joint in a 10.8-μm core, 75-μm o.d. fiber, at 500X magnification. Such joints exhibited losses as low as 11.5 percent.

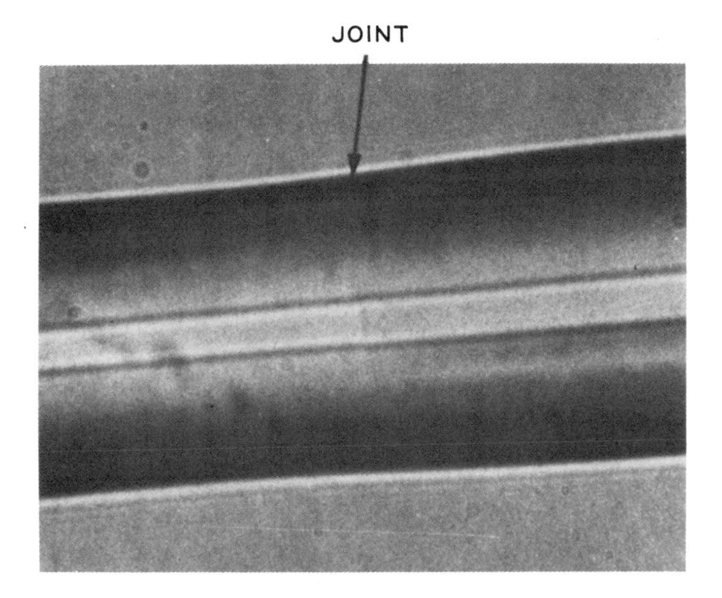

Fig. 5—A thermally fused joint in 10.8-μm core, 75-μm o.d. fiber.

V. CONCLUSIONS

After two fiber ends are carefully broken and properly aligned, they can be fused together by means of a heated wire to give a good mechanical joint with an acceptable amount of loss. Fibers of 10.8-μm core were joined and gave losses as low as 11.5 percent, but with a little more effort it is believed that joints could be made with considerably lower loss than this.

REFERENCES

1. Bisbee, D. L., "Measurements of Loss Due to Offsets and End Separations of Optical Fibers," B.S.T.J., this issue, pp. 3159–3168.

Lösbare Steckverbindung für Ein-Mode-Glasfaserlichtwellenleiter

Es wird eine lösbare Steckverbindung für Ein-Mode-Glasfaserlichtwellenleiter beschrieben, die von Hand ohne großen Justieraufwand bedient werden kann. Die Koppelvorrichtung besteht aus zwei versetzten Bohrungen zur Führung zweier Stifte, in denen die Glasfasern exzentrisch gehaltert sind. Durch entgegengesetztes Drehen der Stifte werden die Faserenden zueinander justiert. Die Anforderungen an die Justiergenauigkeit werden diskutiert. Mit dieser Steckverbindung konnten Koppelverluste von weniger als 0,4 dB erzielt werden.

Detachable Connector for Monomode Glass-fiber Lightwave Guides

A detachable connector for monomode glass-fiber lightwave guides is described, which permits operation by hand with little adjusting effort. The coupling device consists of two offset holes to guide two pins in which the glass-fibers are fastened eccentrically. By rotating the two pins against one another the two fiber ends are adjusted. The demands on the accuracy of adjustment are discussed. With this connector coupling losses smaller than 0.4 dB were achieved.

Es ist vorgesehen, ummantelte Glasfasern als Kabel in einem optischen Nachrichtenübertragungssystem einzusetzen [1]. Die Glasfasern sind so dimensioniert, daß nur die HE_{11}-Grundwelle ausbreitungsfähig ist. Erreicht wird dies dadurch, daß der Durchmesser des Faserkernes nur wenige Lichtwellenlängen beträgt (etwa 2 bis 4 μm). Für die praktische Handhabung eines derartigen Systems ist es von Vorteil, wenn der Sender (Halbleiterlaser) und der Empfänger (Avalanche-Mesaphotodiode) mit einem kurzen Stück Faser zu einer technologischen Einheit zusammengefaßt werden und wenn die Verkopplung der so entstandenen Sender- und Empfängereinheiten als reine Glasfaser-Glasfaser-Kopplung vorgenommen wird. Die Technik der Verbindung zweier Fasern ist also entscheidend. Auf Grund der geringen Abmessungen ist dies jedoch äußerst schwierig und nur mit großem Justieraufwand zu erreichen.

Im AEG-Telefunken Forschungsinstitut wurde ein Faserstecker entwickelt, der robust ist und ohne zusätzliche Hilfsmittel sehr einfach von Hand betätigt werden kann. Zudem läßt sich die Verbindung jederzeit wieder lösen, was beim Auswechseln von defekten Sendern und Empfängern von großem Vorteil ist. Diese lösbare Steckverbindung entspricht also in Funktion und Handhabbarkeit praktisch vollständig einem Koaxialstecker bei Hochfrequenzkabeln. Der prinzipielle Aufbau dieses Fasersteckers ist im Bild 1 skizziert.

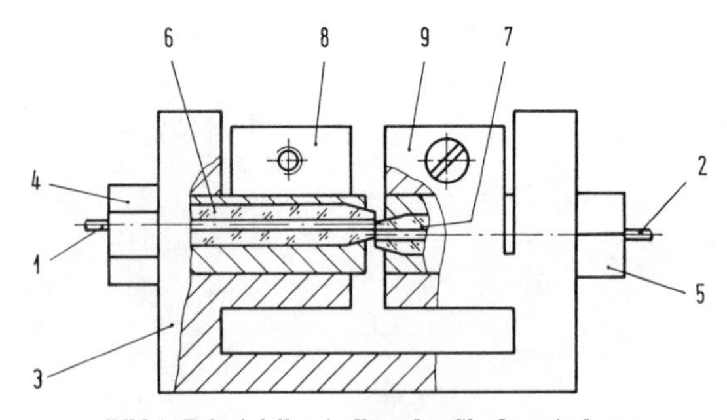

Bild 1. Prinzipieller Aufbau des Glasfasersteckers.

Die Justiervorrichtung besteht aus einem Gehäuse 3, in das zwei gegeneinander versetzte Bohrungen eingebracht sind. Diese Bohrungen dienen zur Führung der Stifte 4 und 5. In den Stiften befinden sich zusätzlich exzentrisch gehalterte Glaskapillaren 6, 7, in die die Fasern eingeklebt sind. Die Gesamtexzentrizität beträgt etwa 100 μm. Die zu verkoppelnden Endflächen müssen plan geschliffen und auf optische Qualität poliert werden. Durch entgegengesetztes Drehen der Stifte werden die Faserenden zueinander justiert. Am Ende des Justiervorganges müssen noch die

Reprinted with permission from *Arch. Elek. Übertragung*, vol. 26, pp. 288–289, June 1972.

Glasfasern in ihrer optimalen Stellung arretiert werden, was durch die Klemmvorrichtungen 8 und 9 erfolgen kann. Im Bild 2 ist ein Labormuster des beschriebenen Fasersteckers zu sehen.

Bild 2. Labormuster des Glasfasersteckers.

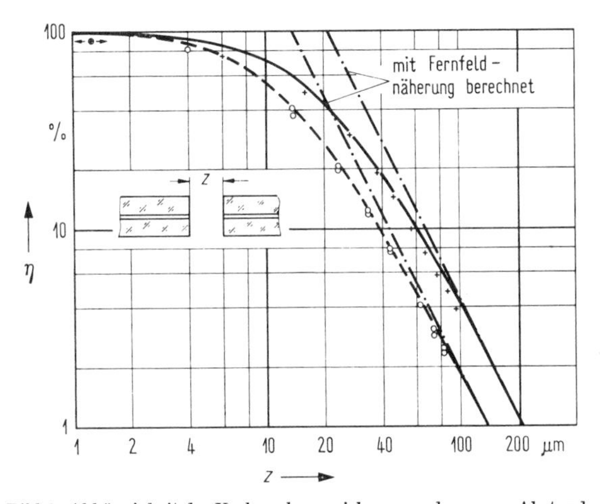

Bild 3. Abhängigkeit des Verkopplungswirkungsgrades η vom Abstand z der Faserenden;
– – – – Zwischenmedium Luft } Rechnung,
——— Zwischenmedium Immersionsöl } Rechnung,
– · – · – Strahlungsfeld der Faser mit Fernfeldnäherung berechnet,
o, × Meßwerte.

Die Anforderungen, die an die Justiergenauigkeit zu stellen sind, gehen aus den Bildern 3, 4 und 5 hervor. Dort ist, für eine Faser mit dem Kerndurchmesser $d = 1{,}98$ µm und den Brechungsindizes $n_1 = 1{,}5311$ für den Kern, $n_2 = 1{,}516$ für den Mantel und einer Lichtwellenlänge $\lambda = 0{,}633$ µm (He-Ne-Laser), der Wirkungsgrad η der Verkopplung als Funktion des Abstandes z [2], [3] (mit Luft und Immersionsöl ($n = 1{,}515$) als Zwischenmedium) und des Versatzes S [3] der beiden Faserenden sowie des Winkels ϑ zwischen den beiden Faserachsen aufgetragen. Bei der Berechnung von η in Abhängigkeit vom Versatz S wurde die Intensitätsveirtelung der anregenden Welle durch eine Gaußkurve angenähert. Der dadurch bedingte Fehler ist sehr klein. Außerdem wurden Reflexionsverluste bei der Berechnung der Kurven vernachlässigt.

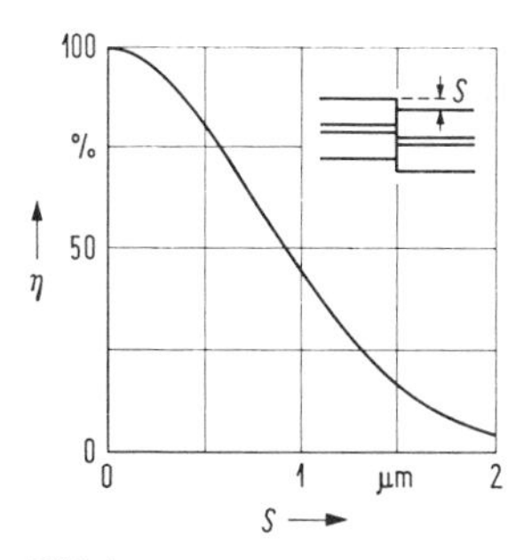

Bild 4. Abhängigkeit des Verkopplungswirkungsgrades η vom Versatz S der beiden Faserenden.

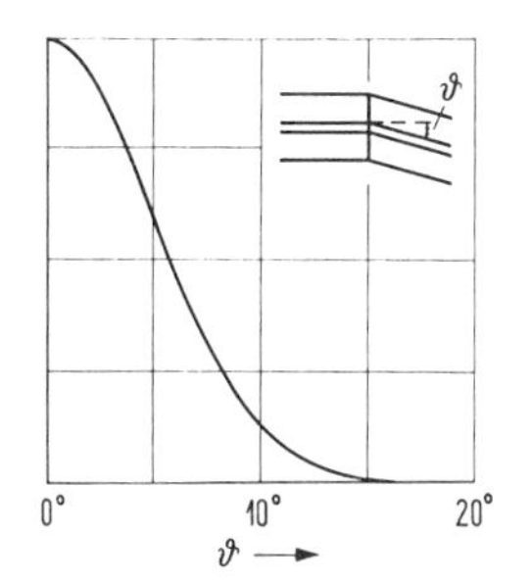

Bild 5. Abhängigkeit des Verkopplungswirkungsgrades η vom Winkel ϑ zwischen den beiden Faserachsen.

Wie den Darstellungen zu entnehmen ist, dürfen, um einen Koppelwirkungsgrad von mindestens 90% zu erreichen, die Abweichungen von der Ideallage nur sehr gering sein. Der Abstand zwischen den beiden Faserenden darf nicht mehr als 3 µm betragen. Außerdem müssen die Faserenden auf Bruchteile eines µm zueinander justiert werden und der zulässige Winkel zwischen den beiden Faserachsen muß kleiner als 2° sein. Trotz dieser hohen Anforderungen ist es möglich, die Steckverbindung von Hand zu justieren. Die gemessenen Koppelwirkungsgrade lagen über 90%, was einem Koppelverlust von weniger als 0,4 dB entspricht.

Wir danken Herrn Dipl.-Phys. A. Jacobsen von der Firma Schott & Gen. in Mainz für die Bereitstellung der Glasfaser und Frau H. Knapp für die mit dem Elektronenrechner ausgeführten Rechnungen.

(Eingegangen am 18.3.1972.)

Dr. Manfred Börner
Dietmar Gruchmann,
Joachim Guttmann
Dr. Oskar Krumpholz
Ing. (grad.) Werner Löffler

i.H. AEG-Telefunken,
Forschungsinstitut
79 Ulm (Donau), Elisabethenstraße 3

Schrifttum

[1] Börner, M., Ein optisches Nachrichtenübertragungssystem mit Glasfaser-Wellenleitern. Wiss. Ber. AEG-Telefunken **44** [1971], 41–44.

[2] Krumpholz, O., Optical coupling problems in communication systems with glass-fiber optical waveguides. Digest of Technical Papers (Las Vegas 1972), S. WB5-1–WB5-4.

[3] Bisbee, D. L., Measurements of loss due to offsets and end separations of optical fibers. Bell Syst. tech. J. **50** [1971], 3159–3168.

Simple, Low-Loss Joints Between Single-Mode Optical Fibers

By C. G. SOMEDA*

(Manuscript received September 21, 1972)

Low-loss joints between single-mode optical fibers have been made without microscopic alignment, without fusing the tips, and without monitoring the transmitted power while the joints are assembled. The fibers are tightly held in an embossed groove; an index-matching liquid is added. Average power coupling efficiencies close to 90 percent in the red and to 85 percent in the infrared have been obtained. Mediocre end faces are acceptable. Realistic discrepancies between the fiber cladding diameters (slightly in excess of twice the core diameter) do not deteriorate the results.

I. INTRODUCTION

Recently, several authors have dealt with techniques for making low-loss joints between multimode[1–4] or single-mode[5] optical fibers. For long-distance communication channels, joints between fiber cables, simply assembled in the field, would be very valuable.

The joints that were made so far with the technique described in this paper connected just one pair of fibers at a time. However, this is intended to be the first step in the development of joints between multifiber cables. Therefore, in the present experiments emphasis has been put on getting repetitive results which are not a consequence of careful laboratory adjustment. This is an important difference with respect to the methods that gave the best results reported so far.[1,5] Nonetheless, the joints between single-mode fibers described here compare favorably with previous ones as far as the losses are concerned. The demonstration with single-mode fibers (core diameters of $\approx 4\ \mu m$) means that the technique applies also to multimode fibers, the larger core diameters of which should make the alignment less critical.

* This work was performed at Bell Telephone Laboratories, Incorporated, Crawford Hill Laboratory, Holmdel, New Jersey, during a six-month internship sponsored by a NATO–CNR fellowship.

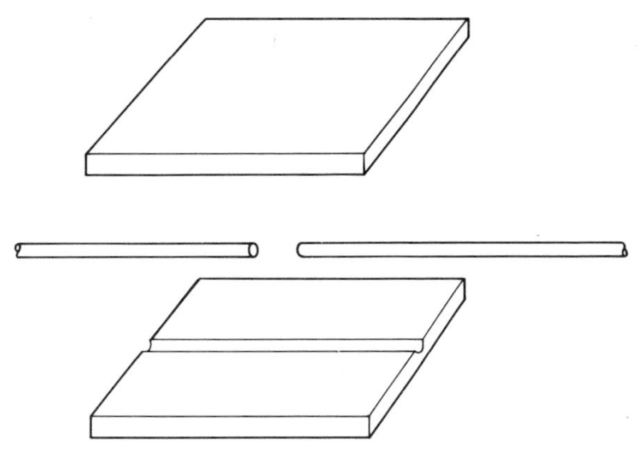

Fig. 1—View of the unassembled parts of a sandwich joint.

A complete description of the technique follows in Section II. As a rough sketch, we may say that the tips of the fibers are aligned in a groove (Fig. 1), embossed in a Plexiglas sheet.* The fiber tips are pushed against each other. An index-matching liquid is added. A "sandwich" is formed with a flat Plexiglas sample, and then squeezed by means of a small vise (Fig. 2). No fusion takes place and no microscopic alignment is needed.

It is obvious that the fiber claddings are aligned in this way and therefore the results depend on the fiber core being coaxial with the cladding. The experiments reported here were performed with single-mode fibers† having a core diameter of 3.7 μm and a cladding diameter of 254 μm; coincidence of the core and cladding axes within ≈ 1 μm had been observed preliminarily.

Experiments were performed at 6328 Å and at ≈ 9000 Å. Repetitive measurements gave average power couplings around 87 percent in the red and around 83 percent in the infrared; best results were 93 percent and 87 percent, respectively. These figures were obtained without paying any particular attention to the quality of the end surfaces of the two fibers.

The experiments were completed by testing the effects of a difference, d, between the cladding diameters, as described in Section IV. It turned out that sandwiches of good mechanical quality exhibit no deterioration of the results for $d \cong 10$ μm (i.e., more than twice the core diameter). For $d \cong 30$ μm, the results were deteriorated, but still showed

* Plexiglas (methyl methacrylate), registered trademark of Rohm and Haas.
† These fibers were manufactured by Corning Glass Works, Corning, N. Y.

Fig. 2—View of an assembled joint. The length of the visible side of the sandwich, perpendicular to the fiber, is ≈ 2 cm.

the presence of an alignment mechanism, which is believed to be due to surface tension in the index-matching liquid.

II. DESCRIPTION OF THE JOINING TECHNIQUE

In order to obtain a joint of the type shown in Fig. 1, the first operation to perform is to emboss a Plexiglas sheet with a groove that fits the dimensions of the fibers to be joined.

A simple and economical technique for embossing grooves in a thermoplastic substrate by means of a glass fiber has been reported by Ulrich *et al.*[6] A further simplification of this procedure proved successful; there is no need to heat the substrate. A rectangular Plexiglas sample and a fiber piece, the length of which slightly exceeds the Plexiglas size, are put between two milled aluminum blocks (flat within ± 1 mil) and then tightly pressed by means of a vise for a few minutes. When the vise is released, the fiber separates from the Plexiglas, leaving a sharply embossed groove. Any other piece of a nominally equal fiber can be introduced into the groove and fits it very closely. Figure 3

Fig. 3—Microscopic picture of a single-mode fiber tip (on the left) lying in an embossed alignment groove. Magnification 200× (1 cm in the picture is 50 μm in actual scale).

is a microscopic picture (magnification 200×) of a fiber tip lying in a groove, embossed with another fiber. Figure 4 is a further enlarged view (magnification 500×) of the groove edge in the vicinity of the fiber end; it shows no defect comparable to the size of the fiber cores to be aligned. It also shows a fairly regular pattern of longitudinal wrinkles due to the compression of the Plexiglas sample.

First, Plexiglas having a nominal thickness of $\frac{1}{16}$ inch was embossed with quartz fibers, 254 μm in cladding diameter. The samples were permanently curved with a remarkable convexity of the embossed side. The cover (Fig. 1) then tended to flip. Using a thicker Plexiglas sample (e.g., $t = \frac{1}{8}$ inch), the curvature was negligible.

There is some evidence that the deformation of the embossed Plexiglas sheets is elastic (at least partially) with a long time constant. Therefore, the joints have to be made with newly embossed samples, or with samples where a fiber has been constantly pressed in.

To assemble a joint, the fibers are put in the groove, a small amount of a suitable index-matching oil is added, and the cover is placed on top. At first, the top is not pressed against the bottom. Next, one fiber is pushed against the other in the axial direction; finally, the sandwich is squeezed with a small vise.

Fig. 4—Microscopic picture of the edge of a groove, with a fiber tip on the left. Magnification 500× (1 cm in the picture is 20 μm in actual scale).

It seems that the index-matching oil is rather slow in wetting the fiber tips, and some time is required for air bubbles to escape. The oil also has the purpose of lubricating the fiber longitudinal motion. Sometimes an excess of it deteriorated the joint performances, probably because an unwanted axial movement of the fiber tips was produced by the oil flux when the sandwich was squeezed. However, the presence of a small amount of liquid is needed in order to get good coupling efficiency (see Section 3.2). Furthermore, the presence of the fluid is believed to be responsible for the excellent behavior in case of unequal diameters of the two fibers (Section IV).

The amount of light transmitted by the joint does not always increase with increasing vise pressure. Often, though, a decrease in the coupling beyond a certain pressure was interpreted as due to some misalignment caused by careless movements accompanying the vise tightening. When the groove was deep enough and the top could not swing or slide, then such a decrease in transmitted power was absent or negligible. A good rule is to press the joint between surfaces that are not too stiff, so that the force is distributed on the whole area of the sandwich.

III. MEASUREMENTS

3.1 *Tests at 6328 Å*

Sandwich joints of the type described in Section II were tested first at the wavelength $\lambda = 6328$ Å. Based on manufacturer's data, the corresponding normalized frequency,[7] $V \cong 2.2$, means that single-mode propagation takes place. Direct observation of the far field at the end of a fiber confirmed that.

The set-up is illustrated in Fig. 5. The light emitted by a commercial He-Ne laser was chopped and then launched into the input fiber by means of a microscope objective (magnification 40×). A curved part of the fiber was lying in an index-matching liquid, to strip off the light launched into the cladding. The length of this mode-stripping section was such that any further addition or small subtraction did not affect the power level at the output of the fiber. This output was detected by means of an Si solar cell and the signal was sent to a lock-in amplifier. Figure 5 shows both arrangements that are needed in order to evaluate the insertion loss of a joint: a "reference" arrangement, where the detector is directly connected to the input fiber, and a "measurement" arrangement, where a joint and an output fiber are added.

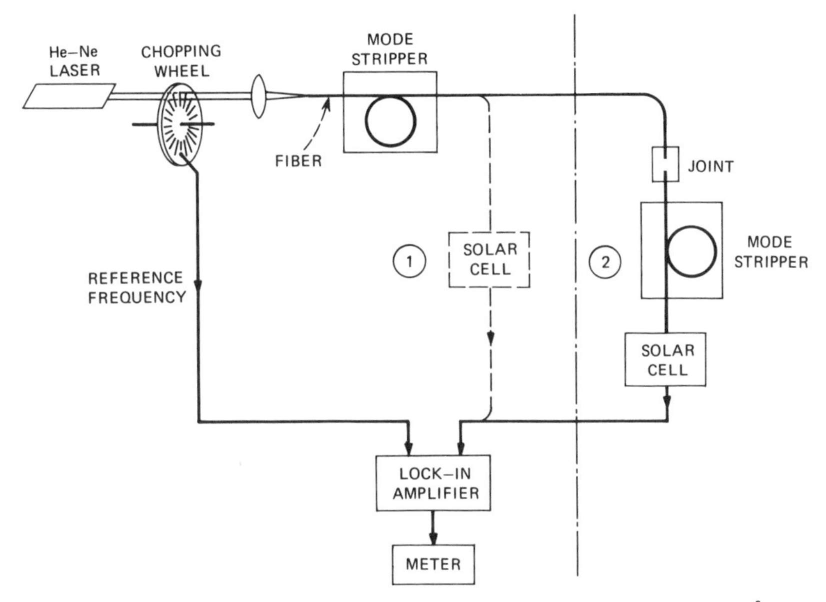

Fig. 5—Block diagram of the set-up used for the measurements at 6328 Å. The broken lines (Part 1) refer to calibration (measurement of a reference level). The solid lines (Part 2) refer to the actual measurement of the light transmitted through a joint.

TABLE I—POWER COUPLING EFFICIENCIES MEASURED AT 6328 Å AND CORRESPONDING INSERTION LOSSES OF THE JOINTS

Sample Number	1	2	3
Best results	92% (0.35 dB)	93% (0.3 dB)	92% (0.35 dB)
Worst results	74% (1.3 dB)	80% (1.0 dB)	81% (0.9 dB)
Average results	80.7% (0.9 dB)	88.7% (0.5 dB)	86.6% (0.6 dB)
Number of measurements	12	6	6
Number of different ends	4	2	2

Attenuations measured in this way result from both the joint insertion loss and the losses in the output fiber. Previous data[8] and direct observation show that the output fiber attenuation is negligible with respect to the loss in the joint, despite the need for a length of the output fiber that could insure an effective stripping of all the light transferred by the joint into the cladding.

This way to perform the measurements was preferred to that where, after taking a reference, the fiber is cut and then joined again, because frequent checks of the reference were needed. Indeed, launching into a single-mode fiber is so critical (the power drops by 3 dB if the fiber tip is misaligned by a few microns), that accidental causes can produce strong shifts in the reference level. For example, in our case the material used to bind the fiber tip on an x-y-z manipulator could deform slowly, under the strength applied by the curved and quite elastic fiber. This effect was compensated for by optimizing the alignment of the fiber with respect to the incoming beam before any meter reading, both in the reference and in the measurement arrangements. Discrepancies between optimized reference levels monitored before and after a measurement were never larger than 3 percent and usually much smaller than that. When the reference levels monitored before and after a measurement were different, the more pessimistic estimate of the joint losses was taken.

Table I contains the best, worst, and average results of three sequences of measurements. As for the "worst" results, they exclude only those instances where clearly identified man-made mistakes took place; i.e., axially separated fibers, one fiber lying out of the groove, and Plexiglas debris shaved by the fibers and accumulated between the two ends. These errors were very rare, and were easily detected because of a large amount of scattered light shining from the sandwich. This point will be discussed further in Section 3.2.

The sequences of Table I refer to assembling, several times, joints that make use of three embossed grooves, one in a $\frac{1}{16}$-inch-thick plate and two in $\frac{1}{8}$-inch-thick plates. All the measurements were independent, in the sense that after any of them the joint was at least disassembled completely and then reassembled. The ends of the fibers were changed often, as shown in Table I. The amount of index-matching oil was changed quite often too.

An important point is that none of the measurements were of a joint where the two coupled end faces resulted from one cut of a fiber; they always resulted from two independent breaks, in order to simulate real situations.

A remarkable advantage of this technique is that there is no need for very accurate flatness of the fiber end faces. As long as there are no lips, which would prevent the two cores from getting close to one another, results are good. Figure 6 is a microscopic picture of one typical pair of ends used in the measurements. Attempts to get end surfaces of this

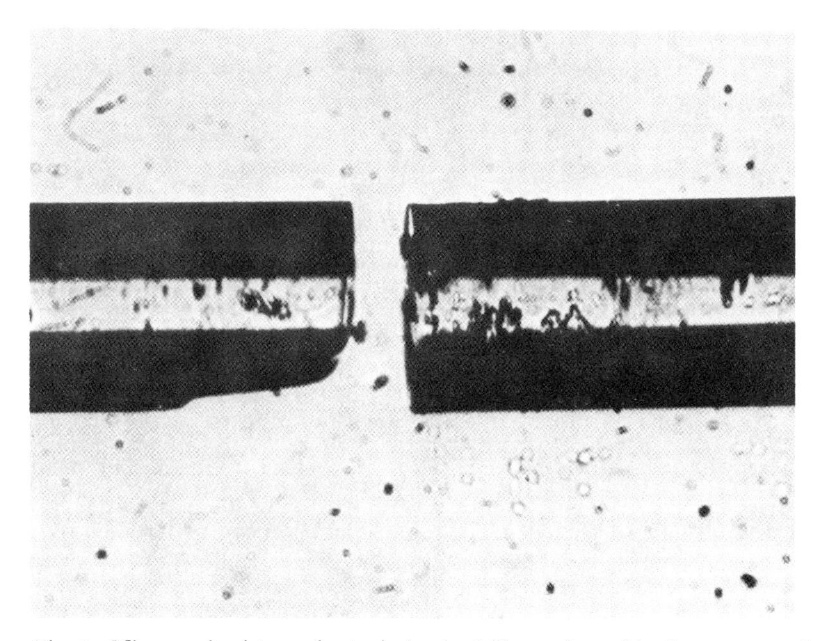

Fig. 6—Microscopic picture of a typical pair of fiber ends used in the sequence of experiments summarized in Table I. The bright zone in the center is not the fiber core, but just a consequence of the illumination.

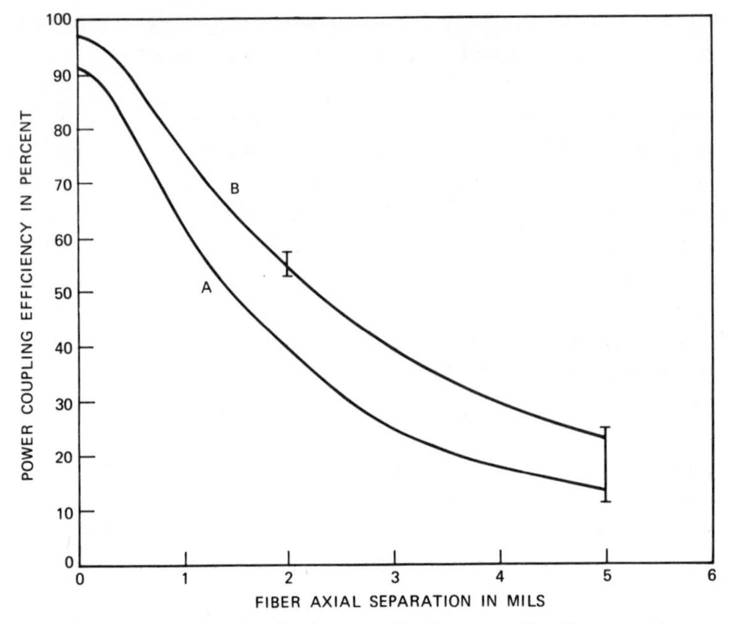

Fig. 7—Effects of the longitudinal separation between the fibers on the coupled power. Lines A and B, from Ref. 2, represent measured values (without and with index-matching oil, respectively) at 6328 Å when the fibers were aligned by means of micromanipulators. The vertical segments represent the range of results obtained when the fibers were held in a sandwich joint.

quality were almost always successful if the fibers were pulled after scoring them with a lathe tool or with an electrical discharge generated by a Tesla coil. Many successful attempts simply consisted of clamping the fibers in a Plexiglas sandwich and then pulling them. Consequences for the multifiber case will be stated in Section V.

Another interesting observation was that when a joint was tightened, bringing the fiber cores into alignment, the forward scattered light on the side surface of the output fiber decreased very remarkably.

Some additional measurements were performed in order to establish the effects of longitudinal separation between the fibers. The results are compared with those obtained by Bisbee[2] in Fig. 7. The very large spread of values observed for a 5-mil separation can be justified by an occasional shortage of index-matching oil, which could cause an air gap sometimes to show up when the fibers were separated; or by deviations of the groove from a rectilinear shape, which would cause unwanted transverse displacements of the fiber tips to accompany the longitudinal ones. The best results ($\eta \cong 56$ percent for a 2-mil separation,

TABLE II—POWER COUPLING EFFICIENCIES MEASURED AT 9000 Å AND CORRESPONDING INSERTION LOSSES OF THE JOINTS

Sample Number	2	3
Best results	87.5% (0.55 dB)	87% (0.6 dB)
Worst results	78.5% (1.05 dB)	80% (1.0 dB)
Average results	82.5% (0.85 dB)	83.7% (0.75 dB)
Number of measurements	11	13
Number of different ends	2	2

$\eta \cong 27$ percent for a 5-mil separation), which are in very good agreement with Bisbee's,[2] were observed more frequently than the bad ones.

In multifiber joints, it is unlikely that the axial separation of each pair of fibers will be as small as it was for the single pairs tested so far. However, one can think of enlarging the waist of the light beam in the fibers, by using a smaller normalized frequency[7]; the confocal length of the beam would then grow as the square of the beam waist. The effect of an axial separation upon the coupling efficiency would be smaller than that shown in Fig. 7.

3.2 *Tests at 9000 Å*

It was pointed out before that, while working with visible radiation, a misalignment or a gap between the fibers was revealed by a large amount of scattered light leaking from the joint. In a multifiber joint, this light will not identify the pair of fibers that form a leaky connection. Hence, it was necessary to check whether the very small number of mistakes in assembling the joints and the high average of their performances were independent of the information that the operator was provided by the scattered light.

A sequence of tests in the near infrared, at $\lambda \cong 9000$ Å (where $V \cong 1.55$), was then performed using a small-area GaAs LED.* The set-up differs from Fig. 5 only because the chopping of the signal is performed by an audio oscillator driving the LED. Launching spatially incoherent radiation into the fiber is much less critical than in the case of the laser; this results in a much smaller long-term drift of the reference level.

* The LED was built and provided by C. A. Burrus.[9]

Tests were performed on joints that made use of the grooved samples referred to in columns 2 and 3 of Table I. Quite repetitive results were obtained. They are summarized in Table II. Comparison with Table I shows a very slight deterioration of the best results (power coupling coefficients from 92–93 percent down to 87–87.5 percent) but even smaller changes of the averages (from 80–81 percent down to 78.5–80 percent) and almost no change in the lower ends of the range. Let us re-emphasize that the tested joints were not assembled while monitoring the transmitted power.

Comparison of both Tables I and II with previously published data on single-mode fiber joints[5] shows the desirability of the present sandwich technique. Best results given by the fusion technique[5] show power transmission efficiencies of ≈ 80 percent, at a normalized frequency $V \cong 2.0$; the average results given by the sandwich technique are better, both above and below that value of V.

The number of times when the attempts to assemble the joints were unsuccessful because of easily identified mistakes in the fiber position increased to three in a sequence of twenty-seven measurements, compared to one in twenty-five measurements at 6328 Å. This value is still low enough to insure that the technique is suitable for further developments in a multifiber system, where the risk of making such mistakes has to be negligible.

Some information on the importance of the index-matching oil has been collected, too: "dry" joints gave best power coupling efficiencies in the order of 50 percent. Comparison with available data[2] shows that the oil produces some other effect besides the elimination of end reflections. Quite likely, it compensates for irregularities of the end faces; also, it provides some alignment mechanism via surface tension. The last point is discussed further in the next section.

IV. EFFECTS OF DISCREPANCIES BETWEEN THE FIBER OUTER DIAMETERS

The usefulness of the sandwich technique would be limited if its performances were sensitive to discrepancies between the diameters of the two joined fibers, a situation that will occur in practice. The results reported in this section show that the sensitivity to this kind of imperfection is small enough to satisfy practical purposes.

A controlled difference, d, between the outer diameters of the fibers was introduced by etching one fiber with commercial hydrofluoric acid. First, a value $d \cong 10\ \mu m$ was chosen, because it would be sufficient to allow more than a complete offset of the two cores if the claddings were

TABLE III—POWER COUPLING EFFICIENCIES OF JOINTS BETWEEN FIBERS WITH A 10-μm DIFFERENCE IN CLADDING DIAMETERS. MEASUREMENTS PERFORMED AT 9000 Å.

Best result	86% (0.65 dB)
Worst result	82% (0.85 dB)
Average result	84.7% (0.7 dB)
Number of measurements	7
Number of different ends	2

aligned along one generatrix; besides which, 10 μm is 4 percent of the cladding diameter, which can be thought of as a realistic tolerance.

A new sequence of tests was performed at $\lambda \cong 9000$ Å, using the embossed sample referring to column 2 in Tables I and II. The results are summarized in Table III. Comparison with Table II shows that there is no appreciable deterioration of the joint performances. On the other hand, it was observed that the transmission was sometimes (not always) sensitive to small changes in the force applied by the vise; e.g., for one of the best joints assembled without monitoring the transmitted power ($\eta \cong 86$ percent), small changes in the tightening could cause a drop of the transmitted power to 82 percent or a raise to 89 percent (insertion loss ≈ 0.5 dB).

In all cases, releasing the vise slowly caused, first, a decrease in transmitted power; then a maximum, close to the figures quoted before, showed up; later, the signal decreased monotonically down to the noise level. The secondary maximum, which was very sensitive to vibrations and shocks, is believed to be entirely due to an alignment caused by the surface tension of the index-matching oil. The same phenomenon is believed to contribute to the excellent alignment under the strongly tightened vise, leading to the figures reported before.

Later, the difference between the cladding diameters was brought to a value $d \cong 30$ μm, i.e., about 12 percent of the outer diameter. The following behavior was observed:

(*i*) There was practically no coupling as long as the joints were not tightened very strongly. This was a remarkable difference with respect to the case of equal diameters, where the simple contact of the two faces, before squeezing the sandwich, could often produce a power transmission in the range 10 to 30 percent.

(*ii*) Several joints, assembled and very tightly squeezed without monitoring the transmitted power, had coupling efficiencies up to 40 to 50 percent and were not very sensitive to vibrations and shocks; however, the worst results ranged down to ≈ 10 percent (very seldom) or 20 percent (more frequently).

(*iii*) Repeated operations of tough tightening followed by partial releasing, while the transmitted power was monitored by a meter, led to coupling efficiencies in the range 80 to 84 percent. These joints were not tightly squeezed, and were very sensitive to vibrations and shocks, which could reduce the coupling all the way to 10 percent. Usually, though, the joints that had been disturbed, once abandoned to themselves, tended to restore spontaneously power coupling efficiencies up to 70 to 75 percent.

These observations and comparison with Bisbee's data[2] show that a remarkable alignment mechanism, due to the presence of the liquid, is still active for such a large difference between the cladding diameters, even if the performances are not any more suitable for practical applications.

V. CONCLUSION

Sandwich joints between single-mode fibers, aligned in embossed grooves, have been made and tested. This technique seems to suit very well the operation of splicing optical waveguides without using a microscope and without fusing their tips together. The average results obtained in this way–power coupling efficiencies ranging from 80 percent to almost 90 percent, i.e., insertion losses under 1 dB and down to 0.5 dB–compete successfully with the best results given by previous techniques.

The facts, that rather mediocre end surfaces of the fibers can be accepted and that cladding diameters can differ at least by a few percent without affecting the performances, induce confidence in the future use of the sandwich technique for splicing multifiber cables in the field. It seems that all the operations could be extended to fiber tapes, dealing with each of them as a whole. A fairly precise alignment of the fibers would be required as the result of the tape manufacturing process, but the final alignment should be provided by a set of precisely embossed grooves on a plate. The fibers ought to be forced into the grooves by means of carefully designed mechanical tools, but without microscopic observation and without dealing with them on an individual scale. The operations to be performed will depend on the ratio between the diameters of the fibers and their spacing in the tape, and on the precision of the preliminary alignment obtained in the manufacturing process. The absence of stringent requirements on the quality of the end faces, proved by the available results, suggests that a simultaneous and coplanar cut of all the fibers (for instance, scoring them by means of a razor blade while pulling along their axis) should be adequate.

VI. ACKNOWLEDGMENTS

This work has been performed during a six-month internship at Bell Laboratories, Crawford Hill Laboratory, Holmdel, New Jersey, under a NATO Fellowship awarded and administered by the Italian National Research Council (C.N.R.) of Rome, Italy. I wish to thank S. E. Miller for arranging this internship in the most useful and pleasant way.

I am very grateful to E. A. J. Marcatili, who suggested the present investigation and stimulated it continuously with very fruitful discussions.

Several other people helped and encouraged me. I wish to thank in particular C. A. Burrus for providing the light-emitting diodes and for his important experimental advice.

REFERENCES

1. Bisbee, D. L., "Optical Fiber Joining Technique," B.S.T.J., *50*, No. 10 (December 1971), pp. 3153–3158.
2. Bisbee, D. L., "Measurements of Loss Due to Offsets and End Separation of Optical Fibers," ibid., pp. 3159–3168.
3. Standley, R. D., and Braun, F. A., "Some Results on Fiber Optic Connectors," unpublished work.
4. Astle, H. W., "Optical Fiber Connector with Inherent Alignment Feature," unpublished work.
5. Dyott, R. B., Stern, J. R., and Stewart, J. H., "Fusion Junction for Glass-Fibre Waveguides," Elec. Ltrs., *8*, No. 11 (June 1, 1972), pp. 290–292.
6. Ulrich, R., Weber, H. P., Chandross, E. A., Tomlinson, W. J., and Franke, E. A., "Embossed Optical Waveguides," Appl. Phys. Ltrs., *20*, No. 6 (March 15, 1972), pp. 213–215.
7. Gloge, D., "Weakly Guiding Fibers," Appl. Optics, *10*, No. 10 (October 1971), pp. 2252–2258.
8. Keck, D. B., and Tynes, A. R., "The Spectral Response of Low-Loss Optical Waveguides," Appl. Optics, *11*, No. 7 (July 1972), pp. 1502–1506.
9. Burrus, C. A., "Radiance of Small-Area High-Current-Density Electroluminescent Diodes," Proc. IEEE, *60*, No. 2 (February 1972), pp. 231–232.

Optical Fiber End Preparation for Low-Loss Splices

By D. GLOGE, P. W. SMITH, D. L. BISBEE, and E. L. CHINNOCK

(Manuscript received May 8, 1973)

Cables made from brittle materials like glass require new techniques of end preparation for the purpose of splicing, especially if such splices are to be made in the field. We report here on a method of breaking fibers in a way which invariably produces flat and perpendicular end faces. We explain the underlying theory and derive optimal parameters that permit the design of a simple breaking tool. Experiments with a tool of this kind show that the tolerances for successful fracture are not critical. Laboratory splices of multimode fibers prepared by this method exhibited losses of less than 1 percent (0.04 dB) when joined in index-matching fluid.

I. INTRODUCTION

With installation and maintenance consuming an ever-larger share of system costs, simple and inexpensive splicing techniques have become a prerequisite for competitive communication systems. One bottleneck in optical fiber cable splicing is the fiber end preparation, as conventional grinding and polishing techniques turn out to be time-consuming and costly, especially in the field. It is well known that glass fibers sometimes break with flat and perpendicular end faces if they are previously scored,[1] and it has thus become common practice in the laboratory to obtain good ends in this way by trial and error. Besides being faster and simpler, this technique has the added advantage of producing perfectly clean surfaces uncontaminated by lossy residues. Such ends were recently used in fiber joining experiments to determine eventual splice losses.[2–5] The lowest losses obtained were about 10 percent for single-mode fibers[4,5] and 3 percent for multimode fibers.[2]

For such laboratory practice to become useful technology, absolute control of the breaking process and utmost reliability in obtaining a successful result are required. We report here on an approach which guarantees this reliability through control of the stress distribution in the fracture zone. The break is initiated by lightly scoring the fiber periphery at the correct point. We explain the underlying theory which allows us to predict the character of the break from the initial stress distribution. By modifying a previous design,[6] we obtained a simple tool that permits us to vary the amount and distribution of stress in the fracture zone. All 130 breaks we have made with this tool have produced the predicted fracture surface. The range within which perfectly flat and perpendicular end faces were obtained was found to be so wide that the eventual construction of a simple hand tool for this purpose should present no problem. The quality of the surfaces obtained makes this method the most promising of all the techniques investigated so far.[7–9] This notion is supported by some fiber-joining experiments which we describe in Section IV of this paper. Low-loss multimode silica glass fibers were prepared by our breaking technique and then joined in an index-matching liquid. With proper alignment, the splice losses were always less than 1 percent. Results on alignment tolerances for multimode fiber splices are also given in Section IV.

II. BRITTLE FRACTURE OF GLASS RODS AND FIBERS

It has been well documented that glass rods tend to break in such a way that the fracture face comprises three regions known as the mirror, the mist, and the hackle zones.[10,11] The mirror zone is an optically

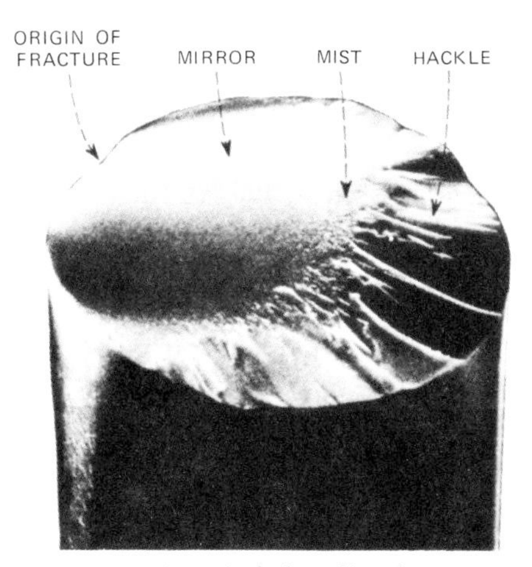

Fig. 1—A typical glass fiber fracture.

smooth surface adjacent to the fracture origin, the hackle zone corresponds to an area where the fracture has forked and the specimen is separated into three or more pieces, and the mist zone is a transition region between these two zones. Such behavior is also observed with glass fibers. Figure 1 shows the fractured end of a 125-μm glass fiber which clearly exhibits these three regions.

It has been experimentally demonstrated[11] that the distance from the origin of fracture to a point on the boundary between the mirror and mist zones, r, is given by

$$Z\sqrt{r} = K\,, \tag{1}$$

where Z is the local stress at the point in question and K is a constant for a given material.

A theoretical justification for eq. (1) can be given. Anderson[12] gives the energy balance equation for a crack of length $2c$ propagating in a brittle isotropic material subject to a plane stress, Z, as

$$\frac{d}{dc}\left(-\pi\frac{Z^2c^2}{E} + \frac{1}{2}k\rho\dot{c}^2\frac{c^2Z^2}{E^2} + 4\gamma c\right) = 0\,. \tag{2}$$

Here E is Young's modulus, ρ is the density, and γ is the surface tension of the material. The parameter k is a geometrical factor which depends on the shape of the crack. The three terms in eq. (2) represent, respectively, the released strain energy, the kinetic energy associated with the moving crack, and the surface energy of the newly created surfaces. As the crack propagates, more and more strain energy is converted into kinetic energy until the crack reaches a limiting velocity, $\dot{c} = v_f$, where v_f is roughly $\frac{1}{3}$ the longitudinal sound velocity for the material (see, for example, Reference 12). At this point the excess energy begins to be taken up by the creation of subsurface cracks (the mist zone). When the released strain energy is sufficient to create four new surfaces, a hackle zone is created. Thus, at the boundary of the mirror and mist zones,

$$\frac{d}{dc}\left(-\pi\frac{Z^2c^2}{E} + \frac{1}{2}k\rho v_f^2\frac{c^2Z^2}{E^2} + 4\gamma c\right) = 0\,. \tag{3}$$

By differentiating, we find

$$Z^2c = \frac{4\gamma E}{2\pi - k\rho v_f^2/E} = \text{a constant}\,, \tag{4}$$

which is of the same form as eq. (1). A similar derivation is given in Reference 11. The value of the constant K in eq. (1) is found experimentally to have the value 6.1 $kg/mm^{\frac{3}{2}}$ for soda-lime-silicate glass and 7.5 $kg/mm^{\frac{3}{2}}$ for fused silica, in reasonably good agreement with the value found[11] from the evaluation of the constant from eq. (4).

In order to break an optical fiber in such a way that the mirror zone extends across the entire fiber, it is necessary to have the stress at all points within the fiber low enough so that $Z\sqrt{r} < K$. The required value of Z at the origin of the fracture depends on the size of the crack or flaw from which the fracture originates.[12] The value of Z cannot be allowed to become zero or negative at any point across the fiber, or the crack will cease to propagate or propagate in a direction which is not perpendicular to the axis of the fiber. Under these conditions, a lip is formed on one fiber end. We see, then, that, to make a reliable clean mirror zone fracture, the stress distribution across the fiber must be suitably adjusted.

III. THE FIBER BREAKING MACHINE

In the preceding section, we have given the conditions necessary to create a mirror zone fracture across an entire fiber end. To determine experimentally the range of stress distributions over which clean mirror zone fractures can be obtained, an apparatus was constructed which could simultaneously bend the fiber and place it under tension. In this way, the stress distribution across the fiber can be varied, as shown in Fig. 2. For a given average tension (force per unit area), T, the stress distribution across the fiber depends on the radius, R, of the form over which the fiber is bent. (We assume no shear friction between the fiber and the form.) In fact, the stress across the fiber, $Z(x)$, is given by

$$Z(x) = T + \frac{E(a - x)}{R}, \qquad (5)$$

where T is the average tension on the fiber, E is Young's modulus, and a is the radius of the fiber.

If $R = \infty$, the maximum diameter, d_M, of fiber that can be fractured with a mirror zone across the entire surface is given by

$$Z'\sqrt{d_M} = K, \qquad (6)$$

where Z' is the stress necessary to initiate the break.

In the experiments to be described later using a diamond or carbide scorer to initiate the break, we find $Z' \approx 25$ kg/mm^2. Thus, for fused silica fibers we find $d_M \approx 100$ μm and for $R = \infty$, when fracturing fused silica fibers with diameters $\gtrsim 100$ μm, we expect hackle to appear.

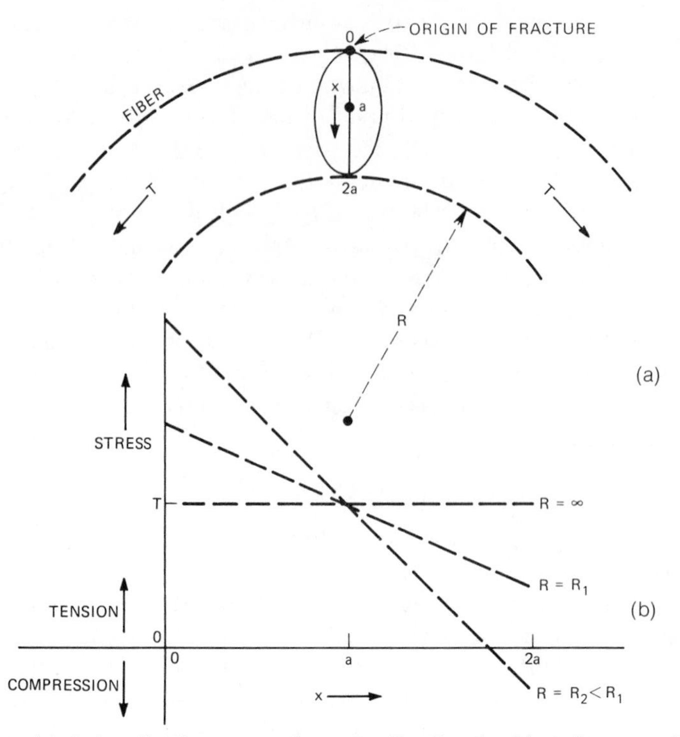

Fig. 2—(a) A glass fiber bent over a form of radius R and subjected to a tension T. (b) The stress as a function of position in the fiber for various bending radii R.

Our experiments showed this to be the case. If we compute the stress from eq. (5), assume T is adjusted so that the stress at $x = 0$ is Z', and select R $(=R_0)$ so that $Z = 0$ at $x = 2a$, we find that the maximum value of $Z\sqrt{r}$ occurs on the surface of the fiber at the position where $r = (\sqrt{4/5})a$, and if we require this product to be $<K$, we find $d_M(R = R_0) = 3.50\ d_M\ (R = \infty)$. Thus, using this technique, fused silica fibers of up to $\sim$350 μm in diameter can be fractured with clean, mirror zone ends.

The fibers used for the experiments reported here were multimode silica glass fibers with an outer diameter of 125 μm and a core diameter of 80 μm. R_0 can be found from eq. (5), letting $Z = 0$ at $x = 125$ μm, assuming the stress necessary to initiate the break, Z', to be equal to the experimentally determined value of 25 kg/mm^2, and using the values $E = 7.2 \times 10^3$ kg/mm^2; $K = 7.5$ kg/mm$^{3/2}$ appropriate for silica glasses. We obtain $R_0 = 3.7$ cm.

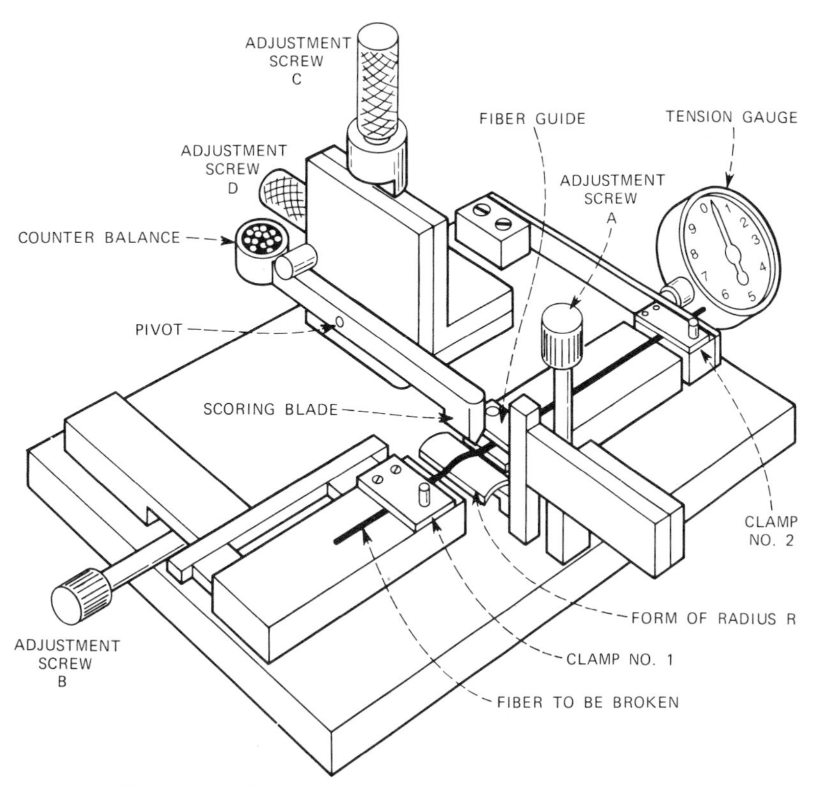

Fig. 3—A semi-schematic view of the fiber breaking machine.

Figure 3 shows a semi-schematic view of the device used to investigate the fracture properties of optical fibers. The fiber to be broken is clamped by clamps No. 1 and No. 2 and slides freely under the Teflon*-coated fiber guide. A Teflon-coated form of suitable radius R can be raised to cause the fiber to conform to the form by adjustment screw A. The tension on the fiber is measured by a tension gauge, which measures the mechanical displacement of a stiff steel bar on which clamp No. 2 is mounted. The tension can be adjusted with the adjustment screw B. A scoring blade can be lowered onto the fiber by adjustment screw C and pulled across the fiber by adjustment screw D. The pressure on the scorer blade can be adjusted by changing the weight in the counterbalance.

---

* Registered trademark of Dupont Co.

IV. EXPERIMENTAL RESULTS

Breaks were made on samples of a low-loss multimode silica glass fiber having a core diameter of 80 μm and a cladding thickness of 22 μm. A wide range of breaking tensions and fiber-bending radii was studied using the fiber-breaking machine described above. We used a variety of scoring techniques and attempted breaks in atmospheres of various relative humidities. The results can be summarized as follows: If the radius of curvature of the form was less than about 2 cm, a lip would be formed. When fractures were made without using a form, i.e., $R = \infty$ or negative, a hackle region was produced. "Good," clean fractures were obtained when a 5.7-cm radius of curvature form was used. These results are illustrated in Fig. 4.

Using the 5.7-cm radius of curvature form, clean fractures with no visible hackle or lip were always produced using breaking tensions in the range of 125 to 175 g and scorer pressures ranging from 1.5 to 7.5 g. The smallest scores were produced when a sharp diamond scorer* was lowered onto the fiber after the tension had been applied. We found no effect on the fracture characteristics when the relative humidity was varied from 7 to 100 percent, or even when water was applied to the point of fracture. In all, a total of 33 fractures were made within this range of conditions. *In no case was there any visible evidence of any hackle or lip.* In the worst case the disturbed region associated with the score extended over a distance of $\sim$22 μm. As the cladding thickness on this optical fiber was 20 μm, this means that in all cases a perfect mirror zone fracture occurred over essentially the entire core region of the fiber.

To establish the minimum splice loss in joining such fiber ends, we used the setup shown in Fig. 5. The joints were made from ends obtained from the same fracture, but rotated with respect to the original fracture position. In this way, the time between fracture and joining was kept at a minimum in order to avoid contamination of the ends. Moreover, utmost accuracy was achieved by comparing the losses immediately before fracture and immediately after joining. This time was typically 10 minutes, while instabilities in the setup caused a power drift at the detector of not more than $\frac{1}{4}$ percent in 30 minutes. Joining adjacent ends, of course, eliminated the possibility of diameter discrepancies which would be encountered in practical splices.

---

* The diamond scorer was supplied by Victory Diamond Tool Co. of East Hanover, N. J.

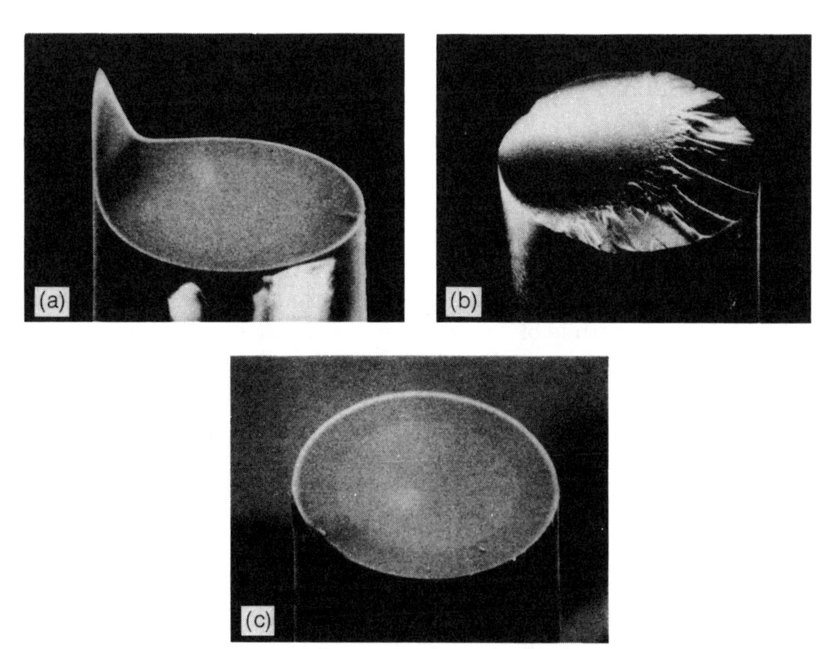

Fig. 4—Electron microscope photographs of 125-μm diameter silicate glass fibers broken using various form radii (R): (a) $R = 0.75$ cm; (b) $R = \infty$ or negative; (c) $R = 5.7$ cm.

Fibers of the type that were used for the splice loss measurements reach a steady-state power distribution after a certain distance independent of the injection conditions. This distribution was measured for the fiber in question at the end of a 1.2-km length. The power distribution in the splice should preferably be the steady-state distribution. Since a sufficient fiber length to achieve such conditions was not available for our measurements, we approximated as well as

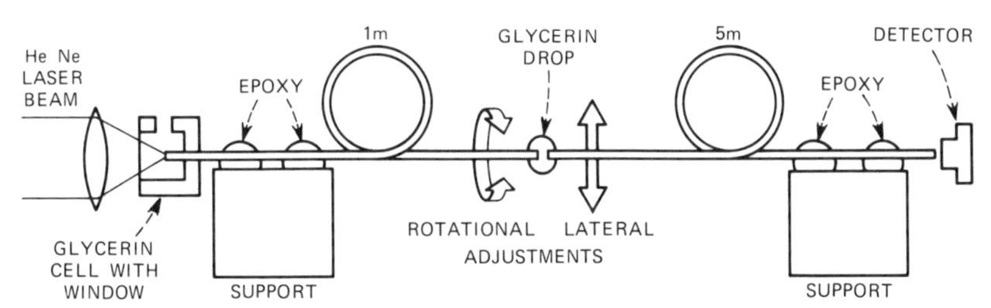

Fig. 5—Schematic of apparatus used to measure laboratory splice loss.

possible the steady-state distribution at the input by properly focusing the input beam onto the fiber front face. Specifically, we made the $1/e^2$ width of the Gaussian field distribution in the input cone equal to the $1/e^2$ width of the steady-state far-field distribution measured at the end of the long sample (0.14 rad half-width). The distance between input and splice was 1 m.

If the splice disturbs the power distribution substantially, a sufficient fiber length must be provided after the splice to allow the distribution to settle, a process which generally is associated with some excess loss. To study the magnitude of this effect, we measured several joints with fiber lengths from 1 to 5 m after the joint. We did not find a consistent increase in loss as the length was increased, although 5 m is admittedly not a sufficient length to reach the steady state. Further study is necessary to estimate the error involved.

To make a good joint, the ends were aligned using a microscope to within a fraction of a degree in angle and within 1 μm laterally, but kept apart by at least 10 μm to avoid damage of the ends by mutual abrasion. A drop of glycerin was then added, which was held between the ends by surface tension. The refractive index of glycerin is 1.473 and almost coincides with that of silica glass ($n = 1.458$). This procedure invariably produced a splice with a loss of less than 1 percent (typically, 0.5 percent). This result was unaffected by rotating one end with respect to the other. Note that no information on the transmitted optical signal was required to achieve this optimal alignment.

To establish the order of magnitude of alignment tolerances permissible in practical splices, we measured the increase in loss as a result of longitudinal or lateral misalignments. The fiber ends could be parted axially by 100 μm (one core diameter) before the losses increased by 1 percent. Lateral displacements were more critical. The loss increased by 1 percent for a 5-μm displacement and by 4 percent for a 10-μm displacement (10 percent of the core diameter).

V. SUMMARY AND CONCLUSIONS

We have presented a theory of glass fiber fracture which allows us to design a machine for reliably producing clean breaks which leave the fiber ends in a suitable condition for splicing. We have built such a machine and demonstrated that, with 125-μm silica glass multimode optical fibers, such breaks are consistently obtained. Laboratory splicing experiments using fibers broken with this machine always produced splices with losses of less than 1 percent (0.04 dB).

VI. ACKNOWLEDGMENTS

We would like to thank R. D. Standley and F. A. Braun for the electron microscope photographs shown in Figs. 1 and 4.

REFERENCES

1. Bisbee, D. L., "Optical Fiber Joining Technique," B.S.T.J., *50*, No. 10 (December 1971), pp. 3153–3158.
2. Bisbee, D. L., "Measurements of Loss Due to Offsets and End Separation of Optical Fibers," B.S.T.J., *50*, No. 10 (December 1971), pp. 3159–3168.
3. Dyott, R. B., Stern, J. R., and Stewart, J. H., "Fusion Junctions for Glass Fiber Waveguides," Elec. Letters, *8*, 11 (June 1, 1972), pp. 290–292.
4. Krumpholz, O., "Detachable Connector for Monomode Glass Fiber Waveguides," Archiv Elektronik Ubertragungstechnik, *26* (1972) pp. 288–289.
5. Someda, C. G., "Simple, Low-Loss Joints Between Single-Mode Optical Fibers," B.S.T.J., *52*, No. 4 (April 1973), pp. 583–596.
6. McCormick, A. R., "Fiber Breaking Technique," unpublished memorandum.
7. Cherin, A. H., Eichenbaum, B. R., and Schwartz, M. I., unpublished memorandum.
8. Saunders, M. J., "Results for the Quality of the Edges of Fibers Cut with a Razor Blade," unpublished memorandum.
9. Gandrud, W. B., "On the Possibility of Two-Dimensional Fiber Splicing," unpublished memorandum.
10. Andrews, A. H., "Stress Waves and Fracture Surfaces," J. Appl. Phys. *30* (May 1959), pp. 740–743.
11. Johnson, J. W., and Holloway, D. G., "On the Shape and Size of the Fracture Zones on Glass Fracture Surfaces," Phil. Mag., *14* (1966), pp. 731–743.
12. Anderson, O. L., "The Griffith Criterion for Glass Fracture," in *Fracture*, B. L. Averbach, et al., eds., New York: Wiley, 1960.

Microlenses for Coupling Junction Lasers to Optical Fibers

L. G. Cohen and M. V. Schneider

Microscopic lenses, fabricated on optical fiber surfaces, have quadrupled the efficiency for coupling astigmatic beams from GaAs junction lasers into 4-μm cores of single-mode fibers. A novel photolithographic technique was used to make hemispherical and hemicylindrical microlenses, with diameters between 4 μm and 10 μm, from commercially available negative type photoresist that is transparent at ir laser wavelengths. Geometrical profiles of photoresist lenses, documented with scanning electron photomicrographs, were remarkably smooth even though their dimensions were more than an order of magnitude smaller than other known lenses.

I. Introduction

Gallium arsenide junction lasers with a stripe geometry have been coupled to optical fibers in the past by aligning the fiber core with the stripe and by optimizing the air gap spacing between the front end of the laser and the fiber.[1] The power coupling efficiencies obtained with this coupling scheme are relatively small, e.g., efficiencies of about 8% have been measured for 4-μm-core single-mode fibers excited by a low-threshold double heterostructure laser. The purpose of this paper is to show that substantially improved coupling efficiencies are obtained by using a sufficiently small lens which is fabricated directly on the fiber surface as shown in Fig. 1(a). The lens, which will be subsequently called a microlens, transforms the strongly astigmatic beam emanating from the junction laser (typically 0.6 μm $\times$ 4 μm) into a beam that more closely matches the circular HE_{11} fiber mode. The dimensions of the lens can be obtained to a first approximation by using simple ray analysis, e.g., a 7-μm-diam lens is needed for a lens material with an index $n = 1.6$ to collimate a laser beam that diverges at $\alpha_y = 30°$ from the junction plane. An alternate technique to couple the laser to a fiber is shown in Fig. 1(b). The lens is fabricated on the front end of the junction laser. The center of curvature must be offset from the surface in order to collimate the beam in air. This more complicated method has not been pursued as yet.

Lenses with diameters and thicknesses on the order of a few micrometers cannot be fabricated by means of conventional polishing or molding techniques. An additional problem is that small lenses cannot be easily mounted and aligned in an optical system. We show in this paper that the fabrication and alignment problems can be solved by making lenses from light-sensitive polymers called photoresist that are crosslinked directly in the optical system. Two types of lenses were made from commercially available photoresist using the following techniques:

(1) Hemispherical microlenses were fabricated by depositing a thin film of photoresist on the front surface of an optical fiber and then exposing the resist to uv light propagating along the fiber core. After removing the unexposed photoresist a crosslinked structure with a hemispherical shape remained over the fiber core.

(2) Hemicylindrical lenses were made by coating the front end of a fiber with resist and then using a slit to project an uv light stripe over the fiber core. After development of the resist a cylindrical lens remained on the fiber surface.

Both types of lenses were used to couple the output of a gallium arsenide injection laser into a single mode optical fiber. Cylindrical lenses were more effective than spherical lenses for collecting light from the laser's rectangular active region.

The following section of this memorandum contains a more detailed description of lens fabrication techniques including physical descriptions of the formed microlenses. Section III contains data on the optical properties of the lenses. They were measured by shining He–Ne laser light (6328 Å) into one end of a fiber and then photographing and electrically scanning the near field image of the light leaving the lens on the other end of the fiber. Finally, in Sec. IV, we describe the power coupling measurements made for GaAs injection lasers (9000 Å) feeding single mode fibers.

The authors are with Bell Laboratories, Crawford Hill Laboratory, Holmdel, New Jersey 07733.

Received 13 June 1973.

Reprinted with permission from *Appl. Opt.*, vol. 13, pp. 89–94, Jan. 1974.

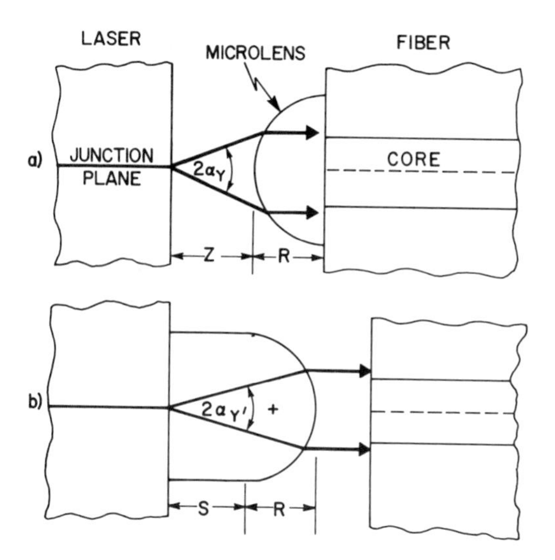

Fig. 1. Arrangement for coupling light power from a junction laser across a small air gap into the core of a single mode fiber. (a) The fiber has a microlens over its core; (b) the laser has a microlens over its active layer.

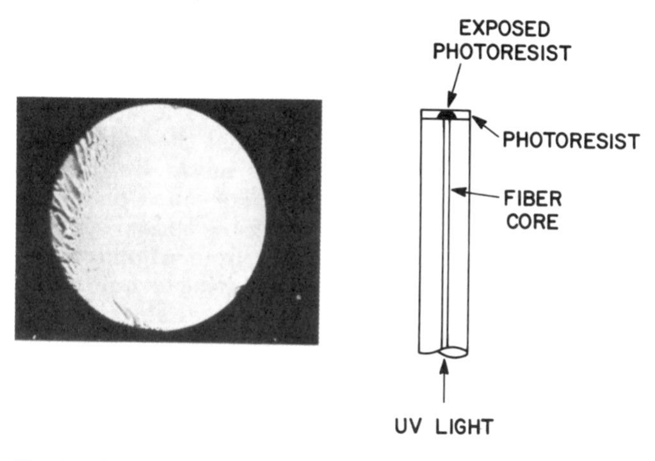

Fig. 2. Experimental arrangement for automatically exposing a hemispherical lens over the core of a single mode fiber. The photograph shows a 7-μm diam microlens in perspective with the fiber cladding diameter of 250 μm.

II. Fabrication of Microlenses

All the lenses were prepared on the surfaces of quartz single mode fibers fabricated by Corning Glass Works with a core diameter of 4 μm and a cladding diameter of 250 μm. A flat surface was obtained by scoring the fiber with a sharp blade and then pulling it apart.[2] The surfaces were carefully cleaned and coated with photoresist (boiling in *J*-100, trichlorethylene, and xylene followed by dipping into photoresist or dropping photoresist onto the fiber; *J*-100 is a photoresist stripping agent manufactured by the Indust-Ri-Chem Laboratory, Richardson, Texas). Undiluted Kodak micronegative resist or Waycoat IC photoresist was used as a photosensitive polymer. The fiber was then prebaked at 80°C for 20 min in order to remove the solvent. Following prebaking, the fiber was exposed to uv light (4000 Å) as described below. The exposed photoresist was developed by dipping the fiber in microneg developer for 30 sec followed by a microneg rinse for 20 sec. The developed photoresist was hardened by postbaking the fiber at 130°C for 30 min.

A. Hemispherical Microlenses

The experimental arrangement used to expose hemispherical lenses is shown in Fig. 2. Ultraviolet light from a mercury arc lamp is coupled into the fiber through the bottom end and is guided by the fiber core to the top end, which is coated with photoresist. Crosslinking occurs in the immediate vicinity of the fiber core. The unexposed photoresist is removed after development. Figure 2 shows a top view of a microlens with a diameter of 7 μm located in the center of a fiber with a diameter of 250 μm. We found that the lens diameter is related to the product of uv light intensity and exposure time, e.g., for a 0.3 mW light intensity a 4-μm wide lens was obtained for an exposure time of 3 min, 7 μm for 10 min, and 14 μm for 25 min. Figure 3 shows scanning electron photomicrographs of two different hemispherical lenses. The top lens has a diameter of 4.2 μm and a thickness of 1.8 μm, and the bottom lens has a diameter of 6.4 μm and a thickness of 2.5 μm. The remarkable smoothly curved lens profiles are obtained because the exposing light intensity has a nearly Gaussian profile and also because the photoresist flows slightly at the postbaking temperature.

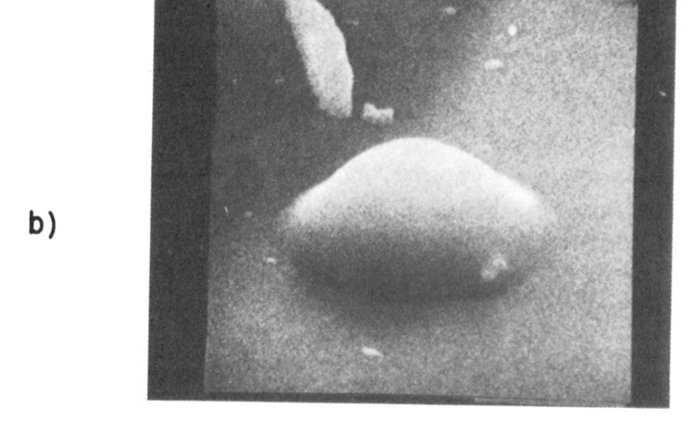

Fig. 3. Scanning electron beam photomicrographs of two different hemispherical lenses. (a) Diameter = 4.2 μm, thickness 1.8 μm; (b) diameter = 6.4 μm, thickness 2.6 μm.

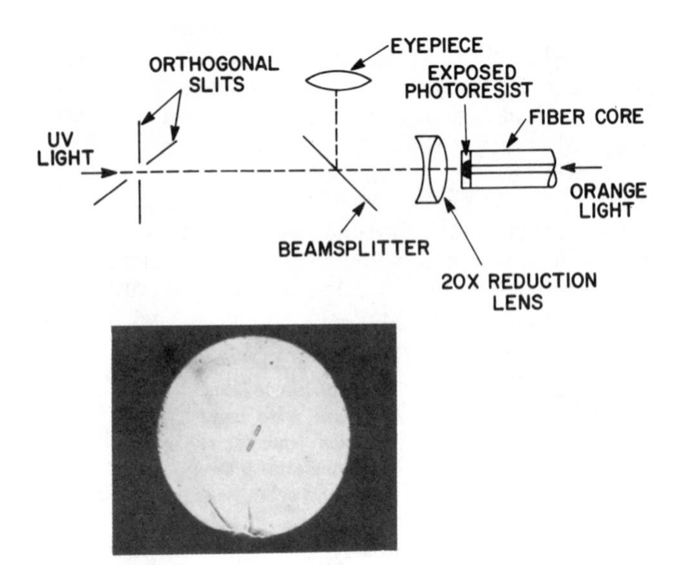

Fig. 4. Experimental arrangement for aligning and exposing hemicylindrical lenses over the core of a single mode fiber. The photograph shows a 7-μm × 35-μm microlens in perspective with the fiber cladding diameter of 250 μm.

Lens quality is estimated, as shown in Fig. 6, by shining He–Ne laser light (6328 Å) into one end of a fiber and using a 40× microscope objective to image the focal plane of the fabricated lens onto a frosted plate. The image at the plane of the frosted plate was magnified to take the photographs shown in Fig. 6. The dimensional scale is the same for all photographs because the distance between the 40× objective lens and the frosted plate was the same for all fibers. The scale was calibrated with the 10-μm wide near field image from a 10-μm multimode fiber. The approximately 5-μm diam large spot in Fig. 6(a) is the near field image of the HE_{11} fiber mode. Figures 6(b) and 6(c) are near field images of the HE_{11} mode transformed by a spherical and a cylindrical lens. Typical 7-μm wide cylindrical lenses transformed a fiber's circular HE_{11} mode into an elliptical image with a 3 to 4:1 aspect ratio. The spherical lens reduced the diameter of the HE_{11} mode by a factor of 3.

B. Hemicylindrical Microlenses

Figure 4 shows the experimental arrangement to expose a cylindrical lens with a controlled length and width. Two orthogonal slits are illuminated by uv light from a mercury arc lamp. The rectangular slit is projected on the fiber surface by means of a 20× reduction objective. A beam splitter and an eyepiece are used to view the image of the crossed slits that is focused on the fiber. The fiber is attached to an x-y-z manipulator in order to align the image with the core. Once aligned, the uv light is used to expose a photoresist stripe with the desired dimensions. The developed photoresist is the cylindrical lens. In order to control the lens curvature, photoresist was exposed in three stages (10 sec, 10 sec, 20 sec) using successively wider uv strips. The final lens has a smooth surface profile because of diffraction effects and also because the photoresist flows slightly at the postbaking temperature.

The photograph shown in Fig. 4 puts a tiny cylindrical lens (7 μm wide × 35 μm long) in perspective relative to the 10-mil fiber cladding. The lens is aligned directly over the fiber core in the center area of the fiber with a diameter of 250 μm. Two cylindrical lenses made with a scanning electron microscope are shown in Fig. 5. Figure 5(a) is a lens with a diameter of 7 μm and length of 25 μm. Figure 5(b) and 5(c) are a side and top view, respectively, of another lens with similar dimensions.

III. Optical Quality of Lenses

Lenses fabricated on fibers, for use in power coupling arrangements, should transform the circular fiber mode into an astigmatic image that more closely matches the laser beam waist. Since the laser beam waist is smaller than the fiber mode, optimal coupling should occur when the laser is aligned in the focal plane of the lens.

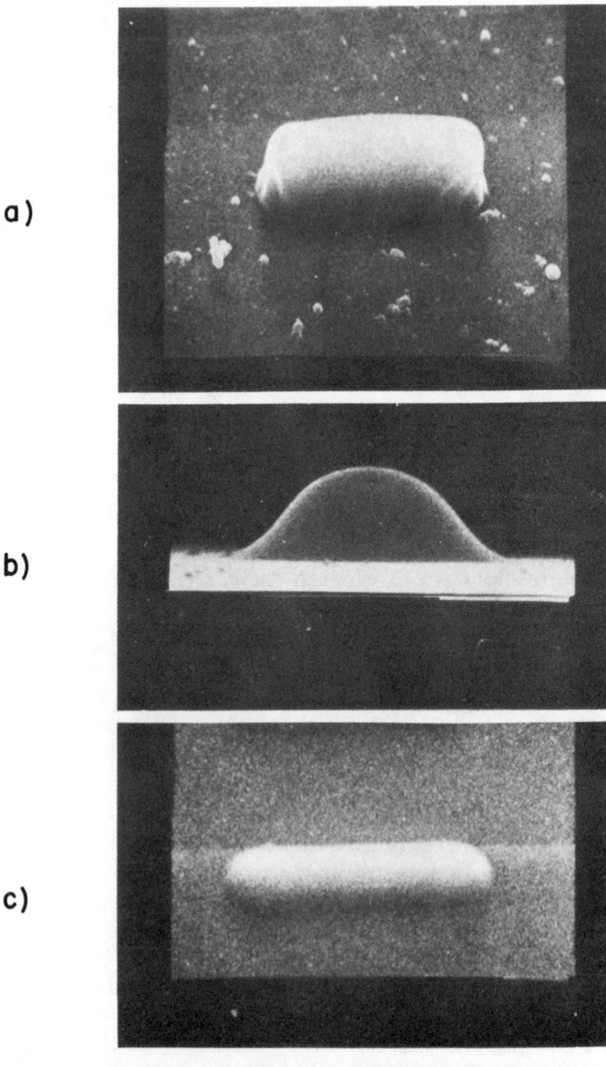

Fig. 5. Scanning electron beam photomicrographs of cylindrical lenses (7-μm diam × 25 μm-length).

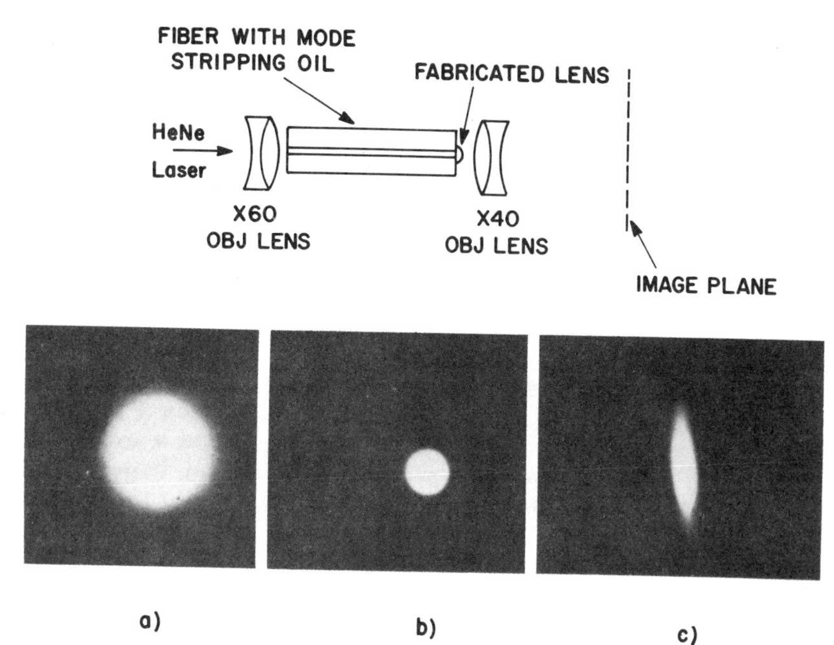

Fig. 6. Experimental arrangement for projecting the near field HE_{11} image, transformed by a lens, onto the plane of a frosted plate. The photographs show near field images emanating from the ends of fibers. (a) HE_{11} image from a fiber end without a lens; (b) image transformed by a 7-μm wide hemispherical reduction lens; (c) elliptical image transformed by a 7-μm × 25-μm long cylindrical lens.

The near field images in Fig. 6 were also focused through a 250-μm pinhole in a mask covering a Si solar cell at a distance of 4 m from the end of the 40× microscope objective. Figures 7(a) and 7(b), recorded with a lockin amplifier, are profiles of light intensity vs pinhole position in the focal plane of a fiber with a spherical lens and with a cylindrical lens. The horizontal scale was calibrated by scanning the 10-μm wide near field image from a 10-μm multimode fiber. Intensity profiles in Fig. 7 are Gaussianlike in shape, consistent with the photographs in Fig. 6.

IV. Power Coupling into Fibers

Light emanating from the rectangular aperture on stripe geometry DH junction lasers was coupled into the 4-μm circular core of a single mode fiber across a small air gap separating the laser from the fiber. (The lasers were fabricated by J. C. Dyment and A. R. Hartman at Bell Laboratories, Murray Hill, N.J.) The coupling technique is diagramed in Fig. 1(a) for a beam that diverges at angle α_y perpendicular to the junction plane. The fiber has a microlens in contact with its core.

Supporting structures for the laser and the fiber were mounted on separate rotatable micromanipulators. The GaAs laser chip was held, by spring contact, with its *p* side face down on a copper block and was driven at room temperature at a 100-Hz repetition rate by negative 100-nsec wide current pulses. The output end of the fiber was immersed in oil to eliminate light reflections, and light power traveling in the cladding was scattered out by immersing several inches of the fiber in oil matched to the index of the cladding. Light power leaving the output end of a fiber was maximized by adjusting the relative axial and transverse position between the fiber core and the active stripe on the GaAs chip. Provision was also made for reducing the tilt angles between the axes of the laser stripe and the axis of the fiber. In addition, the fiber could be rotated about its axis so that the long dimension of a cylindrical lens could be aligned parallel to the laser's junction plane. To facilitate alignment the coupling joint was viewed in two orthogonal planes (top and side) with two microscopes.

To measure coupling efficiency, light power emanating from the end of a fiber was measured and compared with light power radiating from the exposed laser mirror face. [The 4-μm fiber had a rela-

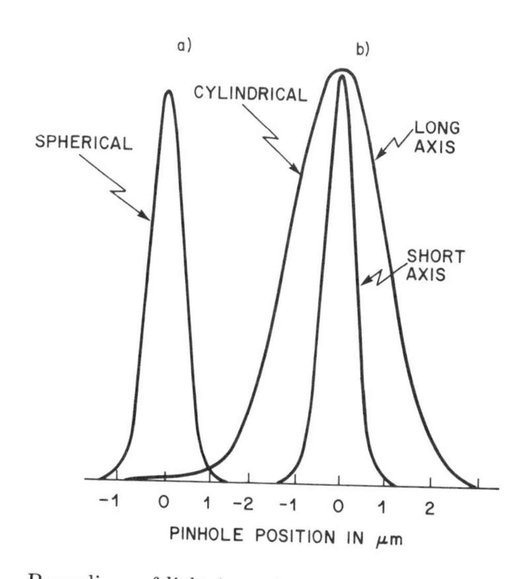

Fig. 7. Recordings of light intensity vs position in the near field of a fiber end. (a) Image transformed by a 7-μm wide hemispherical reduction lens; (b) elliptical image transformed by a 7-μm × 25-μm long cylindrical lens.

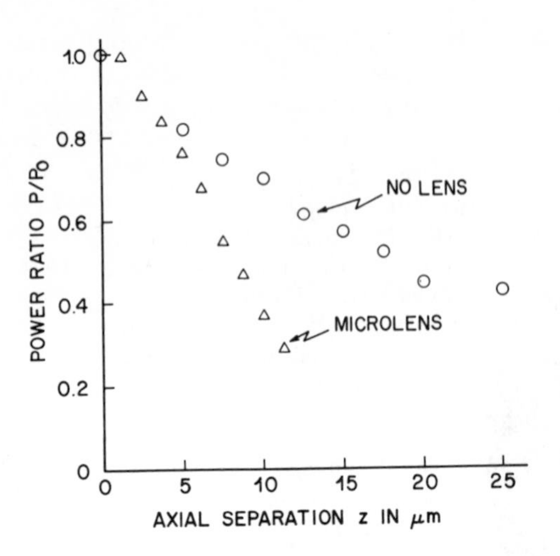

Fig. 8. Normalized power emitted from the end of a fiber is plotted vs axial separation between the laser and the fiber. Circular data points apply to a single-mode fiber without a lens, and triangular data points apply to a fiber with a spherical lens.

tively low transmission loss (20 dB/km). Therefore, the power at an input tip is almost equal to the power leaving the output end of a 25-cm length.] The detectors used were Schottky barrier silicon photodiodes having 8-nsec rise time and sensitive areas equal to 0.99 cm^2. Peak power output was displayed on a sampling oscilloscope.

Transverse misalignments of 3 μm from the optimal position caused a 3-dB loss in coupled power. Axial displacements were less critical. In Fig. 8 coupled power, normalized to its optimal value, is plotted vs z, the separation between fiber and laser. Circular data points apply to the single mode fiber without a lens, and the more rapidly rising triangular points apply to a fiber with a spherical lens.

Table I is a summary of some experiments performed with several lasers from a double-heterostructure batch with proton-bombarded stripes. Measured power coupling coefficients κ are listed for fibers without lenses, with spherical lenses, and with cylindrical lenses fabricated on one end. Lasers were pumped to approximately 15% above threshold, and far field radiation profiles were measured to confirm single-mode behavior parallel and perpendicular to the laser junction. Sample intensity patterns are illustrated in Fig. 9 for one of the tested lasers. The laser properties listed in Table I are: threshold current, I_{th}; far field diffraction angles α_x and α_y to the $(1/e)$-points of radiation profiles; half-width $d/2$ of the GaAs active layer; and % index mismatch Δ_n between the active layer and its surroundings.

It has been shown in Refs. 3 and 4 that a dielectric slab waveguide adequately describes the properties of laser modes perpendicular to the junction plane. Therefore, we determined the beam half-width ω_{0y} by plotting the lowest order TE mode[5] from the characteristic V-number, $V = (2)^{1/2}\pi d/\lambda$ (3.59) (Δ_n),$^{1/2}$ of the slab waveguide from which the laser radiates. The fiber parameters are core radius, $d/2 \approx 2$ μm, index mismatch, $\Delta_n \approx 0.003$, between core and cladding, characteristic V-number $V = (2)^{1/2}\pi d/\lambda\ (1.46)\ (\Delta_n)^{1/2} \approx 1.55$ at 9000 Å, and mode radius $a \approx 3$ μm.[5] The beam half-width ω_{0x} was determined from α_x by assuming Gaussian beam expansion along the junction plane

$$\omega_{0x} = (2)^{1/2}/(\pi\lambda \tan\alpha_x).$$

The virtual beam waist location inside the laser cavity was estimated from

$$Z_{0x} = (L/2)(1/n) \approx 52\mu\text{m},$$

where $L = 375$ μm is the laser cavity length, $n = 3.59$ is the refractive index of the active region, and Z_{0x} is the beam waist distance from a laser mirror.

Theoretical estimates of the power coupling coefficient κ without a lens were computed from[6]

$$\kappa_{th} = \kappa_x\kappa_y = \frac{2/[(\omega_{0x}/a)+(a/\omega_{0x})]}{(1+\{(Z_{0x}\lambda)/[\pi(\omega_{0x}^2+a^2)]\}^2)^{1/2}}\ \frac{2}{[(\omega_{0y}/a)+(a/\omega_{0y})]}$$

in terms of the laser beam waist dimensions (ω_{0x},

Table I. Summary of Measurements

	Laser parameters L-424 1	Laser parameters L-242 2	Fiber parameters
I_{th}(mA)	250	250	
$d/2$ (μm)	0.2	0.2	2
Δ_n (%)	3.6	3.6	0.3
V	1.35	1.35	1.55
α_y (°)	33	30	
α_x (°)	10	6	
ω_{0y}(μm)	0.3	0.3	
ω_{0x}(μm)	2.3	3.8	
a (μm)			3
	Coupling efficiencies		
κ_{th} (%)	12.5	16	
meas κ_n	7	8	
meas κ_{cyl}	30	34	
meas κ_{sph}	23	20	

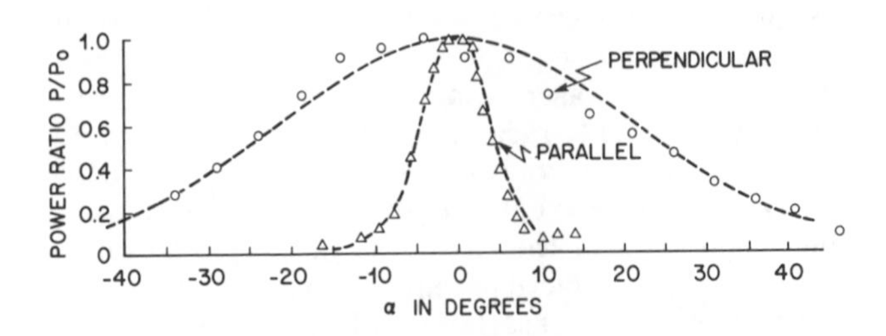

Fig. 9. Far field intensity patterns from a junction laser (L-424A 2); α (parallel) and α (perpendicular) are angles measured in the planes parallel and perpendicular to the junction. The dashed lines are Gaussian fits to the intensity patterns.

ω_{0y}) for a Gaussian laser beam and HE_{11} mode radius a for a Gaussian fiber mode. The factor

$$1/(1+\{(Z_{0x}\lambda)/[\pi(\omega_{0x}^2+a^2)]\}^2)^{1/2}$$

is a measure of the curvature mismatch between the two phase fronts along the junction plane. In Table I the measured power coupling coefficients κ_n are smaller than the theoretical estimates κ_{th}. This is caused by non-Gaussian laser beam behavior especially perpendicular to the junction plane, spontaneous emission from the laser that is not coupled into the fiber core, and uncorrected angular misalignments between the fiber and the laser.

The maximum measured coupling efficiency of a 7-μm diam lens on a 4-μm single mode fiber was 34% for a cylindrical lens, 23% for a spherical lens, and 8% with the lens removed. Rule of thumb estimates for lens effectiveness may be obtained from knowledge of lens magnifying power obtained from measurements in Sec. III. Cylindrical microlenses transform the circular HE_{11} mode into an elliptical image with a 4:1 aspect ratio. The coupling efficiency should increase by a factor of 4 since κ_y is directly proportional to the beam radius mismatch when

$$\omega_{0y}/a \ll 1.$$

Spherical microlenses reduce the diameter of the HE_{11} mode by a factor of 3; $\kappa = \kappa_x\kappa_y$ should increase by approximately 3 times since $\kappa_y \propto (\omega_{0y}/a)$ and because $\kappa_x > 0.6$ when $\omega_{0x}/a > 1/3$.

V. Conclusions

A novel photolithographic technique was used to fabricate hemispherical and hemicylindrical lenses on fiber surfaces. These lenses have significantly increased the power coupling efficiency between GaAs injection lasers and single mode fibers from 8% for a fiber without a lens to 23% for a fiber with a spherical lens and 34% with a cylindrical lens.

All lenses were made from commercially available photoresist, with index $n \approx 1.6$, that is transparent at 9000 Å. No selective etching was necessary. The negative type photoresist we used is a highly stable pure synthetic rubber (polyisoprene). It does not oxidize or absorb moisture, and there are no degradative mechanisms known under atmospheric conditions.[7] The fabrication technique ensures automatic alignment of hemispherical lenses over the fiber core without the aid of a microscope. Spherical lenses were powerful enough to reduce the size of the HE_{11} mode by a factor of 3. Hemicylindrical lenses, fabricated using projection photolithography, were easily, but not automatically, aligned over fiber cores. Cylindrical lenses transformed the fiber's circular HE_{11} mode into an astigmatic image with a 4:1 aspect ratio.

The geometrical profiles of photoresist lenses, documented with scanning electron photomicrographs, were remarkably smooth even though their dimensions were more than an order of magnitude smaller than any other known lens.

The authors thank J. C. Dyment and A. R. Hartmann for supplying the junction lasers and G. W. Kammlott for making the scanning electron-beam photographs. We are also indebted to F. K. Reinhart for helpful comments.

References

1. L. G. Cohen, Bell Syst. Tech. J. **51**, 573 (1972).
2. A. R. McCormick, Bell Labs., Holmdel, N.J.; unpublished work.
3. H. Kressel, J. K. Butler, F. Z. Hawrylo, H. F. Lockwood, and M. Ettenberg, RCA Rev. **32**, 393 (1971).
4. W. O. Schlosser, Bell Labs., Murray Hill, N.J.; unpublished work.
5. D. Marcuse, *Light Transmission Optics* (Van Nostrand Reinhold, New York, 1972), Chap. 8.
6. H. Kogelnik, "Coupling and Conversion Coefficients for Optical Modes," in *Proceedings of the Symposium on Quasi-Optics*, J. Fox, Ed. (Polytechnic Press, Brooklyn, 1964).
7. R. E. Kerwin, Bell Labs., Murray Hill, N.J.; personal communication.

Part VIII
Transmitter, Receiver, and Peripheral Electronics

SMALL-AREA, DOUBLE-HETEROSTRUCTURE ALUMINUM-GALLIUM ARSENIDE ELECTROLUMINESCENT DIODE SOURCES FOR OPTICAL-FIBER TRANSMISSION LINES

C. A. BURRUS
Crawford Hill Laboratory, Bell Telephone Laboratories, Holmdel, New Jersey 07733, USA

and

B. I. MILLER
Bell Telephone Laboratories, Holmdel, New Jersey 07733, USA

Received 2 November 1971

Small-area (50-μm dia) electroluminescent diodes have been fabricated in double-heterostructure configurations of $Al_xGa_{1-x}As/Al_yGa_{1-y}As/Al_xGa_{1-x}As$ grown by liquid-phase epitaxy on a GaAs substrate. The light output has been coupled into multimode optical fibers, and the maximum light output through a short fiber 60 μm in diameter was about 1.7 mW near 0.78-μm wavelength for a bias current of 150 mA dc (7500 A/cm^2). The room temperature operating life (to half output) at this current was at least several thousand hours.

Electroluminescent diodes of GaAs, in the form of small-area devices (50 μm or less in diameter) capable of operation at high current densities ($\approx 10^4$ A/cm^2), can exhibit sufficiently high radiance to serve as sources for multimode optical fiber transmission lines. The fabrication and operating properties of diffused and single-heterostructure diodes of this type have been described elsewhere [1-3]. Here we wish to discuss an adaptation to this application of the double-heterostructure configuration previously used to make room-temperature cw lasers in the 0.77-0.9 μm wavelength region [4,5], and to compare the performance of the resulting diodes with that of the previously-described devices. In general, the advantages to be gained from use of the double-heterostructure material, prepared by liquid-epitaxial growth techniques, are

(1) reliable control of the thickness of the active layer, which allows optimization of the emitting volume;

(2) control of the mole faction y in the $Al_yGa_{1-y}As$ emitting layer, which allows direct-band-gap operation of the resulting light emitters from about 0.75-μm to 0.9-μm wavelength; and

(3) control of the mole fraction x in the $Al_xGa_{1-x}As$ layer surrounding the emitting junction, so that absorption losses between the emitting region and the optical fiber can be reduced.

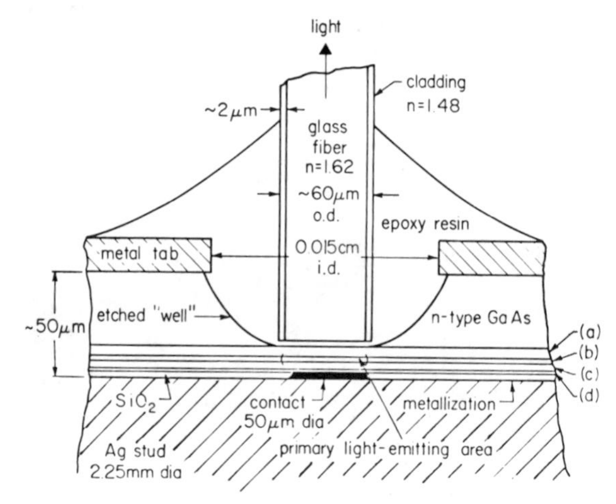

Fig. 1. Cross-sectional drawing (not to scale) of small-area double-heterostructure electroluminescent diode coupled to optical fiber. Layer (a) n-type $Al_xGa_{1-x}As$, 10-μm thick; emitting layer (b), p - $Al_yGa_{1-y}As$, $\approx$1-μm thick; layer (c), p-type $Al_xGa_{1-x}As$, 1-μm thick; layer (d) p±GaAs, $\approx$0.5-μm thick (for contact purposes).

A cross-sectional drawing of a completed diode structure is shown in fig. 1.

The semiconductor layers for this diode were prepared using techniques described elsewhere for the fabrication of near-visible cw lasers [4,5].

Reprinted with permission from *Optics Commun.*, vol. 4, pp. 307-309, Dec. 1971.

Table 1

Characteristics of small-area electroluminescent diodes, 50-μm diameter, feeding 60-μm diameter fiber. Double-heterostructure devices operated near 0.8 μm wavelength; others, near 0.9 μm

Item measured	Double heterostructure [a] (this work)	Single heterostructure (refs. [2,4])	Homo-structure (refs. [1,4])	Units
Realistic operating current, i_{dc} 25 ^{o}C ambient [b]	150	200	300	mA
Current density at i_{dc}	7 500	10 000	15 000	A/cm^2
Output from 30-cm length fiber at i_{dc}	1-1.7	0.9-1.1	0.5-0.7	mW
Input to produce heat saturation, i_s	225-250	250-350	400-500	mA
Rise time, 10-90%, 50-Ω circuit	10-12	8-11	4-6	nsec
2-4 Ω source impedance [c]	1-2	≈ 1	$\lesssim 1$	nsec
Radiance at i_{dc}	60-100	55-75 (3)	30-40 (3)	W/sr cm^2

a) For active-layer thickness ≈ 1 μm.
b) For operating life $\approx 10^4$ h to half output; $i_{dc} \lesssim 0.67\ i_s$.
c) Approximate match to diode resistance.

The GaAs substrate served as a mechanical support for the ternary compound. The composition of the three aluminum-gallium arsenide layers was adjusted so that the band gap of the outer two layers was larger than that of the active center layer. In a laser, this structure provides both electrical and optical confinement in the center layer [6,7]. Its role in the nonlasing diode is not as clearly defined, but it at least provides a low-absorption output path in the fiber-coupled device.

The small-area light-emitting diodes were fabricated from this epitaxial material as described previously [1,2]. The highly absorbing GaAs substrate was completely removed from the area above the contact dot by etching. This left an essentially transparent 10-μm window of $Al_xGa_{1-x}As$ between the primary emitting region and the end of the optical fiber. In this structure, it should be noted that the area of the electrical p-n junction is that of the entire semiconductor chip, while the primary light-emitting area is defined by the region of high current flow through the small contact dot.

The operational performance of this type of diode with and emission peak near 0.8-μm wavelength is summarized in table 1. Values also are listed for comparable homostructure and single-heterostructure devices (0.9-μm wavelength) of the same geometry. In addition to the possibility of fabrication to operate at different wavelengths, the double-heterostructure diodes have provided some advantage in efficiency, output and radiance. However, in heat transfer from the junction and in pulse response time they have not equalled the other structures.

The bias current necessary to produce heat saturation (a measure of heat-transfer efficiency) in the double-heterostructure diodes was found to decrease with increased thickness of the emitting layer when all other variables were held constant. This is shown in table 2, group (b), which lists the correlation between diode behavior and some material properties for three epitaxial growths which were alike except for emitting-layer thickness. As illustrated in table 2, maximum output was achieved with an active-layer thickness near 1 μm. Since we have seen some tendency for the efficiency at very low currents to increase with the emitting layer thickness, we

Table 2
Some properties of typical epitaxial growths and resulting diodes

	Epi growth	Layers a and c: Mole fraction a) of Al, (x)	Layers a and c: Mole fraction a) of Al, (y)	layer b: Dopant and conc. (cm^{-3})	layer b: Thickness (μm)	λ at 100 mA (μm)	Saturation current, i_s(mA)	Max. output at i_s (mW)
a	241 MBP	0.3	0.1	Si, $N_A \approx N_D \approx 10^{19}$	2.1	0.781	≈ 225	1.35
a	493 MB2	0.3	0.1	Sn, $N_D \approx 5 \times 10^{17}$	1.4	0.775	225-250	2.0
a	328 MB2	0.3	0.075	Sn, $N_D \approx 5 \times 10^{17}$	0.95	0.794	225-250	1.5
b	245	0.3	0	Si, $N_A \approx N_D \approx 10^{19}$	3.5	0.9	≈ 175	0.51
b	264	0.3	0	Si, $N_A \approx N_D \approx 10^{19}$	0.9	0.9	225-250	1.1
b	254	0.3	0	Si, $N_A \approx N_D \approx 10^{19}$	0.65	0.9	≈ 250	0.92

a) As estimated from photoluminescence measurements.

believe that the optimum thickness (as measured by maximum output) represents a compromise between emitting efficiency and heat-transfer properties in the structure. The measured pulse rise time in 50-Ω circuits also was affected by the center-layer thickness, increasing from about 7-8 nsec for the 0.65-μm layer to about 30-40 nsec for the 3.5-μm layer.

Our primary interest in output at wavelengths near 0.8-μm wavelength arose from application requirements, and does not in any way represent an optimization of the aluminum-gallium ratio y in the emitting layer. We expect that any composition $0 \leq y \leq 0.21$ should produce light-emitting diodes with essentially the same properties at wavelengths from about 0.9 μm to about 0.75 μm [5]. At higher aluminum concentrations and shorter wavelengths, efficiency would be expected to decrease as the semiconductor takes on the properties of an indirect band-gap material.

We are indebted to R. J. Capik and F. J. Favire for assistance in the materials preparation and device fabrication, respectively, and to R. W. Dawson for the rise-time measurements.

REFERENCES

[1] C. A. Burrus and R. W. Dawson, Appl. Phys. Letters 17 (1970) 97.
[2] C. A. Burrus and E. A. Ulmer Jr., Proc. IEEE 59 (1971) 1263.
[3] R. W. Dawson and C. A. Burrus, Appl. Opt. 10 (1971) 2367.
[4] B. I. Miller, J. E. Ripper, J. C. Dyment, E. Pinkas and M. B. Panish, Appl. Phys. Letters 18 (1971) 403.
[5] B. I. Miller, to be published.
[6] Zh. I. Alferov, V. M. Andreev, E. L. Portnoi and M. K. Truhan, Fiz. i Tekhnika 3 (1969) 1328 [Soviet Phys. Semicond. 3 (1970) 1107.
[7] M. B. Panish, I. Hayashi and S. Sumski, Appl. Phys. Letters 16 (1970) 326.

Pulse Behavior of High-Radiance Small-Area Electroluminescent Diodes

R. W. Dawson and C. A. Burrus

Bell Telephone Laboratories, Inc., Crawford Hill Laboratory, Holmdel, New Jersey 07733.

Received 18 June 1971.

High-radiance small-area electroluminescent diodes can serve as optical-wavelength sources in many applications of optical-fiber transmission lines. The construction of both homostructure (GaAs) and single-heterostructure ($Ga_{1-x}Al_xAs$) diodes for this purpose has been described elsewhere.[1,2] These devices operated at current densities of 10,000–15,000 A/cm² in a room-temperature ambient, delivering cw power of the order of a milliwatt near 0.9-μm wavelength into suitable fibers. A possible advantage of such electroluminescent sources for a communications system is that they can be modulated simply by varying the input current. Direct-modulation methods might employ a pulsed input to the diode, and thus we have undertaken a study of the response-time and heat-saturation characteristics of these devices under pulsed operating conditions.

A cross-sectional drawing of the diodes is shown in Fig. 1. A short optical-fiber coupling was used between the light emitter and the detector. All measurements were made with standard commercial pulse equipment and PIN photodetectors.

The measured diode response times (input to 5% light output and 10–90% rise and fall times) are summarized in Table I.

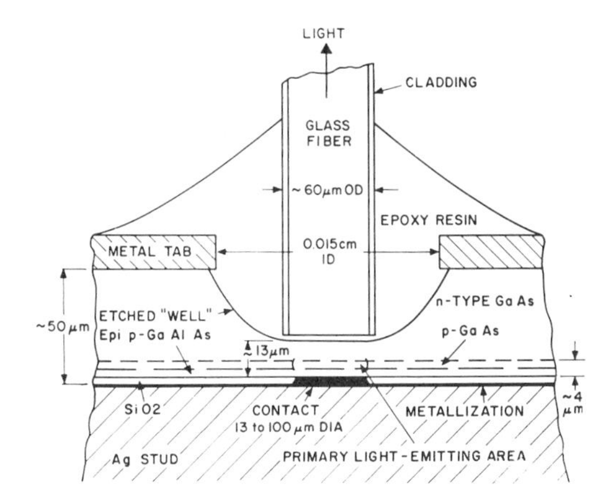

Fig. 1. Construction of single-heterostructure diode for optical-fiber transmission-line source. In homostructure devices, the GaAlAs layer was omitted. Fiber: core refractive index = 1.62, cladding index = 1.48, cladding thickness ≈ 2 μm.

Table I. Measured Light-Output Response Times of GaAs Electroluminescent Diodes Relative to Driving current (~1 A)

Diode type	Source impedance (ohms)	Delay time[c] (nsec)	Rise time (nsec)	Decay time (nsec)
Homostructure	50[a]	0.6–0.8	3.5–5.5	4.0–6.0
	3–5[b]	<0.2	≳1	≳1
Single hetero-structure	50[a]	0.7–1.2	6.5–9.0	6.5–10.0
	3–5[b]	<0.2	~1	~1

[a] Diode matched to circuit with a series resistance.
[b] Approximate match to diode resistance.
[c] Corrected for measured fiber delay.

Diodes driven by a low-impedance pulse generator approximately matched to the diode series resistance (about 5 ohms) exhibited significantly faster rise times than the same units matched to a 50-ohm pulse generator by a series resistance.

The results of varying the amplitude of a narrow, widely spaced input pulse (very low repetition rate) from a 50-ohm generator are illustrated in Fig. 2. The *dc reference* is that dc current at which the diode could be operated in a room-temperature ambient with a life expectancy (to half output) of at least several thousand hours; this *safe* operating point was approximately two-thirds the dc saturation current. The junction temperature at saturation, resulting from dc-induced heating, was estimated from the observed change in output wavelength to be 90–95°C. Sustained operation at saturation, whether dc or pulsed, resulted in rapid, permanent reduction of the optical output. Safe pulse operation was at about 80% of the pulse saturation value.

The first evidence of saturation at very low repetition rates was the appearance of a slope in the detected light pulse as shown in Fig. 3. Measurement of the output wavelength vs pulse width showed the junction temperature at the trailing edge (Fig. 3a) to be in the vicinity of 90°C, with only a few degrees rise at the leading edge. As the driving pulse amplitude was increased, the temperature at the leading edge eventually began to rise and the peak output saturated. This effect was followed

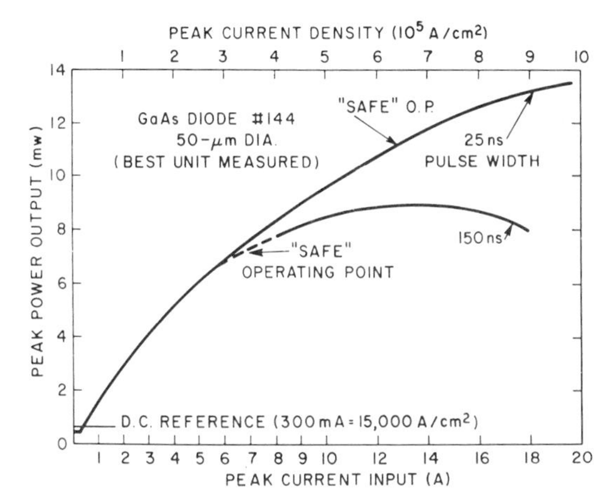

Fig. 2. Measured peak light output vs peak input for isolated short pulses.

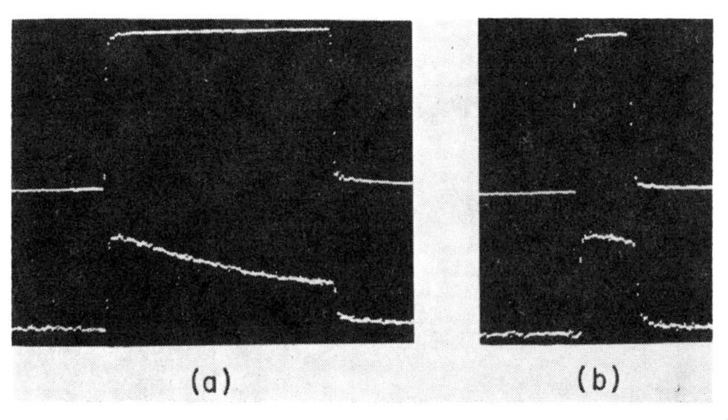

Fig. 3. Driving current (3.5-A peak) and optical output (10-mW peak) pulses showing effects of heating within a single pulse width. Pulse width was (a) 100-nsec, (b) 25 nsec.

Reprinted with permission from *Appl. Opt.*, vol. 10, pp. 2367–2369, Oct. 1971.

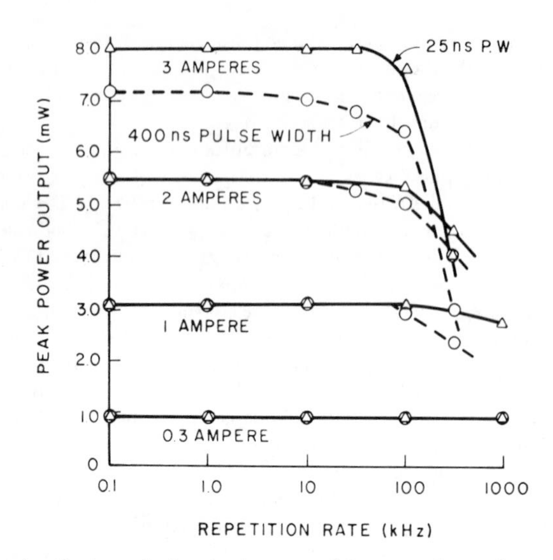

Fig. 4. Peak optical output vs repetition rate for various amplitudes of fixed-width pulses. The diode was No. 128, a typical GaAlAs unit 50 μm in diameter.

quickly by rapid and permanent damage to the diode. For short, single pulses (up to 100 nsec at 100 Hz), typical diodes could be driven safely to produce peak output 12 dB above normal safe dc output. For appreciably wider pulses, this limit was 8–9 dB above dc levels. It was never possible to achieve extremely large peak output by reducing the pulse width to values as low as 12 nsec.

The results of varying the repetition rate of a pulse of constant width are shown in Fig. 4. Saturation here was indicated first by a decrease in peak output as the duty cycle was increased, followed by the appearance of a slope in the individual output pulse.

For a small, flat heat source of radius R on the surface of a heat sink of thermal diffusivity k the ratio R^2/k (appearing as an exponent in a heat transfer equation)[3] can be described as an effective thermal time constant denoting the time from turn-on required for the source to reach 63% of its maximum temperature. We have found this time-constant concept useful in estimating the onset of excessive heating, as a function of repetition rate, when pulses in the range 25–400 nsec are applied to diodes 13–100 μm in diameter mounted on silver studs. Damage-producing heat occurred when the separation between pulses approached the thermal time constant calculated with a value of 1.71 cm^2/sec for the thermal diffusivity of silver. When the period between pulses became less than the time constant (about 3.5 μsec for the 50-μm units, for example), the output peak power declined to the dc reference output level in approximate proportion to the duty cycle. For repetition rates up to the thermal time constant, peak power output 8–9 dB above the safe dc output was possible.

These results indicate that pulses as short as 2–3 nsec (limited by the diode rise time) can be used to drive electroluminescent diodes of this type as directly modulated optical sources. For large duty cycle operation this corresponds to a maximum pulse rate of about 5×10^8/sec, and the peak light output expected would be that corresponding to the dc rating. However, peak output of the order of 10 dB above dc ratings can be achieved with relatively short pulse drive if the repetition rate is restricted to values below about 5×10^4/sec. The peak output exhibits no direct relationship to the average power dissipation, but appears to be limited by heat transfer from the junction area within the time duration of an individual pulse. It is of particular interest to note that diodes degraded permanently and rapidly only when the light-emitting region was heated significantly above 80°C by the applied current, either dc or pulsed. However, diodes operated at somewhat lower currents in ambients to 200°C degraded at rates which suggest the possibility of high-current-density, room-temperature cw operating life (to half output) approaching 10^5 h.

The authors are indebted to D. H. Ring for guidance in this study, to E. A. Ulmer for the heterostructure semiconductor material, and the F. J. Favire for assistance in the diode fabrication.

References

1. C. A. Burrus and R. W. Dawson, Appl. Phys. Lett. **17,** 97 (1970).
2. C. A. Burrus and E. A. Ulmer, Proc. IEEE **59,** 1263 (1971).
3. W. J. Oosterkamp, Philips Res. Rept. **3,** 49 (1948).

Heterostructure Injection Lasers

MORTON B. PANISH

(Invited Paper)

Abstract—**The utilization of the nearly ideal heterojunction that can be achieved between GaAs and $Al_xGa_{1-x}As$ to confine both light and electrical carriers has lead to the evolution of several new classes as injection lasers with very low room-temperature current-density thresholds for lasing ($\lesssim$1000 A/cm²), and structures whose operation can be more readily understood than the earlier homostructure lasers. These are as follows: the single-heterostructure (SH) laser which utilizes one heterojunction to confine light and carriers on one side of the structure; the double-heterostructure (DH) laser in which both carriers and light are confined to the same region; and the separate-confinement-heterostructure (SCH) laser in which the carriers are separately confined to a narrow region within the optical cavity. A state-of-the-art description of these lasers and some of the mode structures encountered in their operation is presented. Recent work is described which permits the growth of low-strain heterostructures with heterojunctions between GaAs and $Al_xGa_{1-x}As_{1-y}P_y$ strain reduction from mismatch and bonding of contacts has resulted in lasers which, while maintaining very low room-temperature current thresholds, also have very long lifetimes (>10^5 h) for continuous operation.**

INTRODUCTION

ONE of the important candidates being considered as the signal generator for optical-communications systems is the heterostructure injection laser. Lasing action by the stimulated recombination of carriers injected across a p-n junction was predicted [1], [2] in 1961 and was achieved [3]–[6] in 1962. These early injection lasers were generally rectangular chips of GaAs containing a p-n junction perpendicular to two polished or cleaved ends of the chip. The structure is illustrated in Fig. 1. The polished or cleaved ends are partial mirrors. Light is generated by the injection of electrons into the p region with subsequent radiative recombination of holes and electrons. This recombination occurs in a volume adjacent to the p-n junction and between the two mirrors. The active region between the mirrors is then an optical cavity. Structures such as that illustrated in Fig. 1 are now generally referred to as homostructure lasers because they are made of a single material such as GaAs, and thus contain no heterojunctions. These types of injection lasers typically have very high room-temperature-threshold current densities (~50 000 A/cm²), because little or no control over the thickness of the recombination region can be achieved as the result of unrestricted diffusion of injected carriers and because that region constitutes a very poor waveguide. The band-edge potential diagram, the refractive index, and the optical-field dis-

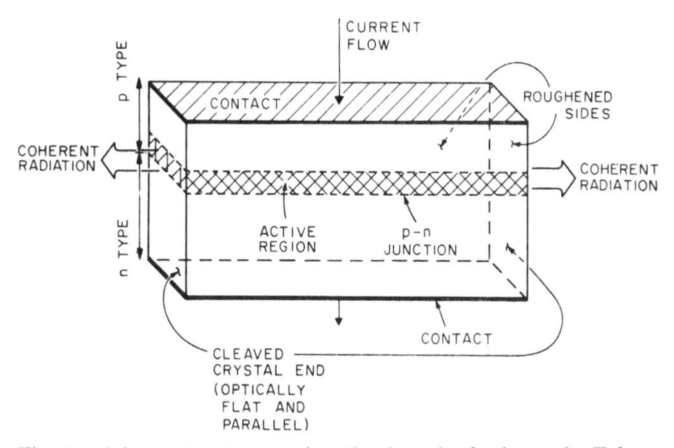

Fig. 1. A homostructure p-n junction laser in the form of a Fabry–Perot cavity. After Panish and Hayashi [21].

Manuscript received March 6, 1974; revised April 25, 1974.
The author is with Bell Laboratories, Murray Hill, N. J. 07974.

Reprinted from *IEEE Trans. Microwave Theory Tech.*, vol. MTT-23, pp. 20–30, Jan. 1975.

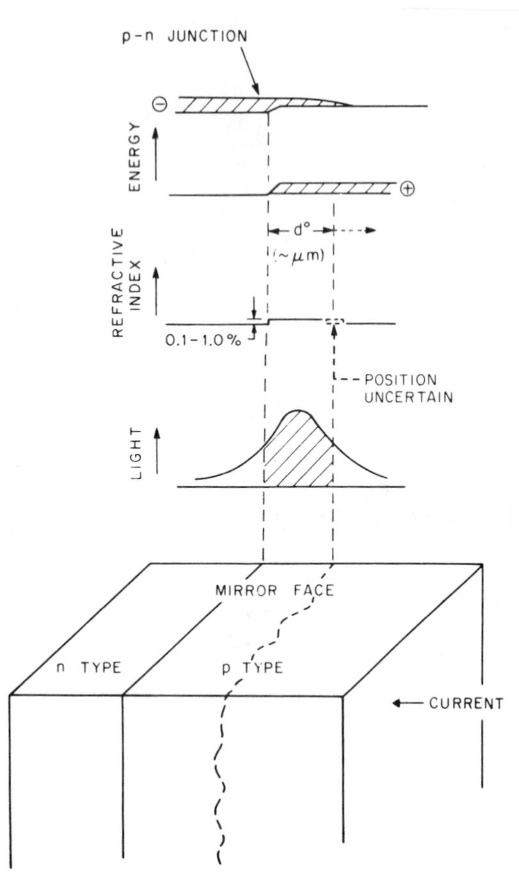

Fig. 2. Schematic representation of the band edges (forward biased), refractive index, and optical-field distribution in diffused homostructure laser diodes. The possibility of uncertainty in the position and regularity of the right boundary of the active region is exaggerated for illustrative purposes. The energy bands are shown with high forward bias and both the n and p regions are degenerately doped. Electrons are injected into the p region but hole injection is negligible. The injected electron population is great enough so that in the region d° the condition that the separation of the quasi-Fermi levels for electrons and holes $F_c - F_v > h\nu$ is satisfied. After Panish and Hayashi [21].

tribution for a homostructure laser are shown schematically in Fig. 2 for a comparison with the more complex structures that will be discussed.

In 1963 Kroemer [7] and Alferov and Kazarinov [8] suggested that the junction laser could be improved by the use of structures with heterojunctions, in which the recombination or active region is bounded by wider band-gap regions. These initial suggestions did not immediately result in injection lasers with lower threshold currents at room temperature [9], [10] because the materials combination suggested at that time, GaAs–$GaAs_xP_{1-x}$, led to the generation of nonradiative traps at lattice-mismatched heterojunctions. In 1967 Rupprecht *et al.* [11] reported the growth by liquid-phase epitaxy (LPE) of layers of $Al_xGa_{1-x}As$ on GaAs. The band gap of $Al_xGa_{1-x}As$ increases with x and is direct to about $x = 0.37$ [12]. Furthermore the replacement of Ga with Al atoms in the lattice has only a very small effect upon the lattice parameter so that heterojunctions between GaAs and $Al_xGa_{1-x}As$ are very nearly lattice matched. It thus appeared that with sufficient control of the layered growth and understanding of the parameters involved, improved junction lasers could be achieved. At about that time, therefore, Alferov [13], [14], Hayashi and Panish [15]–[17], and Kressel [18], all with co-workers, began studies which resulted in injection lasers with reduced room-temperature current thresholds, and eventually lasers capable of operating continuously at room temperature [19], [20].

These lasers, which are designated as heterostructure lasers, fall into three general classes: 1) single-heterostructure (SH) lasers, in which there is only one heterojunction so that light and injected carriers are confined by a heterojunction only at one boundary of the recombination region; 2) double-heterostructure (DH) lasers, in which the carriers and waveguide have the same boundaries on both sides of the recombination region; and 3) separate-confinement heterostructures (SCH), in which the injected carriers are confined to a region within the waveguide. The discussion which follows is a state-of-the-art description of the pertinent physical and chemical properties of GaAs and $Al_xGa_{1-x}As$, of the nature of heterojunctions in the $Al_xGa_{1-x}As$ system, of the properties of several types of heterostructure lasers, and of the degradation of such lasers. A more detailed review is given in [21].

MATERIAL PROPERTIES AND CRYSTAL GROWTH

GaAs and $Al_xGa_{1-x}As$ are semiconductors which crystallize with the zinc-blende lattice structure. The Group III and Group V elements occupy separate sublattices, and thermodynamic studies of $Al_xGa_{1-x}As$ suggest that the Al and Ga are randomly placed on the Group III lattice sites [22], [23]. GaAs is a direct-gap material with a forbidden energy E_g of 1.43 eV at room temperature. The band gap of $Al_xGa_{1-x}As$ increases with x to a transition from a direct to an indirect gap at ~1.92 eV and $x = 0.37$. This is illustrated in the composition-energy diagram of Fig. 3 obtained with data from [12], [24], and [25]. Efficient radiative recombination can readily be achieved in GaAs and in $Al_xGa_{1-x}As$ with compositions in the direct-gap range. Efficient light-emitting diodes have been made with both materials, and minority-carrier diffusion lengths in GaAs have been studied [26]–[28]. These diffusion lengths are long, on the order of several microns for material doped in the 10^{17}–10^{18}-cm^{-3} range. This long diffusion length is an important characteristic because, as will be described later, it is possible to achieve heterostructure lasers with recombination- or active-region widths which are small compared to the diffusion length. Calculations of the optical-field distribution in GaAs–$Al_xGa_{1-x}As$ heterostructure lasers require a detailed knowledge of the refractive index $\bar{n}$ of GaAs and $Al_xGa_{1-x}As$ as a function of wavelength and composition. For GaAs $\bar{n}$ is, in the spectral region of interest, considerably influenced by doping [29]. Detailed new information has recently become available for the refractive index of both GaAs and $Al_xGa_{1-x}As$ [29], [30], and the results have been used in several field calculations which

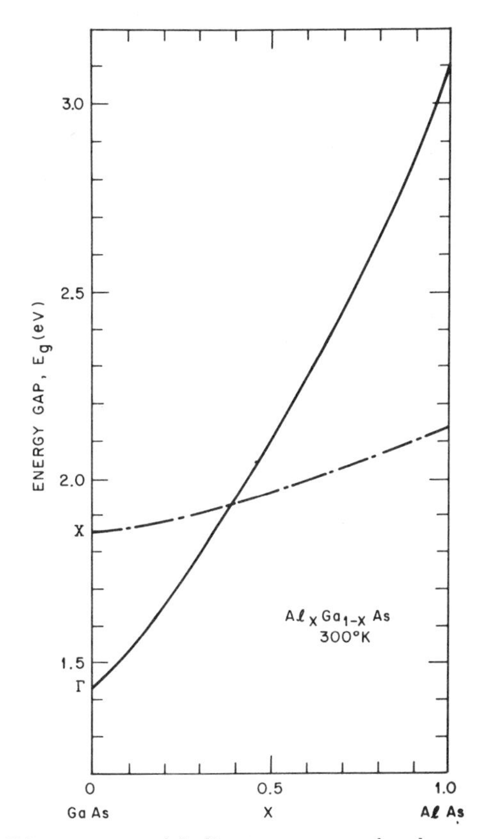

Fig. 3. Direct —— and indirect – – – – band-gap energy of $Al_xGa_{1-x}As$ against composition.

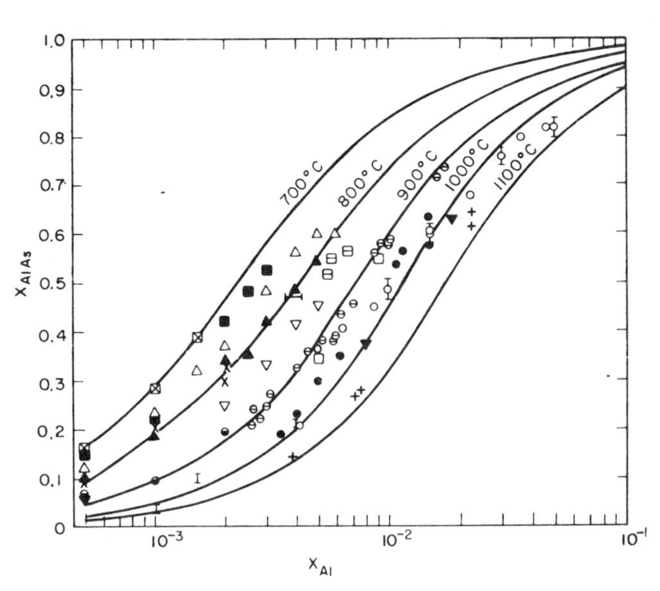

Fig. 5. Solidus isotherms in the Al–Ga–As system. The liquid composition is defined by one elemental composition, x_{Al}, since the liquidus isotherms are defined in Fig. 4. In practice the the liquidus may be considered to be the indicated atom fraction of Al in an Al–Ga solution where it is then saturated with As or GaAs. ⊠, ×, ○, I [34], at 700, 800, 900, and 1000°C, respectively. ⊢⊣, ⊟, ⊖ [33], at 800, 850, and 900°C, ○ [23] at 1000°C, ⏀ [22] at 1000°C. ■ 707°C, △ 756°C, ▲ 802°C, ▽ 852°C, □ 900°C, ● 970°C, ▼ 1020°C, + 1060°C [32]. The solid curves are estimated from the experimental data with a simple solution treatment. After Panish and Hayashi [21].

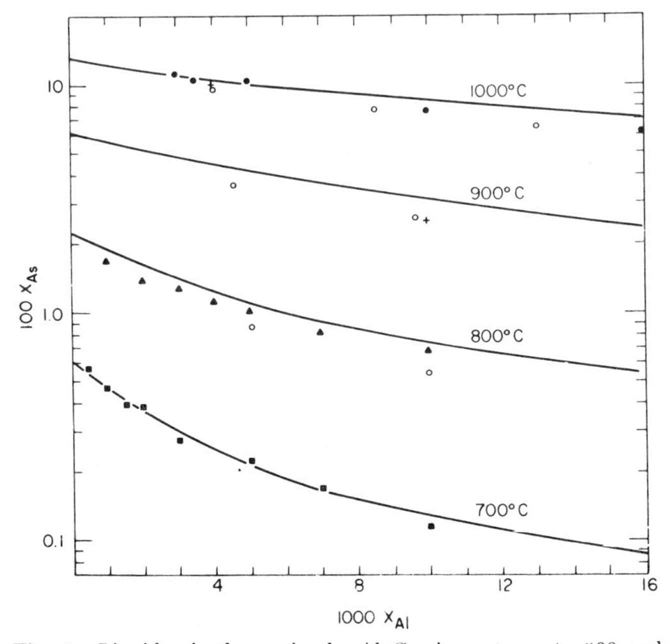

Fig. 4. Liquidus isotherms in the Al–Ga–As system. +: 900 and 1000°C [22]. ○: 800, 900, and 1000°C [23]. ■ 712. ▲ 802. ● 1000°C [32]. The solid curves are estimated from the experimental data with a simple solution treatment. $X_{Ga} + X_{Al} + X_{As} = 1$. After Panish and Hayashi [21].

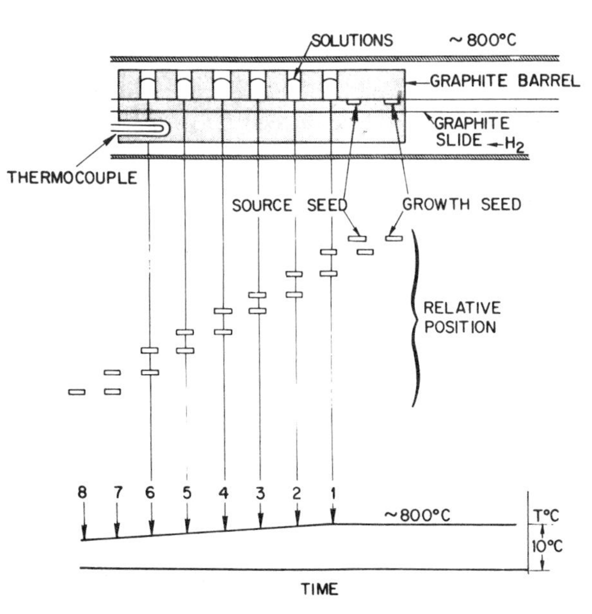

Fig. 6. A recent version of apparatus and procedures used for the growth of DH lasers. The growth boat, the apparatus, and growth and source seed positions relative to solutions and temperature are shown. The procedure results from the work of authors Dawson [71] and Rode [72].

are illustrated in the following discussion of modes and far-field distributions.

At the time of this writing there have been two successful techniques for the preparation of heterostructure wafers. The most frequently used has been LPE, but more recently molecular-beam epitaxy has shown great promise [31]. LPE is a near-equilibrium growth technique in which the successive layers constituting the heterostructure are grown onto a GaAs substrate by successively bringing the substrate into contact with Al–Ga–As or

Ga–As solutions, properly doped, while cooling. It is clear that the composition–temperature relationships for the Al–Ga–As system must be well in hand. A number of phase studies have been done [22], [23], [32]–[34] and the resulting phase diagrams for the liquidus and solidus equilibria at several temperatures are given in Figs. 4 and 5. For lasers, the multilayer growth is generally done near 800°C. The apparatus used recently by this author is shown in Fig. 6. As illustrated, the layers are grown by successively bringing the different solutions into contact with the seed while cooling at 0.05–0.1°C/min. A first seed is generally used to bring the solutions to saturation with respect to arsenic. Solution compositions are chosen from the phase data for the required layer compositions. Dopants are added to the solution.

Molecular-beam epitaxy is a technique based upon the early studies of Arthur [35], which has been further studied and refined by Cho *et al.* [36], [37], for growing multilayers of GaAs and $Al_xGa_{1-x}As$ and which has recently been used to achieve low-threshold DH lasers [31]. The technique consists of impinging controlled beams of Ga, Al, and dopant atoms, and As_2 or As_4 molecules upon a heated substrate. Epitaxy results when the beam contains excess arsenic. This technique is characterized by its potentiality for a very high degree of dimensional control.

THE HETEROJUNCTION IN THE $Al_xGa_{1-x}As$ SYSTEM

The simplest case to discuss is the GaAs–$Al_xGa_{1-x}As$ heterojunction, and the discussion applies also to the more general case of $Al_xGa_{1-x}As$–$Al_yGa_{1-y}As$, in which at least one side of the junction is in the direct-gap composition region. The features of the GaAs–$Al_xGa_{1-x}As$ heterojunction are the wider band gap of $Al_xGa_{1-x}As$, the lack of extensive interface recombination by nonradiative traps, and the fact that the discontinuity in the band-gap energy between GaAs and $Al_xGa_{1-x}As$, while essentially a conduction-band discontinuity when defined in terms of the electron affinity, can be used effectively as a barrier to holes and electrons by adjusting the doping. Schematic diagrams of the band edges for various of the combinations of heterojunctions possible are shown in Fig. 7. A convenient notation which will be used from here on is to use n and p for designating n–GaAs and p–GaAs and N and P to designate n-type $Al_xGa_{1-x}As$ and p-type $Al_xGa_{1-x}As$. For simplicity the Fermi levels of n- and N-type material and of p- and P-type material are taken in Fig. 7 to be at the conduction and valence band edges, respectively. In the lower part of the figure, junctions between materials of different conductivity types are shown at high forward bias so as to permit electron injection (at a p-N heterojunction) or hole injection (at an n-P heterojunction) as shown. At the n-N and N-p heterojunctions the conduction band difference between GaAs and $Al_xGa_{1-x}As$ is expected to cause a discontinuity (shown dotted) in the conduction band. This narrow potential barrier is generally taken to permit passage of

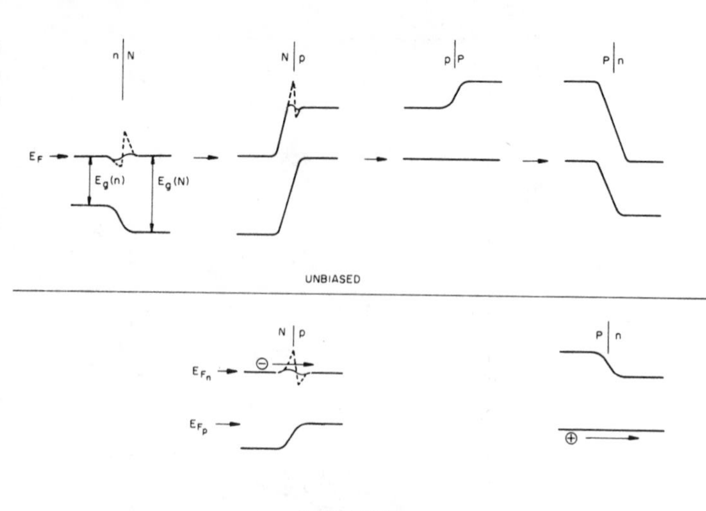

Fig. 7. A schematic drawing of band-edge potential diagrams of n-N, N-p, p-P, and P-n heterojunctions between GaAs and $Al_xGa_{1-x}As$. - - - ungraded junction; —— graded junction. E_F, Fermi level in unbiased case. E_{F_n} and E_{F_p}, Fermi levels in biased case. Fermi levels are taken to coincide with the band edges. E_g shows the forbidden gap.

carriers by tunneling, or to be largely smeared out by composition gradients over relatively short distances [38]. The result of this is, as described for specific structures in the following, that the band-gap difference between GaAs and $Al_xGa_{1-x}As$ is a potential barrier that reflects electrons at p-P and p-N heterojunctions and holes at N-p and N-n heterojunctions. The existence of these potential barriers to minority carriers permits sufficient biasing of the junction devices so that the injected carrier density in the active layer becomes greater than the equilibrium majority carrier density in the emitting layer. Alferov [39] has called this ability to achieve such high injected carrier densities "superinjection."

HETEROSTRUCTURE LASERS

A. Single Heterostructure Lasers

The first reduction of room-temperature current thresholds Jth (300 K) of injection lasers was achieved with a structure which is illustrated in Fig. 8 and is designated in the notation described previously as a n-p-P structure, the so-called single heterostructure. In this structure the p-n junction is located a short distance (about 2 μm usually) from the p-P heterojunction. Under forward bias, electrons are injected across the n-p junction but are then confined by the potential barrier at the p-P junction. Improvement in the establishment of a well-defined active region and optical waveguide over that achieved with the homostructure laser can be qualitatively seen by comparison of Figs. 2 and 8. A plot of Jth (300 K) and Jth (80 K) as a function of the thickness d of the active region is given in Fig. 9. It can be seen that over a considerable range Jth decreases with d. This suggests that the injected electrons are indeed confined by the p-P barrier so that Jth behaves as expected, i.e., decreases with the volume of the active region. At sufficiently small d, however, Jth increases with decreasing d. This can

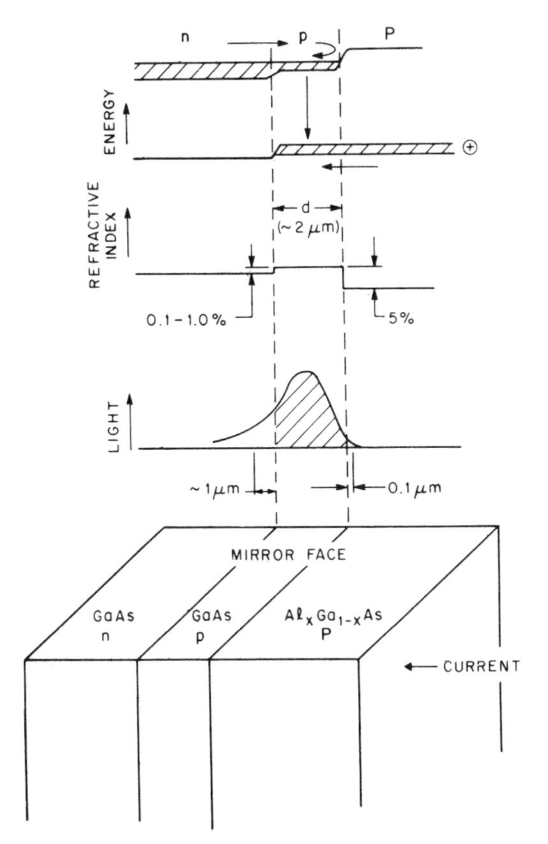

Fig. 8. Band structure, refractive index, optical power distribution, and physical structure of an SH laser. After Panish and Hayashi [21].

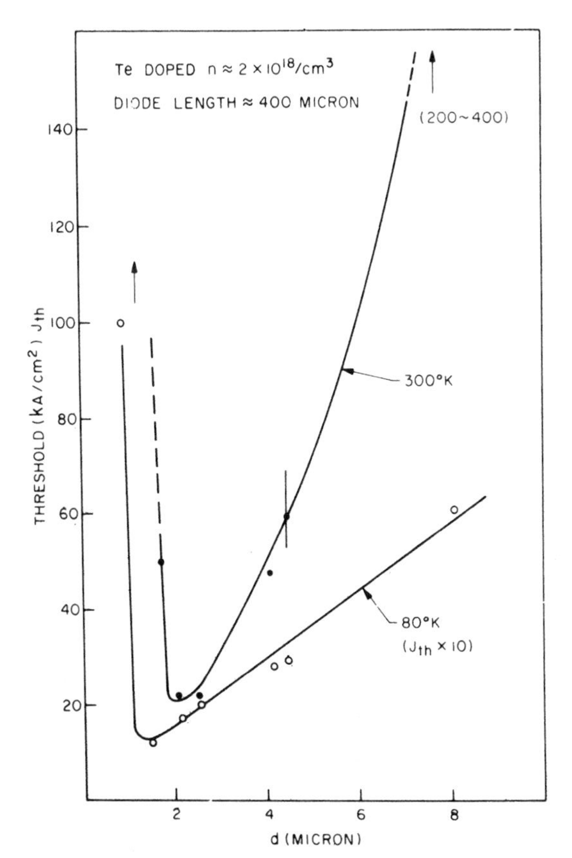

Fig. 9. Jth as a function of d for uncoated (~400 μm-long) Fabry–Perot SH diodes at 300 and 80 K. All diodes used for this plot were made on 2 × 10^{18} n/cm³ Te-doped GaAs. After Hayashi and Panish [17].

result from hole injection when the structure with smaller d must have higher forward bias to reach the lasing electron density because of the greater spread of the optical field out of the active region, and it can also be due to the failure of the optical-field confinement as a result of increasing asymmetry of the structure with decreasing d. With the SH laser, Jth (300 K) in the range of 8000–15 000 A/cm² is readily achieved [15], [18]. These lasers are generally operated with a short duty-cycle current pulse (typically on the order of 0.1 μs at 100–1000 Hz) and have not been run continuously at room temperature. Some typical SH-laser parameters are given in Tables I and II.

B. Double Heterostructure Lasers

The DH laser of GaAs–$Al_xGa_{1-x}As$ is presently the most important heterostructure-laser candidate being developed for use in optical-communications systems. It was the first junction laser capable of continuous operation at room temperature [19], [20] (and even well above [40]), and for which long-lived versions are available [41]. Its layer structure is N-p-P or N-n-P as illustrated in Fig. 10. Minority carriers are effectively confined to the narrow-gap region, the boundaries of which also correspond to the waveguide. The properties of the DH are found to be primarily dependent upon the dimensions of the structure and the composition of the wide-gap regions. At active-region doping levels less than $\sim 10^{18}$ cm^{-3} the DH-laser properties are not strongly dependent upon the doping element or level.

With this structure Jth (300 K) is clearly proportional to the volume of the active region for d greater than about 0.5 μm as is shown for the data for Jth (300 K) against d for a particularly uniform set of diodes in Fig. 11. The thresholds are much lower than those which can be achieved with SH lasers and are achieved at narrower active-region widths. These reduced thresholds result not only from a smaller active volume, reduced loss, and better coupling of the optical field to the active region than in the SH or homostructure laser, but also from greater superinjection. This permits the electron quasi-Fermi level in a p-type active region, for example, to be pushed further into the conduction band with the result that the quasi-Fermi level comes closer to the maximum in the electron distribution in the conduction band [21]. Some typical parameters for DH lasers are given in Tables I and II.

C. Filaments and Modes

One of the difficult problems which must be dealt with in studies of injection lasers is their tendency towards multimode and filamentary operation. A lasing filament is a relatively narrow region of an injection laser which, presumably because of small inhomogenieties in the structure, begins to lase while surrounding regions are still below threshold. These filaments tend to be unstable and cause noise, and with increasing current above threshold multifilament operation is possible in a sufficiently

TABLE I
TYPICAL 300 K HETEROSTRUCTURE LASER PARAMETERS[a]

	SH	Ref	DH	Ref	SCH	Ref
$J_{th}/(A/cm^2)$	8000-15,000	17,18	~1000-6000[c]		575-2500	55,69
$J_{th}/d/(A/cm^2\mu$	~6000-10,000[b]	17	~4000-5000			
gain/(cm/KA)	3-6	17,18				
Loss/ (cm^{-1})	20-40	17,18	~10[d]	67	9-17[d]	55
$g = J^m$,m	1	17,66	~2.8	68	~3	55
efficiency (differential external)	20-50%	17,56	40-50%	68	40-60%	55
efficiency (total external)	~10%	17,56			~40%[e]	55

[a] For pulsed lasers with Fabry–Perot cavities 0.4–1.0 mm long.
[b] $d \sim 1.5$–3.0 μm.
[c] $d \sim 0.2$–1.5 μm.
[d] For active regions doped less than about 10^{18} carriers/cm³.
[e] At four times Jth (300 K), one typical unit tested. Fundamental transverse mode.
[f] Symmetrical SCH. Data for LOC-type structures are given in [52] and [73].

TABLE II
POWER AND CURRENT FOR SEVERAL TYPICAL HETEROSTRUCTURE LASERS

	External Pulsed Peak Power Watts	at	Pulsed Peak Current I/AMPS	External CW Power Watts	at	CW Current I/AMPS
SH	13 (Ref 56)		24	No CW SH lasers		
DH	6[a] (Ref 56)		10	≈ 0.01 (stripe[b])		0.1-0.2
				0.2[a] (broad area[c] Ref 70)		1-2
SCH	3-4[a] (Ref 55)		19	No data available		

[a] Maximum measured.
[b] Typical stripe-laser dimensions are 10 × 400 μm.
[c] Typical broad-area-laser dimensions are 100 × 400 μm.

wide structure. The problem is alleviated by passing current through a narrow-stripe laser rather than a wide area. This stripe which can be defined by a stripe contact [42], proton bombardment [40], doping profile control [43], or a stripe mesa structure [44] is usually about 10 μm wide and can accommodate only a single filament. These various approaches to filament control are illustrated in Fig. 12.

The stripe-geometry lasers, when uniform enough, exhibit mode structures parallel to the junction plane, with the order of the mode increasing with the width of the stripe [45]. These parallel modes have also been observed with stripe-geometry lasers employing the doping profile control [43]. The fundamental mode is obtained for stripe widths of about 12 μm or less so that the criteria for fundamental parallel-mode operation and single-filament operation are the same. Mode confinement parallel to the junction is related to the presence of gain within the stripe confines [46], [47].

In the DH laser the transverse modes perpendicular to the junction plane are generally TE rather than TM because TE modes have a larger reflectivity at the cavity mirror and therefore a lower cavity loss and a lower threshold current density. For the same reason higher order TE modes are preferred because they have a larger angle of incidence and thus lower mirror loss than lower order modes. Mode selection results, however, from a balance between the decrease of mirror loss with increasing mode order and an increase of loss within the volume of the device with increasing order since the higher order modes penetrate farther into the inactive regions which bound the active region. It is usually desirable to maintain the lowest order (fundamental) mode in a practical device. Fortunately, the requirements for fundamental-mode operation and low threshold coincide because of the need for a thin active region in both cases. In structures with sufficiently thick active regions the threshold of a higher order mode can become comparable to that of the fundamental mode so that multimode operation can occur. The elimination of this kind of higher order and multimode operation requires that the gain profile be controlled in such a way as to favor one mode. The SCH

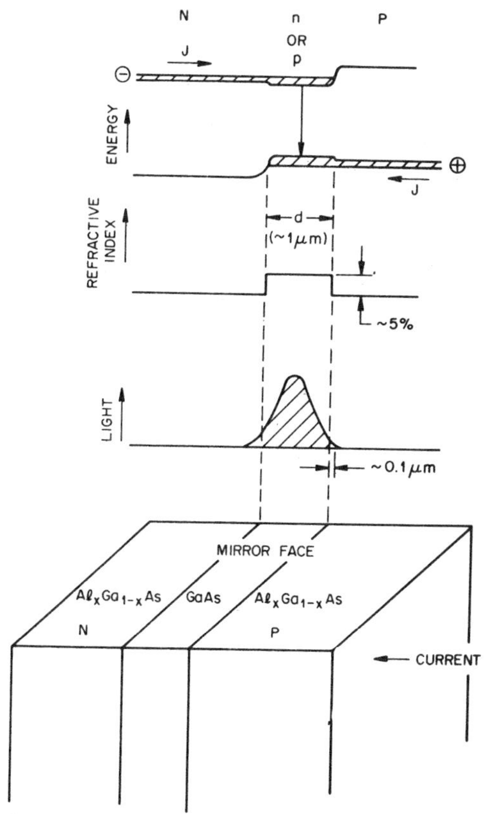

Fig. 10. Schematic representation of the band edges with forward bias, refractive index changes, optical-field distribution, and physical structure of a DH laser diode. After Panish and Hayashi [21].

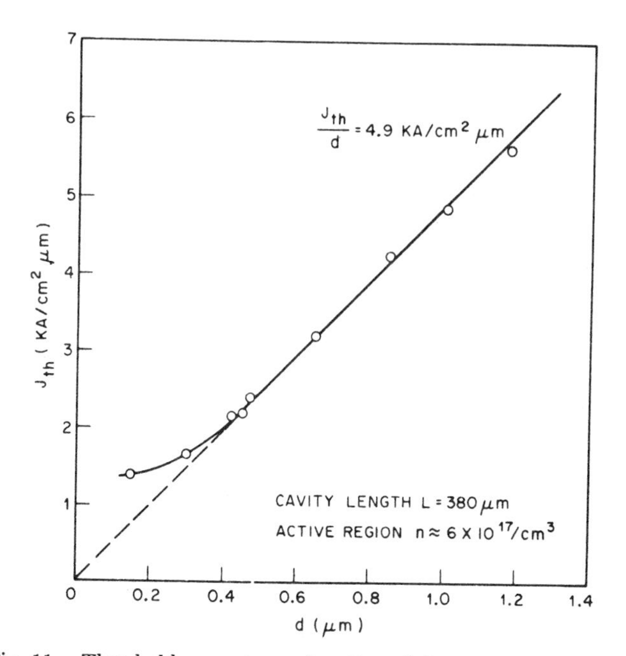

Fig. 11. Threshold current as a function of d at room temperature for a series of similarly doped "uniform" DH lasers. Cavity length is 380 μm; active region $n = 6 \times 10^{17}/\text{cm}^3$. After Pinkas *et al.* [68].

lasers described in the following provide several methods for transverse-mode control. It should be noted here that because of the weaker optical confinement of homostructure and SH lasers, these lasers usually operate in the fundamental transverse mode without further modification.

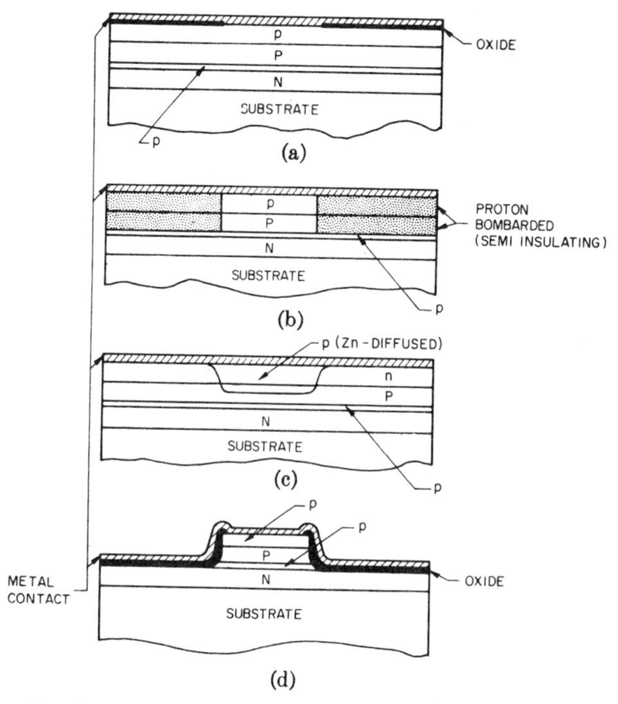

Fig. 12. End (mirror) view of four types of stripe-contact lasers. (a) Stripe contact. (b) Proton-bombardment defined stripe. (c) Doping-profile stripe control. (d) Stripe mesa.

At this point it is useful to consider the optical-field distribution within the structure and in the far field of a DH laser. The laser emission is strongly divergent because of the small size of the source. This divergence is particularly great perpendicular to the junction plane because the source dimensions there are near or less than the lasing wavelength within the structure. The optical-field intensity distribution (E^2) for the fundamental mode perpendicular to the junction plane within the structure is shown for a set of structures with different active-region widths but the same wide-gap Al compositions [48] in Fig. 13. It is interesting to note that for sufficiently narrow structures ($d < \lambda/2$) the field cannot be as effectively confined and spreads rapidly beyond the active region with decreasing d. The effect of the active-layer thickness and the ternary composition upon the far-field distribution of the fundamental mode is clearly seen in Fig. 14. For very small d the effective dimension of the source becomes larger and with decreasing d the divergence perpendicular to the plane of the junction becomes smaller. Unfortunately, the pumped region d is then an increasingly smaller part of the distribution and eventually the lasing-threshold current density must increase with decreasing d. It is also clear that the fundamental transverse-mode operation is restricted to DH lasers with active regions less than about 0.7 μm since the cavity width can support higher order transverse modes at larger d. The far-field intensity distribution for a typical DH laser emitting in the fundamental transverse and parallel modes is shown in Fig. 15. For comparison the far-field intensity distribution perpendicular to the junction plane for a DH laser with a second-order mode is shown as the dashed

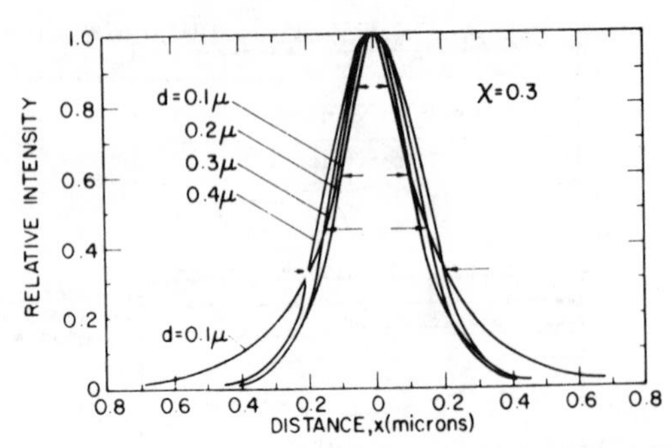

Fig. 13. The square of the electric field as a function of position within a GaAs–Al_xGa_{1-x}As DH laser. The distance is measured from the center of the active region in the direction normal to the plane of the junction. Al concentration is $x = 0.3$. The arrows indicate the active region width d for each curve. After Casey *et al.* [48].

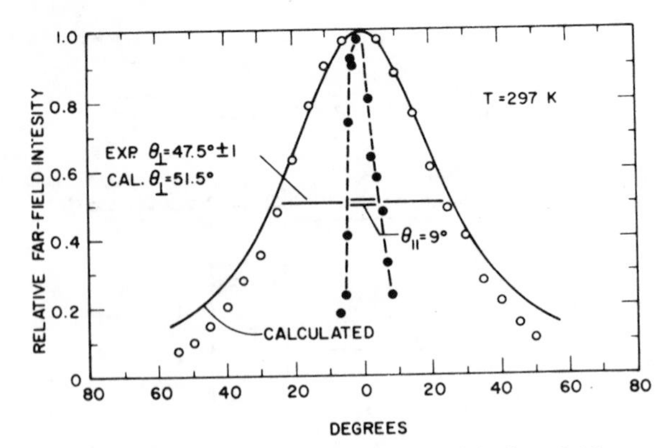

Fig. 15. Far-field pattern for a DH laser with $d = 0.18$ μm and $x = 0.3$. The experimental intensity in the plane of the junction is given by the solid dots ●, and the angle at half intensity is shown by $\theta_{||}$. The experimental intensity perpendicular to the junction plane is given by the open dots ○, and the calculated intensity is shown by the solid line. After Casey *et al.* [48].

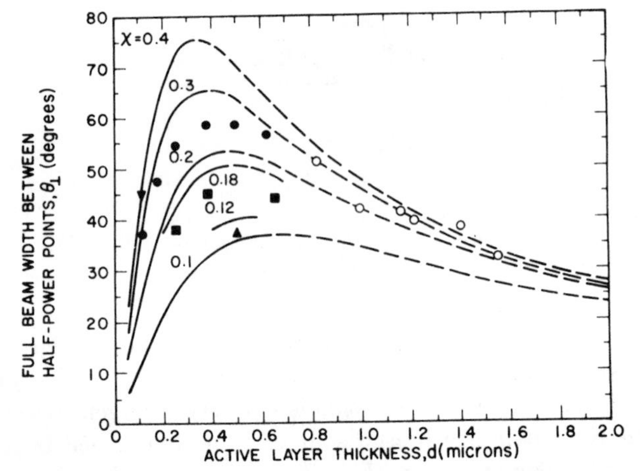

Fig. 14. Full beamwidth at the half-power (3-dB) points as a function of active layer thickness d and composition x for the DH. The solid and dashed curve is the beam divergence calculated for the fundamental TE mode. The dashed portion of the calculated curve represents active-layer thicknesses where higher order modes are possible. The open circles are for SCH lasers (fundamental mode). Data and calculations from [48] and [55].

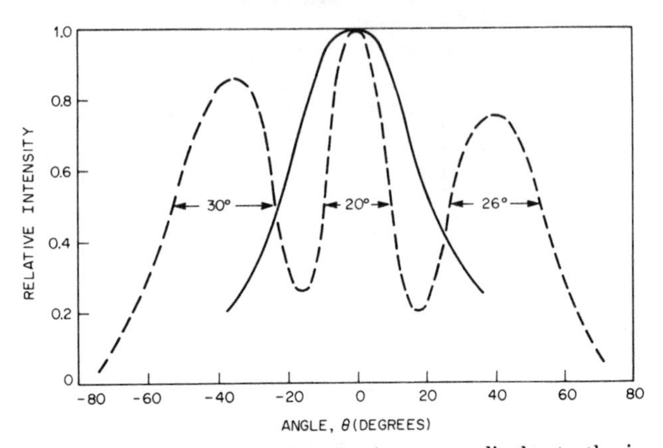

Fig. 16. Far-field intensity distribution perpendicular to the junction for an SCH laser—and a DH laser- - -both with an optical cavity width of 1.05 μm and an Al concentration of $x = 0.3$ in the wide-gap regions. Curves replotted from [55].

curve in Fig. 16. (The zero-order mode also shown in Fig. 16 is for an SCH laser described in the following.)

Not only do the dielectric steps parallel to the junctions, but also the mirror ends act as sources of standing waves. Between these mirrors sets of standing waves constitute the longitudinal modes. These, as all other modes, are separated in frequency. It can be shown that because the gain profile of injection lasers is very wide, about 200 Å, and because the typical laser diode length is about 500 μm, the optical cavity supports many longitudinal modes separated by about 1–3 Å with only slightly differing gain. As illustrated in Fig. 17, many modes may lase simultaneously. A single longitudinal mode is much more difficult to obtain than a single transverse or parallel mode. Rossi *et al.* [49] have achieved single longitudinal-mode operation by coupling an external grating to a homostructure and a SH laser, respectively. The situation should be similar for DH and SCH lasers described in the following but it is doubtful whether an external cavity for mode control will be practical for other than laboratory use.

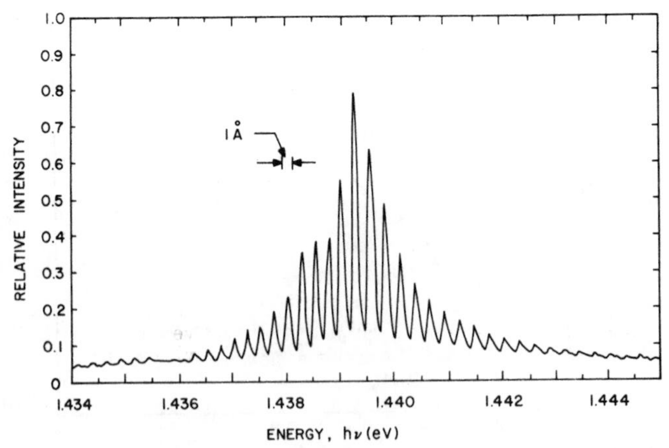

Fig. 17. Lasing spectrum of a single transverse-mode DH laser with multiple longitudinal modes.

D. *Separate-Confinement Heterostructure Lasers*

Several types of structures in which the carriers are confined to a region narrower than the waveguide have been studied. All of these may be classified as SCH lasers although they fall into two somewhat different subclassifications. The simplest of these (N-n-p-P) uses a p–n

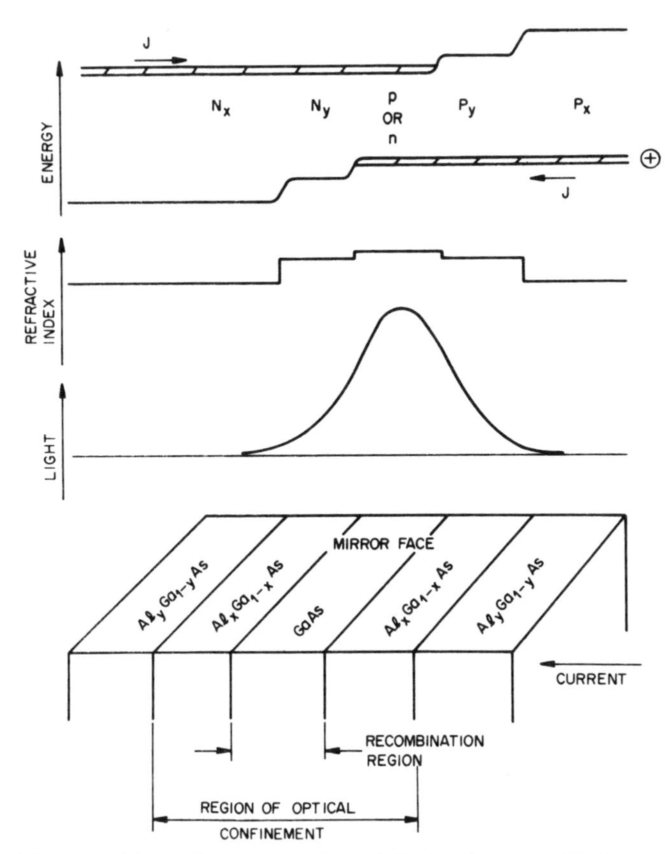

Fig. 18. Schematic representation of the band edges with forward bias, refractive-index changes, optical-field distribution, and physical structure of a symmetrical SCH laser. After Panish and Hayashi [21].

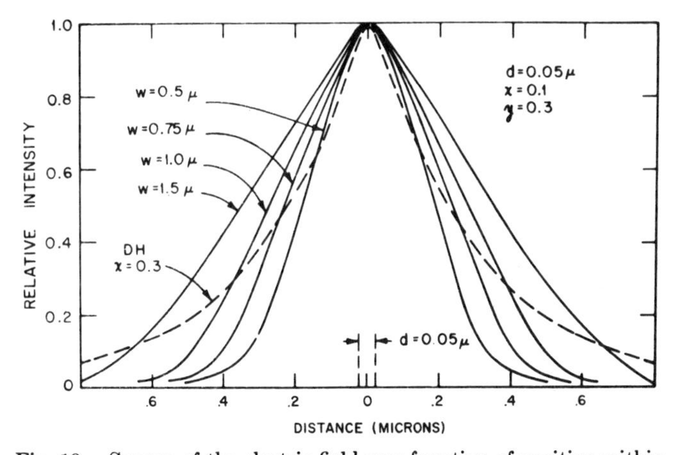

Fig. 19. Square of the electric field as a function of position within SCH and DH lasers with the electrically active region width of both equal to 0.05 μm but at several optical cavity widths for the SCH. After Casey *et al.* [55].

junction within the active region so that as a result of the preference for electron injection in GaAs the gain is limited to only a portion of the active region. The large-optical-cavity (LOC) laser described by Kressel *et al.* [50], [51] is typical of this class. This structure was intended to yield high-power pulsed lasers as a result of its greater efficiency while maintaining a relatively large cavity width to reduce optical-power density. The same structure with the p-n junction properly placed within the optical cavity (a ratio of thickness of the p and n regions of 2:1) may be used to maintain a fundamental transverse mode for a relatively wide waveguide at currents well above threshold [52].

A somewhat more complex structure based upon suggestions by Hayashi [53] and Thompson and Kirkby [54] has also been studied. These authors pointed out that carriers are more easily confined than light by small changes in the band gap at the heterojunction and that the gain in heterostructure lasers depends superlinearly upon current. They suggested that it should be possible to separately confine the carriers to a small fraction of the total optical cavity w while achieving sufficient gain to permit the threshold to decrease as the electrically active layer thickness d decreases.

The notation used previously may be extended to these structures with minor modification. They may be designated N_x-N_y-p-P_y-P_x or N_x-N_y-n-P_y-P_x. Here x and y denote the Al composition in the layers $Al_xGa_{1-x}As$ or $Al_yGa_{1-y}As$ and $x > y$. A typical structure is illustrated in Fig. 18, and a typical set of the optical electric-field distributions for the fundamental mode within the structure for several SCH lasers with different w is compared with a DH laser in Fig. 19. This illustrates that for very narrow electrically active regions the SCH confines the fundamental mode better than the DH, and also that the coupling of the gain region to the fundamental mode is better. In addition, in the symmetrical SCH illustrated in Fig. 18, the fundamental transverse mode is pumped preferentially to all other transverse modes and that mode persists to wide optical-region widths. This is shown for measurements of the half-power beamwidth for several SCH lasers (the open circles) in Fig. 14 and as the solid curve of Fig. 16. The result of the better coupling of the gain to the fundamental mode in SCH lasers than in DH lasers is, as is shown in Table I, that the lowest current-density thresholds have been obtained in these lasers. Furthermore, as is also shown in Table I, the SCH lasers also have somewhat higher efficiencies than DH lasers. The reason for the higher efficiencies is not clear but it has been speculated [55] that this may result from less scatter and less spreading of the optical field out of the structure in SCH lasers. Some typical parameters for SCH lasers are given in Tables I and II.

DEGRADATION OF HETEROSTRUCTURE LASERS

The degradation of injection lasers is a subject which has been characterized by confusion and conflicting information. It is not possible in this brief review to consider all of the earlier experiments and theories. A great deal of progress has been made in the last 18 months or so, and when considering gradual degradation only recent studies will be considered. Degradation during very high pulsed-current operation occurs by a so-called "catastrophic" mode. It is characterized by gross damage (melting, cracking) of the structure. This degradation is apparently related to a power-density limit at the mirrors of 10^6–10^7 W/cm^2. Thus lasers with thin optical

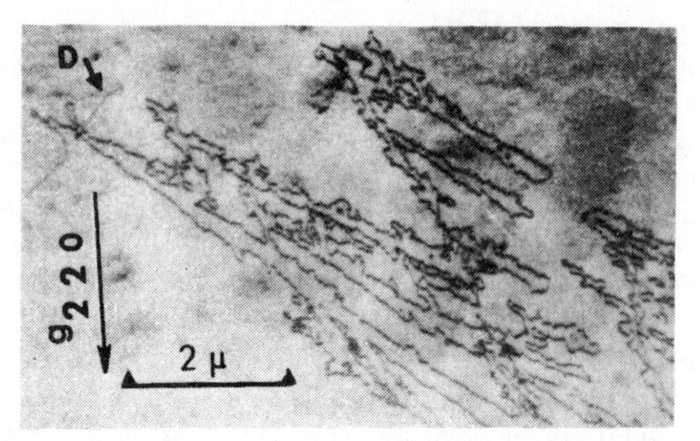

Fig. 20. Bright-field electron micrograph showing a dislocation (*D*) which is at the origin of a dislocation network forming a DLD. After Petroff and Hartman [59].

cavities, such as very low-threshold DH lasers, fail catastrophically at lower pulsed power than do lasers with wide optical cavities or lasers in which the field may spread out of the active region (as with homostructure or SH lasers). Thus lasers intended for high pulsed-power operation have generally been homostructure or SH lasers and low-threshold current is sacrificed for high power at the catastrophic degradation limit. It is worth mentioning here, however, that with the SCH lasers it is possible to make a much better compromise so that relatively low thresholds and relatively high-power thresholds for catastrophic degradation may be achieved with the same laser. A review of catastrophic failure has been given by Kressel [56].

Since lasers used for optical communications will in general be operated as modulated CW devices, we are not primarily concerned with catastrophic degradation, but rather with the second type of degradation—the so-called gradual degradation of injection lasers. In the past year or so an impressive amount of progress has been made towards the understanding and control of gradual degradation. The first DH lasers degraded in a matter of minutes or hours under CW operation to the point at which CW lasing could not be sustained. This rapid degradation was characterized [57], [58] by the appearance of nonluminescent areas called dark-line defects (DLD) crossing the active regions of the lasers during lasing. The DLD has been identified with transmission electron microscopy by Petroff and Hartman [59] as a three-dimensional dislocation network which originates at a dislocation crossing the $Al_xGa_{1-x}As$ and GaAs epitaxial layers. The latter is clearly seen in Fig. 20. The three-dimensional properties of the network have been observed with stereopair photographs. The DLD apparently grows by a climb mechanism during operation of the laser. The source of the point defects required for DLD growth and their driving force are not presently understood. Hartman and Hartman [58] have shown that the lasing lifetime is strongly related to strain, and reduced-strain electrical contacts permitted an increase in laser lifetime from tens of hours to several thousand hours at 20°C [60] and at 30°C [61] ambient temperature. The DLD's are generally not present in operating long-lived lasers.

It has also been established [62], [63] that LPE layers of $Al_xGa_{1-x}As$ on GaAs are compressively stressed at room temperature to levels of 10^8–10^9 dyn cm^{-2} depending on x. This elastic stress can be accounted for by differences in the thermal expansion coefficients of GaAs and $Al_xGa_{1-x}As$ which have the same lattice parameter near the growth temperature. It has been speculated that this stress in heterostructure layers also leads to gradual degradation of DH lasers. Such stress may be removed [64] from the laser structures by the use of $Al_xGa_{1-x}P_yAs_{1-y}$ with $y \approx 0.015$ for the wide band-gap region of the laser. Since P is isoelectronic with As, the small amount present in the quaternary does not materially affect the device electrically or optically, but the relatively small covalent radius of P permits correction of the lattice parameters so that they match at room temperature.

Many DH lasers with GaAs–$Al_xGa_{1-x}P_yAs_{1-y}$ heterojunctions have been studied during CW lasing operations at elevated temperatures [41]. Stable operations with lifetimes greater than 10^5 h are predicted for these lasers when the data are extrapolated to room temperature. Furthermore, these lasers have somewhat lower room-temperature current thresholds [65] than do GaAs–$Al_xGa_{1-x}As$ DH lasers. Although it is believed that the reduced strain is related to both the increased lifetime and reduced threshold of the lasers with the quaternary wide-gap regions, the mechanism by which this occurs is not known.

CONCLUSION

The past few years have been a period of intensive activity in the study of heterostructure lasers. Continuous lasing and vastly reduced room-temperature current thresholds have resulted from the recognition that both optical and carrier confinement can be achieved by utilizing the lattice-matching and band-edge properties at the GaAs–$Al_xGa_{1-x}As$ heterojunction. Detailed studies of the physical and chemical properties have led to the capability for the growth of complex structures and the understanding of the behavior of these structures as lasers. Modification of the composition of the wide-gap region and the development of strain-free fabrication procedures have led to very long-lived lasers. The result of all of this is that the heterostructure laser is a prime candidate to be the optical source in optical-communications systems.

REFERENCES

[1] N. G. Basov, O. N. Krokhin, and Y. M. Popov, *Pis'ma Zh. Eksp. Teor. Fiz.*, vol. 40, p. 1879, 1961; also *Sov. Phys.—JETP*, vol 13, p. 1320, 1961.
[2] M. G. A. Bernard and G. Duraffourg, *Phys. Status Solidi*, vol. 1, p. 699, 1961.
[3] R. N. Hall, G. E. Fenner, J. D. Kingsley, T. J. Soltys, and R. O. Carlson, *Phys. Rev. Lett.*, vol. 9, p. 366, 1962.
[4] M. I. Nathan, W. P. Dumke, G. Burns, F. H. Dill, and G. J. Lasher, *Appl. Phys. Lett.*, vol. 1, p. 62, 1962.
[5] T. M. Quist *et al.*, *Appl. Phys. Lett.*, vol. 1, p. 91, 1962.
[6] N. Holonyak, Jr., and S. F. Beracqa, *Appl. Phys. Lett.*, vol. 1, p. 82, 1962.
[7] H. Kroemer, "A proposed class of heterojunction injection lasers," *Proc. IEEE* (Corresp.), vol. 51, pp. 1782–1783, Dec. 1963.

[8] Zh. I. Alferov and R. F. Kazarinov, Author's certificate 1032155/26-25, USSR, 1963 (as cited in Alferov *et al.* [9]).
[9] Zh. I. Alferov *et al.*, *Fiz. Tverd. Tela*, vol. 9, p. 279, 1967; also *Sov. Phys.—Solid State*, vol. 9, p. 208, 1967.
[10] Zh. I. Alferov, *Fiz. Tekh. Poluprov.*, vol. 1, p. 436, 1967; also *Sov. Phys.—Semicond.*, vol. 1, p. 358, 1967.
[11] H. Rupprecht, J. M. Woodall, and D. G. Pettit, *Appl. Phys. Lett.*, vol. 11, p. 81, 1967.
[12] H. C. Casey, Jr., and M. B. Panish, *J. Appl. Phys.*, vol. 40, p. 4910, 1969.
[13] Zh. I. Alferov, V. M. Andreev, V. I. Korol'kov, E. L. Portnoi, and D. N. Tret'yakov, *Fiz. Tekh. Popuprov.*, vol. 2, p. 1016, 1968; also *Sov. Phys.—Semicond.*, vol. 2, p. 843, 1969.
[14] ——, in *Proc. 9th Int. Conf. Physics of Semiconductors* (1968), p. 504.
[15] I. Hayashi, M. B. Panish, and P. W. Foy, "A low-threshold room-temperature injection laser," *IEEE J. Quantum Electron.* (Corresp.), vol. QE-5, pp. 211–212, Apr. 1969.
[16] M. B. Panish, I. Hayashi, and S. Sumski, "A technique for the preparation of low-threshold room-temperature GaAs laser diode structures," *IEEE J. Quantum Electron.* (Corresp.), vol. QE-5, pp. 210–211, Apr. 1969.
[17] I. Hayashi and M. B. Panish, *J. Appl. Phys.*, vol. 41, p. 150, 1970.
[18] H. Kressel and H. Nelson, *RCA Rev.*, vol. 30, p. 106, 1969.
[19] I. Hayashi, M. B. Panish, P. W. Foy, and S. Sumski, *Appl. Phys. Lett.*, vol. 17, p. 109, 1970.
[20] Zh. I. Alferov *et al.*, *Fiz. Tekh. Poluprov.*, vol. 4, p. 1826, 1970.
[21] M. B. Panish and I. Hayashi, *Applied Solid State Sciences*, vol. 4, R. Wolfe, Ed. New York: Academic, 1974, pp. 235–328.
[22] M. B. Panish and S. Sumski, *Phys. Chem. Solids*, vol. 30, p. 129, 1969.
[23] M. Ilegems and G. L. Pearson, in *1968 Proc. Symp. Gallium Arsenide* (1969), p. 3.
[24] A. Onton, M. R. Lorenz, and J. M. Woodall, *Bull. Amer. Phys. Soc.*, vol. 16, p. 371, 1971.
[25] O. Berolo and J. C. Woolley, *Can. J. Phys.*, vol. 49, p. 1335, 1971.
[26] H. C. Casey, Jr., E. Pinkas, and B. I. Miller, *J. Appl. Phys.*, vol. 44, p. 1281, 1973.
[27] H. Schade, H. Nelson, and H. Kressel, *Appl. Phys. Lett.*, vol. 18, p. 121, 1971.
[28] Zh. I. Alferov, V. M. Andreev, V. I. Morgyin and V. I. Stremin, *Sov. Phys.—Semicond.*, vol. 3, p. 1234, 1970; also *Fiz. Tekh. Poluprov.*, vol. 3, p. 1470 1969.
[29] D. D. Sell, H. C. Casey, Jr., and K. Wecht, *J. Appl. Phys.*, vol. 45, p. 2650, 1974.
[30] H. C. Casey, Jr., D. D. Sell, and M. B. Panish, *Appl. Phys. Lett.*, vol. 24, p. 63, 1974.
[31] A. Y. Cho *et al.*, *Appl. Phys. Lett.*, vol. 25, p. 288, 1974.
[32] Zh. I. Alferov, V. M. Andreev, S. G. Konnikov, V. G. Nitkin, and D. N. Tret'yakov, in *Proc. Int. Conf. Heterojunctions I* (1971), p. 93.
[33] W. G. Rado, W. J. Johnson, and L. I. Crawley, *J. Electrochem. Soc.*, vol. 119, p. 652, 1972.
[34] M. B. Panish and M. Ilegems, *Progress in Solid State Chemistry*, vol. 7, H. Reiss and W. O. McCaldin, Ed. New York: Pergamon, 1972, pp. 39–83.
[35] J. R. Arthur, *J. Appl. Phys.*, vol. 39, p. 4032, 1968.
[36] A. Y. Cho, M. B. Panish, and I. Hayashi, in *Proc. Symp. Gallium Arsenide and Related Compounds* (Aachen, Germany, 1970), p. 18.
[37] A. Y. Cho, *J. Vac. Sci. Technol.*, vol. 8, p. 531, 1971.
[38] J. Womac and R. H. Rediker, *J. Appl. Phys.*, vol. 43, p. 4148, 1972.
[39] Zh. I. Alferov *et al.*, *Fiz. Tekh. Poluprov.*, vol. 5, p. 972, 1971; also *Sov. Phys.—Semicond.*, vol. 5, p. 858, 1971.
[40] a) J. C. Dyment, L. A. D'Asaro, J. C. North, B. I. Miller, and J. E. Ripper, "Proton-bombardment formation of stripe-geometry heterostructure lasers for 300 K CW operation," *Proc. IEEE* (Lett.), vol. 60, pp. 726–728, June 1972.
b) J. C. Dyment, L. A. D'Asaro, and J. C. North, *Bull. Amer. Phys. Soc.*, vol. 16, p. 329, 1971.
[41] R. L. Hartman, private communication.
[42] J. E. Ripper, J. C. Dyment, L. A. D'Asaro, and T. L. Paoli, *Appl. Phys. Lett.*, vol. 18, p. 155, 1971.
[43] H. Yonezu *et al.*, *Japan. J. Appl. Phys.*, vol. 12, p. 1585, 1973.
[44] T. Tsukada, R. Ito, H. Nakashima, and O. Nakada, "Mesa-stripe-geometry double-heterostructure injection lasers," *IEEE J. Quantum Electron.*, (*Special Issue on 1972 IEEE Semiconductor Laser Conference*), vol. QE-9, pp. 356–361, Feb. 1973.
[45] L. A. D'Asaro, *J. Luminescence*, vol. 7, p. 310, 1973.
[46] W. O. Schlosser, *Bell Syst. Tech. J.*, vol. 52, p. 887, 1973.
[47] F. R. Nash, *J. Appl. Phys.*, vol. 44, p. 4696, 1973.
[48] H. C. Casey, Jr., M. B. Panish, and J. L. Merz, *J. Appl. Phys.*, vol. 44, p. 5470, 1973.
[49] J. A. Rossi, S. R. Chinn, and H. Heckscher, *Appl. Phys. Lett.*, vol. 23, p. 25, 1973.
[50] H. F. Lockwood, H. Kressel, H. S. Sommers, and F. Z. Hawrylo, *Appl. Phys. Lett.*, vol. 17, p. 499, 1970.
[51] H. Kressel, H. F. Lockwood, and F. Z. Hawrylo, *Appl. Phys. Lett.*, vol. 18, p. 43, 1971.
[52] T. L. Paoli, B. W. Hakki, and B. I. Miller, *J. Appl. Phys.*, vol. 44, p. 1276, 1973.
[53] I. Hayashi, U. S. Patent 3 691 476, 1972.
[54] G. H. B. Thompson and P. A. Kirkby, "(GaAl)As lasers with a heterostructure for optical confinement and additional heterojunctions for extreme carrier confinement," *IEEE J. Quantum Electron.* (*Special Issue on 1972 IEEE Semiconductor Laser Conference*), vol. QE-9, pp. 311–318, Feb. 1973.
[55] H. C. Casey, Jr., M. B. Panish, W. O. Schlosser, and T. L. Paoli, *J. Appl. Phys.*, vol. 45, p. 322, 1974.
[56] H. Kressel, *Lasers*, vol. 3, A. K. Levine and A. D. Maria, Ed. New York: Marcel Dekka, 1971, p. 49.
[57] B. C. DeLoach, Jr., B. W. Hakki, R. L. Hartman, and L. A. D'Asaro, "Degradation of CW GaAs double-heterojunction lasers at 300 K," *Proc. IEEE* (Lett.), vol. 61, pp. 1042–1044, July 1973.
[58] R. L. Hartman and A. R. Hartman, *Appl. Phys. Lett.*, vol. 23, p. 147, 1973.
[59] P. Petroff and R. L. Hartman, *Appl. Phys. Lett.*, vol. 23, p. 469, 1973.
[60] H. Yonezu *et al.*, presented at the Conf. Solid State Devices, Tokyo, Japan, 1973.
[61] R. L. Hartman, J. C. Dyment, C. J. Hwang, and M. Kuka, *Appl. Phys. Lett.*, vol. 23, pp. 181 and 491, 1973.
[62] G. A. Rozgonyi, C. J. Hwang, and T. J. Ciesielka, *J. Electrochem. Soc.*, vol. 120, p. 333C, 1973.
[63] F. K. Reinhart and R. A. Logan, *J. Appl. Phys.*, vol. 44, p. 3173, 1973.
[64] G. A. Rozgonyi and M. B. Panish, *Appl. Phys. Lett.*, vol. 23, p. 533, 1973.
[65] J. C. Dyment *et al.*, *Appl. Phys. Lett.*, vol. 24, p. 481, 1974.
[66] H. Kressel, H. Nelson, and F. Z. Hawrylo, *J. Appl. Phys.* vol. 41, p. 2019, 1970.
[67] E. Pinkas, B. I. Miller, I. Hayashi, and P. W. Foy, "Additional data on the effect of doping on the lasing characteristics of GaAs-$Al_xGa_{1-x}As$ double-heterostructure lasers," *IEEE J. Quantum Electron.* (*Special Issue on 1972 IEEE Semiconductor Laser Conference*), vol. QE-9, pp. 281–282, Feb. 1973.
[68] ——, *J. Appl. Phys.*, vol. 43, p. 2827, 1972.
[69] G. H. B. Thompson and P. A. Kirkby, *Electron. Lett.*, vol. 9, p. 295, 1973.
[70] B. I. Miller, E. Pinkas, I. Hayashi, and R. J. Capik, *J. Appl. Phys.*, vol. 43, p. 2817, 1972.
[71] L. R. Dawson, to be published.
[72] D. Rode, *J. Growth Cryst.*, vol. 20, p. 13, 1973.
[73] B. W. Hakki and C. J. Hwang, *J. Appl. Phys.*, vol. 45, p. 2168, 1974.

SENSITIVE HIGH SPEED PHOTODETECTORS FOR THE DEMODULATION OF VISIBLE AND NEAR INFRARED LIGHT

H. MELCHIOR
Bell Laboratories, Murray Hill, N.J. 07974, U.S.A.

Photomultipliers, solid state photodiodes and avalanche photodiodes are the preferred detectors for the demodulation of light intensity variations at visible and near infrared wavelengths. The principles of operation of these detectors will be described together with their basic construction and operational characteristics. Photomiltipliers with the recently developed high efficiency photocathodes for visible and near infrared wavelengths and the high gain dynode arrangements with fast speed of response will be mentioned. The various types of silicon and germanium photodiodes with and without internal current gain will be discussed. Tradeoffs involved in the optimization of quantum efficiency, speed of response, internal current gain and sensitivity to weak light signals will be treated in detail for silicon photodiodes operating at wavelengths either in the visible or at the GaAs (~0.7 – 0.9 μm) or YAG (1.06 μm) infrared laser emission lines. The capabilities of germanium avalanche photodiodes for the detection of infrared light at wavelengths up to about 1.6 μm will be mentioned.

1. Introduction

The invention of the laser and its promise as a communications source has greatly stimulated the entire field of photodetection [1–4]. New detection techniques became practical, such as optical mixing or heterodyne detection at longer wavelengths in the infrared. Novel types of photodetectors emerged. A pressing need arose for the development and the perfection of photodetectors with high sensitivity to weak lights signals and response to light intensity modulations at frequencies extending into the microwave region. Great progress has been made, both in the technology of detector materials and in the understanding of their basic device physics. It is now possible to design photodetectors and entire optical receivers with highly optimized performance for various light wavelengths and speed of response combinations.

This paper gives a review and discussion of high speed photodetectors and optical receivers with good sensitivity to weak light signals in the visible and infrared wavelength range between 0.4 and 1.6 μm. After an enumeration of the performance criteria for photodetectors and optical receivers high speed photodetectors will be discussed with particular emphasis on the influence of materials technology and device design on detector performance. Photomultipliers with the new high sensitivity, III-V semiconductor photocathodes and high gain dynodes will be described.

Reprinted with permission from *J. Luminescence*, vol. 7, pp. 390–414, 1973. Published by North-Holland Publishing Company.

The optimization of silicon and germanium photodiodes for various wavelength and speed of response combinations will be described and the use of internal current gain in avalanche photodiodes will be advocated as a means to increase the sensitivity to weak light signals in solid state optical receivers with large signal bandwidths and high speed of response.

2. Performance criteria

Any photodetector with some response at a particular wavelength might be useful for some applications. A well designed optical receiver which consists usually of a photodetector with an amplifier at its output requires however a carefully chosen photodetector and proper optimization of both the photodetector and amplifier [1–8]. The major requirements for a photodetector and optical receiver with high performance include:

(1) large response at the wavelength of the incident optical signal;

(2) sufficient electrical bandwidth, i.e., speed of response, to accommodate the information bandwidth of the incoming signal;

(3) minimum excess noise introduced by the detection and amplification process.

Large response requires a photodetector with a high quantum efficiency at the wavelength of operation and negligible internal shunt conductance for the electrical output signal,

$$\eta = \frac{I_{\text{ph}}}{q} \frac{h\upsilon}{P_{\text{opt}}} \tag{1}$$

and gives the number of photocarriers (I_{ph}/q) collected per unit time in response to the number of photons ($P_{\text{opt}}/h\upsilon$) incident per unit time with energy $h\upsilon$.

Resolution of optical signals with large information bandwidth or short duration requires optical receivers with wide electrical bandwidth and high speed of response. The electrical bandwidth or speed of response of an optical receiver is limited either by carrier transport or multiplication processes within the photodetector or by RC time constants within the photodetector and output amplifier [1–4]. Minimization of both the photodetector (C_I) and amplifier input capacitance (C_A) is mandatory if the photosignal is to develop a high output voltage across the largest possible load or amplifier input resistance. Realization of the risetime T_r or information bandwidth B of an optical signal limits the resistance R_A of the photodetector load or amplifier input to a value lower than

$$R_A \leqslant \frac{T_r}{2.3(C_I + C_A)} \quad \text{or} \quad R_A \leqslant \frac{1}{2\pi(C_I + C_A)B}\,. \tag{2}$$

While high speed of response and adequate quantum efficiency is all that is required for the detection of fast optical signals as long as their intensity is relatively high, the demodulation of weak optical signals requires in addition minimization of the parasitic current and noise sources within the photodetector and output amplifier [1–8]. But even in an ideal optical receiver with unity quantum efficiency and no extraneous noise sources, the sensitivity to weak light signals is limited, by fluctuations in the photoexcitation of carriers [9–11]. Since the number of photocarriers released within a period of time fluctuates statistically, usually in accordance with a Poisson distribution [9–12], there is a mean square noise current

$$\overline{i_{\mathrm{ph}}^2} = 2qI_{\mathrm{ph}}\delta f \tag{3}$$

(with δf = electric bandwidth and q = unit electronic charge) associated with the average photocurrent I_{ph} [1–4]. Assuming an ideal optical receiver and a threshold that counts even single carriers, the average number of photons in an optical pulse has to be at least

$$\bar{n} = 2.3 \log \left(\frac{1}{\text{error rate}} \right) \tag{4}$$

if its presence is to be detected subject to a given error rate [12].

Practical receivers hardly ever reach this so called quantum noise limited sensitivity [1–4]. Especially in receivers with large electrical bandwidth and photodetectors without internal current gain, the sensitivity is much lower and usually limited by the noise of the load or output amplifier [1–4]. A significant improvement in sensitivity is possible if the photocurrent is amplified within the photodetector before it reaches the large noise sources at the output [1–4]. Practically useful current gain mechanisms with relatively low excess noise and high speed of response capability include carrier multiplication in photomultipliers and avalanche photodiodes. Phototransistors and photoconductors too [1–4] exhibit large current gains with low excess noise at somewhat slower speed of response.

Detection of the weakest possible optical signals requires optimization of the photodetector and optical receiver subject to the constraints that either the signal bandwidth and a given signal to noise ratio be maintained or that the presence or absence of the signals which arrive at a given rate be determined with a certain error rate. The power signal to noise ratio at the output of an optical receiver which contains a photodetector with internal current gain and a noisy amplifier is given by [1–9]

$$\frac{S}{N} = \frac{\overline{i_{\mathrm{ph}}^2} M^2 \dfrac{|A(\omega)|^2}{|Y_I + Y_A|^2}}{\displaystyle\int_0^\infty \left\{ 2q(I_{\mathrm{ph}} + I_B + I_D) M^2 F(M) + \frac{\overline{i_I^2}}{\delta f} + \frac{\overline{i_A^2}}{\delta f} + \frac{\overline{v_A^2}}{\delta f} |Y_I + Y_C|^2 \right\} \frac{|A(\omega)|^2}{|Y_I + Y_A|^2} \, \mathrm{d}f} \tag{5}$$

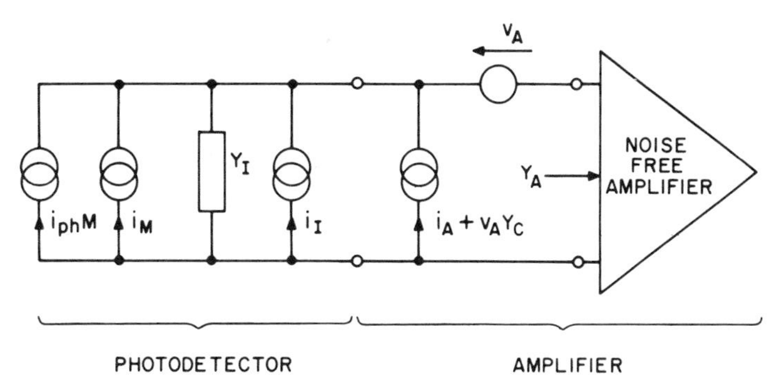

Fig. 1. Equivalent circuit of generalized optical receiver shows the principal signal and noise sources of a photodetector with internal current gain and a noisy output amplifier.

and includes the signal and noise sources of the generalized equivalent circuit of fig. 1. The multiplied signal $i_{ph}{}^2M^2$ of the photodetector with internal current gain M has to override the noise sources of the photodetector and amplifier. The noise sources of the photodetector include the mean square value $\overline{i_M^2} = 2q(I_{ph} + I_B + I_D)M^2F(M)\delta f$ of the multiplied shot noise which is due to the average photocurrent (I_{ph}), the background radiation induced photocurrent (I_B) and the part of the dark current (I_D) that is multiplied within the photodetector. The factor $F(M)$ accounts for the increase in that noise induced by the current gain process within the photodetector. For high gain photomultipliers this noise factor is quite low; F ~ 1.5–3 [1–4]. In avalanche photodiodes the noise factor $F(M)$ increases with carrier multiplication as will be pointed out in more detail later. For well designed phototransistors and photoconductors this noise factor can be as low as $F = 2$ [2–4]. Possible noise due to the non-multiplied dark current and any shunt and load conductance at the output of the photodetector is accounted for by the noise source $\overline{i_I^2}$. The internal admittance of a well designed photodetector Y_I is mainly capacitive (C_I). The output amplifier is represented in fig. 1 by a noise-free amplifier with input admittance $Y_A = G_A + j\omega C_A$ and voltage gain $A(\omega)$ and a set of noise sources which give a complete representation of the noise of the amplifier [13]. In this particular case the amplifier noise is in accordance with Rothe and Dahlke [13] represented by a noise voltage source $\overline{v_A^2}$ in series with its input, an uncorrelated noise current source $\overline{i_A^2}$ in parallel with the input and a noise current source $\overline{v_A^2}|Y_C|^2$ that is fully correlated to the noise voltage source $\overline{v_A^2}$ by the complex correlation admittance $Y_C = ReY_C + jImY_C$.

High sensitivity and a large signal to noise ratio requires low background and dark currents within the photodetector and an output amplifier with the lowest possible noise. Reduction of the influence of the major amplifier noise source $\overline{v_A^2}$, i.e. keeping

$$\int_0^\infty \left(\frac{\overline{v_A^2}}{\delta f}\right) |Y_I + Y_C|^2 \, df \leqslant \int_0^\infty \left(\frac{\overline{i_M^2}}{\delta f} + \frac{\overline{i_i^2}}{\delta f} + \frac{\overline{i_A^2}}{\delta f}\right) df \qquad (6)$$

is possible by using a first amplifier stage with large transconductance and a much higher input resistance R_A than compatible with eq. (2) [6–8]. The full signal bandwidth and pulse shape can be restored by differentiation of the amplified signal at a later stage in the amplifier provided the amplifier is sufficiently fast to follow the voltage rise at the input [7]. As can be seen from eq. (6) minimization of the major noise contribution of the amplifier is greatly aided by low photodetector (Y_I) and cross-correlation admittances (Y_C).

Current gain within the photodetector helps raising the signal level above the amplifier noise in cases where the amplifier is designed in accordance with eq. (2) to pass the full signal bandwidth at its input and in wide bandwidth cases where the amplifier is designed to keep noise extremely small. The highest sensitivity is reached, when the current gain M reaches (or exceeds) an optimal value determined from $\partial\frac{S}{N}/\partial M = 0$, i.e., when as much current gain is used such that the multiplied shot noise of the photodetector reaches (or exceeds) a level comparable to the amplifier noise [1–4].

3. Photomultipliers

As indicated schematically in fig. 2 photomultipliers consist in essence of a photocathode which emits electrons in response to incident optical radiation, a chain of secondary electron emission dynodes for the multiplication of the photo-excited electrons and an anode to couple the multiplied electron current to the output circuit. Major advances in the technology and a better understanding of the operation of photoemitters and secondary electron emission dynodes have lead to the development of new photomultipliers with greatly improved performance. The introduction of a new type of photocathode, the cesiated GaAs photoemitter by Scheer and Van Laar [14] has led to an entire family of new III-V compound photocathodes [15, 16] with substantially improved sensitivities, especially at longer wavelengths in the near infrared. Similarly constructed GaP/Cs dynodes [17] with high secondary electron emission gain made possible photomultipliers with lower excess noise and a smaller number of dynode stages. Special dynode arrangements, which minimize the time of flight differences of the electrons resulted in photomultipliers with electrical bandwidths extending into the GHz region [18, 1–3]. Electrostatically focused photomultipliers are now available with speeds of response as short as 300 ps [19]. These conventional photomultipliers are compact in size, use small high sensitivity photocathodes and have dynodes whose configurations have been chosen with the help of a computer to minimize the time broadening effects on the electron cloud.

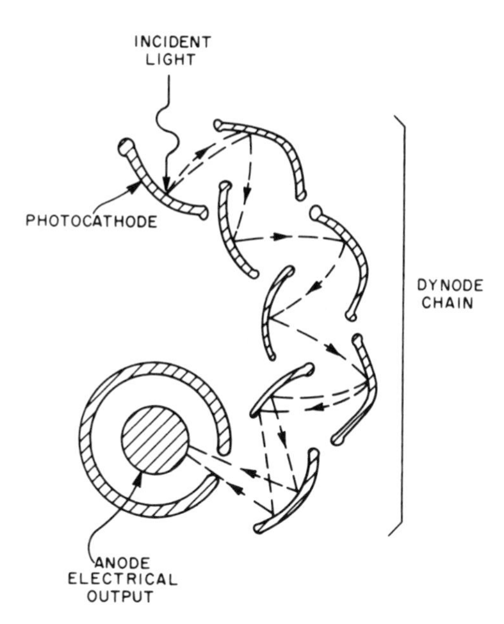

Fig. 2. Schematic view of a photomultiplier with photocathode, conventional electrostatically focused dynode chain and coaxial anode output.

The new III-V compound photocathodes and dynodes basically consist of a heavily doped *p*-type semiconductor whose surface is either coated with a thin monolayer of cesium or with a somewhat thicker Cs_2O layer as indicated in fig. 3. Bandbending near the surface of the semiconductor occurs due to the adsorption of positively charged cesium ions [14–16, 20, 21] or due to the formation of a heterojunction between the III-V semiconductor and the Cs_2O layer [16, 22, 23] and leads to a lowering of the work function and to an effectively negative electron affinity. An effectively negative electron affinity (combined with a sufficiently low interfacial barrier at the heterojunction [23]) allows electrons thermalized at the bottom of the conduction band to escape into the vacuum. The escape depth for electrons is quite large (of order 1 μm) because electrons which diffuse to the band-bending region have sufficient energy for emission into the vacuum. Negative electron affinity photocathodes with low interfacial barriers should thus have a high quantum efficiency for all photon energies a few tenths of an electron volt larger than the bandgap [16]. Narrowing of the bandgap of the photoemitter extends the photoresponse to longer wavelengths in the infrared.

As an indication of the present state of the art, fig. 4 and table 1 compare the spectral response, quantum efficiency and dark current densities of these new III-V semiconductor photoemitters with some of the best conventional photocathodes.

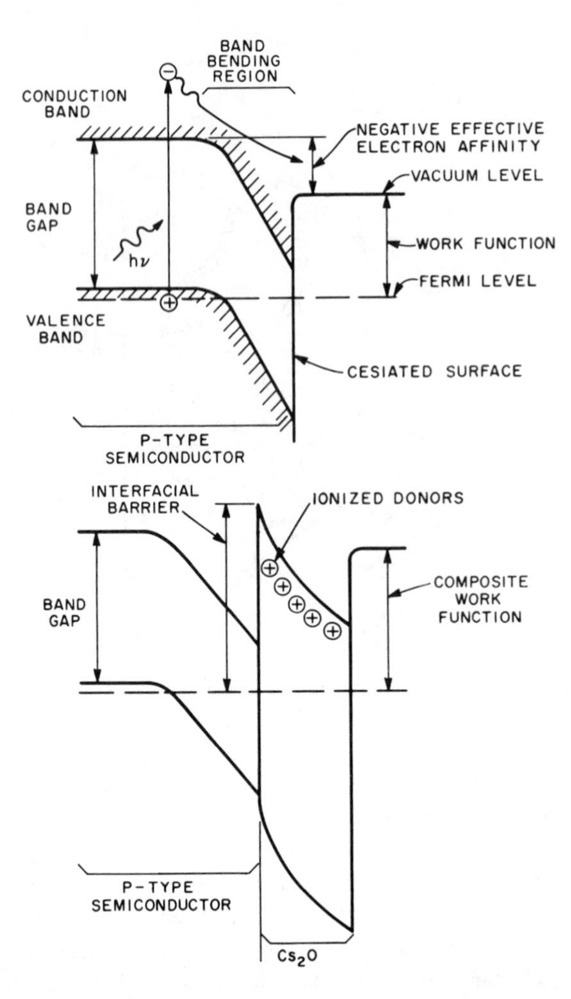

Fig. 3. Idealized energy-band diagrams for cesium (top) and Cs_2O (bottom) covered III-V compound *p*-type semiconductor photocathodes.

The negative electron affinity photocathodes show a higher quantum efficiency, both in the visible and in the infrared. At the GaAs laser wavelength of 0.87 μm the 18% quantum efficiency of a GaAs (Cs) photocathode [20] compares favorably with the 2% efficiency of an extended red sensitive multialkali (ERMA) cathode [24]. At the 1.06 μm YAG laser wavelength, the highest reported quantum efficiencies of 2–3% of the developmental $Ga_{1-x}In_xAs$(Cs) [20] and $InAs_{1-x}P_x$(Cs) [21] photocathodes are more than 10 times larger than for the best S-I infrared photocathodes.

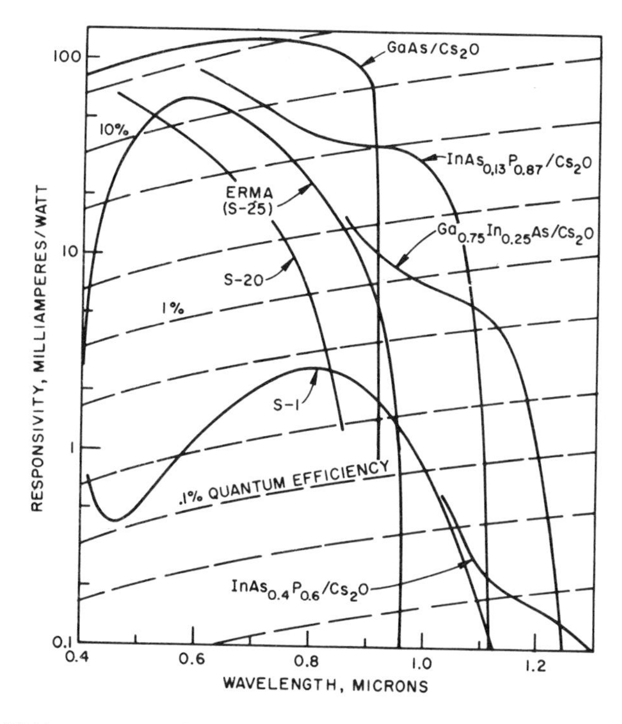

Fig. 4. Responsitivity spectrum and quantum efficiencies of highly developed commercial photocathodes (S-1, S-20, ERMA, $GaAs/Cs_2O$) and developmental infrared sensitive photoemitters ($InAs_{1-x}P_x/Cs-Cs_2O$, $Ga_{1-x}In_xAs/Cs-Cs_2O$).

4. Photodiodes

Solid-state photodiodes, phototransistors [4, 28–30] and avalanche photodiodes contain as an essential element a depleted semiconductor region with a high electric field that serves to separate photoexcited electron-hole pairs. As discussed also by Moss in the preceding paper, these junction photodetectors usually operate in the wavelength range where absorbed photons excite electron-hole pairs through band to band excitation. High speed photodiodes are usually connected to relatively low impedances so as to allow the photoexcited carriers to induce a photocurrent in the load circuit while they are moving through the high field region. Photodiodes for the visible and near infrared range are commonly operated at relatively large reverse bias voltages, since this helps reduce the carrier drift time and lowers the diode capacitance without introducing excessively large dark currents. Operation in the photovoltaic mode, with open circuited terminals is possible, but not suitable for the

Table 1
High-efficiency photocathodes

Cathode material	Quantum efficiency (%)			Dark current density (A/cm^2)	Reference
	0.63 μm	0.87 μm	1.06 μm		
Ag–O–Cs S-1 opaque or semitransparent	0.3	0.7	0.07	$< 10^{-12}$	[15]
Na_2KSb–Cs S-20 semitransparent	6	0.2	–	$< 10^{-15}$	[15]
Extended-red-sensitive-multi-alkali (ERMA) semitransparent	12	2	–	10^{-15}	[24]
GaAs/Cs_2O opaque [semitransparent]	25	18	–	$\sim 10^{-14}$	[14] [25] [16] [20] [26] [21]
$InAs_{0.13}P_{0.87}/Cs_2O$ opaque	16	9	3		[27] [23] [25]
$Ga_{1-x}In_xAs/Cs_2O$ $x \sim 0.14$–0.31	12	8	2		[16] [20]

demodulation of high speed signals. A charge integration mode [28] by which a photodiode or phototransistor is preset by a reverse bias voltage and then open circuited to allow integration of the signal and dark current is often useful, especially in detector arrays [29].

The operation of a reverse biased *p-i-n* [31] photodiode with a load resistance R_L is illustrated in fig. 5. Incident light which is not reflected at the surface penetrates some distance into the photodiode material before it is absorbed and generates photocarriers as indicated by the light absorption or pair generation characteristic of fig. 5. Electrons and holes generated within the high field region (W) of the junction and minority which diffuse from the p and n bulk regions to the junction before recombination are collected across the high field region and contribute to the photocurrent at the output. Carriers generated in the bulk regions, on the average within a diffusion length L_n or L_p respectively from the junction edges can diffuse to the high field region and will be collected.

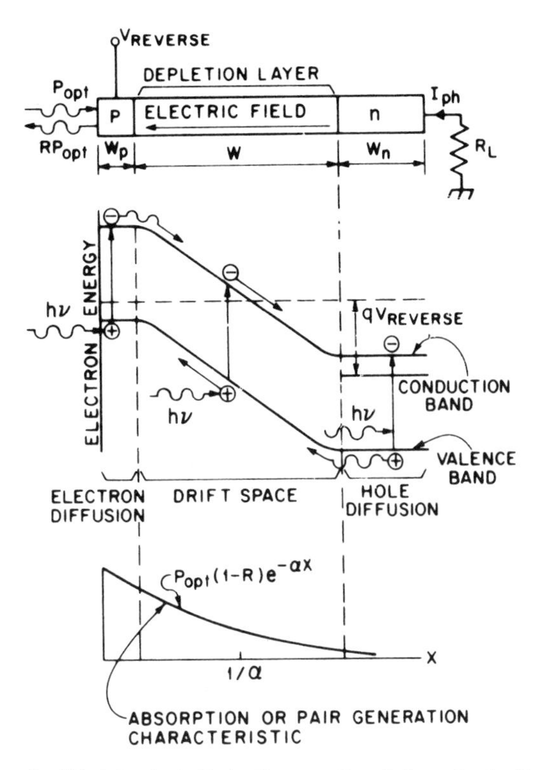

Fig. 5. Operation of solid state photodiode. Cross-sectional view of *p-i-n* diode and energy-band diagram under reverse bias conditions is shown together with optical absorption or pair generation characteristic.

High quantum efficiency requires minimization of the light reflection off the diode surface and placement of the junction in such a way that most of the photoexcited carriers are collected across the junction region. Low reflectivities (R) are usually achieved by means of antireflection coatings as indicated for various diodes

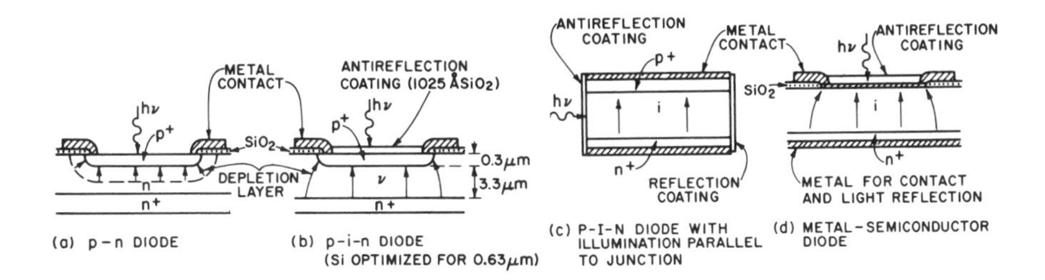

Fig. 6. Construction of high speed photodiodes.

in fig. 6. Most photodiodes are designed for light incidence normal to the junction plane. A large carrier collection efficiency then requires the carrier collection width $W_{coll} = L_p + W + L_n$ or $W_p + W + W_n$, whichever is smaller, to be comparable to or larger than the average penetration depth $(1/\alpha)$ of the light. Specifically, for $\alpha \cdot W_{coll} \geqslant 1.5$ a collection efficiency of 78% or more is achievable.

Unfortunately, minority carrier diffusion is a relatively slow process. It takes carriers a time [32] $T_{Diff} = (Wn, p)^2/2.4\, Dn, p$ (with Dn, p = electron or hole diffusion constant, respectively) to diffuse through a region of thickness Wn, p. Carrier diffusion then limits the frequency response of the photocurrent to $f_{3db} \leqslant 1/2\pi T_{Diff}$ [32]. As an example, electron diffusion through a 5 μm thick *p*-type silicon layer takes 3 ns and leads to a frequency limit of 50 MHz (hole diffusion takes 3 times longer). Although carrier diffusion within the quasi-neutral bulk regions can be speeded up (by factors of 2–6) with fields associated to doping gradients [33], diffusion regions have to be kept narrow in high speed photodetectors. Fast photodiodes are generally so designed, that carriers are mainly excited within the high field region of the junction [31, 34] or so close to it, that diffusion times are shorter or at least comparable to the carrier drift times. At sufficiently high reverse bias voltages, carriers are then collected at scattering limited velocities v_{sat} across the high field region (W) of the junction. Carrier transit times can be as short as $T_{tr} = w/V_{sat}$ and the frequency response limit as high as $f_{3db} = 2.78/2\pi T_{tr}$ [34]. For fields in excess of 2×10^4 V/cm, where scattering limited velocities of order $5 \times 10^6 - 10^7$ cm/sec are reached in silicon and germanium, carrier transit times through thin, 1 μm thick depletion regions can be as short as 10^{-11} sec.

More detailed quantum efficiency and pulse and frequency response calculations for particular diode structures various light penetration depths and different surface boundary conditions can be found in References [31–34] and [36–40].

The actual quantum efficiency and speed of response of photodiodes depends strongly on the wavelength of operation and on the diode material and design. The choice of a particular detector material is primarily determined by the wavelength of operation. Light absorption coefficients for a variety of photodetector materials for the 0.4 –1.6 μm region are shown in fig. 7 [4, 41]. Because of their highly developed technology, silicon photodiodes are preferably used in the near ultraviolet, visible and in the infrared up to about 1 μm. With germanium diodes the response can be extended beyond 1.5 μm. Other materials like GaAs, CdTe, CdSe and CdS are only considered for special applications. As can be seen from fig. 7, the light absorption coefficient of silicon, which is an indirect gap material, changes gradually as a function of wavelength. As a consequence of the strongly varying light penetration depth, silicon photodiodes, as well as phototransistors and avalanche photodiodes, have to be optimized for particular wavelength and speed of response combinations.

The simplest planar p^+-n photodiodes of the type shown in fig. 6a, with space charge layer widths between about 1 and 3 μm, are optimal for light absorption coefficients between 5×10^3 cm^{-1} and 2×10^6 cm^{-1}. For Si this correponds to

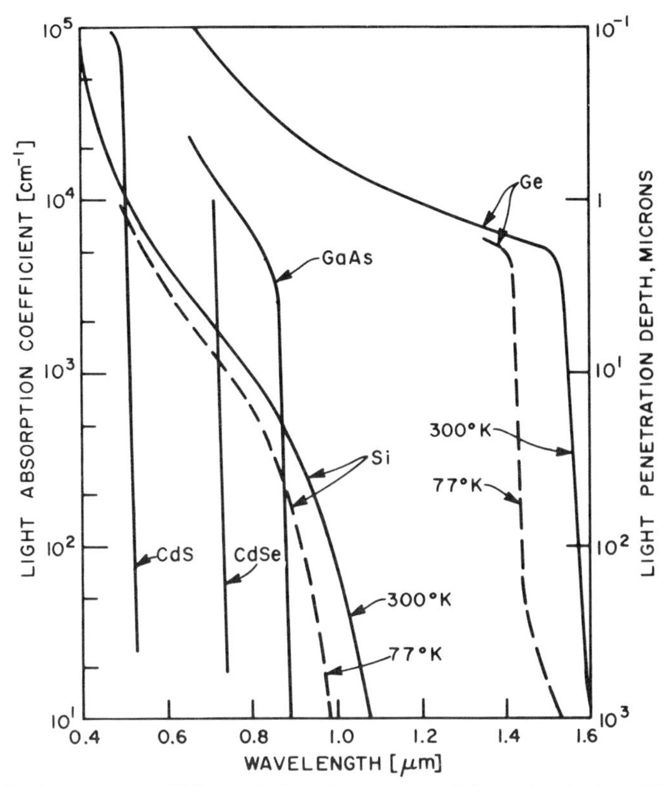

Fig. 7. Light absorption coefficients of photodetector materials for the visible and infrared range between 0.4 and 1.6 μm (Si and Ge data from [41], for other materials see refs. in [4]).

a wavelength range from 0.45 to 0.6 μm and for Ge to a range between 0.95 and 1.5 μm. Fast response in the 100 ps region is possible for small area junctions of this type, as can be seen from table 2. The quantum efficiency of such a Ge n^+ -p diode is 40 to 50% as shown in fig. 8 but could be higher if proper antireflection coatings were used.

At short wavelengths (< 0.5 μm for Si) light is absorbed very close to the semiconductor surface. Recombination of photon-excited carriers at the surface and in the shallow highly doped bulk n^+ or p^+ layer then results in a low quantum efficiency [36, 39] even if the built-in fields do not hamper [39], but help [33] carriers to reach the junction. At short wavelengths, metal-semiconductor junctions with properly antireflection coated semitransparent metal layers are more useful [42, 43]. Extremely high speeds of response and good quantum efficiencies are possible if the depletion regions are small and thin, comparable to the light penetration depth.

Towards longer wavelengths, where light penetrates deeper into the semiconductor material, various front illuminated p^+-i-n^+ and metal-semiconductor structures

Table 2
High speed photodiodes

Diode	Wavelength range (μm)	Peak quantum efficiency (%)	Light sensitive area (cm^2)	Dark current (A)	Capacitance (pF)	Pulse response time	References
Si *p-i-n* with anti-reflection coating	0.5 – 0.7	> 90 at 0.6328 μm	2×10^{-5}	$< 10^{-9}$ at –40 V	< 1 at –5 V	100 ps with 50 ohm load	[44]
Si *p-i-n* with anti-reflection coating	0.4 – 1.1	90 at 0.9 μm	2×10^{-2}	5×10^{-8} at –45 V	3 at –45 V	3ns for –45 V and 50 ohm load	RCA Montreal
		70 at 1.06 μm	2×10^{-2}	5×10^{-8} at –45 V	3 at –45 V	20ns for –200 V and 50 ohm load	
Au-*i-n* Si Schottky barrier with antireflection coating	0.38 – 0.8	> 70	2×10^{-3}	10^{-10} at –10 V	4 at –10 V	5ns for –10 V and 50 ohm load	United Detector Technology
Si *p-n* with side illumination		90 at 1 μm		10 nA at –100 V	1.8 at –300 V	< 500 ps	[47]
Ge n^+-*p* (uncoated) avalanche photodiode	0.6 – 1.65	50	2×10^{-5}	2×10^{-8} at –10 V and 300°K	0.8 at –16 V	120 ps for –10 V and 50 ohm load	[48]
Ge *p-i-n* with side illumination	0.6 – 1.65	60	2.5×10^{-2}	10^{-11} at 77°K 10^{-3} at 300°K	3	25ns at 500 V	[7]

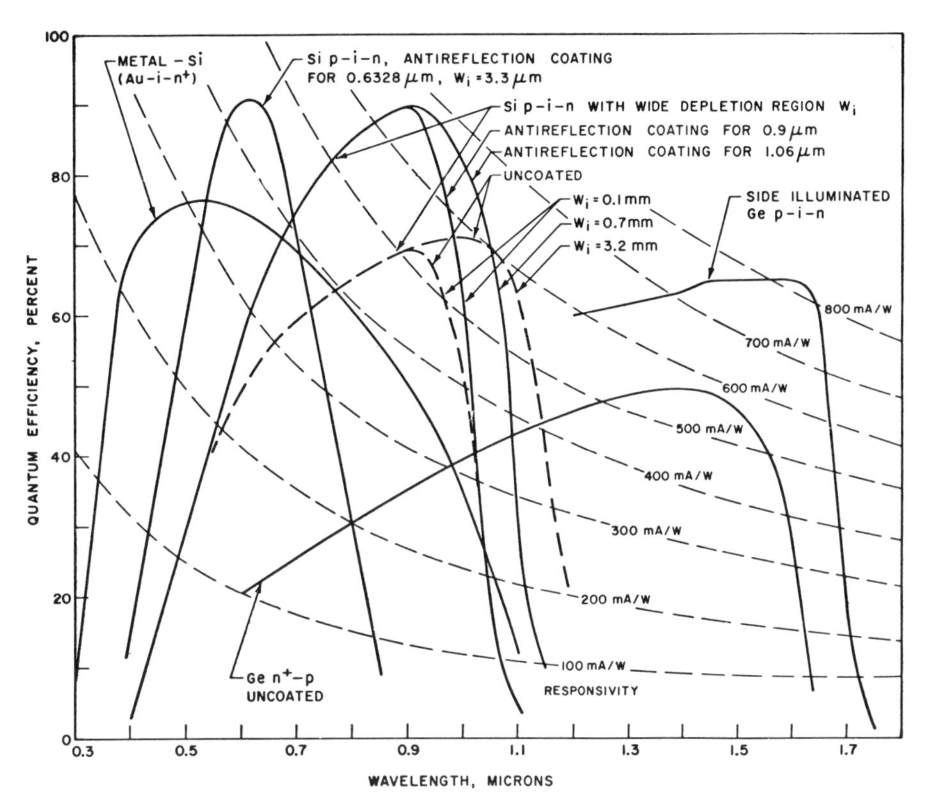

Fig. 8. Quantum efficiency and spectral responsivity of several optimized high speed photodiodes.

with wide depletion regions can be used. Fig. 6b shows such a Si $p^+ - i - n^+$ structure [44] that has been optimized for 0.6328 μm. These diodes have quantum efficiencies exceeding 90% (see fig. 8) and, if mounted in a coaxial header [2, 37, 42], they can resolve optical pulses with 100 ps rise time. Front illuminated Si *p-i-n* diodes [45, 46] with high quantum efficiencies (> 70%) at the GaAs (0.85 – 0.92 μm) and YAG (1.06 μm) wavelengths require depletion layer widths of 20–50 and 500 μm, respectively. With such wide depletion regions, the carrier drift transit times become relatively long, as can be seen from table 3. At wavelengths close to the band edge, a better compromise between quantum efficiency and speed of response can be achieved if light is allowed to penetrate from the side, parallel to the junction, as indicated in fig. 6c. A silicon photodiode of this type has been built [47] and is listed in table 2. Side illuminated Ge *p-i-n* diodes with higher quantum efficiency at wavelengths close to the band edge than front illuminated Ge n^+-p [48] and Ge *p-i-n* [31] diodes have been built [7] and are listed in table 2 and fig. 8.

A different design approach, which stresses highest possible quantum efficiency over an extended wavelength range at the expense of speed of response (especially

Table 3
Avalanche photodiodes

Diode	Wavelength range (μm)	Light sensitive area (cm^2)	Maximum current gain	Noise factor $F \sim M^X$	Current gain bandwidth product (GHz)	Breakdown voltage (V)	Capacitance (pF)	Dark current (A)	Ref.
Si n^+-p	0.4 – 0.8	2×10^{-5}	10^4	$X \sim 0.5$	100	23	0.8 at 23 V	5×10^{-11} at 10 V	[52]
Si n^+-i-p	0.5 – 1.1				Not very high	200 – 2000			[69]
Si n^+-p-π-p^+	0.6 – 1.1	2×10^{-3}	10 – 200	$X \sim 0.3$		200 – 500	2 at 200 V	10^{-7} at 100 V	[68]
Si p-n side illuminated	0.6 – 1.1		~ 200		30	200 – 230	1 at 200 V	10^{-8} at 100 V	[70]
Ge n^+-p	0.8 – 1.65	2×10^{-5}	200 at 300°K 10^4 at 80°K	$X \sim 1$	60	16.3	0.08 at 15 V	10^{-8} at 10 V, 300°K 7×10^{-10} at 10 V, 250°K	[48]
Pt – GaAs Schottky barrier	0.4 – 88	> 100		low	> 50	60			[71]

at short wavelengths) has resulted in metal-semiconductor diodes (Gold-silicon) with wide depletion regions as shown in figs. 5d, 8 and table 2.

As can be seen from fig. 8 and table 2, a number of Si and Ge photodiodes are available with high quantum efficiencies and high speed of response.

These photodiodes, as well as other solid state detectors are relatively small in size. Typical diameters of the light sensitive areas are between 20 and 500 μm. Small size helps keep the diode capacitance and dark currents low.

Dark currents, which limit the sensitivity to weak light signals, originate either from the bulk or from the surface. Surface leakage currents are a problem especially in high resistivity Si devices and in planar, oxide covered Ge diodes. Special surface treatments and various guard ring structures and surface contours are employed to reduce these surface leakage currents. Bulk leakage currents in Si are mainly due to carrier generation within the space charge layer. For carefully processed silicon diodes generation currents as low as $10^{-6} - 10^{-8}$ A per mm^3 of depleted volume have been reached [39, 45, 46]. In germanium diodes the bulk leakage current is mainly due to minority carrier diffusion to the junction.

5. Avalanche photodiodes

Avalanche photodiodes are specially constructed photodiodes which combine the detection of optical signals with internal amplification of the photocurrent. Internal current gain takes place in an avalanche photodiode when carriers gain sufficient energy by moving through the high field region of a highly reverse biased junction to release new electron-hole pairs through impact ionization [49]. High current gains are possible by this process [50], event at microwave frequencies [51, 52, 48].

The typical operation of an avalanche photodiode is illustrated in fig. 9, where the voltage dependence of the dark current of a germanium avalanche photodiode as well as the response to pulses from a phase locked 0.6328 μm He–Ne laser are shown. At low reverse bias voltages, where no carrier multiplication takes place, the diode operates as a regular photodiode. As the reverse bias voltage is increased, carrier multiplication sets in, as indicated by an increase in the current pulse. The highest amplitude of the photocurrent pulse is reached, when the diode is biased to the breakdown voltage. At higher voltages above the breakdown voltage, a self-sustained avalanche current flows which makes the diode less and less sensitive to photon-excited carriers.

The highest gain for the photocurrent is limited either by premature breakdown spots, by spatial inhomogeneities within the avalanche region or more fundamentally by current induced saturation effects [48, 53] and by a current gain bandwidth product [1, 48, 54, 55]. The current gain is influenced by the magnitude of the avalanche current because carriers that emerge from the multiplication region reduce the electric field within the junction and lead to voltage drops across the series and load resistance of the diode. At high light intensities these voltage drops cause the

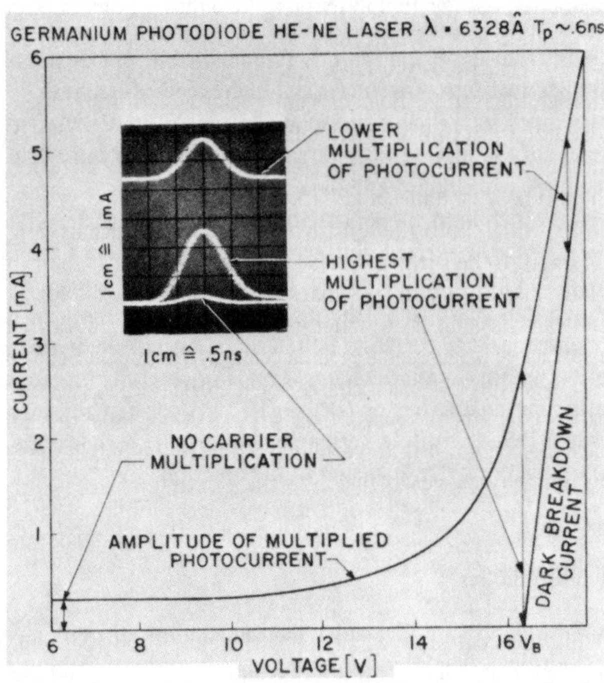

Fig. 9. Current voltage characteristics and pulse response of germanium avalanche photodiode.

multiplied current to increase only as the square root of the photocurrent [48, 53] instead of being proportional to the photocurrent. For low light intensities, the dark current can set a limit on the highest average carrier multiplication that can be reached at low frequencies. The highest carrier multiplications of 200 in germanium avalanche diodes at room temperature are e.g. limited by dark current [48].

For short optical pulses the current gain of silicon and germanium avalanche photodiodes is limited by a current gain bandwidth product, which is inversely proportional to the width of the carrier multiplication region and depends on the ratio between the electron hole ionization coefficients and on the type of carriers that initiates the avalanche [3, 4, 55]. Current gain bandwidth products between 20 – 100 GHz have been reported for various silicon and germanium avalanche photodiodes as can be seen from table 3.

The temporal build-up and decay of the multiplied current flowing through the short circuited output of an avalanche photodiode in response to excitation with a short optical pulse is shown schematically in fig. 10. As indicated the optical pulse generates electron-hole pairs throughout the entire width of the junction. Due to the electric field the electron-hole pairs become separated. The electrons move towards the multiplication region where they free new electron hole-pairs through impact ionization. Since both the secondary electrons and holes give rise to ionizations in

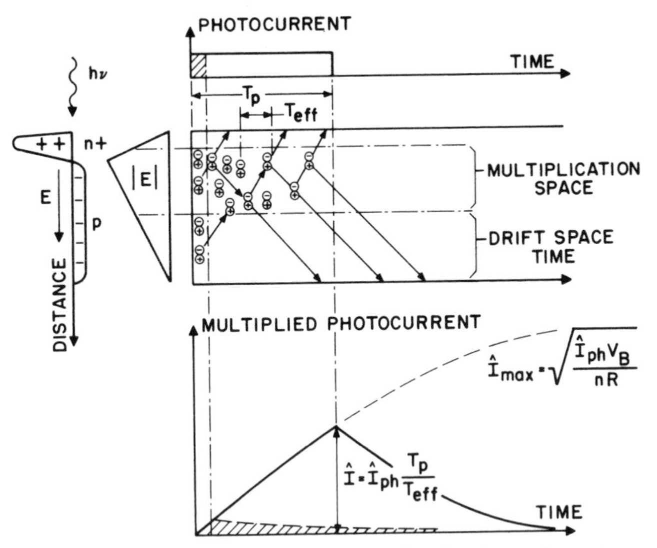

Fig. 10. Schematic view of the build-up and decay of avalanche carrier multiplication and multiplied photocurrent in an avalanche photodiode.

silicon and germanium [4, 49], an avalanche of carriers is initiated, that gives rise to a multiplied photocurrent which persists much longer than the initial excitation. The avalanche of carriers decays at the end of an optical pulse because avalanche photodiodes are usually operated at voltages below the self sustained breakdown region. For short optical pulses, the multiplied photocurrent reaches its peak value

$$\hat{I} = \hat{I}_{\mathrm{ph}} \frac{T_p}{T_{\mathrm{eff}}} \tag{7}$$

at the end of the optical pulse T_p. The effective multiplication time constant T_{eff}, i.e., the mean time of flight of carriers before a new ionization, depends on the width of the carrier multiplication region, on the magnitude of the electric field, on the ratio of the electron and hole ionization coefficients and on the type of carrier excitation [3, 4, 55]. The effective multiplication time constant is related to the current gain bandwidth product by [1, 4, 55]

$$GB = \frac{1}{2\pi T_{\mathrm{eff}}} . \tag{8}$$

As an illustration for the high current gains, that can be reached even for very short optical pulses, fig. 11 shows the pulse response of an n^+-p germanium avalanche photodiode to 80 ps pulses from a 1.06 μm YAG laser. The lower trace of fig. 11 represents the peak photocurrent as a function of the relative optical pulse power for low reverse bias voltages where no carrier multiplication takes place. The upper

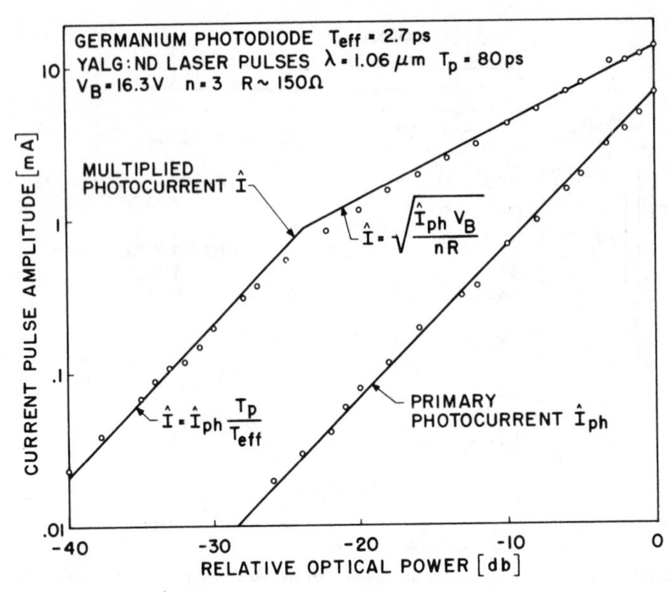

Fig. 11. Pulse response of n^+-p germaium mesa avalanche photodiode as a function of the relative power of short optical pulses from a phase locked YAG laser.

trace shows the maximum pulse amplitudes reached at the breakdown voltage. At moderate light intensities a maximum current gain of 30 is observed for the 80 ps pulses, consistent with a current gain bandwidth product of 60 GHz for these Ge diodes (see table 3). At high light intensities the multiplied photocurrent increases only as the square root of the primary photocurrent due to the current induced saturation effects.

The current gain of an avalanche photodiode fluctuates due to the statistical nature of the carrier multiplication process. Even for spatially uniform avalanche regions the statistical gain variations give rise to excess noise [54, 56] and to a degradation of the Poissonian distribution in the number of photo-generated carriers [57–60]. Measured noise factors $F(M) = \overline{i_m^2}/2qI_{ph}\sigma f$ for various avalanche photodiodes with highly uniform carrier multiplication are shown in fig. 12 and found to be in close agreement with McIntyre's [56] predicted increase of the mean square noise current with average current gain. For the injection of electrons only, the noise factor of a uniformly multiplying avalanche photodiode with an electron to hole ionization rate ratio of $\alpha/\beta = 1/k$ increases as

$$F(M) = M - (1-k)\,\frac{(M-1)^2}{M} \tag{9}$$

with carrier multiplication M [56]. In silicon avalanche photodiodes with a high α/β ratio electron injection which occurs for long wavelength excitation in an $n^{\pm}p$ diode leads to a low noise factor [54, 56, 60, 61]. For germanium avalanche photodiodes

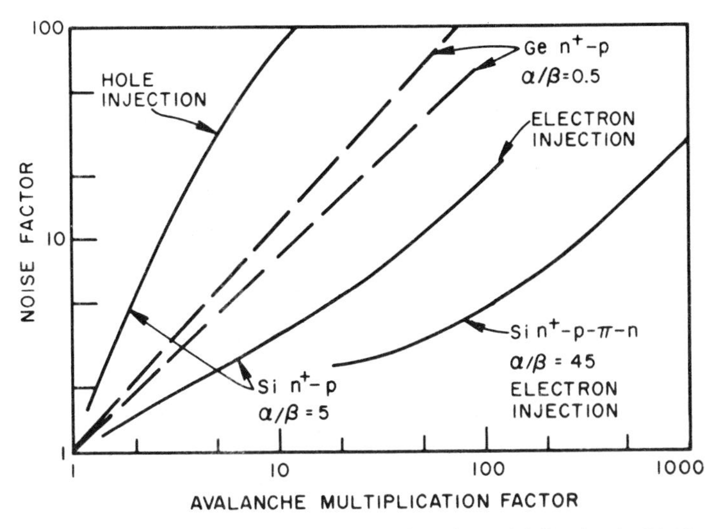

Fig. 12. Measured noise factor $F(M)$ as a function of carrier multiplication for highly uniform silicon [54, 60–63] and germanium [48] avalanche photodiodes.

with almost equal ionization rates for electrons and holes, the noise factor increases proportional to the carrier multiplication, irrespective of the wavelength and type of carrier excitation [48]. If a silicon avalanche photodiode is excited with the wrong type of carriers, i.e., with holes in an $n^{\pm}$ -p diode at short wavelengths, the noise factor increases much steeper with carrier multiplication [61].

The design of avalanche photodiodes requires special precautions to assure spatial uniformity of the carrier multiplication over the entire light sensitive area of the diode [50]. Microplasmas, i.e., small areas with lower breakdown voltages than the remainder of the junction and excessive leakage currents along the junction edges can be eliminated by various guard ring structures and other types of junction contouring as can be seen from fig. 13. Defect-free semiconductor material and cleanliness in processing helps in the fabrication of microplasma free junctions. Highly uniform carrier multiplications in excess of 10^6 have been reached in silicon [50, 52] and in cooled germanium avalanche photodiodes. In large area diodes, that are free of microplasmas, the spatial uniformity of the carrier multiplication is usually limited either by doping inhomogeneities of the starting material or by variations in the doping profile.

The construction and performance characteristics of typical avalanche photodiodes are presented in fig. 13 and table 3. The simplest silicon avalanche photodiodes consist of an n^+-p junction with an n-type guard ring [52] as shown in fig. 13a. The breakdown voltage of these diodes [64, 65] is typically between 20 and 200 V and space charge layer thicknesses between 1.5 and 10 μm render them especially useful for the detection of light in the 0.4 – 0.8 μm range. Since the diameters of the

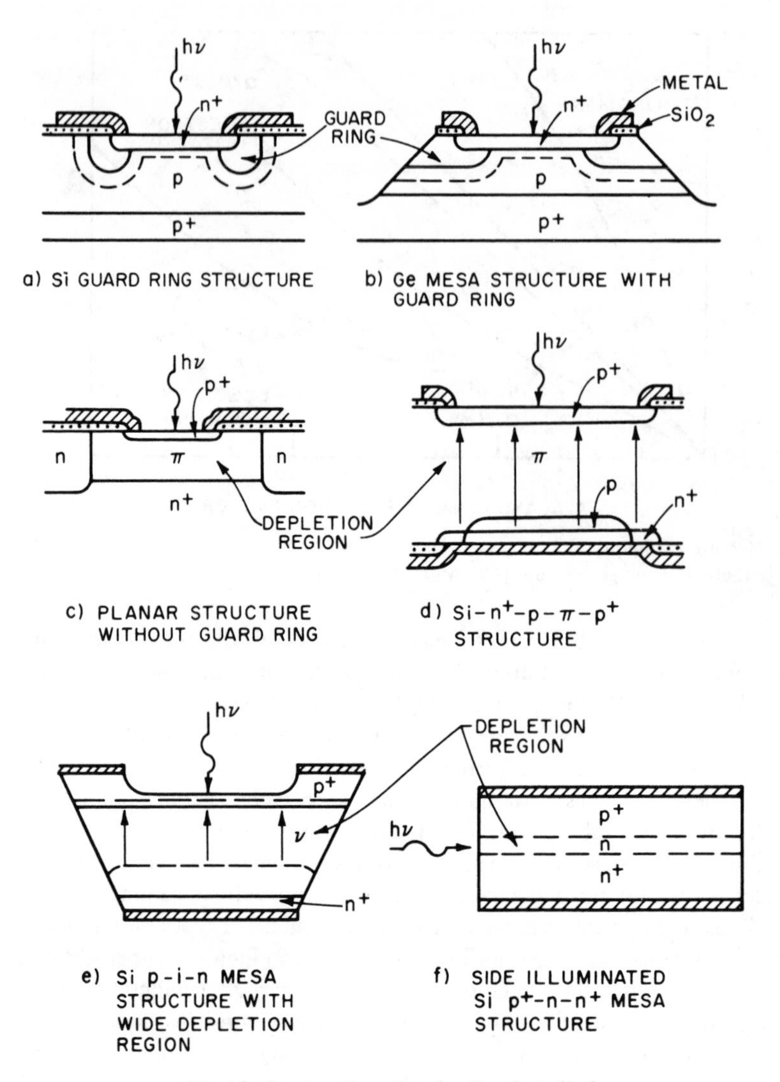

Fig. 13. Construction of avalanche photodiodes.

light sensitive areas are typically very small, between 40 and 200 μm, these diodes have small capacitances at most a few pF. Similarly constructed germanium avalanche photodiodes [48] require mesa etching of the guard ring in order to reduce leakage currents. These n^+-p germanium avalanche photodiodes are the fastest photodetectors with internal current gain in the 0.9 – 1.6 μm wavelength region. A reduction in diode capacitance and elimination of the guard ring [66] is possible in diodes with a burried n^+-p junction that is surrounded by an n ring as shown in fig. 13c [66].

Extension of the response of high speed silicon avalanche photodiodes towards longer wavelengths around 1 μm has been achieved through the use of n^+-p-π-n [67, 68] (fig. 13d) or p-i-n [69] (fig. 13e) structures and for side illuminated p-n diodes (fig. 13) [70]. In the n^+-p-π-n structure of fig. 13d, the multiplication is constrained to the narrow n^+-p region. The wide π region acts mainly as a collection region for the photoexcited electrons. Since the avalanche is initiated with electrons, these photodiodes exhibit high current gain bandwidth products and low excess noise [60, 68].

Silicon avalanche photodiodes with high current gain and low excess noise are thus available for the wavelength range between 0.4 and 0.9 μm. Germanium avalanche photodiodes extend this wavelength range to about 1.6 μm. Despite the excess noise associated with the carrier multiplication process these avalanche photodiodes provide useful gain and significant increase in sensitivity in high speed optical receivers that are otherwise limited by noise of the output amplifier [1, 4, 48, 64]. The highest sensitivity for the detection of weak light signals is reached, when the current gain is adjusted to an optimal value at which the amplified shot noise of the photodiode becomes about equal to the noise of the amplifier [1, 48, 54].

6. Conclusions

It is a fundamental goal of the photodetector design and receiver optimization to reach quantum noise limited operation which results in the highest sensitivities for the detection of weak light signals with a given signal bandwidth or repetition rate. Towards this end, photomultipliers with their large internal current gain have long been established as the best, most sensitive detectors for the ultraviolet, visible and near infrared region to about 0.8 μm. With the appearance of new photomultipliers of much smaller size with more efficient photocathodes (quantum efficiency $>20\%$) and better designed dynode chains even higher sensitivities are possible throughout the visible and especially in the infrared, at present to about 0.9 μm, combined with speeds of response as short as a few hundred pico-seconds. Solid state photodetectors, including photodiodes with and without internal current gain, and to some extent phototransistors and photoconductors have found an ever increasing number of applications throughout the visible and infrared region. Silicon and germanium photodiodes with their highly developed technology can be optimized for various wavelengths (in the 0.4 – 1.6 μm range) and speed of response combinations. They are the simplest, most rugged, cheapest detectors available for the demodulation of optical signals whose intensities are relatively high. Despite their excess noise and relatively large dark currents (at room temperature) silicon avalanche photodiodes start to rival the sensitivity of photomultipliers throughout the visible range and exceed it already in the infrared around 0.9 and 1.06 μm. Germanium photodiodes, which operate in the 0.8 – 1.6 μm range have (as yet) no counterpart in photomultipliers towards the longer wavelengths.

Undoubtedly, as materials technology progresses and device designs are refined, further improvements in quantum efficiency, speed of response, sensitivity and reduction in dark currents and capacitances can be expected. Photocathodes with practically useful quantum efficiencies (at present 3.5%) at 1.06 μm might become available. GaAs photodiodes with internal current gain [71] might outperform silicon devices in the 0.8 – 0.89 μm range. And last, but not least, new devices, like avalanche photodiodes which multiply only one type of carriers might appear with greatly improved speed of response and noise performance.

References

[1] L.K. Anderson and B.J. McMurtry, Proc. IEEE 54 (1966) 1335.
[2] L.K. Anderson, M. DiDomenico and M.B. Fisher, in: *Advances in Microwaves,* vol. 5, Ed. L. Young (Academic Press, New York, 1970).
[3] H. Melchior, M.B. Fischer and F. Arams, Proc. IEEE 58 (1970) 1466.
[4] H. Melchior, in: *Laser Handbook*, Eds. F.T. Arecchi and E.D. Schulz-DuBois (Elsevier/North-Holland Publ. Co., Amsterdam, 1972) 629.
[5] G. Lucovsky and R.B. Emmons, Appl. Opt. 4 (1965) 697.
[6] B.N. Edwards, Appl. Opt. 5 (1966) 1423.
[7] D.P. Mathur, R.J. McIntyre and P.P. Webb, Appl. Opt. 9 (1970) 1842.
[8] R.H. Hamstra and P. Wendland, Appl. Opt. 11 (1972) 1539.
[9] L. Mandel, in: *Progress in Optics,* vol. 2, Ed. E. Wolf (Interscience Publishers, New York, 1963) 183.
[10] B.M. Oliver, Proc. IEEE 53 (1965) 436.
[11] H. Hodara, Proc. IEEE 53 (1965) 696.
[12] T. Curran and M. Ross, Proc. IEEE 53 (1965) 1770.
[13] H. Rothe and W. Dahlke, Proc. IRE 44 (1956) 811.
[14] J.J. Scheer and J. Van Laar, Solid State Commun. 3 (1965) 189.
[15] A.H. Sommer, *Photoemissive Materials* (Wiley, New York, 1968).
[16] R.L. Bell and W.E. Spicer, Proc. IEEE 58 (1970) 1788.
[17] R.E. Simon, A.H. Sommer, J.J. Tietjen and B.F. Williams, Appl. Phys. Letters 13 (1968) 355.
[18] R.C. Miller and N.C. Wittwer, IEEE J. Quantum Electron. QE-1 (1965) 49.
[19] H.R. Krall and D.E. Persyk, IEEE Trans. Nuclear Science NS-19 (1972) 45.
[20] D.G. Fisher, R.E. Engstrom and B.F. Williams, Appl. Phys. Letters 18 (1971) 371.
[21] H. Sonnenberg, Appl. Phys. Letters 19 (1971) 431.
[22] H. Sonnenberg, Appl. Phys. Letters 14 (1969) 289.
[23] R.L. Bell, L.W. James, G.A. Antypas, J. Edgecumbe and R.L. Moon, Appl. Phys. Letters 19 (1971) 513.
[24] P.E. Persyk, Laser Journal, November-December 1969, 21.
[25] B.F. Williams, Appl. Phys. Letters 14 (1969) 273.
[26] Y.Z. Liu, J.L. Moll and W.E. Spicer, Appl. Phys. Letters 17 (1970) 60.
[27] L.W. James, G.A. Antypas, J.J. Uebbing, T.O. Yep and R.L. Bell, J. Appl. Phys. 42 (1971) 580.
[28] G.P. Weckler, IEEE J. Solid State Circuits SC-2 (1967) 65.
[29] IEEE Trans. Electron Devices, Special Issue on Solid-State Imaging, ED-15, April 1968.
[30] F.H. De La Moneda, E.R. Chenette and A. van der Ziel, IEEE Trans. on Electron Devices, ED-18 (1971) 340.

[31] R.P. Riesz, Rev. Sci. Instr. 33 (1962) 994.
[32] D.E. Sawyer and R.H. Rediker, Proc. IRE 46 (1958) 1122.
[33] A.G. Jordan and A.G. Milnes, IRE Trans. on Electron Devices, ED-7 (1960) 242.
[34] W.W. Gaertner, Phys. Rev. 116 (1959) 84.
[35] For refs. see S. Sze, *Physics of Semiconductor Devices* (Wiley-Interscience, New York, 1969) 59.
[36] J.J. Lofersky and J.U. Wysocki, RCA Review 22 (1961) 35.
[37] L.K. Anderson, in: *Proc. Symp. on Optical Masers* (Polytechnic Institute of Brooklyn, April 1963) 549.
[38] O. Krumpholz and S. Maslowski, Telefunkenzeitung 39 (1966) 373.
[39] T.M. Buck, H.C. Casey, J.V. Dalton and M. Yamin, Bell Syst. Tech. J. 47 (1968) 1827.
[40] R.L. Williams, J. Appl. Phys. 52 (1962) 1237.
[41] W.C. Dash and R. Newman, Phys. Rev. 99 (1955) 1151.
[42] M.V. Schneider, Bell System Techn. J. 45 (1966) 1611.
[43] A.J. Tuzzolino, E.L. Hubbard, M.A. Perkins and C.Y. Fan, J. Appl. Pys. 33 (1962) 148.
[44] E. Labate, unpublished.
[45] R.J. McIntyre and H.C. Sprigings, paper presented at Conference on Preparation and Control of Electronics Materials, Boston, Mass., August 1966.
[46] H.C. Sprigings and R.J. McIntyre, paper presented at the International Electron Devices Meeting, Washington, D.C., October 1968.
[47] O. Krumpholz and S. Maslowski, Zeitschr. angew. Phys. 25 (1968) 156.
[48] H. Melchior and W.T. Lynch, IEEE Trans. on Electron Devices ED-13 (1966) 829.
[49] K.G. McKay and K.B. McAfee, Phys. Rev. 91 (1953) 1079.
[50] R.L. Batdorf, A.G. Chynoweth, G.C. Dacey and P.W. Foy, J. Appl. Phys. 31 (1960) 1153.
[51] K.M. Johnson, IEEE Trans. on Electron Devices ED-12 (1965) 55.
[52] L.K. Anderson, P.G. McMullin, L.A. D'Asaro and A. Goetzberger, Appl. Phys. Letters 6 (1965) 62.
[53] K. Nishida, Jap. J. Appl. Phys. 9 (1970) 481.
[54] H. Melchior and L.K. Anderson, paper presented at the International Electron Devices Meeting, Washington D.C., October 1965.
[55] R.B. Emmons, J. Appl. Phys. 38 (1967) 3705.
[56] R.J. McIntyre, IEEE Trans. Electron Devices ED-13 (1966) 164.
[57] R.J. McIntyre, IEEE Trans. Electron Devices ED-19 (1972) 703.
[58] S.D. Personick, Bell Syst. Techn. J. 50 (1971) 167.
[59] S.D. Personick, Bell Syst. Techn. J. 50 (1971) 3015.
[60] J. Conradi, IEEE Trans. Electron Devices ED-19 (1972) 713.
[61] R.D. Baertsch, IEEE Trans. Electron Devices ED-13 (1966) 987.
[62] I.M. Naqvi, C.A. Lee and G.C. Dalman, Proc. IEEE 56 (1968) 2051.
[63] T. Igo and K. Sato, Jap. J. Appl. Phys. 8 (1969) 1481.
[64] J.R. Biard and W.N. Shaunfield, IEEE Trans. on Electron Devices ED-14 (1967) 233.
[65] K. Nishida, Y. Nannichi, T. Uchida and I. Kitano, Proc. IEEE, Correspondence 58 (1970) 790.
[66] W.T. Lynch, IEEE Trans. on Electron Devices ED-15 (1968) 735.
[67] H. Ruegg, IEEE Trans. on Electron Devices ED-14 (1967) 239.
[68] P. Webb and R.J. McIntyre, in: *Proc. Electro-optical Systems Design Conf. 1971,* east, 51
[69] R.J. Locker and G.C. Huth, Appl. Phys. Letters 9 (1966) 227.
[70] O. Krumpholz and S. Maslowski, Wiss. Ber. AEG-Telefunken 44 (1971) 73.
[71] W.T. Lindley, R.J. Phelan, C.M. Wolfe and A.G. Foyt, Appl. Phys. Letters 14 (1969) 197.

An Optical Repeater With High-Impedance Input Amplifier

By J. E. GOELL

(Manuscript received September 14, 1973)

A 6.3-Mb/s repeater for fiber optic communication systems is described which incorporates a high-impedance input amplifier. It is shown that by utilizing an input circuit with a time constant which is long compared to the bit interval and equalizing after the signal has been sufficiently amplified to set the signal-to-noise ratio, thermal noise can be decreased. As a result, a reduction can be realized in the required signal and, with an avalanche detector, in the optimum gain.

The repeater, which was realized in a compact form employing standard integrated circuits, utilizes a GaAs light-emitting diode as its optical source. Other features include automatic gain and threshold controls and recovered timing.

I. INTRODUCTION

Digital communication systems utilizing low-loss optical fibers are presently being investigated. The realization of fibers with losses as low as 4 dB/km[1] has opened the way for numerous applications. System configurations will depend on such factors as fiber dispersion, fiber cost, desired information capacity, and terminal costs. Transmission rates near 6.3 Mb/s are attractive because fiber group delay dispersion is not expected to be a problem even with an incoherent source, and a wide variety of low-cost integrated circuits are applicable. If, as now appears likely, the fiber cost is low, then space multiplex could be an attractive alternative to time multiplex in achieving high capacity.

The repeater described here incorporates a high-impedance input amplifier which is similar in approach to ones that have been used for other applications incorporating a capacitive detector such as nuclear particle counters[2] and television cameras.[3] As a result of the high-input impedance, the power required to achieve a specified error rate is reduced, as is the optimum gain if an avalanche detector is used. The latter advantage is important since it eases the fabrication of the detector diode and increases its thermal stability.

The repeater employs return-to-zero pulses with 50-percent duty cycle. The only word pattern restriction is that an occasional "one" be included so timing can be recovered and signal level can be determined. The optical signal is generated by a gallium arsenide light-emitting diode (LED) operating at 0.9 μ wavelength.[4] Diodes of the type employed have been built with output powers of up to about 5 mW. Optical powers to 1 mW have been coupled from these diodes into a fiber with a 0.63 numerical aperture. The repeater was fabricated in a compact form using standard integrated circuits. Automatic gain and threshold controls were provided so the optical input power could vary over a wide range. Clamping was employed to prevent baseline wander with an unbalanced data content of the signal, and timing was extracted by a phase-locked loop.

II. THEORY

A typical circuit for a photodiode driving an input amplifier is shown in Fig. 1(a) and its equivalent circuit in Fig. 1(b). The current generator i_d is the photo current. R_r is the dc return resistor for the detector, and i_r is the noise generator associated with it. The capacitor C_d is the

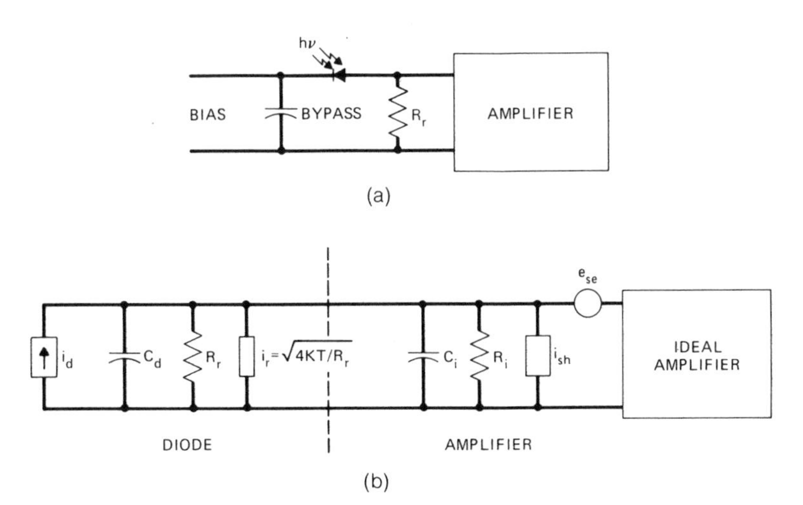

Fig. 1—(a) Input circuit. (b) Equivalent circuit.

output capacitance of the diode, C_i is the input capacitance to the amplifier (excluding feedback effects), and R_i is the input resistance of the amplifier. The quantities e_{se}^2 and i_{sh}^2 are the spectral noise densities of the series voltage and shunt current generators which characterize the noise properties of the amplifier.

A common approach to circuit design has been to set

$$\tau = RC$$
$$< 1/\text{baud rate},$$

where

$$R = \frac{R_r R_i}{R_r + R_i}$$

and

$$C = C_d + C_i$$

to minimize noise while not introducing significant intersymbol interference.

To achieve this, i_r and i_{sh} are often major sources of noise. Therefore, from the standpoint of noise, it is preferable to make R_r very large and employ an amplifier with low i_{sh}, even if $\tau \gg$ baud interval, to amplify the signal sufficiently to set the signal-to-noise ratio, and then to equalize the resulting distortion to eliminate intersymbol interference.

Two possible limitations to the high-impedance approach exist, both related to the low-frequency component of the signal developed across the detector. The difference in voltage between a long string of "ones" and a long string of "zeros" is proportional to the dc load resistance on the diode. Thus, the required dynamic range of the amplifiers preceding the equalizer increases with increasing R. Furthermore, with an avalanche detector this voltage could change the avalanche gain since it is in series with the diode bias. These two factors, which increase in importance as baud rate decreases, ultimately limit the magnitude of the detector load resistance.

In the remainder of this section the relationships between error rate, signal power, and the circuit parameters are discussed for a binary signal with both states equally likely. It is assumed that Gaussian noise statistics apply, that dark current is negligible, and that the amplifiers preceding the equalizer are linear.

Personick[5] has shown that for a pulse of average power p_o with avalanche gain of mean square $\langle g^2 \rangle$, if we assume the optical pulses are distinct, the ratio of the pulse peak to the root mean square thermal noise in the baseband circuit is equal to the ratio of the average current of the received pulse to the square root of the quantity

$$I^2(p_o) = \frac{\eta e^2 p_o f_b \langle g^2 \rangle}{h\nu} + n_t, \tag{1}$$

where n_t, the mean square thermal noise current* weighted to correct for input and output pulse shape, is given by

$$n_t = \left[\left(i_{sh}^2 + \frac{2kT}{R_r} + \frac{e_{se}^2}{R^2}\right) I_1 + (2\pi f_b C)^2 e_{se}^2 I_2\right] f_b. \tag{2}$$

The constants are

T = temperature
h = Plank's constant
ν = optical frequency
η = detector quantum efficiency
e = electron charge
f_b = bit rate.

The weighting functions I_1 and I_2, which take account of pulse shape, are given by

$$I_1 = \frac{f_b}{2\pi} \int_{-\infty}^{\infty} \left|\frac{H_{\text{out}}(\omega)}{H_p(\omega)}\right|^2 d\omega \tag{3}$$

$$I_2 = \frac{1}{(2\pi)^3 f_b} \int_{-\infty}^{\infty} \left|\frac{H_{\text{out}}(\omega)}{H_p(\omega)}\right|^2 \omega^2 d\omega, \tag{4}$$

where $H_{\text{out}}(\omega)$ is the Fourier transform of the output voltage pulse shape, $H_p(\omega)$ is the Fourier transform of the optical power pulse shape, and the pulses have been normalized so that the area of the optical pulse is unity, as is the magnitude of the output pulse at the center of the time slot. The functions I_1 and I_2 can be shown to depend only on pulse shape relative to the time slot length, not on baud rate.

The probability of error can be readily derived from eq. (1), assuming Gaussian noise statistics. For Gaussian-distributed noise, the probability that the noise current will exceed a value D is given by

$$P(D) = \frac{1}{2}\,\text{erfc}\left(\frac{D}{I(p_o)\sqrt{2}}\right),$$

where erfc is the error function complement and I the root mean square

* Note that amplifier shot noise has been included in n_t.

noise current.[6] We assume an ideal regenerator where a "one" is produced if the input exceeds a threshold level D and a "zero" otherwise. Then if a "zero" is transmitted, the condition that the probability will be P_e that the noise will exceed the decision threshold is given by

$$D > QI(0), \tag{5}$$

where

$$Q = \sqrt{2}\ \mathrm{erfc}^{-1}\ (2P_e). \tag{6}$$

Similarly, so that the probability that the noise will not exceed the signal when a "one" is transmitted, the expected value of the signal, given by $p_{\max}\eta e\langle g\rangle/h\nu$, where $\langle g\rangle$ is the mean avalanche gain and $p_{\max}$ the average power for all "ones," must exceed the threshold by Q times the noise current; that is,

$$\frac{p_{\max}\eta e\langle g\rangle}{h\nu} - D > QI_{\max}. \tag{7}$$

From eqs. (1), (5), and (7), the average power required to achieve a specified error probability with avalanche gain is given by

$$p = \frac{h\nu Q}{2\eta}\left[\frac{Q\langle g^2\rangle f_b}{\langle g^2\rangle} + \frac{2}{\langle g\rangle e} n_i^{\frac{1}{2}}\right]; \tag{8}$$

where $\langle g^2\rangle$ is the mean square avalanche gain. (In the case without avalanche gain when thermal noise predominates, the first term in the bracket of eq. (8) can be neglected.) For avalanche photodiodes, it has been found that

$$\langle g^2\rangle = \langle g\rangle^{2+x} \tag{9}$$

and for silicon units $x = 0.5$.

A value of $\langle g\rangle$ exists which optimizes performance.[7] The value which minimizes the power required to achieve a given error rate, found by minimizing eq. (8), is given by

$$g_{\mathrm{opt}} = \frac{2^{4/3}}{(ef_bQ)^{\frac{2}{3}}} n_i^{\frac{1}{3}}. \tag{10}$$

At optimal gain the power required to achieve a specified error rate is given by

$$p = \frac{3h\nu Q}{\eta e g_{\mathrm{opt}}} n_i^{\frac{1}{2}} \tag{11}$$

$$p = \frac{3h\nu Q^{5/3} f_b^{\frac{2}{3}} n_i^{\frac{1}{6}}}{2^{4/3}\eta e^{\frac{1}{3}}}. \tag{12}$$

It is interesting to note that eq. (11) can also be put in the form

$$p = \frac{3}{g_{\mathrm{opt}}} p', \tag{13}$$

where p' is the required power without avalanche gain. From eqs. (8), (10), and (12), the required power without avalanche gain, the optimum avalanche gain, and the required power with optimum avalanche gain are proportional to the second, third, and sixth roots of the thermal noise, respectively.

The series noise source of a junction field effect transistor (FET) is virtually independent of the circuit parameters in the normal range of operation, and the shunt noise source is negligible at 6.3 Mb/s without input tuning. For an FET[8]

$$e_{se}^2 \approx 2kT\left(\frac{0.7}{g_m}\right), \tag{14}$$

where g_m is the transconductance of the device, so the thermal noise referred to the detector is

$$n_t \approx 2kTf_b\left[\left(\frac{1}{R_r} + \frac{0.7}{g_m R^2}\right) I_1 + \frac{0.7(2\pi f_b C)^2}{g_m} I_2\right], \tag{15}$$

assuming all the noise is due to the first stage of the input amplifier. For a good FET the input resistance is virtually infinite, so

$$R = R_r.$$

In a common-source configuration, C is the sum of the drain-gate, gate-source, diode, and gate wiring capacitances.

III. CIRCUITRY

Figure 2 is a block diagram of the repeater, which was constructed in a $5\frac{1}{4} \times 4 \times 1\frac{1}{2}$ inch enclosure. The main signal path is represented by heavy lines and boxes. The signal, which is detected by either a PIN or a silicon avalanche photodiode, is first amplified by a high-impedance input amplifier. Following this, additional gain is provided by an SN52733 integrated video amplifier. Next, the signal is equalized, then further amplified by another SN52733 video amplifier and filtered by a single-section maximally flat LC filter with a 7-MHz bandwidth which, in combination with the other amplifiers, gives a 3-dB point of about 6.3 MHz. From this point the signal is fed to the timing circuits, the automatic gain and threshold circuits, and the regenerator. Finally, the regenerated signal is amplified and applied to the LED.

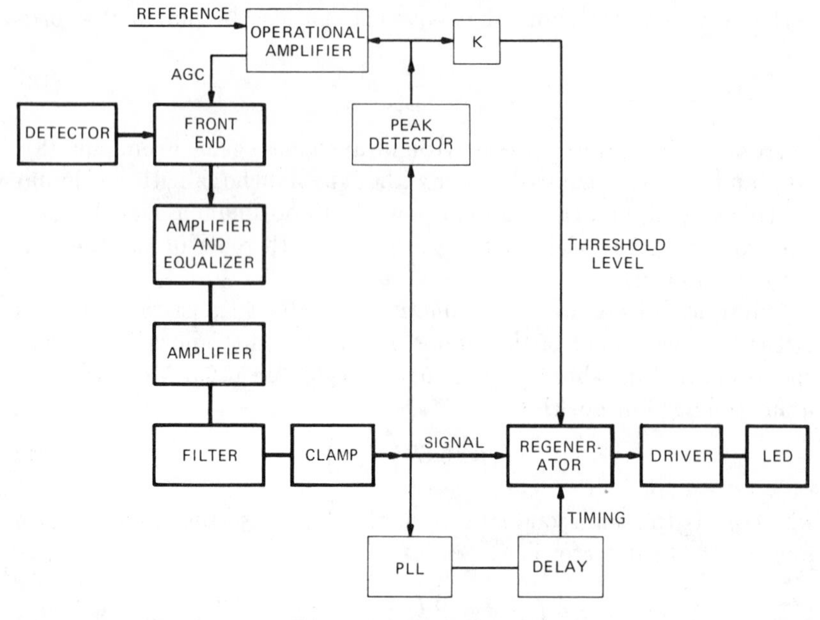

Fig. 2—6.3 Mb/s repeater.

3.1 *Input amplifier*

A 3-stage input amplifier was employed, as shown in Fig. 3. The amplifier consists of a 2N4416 junction field effect transistor followed by a 3N159 tetrode amplifier, and finally a 2N4416 in a source follower configuration. It was found experimentally that input amplifiers with a 2N4416 input stage had an input noise equivalent power about 1 dB less than with the 3N159. However, the tetrode can provide more gain for a single stage because of its low drain to first gate capacitance. Furthermore, the tetrode is well suited to automatic gain control since the g_m of the device is highly dependent on the voltage applied to the second gate. Thus, the configuration of Fig. 3 was chosen. It was found that the input noise dropped by 6 dB when the first gate of the tetrode was shorted to ground, so 75 percent of the thermal noise originates in the first stage. Thus, this configuration is very close to optimum. The source follower was provided to decouple the input amplifier from the subsequent circuits.

3.2 *Equalization*

In order to compensate for the distortion introduced by the long input time constant, the circuit of Fig. 4 was employed. With a source

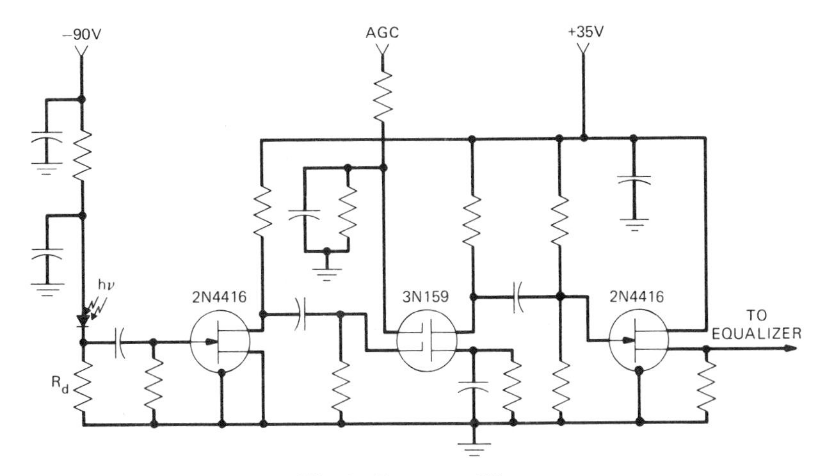

Fig. 3—Input amplifier.

of resistance R_s and a load whose resistance is included in R_2, it can be shown that the transfer function has a pole at

$$s = -\frac{1}{C_1R_1}\left(1 + \frac{R_1}{R_s + R_2}\right)$$

and a zero at

$$s = -\frac{1}{C_1R_1}.$$

Thus, the position of the zero can be adjusted by varying either C_1 or R_1 and, as long as

$$R_1 \gg R_s + R_2,$$

the pole will be above the band of interest and have a negligible effect.

3.3 Clamping and Peak Detection

Since the amplifiers of the repeater are ac-coupled, the dc level would be a function of word pattern unless suitable provisions are made.

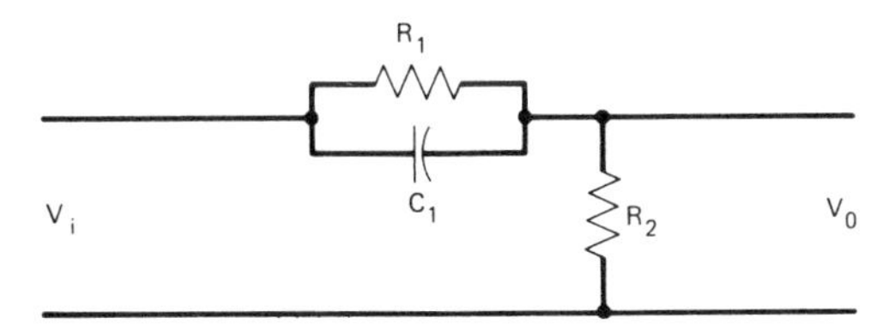

Fig. 4—RC equalizer.

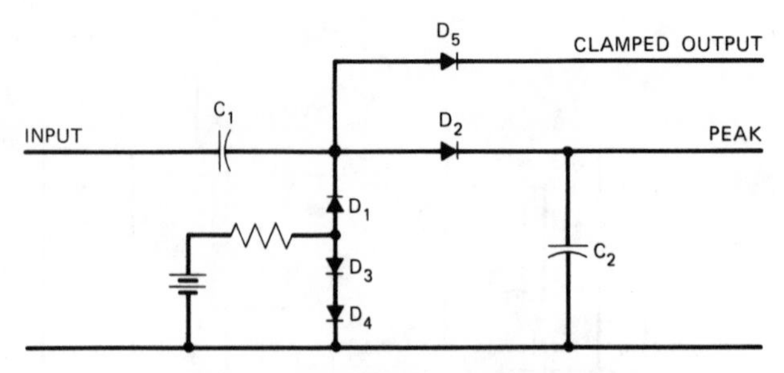

Fig. 5—Clamp and peak detector.

In addition, provision to measure signal level must be incorporated to set the threshold and to control gain. Clamping and peak detection were employed to solve these problems.

By incorporating both automatic threshold and gain controls, the AGC gain need only be high enough to assure that the phase-locked loop will function properly and to prevent compression. This reduces the tendency toward instability because of a high-gain feedback loop.

Figure 5 shows the circuit employed. Diode D_1 and capacitor C_1 serve as the clamp, and diode D_2 and capacitor C_2 serve as the peak detector. The diodes D_3, D_4, and D_5 are included to cancel the diode drops of D_1 and D_2.

3.4 *Digital Circuits*

The timing coincidence and regenerator circuits, Figs. 6 and 7, share an SN72810 dual comparator and an SN7474 dual D flip-flop. The comparator has the property

$$V_0 = 1, \quad V_1 > V_2$$
$$V_0 = 0, \quad V_1 < V_2,$$

where 1 and 0 represent states. For the D flip-flop, the output Q takes

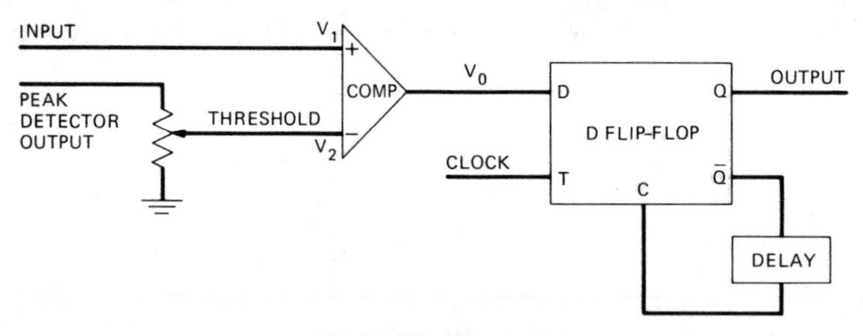

Fig. 6—Regenerator.

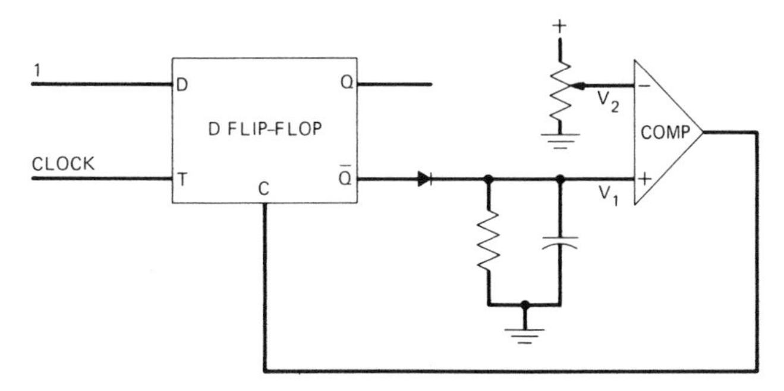

Fig. 7—Timing coincidence.

on the value applied to the D input when the clock input changes from low to high. This value is held until either the clock again shifts from low to high or the clear c is returned to ground. The output $\bar{Q}$ is the complement of Q.

In the regenerator, Fig. 6, the comparator serves as a quantizer and the D flip-flop retimes the signal. On the flip-flop, feedback from $\bar{Q}$ to c is provided so the D flip-flop output will be a pulse of proper duration whenever the clock goes positive and the D input is high.

The circuit shown in Fig. 7 is used to adjust timing coincidence between the phase-locked loop output and the regenerator. The D input to the flip-flop is kept high at all times. Adjusting the input to V_2 of the comparator adjusts the delay between the time when a positive clock pulse is applied and the voltage of the clear (c) drops to a sufficiently low value to clear the flip-flop. The rising edge of the $\bar{Q}$ output is used to trigger the regenerator.

3.5 *LED and driver*

The LED driver consists of a cascade of two emitter followers. The driver is capable of generating 1.5-A pulses into a diode load. For the tests to be described, the diode was driven with 300-mA peak current pulses and generated optical pulses of about 0.3-mW peak power.

IV. RESULTS

Both signal-to-noise ratio and error-rate measurements were made to evaluate the input amplifier and repeater performance. The 2N4416 JFET employed for all the tests had an input capacitance of 5 pF and g_m of 0.006 mho. An additional picofarad of capacitance was added by the diode load circuit. The measurements without gain were performed with an SGD-040A PIN photodetector which had a capacitance of 2 pF and a quantum efficiency of 83 percent. For the measurements with gain, a TIXL56 silicon avalanche photodetector was employed. This diode had a capacitance of 1 pF and a quantum efficiency of 55 percent. For both diodes, the noise calculated from the measured dark current was negligible.

The constants I_1 and I_2 have been evaluated by Personick[5] for rectangular optical pulses and pulses with a raised cosine spectrum and maximum eye opening at the regenerator input. He found $I_1 = 0.6$ and $I_2 = 0.26$. The theoretical curves presented here were obtained using these values.

Measurements of noise equivalent power, that is, optical power to achieve a unity signal-to-noise ratio at the regenerator, were first made to evaluate the performance of the input amplifier. As shown in Fig. 8, the results for the input amplifier of Fig. 3 with equalization closely approximates those predicted from eq. (8) by setting $Q = 1$. The difference between the theoretical and the experimental curve is mainly due to the noise of the second stage, which was about 1 dB with the 1-MΩ diode load resistor.

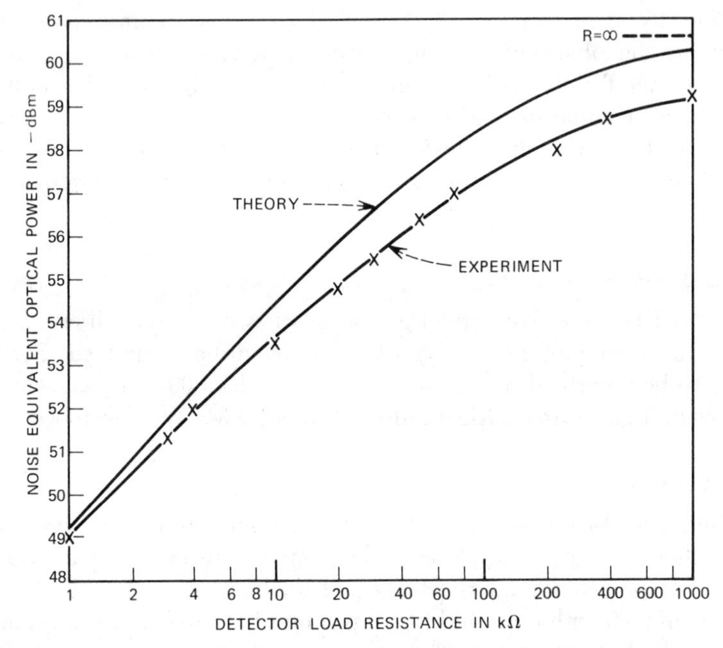

Fig. 8—Noise equivalent power circuit vs. diode load resistance.

Other input amplifiers were also tried. It was found that with a common-drain first stage and a common-source second stage, with a 1-MΩ diode load resistor, the noise equivalent power was about 1 dB higher. With smaller values of diode load resistance, where the resistor noise predominates, the common-drain input amplifier's performance was identical to that of the common-source input amplifier. Similar results were obtained with a cascade configuration using a bipolar transistor for the second stage.

Figure 9 shows "eye" diagrams taken at key points in the receiver with a 1-MΩ diode load resistance. Before equalization, the "eye" is fully closed, as is expected. After equalization, the "eye" is almost fully open. The regenerated pulse was photographed with recovered timing.

The tracking of the threshold level with peak signal is shown in Fig. 10. The tracking in conjunction with the AGC is adequate, as will be apparent from the error performance. With germanium or Schottky barrier diodes in the clamping and peak detection circuit, the tracking could have been held even closer to the ideal, had this been necessary. The AGC also functioned properly. The signal level could be held within about a 20-percent range with the power up to 10 dB above the signal required for 10^{-8} error probability. Greater range could be achieved by cascading tetrode FET stages, if required.

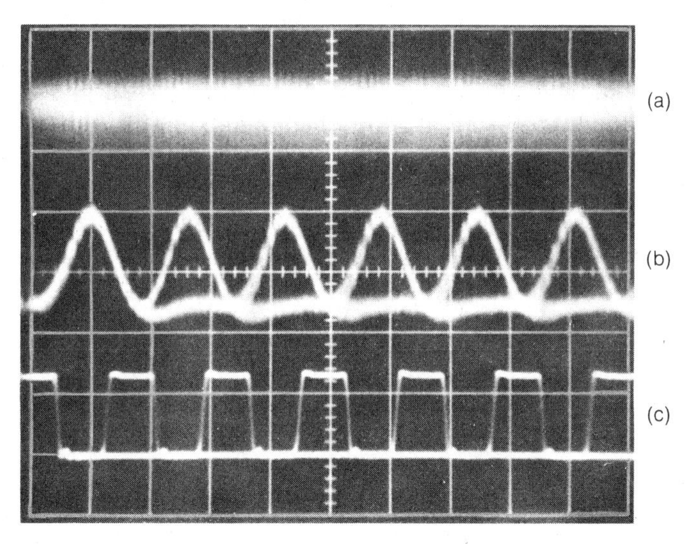

Fig. 9—"Eye" diagrams: (a) Before equalization. (b) After equalization. (c) Regenerated pulse.

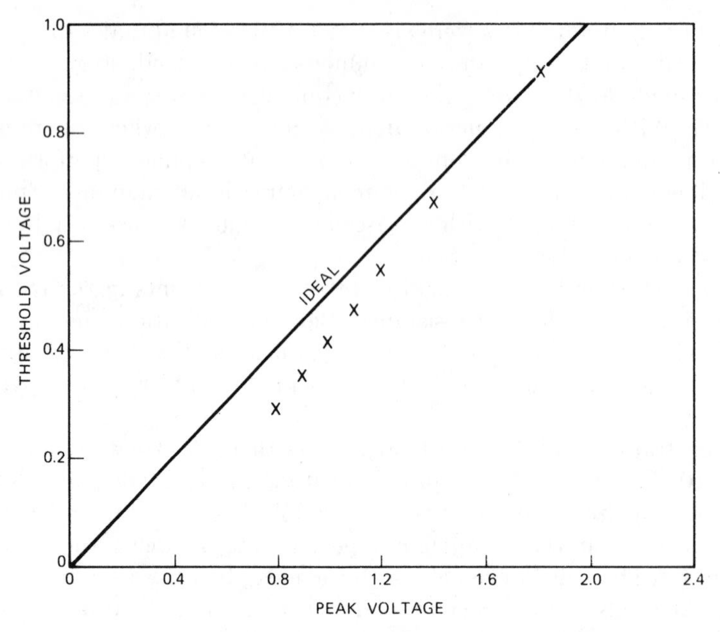

Fig. 10—Threshold voltage vs. peak voltage.

Error probability measurements were made under a variety of conditions. The results are shown in Fig. 11. The signal source was a 2^{15}-1 bit pseudo-random word generator. The signal consisted of 15 bit blocks each separated by a zero. The measurements were made with external timing and performance was optimized at each point, except the points indicated with x's. For these points, which were taken to check the performance of the automatic gain and threshold controls, as well as the timing recovery, the system was optimized with recovered timing and AGC at an error probability of about 10^{-8}. Then all the points were taken without further adjustment to the repeater.

The theoretical curve for a 4-kΩ diode load resistor is shown along with the measured points. These points were taken with the source-follower input amplifier because this amplifier introduced less intersymbol interference with the 4-kΩ diode load resistor. The improvement with a 1-MΩ diode load resistor over a 4-kΩ one was about 8 dB, which is in agreement with the theory.

Measurements of repeater performance were made with a TIXL56 silicon avalanche photodiode. This diode exhibits a significant diffusion tail. A second stage of RC equalization was employed to remove the resulting intersymbol interference.

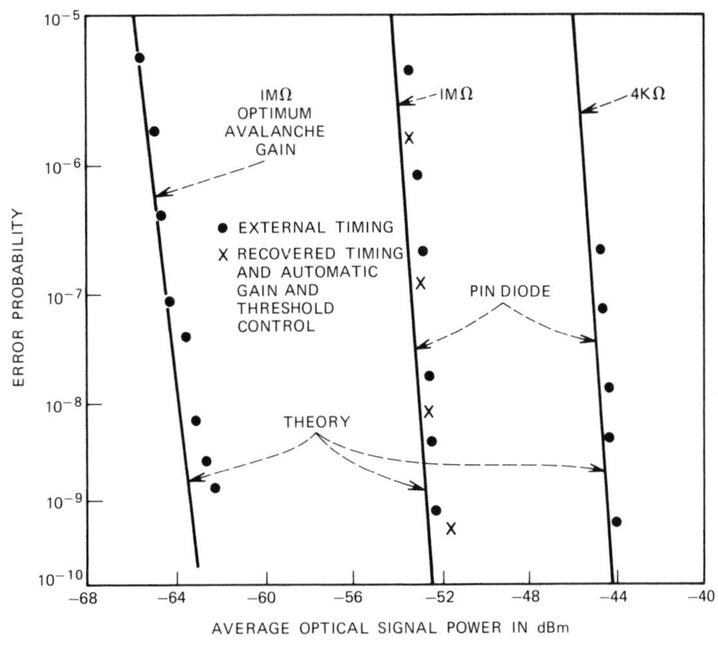

Fig. 11—Error probability vs. average optical signal power.

Equation (10) indicates that the optimum avalanche gain is a function of error rate. However, since it is not practical to optimize the gain at extremely low error probability, the gain was optimized at an error probability of 5×10^{-7} and then held constant for all the points.

Table I shows a comparison of the measured and predicted values of g_{opt} for 4-kΩ and 1-MΩ diode load resistance. In view of the assumption of noise statistics and diode characteristic, the agreement is satisfactory.

V. CONCLUSIONS

It has been demonstrated that a simple compact low-cost* repeater suitable for fiber optic applications can be built which functions close to theory at 6.3 Mb/s. The high-impedance input amplifier and its associated equalizer were realized in a straightforward manner, and compression did not turn out to be a serious problem. A significant reduction in required signal power to achieve a specified error rate with-

* The cost of the active components was about $30, exclusive of the detector and LED. An SGD-040 PIN detector costs about $15 and a TIXL56 avalanche detector $65.

Table I

Detector Load Ohms	Theoretical Gain	Measured Gain
4 K	171	188
1 M	44	62.2

out avalanche gain was achieved. With an avalanche detector, the optimum gain was greatly reduced with high impedance input, as predicted. Thus, the temperature stability will be greatly increased and the diode fabrication requirements eased with an avalanche detector.

VI. ACKNOWLEDGMENT

The author wishes to acknowledge the advice and assistance of W. M. Muska in the fabrication of the repeater.

REFERENCES

1. D. B. Keck, P. C. Schultz, and F. Zimar, "Attenuation of Multimode Glass Optical Waveguide," Appl. Phys. Lett., *21*, No. 5 (September 1, 1972).
2. A. B. Gillespie, *Signals, Noise, and Resolution in Nuclear Counter Amplifiers*, New York: McGraw-Hill, 1953.
3. O. H. Schade, Sr., "A Solid-State Low-Noise Preamplifier and Picture-Tube Drive Amplifier for a 60 mHz Video System," RCA Review, *29*, No. 1 (March 1968), p. 3.
4. C. A. Burrus and R W. Dawson, "Small-Area High-Current GaAs Electroluminescent Diodes and a Method of Operation for Improved Degradation Characteristics," Appl. Phys. Lett., *17*, No. 3 (August 1970), pp. 97–99.
5. S. D. Personick, "Receiver Design for Digital Fiber Optic Communication Systems, Pt. I," B.S.T.J., *52* (July 1973), pp. 843–886.
6. W. R. Bennett and J. R. Davey, *Data Transmission*, New York: McGraw-Hill, 1965, pp. 100, 101.
7. H. Melchior and W. T. Lynch, "Signal and Noise Response of High Speed Germanium Avalanche Photodiodes," IEEE Trans. on Elec. Devices, *ED 13*, No. 12 (December 1966), pp. 829–838.
8. A. van der Ziel, *Noise, Sources, Characterization, Measurement*, Englewood Cliffs, N.J.: Prentice-Hall, 1970, pp. 74–76.

A 274-Mb/s Optical-Repeater Experiment Employing a GaAs Laser

J. E. GOELL

Abstract—**An optical-repeater error-rate experiment is described in which the carrier generator is a directly modulated double heterojunction GaAs laser. It is shown that 274-Mb/s modulation can be achieved with sufficiently high quality that the received signal can be regenerated with error rates in accordance with predictions based on receiver thermal noise.**

This letter describes initial experiments to establish the feasibility of PCM repeaters operating near 270 Mb/s employing a directly modulated optical source. Such repeaters would be far less complex than those employing an external modulator and would be attractive for use in a fiber optic transmission system.

The feasibility of building high-performance compact repeaters using GaAs light-emitting diodes directly modulated at 6.3 Mb/s has already been demonstrated [1]. However, at high bit rates for fiber applications the interrelated problems of dispersion and coupling efficiency point to the use of a single-mode source of narrow spectral width. Room-temperature GaAs diode lasers are potentially attractive candidates.

Using an improved CW AlGaAs double heterojunction injection laser [2] it is shown that a PCM optical signal can be generated at 274 Mb/s with sufficiently high quality that the received signal can be regenerated with error rates in accordance with predictions based on receiver thermal noise. Here a pseudorandom bit stream is employed to evaluate the laser modulation rather than a fixed sequence of ones and zeroes as in previous work [3].

The laser drive circuit is shown in Fig. 1. The driver can supply pulses of up to 150-mA peak to the laser, if required. The laser was prebiased at 130 mA (just below threshold) and was driven with 20-mA non-return-to-zero pulses to give an output of 2.5 mW. Prebias is employed to avoid the turn-on delay when the diode is driven from zero current; this delay is a few nanoseconds [4].

The signal was first observed at high levels with a high-speed silicon p-i-n detector in a 50-Ω mount with a wide-band commercial amplifier to investigate the received pulse shapes. The resulting waveshape for pulses of 1, 2, and 6 consecutive ones is shown in Fig. 2. Considerable ringing, as has been previously observed [4], and a detector diffusion tail are apparent. These effects could seriously degrade system performance. The high-frequency ring was removed by filtering and the diffusion tail by suitable equalization. Fig. 3 shows an eye diagram of the signal (a superimposed recording of all possible outputs) with the 3-dB bandwidth limited to 300 MHz. The eye diagram indicates that the signal applied to the diode can be recovered at the receiver with very little intersymbol interference.

For error-rate measurements a silicon p-i-n diode was mounted directly at the input of a cascode input-stage front end. The signal was then amplified by commercial amplifiers, bandlimited to 300 MHz, and regenerated using a circuit originally designed for a coaxial cable PCM system. A block diagram of the test setup is shown in Fig. 4. For simplicity, regenerator timing is obtained directly from the original clock rather than recovered from the signal as has been done at 6.3 Mb/s [1]. The provision of self-timing can be readily incorporated by conventional techniques.

The eye diagram at the regenerator input is shown in Fig. 5. The degradation is due to the high-frequency rolloff of the front end and detector. It is expected that this degradation will be removed by a filter to be incorporated shortly.

Manuscript received May 30, 1973.
The author is with the Crawford Hill Laboratory, Bell Laboratories, Holmdel, N. J. 07733.

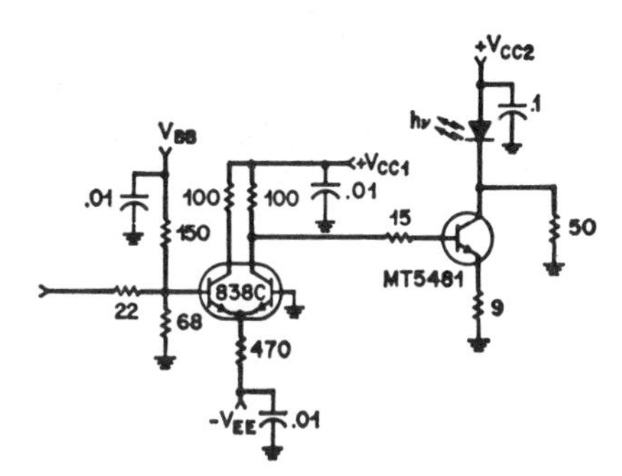

Fig. 1. Laser drive circuit.

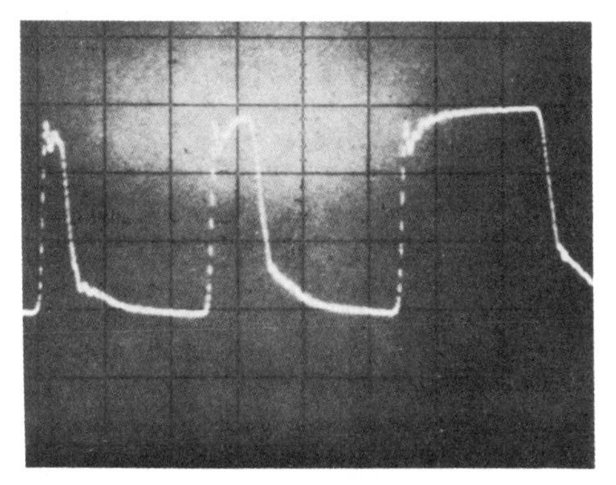

Fig. 2. Received pulse shapes.

Fig. 3. Received eye diagram (1 ns/div).

Reprinted from *Proc. IEEE*, vol. 61, pp. 1504–1505, Oct. 1973.

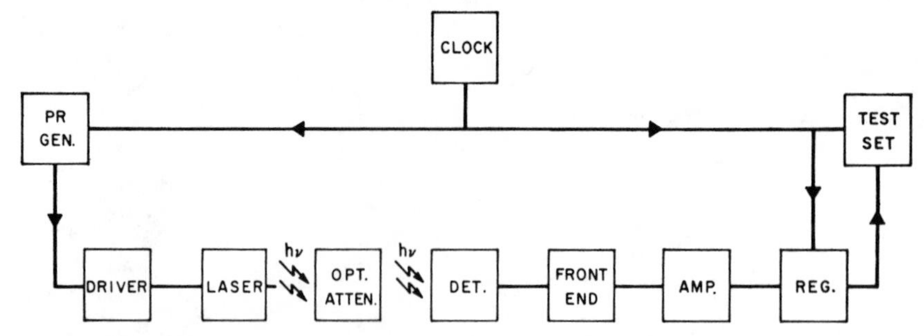

Fig. 4. Block diagram of error rate experiment.

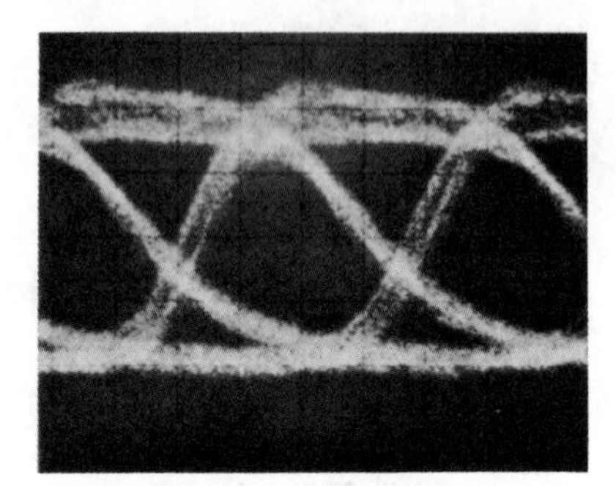

Fig. 5. Eye diagram at input to regenerator (1 ns/div).

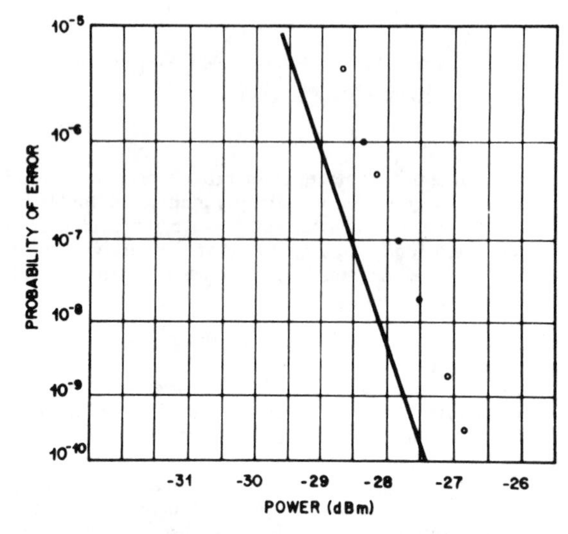

Fig. 6. Probability of error versus input signal power.

The error rate versus optical signal power at the detector is shown in Fig. 6. The data were taken by comparing the output signal with a locally generated pseudorandom word using a TauTron error-rate test set. Burst errors caused by electrical interference which are common in early designs of PCM repeaters that are not well shielded were not included in the data. A theoretical curve based on measured thermal noise is also shown in Fig. 6. This curve is given by [1], [5], [6]

$$P_e = \frac{1}{2} \operatorname{erfc} \frac{kp}{\sqrt{2}\, i_n}$$

where p is the received power, P_e is the probability of error, k is the detector conversion factor which was 0.35 A/W of optical power, erfc the error function complement, and i_n the thermal-noise current referred to the detector output. The latter was inferred from the optical signal required to achieve unity signal-to-noise ratio in the baseband circuit. After optimization, the thermal-noise current at the detector output was inferred to be 0.1 μA. The discrepancy of 0.75 dB between the measured error-rate data and the calculated curve is mainly attributable to the pulse shape degradation introduced by the detector and the input amplifier.

Acknowledgment

The author wishes to thank W. M. Muska for his assistance. He is also indebted to B. C. DeLoach and his colleagues for supplying the laser diode, and to F. D. Waldhauer and his colleagues for the regenerator.

References

[1] J. E. Geoll, "A repeater with high impedance input for optical fiber transmission systems," presented at the Conf. on Laser Engineering and Applications, Washington, D. C., May 30–June 1, 1973.
[2] B. C. DeLoach, "Reliability of GaAs injection lasers," presented at the Conf. on Laser Engineering and Applications, Washington, D. C., May 30–June 1, 1973.
[3] M. Chown, A. R. Goodwin, D. F. Lovelace, G. H. B. Thompson, and P. R. Selway, "Direct modulation of double-heterostructure lasers at rates up to 1 Gb/s," *Electron. Lett.*, vol. 9, pp. 34–36, Jan. 25, 1973.
[4] T. P. Paoli and J. E. Ripper, "Direct modulation of semiconductor lasers," *Proc. IEEE*, vol. 58, pp. 1457–1465, Oct. 1970.
[5] S. D. Personick, "Receiver design for digital fiber optic communication systems," *Bell Syst. Tech. J.*, July–Aug. 1973.
[6] W. R. Bennett and J. R. Davey, *Data Transmission.* New York: McGraw-Hill, pp. 100–101.

Part IX
Applications and Systems Considerations

Military Applications of Fiber Optics and Integrated Optics

R. A. ANDREWS, MEMBER, IEEE, A. FENNER MILTON, AND THOMAS G. GIALLORENZI

Invited Paper

***Abstract*—A general discussion of military applications of integrated optics and fiber optics is presented. Specific applications discussed are: 1) a multiterminal multiplexed data highway for aircraft and shipboard use; 2) optical fibers as tethers; 3) a 10.6-μm heterodyne detector; and 4) integrated optical phased arrays.**

I. Introduction

THE VIRTUES and potential of fiber optics and integrated optics are well known and have been discussed in other papers in this section. This area of optical technology is receiving a lot of attention at the present time from industry and the military. From the military point of view there are a fairly large number of application areas where this technology can possibly offer unique solutions to critical problems. In the remainder of this paper a few possible application areas will be discussed ranging from near-term to long-range possibilities.

Fiber optics is receiving a lot of attention at present due primarily to its potential for the telecommunications industry. Fibers with extremely low losses have been reported and the prospect of ultrahigh bandwidth optical transmission lines with multikilometer repeater spacing is a very real possibility. However, the large-scale use of fibers for telecommunications is probably at least ten years away. The military, on the other hand, has very real problems that can be solved with fibers of shorter length and more modest bandwidth. For example, optical information transfer (OIT) on board aircraft involves fiber lengths of less than 30 m, and bandwidths less than 100 MHz. Since higher losses can be tolerated, these applications can proceed without a large amount of fiber development.

The realization of the full military potential of OIT, and in particular fiber optic systems, will be aided by the development of integrated microoptical circuits [1]. In general, this technology will eliminate many problems inherent in bulk optical devices. The optical processing of information with integrated microoptical circuits will minimize the number of optics–electronics interfaces in OIT systems.

The potential advantages of OIT systems are well understood qualitatively. Generally, the use of fiber optics instead of conventional transmission lines in military hardware should cause the following impact.

1) Reduce, by up to a factor of 5, transmission-line system size and weight for conventional information bandwidths. This will make redundant control systems attractive which will increase reliability and battlefield survivability. As the individual fibers in a fiber bundle break, optical systems are expected to experience a gradual degradation in contrast to the catastrophic failure caused by shorts with conventional technology.

2) Eliminate the problems caused by electromagnetic crosstalk and interference which will allow close spacing of transmission lines and operation near radar transmitters.

Manuscript received June 12, 1973; revised July 19, 1973.
The authors are with the Optical Sciences Division, Naval Research Laboratory, Washington, D. C. 20375.

Reprinted from *IEEE Trans. Microwave Theory Tech.*, vol. MTT-21, pp. 763–769, Dec. 1973.

3) Increase electromagnetic security by eliminating RF emission and making inductive taps impossible.

4) Decrease interface problems, such as ground loops, making modular avionics more attractive.

5) Ultimately lead to ultra broad-band communication bandwidths (1–10 GHz).

It seems probable that integrated optical circuits (IOC) will be used in conjunction with fiber optics transmission systems to overcome some of the difficulties associated with using light to transmit information and may also find an independent role in various forms of optical processing.

The interest in fiber optics and IOC evidenced by the commercial telecommunications industry is clearly motivated by a well-defined eventual requirement for ultra broad-band (1–10-GHz) communication systems. When a high degree of multiplexing becomes common, such a broad-band capability (evidenced by single-mode or SELFOC fibers) will also prove advantageous for future military systems. Over the intermediate term, however, the introduction of this technology will involve applications requiring lower bandwidths where the first four listed advantages are important and bundles of conventional multimode fibers can be used.

II. Areas of Application

An important area that is being worked on now is optical data transmission for aircraft. The first generation of this concept is a collection of point-to-point fiber optic links. More advanced concepts involve optical data highways with several random access terminals. Optical links such as these would carry information between computers, sensors, and systems on board an aircraft. They are attractive because potentially they are small, lightweight, free of EMI, free of impedence matching problems, secure, and could lead to modular avionics.

There is another general area where optical fibers can play a key role. This is the problem of tethers for towed arrays, remotely piloted vehicles, wire guided missiles, etc. The need here is for a lightweight data link over a very long path length. Development of this area will require the lowest loss fibers and/or small repeaters.

The later generations of both these applications will involve integrated microoptical circuits. This technology is farther off than fiber optics but it is progressing steadily and offers a lot of potential. There are also areas where integrated optics alone are important. Efficient two-dimensional acoustooptical scanners for optical data recorders are possible and acoustooptical deflection can be used in an integrated optical circuit along with a linear array of detectors to form a microwave frequency analyzer. Two-dimensional optical signal processing is also possible in IOC form. Other applications are in the IR and in particular at 10.6 μm. These applications are tied to the increased use of CO_2 lasers for military applications. Efficient high-frequency modulators can be made in microoptical form which can be used with modest power levels. Optical receivers using heterodyne detection for laser radars and satellite communications links can also be made in microoptical form. The requirements for frequency tunable local oscillators and complex array geometries make bulk optical heterodyne receivers extremely complex and sensitive to environmental effects. IOC is a possible solution to these problems. Another area is coherent phase front control. An IOC array of phase modulators could conceivably be constructed to correct a laser beam for atmospheric distortion effects and provide some beam steering although the limited power-handling capability of IOC would be a serious constraint.

There are many more areas for IOC application that can be considered such as display, up-conversion, and lightweight atmospheric optical data links. With careful consideration of the state of the art of what can be done in the near future these applications seem further off.

Fiber optics communication systems will be limited to the visible and near IR portion of the spectrum since in that region diode light sources are available, detectors do not have to be cooled, and the most likely fiber materials have their maximum transparency. IOC experimental procedures are usually easier in the visible or near IR, however dimensional tolerances can be relaxed as one proceeds into the IR.

In the remainder of this paper four possible application areas with military relevance are considered in detail—optical data bus, 10.6-μm optical heterodyne detector, tether, and optical phased array antenna. These four applications address real problems and include a rather broad cross section of fiber optics and integrated optics technology.

III. Optical Data Bus

Fiber optics technology offers much promise for data communication systems. Indeed all of the efforts of Bell Labs in this technology are directed towards this application. The military will, however, have special requirements. In general, bandwidths will be smaller than those of the telecommunications industry and lengths will be shorter with more input–output terminals required along a transmission line. The principal military application of fiber optic bundles will involve multiplexed transmission lines of moderate length interconnecting avionics subsystems on aircraft or ships. For military application the emphasis will eventually be on multiterminal systems. Severe volume, weight, and mutual interference problems are now encountered when conventional transmission lines are jammed together on mobile platforms.

For moderately large information capacities, optical transmission lines in avionics systems will offer advantages involving volume, weight, and complete invulnerability to electromagnetic interference and interface problems, compared to conventional RF transmission lines. Fiber optics bundles need be no more than 1 mm in diameter even for large information capacity. For tens of megahertz bandwidths, conventional transmission lines often need diameters to ten times this size. The elimination of ground loops should make modular electronic systems more attractive. The potential advantages of optical transmission systems such as insensitivity to high temperatures, reliability (through redundancy), and lack of fire hazard may also prove important in select cases.

Integrated optical circuits can play a role in such systems by providing sources and modulators for systems which use conventional multimode fibers and additionally by providing couplers (active and passive) and remote optical switches for systems which use single-mode fibers. With the advent of spectral multiplexing techniques, integrated optical circuits can provide wavelength selective couplers and components for single-mode systems. With transmission lines which use a large number of modes IOC will not be able to provide signal processing after the data have passed through the transmission line since efficient coupling from a multimode fiber to an IOC, which contains at most a small number of modes, is not possible.

The first generation of multiplexed optical data transmission systems probably will be configured to be adequate for aircraft avionics and the sensor subsystems of small ships. A main transmission line length of 30–60 m with eight or more major input and output terminals (with multiple access) and a 10–100-Mb/s information capacity should cover both applications. 100 Mb/s is enough to simultaneously transmit one to two TV images. Bandwidth requirements often depend on whether TV imagery is carried on the bus. Hierarchical access to the bus will be required to hold coupling losses down. For these lengths and bandwidths use of a multimode fiber optics bundle with low-cost GaAs light-emitting diodes ($P \sim 1$ mW) and silicon p-i-n photodiode detectors is the most straightforward approach. Nonreciprocal coupling (all the input gets on whereas only a fraction of the throughput is coupled off) to the bus is required to limit terminal loss if a conventional data bus layout is used (see Fig. 1). Such input–output coupling to the main data bus could most easily be provided by nonreciprocal "T" connectors.

A desire for the use of low-cost light sources [1-mW light-emitting diodes (LED's)] and silicon p-i-n diodes sets a maximum transfer loss of ~25 dB for a 100-Mb/s information capacity. Most of this loss will be associated with splitting off from the main line at the major terminals. This loss must be carefully divided up between components. Throughput loss caused by packing fraction problems occurring at each "T" coupler is the major problem with the conventional data bus layout. Alternatively, the "star" layout approach to multiterminal communication systems, developed by Dr. Frank Thiel of the Corning Glass Works, can be used where each terminal has a separate input and output bundle which connects to a central "scrambler."

The principal components which need development are the following.

1) Multimode fiber bundles; 60-m lengths will be required with reasonable packing fraction losses and transmission losses <50 dB/km, so that the loss over 60 m will be less than 3 dB. The low numerical apertures (~0.2) required for the higher bandwidths will require large bundles and special input optics to keep LED input losses reasonable. Higher numerical apertures (NA) should be used with lower bandwidths and/or shorter links. An NA ≥0.4, 80-dB/km fiber would also be very useful. End finishing and repair techniques will have to be worked out for the bundles and high-temperature (300°C) fiber sheathing will need development since the present sheathing melts at temperatures of about 105°C.

2) Optical "T" connectors need development. The approach with the lowest loss is to essentially use bundle fractionation along with a number of scramblers to ensure that all the information is carried on each fiber before the next terminal is reached. This approach is nonreciprocal in the sense that all the information is coupled onto the rod, whereas only a fraction is coupled off.

If a light pipe integrator (clad glass rod) is used as the scrambler packing, fraction losses will be limiting. Present low-loss multimode optical fibers lead to high packing fraction losses (5 dB) unless the cladding is removed.

An optical "T" connector currently being worked on at NRL [2] is shown in Fig. 2. Here a fiber bundle is terminated and contacted with a solid glass cylinder. Light from any single fiber in the bundle illuminates the entire exit face of the cylinder. Hence this element acts to integrate or "scamble" the incoming signal over the entire aperture. The technique shown in Fig. 2 uses two cylinders with the input–output

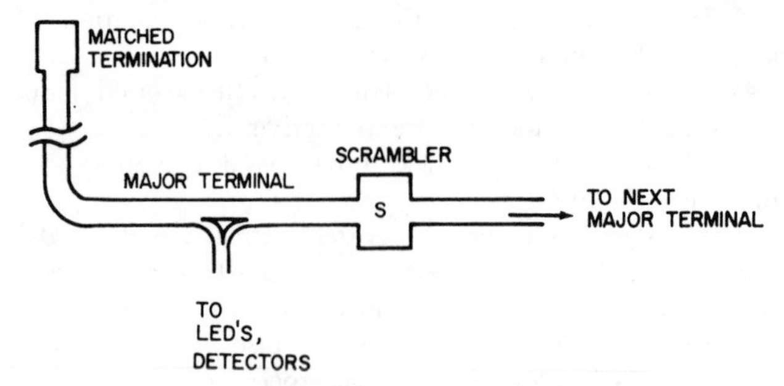

Fig. 1. Optical data bus with fractionation at each major terminal, and scramblers.

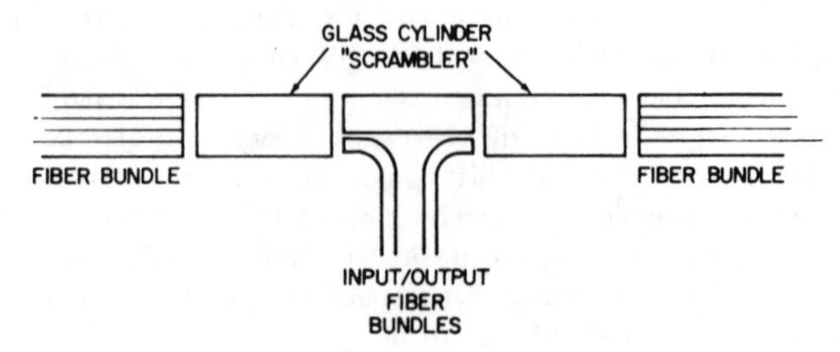

Fig. 2. A schematic diagram of a scrambler–coupler technique being developed at NRL for use with an optical data highway.

coupling to the bundle between them. The central portion is a third glass cylinder with a flat ground on one side. This area serves to allow a small fiber bundle to couple to the scrambler for input–output functions. This "T" coupler is nonreciprocal, self-contained, and needs only suffer one packing fraction loss on throughput.

3) High radiance LED's and silicon photodiode p-i-n structures which meet environmental specification will have to be put together. For modulation frequencies of 50 MHz and above some LED development is necessary. Otherwise, satisfactory components are commercially available.

Growth versions of the multiplexed data bus will be appropriate for large ships. Here 300-m lengths will be needed with up to 100 terminals. 300-Mb/s information capacity should suffice. Transmission losses and terminal bypass losses will now dominate input–output losses so that lower loss fiber bundles, higher power sources, and more sensitive detectors will be needed. Spectral multiplexing should be used to reduce bandwidth requirements. Repeaters will be needed and laser sources and avalanche diodes should be considered with low numerical aperture bundles.

An alternate approach is to use single-mode or SELFOC fibers. Spectral multiplexing might be useful but would not be required, since with these fibers the transmission line will not limit the bandwidth. Laser sources will be required. Diode lasers can operate CW at room temperature, but the lifetime is still limited. Efficient coupling into a bundle of single-mode or SELFOC fibers from a single source is not straightforward. Use of a single fiber instead of a bundle loses the advantages of redundancy and the advantage of fractionation. However, single-mode operation (or operation with a limited number of modes) is more advantageous to IOC device techniques and coupling to the main bus could be performed by using IOC. If single-mode operation is used, coupling will have to be achieved by the use of active switches which change the coupling with time. A bidirectional repeater will require the development of an optical circulator which will require magnetic optical devices. With single-mode operation optical switches at the terminals can be made in IOC form (see Fig. 3).

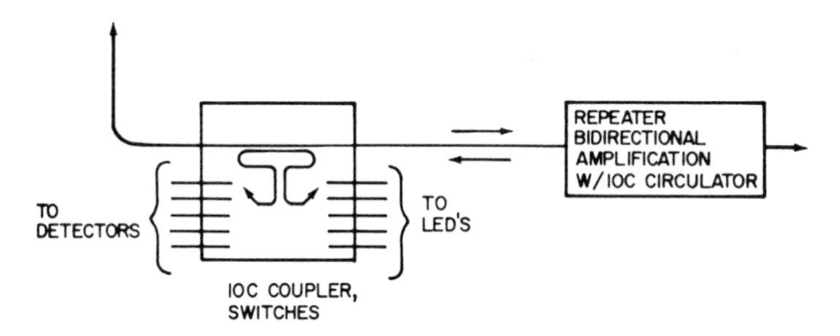

Fig. 3. Single-mode fiber optical data bus using IOC terminals and IOC bidirectional repeaters.

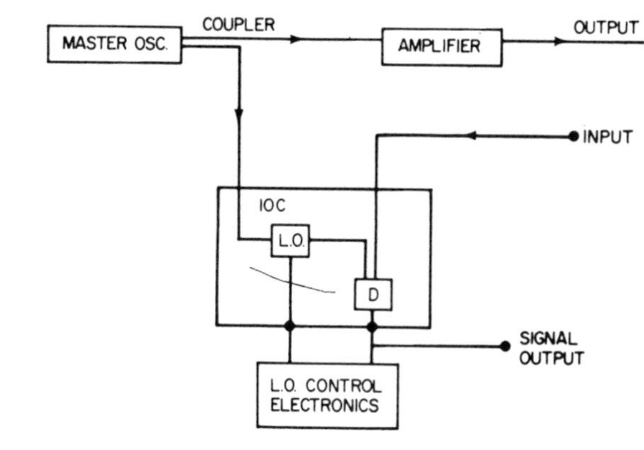

Fig. 4. Block diagram of an optical heterodyne detector where the LO signal is obtained by Doppler shifting a portion of the master oscillator power (see text).

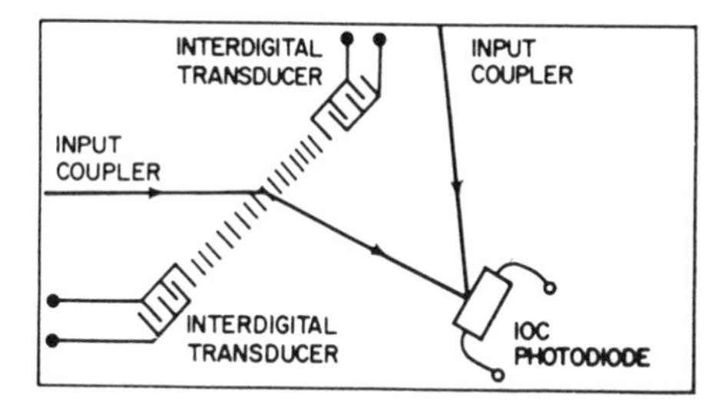

Fig. 5. Basic optical heterodyne unit with Bragg scatterer for LO generation and IOC photodiode.

IV. 10.6-μm Optical Heterodyne Detector

In the infrared, the heterodyne detection of laser radiation offers significant advantages over direct detection. When a heterodyne receiver is operated with enough local oscillator power, dark current noise and receiver thermal noise can be suppressed even for very wide bandwidth operation. The signal-to-noise ratio (SNR) can thereby be increased to the signal photon noise limit. Indeed the narrow spectral acceptance of the heterodyne system ($\Delta\lambda/\lambda < 10^{-3}$) is the only way to achieve this result near 10 μm due to the large thermal backgrounds which are usually present at these wavelengths.

Both the large thermal backgrounds and the absence of high-speed detectors for $\lambda = 10.6$ μm with internal current gain make heterodyne detection particularly attractive for CO_2 laser systems. Heterodyne techniques would also be advantageous for laser systems operating in the 3–5-μm window (DF or CO) since thermal backgrounds are still troublesome and no high-efficiency photocathodes exist for those wavelengths.

Heterodyne detection for optical communication systems, optical radar systems, and laser-line and raster scanners can significantly reduce transmitter power requirements; however, at this time this detection technique is complicated and expensive. The primary reason is that a laser local oscillator must be provided and careful alignment maintained.

Master oscillator fluctuations and Doppler shifts due to moving targets and receivers will change the frequency of the received radiation. Unless the local oscillator tracks these shifts the detector must be operated with a very wide bandwidth with bandwidth reduction accomplished after the IF channel. Since Doppler shifts for space targets get as high as 1.5 GHz at 10.6 μm (150 MHz for airplane targets) detector frequency response and electrical crosstalk problems are serious, and large LO power and expensive electronics are necessary with a fixed-frequency LO approach.

A far more elegant solution is to frequency shift radiation from the transmitter master oscillator for the local oscillator and make it frequency track the received signal. Indeed this would in many cases be the only way to use heterodyne detection in the mid-IR, since Doppler shifts get larger at shorter wavelengths. Even at 10.6 μm severe problems have been encountered in developing frequency shifters of sufficiently high bandwidth using bulk electrcoptical and acoustooptical interactions [3], [4]. Interactions which are confined to an optical waveguide may offer the best approach to constructing a practical frequency shifter since higher efficiency can be obtained due to the absence of diffraction effects. The alternative of using $Pb_{1-x}Sn_xTe$ diode lasers for tunable LO power does work but problems in obtaining sufficient output power in a single mode remain [5].

Integrated optics offers potential for the heterodyne detector. With IOC the detector diode, local oscillator source, and frequency shifter are an integral unit. This unit is small, lightweight, and free of environmental effects such as vibrations. Frequency shifting of the LO should be very efficient with IOC. Packaged in this manner the heterodyne detector could be stacked into arrays for IR image detection. Present technology concepts for matrix heterodyne receivers, to be used with pulsed imaging laser radars, have serious image disection problems which could be helped by the introduction of 10.6-μm optical waveguides. Photodiodes have not yet been fabricated to be compatible with optical waveguides. There are potential materials compatibility problems and fabrication technique problems, but there are no reasons to assume these are unsolvable. Further there has been no demonstration of an IOC local oscillator at 10.6 μm. The easiest approach to this problem is an external source frequency shifted by an appropriate technique.

A basic heterodyne detector is shown in Fig. 4. A master oscillator–amplifier combination provides the output signal for the system. A part of the master oscillator signal is tapped off to provide an LO source. The LO–detector (D) combination is an IOC. The LO is obtained by Doppler shifting the master oscillator input. The Doppler-shifted frequency is controlled by the electronics package which monitors the intermediate frequency at the detector.

The general problem areas associated with the IOC portion of this detector are: 1) low-loss 10-μm planar waveguides; 2) photodiodes compatible with IOC; 3) 10-μm planar waveguide output couplers; and 4) acoustooptic capability in 10-μm waveguide.

The basic optical heterodyne unit is shown in Fig. 5. "End-fire" tapered couplers are used to couple 10.6-μm radiation into the IOC. They must efficiently (>90 percent) accept the collimated input radiation when incident in a given field of view. The local oscillator signal is generated by using the Doppler-shifted light from Bragg scattering. The scattering is due to an acoustic surface wave generated by interdigital transducers. LO and signal are incident in a photodiode fabricated on the IOC. The bandwidth of this device should be 1 GHz. Some applications will require the development of a 2-D array of these basic elements.

V. Tether

Tethers are used to tow acoustic detector arrays behind surface ships and submarines. The basic idea is to have the listening device as far away from the noise of the towing ship as possible. Distances in excess of 6000 yd are sometimes used. The cables that are used for this application must be strong enough to overcome the tremendous drag on itself and the detectors. Hence they are typically 1-in diameter cables and when rolled up on the stern of a surface ship are very conspicuous and heavy. Besides providing the towing function for the detectors the cable must provide electrical conductors for power signal transmission. Since the loss per unit length of coaxial cable is a function of its diameter, this accounts for the large diameter of the tether. Further, to extend the length of the cable the diameter of the coaxial line must increase to maintain the signal. This requires the diameter of the entire cable to increase and therefore increase the drag. Hence it must be made stronger. The effect is not linear and represents a major problem. In short, the major problem areas are weight and size of the cable necessary for sufficient data rate handling capability.

Tethers are also used for guided ordnance delivery. A tether is the only possible guidance approach for a long-range torpedo. Midcourse guidance must be provided until the target is within range of the sensitivity limited acoustic sensors carried by the torpedo.

Above-the-water missiles also use a tether as a guidance command link. A wire guided approach has proven far more reliable than the free-space IR optical links. Tethered links are, of course, not susceptible to jamming. For ordnance applications the data rates are small ~10 kb/s (unless images are transmitted back to the launcher), but the required lengths are several miles. Any new technology which reduces the size and weight of tethers could open up new realms of performance for such ordnance.

OIT is a possible solution to the tether problems. Glass fiber waveguides are not only lighter than coaxial cables, but smaller in diameter. The fiber bundle itself can be less than 1 mm in diameter. This means that the tether can have a smaller diameter for less drag and a smaller weight per unit length. The reduction in overall diameter of an armored cable should be at least a factor of 2. The total volume of reeled cable would be smaller and less conspicuous. The introduction of this technology for tether application must await the development of manufacturing techniques to make truly long, low-loss fiber bundles.

This application of OIT is basically a very long point-to-point communication problem. For acoustic arrays the information capacity requirement is 10 Mb/s at maximum. Signals for N sensors must be processed and converted to optical signals. At the other end the signals must be detected and demultiplexed into N channels of information. This communication link requires a low-loss very long multimode optical fiber bundle with low numerical aperture to avoid geometrical dispersion effects which limit bandwidth for long lengths. A numerical aperture of ~0.1 would be compatible with laser diode sources and would allow for a bit rate of 10 Mb/s for lengths up to 3000 m. Laser sources would be needed for efficient input coupling (diode lasers operating CW at room temperature exist at present on a laboratory basis). Assuming a transmission loss of 25 dB/km (20 dB/km has already been observed in short lengths of fiber with NA = 0.1), a length of 3000 m represents an overall loss of 75 dB. Use of avalanche diode detection would place a minimum requirement on the diode laser of ~200-mW output power for a 10-Mb/s capacity. This should be feasible with development; however, for first-generation systems a bandwidth of 500 kHz will ease the source requirements to the 10-mW region.

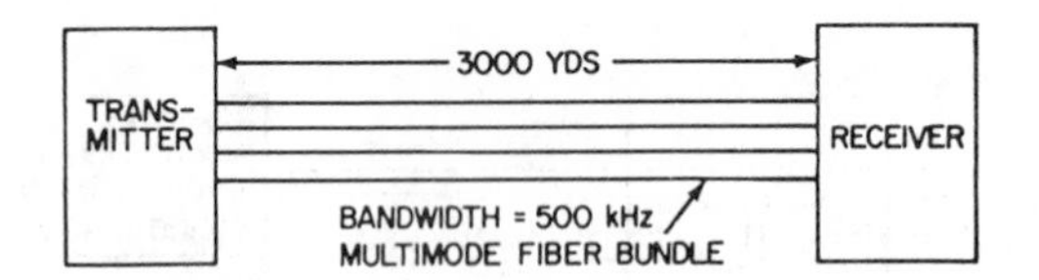

Fig. 6. First-generation tether demonstration.

If losses are increased, spectral multiplexing will be required to reduce the bandwidth and retain the information capacity. An alternative to spectral multiplexing is the use of a repeater or repeaters along the line to maintain the signal. The repeater could use IOC components to avoid a large package and the extra drag and weight. Also the spectral multiplexing would be accomplished either by separate laser sources into different parts of the fiber bundle immediately followed by a "scrambler" or an IOC with several sources, a spatial mixer, and IOC–fiber coupler. Similar concepts would apply at the receiver end of the line. The components and details of a first-generation approach to this problem are given below.

The first-generation system is shown in Fig. 6. It is a point-to-point optical fiber communication link with 500-kHz bandwidth ($S/N = 8$). There is a total length of 3000 yd. The transmitter consists of a single laser diode end coupled to the fiber bundle. The receiver is a single avalanche diode. The goals required for this demonstration are: 1) continuous low-loss fiber bundle (<25 dB/km) of 3000-m length and numerical aperture of <0.2; and 2) minimize the input and output coupling losses by design of an appropriate mechanical coupler. A laser diode with 10 mW of output power could suffice.

VI. Optical Phase Front Control

Electromagnetic phase front control has been extensively used in the microwave region to provide electronically controlled beam steering and beam forming. Indeed the phased array antenna has provided a beam-steering agility which would be impossible by mechanical means. The application of such techniques at optical frequencies could alleviate many pointing and deflection problems associated with the use of laser beams.

The usual phased array antenna consists of an $N \times N$ array of radiating apertures, each $1/2\ \lambda$ apart and whose phases and amplitudes are electronically controlled. The minimum beamwidth from such an antenna is approximately π/N rad which can be scanned over π rad. IOC with batch processing could provide the high density of phase-controlled apertures required at optical frequencies; but no matter how the phased array was configured beam steering would be limited to $N \times N$ resolutions elements. There is a strong tradeoff between beam size and maximum deflection angle. Even for a large number of phase-controlled apertures (10^4) a milliradian beam could only be deflected over 5°.

For high-resolution scanning applications, with larger numbers of resolution elements, one would like to design a nearly continuous antenna. The continuous antenna must have a capability of varying the phase across the beam in any desired fashion.

Using IOC technology, an optical antenna that will act as a nearly continuous antenna (the number of elements approaching infinity) is shown in Fig. 7. An input laser is coupled into a waveguide which has electrooptic capabilities. On this waveguide are placed 10 or more electrodes of precisely determined shapes. These shapes are designed to modulate the beam in a precise fashion. In particular, the modulation defined by each electrode is described by a function that is approximately orthogonal to the other nine. By varying the voltage across each of the electrodes, a quasi-continuous phase shift may be impressed on the beam. The IOC will thus act as a nearly continuous antenna, i.e., one which has an extremely large number of antenna elements. Complete phase front control will be limited in this case by the closeness by which the 10 functions chosen approach a complete set. High-resolution beam steering, however, can be accomplished with two modulators.

Acoustooptic scattering of light in IOC is an alternate means available for beam scanning. Through Bragg scattering, the optical beam in the IOC may be directed into the Bragg angle which is (acoustic) frequency dependent. Thus, with careful design, continual scanning could be available through this technique. However, there are several disadvantages of this technique. 1) There are problems in launching acoustic-surface waves with varying frequencies in IOC. Very high frequencies are needed for appreciable deflection of visible radiation (this problem is eased for 10.6 μm). Good broadband interdigital transducers are not readily available. 2) The angle between the incident light wave and the acoustic wave must be varied as the acoustic frequency is changed for efficient Bragg scattering. 3) The electronics associated with acoustic-wave generation is quite bulky and conversion of acoustic power to surface waves is generally inefficient. 4) Electrooptic switching offers much faster switching capabilities. 5) Electrooptically controlled antenna offer better resolution, more freedom for tailoring the output beam to the desired pattern, and more compact packaging.

The technology necessary to construct the required IOC is or will shortly be available. Waveguides (planar) for use in IOC presently have losses of about 1 dB/cm. This magnitude of loss is acceptable for the antenna array. Fabrication of waveguide by ion implantation, ion exchange, sputtering, and liquid epitaxy has been demonstrated. Passive thin-film optical elements such as lenses, prisms, etc., have been fabricated and will find use in the present system (especially beam splitters). Construction of waveguides with 1-μm resolution has been demonstrated, and this will enable the construction of the electrode shapes needed to define the modulation function. However, good endfire couplers have yet to be demonstrated and thin-film lasers and amplifiers have only been

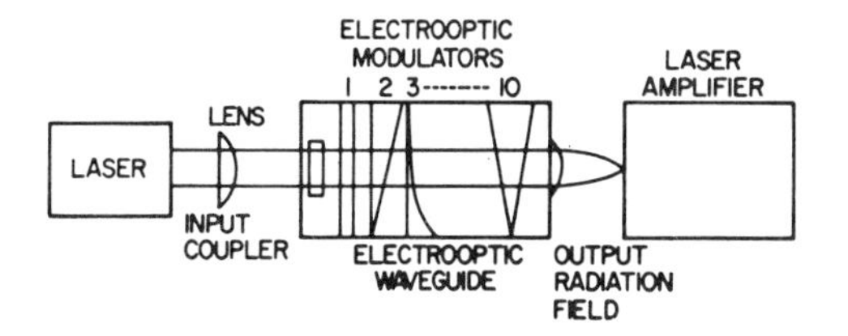

Fig. 7. Schematic diagram of first-generation linear phased array.

developed in the visible. Electrooptic modulators have been developed and demonstrated but these do not have the required degree of sophistication. These shortcomings, however, are well within the scope of present technology.

The main use for optical antenna arrays is the control of the radiation field they produce. Two important areas that will use precise beam control are the following.

A. Display or Scanning Applications

Since the beam position is controlled by the element-to-element phase shift, the beam maximum can be scanned by electrooptically shifting the phase of the wavefront. This allows scanning without changing the mechanical orientation of the antenna, a practical consideration especially important in applications where vibrations are present. The antenna in this case can be rigidly mounted, thus avoiding misalignment problems. Another important characteristic of this system is that the number of resolution spots attainable from this type of system, i.e., continuous-type antenna, is quite high because the field may be extremely well defined and controlled. TV quality displays require more than 10^5 resolution elements so that the continuous antenna approach is the only one possible.

B. Beam Control

By varying the phases across the beam in the IOC it is possible to tailor the resulting output beam to any desired form. This is important in beam propagation applications where atmospheric distortion of the propagating beam destroys the beam quality. Atmospheric turbulance limits the beam size to 0.5×10^{-4} rad at sea level regardless of beam aperture if phase front control is not employed. Beam control is an important long-range goal for many military laser systems where the effectiveness of the system is reduced by beam distortion. The phased array optical antenna concept, in conjunction with beam feedback, makes it possible to continually correct for atmospheric disturbance, thereby permitting the transmission of optical beams with the desired optical quality ($\sim 10^{-5}$ rad). Some beam-steering capability will also be available in a beam-control antenna system. Return beams can also yield image resolution finer than the turbulence limit if phase front control is introduced. Approximately 100 phase-controlled antennas (10×10) should be able to obtain the order of magnitude better beam quality than is generally desired ($0.5 \times 10^4 \rightarrow 0.5 \times 10^5$ rad). An IOC approach with discreet modulators is also possible; however, the optimum approach may be to use a continuous antenna with the modulators carefully designed to correct for typical distortions. Power-handling problems in thin films will limit the use of IOC to portions of the transmitter chain where average power is still modest.

VII. Summary

As seen from the above discussion there are a wide variety of military applications for fiber optics and integrated optics technology. The fiber optics application are certainly furthest along in terms of practical development. However, future generations of applications will almost certainly have integrated optical circuits. Only then will the full potential advantages of this technology be realized.

This technological area is not without problems. There are serious materials fabrication problems. Radiation sensitivity of these optical materials is a potentially serious problem area that has not been discussed.

It is probably fair to conclude that the military-related applications of this technology are nearer term than those of the telecommunication industry. Further, the impetus provided by military needs will lead to successful applications of fiber optics and integrated optics in the near future.

References

[1] R. A. Andrews, "Optical waveguides and integrated optics technology," NRL Rep. 7291, Aug. 1971.

[2] A. F. Milton and L. W. Brown, "Nonreciprocal access to multiterminal optical data highways," in *Dig. Tech. Papers IEEE/OSA Conf. Laser Engineering and Applications* (Washington, D. C.), May 30, 1973, pp. 24–25.

[3] J. E. Keifer, J. A. Nussmeier, and F. E. Goodwin, "Intracavity CdTe modulators for CO_2 lasers," *IEEE J. Quantum Electron.* (*Special Issue on 1971 IEEE/OSA Conf. Laser Engineering and Applications*), vol. QE-8, pp. 173–179, Feb. 1972.

[4] F. S. Chen, "Modulators for optical communications," *Proc. IEEE* (*Special Issue on Optical Communication*), vol. 58, pp. 1440–1457, Oct. 1970.

[5] J. Dimmock, Lincoln Laboratory, private communication.

CATV AND SUBSCRIBER NETWORKS USING OPTICAL FIBRES

B.S. HELLIWELL
Plessey Telecommunications Research,
Taplow, Maidenhead,
England.

SUMMARY

Progress in the field of optical fibre transmission has been rapid over the last few years. It is now possible to consider practical application of these techniques and this paper considers how they might be applied to CATV type networks. Problem areas requiring further study are discussed.

Basic Design Considerations

Optical fibre transmission systems have several possible advantages as a transmission medium for carrying wideband signals to the subscriber[1]. The following can be listed:-

1. Because of their small physical size, optical cables should incur lower civil engineering costs in terms of duct space than conventional cables offering an equivalent service.

2. The potential low attenuation of optical fibre combined with good semiconductor source and detector performance largely eliminates the need for intermediate amplifiers except for very long subscriber loops.

3. The cost of copper and aluminium for cables is likely to continue to escalate.

4. The demand for the wideband services by the business user at least may well be hastened by the energy supply situation.

5. The semiconductor costs should follow the downward trends displayed by most devices to date.

With a multimode fibre there are several transmission parameters which can be varied to alter the performance and hence meet system design requirements. Correspondingly, the fibre parameters lead us to transmit and receive device specifications. Let us therefore summarise the present state of the fibre and device technology.

As described elsewhere, fibre attenuation[2], has been reduced to a few dB's per km. For CATV type applications low cost will obviously be of great importance, and therefore we may need to set a loss figure somewhat above the minimum that could be achieved. It is assumed that a form of multimode fibre will be used in early systems in order to retain choice of light source and to avoid manufacturing and jointing difficulties. If we assume a figure of 8 dB/km it will be shown later that we can achieve wideband transmission over distances of several kms - quite sufficient for many applications. The available bandwidth is mainly a function of the refractive index of the glass and the refractive index profile across the core[3,4,5]. The shape of the profile and whether the type of construction performs the function of mode mixing are the important factors.

Fig. 1 illustrates the bandwidth of three types of multimode fibre when fully filled as from a LED source and in which no mode mixing is assumed. Equally important is the amount of light one can get into the fibre as shown in Fig. 2 overleaf. This amount is a function of the fibre N.A. and the core size[6]. We have to make a choice of core size and N.A. which accords with overall system needs including mechanical problems such as fibre jointing and manufacturing tolerances. Present thoughts are towards an N.A. of ~ 0.2 and diameter 90 μm. This gives an input power of about +12 dB relative to 1 μW using a $1 W/cm^2$/ sterad LED source, or 300-500 μW from an available LED. If the fibre exhibits mode mixing and hence shows an inverse square root of length-proportionality against bandwidth, then the bandwidth could be around 40 MHz for a 1 km length of fibre or 30 MHz over 2 km. These bandwidth figures relative to the TV bandwidths we wish to carry, suggest that we must consider the possibilities of baseband analogue transmission. Let us therefore consider the LED as a source.

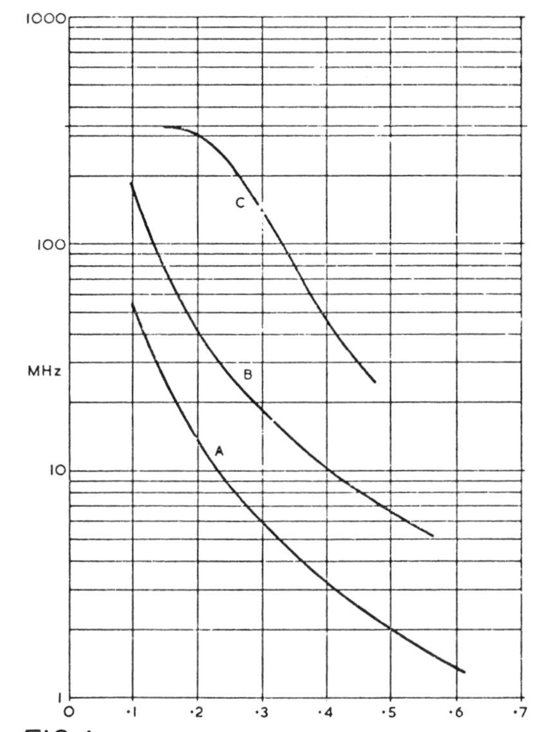

FIG. 1.

BANDWIDTH FOR 1km LENGTH OF VARIOUS MULTIMODE FIBRES

$$N.A. = n_o \sqrt{2\Delta}$$

A Ungraded index fibre $n = n_o(1 + \Delta)$

B Ring shaped parabolic index

$$n = n_o\left(1 - \Delta\frac{\{r - R_o\}^2}{a^2}\right)$$

C Graded index fibre $n = n_o\left(1 - \frac{\Delta r^2}{a^2}\right)$

The properties of main interest are:-

1. Radiance and efficiency

Reprinted from *IEEE INTERCON Tech. Program Papers*, session 12, paper 5, Mar. 26-29, 1974, pp. 1-6.

2. Linearity of light O/P with drive current.

3. Rise time of light output, including bandwidth.

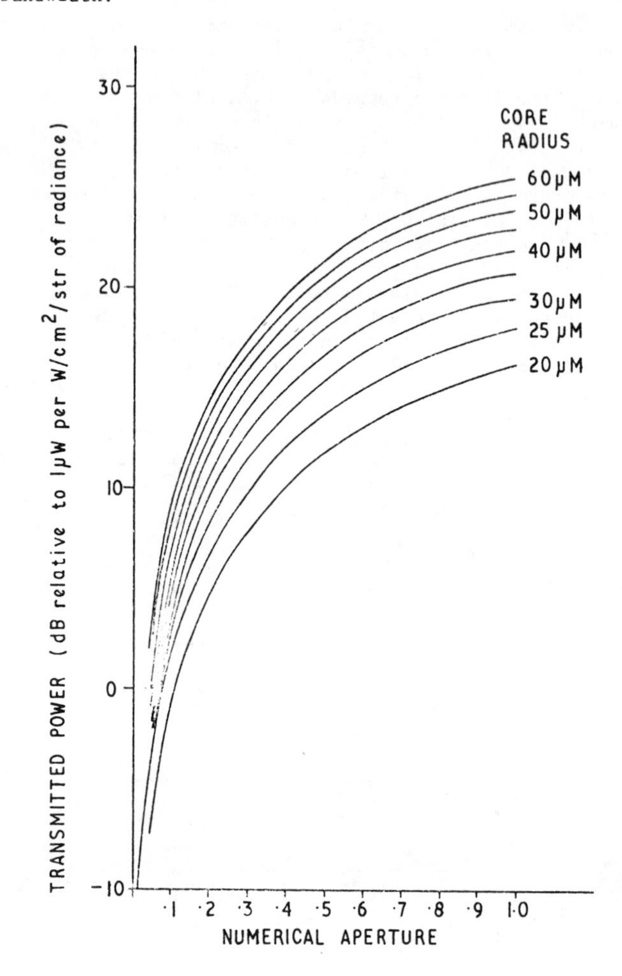

FIG. 2 TRANSMITTED POWER INCOHERENT LIGHT

Radiance & Efficiency

High radiance diodes have been made at Bell Laboratories and elsewhere[7,8,9,10]. Figures as high as 100W/cm^2/steradian have been quoted by Bell for a source spectral width of 45 nm. at a mean wavelength of 0.9 μm. We have achieved lower total radiance of 15W/cm^2/steradian in a GaAs homostructure device, but with a reduced spectral width of 25 nm, thus reducing material dispersion effects. The area of the brightly emitting surface is not very sharply defined and visual inspection of a typical diode gives the emitting area diameter as 50 μm to the half brightness points with a 10% tolerance.

Some of the light is scattered out of the LED chip from regions remote from the desired central emitting area, and is therefore of no value for transission down the fibre. Our results show that approximately 50% of the total light output is lost in this way.

The system designer will wish to optimise the input light power into the fibre and therefore match the source diameter to the fibre. Other factors which have to be taken into account are the problems of the LED design for a given emitting area and also the LED efficiency.

At present in order to launch about 250-350 μW peak power, a typical Plessey LED requires to be driven by about 300 mA current at 1.8V for a good linear performance in analogue systems. The efficiency calculated as the light into the fibre/input d.c. power is only of the order of 0.05%. More important is the fact that we are dissipating 0.54W in the diode. This can lead to non-linearity effects due to heating of the LED chip, and also a heat dissipation problem if large quantities of diodes are used in a terminal building or concentrator. We would therefore wish to improve the efficiency of the LED, and recent work reported by RCA using laser-striped geometry for LEDs with improved efficiency is encouraging.[11]

Linearity

The light output against d.c. drive current is shown in Fig. 3 for a prototype Plessey diode. As the bias current exceeds 150 mA the efficiency of the LED decreases due to heating of the chip and the device starts to become non-linear. The time-constant of the heating effect was measured by applying a fast rising flat-topped current pulse and observing the light output. The time-constant is of the order of 10 μsec.

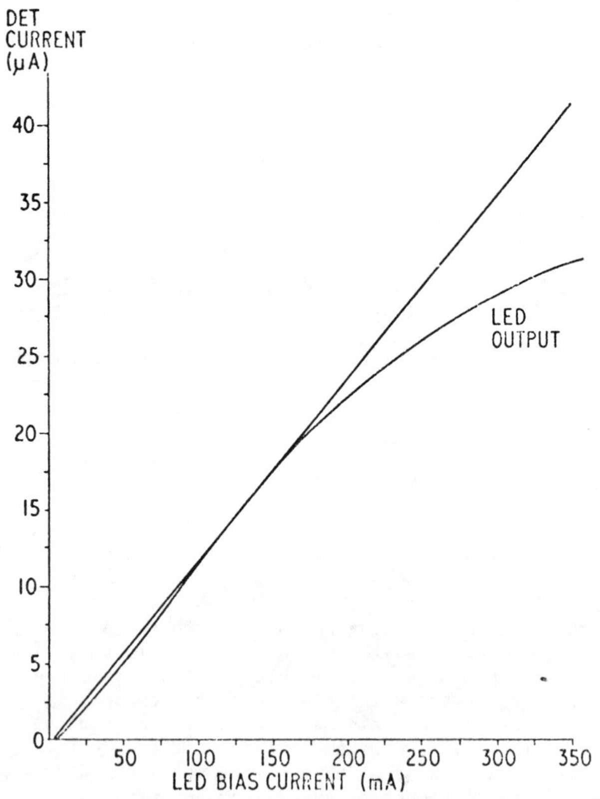

FIG. 3 STATIC CHARACTERISTIC: PLESSEY HR LIGHT EMITTING DIODE

LED Rise Time

Fig. 4 shows the output of a Plessey LED biased by a current of 50 mA and driven by pulses of 100 mA with 1 msec rise time and 70 nsec duration. The output was coupled by a short length multimode fibre to a PIN diode and associated amplifier with 1.5 nsec rise time. The fast initial portions of the rising and falling edges are attributed to the LED and are about 5.nsec. We note that the maximum analogue rate of modulation of this particular combination is about 50 MHz.

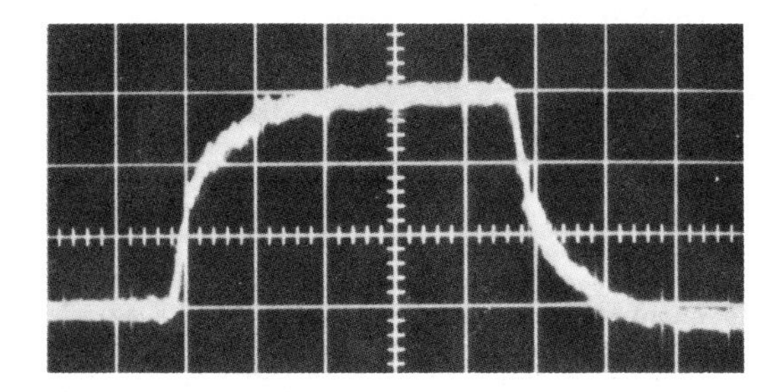

FIG. 4 LIGHT OUTPUT PULSE RESPONSE - PLESSEY H.R.L.E.D.

Detectors

There are a number of devices that can be used as photo detectors including avalanche diodes, photo FETs, photo transistors and PIN diodes[12]. The operation of all of them is similar, but the avalanche diode provides signal gain before encountering high level Johnson- and flicker-noise from the amplifier thus giving an improved signal to thermal noise. Noise from other sources now becomes significant and Hubbard has listed six types of noise[13]. They are in order of magnitude:

1. Thermal noise of the input device of the succeeding amplifier.

2. Quantum noise in the detector.

3. Dark current noise that is subject to avalanche gain.

4. Leakage current noise not subjected to gain.

5. Noise created by incoherent background radiation.

6. Noise due to beats between spectral components within the spectral width of the source.

At present commercially-available avalanche diodes suffer from the disadvantage that they require bias voltages of the order of 150-200V and it is difficult to couple light into them. Work is however going on to develop devices suitable for lower voltage workings.

In the meantime, PIN diodes are available giving a noise performance only a few dBs worse than avalanche diodes for baseband TV as shown in Table 1:

Table 1

	DEVICE PARAMETERS			
	AVALANCHE DIODE		PIN DIODE + AMP	
	NOW	SOON	NOW	SOON
QUANTUM EFFICENCY	60%	90%	70%	90%
RISETIME	0·5 nS	0·5 nS	0·3 nS	0·5 nS
CAPACITY	1·5 pF	<1 pF	2 pF	< 1 pF
SUPPLY VOLTAGE	100 - 200V (WELL REGULATED)	30 - 40V	10 V	10V
RECEIVED POWER	DETECTOR PERFORMANCE INC. MATCHING AMP			
120 Mb/s PCM (20 dB S/N)	6×10^{-8} W	2×10^{-8} W	10^{-6} W	2×10^{-7} W
TV (46 dB S/N)	3×10^{-7} W	2×10^{-7} W	10^{-6} W	3×10^{-7} W
TELEPHONY (70dB S/N) (1 VOICE CH.)	4×10^{-8} W	$2{\cdot}5 \times 10^{-8}$ W	4×10^{-8} W	$2{\cdot}5 \times 10^{-8}$ W

DETECTOR PERFORMANCE LIMITS ±20%

Let us now put the fibre and devices to use by considering a possible scheme for CATV type applications employing analogue transmission techniques.

Performance of an Analogue Optical Link

The diode frequency response, linearity and power output which have been discussed, when considered with the probable fibre and detector performance, were adequate to enable an experimental TV link to be produced, where the colour TV video signal is transmitted at baseband frequencies. Such a link has been designed and built. The performance is given in Table 2 and illustrated in Figs. 5 & 6.

Table 2

Input to link:	TV baseband video input signal level (composite). (bandwidth 10 Hz to 5.5 MHz)	1V
	+	
	FM sound channel (carrier 6 MHz ±25 kHz deviation)	100 mV pk-pk
Output from link:	nominal levels as above (loss - 24 dB)	
	harmonic distortion	< 1%
	frequency response 10 Hz - 7 MHz (3 dB down)	
	Output signal to noise ratio (pk-pk signal/R.M.S. noise) (unweighted)	44 dB

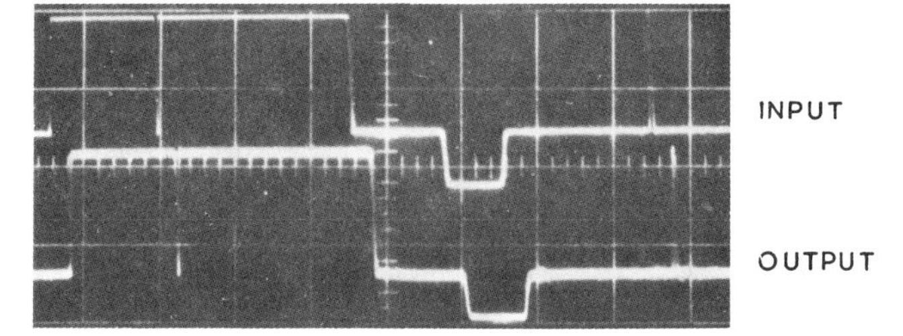

FIG. 5 SIN SQUARED PULSE AND BAR RESPONSE - T PULSE

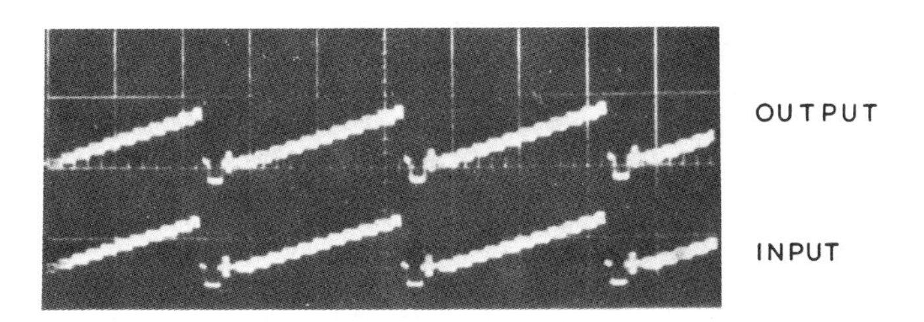

FIG. 6 GREY SCALE RESPONSE

These results are sufficiently encouraging to enable a CATV type network to be considered.

A Suggested Connection Plan

The connection plan to be described aims to provide the means by which a large number of services, many of which are wideband, may be distributed to any subscriber. Modifications to it are easily made to provide an overlay to existing networks, so as to give the video capacity. Analogue transmission is mainly used, but digital data may also be transmitted as and when required. It is expensive and inconvenient to modulate the video signal on to an r.f. carrier as is common at present, but we have demonstrated that this is unnecessary with the experimental link. One fibre for each broadcast TV or videophone terminal is proposed using video baseband transmission. The additional services requiring less bandwidth are then placed by a frequency division multiplex in the band above the TV signals. The proposed use of the available bandwidth is shown in Fig. 7 in which a system bandwidth of 10 MHz is suggested, with four fibres per subscriber. Such a bandwidth appears to be well within the performance limits of both devices, and a low-loss multimode fibre.

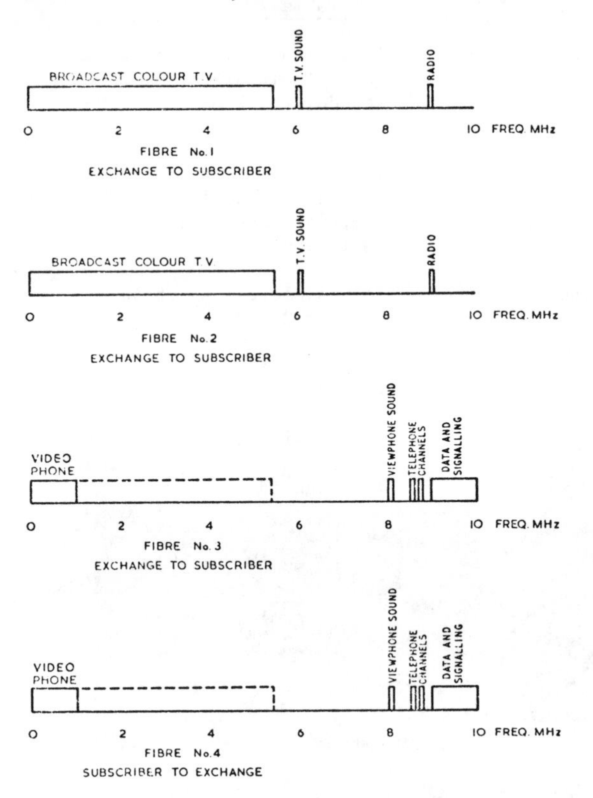

FIG. 7 FIBRE-OPTIC DISTRIBUTION SCHEME FREQUENCY ALLOCATIONS

TV programme choice is made by means of remote switching apparatus. This has the advantage that the services to individual subscribers can be limited to those requested and that Pay TV and similar services requiring special charges are simple to provide. Two fibres have been shown allowing two TV receivers to be used independently. Remote switching could also be used to allow selection of radio programmes, though it may be preferred to provide a limited number of them in an FDM multiplex. Fibres 3 and 4 contribute a full two-way fully switched system which can have access to the telecommunications network. The videophone uses the low end of the spectrum 1 MHz bandwidth being indicated but obviously 5 MHz for a full standard TV signal is available. Telephone, data and all signalling channels are shown as provided in these fibres. Where a limited two-way facility only is required, Fibre 3 can be omitted when one of the TV equipments would be used instead of the videophone receiver. Fibre 4 controls the programme selection for Fibres 1 & 2. Fig. 8 is a block schematic of the proposal showing the remote switches. Because we are proposing an all-optical link, surge and pick-up problems are largely eliminated and solid state switches may be used[14,15].

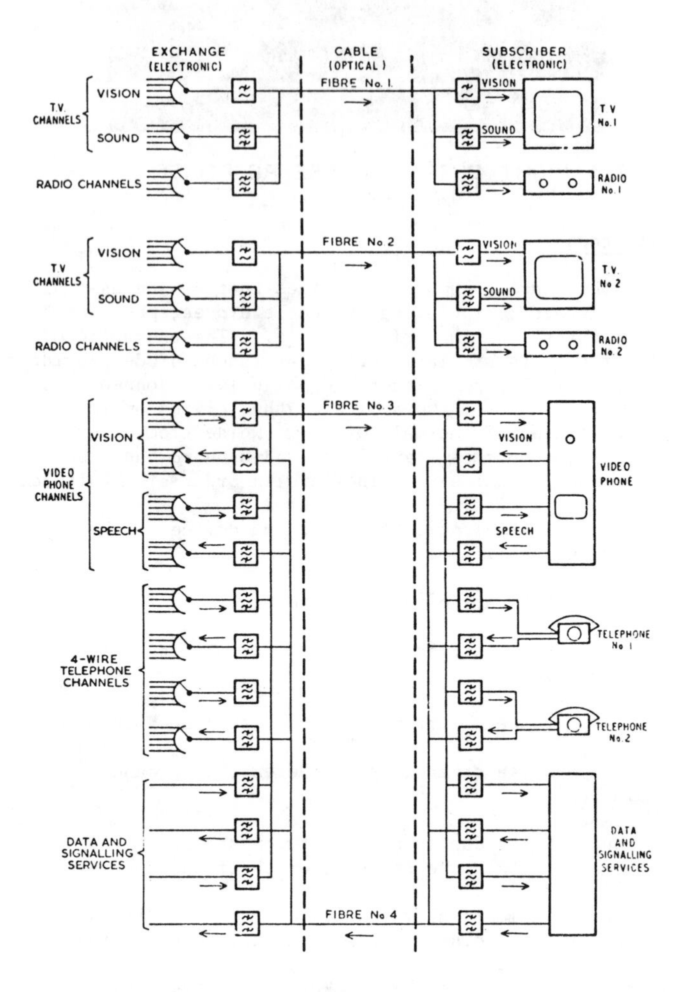

FIG. 8 FIBRE-OPTIC DISTRIBUTION SCHEME BLOCK SCHEMATIC

It will be necessary in a practical system to consider cable joints and the losses of these must be allowed for in the cabling proposal, plus an allowance for subsequent accidental cable damage. Present practice in the U.K. for local area telephone distribution plans utilises special cabinets or distribution points; such cabinets are useful for cable splitting and jointing purposes and a typical connection to a subscriber could be envisaged as shown in Table 3 on page 5.

There will be, however, a need to transmit services even in a local network scheme over distances longer than 2-3 km and hence repeaters will have to be used. Most schemes could be arranged so that main trunks are used for the longer distances, to connect, for example, the common antenna to the main distribution centre and on to subsidiary distribution centres. Once repeaters are required there is much to be said for the use of digital transmission methods, at rates of 70-120 Mbit/s for video signals and 2-8 Mbit/s for the narrow-band services. Wherever a signal tapping point is to be made it is necessary at present to use a repeater, with the signal removed in electrical form.

Table 3

Exchange - 1st jointing point (1 km typical)	8 dB	
Loss at joint	1 dB	
1st point - cabinet (up to 1 km)	8 dB	
Loss at joint	1 dB	
Cabinet - subscriber (100m-300m)	2 dB	
Loss at joint	1 dB	
Sub-total for losses as installed	21 dB	
Allowances for tolerances, subsequent extra joints, repairs to cables etc.	6 dB	
Total system loss	27 dB	
Light coupled into fibre (500 μW)	-3 dBm	
Received light power required	-30 dBm	(to give a 33 dB signal to noise ratio at TV receiver)

Light source :	high radiance LED
Detector :	PIN diode
Fibre bandwidth :	better than 20 MHz/km
Fibre loss :	< 8 dB/km

Why not digital methods right to the subscriber if we are suggesting digital trunks? Full digital transmission would require every subscriber to have a decoder, possibly a demultiplexer and certainly a coder and multiplexer in addition if a full two-way service is to be provided. The fibre and device requirements are also much more severe. The costs of such a scheme are likely to be prohibitive and the final distribution should be analogue.

Problem Areas

Whilst big advances have been made in the laboratories over the past few years, problems remain to be solved before optical fibre systems can take their place in networks carrying live traffic. Some of these problems are:

1. The manufacture of fibres in long lengths to specified optical and mechanical tolerances meeting the system design requirements.

2. The making up of fibres into cables and their installation with no breakage or at least an acceptable level of wastage.

3. Reliable jointing methods have to be developed. Adequate joints have been made in the laboratory with under 0.5 dB loss[16,17]. Techniques for joints in fibre cables in field conditions are being studied.

4. Practical CATV networks may well require concentrators and the use of video solid state switches. This area requires much more study.

5. Improved devices, particularly LEDs are desirable and their life time requires study. LEDs with reasonable lifetimes for practical systems have been made. However, attempts to increase radiance and or efficiency have shown that the resulting life time may be inadequate. This problem of life expectancy is almost certainly related to the problem of GaAs laser degradation[18] though less severe. The main parameters affecting degradation are mechanical strain in the LED chip and the current density.

6. Fibre systems must be able to compete economically with other possible means of providing services such as coaxial cables. At present it is difficult to predict large-scale production costs of low loss fibre cables and this is the key to the system economics. It seems possible that a range of fibre cables will be available just as we have a range of coaxial and pair cables now.

Acknowledgements

The Author wishes to thank his many colleagues at Plessey Telecommunications Research, Taplow and the Allen Clark Research Laboratory, Plessey Caswell for their assistance in the writing of this paper and the Directors of the Plessey Company for permission to publish.

REFERENCES

1. Cher.ˀ A.H. 'Fibre Optics : An Overview of a Future ˀransmission Medium' Wire and ˀable Symposium, Atlantic City, December 1ˀ72.

2. Maurer R.D. ˀlass Fibers for Optical Communications'. Proc. IEEE Vol. 61, No. 4, pp 452-462, April ˀ973.

3. Gloge D. 'Weakly Guiding Fibers' Applied Optics, Vol. 10, No. 10, October 1971 pp 2252-2258.

4. Gloge D. and Marcatili E.A.J., 'Impulse Response of Fibers with Ring-Shaped Parabolic Index Distribution'. B.S.T.J. Vol. 52, No. 7, September 1973, pp 1161-1168.

5. Marcuse D. 'The Impulse Response of an Optical Fiber with Parabolic Index Profile'. B.S.T.J. Vol. 52, No. 7, September 1973, pp 1169-1174.

6. Timmerman C.C. 'Launching Efficiency of Incoherent Light Coupled into Optical Fibres'. AEU, Vol. 27 No. 3, March 1973, pp 150-152.

7. Burrus C.A. and Ulmer E.A. 'Efficient Small-Area GaAs-$Ga_{1-x}Al_x$ As Heterostructure Electroluminescent diodes Coupled to Optical Fibres'. Proc. IEEE (lett) Vol. 59, pp 1263, 1264. August 1971.

8. Burrus C.A. 'Radiance of small-area high-current density electroluminescent diodes'. Proc. IEEE (lett). Vol. 60, pp 231-232. Feb. 1972.

9. Goodfellow R. 'High Radiance, Small Area, GaAs Lamps', Institute of Physics, Semiconductor Device Conference, 1973. Nottingham, U.K.

10. 'Diodes Challenge Light-Link Lasers' Electronics, September 27th 1973, pp 9E 10E.

11. Ettenberg M., Hudson K.C., Lockwood H.F. 'High Radiance Light-Emitting Diodes'. IEEE Journal of Quantum Electronics Vol. QE-9 No. 10, October 1973, pp 987-991.

12. Rokos G.H.S. 'Optical Detection using Photodiodes' Opto-Electronics 5 (1973) pp 351-366.

13. Hubbard W.M. 'Utilization of Optical Frequency Carriers for Low and Moderate Bandwidth Channels'. B.S.T.J. Vol. 52, No. 5, May-June 1973 pp 731-765.

14. Flood J.E. and Deller W.B. 'The p-n-p-n Diode as a Cross-point for Electronic Telephone Exchanges'. Inst. of Elec. Eng. Conference on Electronic Telephone Exchanges', 22-29th Nov. 1960. (Published in Vol. 108, Proc. IEE Part B, 1961, pp 291-303).

15. Bhatt J., Hawkins R., Winslow T. 'The Advent of Semiconductor Crosspoint Networks' Telesis Vol.2, No. 2, pp 2-7, Summer 1971.

16. Someda C.G. 'Simple, Low-loss Joints between Single-Mode Optical Fibers'. B.S.T.J. 52, No. 4, (April 1973) pp 583-596.

17. Gloge D., Smith P.W., Bisbee D.L., and Chinnock E.L. 'Optical Fiber End Preparation for Low Loss Splices'. B.S.T.J. Vol. 52, No. 9, Nov. 1973, pp 1579-1588.

18. De Loach Jr., B.C. 'Reliability of GaAs Injection Lasers', 1973 IEEE/OSA Conference on Laser Engineering and Applications. (Digest of Technical Papers, 12.9, p. 70).

Receiver Design for Digital Fiber Optic Communication Systems, I

By S. D. PERSONICK

(Manuscript received January 15, 1973)

This paper is concerned with a systematic approach to the design of the "linear channel" of a repeater for a digital fiber optic communication system. In particular, it is concerned with how one properly chooses the front-end preamplifier and biasing circuitry for the photodetector; and how the required power to achieve a desired error rate varies with the bit rate, the received optical pulse shape, and the desired baseband-equalized output pulse shape.

It is shown that a proper front-end design incorporates a high-impedance preamplifier which tends to integrate the detector output. This must be followed by proper equalization in the later stages of the linear channel. The baseband signal-to-noise ratio is calculated as a function of the preamplifier parameters. Such a design provides significant reduction in the required optical power and/or required avalanche gain when compared to a design which does not integrate initially.

It is shown that, when the received optical pulses overlap and when the optical channel is behaving linearly in power,[1] *baseband equalization can be used to separate the pulses with a practical but significant increase in required optical power. This required power penalty is calculated as a function of the input and equalized pulse shapes.*

I. INTRODUCTION

The purpose of this paper is to provide insight into a systematic approach to designing the "linear channel" of a repeater for a digital fiber optic communication system.

In particular, we are interested in how one properly chooses the biasing circuitry for the photodetector; and how the required power to achieve a desired error rate varies with the bit rate, the received optical pulse shape, and desired baseband output pulse shape.

Throughout this paper, performance will be measured in terms of signal-to-noise ratios. Efforts to calculate exact error rates and bounds to error rates are difficult to carry out, and, in the past, the results of such efforts have shown little deviation (for practical design purposes) from calculations of error rates using the signal-to-noise ratio (Gaussian approximation) approach. (See Refs. 2 through 5 and Appendix A.)

II. INPUT–OUTPUT RELATIONSHIPS FOR AN AVALANCHE DETECTOR

An avalanche photodiode is the device of interest in fiber applications for converting optical power into current for amplification and equalization, ultimately to produce a baseband voltage for regeneration.

In order to appreciate its performance in practical optical systems, we have to characterize the avalanche photodiode from three points of view: the physical viewpoint, the circuit viewpoint, and the statistical viewpoint.

When we study the device from the physical viewpoint, we ask how does it operate, how do we develop circuit and statistical models of its operation, and what are the limitations of the models.

From the circuit viewpoint, we investigate how to design a piece of equipment in which the device will perform some function.

From the statistical viewpoint, we investigate the probabilistic behavior of the device to allow us to quantify its performance in a circuit.

2.1 *The Physical Viewpoint*

The avalanche photodiode is a semiconductor device which is normally operated in a backbiased manner–producing a region within the device where there is a high field (see Fig. 1). Due to thermal agitation and/or the presence of incident optical power, pairs of holes and electrons can be generated at various points within the diode. These carriers drift toward opposite ends of the device under the influence of the applied field. When a carrier passes through the high-field region, it may gain sufficient energy to generate one or more new pairs of holes and electrons through collision ionization. These new pairs can in turn generate additional pairs by the same mechanism. Carriers accumulate at opposite ends of the diode, thereby reducing the potential across the device until they are removed by the biasing and other circuitry in parallel with the diode (see Fig. 2). The chances that a carrier will generate a new pair when passing through the high-field region depends upon the type of carrier (hole or electron), the material out of which the diode is constructed, and the voltage across the device. To the extent that carriers do not accumulate to significantly modulate the

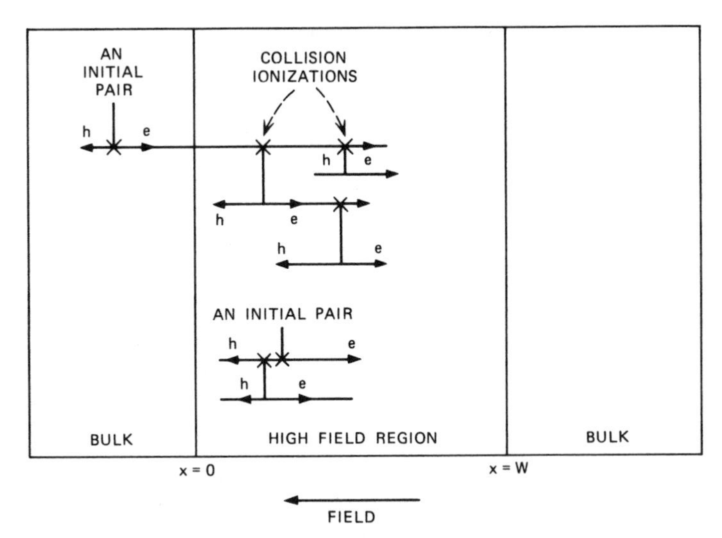

Fig. 1—Avalanche detector.

voltage across the device, it can be assumed that all ionizing collisions are statistically independent. This assumption also requires that the mean time between ionizing collisions be large compared to the time it takes for a carrier in the high-field region to randomize its momentum.

2.2 *The Circuit Viewpoint*

From the discussion above, and of course more detailed investigations,[3,6–8] it has been concluded that a reasonable small-signal model of an avalanche photodiode with a biasing circuit shown in Fig. 2 is the equivalent circuit of Fig. 3. In Fig. 3, C_d is the junction capacitance of the diode† across which voltage accumulates when charges produced within the device separate under the influence of the bias field. The current generator $i(t)$ represents the production of charges (holes and electrons) by optical and thermal generation and collision ionization in the diode high-field region. In order to use the photodiode efficiently,

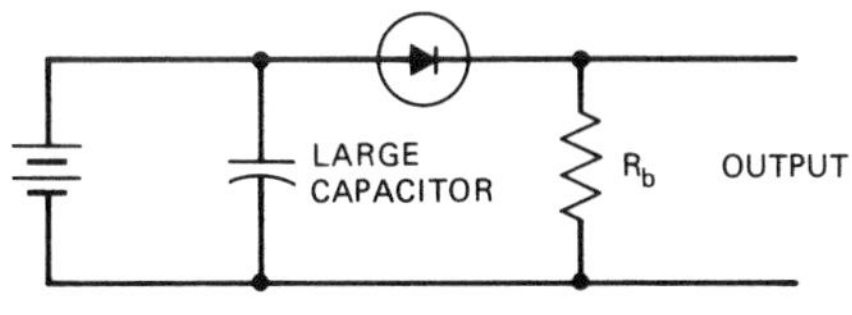

Fig. 2—Detector biasing circuit.

† Not to be confused with the large power supply bypass capacitor of Fig. 2.

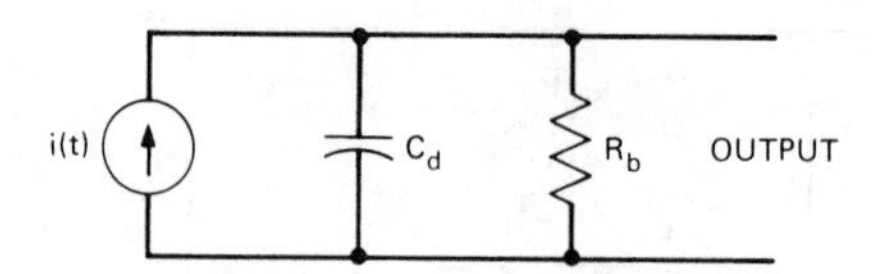

Fig. 3—Equivalent circuit of biased detector.

we must design a circuit which will respond to the current $i(t)$ with as little distortion and added noise as possible.

In order to derive information from the circuit responding to $i(t)$, we must understand the statistical relationship between $i(t)$ (the equivalent current generator) and the incident optical power $p(t)$.

2.3 *The Statistical Viewpoint*

In Fig. 3, the current source $i(t)$ can be considered to be a sequence of impulses corresponding to electrons generated within the photodiode due to optical or thermal excitation or collision ionization. We shall now specify, in a statistical way, how many electrons are produced and when they are produced.

From various physical studies,[3,7,9] it has been concluded that for cases of current interest the electron production process can be modeled as shown in Fig. 4.

Let the optical power falling upon the photon counter be $p(t)$.† In response to this power and due to thermal effects, the photon counter of Fig. 4 produces electrons at average rate $\lambda(t)$ per second where

$$\lambda(t) = [(\eta/\hbar\Omega)p(t)] + \lambda_0, \tag{1}$$

where

$$\eta = \text{photon counter quantum efficiency}$$
$$\hbar\Omega = \text{energy of a photon}$$
$$\lambda_0 = \text{dark current ``counts'' per second.}$$

$\lambda(t)$ is only the average rate at which electrons are produced. In any interval T seconds long, the probability that exactly N counts are produced is given by

$$P[N, (t_0, t_0 + T)] = \frac{\Lambda^N e^{-\Lambda}}{N!},$$

where

$$\Lambda = \int_{t_0}^{t_0+T} \lambda(t)dt. \tag{2}$$

† The reader is cautioned not to confuse $p(t)$, the optical power, with the probability densities (e.g., $P[N, \{t_k\}]$) in this paper.

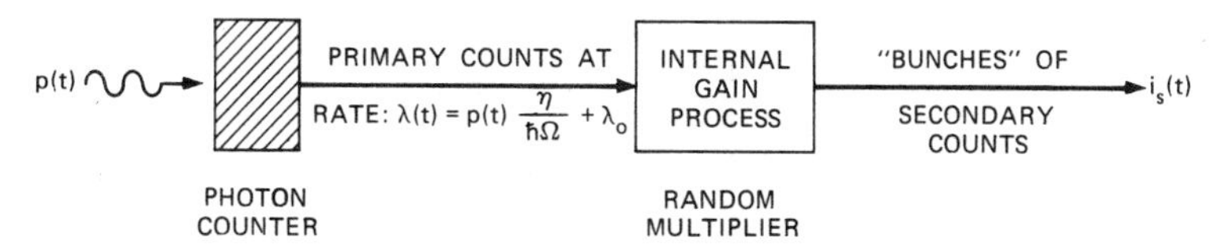

Fig. 4—Model of $i_s(t)$ generation process.

Given $p(t)$, the number of electrons produced in any interval is statistically independent of the number produced in any other disjoint interval.

A process of impulse (electron) production satisfying (2) and the above independent increments condition is said to be a "Poisson impulse process" with arrival rate $\lambda(t)$.[10]

A useful equivalent description of the above process follows.

If T is an interval, the probability that exactly N electrons will be produced at the (approximate) times $t_1 \pm \frac{1}{2}\Delta$, $t_2 \pm \frac{1}{2}\Delta$, $\cdots$ $t_N \pm \frac{1}{2}\Delta$ where the widths Δ are very small is

$$P[N, \{t_k\}] = \{e^{-\Lambda} \prod_1^N [\lambda(t_k)\Delta]/N!\} + o(\Delta), \tag{3}$$

where Λ is defined in (2) and $o(\Delta)$ is a term such that

$$\lim_{\Delta \to 0} \frac{o(\Delta)}{\Delta} = 0.$$

It is *important* to note that in (3) the times $\{t_k\}$ are *not* in order, that is, in (3) it is *not* necessarily true that $t_1 < t_2$, etc.

Each of the "primary" impulses (electrons) produced by the photon counter enters a random multiplier where, corresponding to collision ionization, it is replaced by g contiguous "secondary" impulses (electrons). The number g is governed by the statistics of the internal gain mechanism of the photodiode. Each primary impulse (electron) is "multiplied" in this manner by a value g which is statistically independent of the value g assigned to other primaries.

Thus the current leaving the photodiode consists of "bunches" of electrons, the number of electrons in the bunch being a random quantity having statistics to be described below. For applications of interest here, it will be assumed that all electrons in a bunch exit the photodiode at the time when the primary is produced. This implies that the duration of the photodiode response to a single primary hole-electron pair is very short compared to the response times of circuitry to be used with the photodiode.

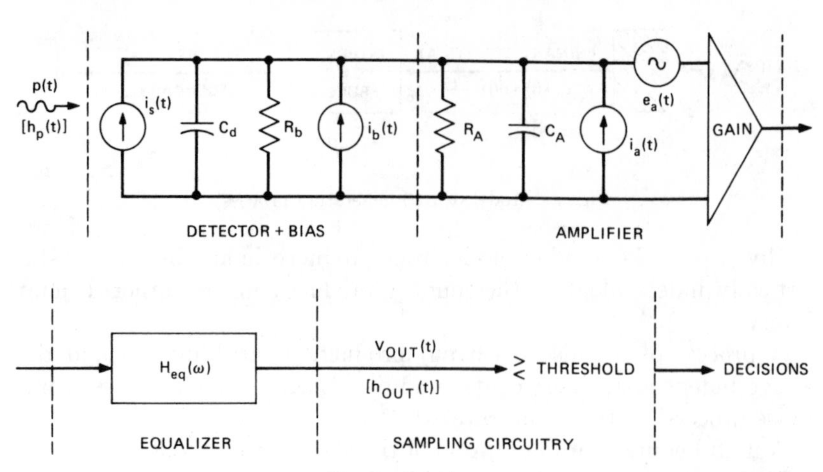

Fig. 5—Receiver.

Different avalanche photodiodes have different statistics governing the number of electrons in a bunch, i.e., the gain. For applications below, we will only need to know the mean gain $\langle g \rangle$ and the mean square gain $\langle g^2 \rangle$. For a large class of avalanche photodiodes of interest, it has been found that[3,7]

$$\langle g^2 \rangle \cong \langle g \rangle^{2+x}, \tag{4}$$

where $\langle g \rangle$ is determined by the applied bias voltage and x, a number usually between 0 and 1, depends upon the materials out of which the diode is constructed. For germanium photodiodes, $x \cong 1$; for well-designed silicon photodiodes, $x \approx 0.5$.

III. AN OPTICAL RECEIVER

Figure 5 shows a fairly typical receiver, in schematic form, consisting of an avalanche photodiode, an amplifier, and an equalizer.

The amplifier is modeled as an ideal high-gain infinite-impedance amplifier with an equivalent shunt capacitance and resistance at the input and with two noise sources referred to the input. For the purposes of this paper, the noise sources will be assumed to be white, Gaussian, and uncorrelated. Extensions to other amplifier models will be straightforward when the techniques of this paper are understood.†

It is assumed that the amplifier gain is sufficiently high so that noises introduced by the equalizer are negligible.

† With this model, the noise sources of the amplifier do not change when the input and output load circuitry changes.

The power falling upon the detector will be assumed to be of the form of a digital pulse stream

$$p(t) = \sum_{-\infty}^{\infty} b_k h_p(t - kT), \tag{5}$$

where b_k takes on one of two values for each integer value of k, T = the pulse spacing, $h_p(t - kT)$ = pulse shape and is positive for all t. We shall assume $\int_{-\infty}^{\infty} h_p(t - kT)dt = 1$, therefore b_k is the energy in pulse k. The assumption that the received power will be in the form (5) appears reasonable for intensity modulation and fiber systems of interest.[1]

From (1) we have the average detector output current $\langle i_s(t)\rangle$ given by

$$\langle i_s(t)\rangle = \frac{\eta\langle g\rangle e p(t)}{\hbar\Omega} + \langle g\rangle e\lambda_0,$$

where

$\langle g\rangle$ = average detector internal gain

e = electron charge

λ_0 = dark current electrons per second

$\frac{\eta}{\hbar\Omega} p(t)$ = average optical primary electrons per second.

Therefore, the average voltage (neglecting dc components) at the equalizer output is

$$\langle v_{\text{out}}(t)\rangle = \frac{A\eta\langle g\rangle e p(t)}{\hbar\Omega} * h_{\text{fe}}(t) * h_{\text{eq}}(t), \tag{6}$$

where "*" indicates convolution and A is an arbitrary constant.

$$h_{\text{fe}}(t) = F\left\{\frac{1}{\frac{1}{R_T} + j\omega(C_d + C_A)}\right\}$$

= amplifier input circuit current impulse response,

$$R_T = \left[\frac{1}{R_b} + \frac{1}{R_A}\right]^{-1} = \text{total detector parallel load resistance},$$

and $h_{\text{eq}}(t)$ = equalizer impulse response.

Clearly, $\langle v_{\text{out}}(t)\rangle$ is of the form

$$\langle v_{\text{out}}(t)\rangle = \sum_{-\infty}^{\infty} b_k h_{\text{out}}(t - kT) \tag{7}$$

and $v_{out}(t)$ is of the form

$$v_{out}(t) = \sum_{-\infty}^{\infty} b_k h_{out}(t - kT) + n(t),$$

where $n(t)$ represents *deviations* (or noises) of $v_{out}(t)$ from its average.

The fundamental task ahead is to pick R_b (the bias circuit resistor) and $h_{eq}(t)$ so that a system which samples $v_{out}(t)$ at the times $\{kT\}$ can make decisions as to which value b_k has assumed (by comparing the sample to a threshold) with minimum chance of error.

IV. CALCULATING SIGNAL–TO–NOISE RATIO IN TERMS OF THE EQUALIZED PULSE SHAPE

Having defined the receiver and its statistics in the above sections, we can now calculate the variance of $n(t)$, the noise portion of the output $v_{out}(t)$ of the system of Fig. 5, defined as follows:

$$N = \langle (n(t))^2 \rangle = \langle v_{out}^2(t) \rangle - \langle v_{out}(t) \rangle^2. \tag{8}$$

The noise, N, of (8) above depends upon the coefficients $\{b_k\}$ defined in (5) and upon the time t.

We shall first of all restrict consideration to the set of times $t = \{kT\}$ when a decision as to the values $\{b_k\}$ will be made by sampling $v_{out}(t)$. We shall next assume that the equalized pulses satisfy

$$\begin{aligned} h_{out}(0) &= 1 \\ h_{out}(t) &= 0 \qquad \text{for} \qquad t = kT, \qquad k \neq 0. \end{aligned} \tag{9}$$

That is, we shall assume that the equalized pulse stream has no intersymbol interference at the sampling times kT.† Therefore,

$$v_{out}(kT) = b_k + n(kT). \tag{10}$$

In eq. (10) the noise, $n(t)$, still depends upon all the $\{b_k\}$ and the time t. This is a property which distinguishes fiber optic systems from many other systems where the noise is signal-independent and stationary (not time-dependent). Consider, without loss of generality, the output, $v(t)$, at $t = 0$. We define the worst-case noise, $NW(b_0)$, for each of the two possible values of b_0 as follows:

$$NW(b_0) = \max_{\{b_k\}, k \neq 0} [\langle v_{out}^2(0) \rangle - \langle v_{out}(0) \rangle^2], \tag{11}$$

where in (11) the maximization is over all possible sets $\{b_k\}$ for $k \neq 0$, and where b_0 can take on either of two values as previously stated. The

† The limitations imposed by this assumption are discussed in Section VII.

quantity $NW(b_0)$ shows, for the two possible values of b_0, what the noise for the worst combination of the other symbols is.

We shall next calculate $\langle v_{\text{out}}^2(t)\rangle - \langle v_{\text{out}}(t)\rangle^2$ as a function of the set $\{b_k\}$.

Examine Fig. 5. We shall define the two-sided spectral density of the amplifier-current noise source $i_a(t)$ as S_I and the two-sided spectral height of the amplifier-voltage noise source $e_a(t)$ as S_E. The two-sided spectral density of the Johnson-current noise source $i_b(t)$ associated with R_b is $2\text{k}\theta/R_b$, where k is Boltzmann's constant and θ is the absolute temperature.

We can write the output noise as follows:

$$v_{\text{out}}(t) - \langle v_{\text{out}}(t)\rangle = n_S(t) + n_R(t) + n_I(t) + n_E(t), \tag{12}$$

where

$n_S(t)$ is the output noise due to the random multiplied Poisson process nature of the current $i_s(t)$ produced by the detector,

$n_R(t)$ is the output noise due to the Johnson noise current source of the resistor R_b,

$n_I(t)$ is the output noise due to the amplifier input current noise source $i_a(t)$, and

$n_E(t)$ is the output noise due to the amplifier input voltage noise source $e_a(t)$.

We have

$$\begin{aligned}
&\langle v_{\text{out}}^2(t)\rangle - \langle v_{\text{out}}(t)\rangle^2 \\
&\quad = \langle (v_{\text{out}}(t) - \langle v_{\text{out}}(t)\rangle)^2\rangle \\
&\quad = \langle n_S^2(t)\rangle + \langle n_R^2(t)\rangle + \langle n_I^2(t)\rangle + \langle n_E^2(t)\rangle \\
&\quad = \langle n_S^2(t)\rangle + (2\text{k}\theta/R_b)\frac{1}{2\pi}\int_{-\infty}^{\infty}\left|H_{\text{eq}}(\omega)\frac{1}{\dfrac{1}{R_b}+\dfrac{1}{R_A}+j\omega(C_d+C_A)}\right|^2 d\omega \\
&\qquad\quad + (S_I)\frac{1}{2\pi}\int_{-\infty}^{\infty}\left|H_{\text{eq}}(\omega)\frac{1}{\dfrac{1}{R_b}+\dfrac{1}{R_A}+j\omega(C_d+C_A)}\right|^2 d\omega \\
&\qquad\qquad + (S_E)\frac{1}{2\pi}\int_{-\infty}^{\infty}|H_{\text{eq}}(\omega)|^2 d\omega.
\end{aligned} \tag{13}$$

In (13), the last three terms were evaluated using the well-known formula for the average-squared output of a filter driven by white noise. We must now calculate the "shot noise" term $\langle n_S^2(t)\rangle$.

Recall that $i_s(t)$ consists of impulses of random charge corresponding to "bunches" of electrons with a random number g per bunch, this number being independent from bunch to bunch.

Consider a finite interval of duration L. Let g_k be the number of electrons in bunch k in the interval; where the bunches are labeled *not* in order of time but at random. Let t_k be the arrival time of bunch k. Let $h_I(t)$ be the response of the RC circuit, amplifier, equalizer combination to a current impulse from $i_s(t)$. Then the output $v_{\text{out}}(t)$ *just due to the current* $i_s(t)$ *in the interval* L is

$$v^L_{\text{out}}(t) = \sum_1^N eg_k h_I(t - t_k), \tag{14}$$

where N is the number of bunches.

Recall that the probability density of N bunches at the times $\{t_k\}$ is

$$p[N, \{t_k\}] = \frac{e^{-\Lambda} \prod_1^N \lambda(t_k)}{N!}, \tag{15}$$

where

$$\lambda(t) = p(t) \frac{\eta}{\hbar\Omega} + \lambda_0.$$

Thus combining (14) and (15) and leaving out some tedious algebra we obtain[10]

$$\langle v^L_{\text{out}}(t) \rangle = \int_{\text{interval } L} e\langle g \rangle (p(t')\eta/\hbar\Omega + \lambda_0) h_I(t - t')dt'. \tag{16}$$

In a similar manner, we obtain

$$\langle (v^L_{\text{out}}(t))^2 \rangle - \langle v^L_{\text{out}}(t) \rangle^2 = \int_{\text{interval } L} e^2 \langle g^2 \rangle \left(p(t') \frac{\eta}{\hbar\Omega} + \lambda_0 \right) h_I^2(t - t')dt,$$

where $\langle g \rangle$ is the mean internal gain of the detector and $\langle g^2 \rangle$ is the mean-squared internal gain.

We therefore obtain, letting $L \to \infty$, the result

$$\begin{aligned} \langle n_S^2(t) \rangle &= \lim_{L \to \infty} [\langle [v^L_{\text{out}}(t)]^2 \rangle - \langle v^L_{\text{out}}(t) \rangle^2] \\ &= \int_{-\infty}^{\infty} e^2 \langle g^2 \rangle \left\{ [\sum b_k h_p(t' - kT)] \frac{\eta}{\hbar\Omega} + \lambda_0 \right\} h_I^2(t - t')dt'. \end{aligned} \tag{17}$$

Further,

$$H_I(\omega) = F\{h_I(t - t')\} = H_{\text{eq}}(\omega) \frac{1}{\dfrac{1}{R_b} + \dfrac{1}{R_A} + j\omega(C_d + C_A)}. \tag{18}$$

Thus we have the remaining term in (13) in terms of the input optical pulse, the equalizer response, and the RC circuit at the amplifier input.

Converting everything to the frequency domain and recalling that we have normalized the equalized output pulse $h_{\text{out}}(t)$ to unity at $t = 0$, we obtain

$$NW(b_0) \underset{\{b_k, k\neq 0\}}{\overset{=}{\max}} \Bigg[\left(\frac{1}{2\pi}\right)^2 \int_{-\infty}^{\infty} \frac{\langle g^2 \rangle}{\langle g \rangle^2} \frac{\hbar\Omega}{\eta} H_p(\omega) \left(\sum_{-\infty}^{\infty} b_k e^{j\omega kT} \right)$$

$$\times \left(\frac{H_{\text{out}}(\omega)}{H_p(\omega)} * \frac{H_{\text{out}}(\omega)}{H_p(\omega)} \right) d\omega$$

$$+ \frac{(\hbar\Omega/\eta)^2}{2\pi\langle g\rangle^2 e^2} \left(\frac{2\text{k}\theta}{R_b} + S_I + e^2\langle g^2\rangle\lambda_0 \right) \int_{-\infty}^{\infty} \left| \frac{H_{\text{out}}(\omega)}{H_p(\omega)} \right|^2 d\omega$$

$$+ \frac{(\hbar\Omega/\eta)^2}{2\pi\langle g\rangle^2 e^2} S_E \int_{-\infty}^{\infty} \left| \frac{H_{\text{out}}(\omega) \left(\frac{1}{R_b} + \frac{1}{R_A} + j\omega(C_d + C_A) \right)}{H_p(\omega)} \right|^2 d\omega \Bigg], \quad (19)$$

where

$H_p(\omega) = F\{h_p(t)\}$ = input power pulse transform,

$H_{\text{out}}(\omega) = F\{h_{\text{out}}(t)\}$ = output pulse transform,

"$*$" = convolution,

b_0 = coefficient multiplying zeroth input pulse,

and

$$\frac{1}{2\pi} \int_{-\infty}^{\infty} H_{\text{out}}(\omega) d\omega = 1. \quad (20)$$

In principle, we wish to minimize $NW(b_0)$ by choosing R_b and $H_{\text{eq}}(\omega)$ for the worst-case combination of symbols $\{b_k\}$, subject to the zero intersymbol interference condition on the ouput pulse stream $v_{\text{out}}(t)$ [recognizing that we have normalized $h_{\text{out}}(t)$ and $H_{\text{out}}(\omega)$ as given in (9) and (20) above].

4.1 *Comments*

(*i*) One observation, which follows regardless of the choice of $H_{\text{out}}(\omega)$, is that the noise is always made smaller when R_b is increased. Therefore, subject to practical constraints and for a fixed amplifier and a fixed desired output pulse shape (which is determined by the equalizer and R_b), it is always best to make R_b, the bias circuit resistor, as large as possible.

(*ii*) It is also clear, from (17) and the fact that the input pulse $h_p(t)$ is positive for all t, that the worst-case noise occurs when all the b_k (except b_0) assume the larger of the two possible values. Recall that we are interested in the noise for both values of b_0.

(*iii*) Furthermore, for a given S_E and S_I and a given output pulse shape, it is desirable that the amplifier input resistance be as large as possible and that the amplifier shunt capacitance be as small as possible.

(*iv*) It is desirable that the diode shunt capacitance be as small as possible.

V. CHOOSING THE EQUALIZED PULSE SHAPE

In principle, using (19) and given $H_p(\omega)$, $\langle g\rangle$, $\langle g^2\rangle$, S_I, S_E, R_b, R_A, C_d, and C_A one can find the equalized pulse shape $H_{\text{out}}(\omega)$ for each value of b_0 that minimizes the worst-case noise.

In practice, other considerations in addition to the noise are also of interest. In particular, it is important not only that the intersymbol interference be low at the nominal decision times kT, but that it be sufficiently small at times offset from $\{kT\}$ to allow for timing errors in the sampling process.

Therefore, rather than seeking the equalized pulse shape that minimizes the noise, we shall consider various equalized pulse shapes to see how the noise trades off against eye width.

Before proceeding, it is helpful to perform some normalizations upon (19) to reduce the number of parameters.

Make the following definitions:

$$R_T = \left(\frac{1}{R_b} + \frac{1}{R_A}\right)^{-1} = \text{total detector parallel load resistance,} \tag{21}$$

$$C_T = C_d + C_A = \text{total detector parallel load capacitance,}$$

$$b_{\max} = \text{larger value of } b_k,\ b_{\min} = \text{smaller value of } b_k,$$

$$H_p'(\omega) = H_p\left(\frac{2\pi\omega}{T}\right),$$

$$H_{\text{out}}'(\omega) = \frac{1}{T} H_{\text{out}}\left(\frac{2\pi\omega}{T}\right).$$

In this normalization, the functions $H_p'(\omega)$ and $H_{\text{out}}'(\omega)$ depend only upon the *shapes* of $H_p(\omega)$ and $H_{\text{out}}(\omega)$, not upon the time slot width T. The previous normalizing conditions on $H_p(\omega)$ and $H_{\text{out}}(\omega)$ imply conditions on $H_p'(\omega)$ and $H_{\text{out}}'(\omega)$

$$H_p(0) = 1 \Rightarrow H_p'(0) = 1 \tag{22}$$

which implies

$$\int_{-\infty}^{\infty} h_p(t)dt = 1.$$

Also,

$$h_{\text{out}}(0) = 1 \Rightarrow \frac{1}{2\pi}\int_{-\infty}^{\infty} H_{\text{out}}(\omega)d\omega = 1 \Rightarrow \int_{-\infty}^{\infty} H'_{\text{out}}(f)df = 1.$$

With the above normalizations, (19) becomes

$$NW(b_0) = \left(\frac{\hbar\Omega}{\eta}\right)^2 \left\{ \frac{\langle g^2\rangle}{\langle g\rangle^2}\frac{\eta}{\hbar\Omega}\underbrace{[b_0 I_1 + b_{\max}[\Sigma_1 - I_1]]}_{\text{SHOT NOISES}} + \frac{T}{(\langle g\rangle e)^2}\left[S_I + \frac{2k\theta}{R_b} + \underbrace{\langle g^2\rangle e^2\lambda_d}_{\text{SHOT NOISE}} + \frac{S_E}{R_T^2}\right] I_2 + \frac{(2\pi C_T)^2 S_E I_3}{T(\langle g\rangle e)^2}\right\}, \tag{23}$$

(In the printed equation, arrows label the two terms $b_0 I_1$ and $b_{\max}[\Sigma_1 - I_1]$ as SHOT NOISES, $\langle g^2\rangle e^2\lambda_d$ as SHOT NOISE, and $\frac{2k\theta}{R_b}$, $\frac{S_E}{R_T^2}$ and the I_3 term as THERMAL NOISES.)

where

$$I_1 = \int_{-\infty}^{\infty} H_p'(f)\left[\frac{H'_{\text{out}}(f)}{H_p'(f)} * \frac{H'_{\text{out}}(f)}{H_p'(f)}\right]df$$

$$\Sigma_1 = \sum_{k=-\infty}^{\infty} H_p'(k)\left[\frac{H'_{\text{out}}(k)}{H_p'(k)} * \frac{H'_{\text{out}}(k)}{H_p'(k)}\right]$$

$$I_2 = \int_{-\infty}^{\infty}\left|\frac{H'_{\text{out}}(f)}{H_p'(f)}\right|^2 df$$

$$I_3 = \int_{-\infty}^{\infty}\left|\frac{H'_{\text{out}}(f)}{H_p'(f)}\right|^2 f^2 df.$$

In (23), the first shot-noise term is due to the pulse in the time slot under decision, the second term being shot noises from the other pulses which are assumed to be all "on." From this normalized form of (19), we see that for a fixed input pulse *shape* and a fixed output pulse *shape* and with fixed R_b, R_A, C_T, S_E, and S_I, the noise decreases as the bit rate, $1/T$, increases (a consequence of the square-law detection) until the term involving I_3 dominates. After that, the noise increases with increasing bit rate (due to the shunt capacitance C_T).

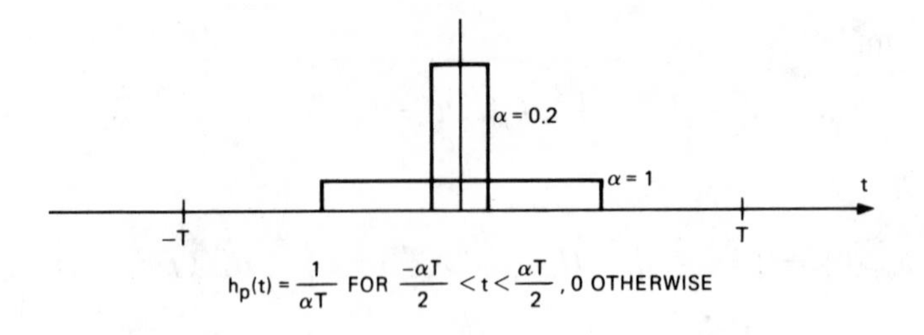

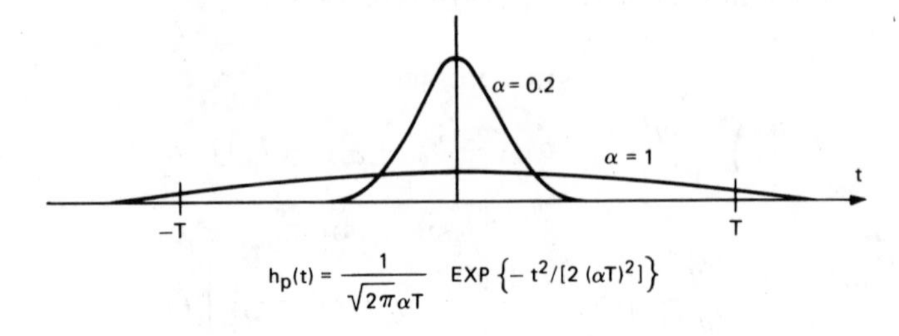

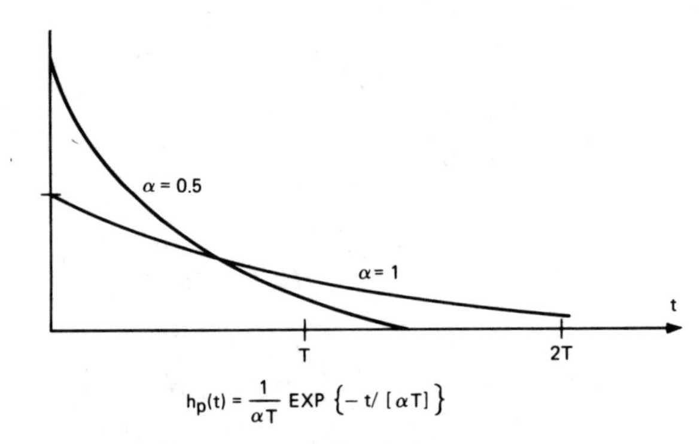

Fig. 6a—Input pulse families.

Example of Normalization:

Suppose the input optical pulse is a rectangular pulse of unit area having width equal to one-half a time slot T; then

$$
\begin{aligned}
H_p(\omega) &= \int_{-\frac{1}{4}T}^{\frac{1}{4}T} \frac{2}{T} e^{i\omega t} dt \\
&= \frac{1}{i\omega}\left(\frac{2}{T}\right)\left(e^{i\omega t/4} - e^{-i\omega T/4}\right) \\
&= \sin\frac{(\omega T/4)}{\omega T/4}\cdot \qquad (24)
\end{aligned}
$$

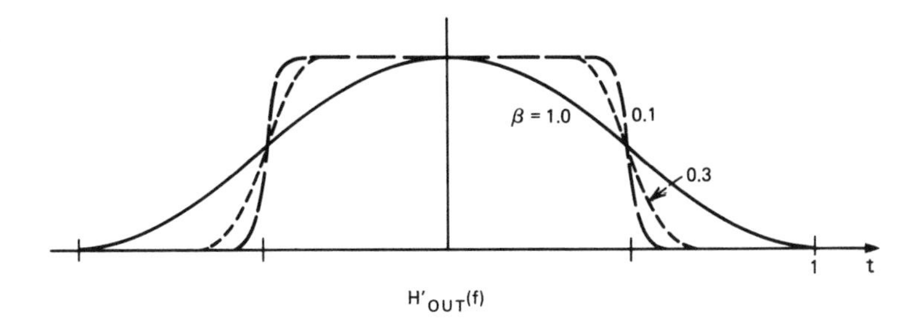

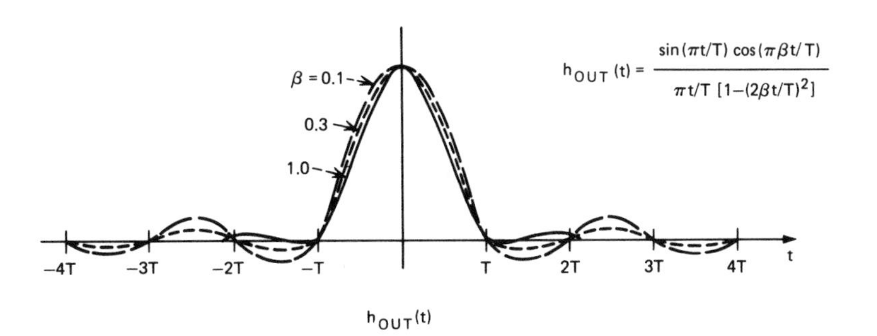

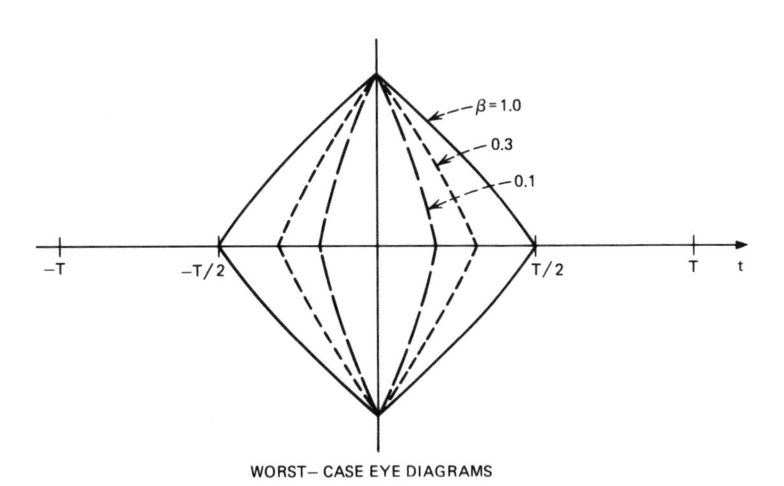

Fig. 6b—Frequency domain, time domain, and eye diagram representations of raised cosine family.

Therefore,

$$H_p'(f) = H_p\left(\frac{2\pi f}{T}\right) = \frac{\sin(\pi f/2)}{\pi f/2}.$$

As expected, the normalized pulse spectrum $H_p'(f)$ is independent of the time slot width T and merely reflects the fact that the pulse $h_p(t)$ is a rectangular pulse with width equal to half a time slot.

In order to obtain the noise for various input and output (equalized) pulse shapes, one needs to calculate the three integrals I_1, I_2, and I_3 and the sum $\sum_1$.

Consider the following three families of input pulse shapes (see Fig. 6a) and single family of output pulse shapes (see Fig. 6b).

(*i*) Rectangular input pulses:

$$h_p(t) = \frac{1}{\alpha T}, \qquad -\frac{\alpha T}{2} < t < \frac{\alpha T}{2}, \qquad 0 \text{ otherwise} \tag{25}$$

$$H_p'(f) = \frac{\sin(\alpha\pi f)}{\alpha\pi f}.$$

(*ii*) Gaussian input pulses:

$$h_p(t) = \frac{1}{\sqrt{2\pi}\alpha T} e^{-[t^2/2(\alpha T)^2]}$$

$$H_p'(f) = e^{-(2\pi\alpha f)^2/2}.$$

(*iii*) Exponential input pulses:

$$h_p(t) = \frac{1}{\alpha T} e^{-t/\alpha T}$$

$$H_p'(f) = \frac{1}{1 + j2\pi\alpha f}.$$

(*iv*) "Raised cosine" output pulses:

$$h_{\text{out}}(t) = \left[\sin\left(\frac{\pi t}{T}\right)\cos\left(\frac{\pi\beta t}{T}\right)\right]\left[\frac{\pi t}{T}\left(1 - \left(\frac{2\beta t}{T}\right)^2\right)\right]^{-1}$$

$$H_{\text{out}}'(f) = 1, \qquad \text{for} \qquad 0 < |f| < \frac{(1-\beta)}{2}$$

$$= \frac{1}{2}\left[1 - \sin\left(\frac{\pi f}{\beta} - \frac{\pi}{2\beta}\right)\right], \qquad \text{for } \frac{1-\beta}{2} < |f| < \frac{1+\beta}{2}$$

$$= 0 \text{ otherwise.}$$

(Time, frequency, and eye diagram representations of the raised cosine family are shown as a function of β in Fig. 6b.[11])

In Figs. 7 through 18 calculations of I_1, I_2, I_3, and I_4 are given graphically for each input pulse family as a function of α and β.

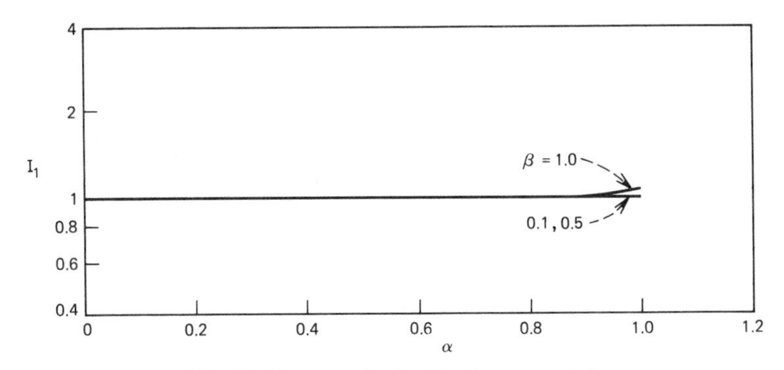

Fig. 7—Rectangular family I_1 vs α and β.

5.1 *Comments on the Numerical Results*

For the rectangular input pulses with widths between 0.1 and 1 time slot, I_1, $\sum_1$, I_2, and I_3 vary very little. Thus, if one expects to receive rectangular optical pulses which are fixed in energy, the required energy per pulse is insensitive to the pulse width for widths up to 1 time slot.

The curves for Gaussian-shaped input pulses imply very strong sensitivity of required energy per pulse to pulse width. This is a consequence of the rapid falloff of the spectrum of a Gaussian pulse with frequency. It is suspected that, although for certain fiber systems the received pulses may appear approximately Gaussian in the time domain, the frequency spectrum will not suffer such a rapid falloff. The results for the exponential-shaped input pulses seem much more realistic.

For exponential-shaped optical pulses we notice, from Figs. 15 and 16, that the shot noise coefficients I_1 and $\sum_1$ are sensitive to the optical

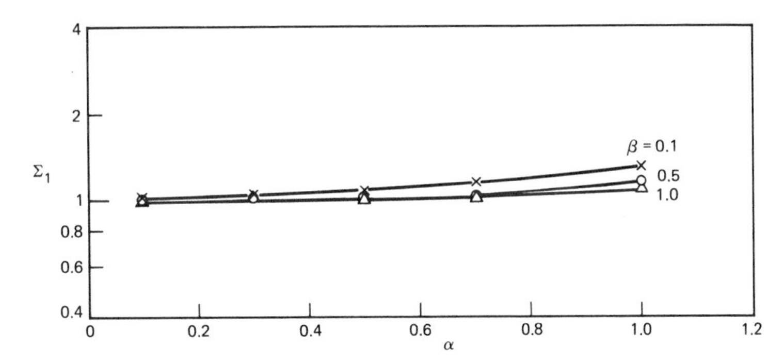

Fig. 8—Rectangular family Σ_1 vs α and β.

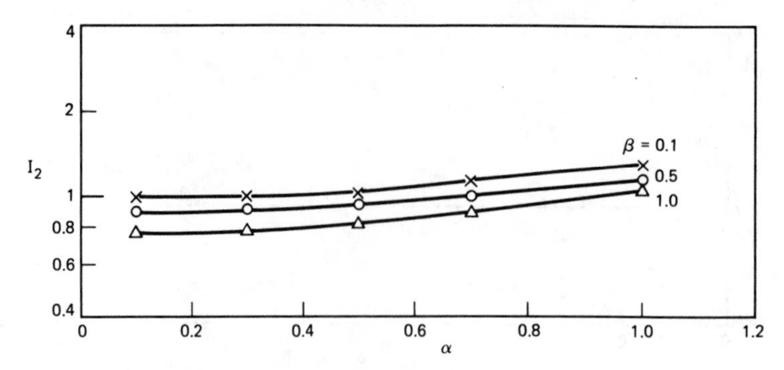

Fig. 9—Rectangular family I_2 vs α and β.

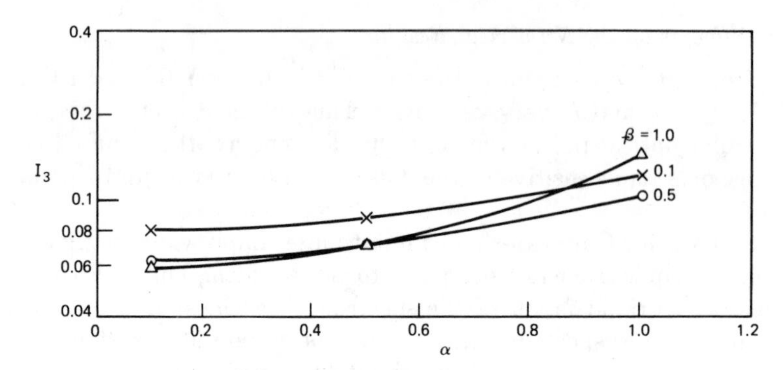

Fig. 10—Rectangular family I_3 vs α and β.

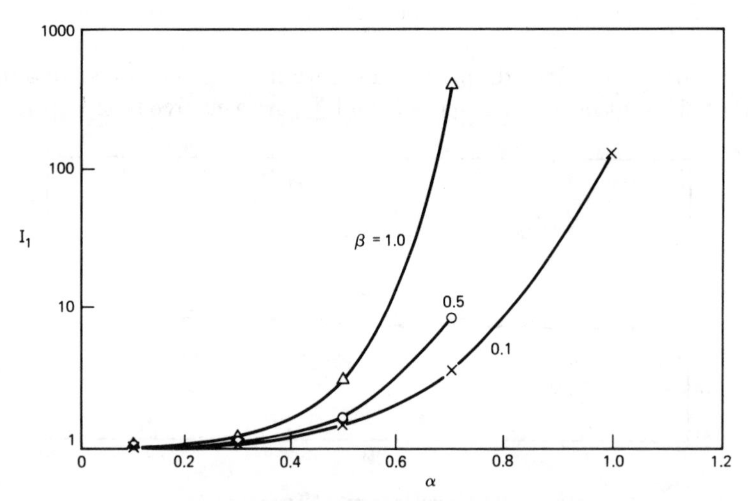

Fig. 11—Gaussian family I_1 vs α and β.

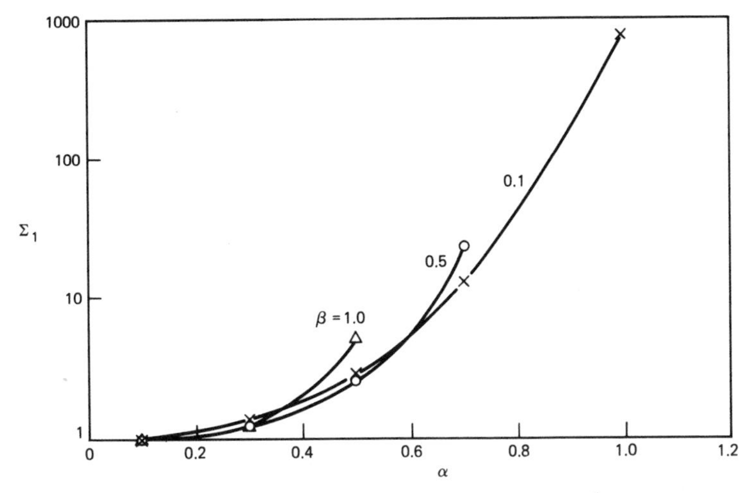

Fig. 12—Gaussian family Σ_1 vs α and β.

pulse width, but that these sensitivities imply a practically useful tradeoff in required optical power vs allowable bit rate. That is, one might take a certain power penalty to allow equalization which can substantially increase the usable bit rate on a channel having a fixed optical output pulse width. The sensitivity of I_2 and I_3 to the optical pulse width is similar to that of $\sum_1$ and less significant because in-

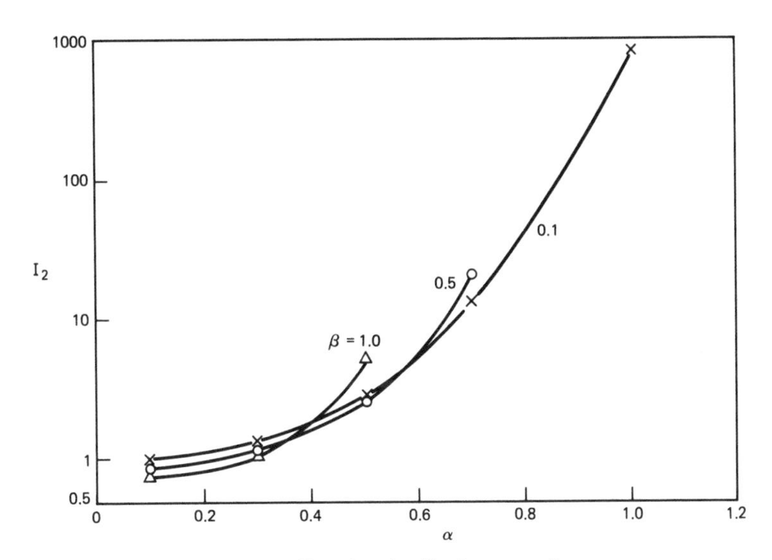

Fig. 13—Gaussian family I_2 vs α and β.

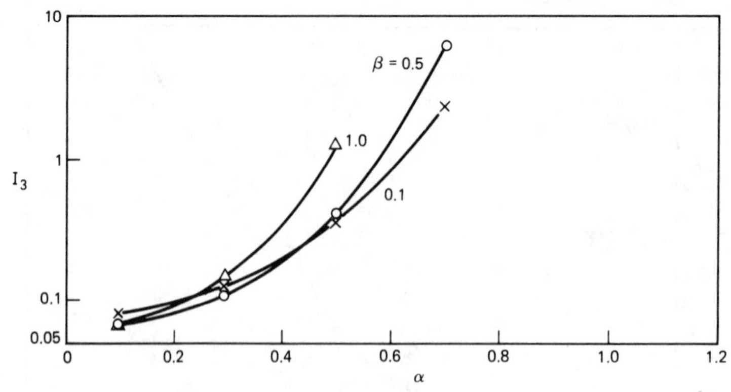

Fig. 14—Gaussian family I_3 vs α and β.

creases in the thermal noises of the receiver are for the most part compensated for by adjustment of the avalanche gain, with only a small penalty in excess shot noise. The above statements will be made quantitative in Section VI.

VI. OBTAINING THE RELATIONSHIPS FOR FIXED ERROR RATE BETWEEN THE REQUIRED ENERGY PER PULSE, OPTIMAL AVALANCHE GAIN, AND OTHER PARAMETERS

Suppose that in (23) all parameters are fixed except $\langle g \rangle$, $\langle g^2 \rangle$, $b_{\min}$, and $b_{\max}$.

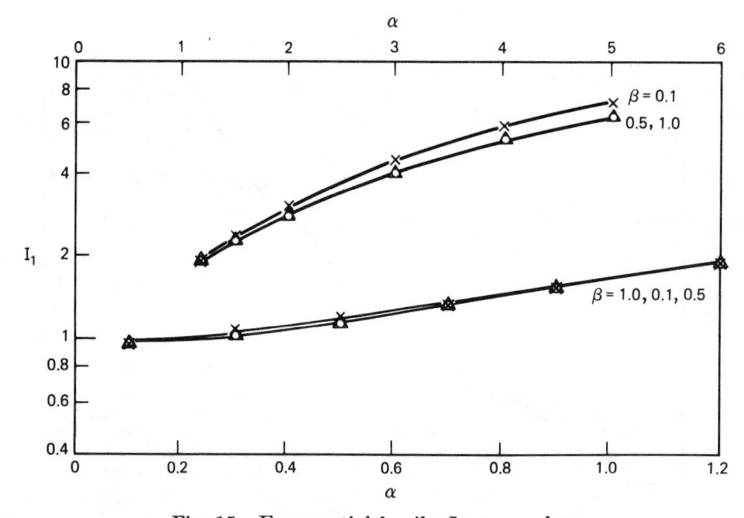

Fig. 15—Exponential family I_1 vs α and β.

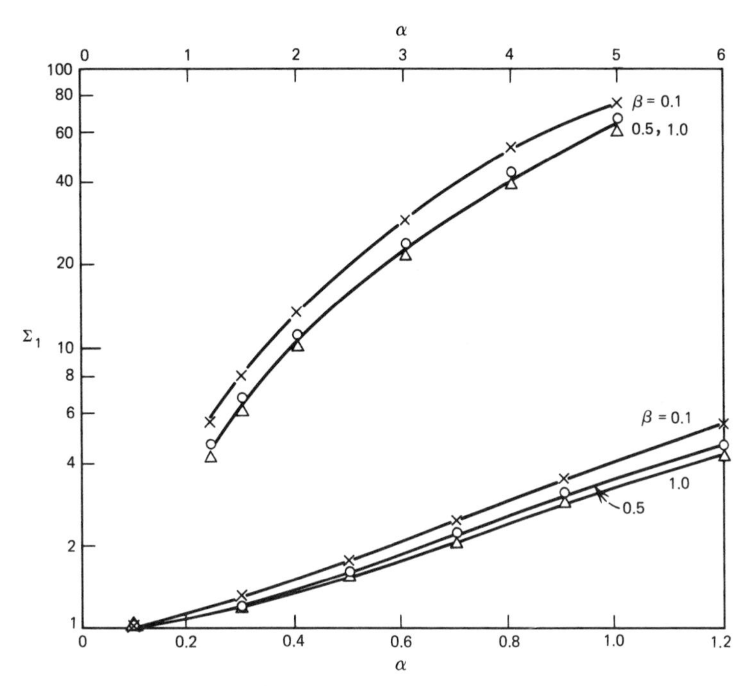

Fig. 16—Exponential family Σ_1 vs α and β.

The receiver equalized output at the sampling time, due to an optical pulse of energy b_0, is b_0.

When $b_0 = b_{\min}$, we must be sure that the probability that noise drives the receiver output $v_{\text{out}}(t)$ at the sampling time above the threshold D is less than 10^{-9}.† Using the signal-to-noise ratio approximation,‡ we require the noise variance, $NW(b_{\min})$, to be less than $\{\frac{1}{6}[D - b_{\min}]\}^2$.

Therefore, we require that

$$NW(b_{\min}) \leqq \frac{1}{36}[D - b_{\min}]^2. \tag{26}$$

Furthermore, when $b_0 = b_{\max}$ we must be sure that the probability that the noise drives the receiver output below the threshold is less than 10^{-9}. Therefore, we require that

$$NW(b_{\max}) \leqq \frac{1}{36}[b_{\max} - D]^2. \tag{27}$$

† An error rate of 10^{-9} is arbitrarily chosen here. Dependence of required optical power on error rate is discussed in Part II of this paper.

‡ See Appendix A.

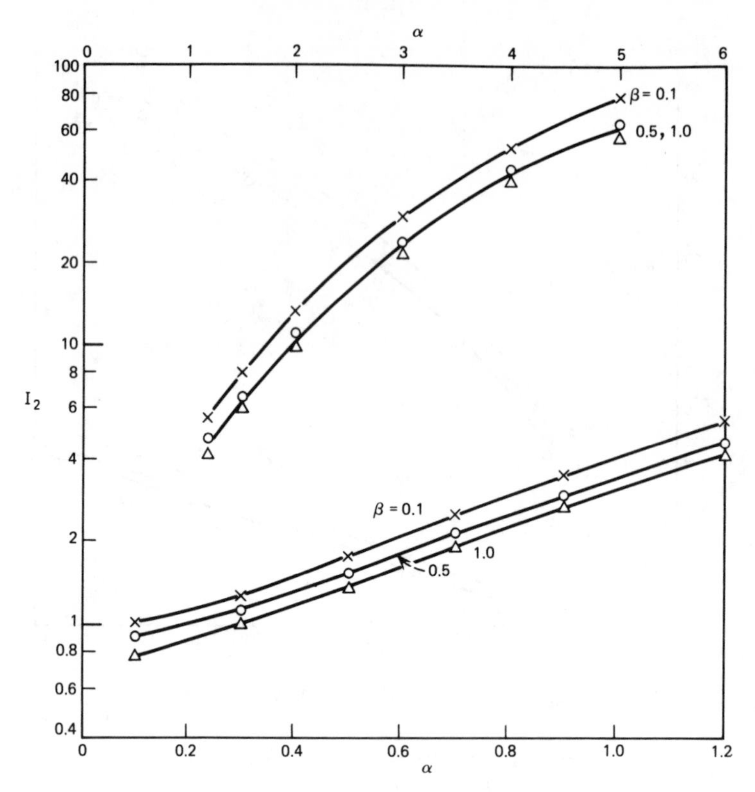

Fig. 17—Exponential family I_2 vs α and β.

Using equality in (26) and (27), we require for a 10^{-9} error rate

$$\sqrt{NW(b_{\max})} + \sqrt{NW(b_{\min})} = \tfrac{1}{6}(b_{\max} - b_{\min}). \tag{28}$$

Very often we have a fixed ratio $(b_{\min}/b_{\max}) = \rho$.

Rearranging (28) we obtain

$$b_{\max} = \frac{6}{1-\rho}\left[\sqrt{NW(b_{\max})} + \sqrt{NW(\rho b_{\max})}\right]. \tag{29}$$

In order to obtain numerical results, we shall make the following reasonable assumptions. Let the dark current be negligible and let $b_{\min}/b_{\max}$ be much less than unity. Therefore we shall set $\lambda_0 = 0$, $b_{\min} = 0$.† We obtain from (23)

† Quantitative discussion of the consequences of these approximations are given in Part II.

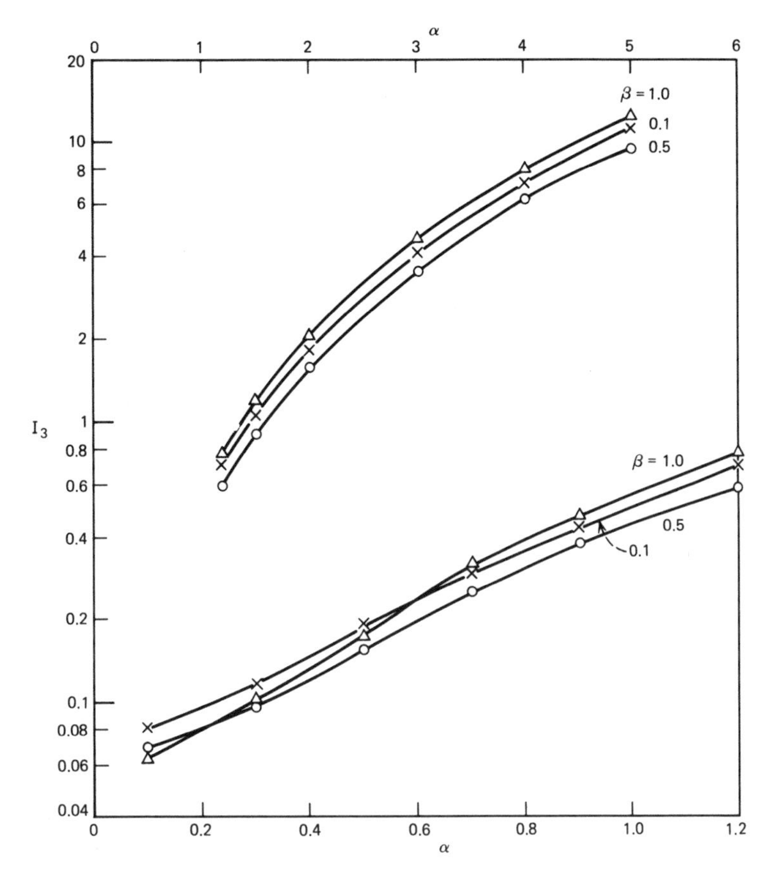

Fig. 18—Exponential family I_3 vs α and β.

$$NW(b_0) \cong \left[\frac{\hbar\Omega}{\eta}\right]^2 \left\{\frac{\langle g^2\rangle}{\langle g\rangle^2}\frac{\eta}{\hbar\Omega}\left[b_0 I_1 + b_{\max}(\textstyle\sum_1 - I_1)\right] + \frac{1}{\langle g\rangle^2}[Z]\right\},$$

where

$$Z \triangleq \left\{\frac{T}{e^2}\left[S_I + \frac{2k\theta}{R_b} + \frac{S_E}{R_T^2}\right] I_2 + \frac{(2\pi C_T)^2}{Te^2} S_E I_3\right\}. \tag{30}$$

In (30), Z includes all the thermal noise terms of (23).

From (29), taking the limit as $\rho \rightarrow 0$ ($b_{\min} \rightarrow 0$), we obtain the conditions to achieve a 10^{-9} error rate as follows.

Case I: Thermal noise (Z) dominates (i.e., little or no avalanche gain).

$$b_{\max} = \frac{12\hbar\Omega}{\eta\langle g\rangle} Z^{\frac{1}{2}}. \tag{31}$$

Case II: Optimal gain (i.e., $\langle g \rangle$ adjusted to minimize the required optical energy in an "on" pulse $b_{\max}$).

Let the relationship between $\langle g^2 \rangle$ and $\langle g \rangle$ be specified in the usual way:

$$\langle g^2 \rangle = \langle g \rangle^{2+x}, \tag{32}$$

where x depends upon the type of detector. We obtain the following formula for the optimal gain:

$$\langle g \rangle_{\text{optimal}} = (6)^{-1/(1+x)}(Z)^{1/(2+2x)}(\gamma_1)^{1/(2+2x)}(\gamma_2)^{-1/(1+x)}, \tag{33}$$

where defining $I_5 = \sum_1 - I_1$ [see eq. (23)]

$$\gamma_1 \triangleq \frac{-[\sum_1 + I_5] + \sqrt{(\sum_1 + I_5)^2 + \frac{16(1+x)}{x^2} \sum_1 I_5}}{2 \sum_1 I_5} \tag{34}$$

$$\gamma_2 \triangleq \sqrt{1/\gamma_1 + I_5} + \sqrt{1/\gamma_1 + \textstyle\sum_1}.$$

We obtain the following formula for $b_{\max}$:

$$b_{\max} = \frac{\hbar\Omega}{\eta}(6)^{(2+x)/(1+x)}(Z)^{x/(2+2x)}(\gamma_1)^{x/(2+2x)}(\gamma_2)^{(2+x)/(1+x)}. \tag{35}$$

That is,

$$b_{\max} \propto [Z]^{x/(2+2x)}. \tag{36}$$

We therefore see that for these assumptions and $x = 0.5$ corresponding to a silicon avalanche detector the minimum required energy per pulse varies as the one-half power of the thermal noise term, Z, without avalanche gain, and as the one-sixth power of the thermal noise term, Z, with optimal gain.

However, this does not mean that at optimal gain the value of Z is unimportant. By reducing Z (the thermal noise terms) through proper choice of biasing and amplifier circuitry, we still minimize the optimizing avalanche gain [see (33)] and obtain some reduction in the required energy per pulse (see Part II).

6.1 *Example*

From eqs. (23), (30), (34), and (35) we can calculate, for various shaped optical pulses, the effect of intersymbol interference on the required energy per "on" pulse ($b_{\max}$) and therefore on the required average optical power needed for a 10^{-9} error rate.† We shall assume

† That is, if pulses are "on" half the time, the required optical power equals $b_{\max}/2T$.

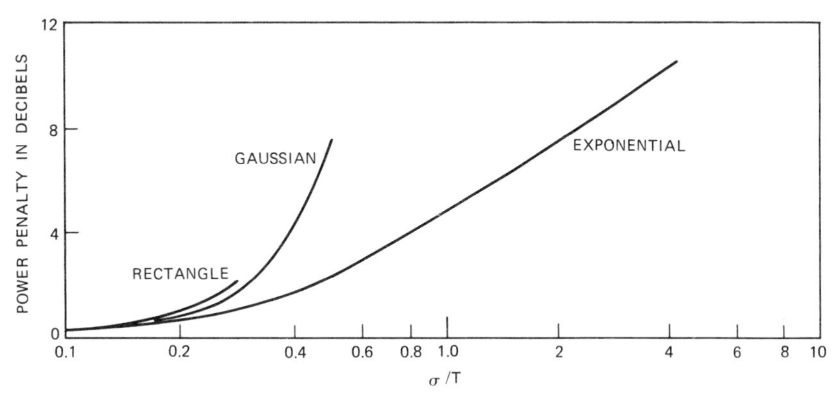

Fig. 19—No avalanche gain.

that the detector amplifier shunt resistance R_T is sufficiently large so that the term $((2\pi C_T)^2/Te^2)S_E I_3$ dominates the thermal noise in (23) and (30).

The minimal required optical power is obtained for very narrow optical input pulses.† For other pulse shapes, the *excess* required optical power can be defined as a *penalty* in dB for not using narrow pulses. This penalty is plotted in Figs. 19 and 20 for the case of no avalanche gain and optimal avalanche gain using the pulse shapes of (25), assuming a silicon detector ($x = 0.5$). In those figures, the abscissa is the normalized rms optical pulse width defined as follows:

$$\frac{\sigma^2}{T^2} = \frac{\left(\int t^2 h_p(t)dt\right) - \left(\int th_p(t)dt\right)^2}{T^2}, \tag{37}$$

where T = time slot width.

VII. CONCLUSIONS

7.1 *Conclusion on Choosing the Biasing Circuitry*

From the results of Sections IV and VI, and from (23), it is clear that, to minimize the thermal noise degradations introduced by the amplifiers following the detector, it is necessary to make the amplifier input resistance and the biasing circuit resistance sufficiently large so that the amplifier series noise source dominates the Johnson noise of these parallel resistances. When designing the amplifier, one should keep in mind that for a silicon avalanche detector the required optical

† See Appendix B.

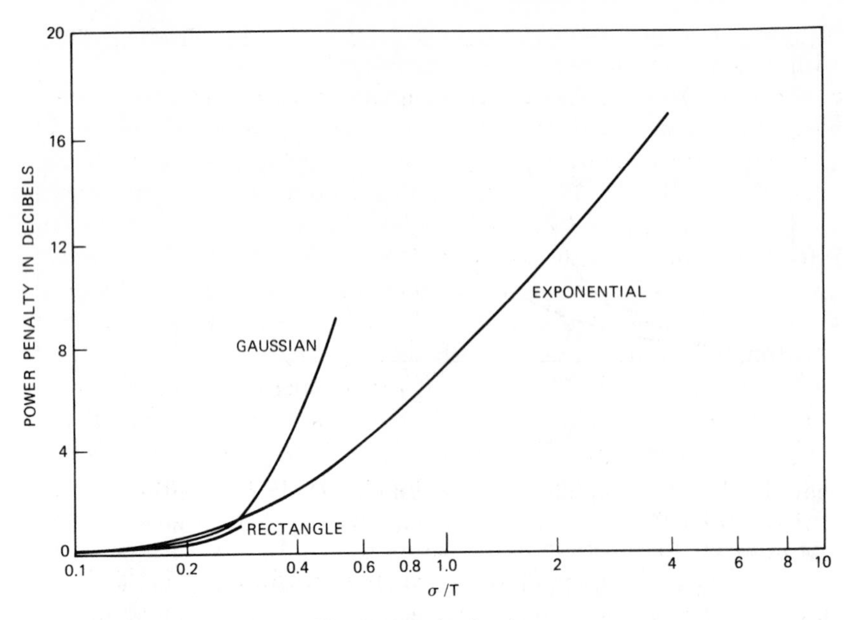

Fig. 20—Optimal gain.

energy per pulse at optimal gain varies roughly as the one-sixth power of the thermal noise variance at the receiver output, and therefore it is not wise to spend too much money on thermal noise reduction. On the other hand, if one is not using avalanche gain, the required energy per pulse varies roughly as the one-half power of the thermal noise variance at the receiver output.

In order to minimize the effects of the thermal noise, the total capacitance shunting the detector should be as small as possible and the equivalent series thermal noise source of the amplifier should also be as small as possible.

7.2 *The Effect of Bit Rate on Required Energy Per Pulse*†

The effect of bit rate on the required energy per pulse is small if the received pulses remain well confined to a time slot. In (23), assume I_1, Σ_1, I_2, and I_3 are fixed corresponding to a fixed received pulse width relative to a time slot. Then the shot noise terms due to the signal are independent of the bit rate $1/T$, and the shot noise due to the dark current decreases with increasing bit rate. If the series noise from the amplifier dominates, then the thermal noise increases with increasing

† This subject will be discussed in more detail in Part II.

bit rate, but is for the most part compensated for by the avalanche gain with little penalty in required energy per pulse.

If considerable equalization is being used, then the required energy per pulse increases with the bit rate (because a higher bit rate necessitates greater equalization). For the equalization assumed above, where the equalized pulses are forced to go to zero at all sampling times except one, the required energy per pulse is a strong function of the bit rate. For example, with exponential-shaped received pulses, the required optical power at optimal avalanche gain was roughly 6 dB higher for a pulse 1 time slot wide to the $1/e$ point compared to a pulse only 0.25 time slot wide to the $1/e$ point (see Fig. 20).

On the other hand, it is clear that zero-forcing-type equalization is not optimal, particularly for received pulses whose spectra fall off rapidly with frequency. It is more likely that some compromise between eye opening and output noise variance results in minimum required energy per pulse.

For the assumed zero-forcing equalization, we still can conclude that a usable tradeoff exists between required energy per pulse and bit rate, and this will allow some extension of the usable rate on "dispersion-limited" fibers.

7.3 *Comments on Previous Work*

The purpose of this paper has been to illustrate the application of the "high-impedance" front-end design to optical digital repeaters, to take into account precisely the input pulse shape and the equalizer-filter shape, and to obtain explicit formulas for the required optical power to achieve a desired error rate as a function of the other parameters.

Previous authors[12,13] working in the areas of particle counting and video amplifier design have recognized that a high-impedance front end followed by proper equalization in later stages provides low noise and adequate bandwidth. However, optical communication theorists[5,6,14,15] have in the past often used the criterion "$RC \leqq T$"–loading down the front-end amplifier so as to have adequate bandwidth without equalization–therein incurring an unnecessary noise penalty. Some optical experimenters[16,17] have recognized the high-impedance design for observing isolated pulses or single frequencies, but failed to recognize the use of equalization.

Many previous authors[3,4,6,15] have used simple formulas (which usually assume isolated rectangular input pulses and a front-end bandwidth of the reciprocal pulse width) to obtain the required power in optical communication systems for a desired signal-to-noise ratio. Often these formulas average out the signal-dependent nature of the shot noise. If modified to include the high-impedance design concept, such formulas are very useful for obtaining "ball park" estimates of optical power requirements. Such formulas are, in general, special cases of the formulas described here.

7.4 *Experimental Verification*

In work recently reported,[18] J. E. Goell has shown that, in a 6.3-Mb/s repeater operating at an error rate of 10^{-9}, agreement of experimentally determined power requirements and the above theory were within 1 dB (0.25 dB in cases without avalanche gain). In particular, using an FET front end and the "high-impedance" design, the optical power requirement without avalanche gain was 8 dB less than with the front end loaded down to the "$RC = T$" design.

APPENDIX A

Signal-to-Noise Ratio Approximation

In this paper we have calculated the mean voltage ($b_{\max}$ or $b_{\min}$) and the average-squared deviation from the mean voltage ($NW(b_{\max})$ or $NW(b_{\min})$) at the receiver output at the sampling times. In order to calculate error rates, we shall assume that the output voltage is approximately a Gaussian random variable. This is the signal-to-noise ratio approximation. Thus if the threshold, to which we compare the output voltage, is D, and if the desired error probability is P_e, we have

$$\frac{1}{\sqrt{2\pi\sigma_o^2}} \int_D^\infty \exp\left[-(v - b_{\min})^2/2\sigma_o^2\right]dv = P_e, \tag{38}$$

where

$$\sigma_o^2 = NW(b_{\min})$$

and

$$\frac{1}{\sqrt{2\pi\sigma_1^2}} \int_{-\infty}^D \exp\left[-(v - b_{\max})^2/2\sigma_1^2\right]dv = P_e,$$

where

$$\sigma_1^2 = NW(b_{\max}).$$

Changing the variables of integration we obtain the following expressions, equivalent to (38):

$$\frac{1}{\sqrt{2\pi}} \int_Q^\infty e^{-x^2/2}dx = P_e, \tag{39}$$

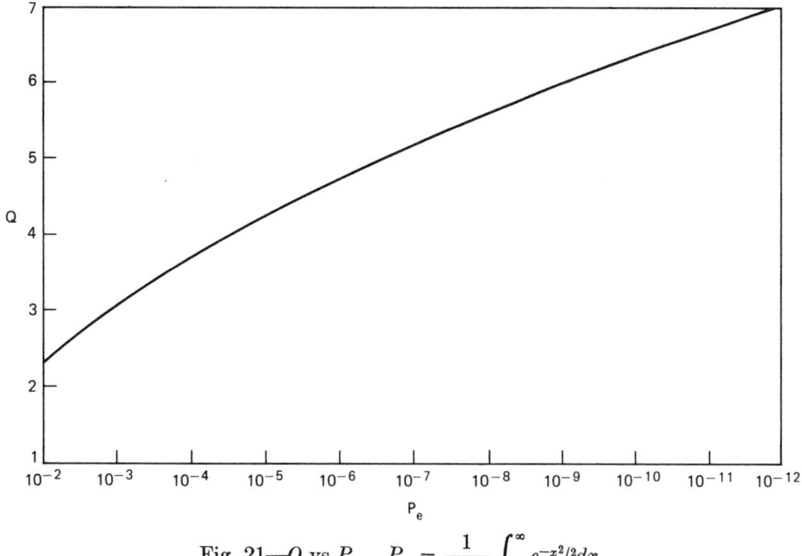

Fig. 21—Q vs P_e, $P_e = \frac{1}{\sqrt{2\pi}} \int_Q^\infty e^{-x^2/2} dx.$

where

$$Q = (D - b_{\min})/\sigma_o$$

and also

$$Q = (b_{\max} - D)/\sigma_1.$$

Thus we must have

$$\sigma_o = \sqrt{NW(b_{\min})} = (D - b_{\min})/Q$$

and

$$\sigma_1 = \sqrt{NW(b_{\max})} = (b_{\max} - D)/Q.$$

Therefore we must also have (eliminating D)

$$\sqrt{NW(b_{\max})} + \sqrt{NW(b_{\min})} = (b_{\max} - b_{\min})/Q.$$

The value of Q is determined by the error rate through (39) above. Figure 21 shows a plot of Q vs P_e which can be obtained from standard tables.

Equation (39) states that the threshold must be Q standard deviations (of the noise at $b_{\min}$) above $b_{\min}$, and also must be Q standard deviations (of the noise at $b_{\max}$) below $b_{\max}$ to insure the desired error rate. For an error rate of 10^{-9} ($P_e = 10^{-9}$) Q is roughly 6 (5.99781).

APPENDIX B

Optimal Input Pulse Shape

We now wish to show that the optimal input pulse, $h_p(t)$ shape (which minimizes the required average optical power) is ideally an impulse; and for practical purposes a pulse which is sufficiently narrow so that its Fourier transform is almost constant for all frequencies passed by the receiver. To do this, we shall show that such a narrow pulse minimizes the noises $NW(b_{\max})$ and $NW(b_{\min})$ defined in (23).

We begin with the already established condition that the area of $h_p(t)$ is equal to unity and that $h_p(t)$ is positive (power must, of course, be positive).

$$\int h_p(t)dt = 1; \qquad h_p(t) > 0. \tag{40}$$

These conditions imply the following weaker condition:

$$|H_p'(f)| = \left| \int h_p(t)e^{-i2\pi ft/T}dt \right|$$
$$\leqq \int |h_p(t)|\,|e^{-i2\pi ft/T}|\,dt = \int h_p(t)dt = H_p'(0) = 1. \tag{41}$$

Consider first the thermal noise terms of (23) involving the integrals I_2 and I_3:

$$I_2 = \int \frac{|H_{\text{out}}'(f)|^2}{|H_p'(f)|^2}\,df; \qquad I_3 = \int \frac{|H_{\text{out}}'(f)|^2}{|H_p'(f)|^2}\,f^2 df. \tag{42}$$

Using (41) in (42) we see that these terms I_2 and I_3 are minimized for any desired output pulse $H_{\text{out}}'(f)$ by setting $|H_p'(f)| = H_p'(0) = 1$ for all frequencies, f, for which $|H_{\text{out}}'(f)| > 0$. Thus, ideally, to minimize I_2 and I_3, $h_p(t)$ is an impulse of unit area which also satisfies the conditions (40).

We must now show that the shot noise terms of (23), I_1 and $\Sigma_1 - I_1$, are minimized by a very narrow pulse $h_p(t)$.

First recall that $(\Sigma_1 - I_1)b_{\max}(\hbar\Omega/\eta)\langle g^2\rangle/\langle g\rangle^2$ is the worst-case, mean-square shot noise at the sampling time due to all other pulses except the one under decision, and assuming all of those pulses are "on" ($b_k = b_{\max}$ for $k \neq 0$). Thus, from (17), we obtain

$$\Sigma_1 - I_1 \geqq 0, \tag{43}$$

where

$$\Sigma_1 - I_1 = \int \Big(\sum_{k\neq 0} h_p(t' - kT)\Big)h_I^2(-t')dt'$$

and where $h_I(t)$ is the overall receiver impulse response relating $h_p(t)$ to $h_{\text{out}}(t)$.

Now let the optical pulse $h_p(t)$ be an impulse of unit area. Then the overall impulse response h_I must be equal to $h_{\text{out}}(t)$ and therefore using (43) and (9) we obtain

$$\Sigma_1 - I_1 = \sum_{k \neq 0} h_{\text{out}}^2(-kT) = 0 \quad \text{(for } h_p(t) = \delta(t)\text{).} \tag{44}$$

Because condition (9) requires zero-crossing equalization, we have shown that an impulse shape for $h_p(t)$ minimizes (removes) the shot noise from pulses other than the one under decision.

Finally, consider the shot noise from the pulse under decision given by $I_1(\hbar\Omega/\eta)b_o\langle g^2\rangle/\langle g\rangle^2$ where

$$I_1 = \int h_p(t')h_I^2(-t')dt' > 0. \tag{45}$$

We already have the condition (9)

$$h_{\text{out}}(0) = \int h_p(t')h_I(-t')dt' = 1. \tag{46}$$

We can next use the Shwarz inequality on (46)

$$(h_{\text{out}}(0))^2 = 1 = \left(\int h_p^{\frac{1}{2}}(t)h_p^{\frac{1}{2}}(t)h_I(-t)dt\right)^2 \leq \int h_p(t)dt \int h_p(t)h_I^2(-t)dt. \tag{47}$$

Since $\int h_p(t)dt = 1$, we have from (47) and (45)

$$I_1 \geqq 1. \tag{48}$$

Now set $h_p(t)$ equal to a unit area impulse. It then must follow from (46) that $h_I(0) = 1$. We finally obtain

$$\int h_p(t)h_I^2(t)dt = h_I^2(0) = 1. \tag{49}$$

From (48) and (49) we see that an impulse-shaped $h_p(t)$ makes I_1 achieve its minimum value of unity.

Summarizing, an impulse-shaped optical input pulse $h_p(t)$ (for practical purposes a sufficiently narrow pulse so that its Fourier transform is approximately constant for all frequencies passed by the receiver) minimizes all the pulse-shape-dependent coefficients (I_1, $\Sigma_1 - I_1$, I_2, and I_3) in the noise expression (23) and thereby minimizes the required average optical power to achieve a desired error rate (using the signal-to-noise ratio approximation of Appendix A).

REFERENCES

1. Personick, S. D., "Baseband Linearity and Equalization in Fiber Optic Communication Systems," to appear in B.S.T.J., September 1973.
2. Personick, S. D., "Statistics of a General Class of Avalanche Detectors with Applications to Optical Communication," B.S.T.J., *50*, No. 10 (December 1971), pp. 3075–3096.
3. Melchior, H., et al., "Photodetectors for Optical Communication Systems," Proc. IEEE, *58*, No. 10 (October 1970), pp. 1466–1486.
4. Pratt, W. R., *Laser Communication Systems*, New York: John Wiley and Sons, 1969.
5. Hubbard, W. M., "Comparative Performance of Twin-Channel and Single-Channel Optical Frequency Receivers," IEEE Trans. Commun. *COM20*, No. 6 (December 1972), pp. 1079–1086.
6. Anderson, L. K., and McCurtry, B. J., "High Speed Photodetectors," Proc. IEEE, *54* (October 1966), pp. 1335–1349.
7. McIntyre, R. J., "Multiplication Noise in Uniform Avalanche Diodes," IEEE Trans. Electron Devices, *ED-13*, No. 1 (January 1966), pp. 164–168.
8. Melchior, H., and Lynch, W. T., "Signal and Noise Response of High Speed Germanium Photodiodes," IEEE Trans. Electron Devices, *ED-13* (December 1966), pp. 829–838.
9. Klauder, J. R., and Sudarshan, E. C. G., *Fundamentals of Quantum Optics*, New York: W. A. Benjamin, Inc., 1968, pp. 169–178.
10. Parzen, E., *Stochastic Processes*, San Francisco: Holden-Day, 1962, p. 156.
11. Figure 5b is from *Transmission Systems for Communication*, Bell Telephone Laboratories, 1970, p. 651.
12. Gillespie, A. B., *Signal, Noise, and Resolution in Nuclear Particle Counters*, New York: Pergamon Press, Inc., 1953.
13. Schade, O. H., Sr., "A Solid-State Low-Noise Preamplifier and Picture-Tube Drive Amplifier for a 60 MHz Video System," RCA Rev., *29*, No. 1 (March 1968), p. 3.
14. Chown, M., and Kao, K. C., "Some Broadband Fiber-System Design Considerations," ICC 1972 Conf. Proc., June 19–21, 1972, Philadelphia, Pa., pp. 12-1, 12-5.
15. Ross, M., *Laser Receivers*, New York: John Wiley and Sons, 1967, p. 328.
16. Edwards, B. N., "Optimization of Preamplifiers for Detection of Short Light Pulses with Photodiodes," Appl. Opt. *5*, No. 9 (September 1966), pp. 1423–1425.
17. Mathur, D. P., McIntyre, R. J., and Webb, P. P., "A New Germanium Photodiode with Extended Long Wavelength Response," Appl. Opt., *9*, No. 8 (August 1970), pp. 1842–1847.
18. Goell, J. E., work to be presented at the Conference on Laser Engineering and Applications (CLEA) Washington, D.C., May 30–June 2, 1973.

Receiver Design for Digital Fiber Optic Communication Systems, II

By S. D. PERSONICK

(Manuscript received January 15, 1973)

This paper applies the results of Part I to specific receivers in order to obtain numerical results. The general explicit formulas for the required optical average power to achieve a desired error rate are summarized. A specific receiver is considered and the optical power requirements solved for. The parameters defining this receiver (e.g., bit rate, bias resistance, dark current, etc.) are then varied, and the effects on the required optical power are plotted.

I. INTRODUCTION

This paper will apply the theory of Part I to illustrate in detail how the required received optical power in a digital fiber optic repeater varies with the parameters such as the desired error rate, the thermal noise sources, the bit rate, detector dark current, imperfect modulation, etc. We shall begin by first applying the formulas of Part I to a specific realistic example to obtain reference point. We shall then derive curves of how the required power varies around this point as we vary the system parameters.

II. REVIEW OF RESULTS OF PART I

In Part I we derived explicit formulas for the required optical power at the input of a digital fiber optic communication system repeater to achieve a desired error rate. One formula was applicable when little or no internal (avalanche) detector gain was used, so that thermal noise from the amplifier dominated. The other formula was applicable when optimal gain was being used. These formulas are repeated below:

$$p_{\text{required}} = \frac{QZ^{\frac{1}{2}}}{GT}\frac{\hbar\Omega}{\eta}, \qquad \text{(Thermal Noise Dominates)} \qquad (1)$$

Reprinted with permission from *Bell Syst. Tech. J.*, vol. 52, pp. 875–886, July–Aug. 1973.

where

$$Z = \left\{ \frac{T}{e^2} \left[S_I + \frac{2k\theta}{R_b} + \frac{S_E}{R_T^2} \right] I_2 + \frac{(2\pi C_T)^2 S_E I_3}{Te^2} \right\}; \tag{1a}$$

$$p_{\text{required}} = \frac{1}{2T} (Q)^{(2+x)/(1+x)} \left[(Z)^{x/(2+2x)} (\gamma_1)^{x/(2+2x)} (\gamma_2)^{(2+x)/(1+x)} \right] \frac{\hbar\Omega}{\eta},$$

(Optimal Avalanche Gain) (2)

where

$$G_{\text{optimal}} = (Q)^{-1/(1+x)} \left[(Z)^{1/(2+2x)} (\gamma_1)^{1/(2+2x)} (\gamma_2)^{-1/(1+x)} \right],$$

where (referring to Fig. 1)

$\eta/\hbar\Omega$ = detector quantum efficiency/energy in a photon
T = interval between bits = 1/bit rate
G = average detector internal gain
G^x = detector random internal gain excess noise factor
Q = number of noise standard deviations between signal and threshold at receiver output. $Q = 6$ for an error rate of 10^{-9}. (See Fig. 21 in Part I for a graph of error rate vs Q.)
e = electron charge
$k\theta$ = Boltzman's constant·the absolute temperature
R_T = total parallel resistance in shunt with the detector including the physical biasing resistor and the amplifier input resistance
R_b = value of physical detector biasing resistor
C_T = total shunt capacitance across the detector including the shunt capacitance of the detector and that of the amplifier
S_I = amplifier shunt noise source spectral height (two-sided) in amperes^2/Hz
S_E = amplifier series noise source spectral height (two-sided) in volts^2/Hz.

I_2, I_3, γ_1, and γ_2 are functions only of the shapes of the input optical pulses and the equalized repeater output pulses, where the length of a time slot has been scaled out. These functions are defined in eqs. (23) and (34) of Part I.

Formulas (1) and (2) neglect dark current and assume perfect modulation (received optical pulses completely on or off). We shall investigate deviations from these idealizations later in the paper. For silicon detectors and bit rates above a few megabits per second, these idealizations are reasonable approximations.

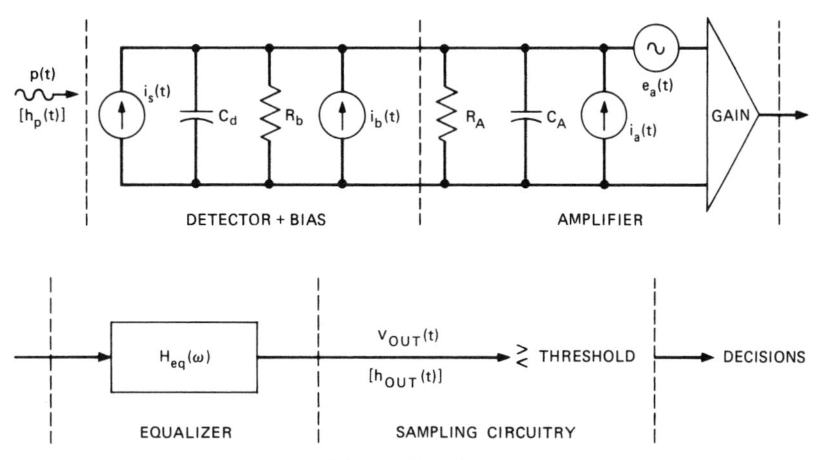

Fig. 1—Receiver.

III. A TYPICAL OPTICAL REPEATER

Consider the following practical optical repeater, operating at a bit rate of 2.5×10^7 bits per second and an error rate of 10^{-9}. The detector is a silicon device with excess noise exponent $x = 0.5$, quantum efficiency 75 percent, dark current before avalanche gain of 100 picoamperes, and an operating wavelength of 8500 angstroms. The front-end amplifier is a field-effect transistor in a common-source configuration. The total shunt capacitance across the detector is 10 pF. The detector biasing resistor is 1 megohm. The amplifier input resistance is 1 megohm. The amplifier shunt-current noise-source spectral height is equal to the thermal noise of a 1-megohm resistor. The amplifier series-voltage noise-source spectral height is equal to the thermal noise of a conductance with a value equal to the transistor transconductance, g_m, which is 5000 micromhos. The received optical pulses are half-duty-cycle rectangular pulses. The desired equalized output pulse is a raised cosine pulse [see Part I, eq. (25)] with parameter $\beta = 1$.

We must first calculate the value of Q which depends only upon the desired error rate. From Part I, Fig. 21, we see that for an error rate of 10^{-9}, $Q = 6$.

Next we must obtain the constants I_2, I_3, γ_1, and γ_2. These depend only upon the input optical pulse shape and the equalized output pulse shape. From (23) and (34) of Part I we obtain

$$I_2 = 0.804046, \qquad I_3 = 0.071966, \qquad \gamma_1 = 21.4106, \qquad \gamma_2 = 1.25424. \tag{3}$$

Using the above data we obtain the thermal noise parameter Z as follows:

$$Z = \left\{ \frac{\overset{T}{4 \times 10^{-8}}}{(\underset{e}{1.6 \times 10^{-19}})^2} \left[\overset{2k\theta}{8.28\ 10^{-21}} \left(10^{-6} + 10^{-6} + \frac{4 \times 10^{-12}}{5 \times 10^{-3}} \right) \overset{I_2}{0.804046} \right] + \frac{(2\pi \times 10^{-11})^2 \left(\frac{8.28 \times 10^{-21}}{5 \times 10^{-3}} \right) (\overset{I_3}{0.071966})}{(1.6 \times 10^{-19})^2 (4 \times 10^{-8})} \right\} = 4.8027 \times 10^5. \quad (4)$$

From the data we have $\hbar\Omega/\eta = 3.117 \times 10^{-19}$ joules.

We obtain from (1), at unity internal gain (no avalanche), $p_{\text{required}} = 3.25 \times 10^{-8}$ watts $= -44.89$ dBm (no gain).

We obtain from (2), at optimal avalanche gain, $p_{\text{required}} = 1.6409 \times 10^{-9}$ watts $= -57.85$ dBm, $G_{\text{optimal}} = 56.89$.

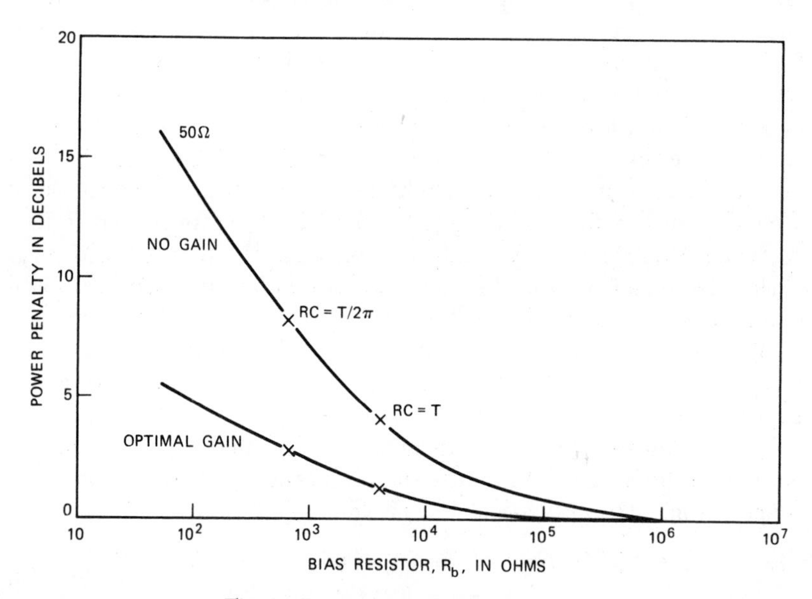

Fig. 2—Required power penalty vs R_b.

We therefore observe that optimal avalanche gain buys a 13-dB reduction in required optical power in this example. Before proceeding, we can check the validity of neglecting dark current. The average number of primary photoelectrons produced by the signal per pulse interval T is the required optical power multiplied by $\eta T/\hbar\Omega$. When shot noise is important (with avalanche gain) this number is 210 primary signal counts per interval T. The number of dark current counts per interval T is the dark current in amperes multiplied by T/e, which in this example is 25 primary dark current counts. Thus, the shot noise due to the dark current is about 10 percent of the signal shot noise. It is therefore a reasonable approximation to neglect this dark current noise. In Section VI we shall calculate precisely the effect of dark current upon the required optical power.

IV. VARYING THE PARAMETERS

In this section, we shall calculate the effect of varying parameter values used in the example of Section III.

4.1 *Biasing Resistor Value*

It was pointed out in Part I that the biasing resistor R_b should be sufficiently large so that the amplifier series noise source S_E dominates in the expression for Z of (1a). This was in fact the case in the example of Section III. We can calculate the penalty in required optical power for using a smaller biasing resistor. This penalty is plotted in Fig. 2 in dB with zero dB being the penalty associated with an infinitely large biasing resistor. The exact penalty of Fig. 2 is applicable with the other relevant parameters (which make up Z) given in the example above. However, the qualitative conclusions are that significantly more optical power is needed if one adheres to the "$RC = T$" design rather than the "large R" (high-impedance) design, in the absence of avalanche gain. Figure 3 shows how the optimal avalanche gain varies when R_b is changed. The qualitative conclusion is that the "$RC = T$" design requires significantly more avalanche gain that the "large R" (high-impedance) design. It should be pointed out that, for lower bit rates and/or a smaller capacitance C_T, the improvement associated with use of a large R_b rather than a value to keep $R_bC_T = T$ is more pronounced.

4.2 *Desired Error Rate*

As mentioned before, the error rate is coupled to the parameter Q in (1) and (2). Figures 4a and 4b show plots of the variation in the re-

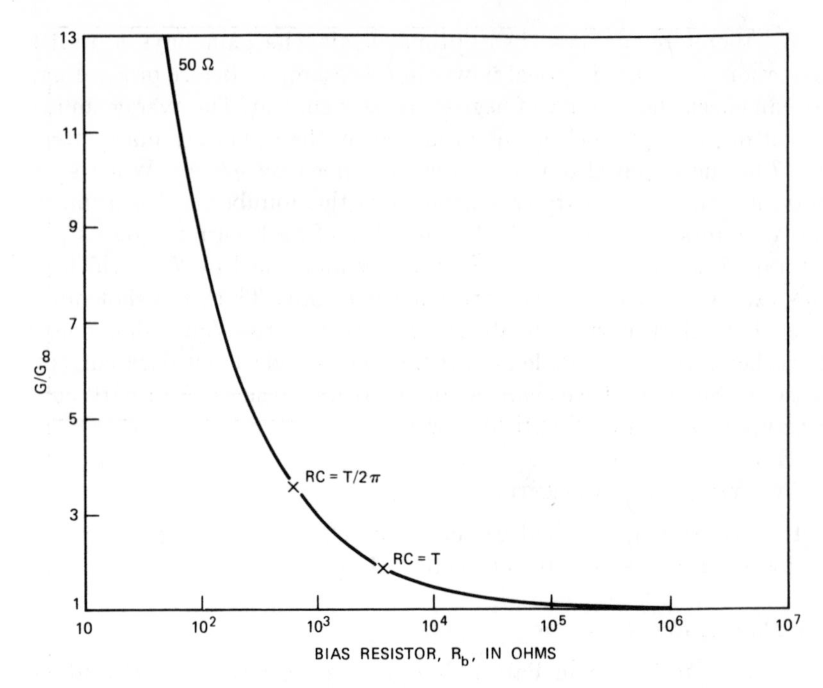

Fig. 3—Optimal gain penalty vs R_b, G_∞ = optimal gain at $R_b = \infty$.

quired power in dB with the desired error rate without gain and with optimal gain. The absolute power in dBm is only applicable to the example of Section III above. However, the difference in required power in dB between any two error rates is applicable in general, as should be apparent from (1) and (2), provided a silicon detector ($x = 0.5$) is being used.

4.3 *Bit Rate* $(1/T)$

As mentioned before, the pulse spacing T is scaled out of I_2, I_3, γ_1, and γ_2. These numbers depend only upon the input and output pulse *shapes* (e.g., half-duty-cycle rectangular input pulse, raised-cosine equalized output pulse). Therefore, the effect of the parameter T is explicitly given in (1) and (2) without any hidden dependencies. (This of course assumes that the input pulse shape is not limited by dispersion in the transmission medium and can therefore be held to a half-duty-cycle rectangle.) If we assume that the high-impedance design is being used and that this dominance of the term proportional to $1/T$ in Z of (1a) can be maintained as the bit rate is varied (becomes difficult

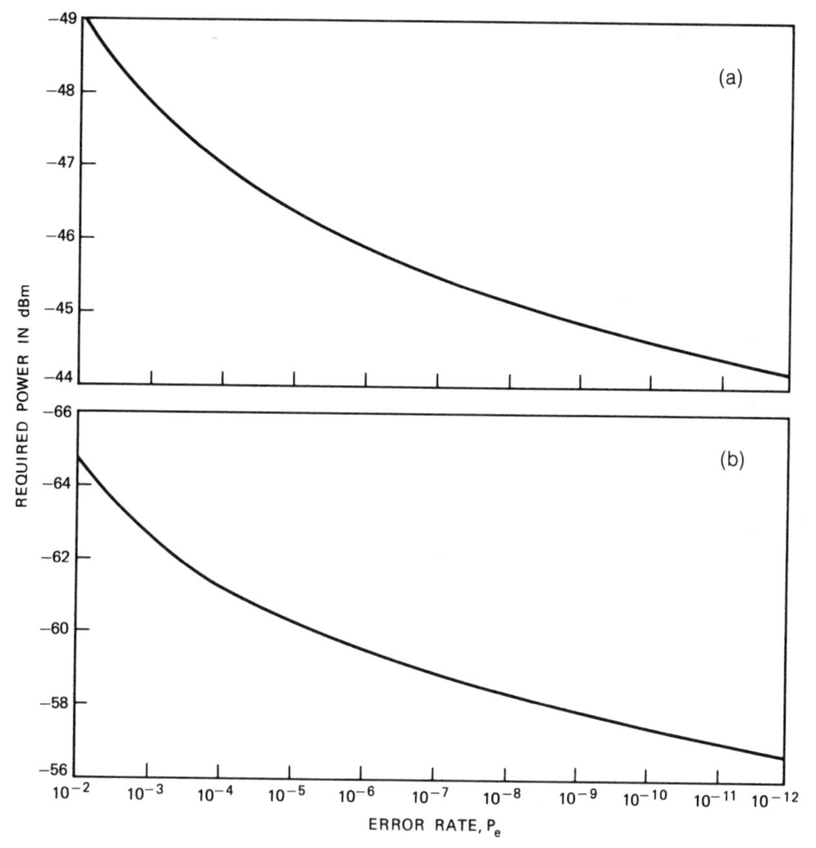

Fig. 4—(a) Required power vs error rate (no avalanche gain). (b) Required power vs error rate (optimal gain).

at low bit rates), then we have the following dependence of the required optical power upon the bit rate $1/T$ without gain and with optimal gain:

$$p_{\text{required}} \propto T^{-\frac{3}{2}} \quad \text{(no gain)} \quad \text{(4.5 dB/octave of bit rate)} \tag{5a}$$

$$p_{\text{required}} \propto T^{-7/6} \quad \text{(optimal silicon gain)} \quad \text{(3.5 dB/octave of bit rate)} \tag{5b}$$

$$G_{\text{optimal}} \propto T^{-\frac{1}{3}} \quad \text{(1 dB/octave of bit rate).}$$

One should be careful extrapolating (5a) and (5b) to very low bit rates. First, the shot noise is no longer negligible compared to the thermal noise at bit rates where the optimal gain is low. Thus (5a) loses validity at very low bit rates. Further, (5b) is only valid for optimal gains greater than unity. Near unity optimal gain, the silicon excess noise factor departs from $G^{.5}$. In addition, at low bit rates, dark current may not be negligible. It is reasonable to use (5a) and (5b) to extrapolate the results of the example in Section III to bit rates between 5 and 300 Mb/s.

V. THE EFFECT OF IMPERFECT MODULATION

The above formulas (1) and (2) assume that there is perfect modulation. That is, it was assumed that each optical pulse is either completely on or completely off. In this section we shall investigate two versions of imperfect modulation.

Case 1: Pulses Not Completely Extinguished

This case is illustrated in Fig. 5. In each time slot the optical pulse is either completely or partly on. The partly on pulse has the same shape as a completely on pulse, but has area EXT times the area of a completely on pulse. This may correspond to an externally modulated mode-locked laser source. Thus the ratio of the power received when a sequence of all "off" pulses is transmitted to the power received when a sequence of all "on" pulses is transmitted is EXT. Using the results of Part I eqs. (23) and (34), we obtain the following power requirements which are modifications of (1) and (2) above:

$$p_{\text{required}} = \left(\frac{1 + EXT}{1 - EXT}\right) \frac{QZ^{\frac{1}{2}}}{GT} \frac{\hbar\Omega}{\eta}, \qquad \text{(Thermal Noise Dominates)} \quad (6)$$

$$p_{\text{required}} = \frac{1 + EXT}{2T} \left(\frac{Q}{1 - EXT}\right)^{(2+x)/(1+x)} \times \left[(Z)^{x/(2+2x)} (\gamma_1')^{x/(2+2x)} (\gamma_2')^{(2+x)/(1+x)}\right] \frac{\hbar\Omega}{\eta}; \qquad \text{(Optimal Gain)} \quad (7)$$

where defining from Part I (23) and (34)

$$I_6 = \Sigma_1 - (1 - EXT) I_1$$

we have

$$\gamma_1' = \frac{-(\Sigma_1 + I_6) + \sqrt{(\Sigma_1 + I_6)^2 + \dfrac{16(1 + x)}{x^2} \Sigma_1 I_6}}{2 \Sigma_1 I_6}$$

$$\gamma_2' = \sqrt{1/\gamma_1' + \Sigma_1} + \sqrt{1/\gamma_1' + I_6}.$$

[Compare (6) and (7) to (1) and (2).]

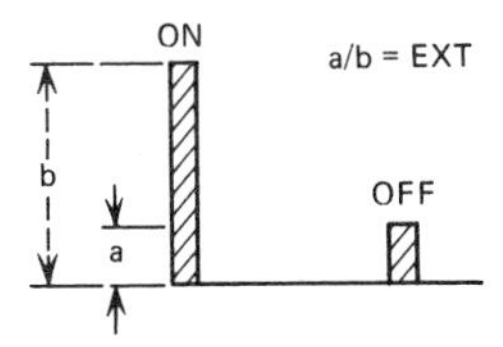

Fig. 5—Imperfect modulation, pulses not completely extinguished.

Using (1), (2), (7), and (8), we can calculate the extra required optical power due to a nonzero value of EXT with and without avalanche gain. When avalanche gain is being used, this power penalty depends upon the input and output pulse shapes. We plot in Fig. 6 the power penalty vs EXT, assuming the pulse shapes of the example in Section III above for the avalanche gain case.

Case 2: Pulses on a Pedestal

This case is illustrated in Fig. 7. The received optical pulses arrive on a pedestal, which may correspond to inability to completely extinguish the light from a modulated source which is not in a pulsing (mode-locked) condition. We set the ratio of average received optical power when all pulses are "off" to average received optical power when all

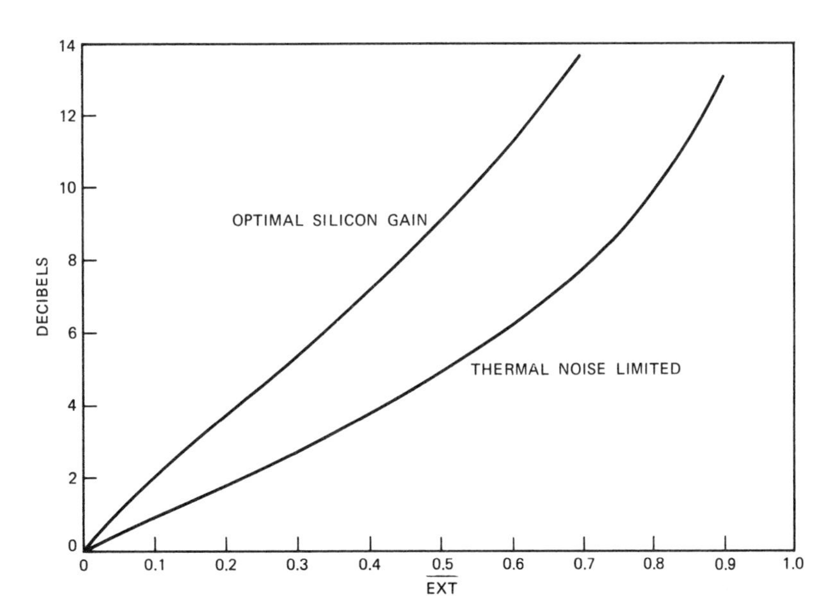

Fig. 6—EXT penalty (dB) vs EXT.

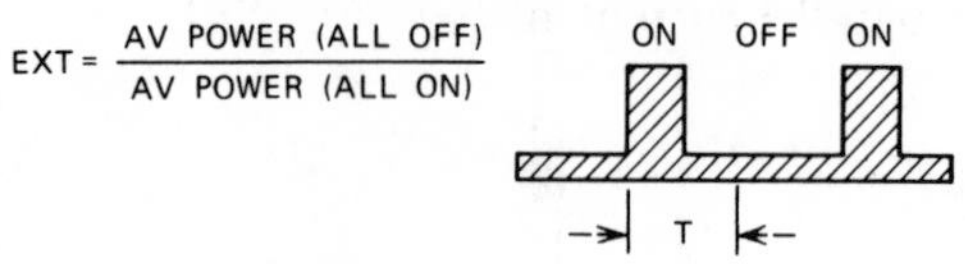

EXT = AV POWER (ALL OFF) / AV POWER (ALL ON)

ON OFF ON

T

Fig. 7—Imperfect modulation, pulses on pedestal.

pulses are "on" to be EXT in analogy to Case 1 above. This ratio will remain fixed if the pulse changes in propagation from transmitter to receiver. Using the results of Part I we obtain the following formulas for the required optical power:

$$p_{\text{required}} = \frac{1 + EXT}{1 - EXT} \frac{QZ^{\frac{1}{2}}}{GT} \frac{\hbar\Omega}{\eta}, \qquad \text{(Thermal Noise Dominates)} \qquad (8)$$

$$p_{\text{required}} = \frac{1 + EXT}{1 - EXT} \frac{(Q)^{(2+x)/(1+x)}}{2T} \times [(Z)^{x/(2+2x)}(\gamma_1'')^{x/(2+2x)}(\gamma_2'')^{(2+x)/(1+x)}] \frac{\hbar\Omega}{\eta}, \qquad \text{(Optimal Gain)} \qquad (9)$$

where defining from Part I (23) and (34)

$$\Sigma_1' = \Sigma_1 + \left(\frac{EXT}{1 - EXT}\right) I_2$$

$$I_7 = \Sigma_1 - I_1 + \left(\frac{EXT}{1 - EXT}\right) I_2$$

we have

$$\gamma_1'' = \frac{-(\Sigma_1' + I_7) + \sqrt{(\Sigma_1' + I_7)^2 + \frac{16(1+x)}{x^2}(\Sigma_1')I_7}}{2(\Sigma_1')I_7}$$

$$\gamma_2'' = \sqrt{1/\gamma_1'' + \Sigma_1'} + \sqrt{1/\gamma_1'' + I_7}.$$

Once again we can use (1), (2), (8), and (9) to calculate the penalty for nonzero extinction. This penalty is plotted in Fig. 8 vs EXT where we assume the input and output pulse shapes of the example in Section III when there is optimal avalanche gain.

VI. THE EFFECT OF DARK CURRENT

In order to allow for dark current, we must solve the following set of simultaneous equations which treat the dark current as an equivalent pedestal-type nonzero extinction. (When thermal noise dominates, dark current is either negligible or its shot noise can be added trivially to the amplifier parallel current noise source S_I.)

$$p_{\text{required}} = \frac{Q^{(2+x)/(1+x)}}{2T} (Z)^{x/(2+2x)} (\gamma_1''')^{x/(2+2x)} (\gamma_2''')^{(2+x)/(1+x)} \frac{\hbar\Omega}{\eta},$$

(At optimal avalanche gain) (10)

where defining from Part I (23) and (34)

$$\Sigma_1'' = \Sigma_1 + \delta I_2$$
$$I_8 = \Sigma_1 - I_1 + \delta I_2$$

we have

$$\gamma_1''' = \frac{-(\Sigma_1'' + I_8) + \sqrt{(\Sigma_1'' + I_8)^2 + \frac{16(1+x)}{x^2} \Sigma_1'' I_8}}{2 \Sigma_1'' I_8}$$

$$\gamma_2''' = \sqrt{1/\gamma_1''' + \Sigma_1''} + \sqrt{1/\gamma_1''' + I_8}$$

$$i_d = (2p_{\text{required}}) \frac{\eta e \delta}{\hbar\Omega}, \tag{11}$$

where i_d = primary dark current in amperes.

There are various ways to solve (10) and (11) simultaneously. One way is to solve (10) first with $\delta = 0$ for p_{required}. Then one can solve

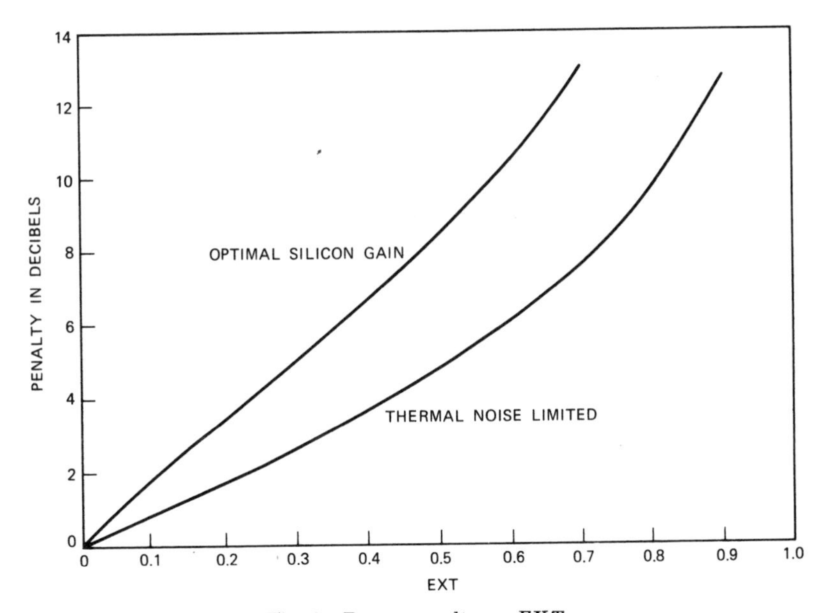

Fig. 8—Power penalty vs EXT.

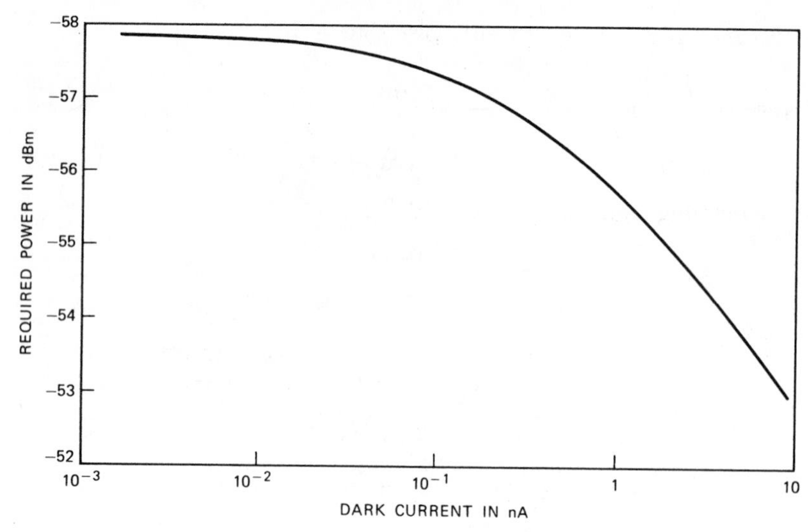

Fig. 9—Required power vs dark current.

(11) for a new value of δ. One then resolves (10) and then (11), etc.–repeating the iterations until satisfactory convergence is obtained. Figure 9 shows a plot of the required power in dBm for the example in Section III vs dark current in nanoamperes. We see that a dark current of 100 picoamperes results in an optical power requirement which is about 0.5 dB more than that which would be required with zero dark current. Thus it was reasonable to neglect dark current when calculating the required power in Section III. Dark current will result in even less of a penalty at higher bit rates. Although the curve of Fig. 9 is applied to the specific example of Section III, it is apparent in general that, at bit rates above a few megabits per second and with primary dark currents less than 0.1 nanoampere, dark current will have a small effect upon the required optical power.

Author Index

Subject Index

Boldface numbers indicate entries of major importance.

X

Y

Z

Editor's Biography

Detlef Gloge (M'66–SM'75) was born in Breslau, Silesia, Germany, on February 2, 1936. He received the degrees of Diplom Ingenieur and Doktor Ingenieur from the Technische Universität Braunschweig, Braunschweig, Germany, in 1961 and 1964, respectively.

He joined the staff of Bell Laboratories, Murray Hill, NJ in 1965. Presently, he is Head of the Optical System Research Department at Bell Laboratories, Holmdel, NJ. His research in the optical communications field has involved work on various optical transmission media and optical components. He is currently engaged in research relating to optical fiber communication systems.

Dr. Gloge is the author of 50 technical papers and holds 12 patents in the optical communications field. He is a member of the Optical Society of America.